Synoptic–Dynamic Meteorology in Midlatitudes

Synoptic–Dynamic Meteorology in Midlatitudes

VOLUME II
Observations and Theory of Weather Systems

HOWARD B. BLUESTEIN

New York Oxford
OXFORD UNIVERSITY PRESS
1993

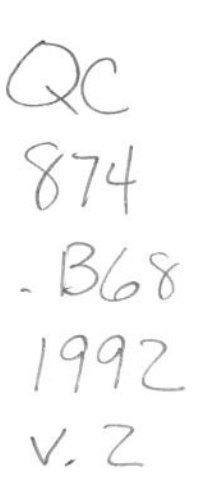

Oxford University Press

Oxford New York Toronto
Delhi Bombay Calcutta Madras Karachi
Kuala Lumpur Singapore Hong Kong Tokyo
Nairobi Dar es Salaam Cape Town
Melbourne Auckland Madrid

and associated companies in
Berlin Ibadan

Published by Oxford University Press, Inc.,
200 Madison Avenue, New York, New York 10016

Oxford is a registered trademark of Oxford University Press

Library of Congress Cataloging-in-Publication Data
Bluestein, Howard.
Synoptic–dynamic meteorology in midlatitudes.
Includes bibliographical references and index.
Contents: v. 1. Principles of kinematics and dynamics—v.
2. Observations and theory of weather systems.
ISBN 0-19-506268-X
1. Synoptic meteorology.
2. Dynamic meteorology.
I. Title.
QC874.B68 1993 551.5 91-7347

9 8 7 6 5 4 3 2 1

Printed in the United States of America
on acid-free paper

DEDICATION

To my mother and my father, and particularly to my maternal grandfather, Isaac Springer, who in his native Russia, at the turn of the century, saw no need for weather forecasting. (Pa wondered what all the fuss was while circumnavigating fallen trees and electrical wires during the great 1938 New England hurricane.) Finally, in the words of the Shirelles, this is dedicated to the one I love . . . my wife Kathleen Welch.

ACKNOWLEDGMENTS

This is the second volume of a textbook I have written in an attempt to provide a modern, updated, replacement to the two-volume work by Sverre Petterssen, which was last published over 30 years ago. I am grateful to Fred Sanders (Olde Dad), my thesis advisor at the Massachusetts Institute of Technology (M.I.T.), whose course notes formed the seminal roots of this textbook. His charm, wit, and wisdom have been the source of many years of inspiration. Courses taught at M.I.T. in the early 1970s by Ed Lorenz and Norm Phillips have also influenced the writing of this text. I am particularly grateful to many undergraduate and graduate students at the University of Oklahoma (OU), who between 1976 and 1991 endured many imperfect versions of my "notes," and who did not hesitate to suggest improvements. Lance Bosart's friendship and continued encouragement sparked me to forge on and complete this almost never-ending project. A substantial portion of the book was written while I was on sabbatical leave at the National Center for Atmospheric Research (NCAR) from 1984 to 1985, and at NCAR during the summers of 1988, 1989, and 1990 and at the University of Washington during the summer of 1986. A summer colloquium and workshop on synoptic meteorology and its instruction at NCAR in 1988 provided further opportunity to update the text. Criticism from Fred Carr, Morris Weisman, Rich Rotunno, Cliff Mass, Dan Keyser, Glen Lesins, Earle Williams, John Brown, and Mohan Ramamurthy is gratefully acknowledged. Don Fred provided some early editing. I thank the editors at Oxford University Press for their expert work. I thank former secretaries at OU for typing early versions of the notes. I owe particular gratitude to Ginger Rollins and Susan Hovis for painstakingly making revision after revision on the word processor. Neal Shores assisted with some of the climatological analyses. Paul Mulder (NCAR) provided some of the climatological data. Maria Neary expertly drafted many of the figures. Sue Weygandt assisted with many of the figures. Hope Hamilton and Graphic Services at NCAR also assisted with some figures. I acknowledge Tim Hughes, Keith Brewster, and Meta Sienkiewicz for helping me generate some figures on the Geosciences Computer Network (GCN) at OU. I thank Mary desJardins for developing and providing GEMPAK, a general meteorological software package. Instructional Services at OU drafted many of the figures for the problems at the end of each chapter. Many friends and colleagues provided figures from their publications. My residence in Miami at the National Hurricane Research Laboratory (currently the Hurricane Research Division) allowed me to nurture my interest in tropical convection. Ed Kessler and Rex Inman made it possible for me to explore convective phenomena in Oklahoma. I thank Bob Burpee, Jeff Kimpel, Claude Duchon, the National Science Foundation, and the COMET secretaries for their support. Finally, I am indebted to the atmosphere for inspiring me with impressive, though not frequent enough, episodes of tornadoes, waterspouts, funnel clouds, strong straight-line winds, wall clouds, large hail, downpours, lightning, heavy snow, ice pellets, freezing rain, biting cold, and spectacular clouds.

Contents

Synoptic–Dynamic Meteorology in Midlatitudes

1

The Behavior of Synoptic-Scale, Extratropical Systems

It was a quiet seeming Day—
There was no harm in earth or sky—
Till with the setting sun
There strayed an accidental Red
A Strolling Hue, one would have said
To westward of the Town—

But when the Earth began to jar
And Houses vanished with a roar
And Human Nature hid
We comprehended by the Awe
As those that Dissolution saw
The Poppy in the Cloud.

EMILY DICKINSON

Much of this chapter details techniques used to diagnose the physical mechanisms responsible for height and pressure changes, and for vertical motion, using quasigeostrophic reasoning. The following three techniques are described in varying degrees of detail: idealized conceptual modeling, analytical modeling, and "diagnostic" studies of real data. We will draw heavily upon material discussed in Vol. I; it will be assumed that the reader has access to it.

1.1. THE FORMATION OF SURFACE PRESSURE SYSTEMS

The processes of surface cyclogenesis and anticyclogenesis, the formation of surface cyclones and anticyclones, will be treated equally, even though cyclogenesis is more exciting, more violent, and hence more often discussed. A nocturnal freeze, owing to radiational cooling in a dry, calm air mass under a high at temperate latitudes in a citrus-growing area is, in a sense, more destructive behavior than that exhibited by many cyclones. Furthermore, cyclogenesis and anticyclogenesis are usually dependent upon each other. We will therefore not neglect anticyclones!

Unfortunately, we cannot use the quasigeostrophic height-tendency equation to explain cyclogenesis and anticyclogenesis at the surface: The height tendency at the surface is a lower boundary condition for the height-tendency equation. That is, the height tendency at the surface, which we are trying to infer, is itself necessary in order to solve the height-tendency equation. To determine the height-tendency field at the surface, we can use the quasi-geostrophic ω equation evaluated above the surface (three dimensionally) and the vorticity equation evaluated at the surface. If there is rising motion above a level surface, then $\omega < 0$, and hence from the equation of continuity

$$\frac{\partial \omega}{\partial p} = -\delta > 0 \tag{1.1.1}$$

at the surface. It follows from the quasigeostrophic vorticity equation that the effect of this convergence at the surface is to make vorticity more cyclonic locally (i.e., to increase the geostrophic vorticity locally in the Northern Hemisphere), since

$$\frac{\partial \zeta_g}{\partial t} = -\delta f_0 > 0. \tag{1.1.2}$$

(We consider the effect of convergence alone, and hence neglect the effects of vorticity advection and friction). Because $\zeta_g = (1/f_0)\nabla_p^2\Phi$, it follows from Eq. (1.1.2) that

$$\frac{\partial}{\partial t}\nabla_p^2\Phi = \nabla_p^2\left(\frac{\partial \Phi}{\partial t}\right) > 0. \tag{1.1.3}$$

Therefore the height falls locally, because $\nabla_p^2(\partial\Phi/\partial t)$ and $\partial\Phi/\partial t$ tend to have opposite signs.

Let us now determine the relationship between the local height tendency and the local pressure tendency. Consider the total differential of pressure, which is in general a function of x, y, z, and t:

$$dp = \left(\frac{\partial p}{\partial x}\right)dx + \left(\frac{\partial p}{\partial y}\right)dy + \left(\frac{\partial p}{\partial z}\right)dz + \left(\frac{\partial p}{\partial t}\right)dt. \tag{1.1.4}$$

Substituting the hydrostatic equation into Eq. (1.1.4), using the definition of geopotential, and rearranging terms, we find that

$$d\Phi = -\frac{dp}{\rho} + \frac{1}{\rho}\left(\frac{\partial p}{\partial x}\right)dx + \frac{1}{\rho}\left(\frac{\partial p}{\partial y}\right)dy + \frac{1}{\rho}\left(\frac{\partial p}{\partial t}\right)dt. \tag{1.1.5}$$

The total differential of geopotential, which is also in general a function of x, y, z, and t, can be expressed as

$$d\Phi = \left(\frac{\partial \Phi}{\partial x}\right)dx + \left(\frac{\partial \Phi}{\partial y}\right)dy + \left(\frac{\partial \Phi}{\partial p}\right)dp + \left(\frac{\partial \Phi}{\partial t}\right)dt. \tag{1.1.6}$$

Substituting the hydrostatic equation into Eq. (1.1.6), using the ideal gas law, and rearranging terms, we find that

$$d\Phi = -\frac{dp}{\rho} + \left(\frac{\partial \Phi}{\partial x}\right)dx + \left(\frac{\partial \Phi}{\partial y}\right)dy + \left(\frac{\partial \Phi}{\partial t}\right)dt. \tag{1.1.7}$$

We see then from Eqs. (1.1.5) and (1.1.7) that

$$\frac{1}{\rho}\left(\frac{\partial p}{\partial t}\right)_z = \left(\frac{\partial \Phi}{\partial t}\right)_p. \tag{1.1.8}$$

Therefore, if the height of the pressure surface that had been at ground level falls, then the pressure at ground level also falls. [This is obvious even without considering Eqs. (1.1.4)–(1.1.8), since pressure decreases with height!]. Thus, surface cyclones (low-pressure areas) form in areas of convergence at the surface underneath regions of rising motion. Similarly, surface anticyclones (high-pressure areas) form in areas of divergence at the surface underneath regions of sinking motion. (If the ground is not level, then ω may not necessarily be zero at the ground, and hence our analysis is not applicable. For example, we will soon see how sinking motion can lead to lee cyclogenesis!)

The four quasigeostrophic forcing functions that are associated with rising motion and pressure falls at the surface are as follows:

1. Vorticity advection becoming more cyclonic (or less anticyclonic) with "height" (i.e., with respect to $-p$);
2. A local maximum in temperature advection;
3. The vertical component of the curl of the frictional force becoming more cyclonic (or less anticyclonic) with "height";
4. A local maximum in diabatic heating.

We see from Eq. (1.1.8) that the surface pressure change

$$\left(\frac{\partial p}{\partial t}\right)_{z=0} = \rho\chi_0. \tag{1.1.9}$$

Formally, an equation for $\partial p/\partial t$ at the surface ($p = p_0$) may be obtained from the frictionless form of the quasigeostrophic vorticity equation

$$\frac{\partial \zeta_g}{\partial t} = -\mathbf{v}_g \cdot \nabla_p(\zeta_g + f) - \delta f_0.$$

It follows from the equation of continuity and the relationship between vorticity and the geopotential–height field that

$$\nabla_p^2\chi_0 = f_0[-\mathbf{v}_g \cdot \nabla_p(\zeta_g + f)]_{p_0} + f_0^2\left(\frac{\partial \omega}{\partial p}\right)_{p_0}, \tag{1.1.10}$$

where

$$\chi_0 = \left(\frac{\partial \Phi}{\partial t}\right)_{p_0}. \tag{1.1.11}$$

If $\omega = 0$ at the surface, then

$$\nabla_p^2\chi_0 = f_0[-\mathbf{v}_g \cdot \nabla_p(\zeta_g + f)]_{p_0} - f_0^2\left(\frac{\omega_{p'}}{p_0 - p'}\right), \tag{1.1.12}$$

where $\omega_{p'}$ is the vertical velocity at $p = p'$, for $p' < p_0$. Therefore, from Eq. (1.1.9) or (1.1.11) we find that surface height changes (χ_0) are due to vorticity advection (χ_V) and divergence (or convergence) associated with vertical motions due to differential vorticity advection (χ_{ω_V}), temperature advection (χ_{ω_T}), differential friction (χ_{ω_F}), and diabatic heating (χ_{ω_Q}):

$$\chi_0 = \chi_V + \chi_{\omega_V} + \chi_{\omega_T} + \chi_{\omega_F} + \chi_{\omega_Q}. \tag{1.1.13}$$

1.1.1. The Effects of Differential Vorticity Advection

Let us first consider the effects of differential vorticity advection alone. (The effect of each forcing function in the quasigeostrophic ω equation may be discussed as if each were operating alone, even though each is dependent upon the others; the net effect of all forcing functions is additive.) A typical wavetrain in the baroclinic westerlies aloft is depicted in Fig. 1.1. There is cyclonic vorticity advection (CVA) downstream from the maxima in absolute vorticity, which are located along the trough axes. Ordinarily, vorticity advection aloft is larger in magnitude than it is at the surface, where pressure (and height) systems tend to be circular. Therefore vorticity advection is more cyclonic with height; there is thus rising motion downstream from upper-level troughs. It follows that the effect of differential vorticity advection is to make the surface pressure fall, and hence contribute to the formation of a surface cyclone or trough. Similarly, downstream from a ridge aloft, the effect of differential vorticity advection is to make the surface pressure rise and contribute to the formation of a surface anticyclone or ridge.

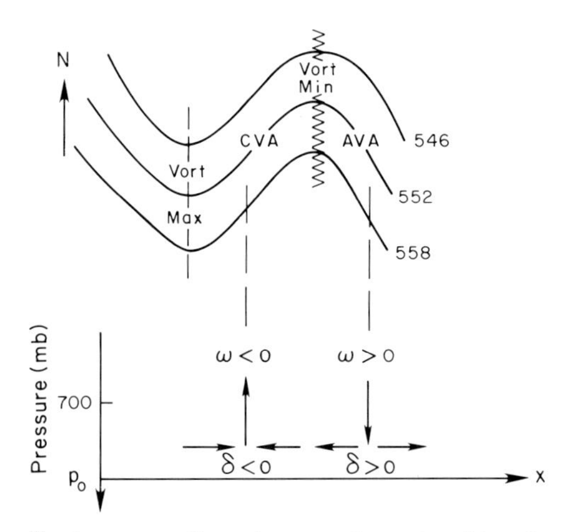

Figure 1.1 The effects, according to quasigeostrophic theory, of differential geostrophic vorticity advection in a typical wavetrain in the baroclinic westerlies. Height contours at 500 mb in dam (solid lines); CVA (cyclonic vorticity advection) and AVA (anticyclonic vorticity advection) located downstream from vorticity maxima and minima, respectively, at 500 mb (top). Below 500 mb, vorticity advection is assumed to be relatively weak. Consequently the vorticity advection is more cyclonic with height under the CVA region and more anticyclonic with height under the AVA region. The corresponding vertical velocity (ω_v) and horizontal divergence (δ) patterns are depicted (bottom) in the following.

Surface cyclones that form mainly in response to differential vorticity advection associated with a pre-existing finite-amplitude upper-level disturbance have been named *Type B* cyclones by Petterssen and Smebye. Although there may be little surface temperature advection initially, if often becomes substantial as the cyclone deepens.

1.1.2. The Effects of Temperature Advection

We now consider the effects of temperature advection alone. Although surface cyclogenesis may occur along frontal zones, it does not happen in response to localized, intense, warm advection *alone*: If there were either a jet or a local concentration of temperature gradient, then there would also be pre-existing differential vorticity advection.

Bands of warm advection are often found along warm frontal zones. Since these bands represent relative maxima in temperature advection, there is a band of rising motion, which results in the intensification of the pre-existing frontal trough (Fig. 1.2). Pressure ridges can intensify similarly in response to strong surface cold advection (Fig. 1.2) behind cold frontal zones. The Polar-Front theorists, whose work will be discussed later, expounded during the early part of this century on the importance of surface baroclinic zones in the formation of cyclones.

Petterssen and Smebye have classified surface cyclones that form along frontal waves in the absence of a pre-existing finite-amplitude upper-level disturbance as *Type A* cyclones. (It is not known whether or not these cyclones have *Type A* personalities, even if they form near "hyper"–baroclinic zones!) It is of historical interest that Polar-Front theory was developed because it was thought that Type A cyclones dominated the region under study. Because

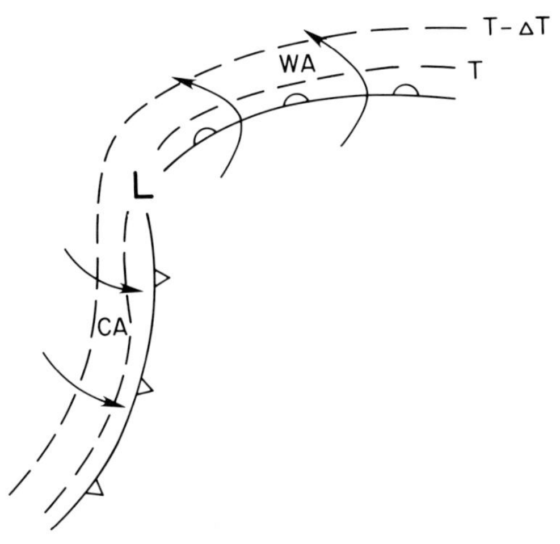

Figure 1.2 Illustration of temperature–advection patterns associated with surface fronts in a typical extratropical cyclone. Quasigeostrophic theory argues that warm advection (WA) poleward of the warm front can form a trough at the surface, and cold advection (CA) to the rear of the cold front can form a ridge at the surface.

upper-air data were not available, it was not possible to see that vorticity advection aloft could have also played a significant role.

1.1.3. The Effects of Surface Friction

The effects of differential friction alone are now considered. In a cyclone whose geostrophic vorticity is independent of height in the friction layer, the vertical component of the curl of the frictional force is anticyclonic (Fig. 1.3) and becomes zero at the top of the friction layer. There is therefore rising motion in and just above the friction layer (Ekman pumping), while in the friction layer there is convergence; the effect of convergence is to intensify the cyclone, while the direct effect of friction is to dissipate energy and weaken the cyclone. However, just above the top of the friction layer there is divergence and no friction, and hence the cyclone weakens: The effects of friction below have been "communicated" to the "free" atmosphere by the secondary circulation (see Vol. I). In the region of a surface anticyclone, there is similarly sinking motion in the friction layer only, while at the surface there is divergence; just above the friction layer there is convergence, and the anticyclone weakens.

1.1.4. The Effects of Diabatic Heating

Diabatic heating alone may be responsible for the formation of a surface cyclone, if the region of diabatic heating represents a local maximum. For

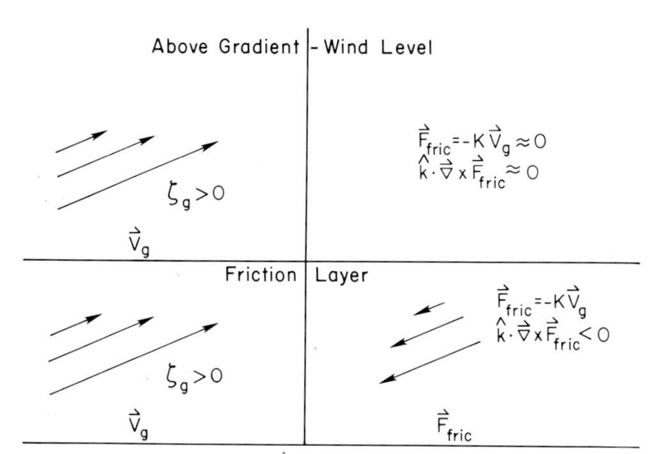

Figure 1.3 Illustration of the quasigeostrophic effects of surface friction. Above gradient-wind level friction is negligible, and hence the vertical component of the curl of the friction force is zero. Within the friction layer, where vorticity is independent of height, the vertical component of the curl of the friction force is opposite in sign to that of the vertical component of vorticity. In the example shown there is cyclonic vorticity; the vertical component of the curl of the friction force is anticyclonic. The vertical component of the curl of the friction force becomes less anticyclonic with height at gradient-wind level. Consequently there is rising motion.

example, intense localized diabatic heating owing to latent heat release from cumulus convection that acts for a time long enough so that the Earth's rotation is dynamically important may form a low or a trough. The upward flux of heat and water vapor from the ocean surface and the vertical distribution of latent heat release through cumulus convection is thought to be responsible for tropical cyclogenesis. It is difficult, however, to model the effect of small-scale cumulus convection on the larger scale in a realistic manner. On the other hand, a region of diabatic cooling owing to evaporation of rain in an unsaturated layer below may be responsible for the formation of a high or a ridge.

However, diabatic heating or cooling associated with cumulus convection typically acts on time scales too short for the effects of Earth's rotation to be felt. In this case lows, troughs, highs and ridges still form as a consequence of the hydrostatic equation; however, their associated wind fields are not geostrophic, as, for example, in thunderstorm gust fronts or in the stratiform-precipitation area of some squall lines (see Chap. 3).

Diabatic heating and cooling from effects other than cumulus convection can also be important. For example, when cold, continental air streams out over relatively warm water, sensible heat is vigorously transferred to the air and may contribute to the formation of a trough or a low. This happens, for example, during the winter over the Gulf Stream and Gulf of Mexico. Relatively warm air flowing over a cold, icy, or snowy surface, or a body of water that is much colder, may contribute to the formation of an anticyclone. This happens, for example, over the Great Lakes or off the New England coast during the summer. Local, intense heating owing to solar radiation, such as happens over a desert, contributes to the formation of a "thermal low."

1.1.5. The Effects of Upslope and Downslope Motions

Even in the absence of any quasigeostrophic forcing, cyclogenesis and anticyclogenesis can occur if the surface is sloped. Downslope motion over a mountain range results in the formation of a *lee trough* (Fig. 1.4). Subsiding air in the lee of a mountain range warms as it is compressed, and therefore acts like the warm-advection or diabatic heating forcing functions. The atmosphere's response is rising motion, which is accompanied by convergence at the surface, and hence a drop in height and pressure. Note that the "forcing" is sinking motion that is *not* dynamically induced; the response is rising motion, and the net vertical motion is the sum of the two. (Another way to view lee-trough formation is that downslope motion in a statically stable atmosphere in which θ surfaces are initially horizontal acts to push θ surfaces downward on the lee slope. The air column becomes warmer on the lee slope, and therefore as a hydrostatic consequence must have lower pressure than air at the same height away from the slope.)

Mathematically, the quasigeostrophic ω equation (see Vol. I, Chap. 5) in the absence of forcing other than that associated with orography (the

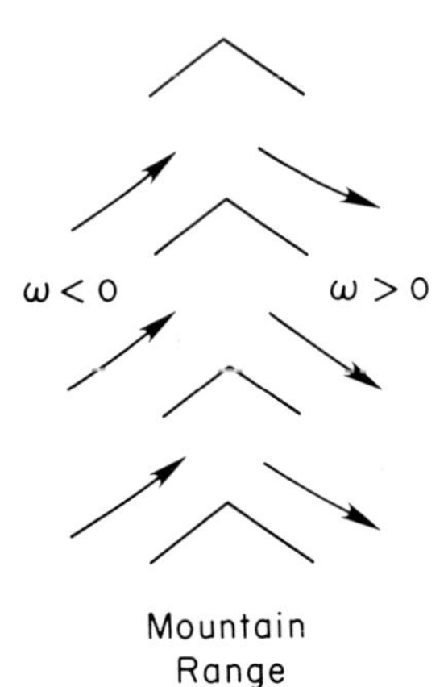

Figure 1.4 The effects of a mountain range on the lower boundary condition on ω. On the windward side, air ascends ($\omega < 0$); on the lee side, air descends ($\omega > 0$).

homogeneous form) can be expressed at the surface as

$$\nabla_p^2 \omega_0 = -\frac{f_0^2}{\sigma}\frac{\partial^2 \omega_0}{\partial p^2}, \tag{1.1.14}$$

where σ, the static stability parameter, is $(-RT/p)/(\partial \ln \theta/\partial p)$. If $\omega_0 > 0$ owing to downslope motion over a mountain range, then

$$\nabla_p^2 \omega_0 < 0$$

and hence in a statically stable atmosphere

$$\frac{\partial^2 \omega_0}{\partial p^2} > 0. \tag{1.1.15}$$

In other words, from the continuity equation and Eq. (1.1.15) we see that

$$-\frac{\partial \delta}{\partial p} > 0;$$

that is, divergence increases with height. Hence, if $\delta = 0$ aloft where the atmosphere does not "feel" the effects of the mountains, there must be convergence ($\delta < 0$) at the surface. This convergence increases the surface vorticity and drops the surface height and pressure.

Alternatively, we may use Eq. (1.1.10) in the absence of dynamical forcing:

$$\nabla_p^2 \chi_0 = f_0^2 \left(\frac{\partial \omega}{\partial p}\right)_{p_0}. \tag{1.1.16}$$

If $\omega = 0$ aloft, where the atmosphere does not "feel" the effects of the mountains, but $\omega > 0$ at $p = p_0$ owing to downslope motion, then $(\partial \omega/\partial p)_{p_0} > 0$, and hence

$$\nabla_p^2 \chi_0 > 0.$$

It follows that χ_0 and $(\partial p/\partial t)_{z=0}$ are "less than zero."

Other interesting phenomena such as lee waves and their sometimes

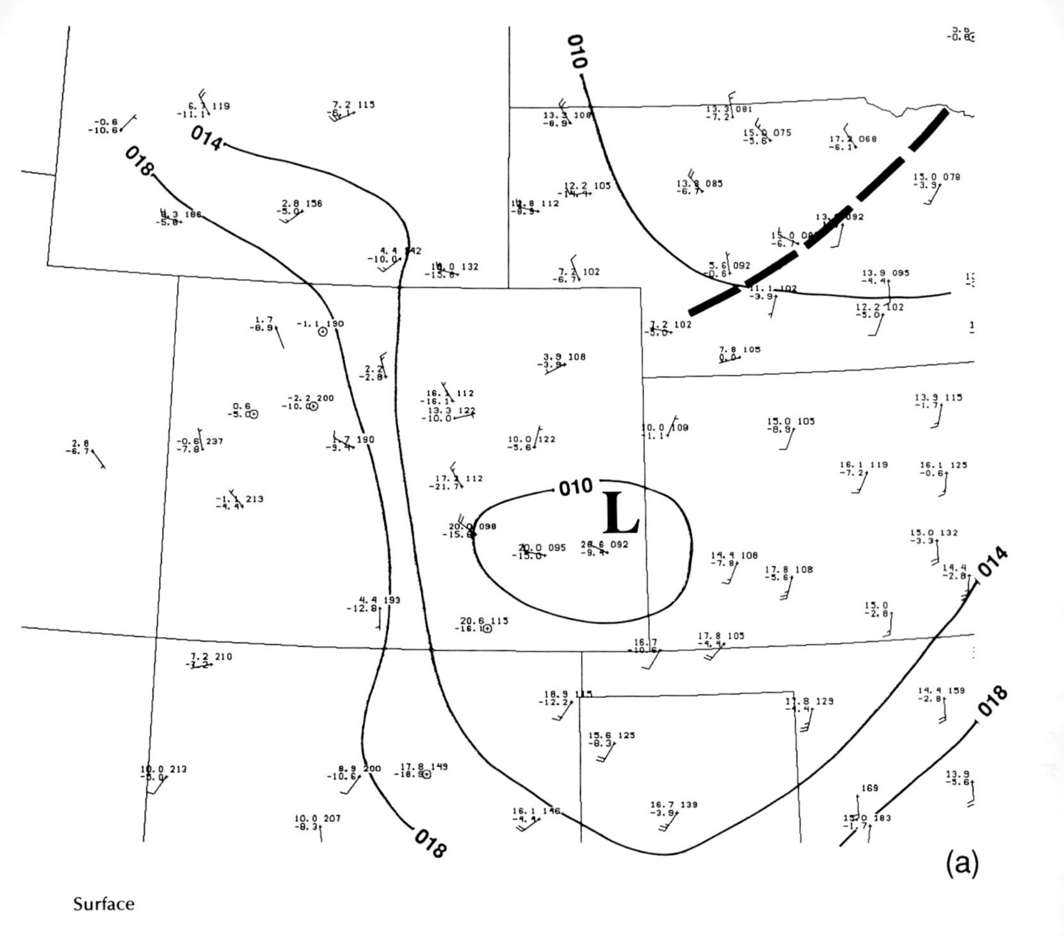

Figure 1.5 Structure of a lee cyclone (a) for which there is very weak synoptic-scale forcing; at 2100 UTC, November 30, 1990; surface wind field (whole barb = 5 m s^{-1}; half barb = 2.5 m s^{-1}), temperature and dewpoint (°C), and altimeter setting in ×10 mb, without the leading "10"; isobars in mb without the leading "1" (solid lines); note that the cyclone in southeastern Colorado is nearly co-located with the area of warmest temperatures (of 20°C or higher), owing to a westerly component of downslope motion; temperatures in southwestern Nebraska and northeastern Colorado are locally much colder (below 10°C), owing to snow cover; (b) NMC analysis at 700 mb for 0000 UTC, December 1, 1990; winds are relatively weak and 12-h height falls are small in eastern Colorado; and (c) for the much more common situation in which there is strong quasigeostrophic forcing owing to a short-wave trough aloft; at 1200 UTC, February 22, 1977; surface wind field [as in (a)]; NMC sea-level pressure field (solid lines in tens of mb, without the leading "10" or "9"; temperature and dew point in °F; note the locally warm air (60°F near the low in southeastern Colorado) owing to downslope motion. (d) NMC analysis at 700 mb for 1200 UTC, February 22, 1977; winds are strong through northern New Mexico, and flow curvature is cyclonic in Colorado, where 12-h height falls are in excess of 10 dam. There is strong cyclonic vorticity advection over southeastern Colorado. In the United States lee troughs are sometimes found in response to easterly downslope along the California coast when there is a cold anticyclone to the northeast. The dry downslope winds that occur are referred to as *Santa Ana conditions*. Lee troughs are also common east of the Appalachians and east of the Northern Rockies in Wyoming and Montana.

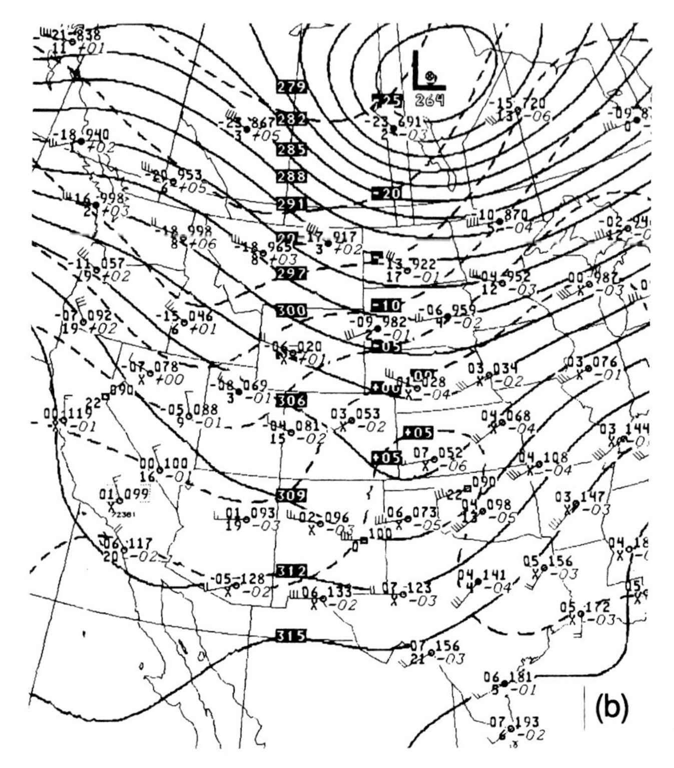

700 mb 12/1/90 0000 UTC

Figure 1.5 (*cont.*)

attendant wind storms are not quasigeostrophic, and therefore will not be discussed here. When the downslope motion field is organized into a jetlike configuration, or when there is uniform downslope motion over terrain in which there is a localized region of steep dropoff in elevation, lee cyclogenesis may occur (Fig. 1.5). (In this context lee cyclogenesis refers to localized downslope-induced surface pressure falls; "lee cyclogenesis" is often used to described *any* surface cyclogenesis in the lee of mountains, regardless of the mechanism.) However, other factors, which are not quasigeostrophic (e.g., internal gravity wave motion), also may have a significant effect on lee cyclogenesis. Since the downslope motion is strongest near the ground, lee cyclones are most intense near the ground, where most of the downslope warming has occurred, and hence the lee cyclone is a "warm-core" feature. Similarly, "windward ridges" may form as air moves upslope over a mountain range. (Windward-ridge formation may also be viewed as a hydrostatic consequence of θ surfaces that are pushed upward on the windward slope.) This may happen east of the Appalachians or Rockies under easterly low-level. However, since the kinetic energy required to overcome the potential energy associated with the weight of the air column above may not be available, air may not make it all the way over the mountain range. A surface ridge may

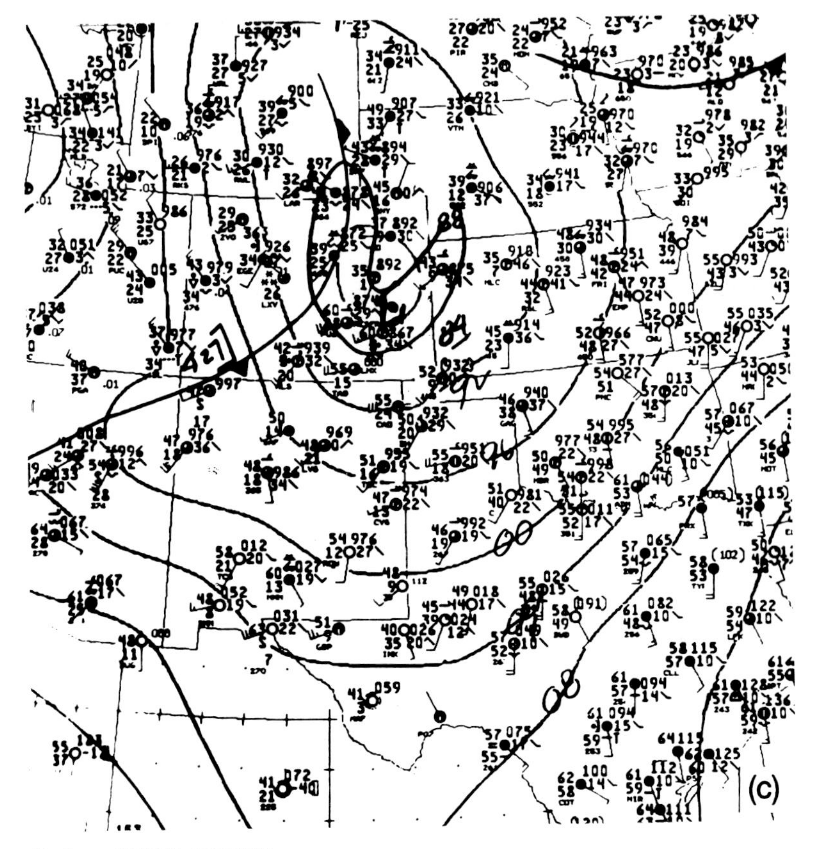

Surface 2/22/77 1200 UTC

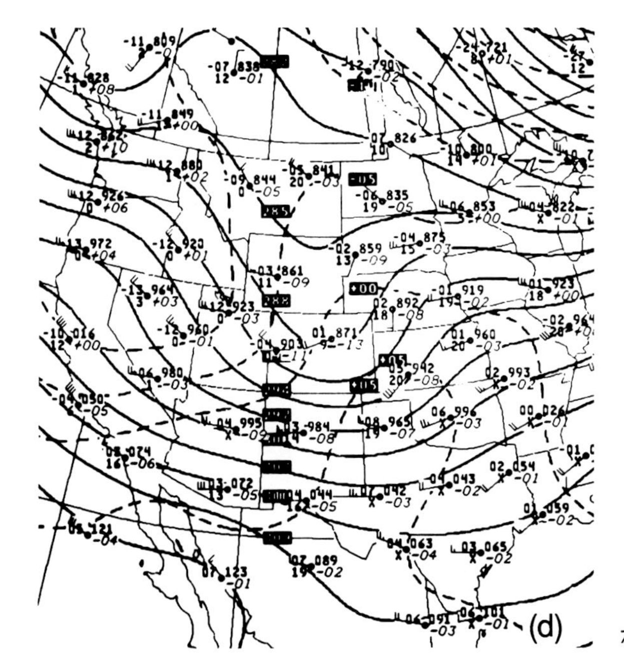

700 mb 2/22/77 1200 UTC

Figure 1.5 (*cont.*)

form simply because a shallow cold air mass is unable to make it over the mountains.

1.1.6. The Effects of Static Stability

Since the forcing functions in the ω equation are inversely proportional to the static stability parameter σ, cyclogenesis and anticyclogenesis are hastened by low values of static stability, and are slowed down by high values. Sometimes cyclogenesis does not occur until an upper-level disturbance moves into a region of lower static stability, for example when a cold upper-level trough moves over a relatively warm ocean during the winter.

1.1.7. The Combined Effects of the Quasigeostrophic Forcing Functions

Sometimes one forcing function "preconditions" or "ripens" the atmosphere for another so as to enhance cyclogenesis or anticyclogenesis. Other times the forcing functions in the ω equation oppose each other, thereby inhibiting cyclogenesis and anticyclogenesis. We will now consider examples of both the latter and former.

Differential vorticity advection and temperature advection. Figure 1.6 shows a region of cyclonic differential vorticity advection downstream from an upper-level trough, which is coincident with an area of strong cold advection below. Thus, cyclogenesis is inhibited. [Using the Trenberth formulation of the quasigeostrophic ω equation, we see that advection of geostrophic vorticity by the thermal wind is negligible (Fig. 1.7), and hence the cyclone already present does not develop any further. Sutcliffe in 1947 presented a similar argument using his "development" equation. The Q-vector formulation gives the same result (Fig. 1.8).] Similarly, suppose there is a region of anticyclonic differential vorticity advection downstream from an upper-level ridge, which is coincident with an area of strong warm advection below: Anticyclogenesis at the surface is inhibited.

However, if an upper-level trough passes from the region of cold advection west of a surface front to the region beyond that of reduced cold advection or neutral temperature advection, to the region of warm advection east of the surface front, differential vorticity advection and temperature advection act together. (In other words, there is cyclonic advection of geostrophic vorticity by the thermal wind, and cyclogenesis occurs.) Similarly, as the region of anticyclonic differential vorticity advection upstream from the trough moves over the area of strong cold advection, anticyclogenesis occurs.

Temperature advection and diabatic heating. When cold air flows out over warm water (cold advection over the water area), surface pressure falls associated with the diabatic heating (turbulent sensible heat transport) may be inhibited by the cold advection (Fig. 1.9). This may happen over the Gulf of Mexico during the winter for example. Cyclogenesis may be delayed until

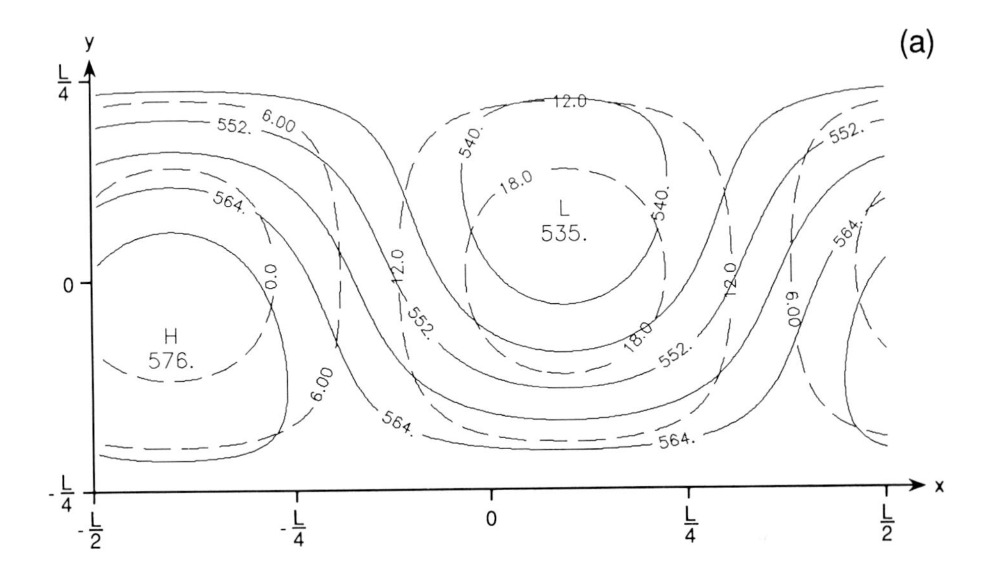

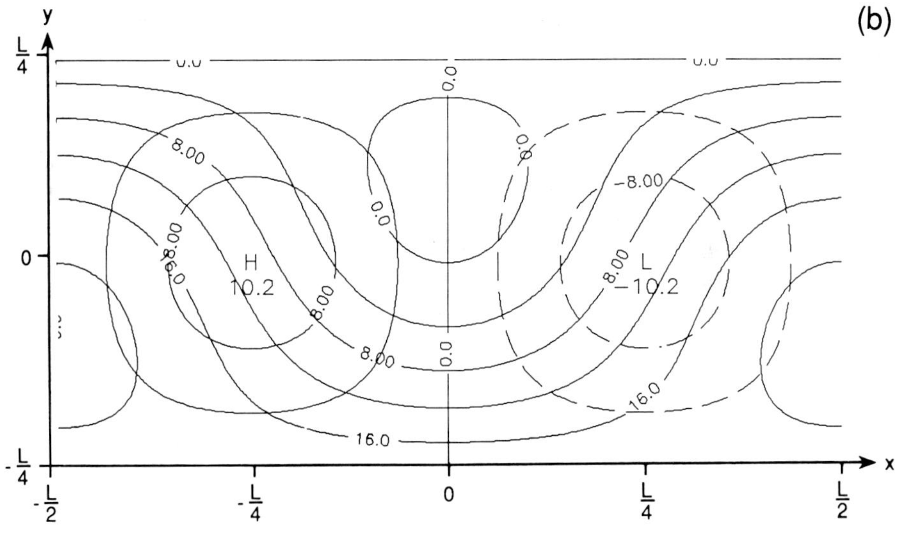

Figure 1.6 Illustration of the cancellation of the forcing-function terms in the traditional formulation of the quasigeostrophic ω equation. (a) Vorticity advection becoming more cyclonic with height contributes to rising motion downstream from the 500-mb trough; 500-mb height in dam (solid lines); geostrophic absolute vorticity in $10^{-5}\,s^{-1}$ (dashed lines); the absolute vorticity is unrealistically negative in the vicinity of the ridges; (b) low-level cold advection contributes to sinking motion southwest of the 1000-mb low, and low-level warm advection to rising motion northeast of the low; 1000-mb height in dam (solid and dashed lines for positive and negative values, respectively); 1000-mb temperature in °C (solid lines). In the analytic temperature and wind field illustrated here, the horizontal wavelength is 3500 km, and the trough in the 1000-mb temperature field lags the trough in the height field by one-quarter of a wavelength.

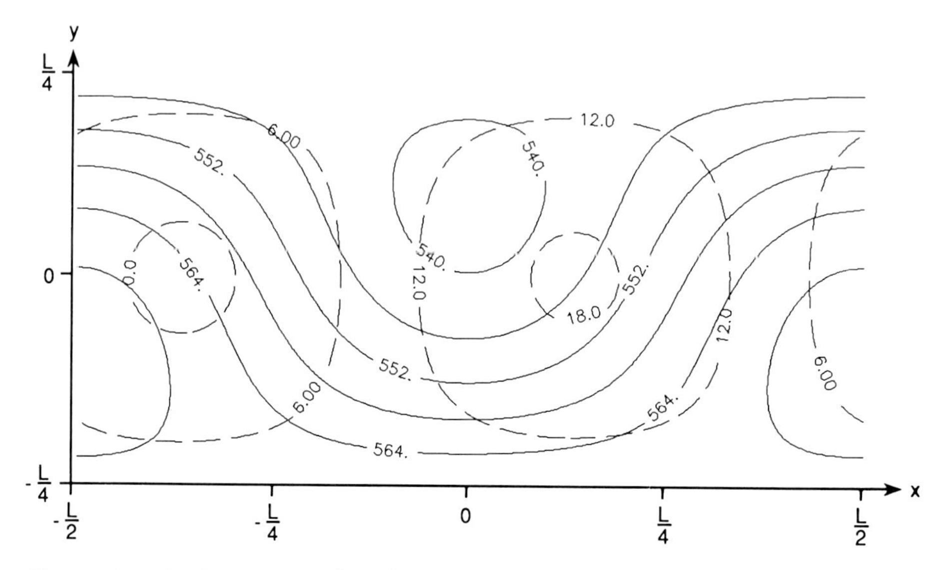

Figure 1.7 As in Fig. 1.6, but for the Trenberth formulation of the quasigeostrophic ω equation; 1000–500 mb thickness in dam (solid lines); geostrophic absolute vorticity at 700 mb in $10^{-5}\,s^{-1}$ (dashed lines); there is either little advection or anticyclonic advection of 700-mb vorticity by the 1000–500 mb thermal wind just downstream from the 500-mb trough [e.g., at $x = L/8$, $y = 0$; see Fig. 1.6(a)]; farther downstream (e.g., at $x = L/4$, $y = 0$) there is cyclonic advection of 700-mb vorticity by the 1000–500 mb thermal wind.

there is additional forcing from a trough aloft approaching from the west. Similarly, diabatic cooling over an ice field or snow field or relatively cold water is strongest when there is warm advection, and surface anticyclogenesis may be inhibited by the warm advection until a ridge aloft approaches from the west. This can happen, for example, over the northwest section of Canada during the winter as mild Pacific air flows over the snow-covered continent.

Differential vorticity advection, temperature advection, and diabatic heating. Suppose that cyclogenesis at the surface occurs in response to the combined effect of cyclonic differential vorticity advection and warm advection. Cumulus convection may ensue if the atmosphere is conditionally or convectively unstable, which results in the release of latent heat into the air. It is believed that the effect of the latent-heat release is to hasten cyclogenesis.

When surface anticyclogenesis occurs in response to anticyclonic differential vorticity advection and cold advection, the surface wind near the center becomes calm, and if the surface is snow covered, radiational cooling is very efficient and can enhance the rate of surface anticyclogenesis.

Friction and diabatic heating. Although friction by itself is dissipative, and the secondary circulation is also dissipative above the friction layer (in the so-called *interior*), friction may act together with latent-heat release to produce cyclogenesis at the top of the friction layer and also below (Fig. 1.10). Charney

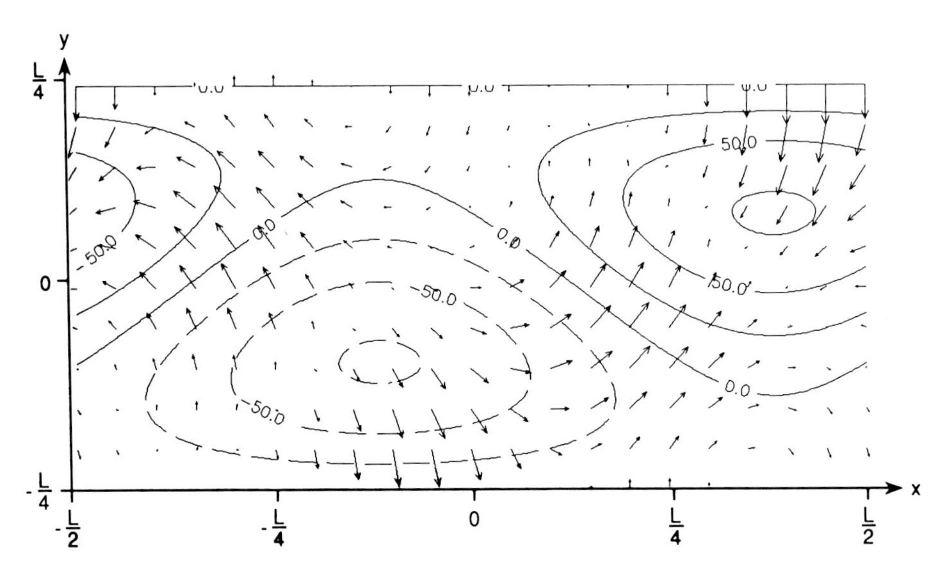

Figure 1.8 As in Fig. 1.6, but for the Q-vector formulation of the quasigeostrophic ω equation; Q vectors at 700 mb represented by arrows; largest vector has magnitude of 1.84×10^{-9} kPa s^{-1} m^{-1}; $-2\nabla_p \cdot \mathbf{Q}$ at 700 mb, which also represents the sum of the forcing functions in the traditional formulation (less the small β term) or the forcing function in the Trenberth formulation (less the small deformation terms), in 10^{-16} kPa s^{-1} m^{-2}. The forcing is near zero just downstream from the 500 mb trough [e.g., at $x = L/8$, $y = 0$; Fig. 1.6(a)]; the forcing is positive farther downstream, where the effects of warm advection at 1000 mb [Fig. 1.6(b)] and vorticity advection becoming more cyclonic with height [Fig. 1.6(a)] are in the same sense, and therefore combine (e.g., at $x = 3L/8$, $y = L/8$).

and Eliassen in 1964 termed this process "conditional instability of the second kind" (CISK) (Fig. 1.11). The Ekman pumping associated with a cyclone in the friction layer may trigger or enhance cumulus convection if the atmosphere is conditionally unstable and if there is an adequate supply of moisture. The latent-heat release itself induces additional rising motion aloft and convergence

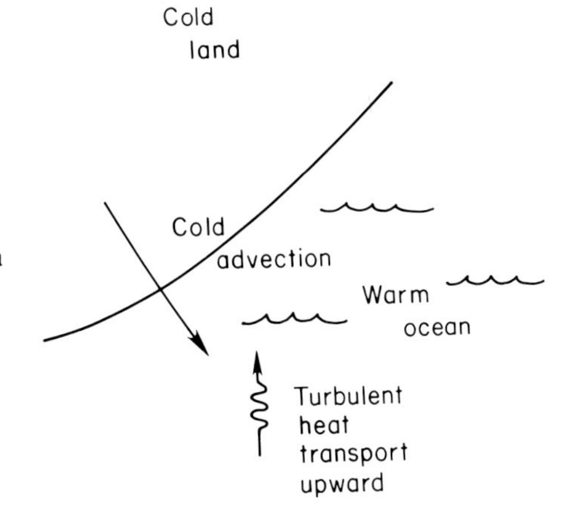

Figure 1.9 Illustration of how flow off a cold land surface onto a warm water surface is responsible for both cold advection and diabatic heating. Cold advection and diabatic heating are associated quasigeostrophically with sinking motion and rising motion, respectively.

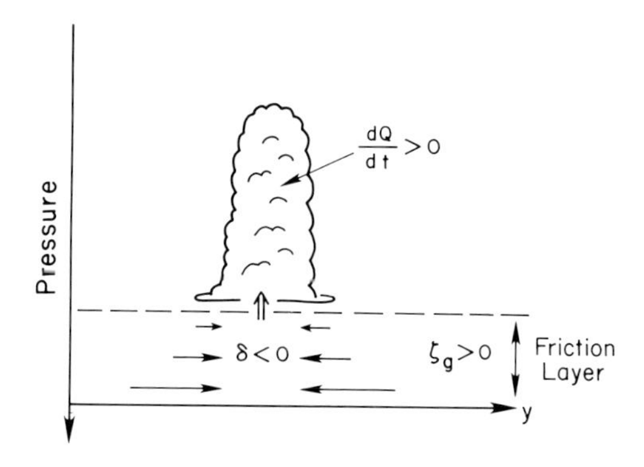

Figure 1.10 Cooperative action between friction and diabatic heating. Suppose there is cyclonic geostrophic vorticity in the friction layer. Quasigeostrophic theory requires that there be rising motion at the top of the friction layer. Cumulus convection may be triggered in response to this rising motion. Diabatic heating associated with the condensation of water vapor induces more rising motion according to quasigeostrophic theory. Surface convergence associated with the rising motion induces an increase in cyclonic vorticity and a drop in pressure at the surface.

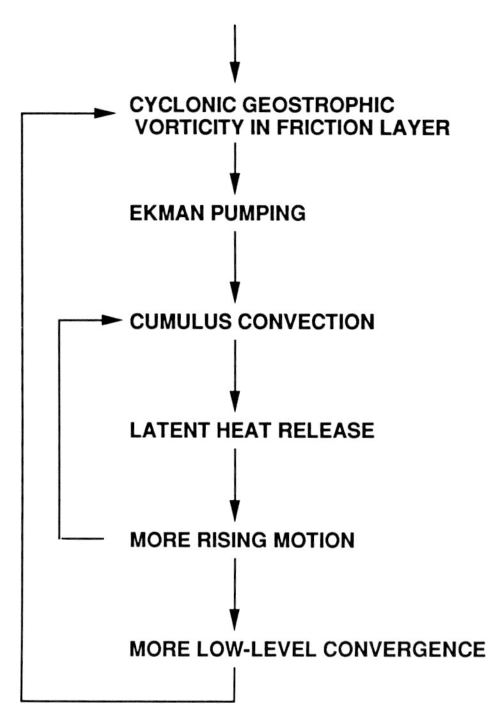

Figure 1.11 Schematic diagram of CISK—conditional instability of the second kind.

at the surface, and falling pressures at the surface. Convergence acting on the existing cyclonic vorticity produces more vorticity, which induces more Ekman pumping, and so on.

In a similar manner, the frictional inflow of air toward the center of a warm-core, maritime, surface low results in diabatic heating owing to turbulent sensible heat transport upward from the sea surface. This diabatic heating is concentrated around the low and leads to additional upward motion, surface convergence, a drop in surface pressure, and stronger surface winds. The sensible flux is thus enhanced, and cyclogenesis proceeds. The latent heat flux is also enhanced and provides a source of water vapor for cumulus convection, which redistributes the sensible and latent heat vertically. Tropical cyclogenesis has been attributed to this "air–sea interaction" instability (by Emanuel and Rotunno) (Fig. 1.12) and to CISK. It may also be responsible for some oceanic cyclogenesis in the midlatitudes. In the case of the latter, the transport of sensible heat (temperature) is dominant, while in the case of the former, the transport of latent heat (water vapor) is dominant. It is not necessary for the atmosphere to have any prior conditional instability for this process to occur.

Downslope or upslope motions and differential vorticity advection. Downslope motion in the lee of the Colorado Rockies, in advance of a ridge aloft, usually results in the formation of a lee trough or low; the effect of downslope warming must be overwhelming the effect of vorticity advection becoming more anticyclonic with height. However, it is unusual to find cases of lee cyclogenesis in which the lee cyclone moves away from the mountains in the absence of

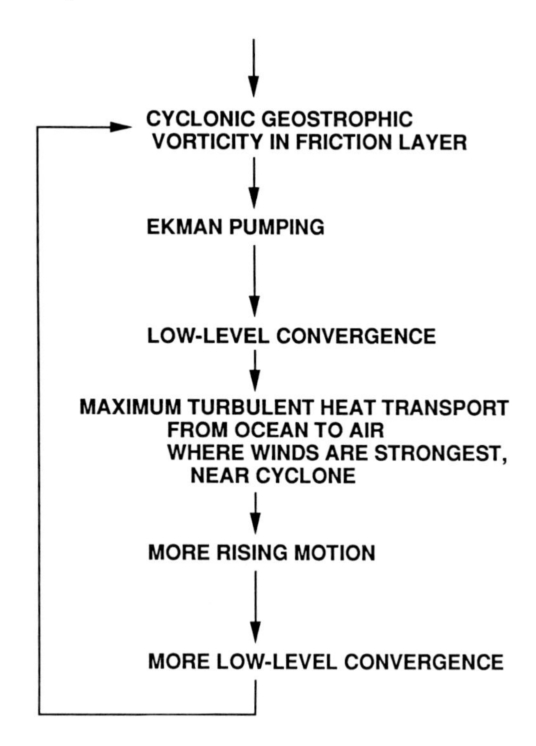

Figure 1.12 Schematic diagram of air–sea interaction instability.

(vorticity advection becoming more cyclonic with height in association with) a mobile upper-level trough. If the mountains were not present, would a cyclone form? This raises the following question: Is lee cyclogenesis just ordinary cyclogenesis modified by orography, or is it indeed special? We leave this as an unanswered question.

1.1.8. Climatology of Cyclogenesis and Anticyclogenesis at the Surface over North America

We turn our attention to the synoptic climatology of cyclogenesis and anticyclogenesis over North America and the surrounding bodies of water in

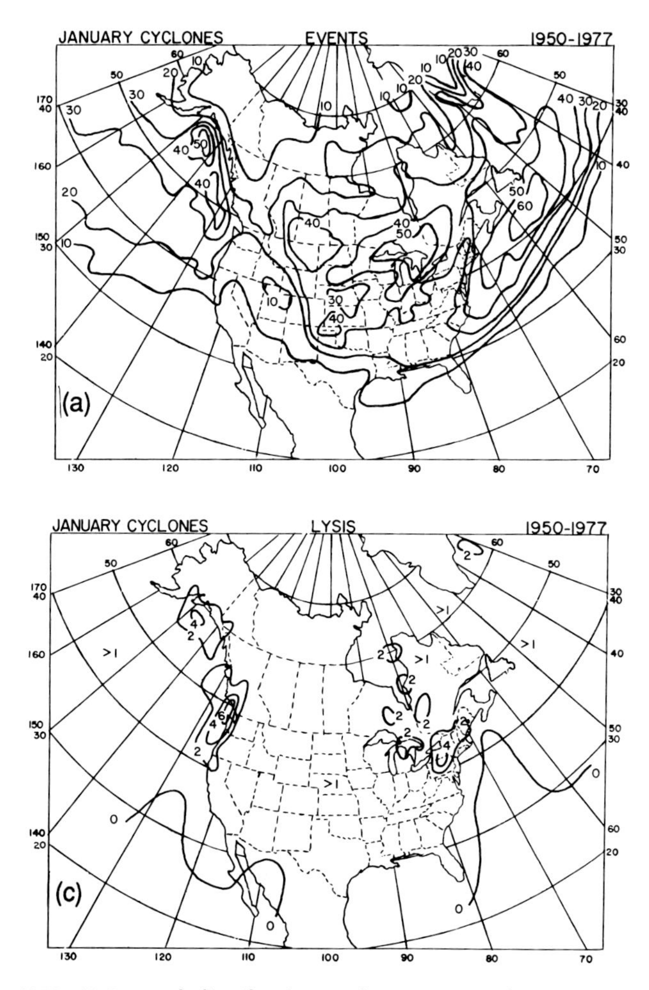

Figure 1.13 1950–1977 areal distributions of (a) events, (b) genesis, (c) lysis (i.e., demise of; opposite of genesis), and (d) relative variability with preferred propagation tracks superimposed for January cyclones. Values represent 28-year totals. Areas in (b) and (c) in which individual quadrangles contain a frequency greater than 1, but do not form centers defined by three contiguous quadrangles

order to illustrate further the most common mechanisms of surface cyclogenesis and anticyclogenesis. (The author's intent is *not* to discuss the climatology of the world in detail. The author apologizes to readers from Europe and Asia and the Southern Hemisphere for appearing to be so provincial!) A *surface* climatology is presented here first, because we all experience surface weather; a discussion of upper-level climatology is delayed until later.

Cyclogenesis at the surface. There are three regions of relatively frequent cyclogenesis in the United States during the winter (Fig. 1.13):

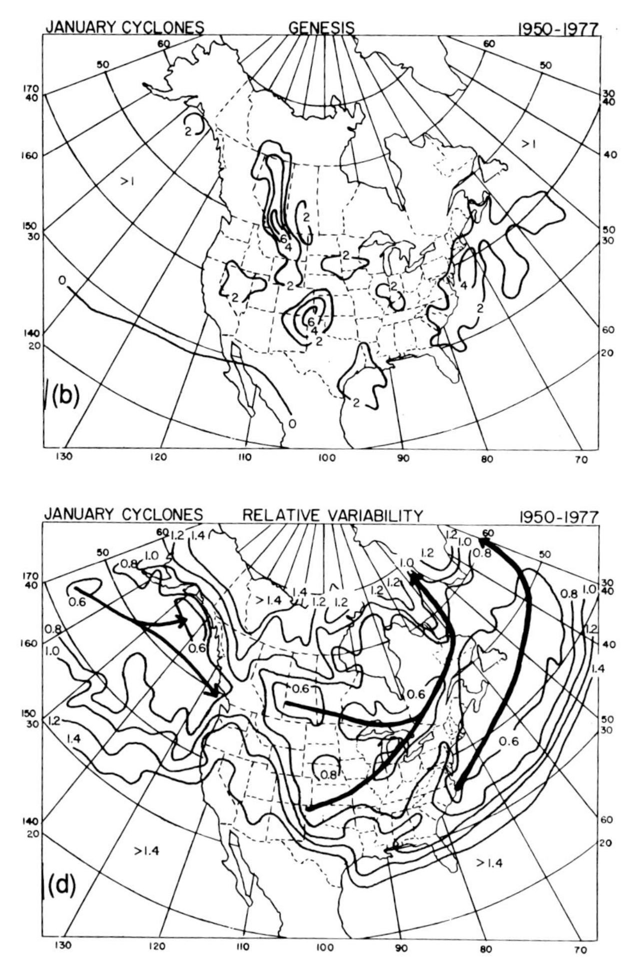

Figure 1.13 (*cont.*)
are designated by ">1." Relative variability represents the mean absolute difference of individual year frequencies from the 28-year mean divided by the mean (from Zishka and Smith, 1980). (Courtesy of the American Meteorological Society)

1. The lee of the northern Rocky Mountains from Alberta through Montana (where "Alberta clippers" are spawned);
2. The lee of the southern Rocky Mountains, including southeastern Colorado, western Kansas, northeastern New Mexico, the Oklahoma Panhandle, and the northern Texas Panhandle (where "Colorado lows" are born);
3. Offshore from the Mid-Atlantic states and southern New England (where "Hatteras lows" form).

Although lee cyclogenesis often occurs in the former two areas, the "Alberta clippers," e.g., may actually be Pacific cyclones that have been weakened on the upslope side of the cordillera of western Canada. The latter area (3), is located over relatively warm water, and surface lows in this region may be associated with diabatic heating, strong warm advection in association

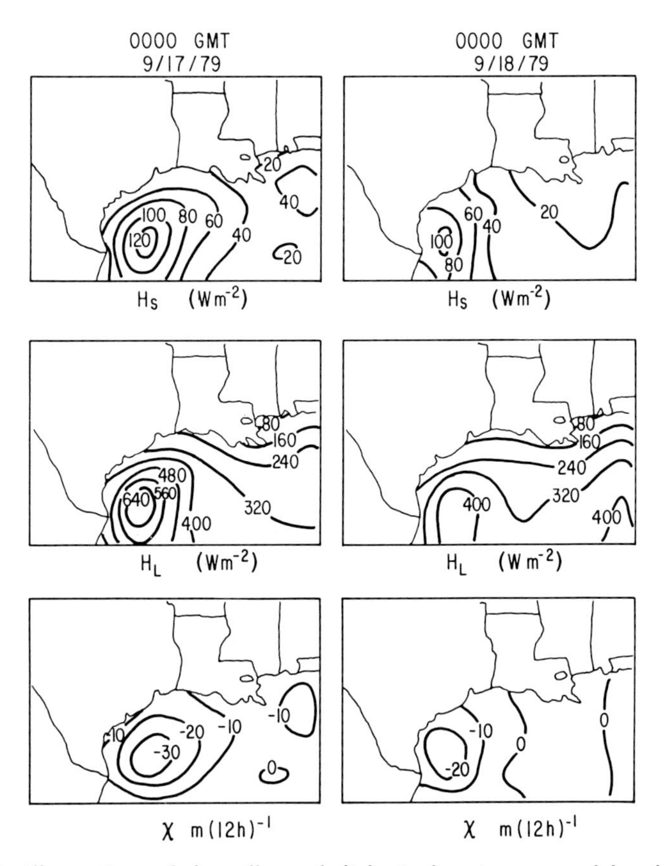

Figure 1.14 Illustration of the effect of diabatic heating caused by the sensible heat flux from a warm ocean (Gulf of Mexico) to colder overlying air. Sensible heat flux H_S (top), latent heat flux H_L (middle), and 1000-mb height tendency ($\chi = \partial z/\partial t$) associated with sensible heating (bottom) for 0000 UTC September 17 (18), 1979, left (right). Units are W m^{-2} except m (12 h)$^{-1}$ for the 1000-mb height tendency (from Bosart, 1984). (Courtesy of the American Meteorological Society)

with air that has been warmed by the sea surface, the reduction in static stability encountered by upper-level troughs that during the winter move off the relatively cold continent and over the warm ocean, or a combination of all the aforementioned. "Hatteras lows" usually become deeper than the Colorado lee cyclones during the winter.

Another area of relatively frequent cyclogenesis is in the western Gulf of Mexico, for perhaps the same reason cyclones form off the Mid-Atlantic coast. In addition, it should be noted that the shapes of the western Gulf of Mexico and Mid-Atlantic coastal regions are similar; they are both convex. Thus, diabatic heating from the relatively warm ocean surface is a relative maximum when there is offshore flow of cold air, and hence χ_{ω_Q} is negative (Fig. 1.14). During the summer the regions of most frequent cyclogenesis move northward, except for the Colorado-low region, which nearly disappears (Fig. 1.15) as the westerlies weaken and move poleward.

Anticyclogenesis at the surface. The area of most frequent anticyclogenesis during the winter stretches from the South Plains of Texas northward into western Kansas (Fig. 1.16). These anticyclones form in response to strong cold advection at the surface west of developing Colorado lows. The Northwestern States is another area of frequent anticyclogenesis, though not as frequent as the former. The anticyclones which form here are generated in part by the diabatic cooling of relatively warm Pacific air brought over the mountains onto snow and ice fields, and by cold advection west of developing Alberta clippers. The highs in the High Plains are usually shallow, fast moving, and originate poleward of, or on the "cold" side of, the westerlies. As the "cold" or polar highs plunge equatorward, subsidence aloft makes the layer of shallow, cold air even shallower, and surface heating can ultimately result in the mixing of the warm air aloft with the air mass below, resulting in the death of the surface anticyclone.

Sometimes the "cold" highs later become transformed into "warm" highs owing to subsidence-induced warming forced by vorticity advection becoming more anticyclonic with height downstream from a strong upper-level ridge. The "warm" anticyclones are usually deep, slow moving, and appear on the "warm" side of the westerlies aloft.

During the summer, anticyclogenesis occurs most frequently in southern Alberta (Fig. 1.17). This area is often influenced by cold advection west of developing Alberta cyclones.

In general there are twice as many cyclones as anticyclones, a happy state of affairs for storm aficionadoes! (The reader is challenged to explain this curious statistic.)

Cyclolysis and anticyclolysis at the surface. Cyclones most frequently decay in the Gulf of Alaska and off the northwest coast of the United States and Canada during the winter (Fig. 1.13). They also often decay in the interior of the northeastern part of the United States. Cyclones in the summer frequently decay in the Gulf of Alaska and along the Pacific coast of Canada, and between the extreme northeastern portion of Canada and southern Greenland (Fig. 1.15).

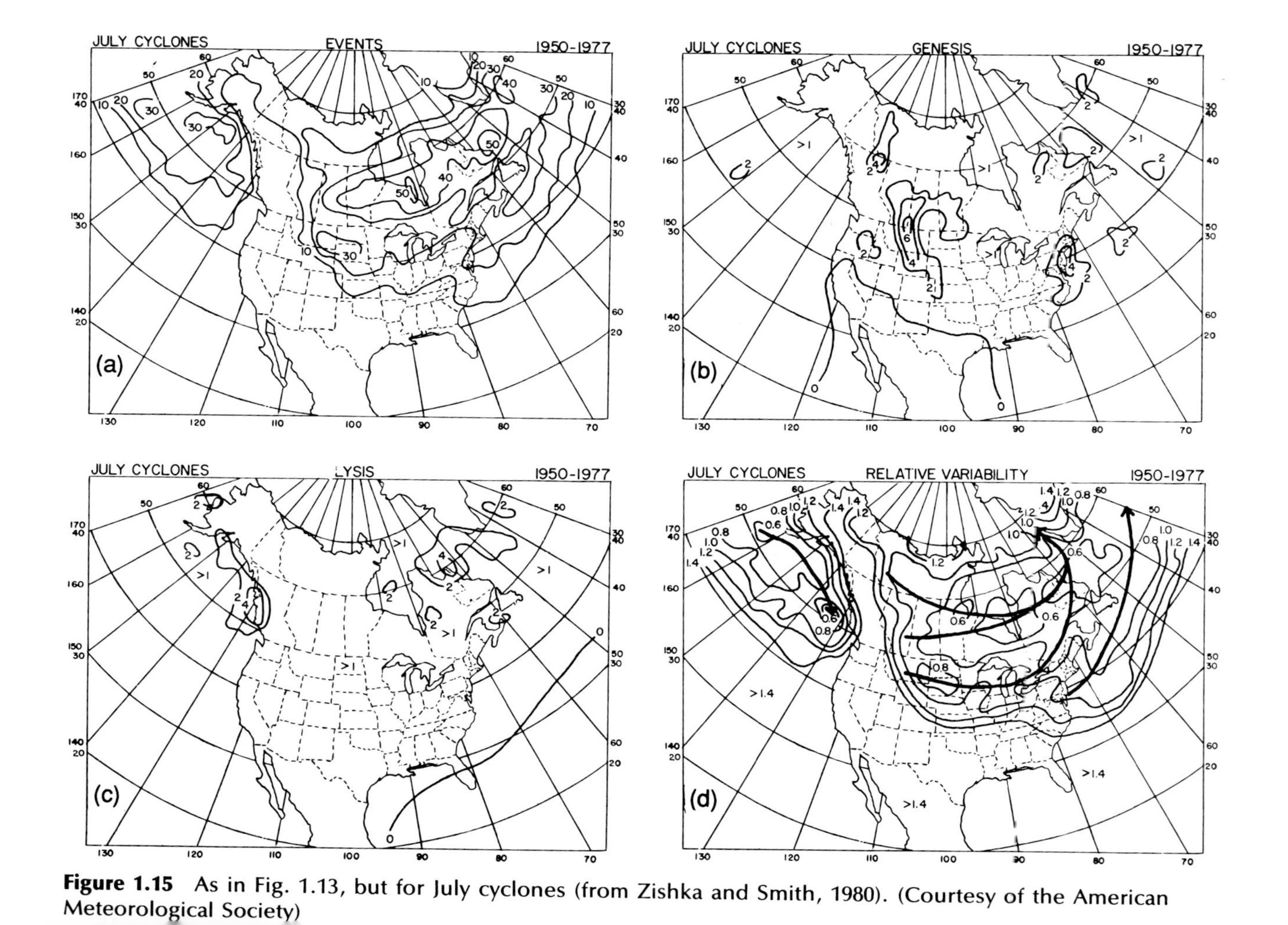

Figure 1.15 As in Fig. 1.13, but for July cyclones (from Zishka and Smith, 1980). (Courtesy of the American Meteorological Society)

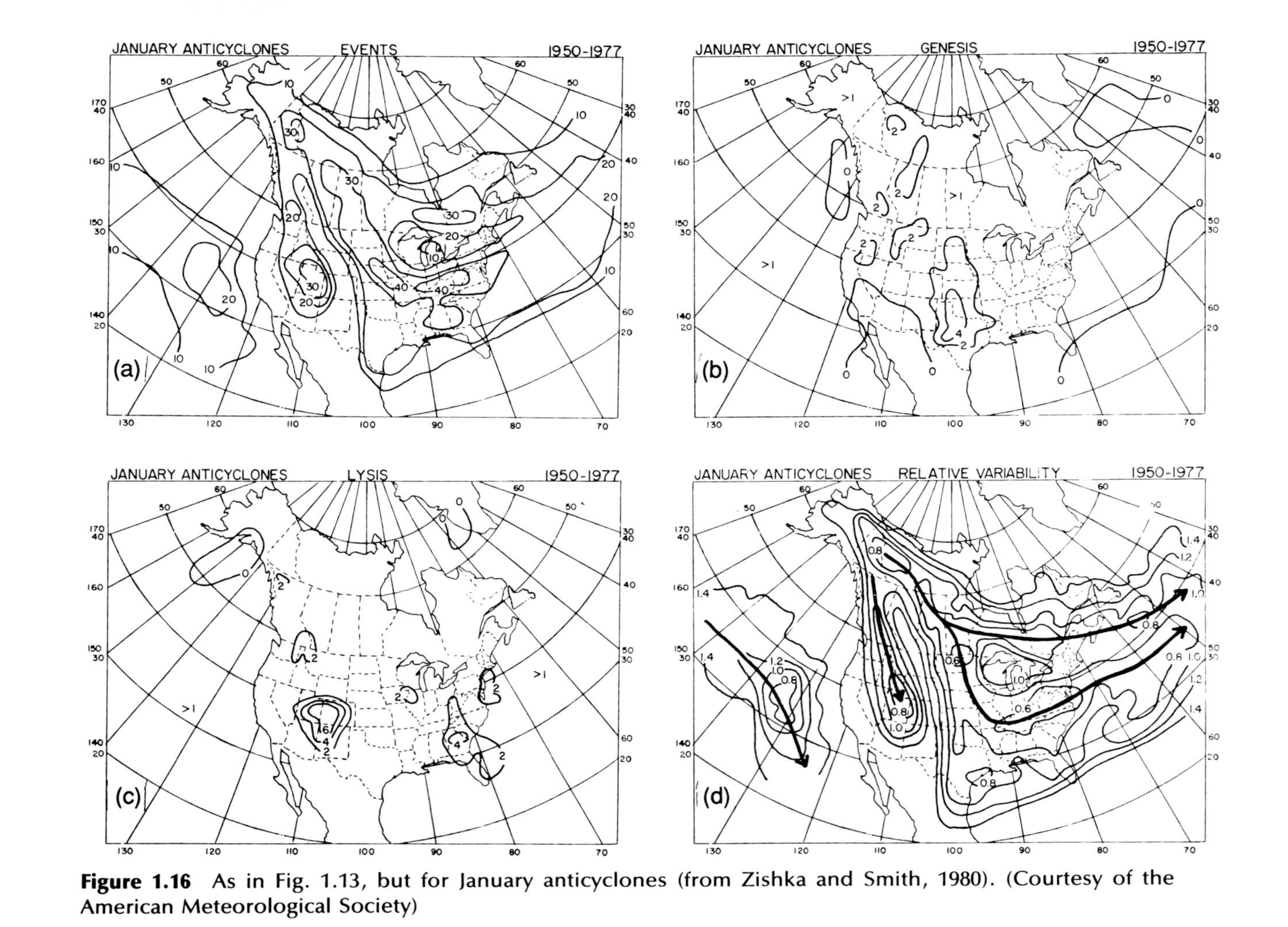

Figure 1.16 As in Fig. 1.13, but for January anticyclones (from Zishka and Smith, 1980). (Courtesy of the American Meteorological Society)

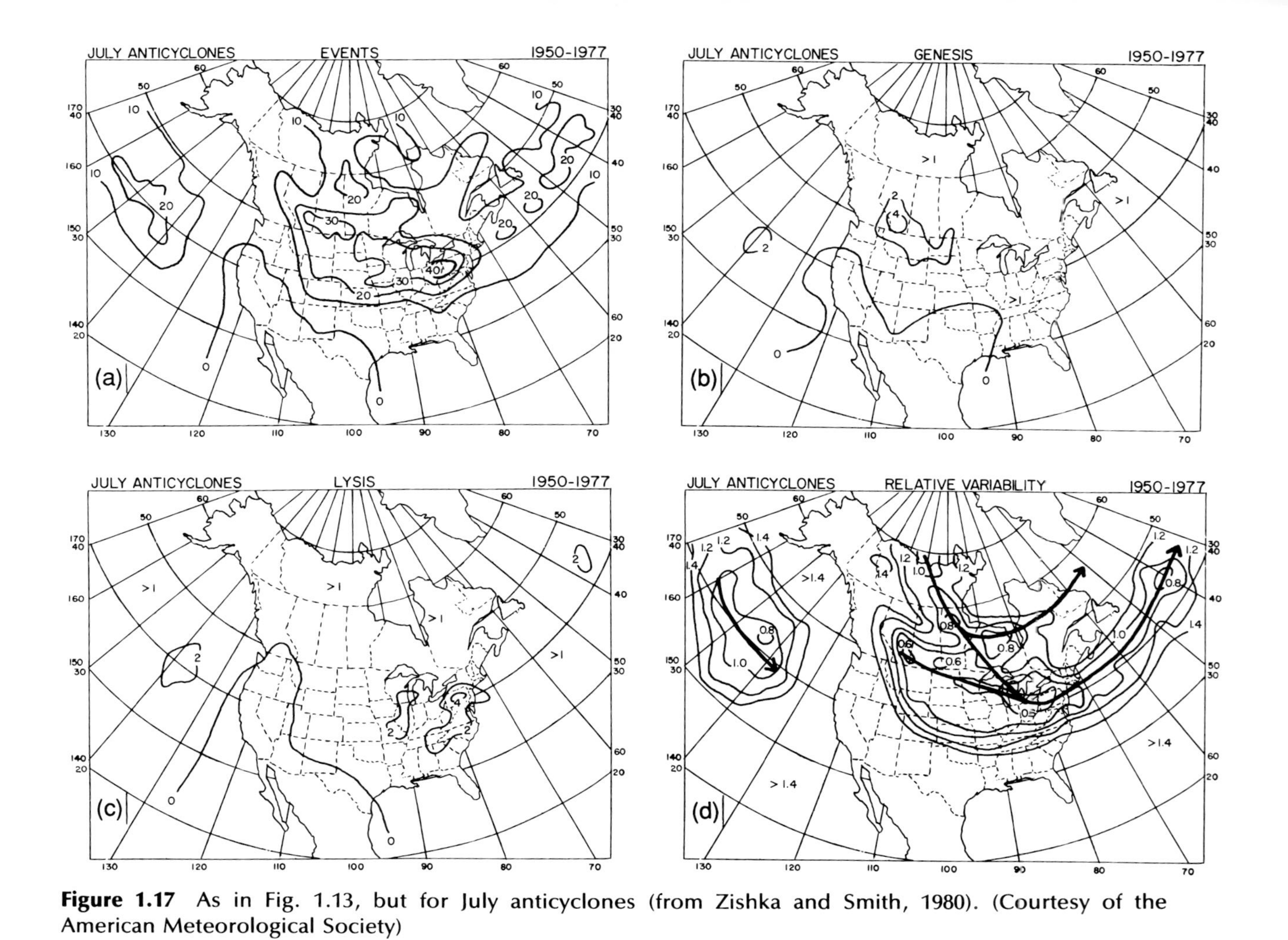

Figure 1.17 As in Fig. 1.13, but for July anticyclones (from Zishka and Smith, 1980). (Courtesy of the American Meteorological Society)

Anticyclones during the winter tend to decay most often near the "four-corners" region of Utah, Colorado, Arizona, and New Mexico, and over the southeastern portion of the United States (Fig. 1.16). During the summer anticyclones most frequently decay around the Appalachians (Fig. 1.17).

1.2. THE MOVEMENT OF SURFACE PRESSURE SYSTEMS

In Vol. I we discussed the kinematics of pressure systems: A surface high-pressure area, an anticyclone, moves from a region of pressure falls toward a region of pressure rises, that is, in the direction of the isallobaric gradient. In Sec. 1 of this chapter, we argued that a region of pressure falls at the surface is also a region of height falls near the ground and convergence. (Vorticity advection is neglected because surface systems are usually relatively circular and thus are associated with little vorticity advection.) Similarly, a region of pressure rises at the surface is also a region of height rises near the ground and divergence.

On a level surface a region of convergence is accompanied by rising motion aloft owing to continuity, and a region of divergence is accompanied by sinking motion aloft. Therefore the following rules govern the motion of surface pressure systems:

1. A surface anticyclone moves away from a region of rising motion toward a region of sinking motion [Fig. 1.18 (top)].
2. A surface cyclone moves away from a region of sinking motion toward a region of rising motion [Fig. 1.18 (bottom)].

1.2.1. The Effects of the Forcing Functions in the Quasigeostrophic ω Equation

According to quasigeostrophic theory, surface anticyclones on level surfaces have the following motion characteristics:

1. From regions of geostrophic absolute vorticity advection becoming more cyclonic with height toward regions of geostrophic absolute vorticity advection becoming more anticyclonic with height;
2. From regions of geostrophic warm advection toward regions of geostrophic cold advection;
3. From regions of diabatic heating toward regions of diabatic cooling.

By similar reasoning, cyclones on level surfaces have the following motion characteristics:

1. From regions of geostrophic absolute vorticity advection becoming more anticyclonic with height toward regions of geostrophic absolute vorticity advection becoming more cyclonic with height;
2. From regions of geostrophic cold advection toward regions of geostrophic warm advection;
3. From regions of diabatic cooling toward regions of diabatic heating.

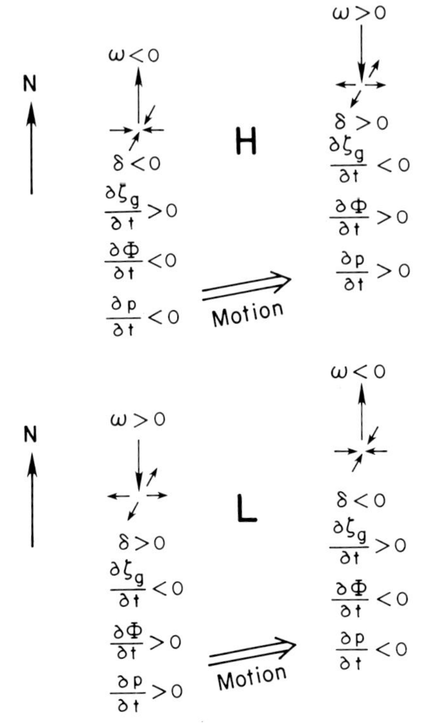

Figure 1.18 How the motion of (top) high-pressure areas and (bottom) low-pressure areas is related to the vertical-motion field: Rising motion is associated with surface convergence, a cyclonic tendency in surface geostrophic vorticity, a drop in geopotential height at the lowest pressure level, and a drop in surface pressure; sinking motion is associated with surface divergence, an anticyclonic tendency in surface geostrophic vorticity, an increase in geopotential height at the lowest pressure level, and an increase in surface pressure.

The actual movement of cyclones and anticyclones over a level surface is determined by the sum of these effects, which in turn is controlled by the geometry of the height field.

There is usually cold advection equatorward and eastward of, and warm advection poleward and westward of, surface anticyclones. Thus the effect of temperature advection is usually to "move" surface anticyclones toward the southeast (in the Northern Hemisphere). Since surface anticyclones often form upstream from a trough and downstream from a ridge in the westerlies, the flow aloft is also toward the southeast (Fig. 1.19).

There is usually warm advection poleward and eastward of, and cold advection equatorward and westward of, surface cyclones. Thus the effect of temperature advection is usually to "move" surface cyclones toward the northeast (in the Northern Hemisphere). Since surface cyclones often form downstream from a trough and upstream from a ridge in the westerlies, the flow aloft is also toward the northeast (Fig. 1.19).

The effects of differential vorticity advection cannot be as easily generalized. If the surface low is co-located with the region of maximum vorticity advection becoming more cyclonic with height, then the cyclone will deepen, but will not move as a result of differential vorticity advection alone. The effect of differential vorticity advection is to retard (hasten) the eastward component of motion if the region of maximum vorticity advection becoming more cyclonic with height is located upstream (downstream) aloft from the surface low.

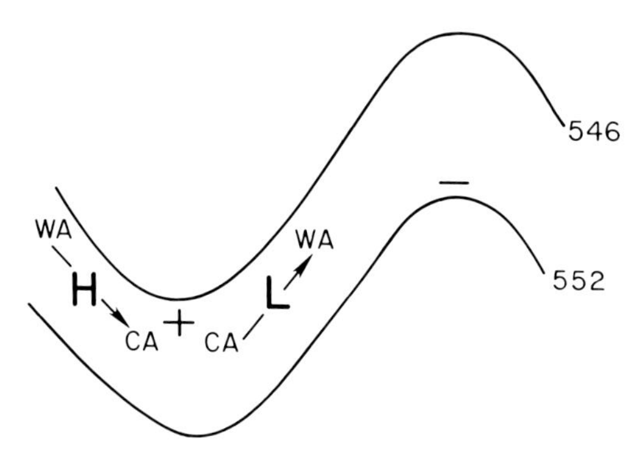

Figure 1.19 Illustration of the relationship between the upper-level flow pattern and the motion of surface systems. 500-mb height contours in dam (solid lines); surface high (H) and low (L); areas of low-level warm advection (WA) and cold advection (CA). Location of maximum cyclonic geostrophic vorticity (+) and maximum anticyclonic geostrophic vorticity (−) at 500 mb. Approximate direction of motion of lows and highs indicated by arrows.

It is apparent from the geometry of the three-dimensional height field associated with a typical cyclone or anticyclone that the apparent motions of surface cyclones and anticyclones to a large extent follow the flow aloft (Fig. 1.19). The tendency of surface pressure systems to follow the flow aloft is called *steering*. However, surface cyclones and anticyclones are not solid objects being carried along by the flow aloft, and their movement is really by propagation of the low- and high-pressure centers.

1.2.2. The Effects of Orography

Orography can also have a significant effect on the motion of surface systems. Suppose that a cyclone is on a surface whose elevation slopes upward toward the west (Fig. 1.20). There is therefore upslope motion and pressure rises poleward of the cyclone, and downslope motion and pressure falls equatorward of the cyclone. In the absence of any other forcing functions, the cyclone moves equatorward. Using the same analysis technique, we find that an anticyclone on the same surface also moves equatorward (Fig. 1.21). In general, pressure systems at the surface in the Northern Hemisphere move with higher elevation to their right.

This behavior is analogous to that of a "Rossby wave," with the sloping surface playing the role of the variation of f with lattiude: Consider an air parcel in the Northern Hemisphere that is given a northward push (Fig. 1.22). Suppose the atmosphere is barotropic, so that geostrophic absolute vorticity is conserved. Because f at the parcel increases as it moves northward, the parcel's geostrophic vorticity must decrease. The parcel turns anticyclonically to the right. In the absence of friction, the parcel has enough momentum to turn southward eventually. Now f at the parcel decreases, and hence the parcel's geostrophic vorticity must increase. The parcel turns cyclonically to

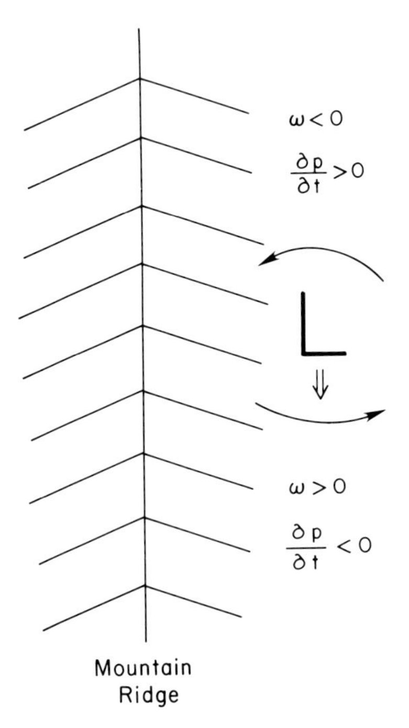

Figure 1.20 Relationship between surface low-pressure area (L), the cyclonic flow around it (arrows) over the sloping surface of a mountain range, and the vertical motion field at the surface (ω) and the pressure-tendency field at the surface ($\partial p/\partial t$). In this example the cyclone propagates southward.

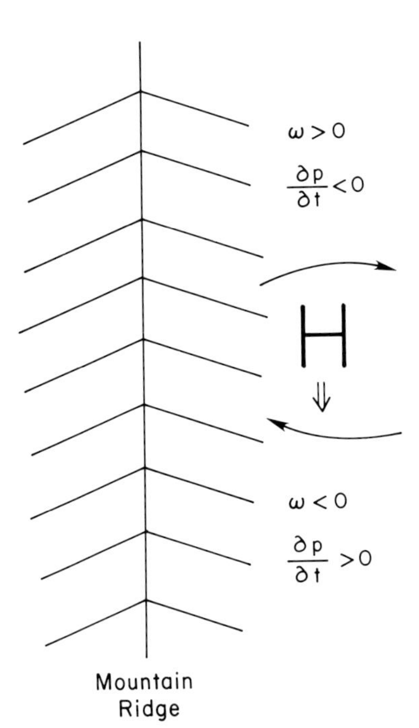

Figure 1.21 As in Fig. 1.20, but for a surface high-pressure area (H).

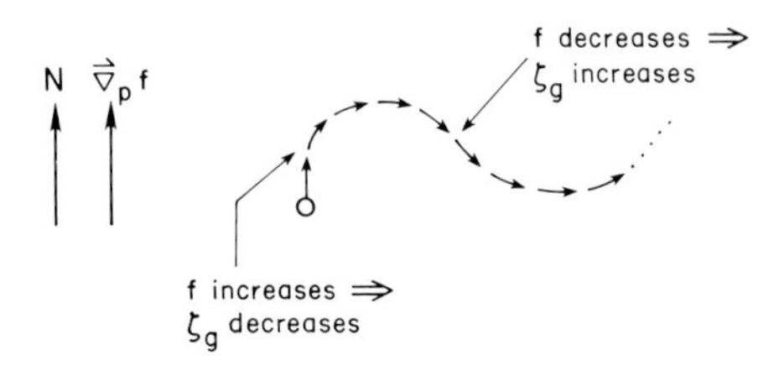

Figure 1.22 Illustration of the wavelike trajectory of an air parcel in the Northern Hemisphere (in a barotropic, frictionless atmosphere) that is given an initial push toward the north.

the left and eventually turns northward, and so on. Rossby waves will be discussed in more detail later on.

Now, consider an air column that is given a push up a sloping surface (Fig. 1.23). The air column shrinks, and hence its absolute vorticity must decrease. An air parcel in the column turns anticyclonically (if changes in f are ignored) to the right. In the absence of friction, the parcel has enough momentum to turn and move in the downslope direction. Now the air column is stretched, and hence its vorticity must increase. The parcel turns cyclonically to the left and eventually turns upslope, and so on. The former wavelike motion is a *Rossby wave*; the latter is a *topographic Rossby wave.*

Colorado lows in the lee of the Rockies, for example, usually move southward initially, since the elevation of the ground increases toward the west. As warm advection becomes significant east of the low, the cyclone will move towards the southeast, and finally eastward and northeastward as the effects of orography diminish and the quasigeostrophic forcing functions become dominant. Similarly, anticyclones in the lee of the Rockies "smell the Gulf" and initially plunge southward.

Climatological cyclone and anticyclone tracks for January and July are shown in Figs. 1.13, 1.15, 1.16, and 1.17. Cyclones (the "Aleutian lows") enter British Columbia in January from the Pacific and may redevelop into Alberta cyclones. The July pattern is shifted northward. In July, the tracks of Alberta and Colorado anticyclones seen in the January charts are shifted northward.

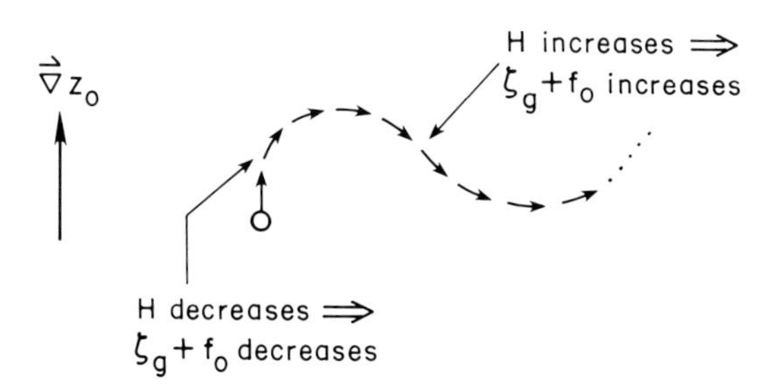

Figure 1.23 As in Fig. 1.22, but for flow at the surface, whose height (z_0) increases towards the north. The depth of the air parcel is H. In this case we neglect variations in f.

1.2.3. Some Illustrations of the Application of Quasigeostrophic Theory

Descartes, of Cartesian-coordinates fame, left out elements of a proof in his *Geometry* so that the reader (read "student" here) could have the fun of discovering it for him(her)self.[1] As preparation for working out the problems at the end of this chapter, several illustrative problems (à la Descartes) and their solutions are now presented. Although the situations presented are physically plausible, they are not commonly observed, and might even appear to the keen observer to be bizarre. The intent here is to present cases for which the diagnosis is not known from experience; hence, the student must apply quasigeostrophic reasoning completely, and not rely on any preconceived notions. The student is then referred to the problems at the end of this chapter to satisfy the Cartesian desire for self-fulfillment.

The convergence conundrum. Consider the 500-mb height field (in the Northern Hemisphere) sketched in Fig. 1.24. The geostrophic wind at the surface, which is level, is calm. Neglect friction, diabatic heating, and water vapor. At what locations (number designation) is there surface convergence?

Surface convergence is found under a region of rising motion, owing to continuity and an $\omega = 0$ kinematic lower boundary condition. We can use the quasigeostrophic ω equation to diagnose ω qualitatively (the *sign* of ω). We neglect the friction and diabatic heating forcing functions.

First, let us consider the temperature-advection forcing function: If the geostrophic wind is calm at the surface, then the pressure is a constant there, and hence the height of the pressure surface at the surface is also uniform. Then the thickness contours that depict the difference between the 500-mb height and the surface height are parallel to the 500-mb height contours. Since the mean temperature in the 500-mb–to–surface layer is proportional to its thickness, we infer that isotherms are in a vertically averaged sense parallel to the vertically averaged geostrophic wind field, and hence there is no mean temperature advection.

Alas, the only forcing left to consider is that of differential vorticity advection. Since the horizontal scale of the height field is relatively small, we can neglect geostrophic advection of Earth's vorticity. At the surface there is no geostrophic vorticity advection because the geostrophic wind is calm. Since rising motion is associated with vorticity advection becoming more cyclonic with height, rising motion in this case is associated with regions of cyclonic vorticity advection at 500 mb. Look at the figure carefully to see that minima in geostrophic vorticity, owing mostly to curvature, are located at 1 and 5. Hence

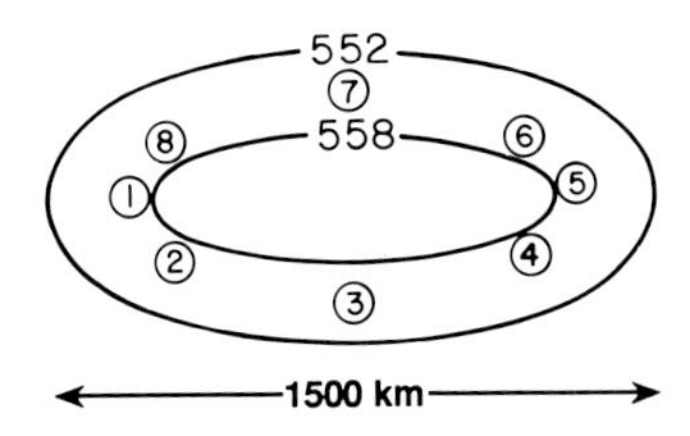

Figure 1.24 500-mb height contours in dam (solid lines). Numbers refer to specific locations discussed in the text.

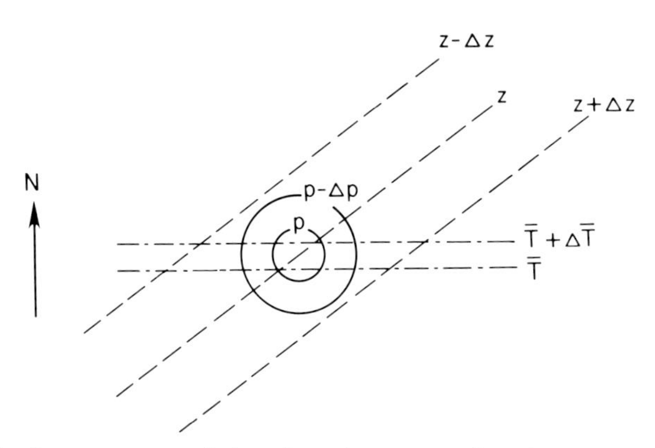

Figure 1.25 Surface pressure (p) reduced to some height z (solid lines); mean temperature ($\bar{T}$) in the 1000–500 mb layer (dashed–dotted lines); contours of elevation above sea level (dashed lines).

there is CVA at 2 and 6, which are upstream, with respect to the geostrophic wind, from the minima. (At 3 and 7 there is no vorticity advection because $\nabla_p \zeta_g = 0$.) We therefore infer that there is surface convergence at 2 and 6.

The quasigeostrophic query. Consider the pattern of isobars (reduced to height z) and of mean temperature in the 1000–500 mb layer shown in Fig. 1.25. The surface topography is as depicted by the contours of elevation above sea level. Neglect diabatic heating and friction. Which way will the circular pressure pattern move?

After taking a deep breath and avoiding sudden panic, we note that the closed height contours depict a high. The high will move from a region of height falls toward a region of height rises. Surface height falls and rises are associated with local increases and local decreases in geostrophic vorticity, respectively. Local increases and local decreases in geostrophic vorticity are associated (*cf.* the quasigeostrophic vorticity equation) with convergence and divergence, respectively. Convergence and divergence are associated with rising and sinking motion aloft. We are thus led to consider the quasi-geostrophic ω equation again: The surface high will move from a region of rising motion toward a region of sinking motion.

Since diabatic heating and friction are neglected, we need to determine qualitatively only the differential vorticity advection and temperature-advection forcing functions. Let us evaluate the latter first, because in this case it is the easier one. East of the high at z the geostrophic wind is northerly; hence there is warm advection and rising motion. West of the high the geostrophic wind is southerly; hence there is cold advection and sinking motion. Therefore temperature advection by itself makes the high move westward.

We now evaluate the differential vorticity advection forcing function aloft. First we consider the geostrophic advection of geostrophic vorticity. At z, there is no geostrophic vorticity advection: Since the height contours are circular, the isopleths of geostrophic vorticity are parallel to the height contours. There is therefore a contribution to rising motion and sinking motion aloft where there is CVA and AVA aloft, respectively. What is the vorticity–advection pattern aloft? We add the thermal-wind vector, which in this case is easterly, to the surface geostrophic wind vectors, to find that there is a ridge in easterly flow aloft. Hence there is CVA aloft east of the ridge (and surface high), and AVA aloft west of the ridge. What about the geostrophic advection of Earth's vorticity? At the surface there is a maximum in CVA east of the high in northerly geostrophic flow, and a maximum in AVA west of the high in southerly geostrophic flow. Aloft the northerly winds at the surface east of the high veer to northeasterly (indicative of warm advection), while the southerly winds west of the high back to southeasterly (indicative of cold advection); hence the CVA and AVA aloft are weaker. Therefore east and west of the high Earth's vorticity advection becomes less cyclonic and less anticyclonic with height, respectively. Hence, the differential advection of geostrophic vorticity and the differential advection Earth's vorticity are of opposite sign both east and west of the high. North and south of the high there is no advection of Earth's vorticity, owing to the zonal winds. It is therefore not clear how the differential vorticity advection forcing function would contribute toward motion, unless we actually made a quantitative calculation. It is likely, however, that the scale of the high is not too large (information on the dimension of the high is not given), so that relative vorticity advection dominates over Earth's vorticity advection.

Finally, the effects of upslope and downslope motion must be considered. Maximum downslope motion and height falls occur southwest of the high; maximum upslope motion and height rises occur northeast of the high. The effect of topography is to make the high track northeastward.

(Determining the *net* motion of the circular pressure pattern demands a *quantitative* calculation of the components of motion from each effect, and cannot be done without further information.)

The vorticity vexation. Consider the 500-mb height field shown in Fig. 1.26. The atmosphere is equivalent barotropic, adiabatic, and inviscid. At what locations (letter designation) does the height rise for zonal wavelength $\ll 6000$ km? for zonal wavelength $\gg 6000$ km?

To find χ we use the quasigeostrophic height-tendency equation. Since the atmosphere is equivalent barotropic, there is no temperature advection, and hence the differential temperature advection forcing function is zero. The differential diabatic heating forcing function and friction forcing function are also zero.

We need to consider only the vorticity-advection forcing function. (1) For zonal wavelength $\ll 6000$ km, geostrophic vorticity advection in a wavetrain dominates over Earth's vorticity advection. Relative maxima in cyclonic vorticity are located at D and C; weaker relative maxima in anticyclonic vorticity are located at A and F. Therefore AVA and height rises are found upstream from D at M, upstream from C at L, downstream from A at J, and downstream from F at O.

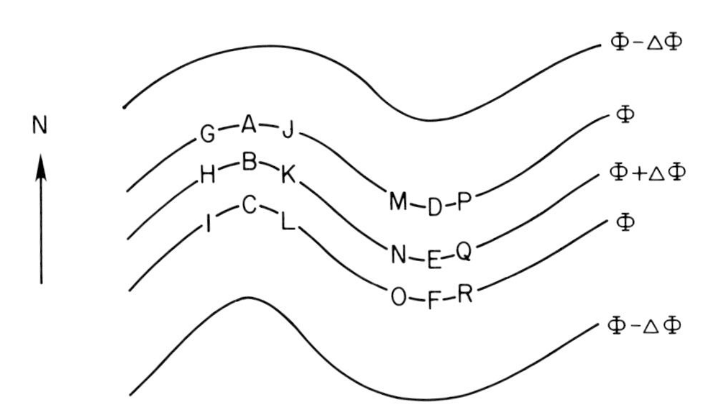

Figure 1.26 500-mb geopotential height field (Φ) given by solid lines; letters refer to locations discussed in the text.

(2) For zonal wavelength ≫6000 km, Earth's vorticity advection in a wavetrain dominates over geostrophic vorticity advection. Therefore AVA and height rises are located wherever there is a southerly component to the geostrophic wind (i.e., at G, P, O, and L).

1.2.4. An Analytic Model for the Quasigeostrophic Diagnosis of Synoptic-Scale Systems

J. M. Austin in 1947 pointed out that surface cyclones (like the politically conservative) often move to the right of the flow and thermal wind, above 850 mb, not exactly along with it. However, on some occasions surface cyclones move to the left of the flow. In a *quantitative* sense, what controls the movement of surface systems? The motion of surface systems can be easily studied for a wide range of parameters without truncation error, and without great expense, with an analytic model. We will use an analytic model first described by F. Sanders in 1971. Although this quasigeostrophic model is being introduced here so that we can discuss more quantitatively the motion of surface systems, it will be used later on again, especially in the context of baroclinic instability. It is ironic that although quasigeostrophic theory was formulated in the late 1940s, quasigeostrophic diagnostic studies of *typical* midlatitude systems are not often done. Most recent diagnostic case studies have focused on extreme events. We offer the analytic model as a tool for diagnosing the typical system.

Specification of the height field. The simple 1000-mb geopotential height field

$$\Phi(x, y, p = 1000\,\text{mb}) = \hat{\Phi}_0 \cos\frac{2\pi}{L}(x + \lambda)\cos\frac{2\pi}{L}y \tag{1.2.1}$$

is an infinite checkerboard of highs and lows, a small portion of which is shown in Fig. 1.27, and the temperature field (Fig. 1.28)

$$T(x,y,p) = T_m(p) - \left(1 - \alpha\ln\frac{1000}{p}\right)\left(ay + \hat{T}\cos\frac{2\pi}{L}x\cos\frac{2\pi}{L}y\right) \tag{1.2.2}$$

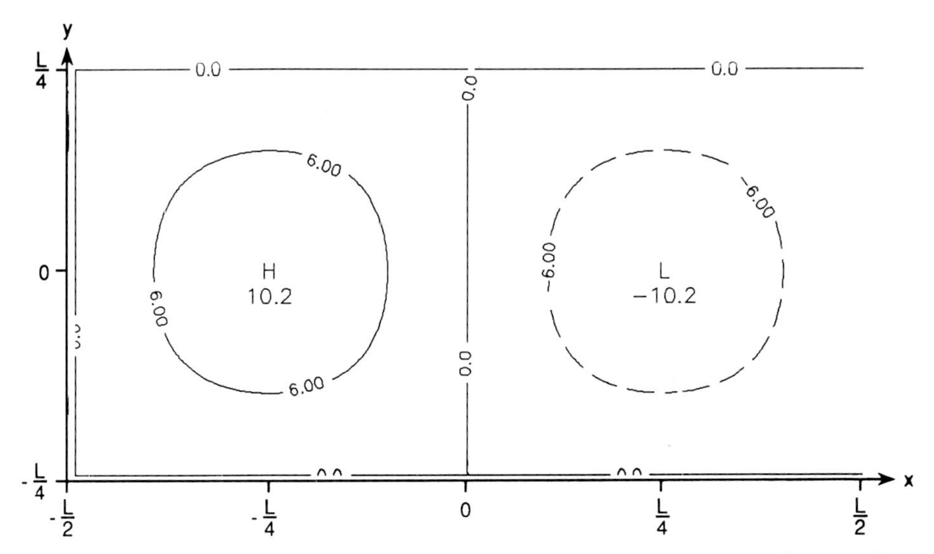

Figure 1.27 Isopleths of 1000-mb height at intervals of 6 dam in Sanders' analytic model. $\hat{\Phi}_0 = 1020\ \mathrm{m^2\,s^{-2}}$, and $\lambda = L/4$. The abscissa is the x direction; the ordinate is the y direction.

is an infinite checkerboard of warm and cold perturbation centers $-[\hat{T}\cos(2\pi/L)x\cos(2\pi/L)y]$ superimposed upon a field of uniform meridional temperature gradient $(-ay)$. The x direction points toward the east, and the y direction points toward the north. The parameter L is the (isotropic) horizontal wavelength, and λ is the phase lag of the 1000-mb temperature field

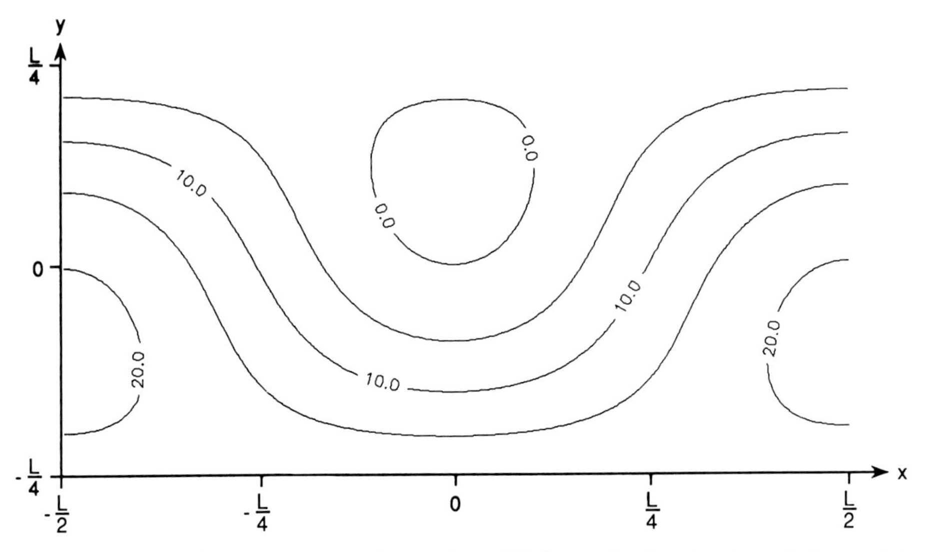

Figure 1.28 Isotherms at 1000 mb in °C (solid lines) in Sanders' analytic model. $a = 1.0 \times 10^{-5}\,°\mathrm{C\,m^{-1}}$, $\hat{T} = 10°\mathrm{C}$, and $L = 3500$ km. The abscissa is the x direction; the ordinate is the y direction.

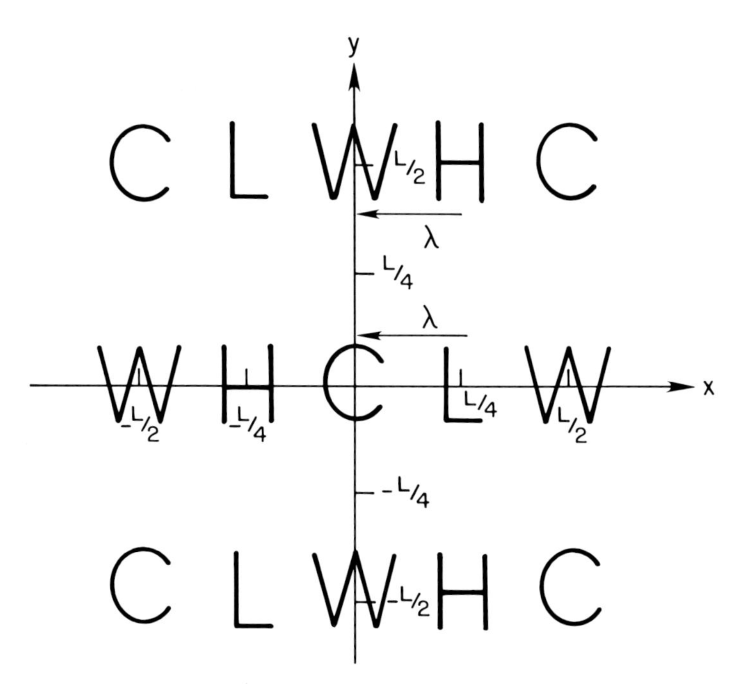

Figure 1.29 Locations of high (H) and low (L) centers of 1000-mb geopotential height, and warm (W) and cold (C) centers at 1000 mb in Sanders' analytic model. λ indicates the phase lag of the temperature field relative to the geopotential field.

relative to the geopotential field (Fig. 1.29). The function $T_m(p)$ is the "standard-atmosphere" profile and $1-\alpha \ln(1000/p)$ represents the vertical variation of the amplitude of the horizontal temperature field. The value of α is chosen so that at the tropopause, where $p = p_{\text{trop}}$,

$$1 - \alpha \ln \frac{1000}{p_{\text{trop}}} = 0. \tag{1.2.3}$$

Since the atmosphere is hydrostatic,

$$d\Phi = \frac{-RT}{p}\, dp. \tag{1.2.4}$$

Integrating Eq. (1.2.4), we find that

$$\int_{\Phi(p=1000\text{ mb})}^{\Phi(p)} d\Phi' = -R \int_{p'=1000\text{ mb}}^{p} \Bigg[T_m(p') - \left(1 - \alpha \ln \frac{1000}{p'}\right)\left(ay + \hat{T} \cos \frac{2\pi}{L} x \cos \frac{2\pi}{L} y\right)\Bigg] \frac{dp'}{p'}, \tag{1.2.5}$$

and so

$$\Phi(x,y,p) = \Phi_{\mathrm{m}}(p) + \hat{\Phi}_0 \cos\frac{2\pi}{L}(x+\lambda)\cos\frac{2\pi}{L}y$$
$$+ R\left(ay + \hat{T}\cos\frac{2\pi}{L}x\cos\frac{2\pi}{L}y\right)\left[\ln\frac{p}{1000} + \frac{\alpha}{2}\left(\ln\frac{p}{1000}\right)^2\right], \quad (1.2.6)$$

where

$$\Phi_{\mathrm{m}}(p) = -R\int_{p'=1000\,\mathrm{mb}}^{p} T_{\mathrm{m}}(p')\frac{dp'}{p'}. \quad (1.2.7)$$

The 500-mb geopotential field for $\lambda = L/4$ is shown in Fig. 1.30.

The static stability. The static stability parameter σ used in this model is

$$\sigma = \frac{RT_{\mathrm{m}}(p)}{p}\left(\frac{\kappa}{p} - \frac{\partial \ln T_{\mathrm{m}}}{\partial p}\right) = \frac{RT_{\mathrm{m}}(p)}{p^2}\gamma(p), \quad (1.2.8)$$

where

$$\gamma(p) = \kappa - p\frac{\partial \ln T_{\mathrm{m}}}{\partial p}. \quad (1.2.9)$$

Quasihorizontal variations in σ are thus neglected. A dryadiabatic lapse rate corresponds to $\gamma = 0$, while an isothermal lapse rate corresponds to $\gamma = \kappa = R/C_p$. Since $T_{\mathrm{m}}(p)$ varies only about 20 percent from the surface to the tropopause, we can replace it where it is an undifferentiated coefficient

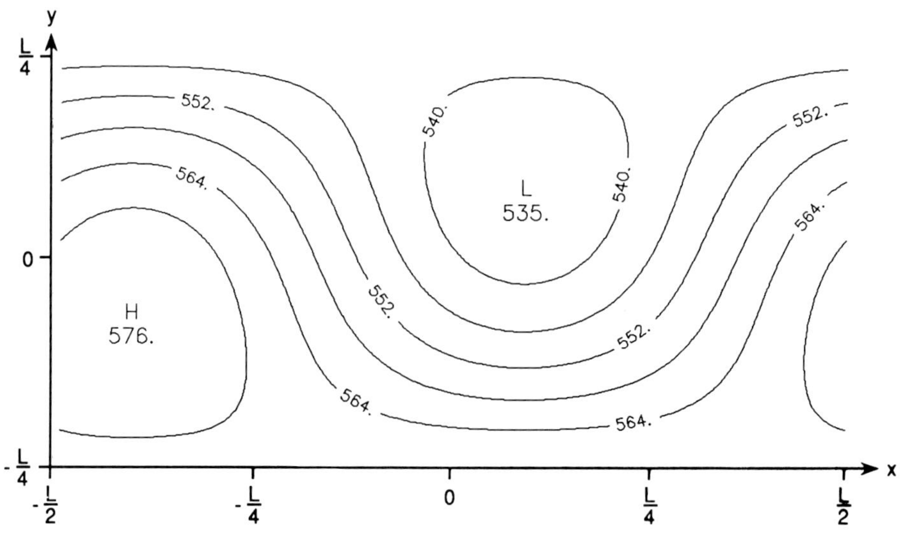

Figure 1.30 500-mb height at intervals of 6 dam (solid lines) in Sanders' analytic model. The value of α is 0.722, and Φ_{m} (500 mb)/g = 5555 m. The abscissa is the x direction; the ordinate is the y direction.

by T_0 without introducing any serious error, so that

$$\sigma = \frac{R}{p^2} T_0 \gamma(p). \tag{1.2.10}$$

Limitations of the model. This analytic model will be used to infer properties of the real atmosphere, which one hopes bear resemblance to those of our model. However, at the outset, we note several limitations to the model:

1. Its cyclones and anticyclones are equally intense. In the real atmosphere cyclones are more intense than anticyclones.
2. Its "basic" temperature gradient (i.e., a) is meridional and independent of latitude. In the atmosphere most of the horizontal temperature gradient is concentrated within a latitude belt and is not necessarily meridional.
3. The temperature perturbation centers are always at the same latitude as the surface highs and lows. This is not always so in the real atmosphere. (Surface highs and lows themselves are not always at the same latitude.)
4. Horizontal temperature gradients are strongest in the lower part of the troposphere. This is unrealistic over oceanic areas, where sensible heat fluxes from the sea surface tend to destroy horizontal temperature gradients.
5. The model produces grossly unrealistic results above 50 mb. What goes on in the stratosphere, however, probably does not ordinarily affect what goes on in the troposphere (and vice versa) on "synoptic" time scales because the static stability in the stratospheric is so high. [In a study of cyclogenesis over the ocean, Sanders and Gyakum (1980) used a similar analytic model in which there is no variation of horizontal temperature gradient with height; that is, $\alpha = 0$. However, ω was constrained to be zero at the tropopause.)
6. The model does not incorporate the complicating effects of friction, diabatic heating, or topography.

Despite all the limitations of the model, the most important features of the troposphere (i.e., wavelike structure aloft and checkerboard structure at the surface, and pole-to-equator temperature gradient) are in fact modeled reasonably well, and hence we will use the model to diagnose certain aspects of the behavior of synoptic-scale systems in midlatitudes.

Outline of the procedure used to find the motion of surface systems. The motion of surface cyclones and anticyclones ($\mathbf{c} = c_x\hat{\mathbf{i}} + c_y\hat{\mathbf{j}}$) is determined quantitatively from the following relations:

$$c_x = -\frac{\partial \chi}{\partial x} \bigg/ \frac{\partial^2 \Phi}{\partial x^2} \tag{1.2.11}$$

$$c_y = -\frac{\partial \chi}{\partial y} \bigg/ \frac{\partial^2 \Phi}{\partial y^2} \tag{1.2.12}$$

evaluated at $p = p_0$. The geopotential-height tendency at p_0 (χ_0) is computed from Eq. (1.1.10). To solve Eq. (1.1.10), we need to compute $(\partial\omega/\partial p)_{p_0}$, and hence we need to find ω as a function of x,y, and p.

(The local geopotential height tendency χ at any arbitrary pressure p is found by using Eq. (1.1.10) evaluated at pressure level p, not at $p = p_0$. The motion of features in the height field such as lows, highs, troughs, and ridges can be computed from the field of geopotential height itself, and from the geopotential-tendency field using Eqs. (1.2.11) and (1.2.12).)

The solution to the quasigeostrophic ω equation subject to the simple boundary conditions

$$\omega(p = 1000\text{ mb}) = 0 \tag{1.2.13}$$

$$\omega(p = 0\text{ mb}) = 0 \tag{1.2.14}$$

is found analytically, given forcing functions determined by $\Phi(x,y,p)$ and $\sigma(p)$.

The solutions to the quasigeostrophic ω equation that are used in Eq. (1.1.10) may be easily interpreted in the following manner: Let

$$\mathbf{v}_g(x,y,p) = \mathbf{v}_0(x,y) + \mathbf{v}_M(p) + \mathbf{v}'(x,y,p), \tag{1.2.15}$$

where $\mathbf{v}_0$ is the geostrophic wind vector at 1000 mb, $\mathbf{v}_M$ is the mean zonal part of the thermal wind vector, $\mathbf{v}_g(p) - \mathbf{v}_g(p = 1000\text{ mb})$, owing to the term $-ay$, and $\mathbf{v}'$ is the perturbation part of the thermal wind vector, $\mathbf{v}_g(p) - \mathbf{v}_g(p = 1000\text{ mb})$, owing to the term $-\hat{T}\cos(2\pi/L)x\cos(2\pi/L)y$. (Note that subscript M denotes the thermal wind vector owing to the mean meridional temperature gradient, while m denotes the standard atmosphere vertical profile.) Then it follows that

$$\zeta_M = 0, \tag{1.2.16}$$

since a is independent of x and y, and so

$$\zeta_g(x,y,p) = \zeta_0(x,y) + \zeta'(x,y,p). \tag{1.2.17}$$

Also

$$\nabla_p T_m(p) = 0, \tag{1.2.18}$$

and so

$$\nabla_p T(x,y,p) = \nabla_p T_M(p) + \nabla_p T'(x,y,p), \tag{1.2.19}$$

where T_M is the deviation of the temperature from its mean at a given pressure level owing to the mean meridional temperature gradient, and T' is the deviation of the temperature from its mean at a given pressure level owing to the perturbation part of the temperature field. The quasigeostrophic ω equation, without friction or diabatic heating, may then be written as

$$\left(\nabla_p^2 + \frac{f_0^2}{\sigma}\frac{\partial^2}{\partial p^2}\right)\omega = -\frac{f_0}{\sigma}\frac{\partial}{\partial p}[(-\mathbf{v}_M \cdot \nabla_p \zeta_0)_{\text{I}} + (-\mathbf{v}_M \cdot \nabla_p \zeta')_{\text{II}} + (-\mathbf{v}' \cdot \nabla_p f)_{\text{III}}]$$

$$-\frac{R}{\sigma p}\nabla_p^2[(-\mathbf{v}_0 \cdot \nabla_p T_M)_{\text{IV}} + (-\mathbf{v}_0 \cdot \nabla_p T')_{\text{V}}]. \tag{1.2.20}$$

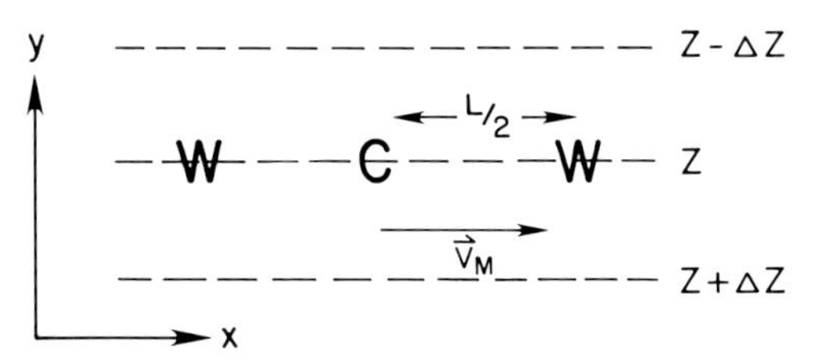

Figure 1.31 Mean temperature (thickness) isotherms Z (dashed lines); cold (C) and warm (W) centers. Cold and warm centers associated with cyclones and anticyclones, respectively, in the perturbation thermal-wind field; thermal wind associated with the mean meridional temperature gradient ($\mathbf{v}_M$) is westerly in the example. Advection of perturbation thermal vorticity by the thermal wind associated with the mean meridional temperature gradient is cyclonic east of C, and anticyclonic west of C (east of W).

The solutions to each of the five forcing functions in Eq. (1.2.20) are designated as follows:

- ω_{11} is due to advection of perturbation thermal vorticity by the thermal wind associated with the mean meridional temperature gradient (II) (Fig. 1.31);
- ω_{12} is due to advection of Earth's vorticity by the perturbation thermal wind (III) (Fig. 1.32);
- ω_{21} is due to advection of geostrophic vorticity at 1000 mb by the thermal wind associated with the mean meridional temperature gradient (I) (Fig. 1.33);
- ω_{22} is due to advection of the mean meridional temperature field by the geostrophic wind at 1000 mb (IV) (Fig. 1.34);
- ω_3 is due to advection of perturbation temperature by the geostrophic wind at 1000 mb (V) (Fig. 1.35).

The solutions to the quasigeostrophic ω equation using Eqs. (1.2.6), (1.2.8), (1.2.13), and (1.2.14) are given in Appendix 1.

Interpretation of the solutions to the ω equation. Vertical-profile factors of various contributions to ω are seen in Fig. 1.36. Advection of Earth's vorticity by the perturbation thermal wind (ω_{12}) is relatively insignificant. The fields of vertical motion ω_1 and ω_2 are symmetric about the line connecting centers of

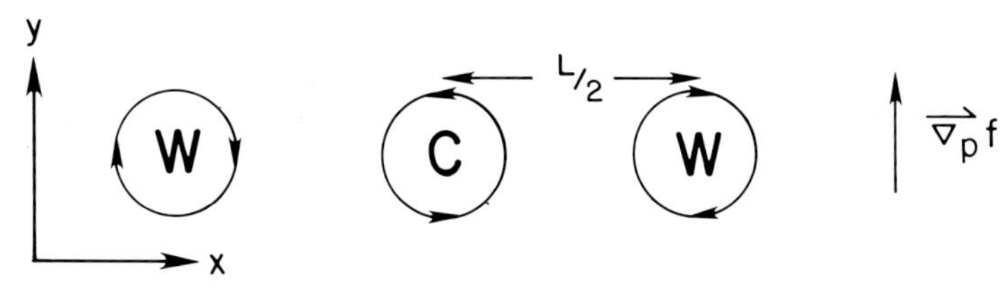

Figure 1.32 Perturbation thermal wind field (streamlines) associated with cold (C) and warm (W) centers. Advection of Earth's vorticity by the perturbation thermal wind is cyclonic east of W, west of C, and anticyclonic east of C, west of W.

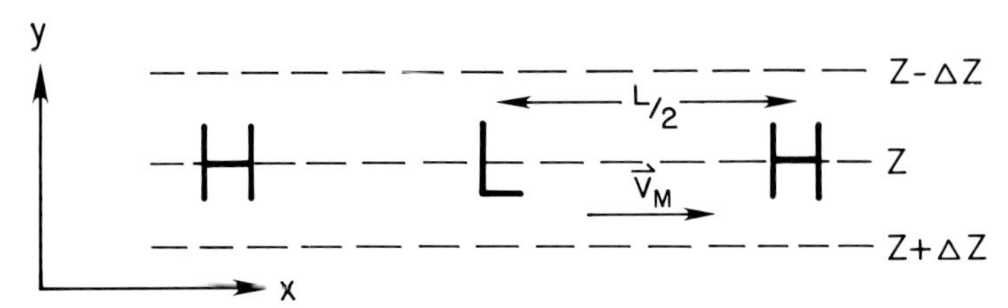

Figure 1.33 1000-mb highs (H) and lows (L), which are locations of minima and maxima in cyclonic vorticity, respectively. Mean temperature (thickness) isotherms Z (dashed lines). Thermal wind associated with the mean meridional temperature gradient ($\mathbf{v}_M$) is westerly in this example. Advection of relative vorticity at 1000 mb by the thermal wind associated with the mean meridional temperature gradient is cyclonic east of L, west of H, and anticyclonic east of H, and west of L.

temperature perturbation. Thus, these effects are responsible for the zonal movement of surface cyclones and anticyclones. The field of vertical motion induced by the advection of perturbation temperature by the geostrophic wind at 1000 mb (ω_3) is antisymmetric about the line connecting centers of temperature perturbation. Rising motion is found poleward of the lines, and sinking motion is found equatorward. The "ω_3 effects" in general are responsible for the rising motion (and "bad" weather) poleward of cyclone tracks, and sinking motion (and "good" weather) equatorward of cyclone tracks. The advection of perturbation temperature by the geostrophic wind at 1000 mb is responsible for the meridional movement of surface cyclones and anticyclones. The fields of ω_1 and ω_2 may counteract ω_3. The reader should note that, when the temperature and height fields at 1000 mb are in phase ($\lambda = 0$) or completely out of phase ($\lambda = L/2$), there is no "ω_3 effect." We will not consider the unusual cases in which

$$L/2 < \lambda < L, \qquad (1.2.21)$$

for which the "ω_3 effect" is reversed, because the height field is rarely observed in such a position relative to the temperature field.

The low-level (800 mb) cyclones track to the northeast for $\lambda = L/4$ (Fig. 1.37). This motion is to the right of the geostrophic wind at 500 mb (cf. Fig. 1.30). The low-level anticyclones track to the southeast, to the left of the

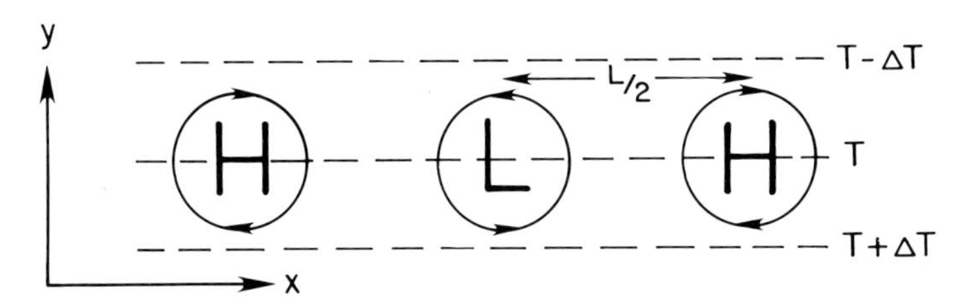

Figure 1.34 1000-mb high (H) and low (L) centers and their associated anticyclones and cyclones (streamlines); mean meridional isotherms (dashed lines). Advection of mean meridional temperature by the geostrophic wind at 1000 mb is warm east of L, west of H, and cold west of L, east of H.

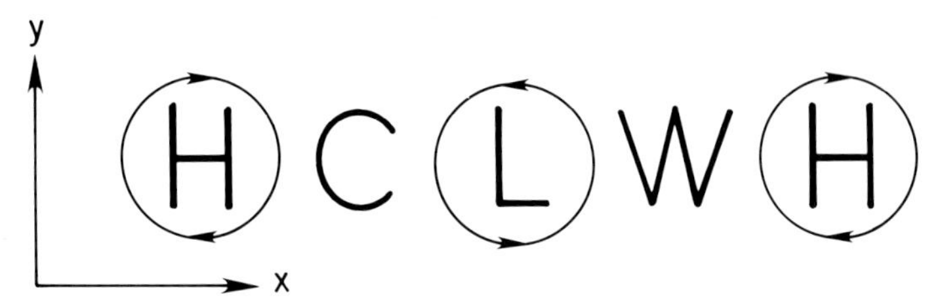

Figure 1.35 1000-mb high (H) and low (L) centers and their associated anticyclones and cyclones (streamlines); cold (C) and warm (W) centers. Advection of perturbation temperature by the geostrophic wind at 1000 mb is warm north of the line connecting the centers, and cold south of the line connecting the centers.

500-mb wind vectors. Since ω_3 is antisymmetric about the line connecting centers of temperature perturbation, whereas ω_1 and ω_2 are symmetric about the line, the effects of ω_1 and ω_2 are responsible for the right-moving (with respect to the 500-mb flow) character of cyclones, which would otherwise move in the poleward direction, to the left of the 500-mb flow. The effects of ω_1 and ω_2 are similarly responsible for the left-moving character of anticyclones, which would otherwise move in the equatorward direction, to the right of the 500-mb flow.

We use Eqs. (1.2.15), (1.2.17), the solution to the ω equation (Appendix

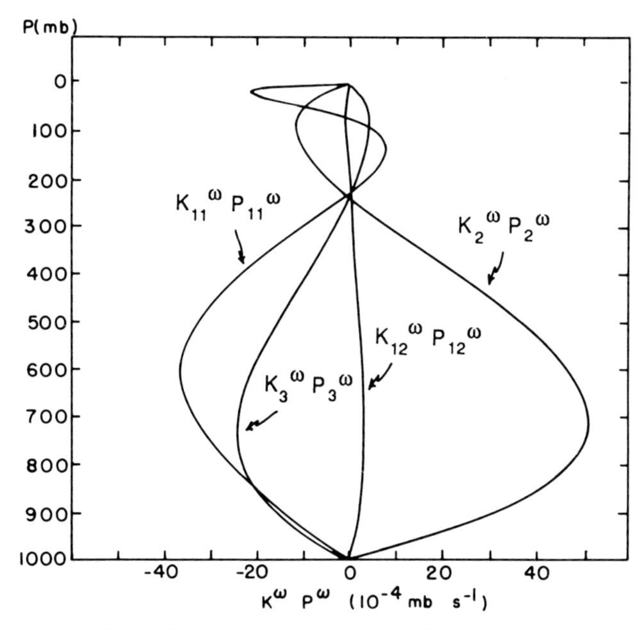

Figure 1.36 Vertical-profile factors (see Appendix 1) of various contributions to vertical motion (ω_1 at $x = L/4$, $y = 0$; ω_2 at $x + \lambda = L/4$, $y = 0$; ω_3 at $y = L/8$) in Sanders' analytic model. $\alpha = 0.114$, $f_0 = \eta_0 = 0.92 \times 10^{-4}\ \mathrm{s}^{-1}$, $T_0 = 250$ K, $L = 2900$ km, and $\lambda = L/4$ (from Sanders, 1971). (Courtesy of the American Meteorological Society)

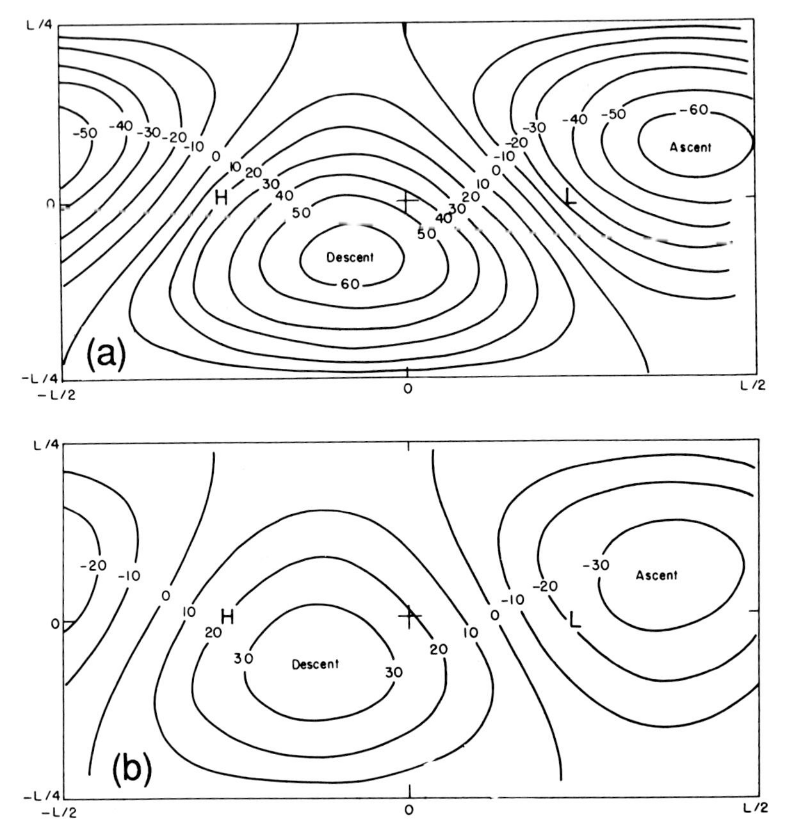

Figure 1.37 Vertical motion at (a) 800 mb and (b) 400 mb in 10^{-4} mb s^{-1}. The positions of the 1000-mb lows (L) and highs (H) are indicated. Lows track from region of maximum descent to region of maximum ascent. The abscissa is the x direction; the ordinate is the y direction (from Sanders, 1971). (Courtesy of the American Meteorological Society)

1), and Eq. (1.1.10), and find that

$$\nabla_p^2\chi = f_0[(-\mathbf{v}_M \cdot \nabla_p\zeta_0)_V + (-\mathbf{v}_M \cdot \nabla_p\zeta')_I + (-\mathbf{v}_0 \cdot \nabla_p f)_{VI} + (-\mathbf{v}' \cdot \nabla_p f)_{III}]$$
$$+ f_0^2\left[\left(\frac{\partial\omega_{11}}{\partial p}\right)_{II} + \left(\frac{\partial\omega_{12}}{\partial p}\right)_{IV} + \left(\frac{\partial\omega_2}{\partial p}\right)_{VII} + \left(\frac{\partial\omega_3}{\partial p}\right)_{VIII}\right]. \qquad (1.2.22)$$

This equation is essentially a height-tendency equation valid at the "surface," that is, on a pressure surface at the ground.

The solutions to each of the eight forcing functions in Eq. (1.2.22) are designated as follows:

χ_{11} is due to advection of perturbation thermal vorticity by the thermal-wind component associated with the mean meridional temperature gradient (I):

χ_{11}^{δ} is due to the divergence associated with ω_{11}, that is, to vertical motion related to the advection of perturbation thermal vorticity by the thermal-

wind component associated with the mean meridional temperature gradient (II);
χ_{12} is due to advection of Earth's vorticity by the perturbation thermal wind (III);
χ_{12}^{δ} is due to the divergence associated with ω_{12}, that is, to vertical motion

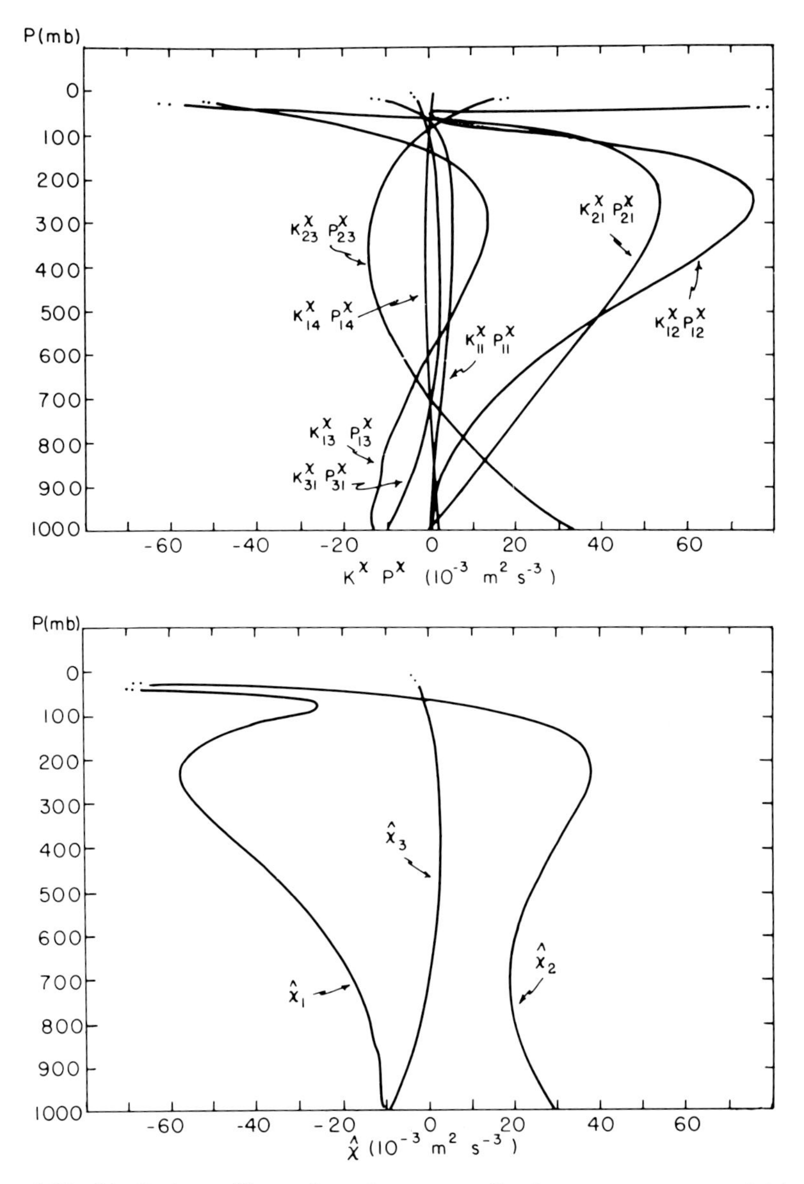

Figure 1.38 Vertical profiles of various contributions to geopotential-height tendency (see Appendix 1). χ_{11} is associated with $K_{11}^x P_{11}^x$, χ_{11}^{δ} with $K_{13}^x P_{13}^x$, χ_{12} with $K_{12}^x P_{12}^x$, χ_{12}^{δ} with $K_{14}^x P_{14}^x$, χ_{21} with $K_{21}^x P_{21}^x$, χ_{23}^{δ} with $K_{23}^x P_{23}^x$, and χ_3^{δ} with $K_{31}^x P_{31}^x$. $\hat{\chi}_1 = K_{11}^x P_{11}^x - K_{12}^x P_{12}^x + K_{13}^x P_{13}^x - K_{14}^x P_{14}^x$, $\hat{\chi}_2 = K_{21}^x P_{21}^x - K_{22}^x + K_{23}^x P_{23}^x$, $\hat{\chi}_3 = K_{31}^x P_{31}^x$ (from Sanders, 1971). (Courtesy of the American Meteorological Society)

related to the advection of Earth's vorticity by the perturbation thermal wind (IV);

χ_{21} is due to advection of 1000-mb geostrophic vorticity by the mean thermal-wind component associated with the mean meridional temperature gradient (V);

χ'_{21} is due to advection of Earth's vorticity by the geostrophic wind at 1000 mb (VI);

χ^{δ}_{23} is due to the divergence associated with ω_2, that is, to vertical motion related to the advection of relative vorticity at 1000 mb by the thermal-wind component associated with the mean meridional temperature gradient and advection of the mean meridional temperature field by the geostrophic wind at 1000 mb (VII);

χ^{δ}_{3} is due to the divergence associated with ω_3, that is, to vertical motion related to the advection of perturbation temperature by the geostrophic wind at 1000 mb (VIII).

The (first) subscripts in the χ's, 1, 2, and 3, are associated with the $\sin(2\pi/L)x\cos(2\pi/L)y$, $\sin(2\pi/L)(x+\lambda)\cos(2\pi/L)y$, and $\sin(4\pi/L)y$ distributions in x and y, respectively (see Appendix 1).

The complete solutions to the height-tendency equation at the "surface" Eq. (1.1.10) are given in Appendix 1.

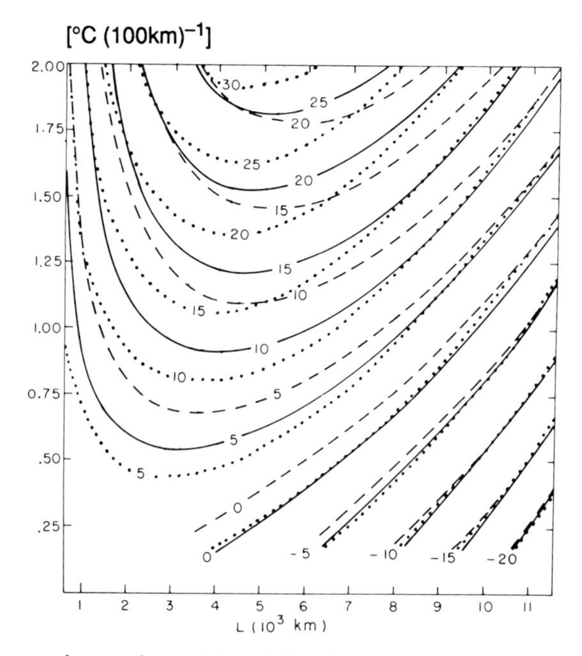

Figure 1.39 Eastward speed c_x of the 1000-mb center as a function of wavelength and meridional temperature gradient for selected values of the vorticity-stability parameter $(\zeta_g + f_0)/T_0\gamma$. The dashed, solid, and dotted lines are for values of 1.4, 2.8, and $5.6 \times 10^{-6}\,\mathrm{s}^{-1}\,\mathrm{K}^{-1}$, respectively. The intermediate value most nearly represents average atmospheric conditions. The isotachs are labeled in m s^{-1} (from Sanders, 1971). (Courtesy of the American Meteorological Society)

Interpretation of the solutions to the height–tendency equation. In Fig. 1.38 we see that at 1000 mb χ_{11}^{δ} (see $K_{13}^{x}P_{13}^{x}$), χ_{3}^{δ} (see $K_{31}^{x}P_{31}^{x}$), and χ_{23}^{δ} (see $K_{23}^{x}P_{23}^{x}$) are most important. These are three of the four solutions that depend upon "surface" divergence and convergence associated with vertical motion. The fourth (χ_{12}^{δ}, see $K_{14}^{x}P_{14}^{x}$), which is dependent upon the advection of Earth's vorticity, is insignificant. The Sanders analytic model therefore confirms our earlier qualitative arguments that divergence and convergence at the surface are the main mechanisms responsible for the motion of surface cyclones and anticyclones.

The analytic expressions for c_x and c_y are given in Appendix 1. For $\lambda = L/4$,

$$c_x(p = 1000\text{ mb}) = c_{x2}. \tag{1.2.23}$$

The eastward component of the motion of surface cyclones increases with the magnitude of the meridional temperature gradient a, and attains a maximum value at some wavelength (Fig. 1.39). The β effect [see Eq. (A.1.54)] in general slows down the eastward movement, though not very much for most cyclones, which have a relatively small scale. The poleward component of motion

$$c_{y3}(p = 1000\text{ mb}) > 0, \qquad \text{for } 0 < \lambda < L/2 \tag{1.2.24}$$

$$c_{y3}(p = 1000\text{ mb}) < 0, \qquad \text{for } -L/2 < \lambda < 0. \tag{1.2.25}$$

The poleward component is proportional to the amplitude of the horizontal temperature fluctuation $\hat{T}$, which is shown in Fig. 1.40 for $\lambda = L/4$. It decreases with horizontal wavelength[2] L and increases with the vorticity-stability parameter

$$\frac{\zeta_g + f_0}{T_0\gamma} \equiv \frac{\eta_0}{T_0\gamma}. \tag{1.2.26}$$

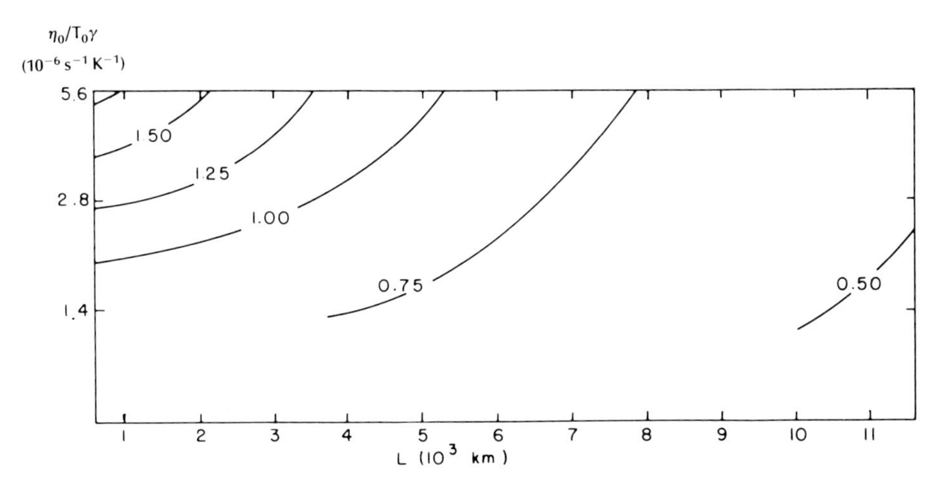

Figure 1.40 Northward speed c_{y3} of the 1000-mb center as a function of wavelength and vorticity-stability parameter for $\hat{T} = 1°\text{C}$. Isotachs labeled in m s^{-1} (from Sanders, 1971). (Courtesy of the American Meteorological Society)

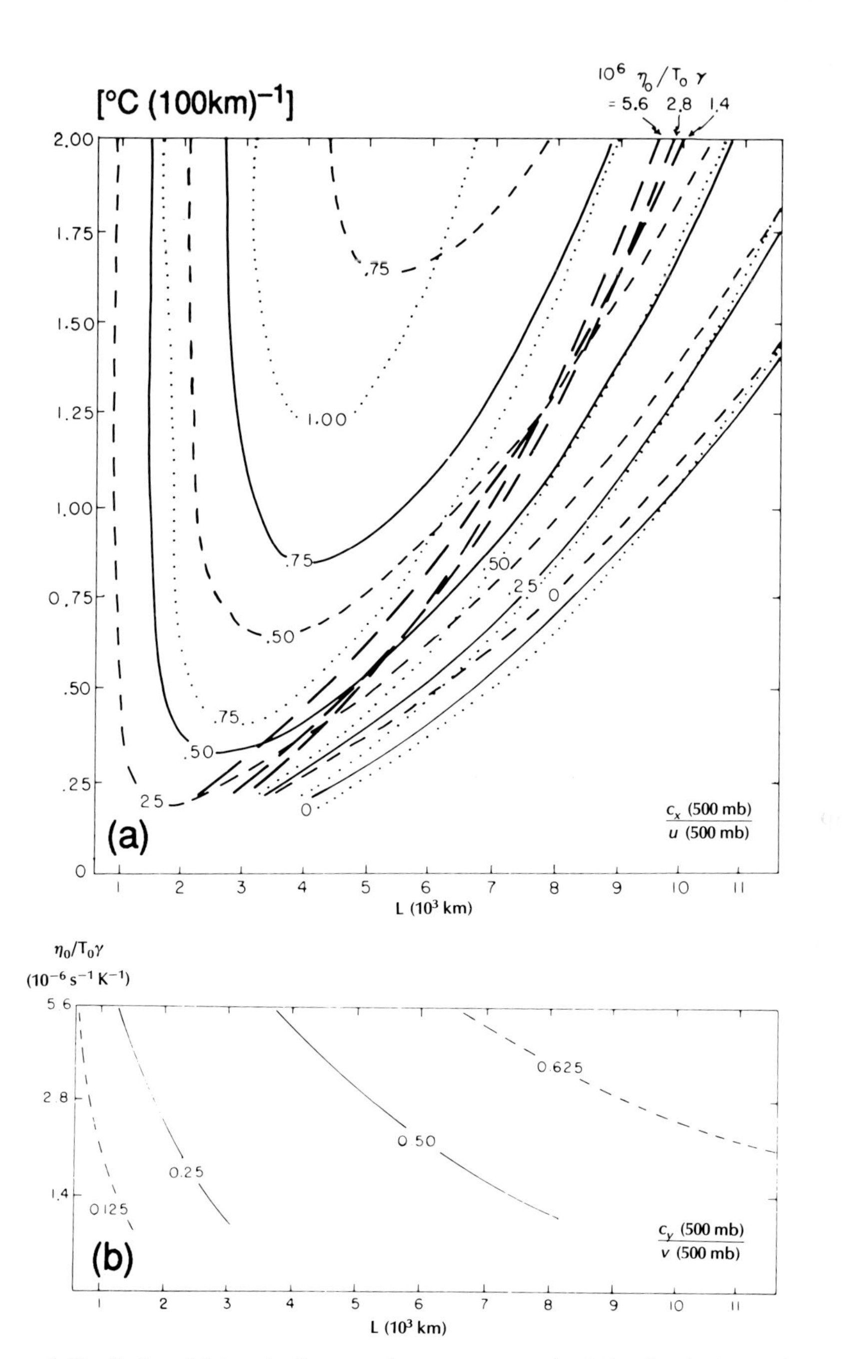

Figure 1.41 Ratio of (a) c_x to the zonal component of a 500-mb wind as a function of wavelength and meridional temperature gradient and (b) c_y to the meridional component of a 500-mb wind as a function of wavelength and vorticity-stability parameter. No isopleths are shown where the ratios are negative. In (a), the dashed, solid, and dotted lines are for values of the vorticity-stability parameter as explained in Fig. 1.39. The heavy dashed lines are loci of maximum wavelength for which the 1000-mb low moves to the right of the upper level flow. (Courtesy of the American Meteorological Society)

Surface cyclones track to the right of the flow at 500 mb when

$$\frac{c_y(p=1000\text{ mb})}{c_x(p=1000\text{ mb})} < \frac{v(p=500\text{ mb})}{u(p=500\text{ mb})}; \tag{1.2.27}$$

that is, for $\lambda = L/4$ (Fig. 1.41) when

$$\frac{c_{y3}}{c_{x2}} < \frac{v(p=500\text{ mb})}{u(p=500\text{ mb})}. \tag{1.2.28}$$

A similar analysis for the motion of surface anticyclones at

$$x = L - \lambda \pm nL, \qquad n = 0, 1, 2, \ldots \tag{1.2.29}$$

yields similar results. [Surface cyclones track to the left of the flow at 500 mb when the inequality in Eq. (1.2.27) is reversed.]

1.2.5. The Climatology of 1000-mb Height and 1000-mb Temperature

Northern Hemisphere Climatology. Winter and summer averages of 1000-mb height are shown in Figs. 1.42 and 1.43, and winter and summer season averages of 1000-mb temperature are shown in Figs. 1.44 and 1.45. The 1000-mb height and temperature fields represent the surface pressure and temperature fields. A high frequency of occurrence of lows (highs) is not necessarily reflected in the mean pressure field because the occurrence of lows (highs) may alternate with the occurrence of highs (lows). For example, although lows are often found in Southeast Colorado during the winter (Fig. 1.13), the mean 1000-mb height is not relatively low (Fig. 1.42), owing to the relatively high frequency of occurrence of highs (Fig. 1.16).

During the winter, many Hatteras lows end up near Iceland (and Greenland) and account for the "Icelandic lows." Lows that form east of the Asian continent end up near the Aleutian Islands and account for the "Aleutian lows." (The central sea-level pressure in Icelandic and Aleutian lows is sometimes as low as 950 mb or less.) High-pressure areas are located in the mean near 30°N over the Atlantic and eastern Pacific, and over Siberia. A secondary area of low pressure is found near Italy as a result of cyclogenesis in the lee of the Alps. A secondary area of high pressure is found over the snow fields of Northwest Canada. (The central sea-level pressure in highs over Siberia and Northwest Canada is sometimes as high as 1050 mb or greater.)

[The mean 1000-mb height distribution in spring (not shown) is qualitatively similar to that during the winter. The mean cyclones near the Aleutians, Iceland and Greenland, and Italy are weaker than they are during the winter. The mean high over Siberia is weaker, and lows are now apparent over India and North Africa. The secondary high over northwest Canada during the winter is now more prominent and is located over extreme Northern Canada.]

During the summer the high-pressure areas in the Atlantic and eastern

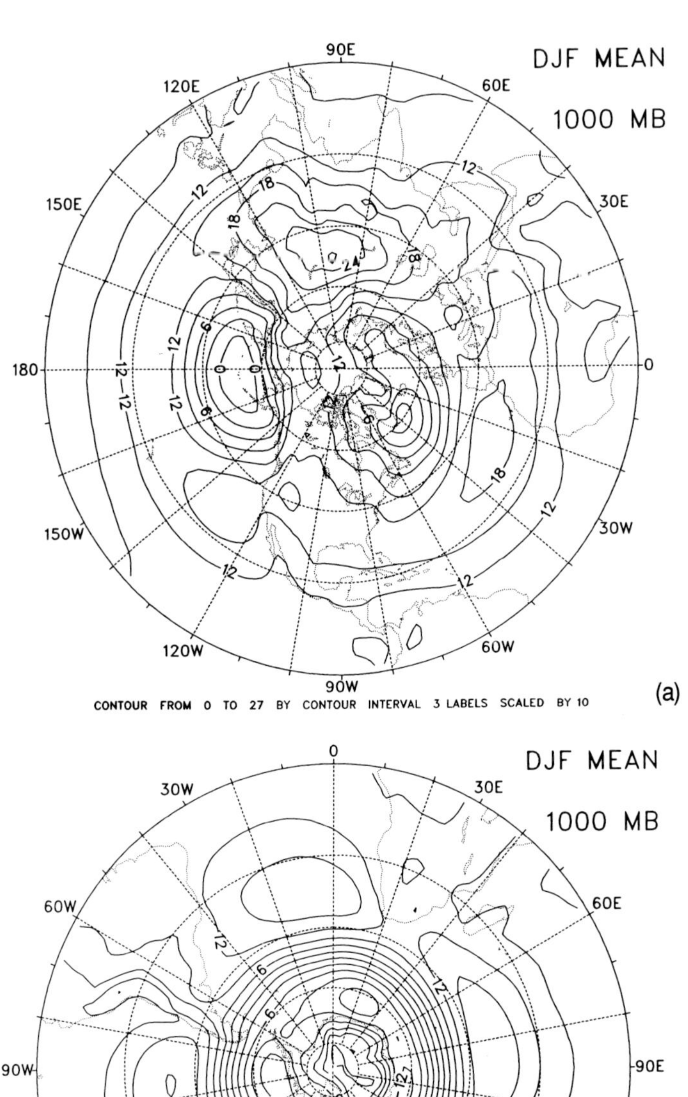

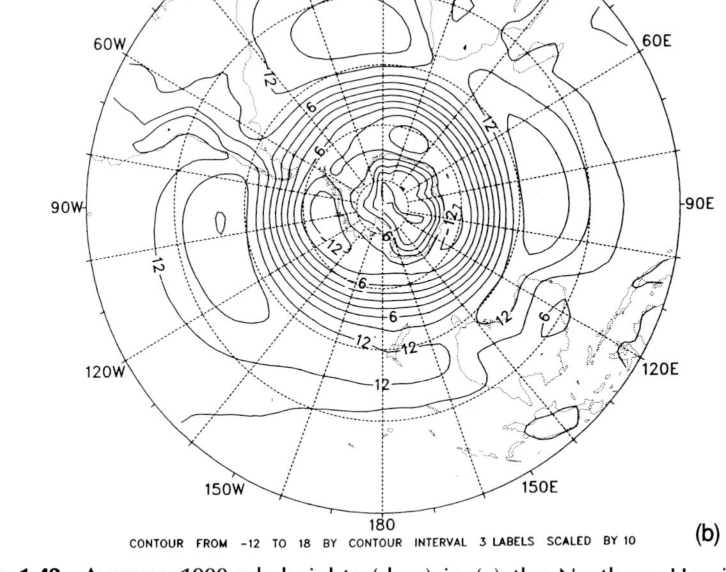

Figure 1.42 Average 1000-mb heights (dam) in (a) the Northern Hemisphere for the winter and (b) the Southern Hemisphere for the summer (December, January, and February) (from ECMWF data, 1979–1988; courtesy Kevin Trenberth and Amy Solomon, NCAR).

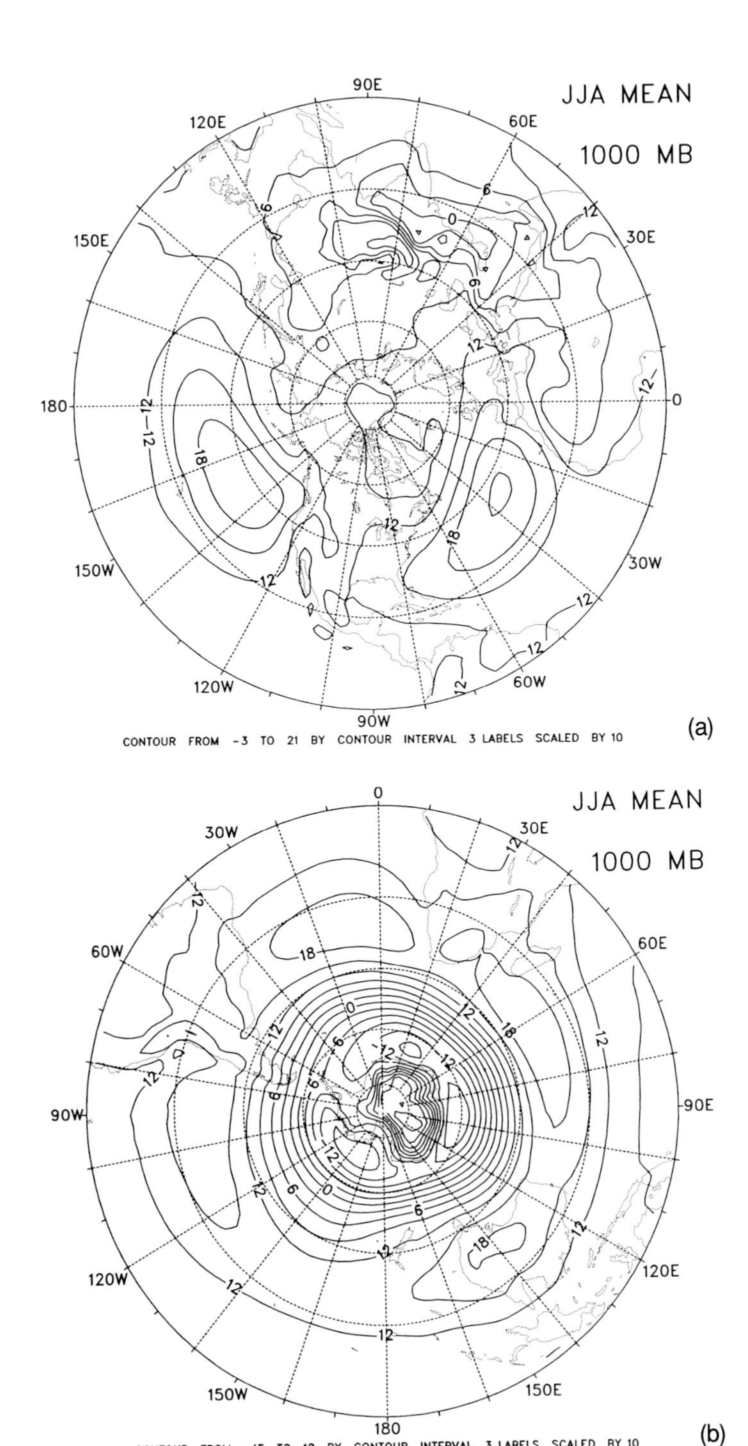

Figure 1.43 As in Fig. 1.42, but for the Northern-Hemisphere summer and Southern-Hemisphere winter (June, July, and August). (Courtesy Kevin Trenberth and Amy Solomon, NCAR)

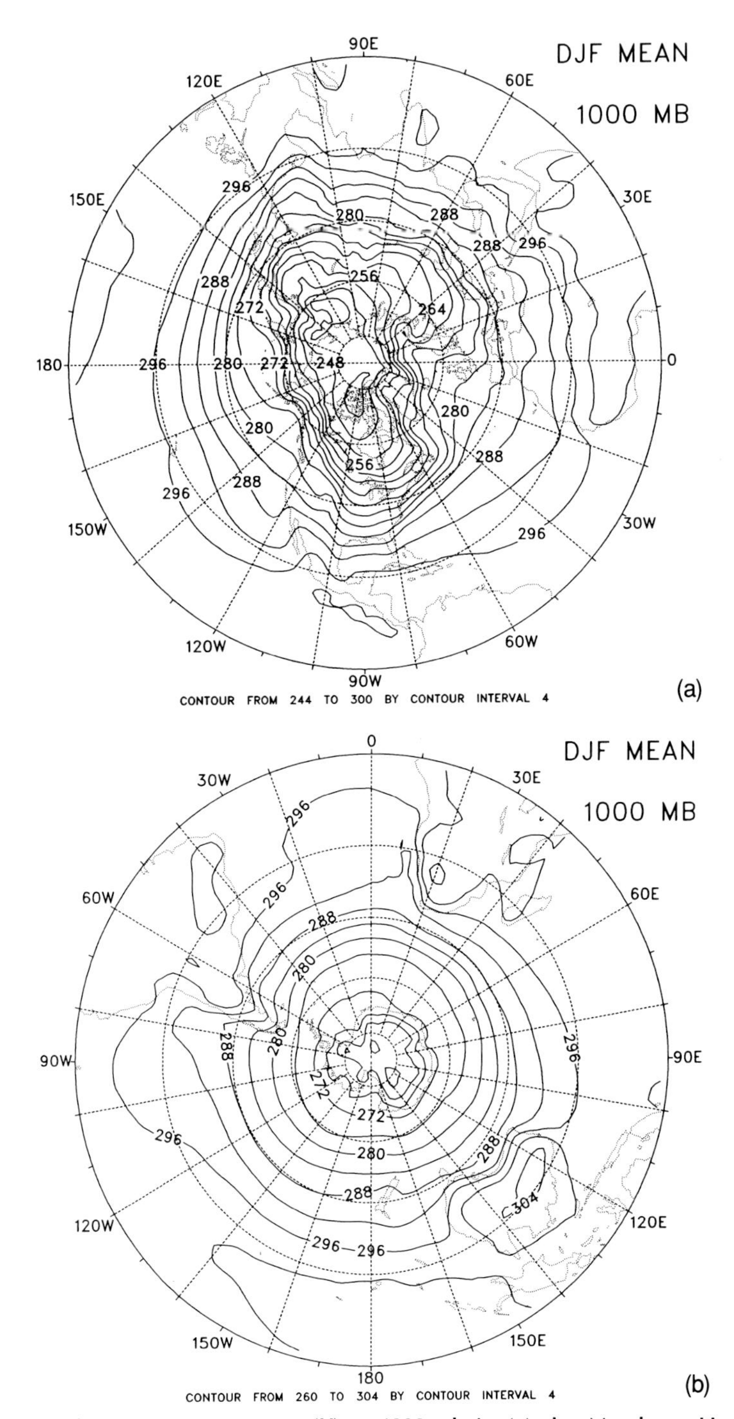

Figure 1.44 Average temperature (K) at 1000 mb in (a) the Northern Hemisphere for the winter and (b) the Southern Hemisphere for the summer (December, January, and February) (from ECMWF data, 1979–1988; courtesy Kevin Trenberth and Amy Solomon, NCAR).

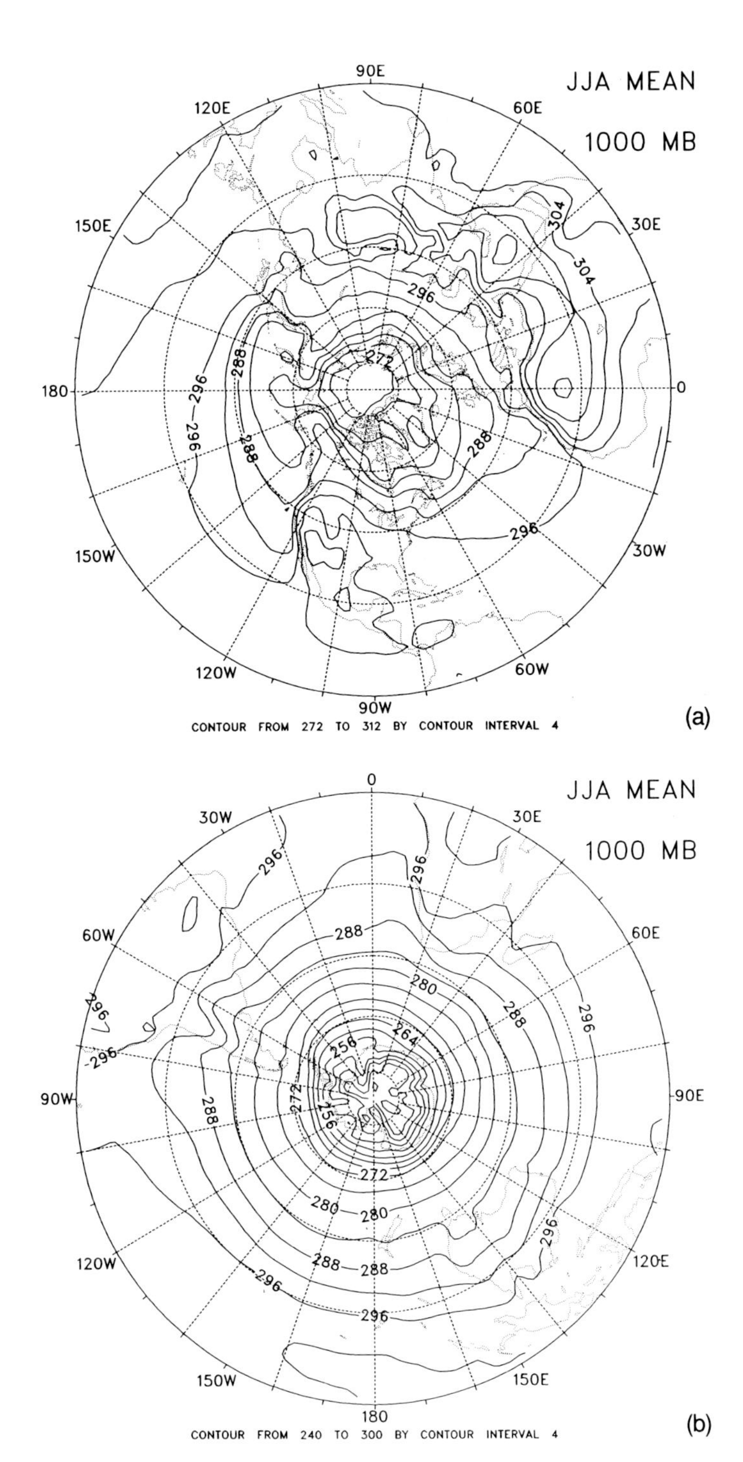

Figure 1.45 As in Fig. 1.44, but for the Northern-Hemisphere summer and Southern-Hemisphere winter (June, July, and August). (Courtesy Kevin Trenberth and Amy Solomon, NCAR)

Pacific are stronger than they are in any other season. They are located near 35°N, farther north than in any other season. The Aleutian and Icelandic lows and the Siberian high are nonexistent. Zonally oriented troughs of low pressure are found extending from northern India into Saudia Arabia and North Africa. A secondary trough of low pressure is found over Mexico and the southwest United States, while a secondary high is found over the North Pole.

[During the fall (not shown) the Aleutian and Icelandic lows and the Siberian high reappear. The eastern Pacific and Atlantic highs persist around 30–35°N, but are weaker than they are during the summer. A secondary trough of low pressure remains over the southwest United States and Mexico, while a secondary trough reappears west of Italy. The Asian and African tropical troughs are no longer evident.]

In general, surface temperatures decrease with latitude. The coldest surface temperatures are found in the vicinity of the Siberian and Northwest Canada highs and over Greenland in winter. (Cold temperatures are also found over the Tibetan Plateau.) The strongest surface temperature gradients are located over North America and Asia.

The surface temperature gradients are considerably weaker during the summer in comparison with those in other seasons. Prominent warm centers are located over the southwest United States and west North Africa where there are nearly co-located regions of low pressure. Coldest surface temperatures are found over Greenland, the Mexican mountains, and the Tibetan Plateau.

Southern Hemisphere climatology. In the Southern Hemisphere winter, areas of low pressure are found in the mean just beyond the edge of the Antarctican continent. A high is found over Antarctica itself. Highs are also found over the subtropical oceans west of South America, in the Atlantic, and in the Indian Ocean, and also over Australia.

During the summer in the Southern Hemisphere, lows are also found along the edge of the Antarctican continent; a high is also located over the continent itself. Subtropical highs are still located west of South America, in the Atlantic, and in the Indian Ocean. However, the winter high over Australia is replaced by a low, which is especially prominent over the northwest. A weak trough is found just off the west coast of South America.

In the Southern Hemisphere winter, there is an intense temperature gradient around the edge of the Antarctican continent, with the coldest air over the continent itself. Owing to the larger area of ocean, the temperature gradients are weaker in middle latitudes in the Southern Hemisphere than they are in the Northern Hemisphere. The temperature field is also more symmetrical about the South Pole than it is about the North Pole.

During the summer in the Southern Hemisphere, the coldest temperatures are found over Antarctica, with the strongest temperature gradients found along the edge of the Antarctican continent. The highest temperatures are found over western Australia, South Africa, and eastern South America.

1.3. THE FORMATION OF UPPER-LEVEL SYSTEMS

The term *upper-level systems* refers to height or pressure systems such as troughs, ridges, lows, and highs that are below the tropopause, but at and above the level at which roughly half of the mass of the troposphere lies above and below. In this sense, the troposphere is divided simply into an upper half and a lower half. In a more physical sense, "upper level" refers to the layer within the troposphere that is not significantly affected by friction. This usually includes the 700-mb level on upwards to the tropopause. (Of course, in mountainous areas "upper level" means "well above mountain level," which could be above 700 mb.) In this text we will use the latter terminology. The expression *upper air* usually refers to any level above the ground, and includes the 850-mb level, which is usually near the top of the friction layer.

From an analysis of nine seasons, F. Sanders has found that upper-level troughs in the Northern Hemisphere are born most frequently over and east of the Rocky Mountains and the highlands of central Asia, the two major mountain masses of the North Hemisphere, and also within northwesterly flow. They disappear most frequently over the eastern portions of the oceans, approximately 1000 km upwind from major masses of elevated terrain, and also in southwesterly flow. Topography and the planetary-scale flow pattern must therefore play some role in the formation of upper-level systems. It might be that since vorticity decreases (increases) while air ascends (descends) over elevated terrain, trough formation and trough dissipation is masked by the effects of terrain. Furthermore, it might be that trough formation and dissipation is masked by the effects of strong vorticity associated with high-amplitude planetary-scale waves.

Recall the quasigeostrophic height-tendency equation discussed in Vol. I for the case in which σ is treated as height independent. Aloft, where friction is negligible, it is expressed as follows if there is no diabatic heating:

$$\left(\nabla_p^2 + \frac{f_0^2}{\sigma}\frac{\partial^2}{\partial p^2}\right)\chi = f_0[-\mathbf{v}_g \cdot \nabla_p(\zeta_g + f)] - \frac{f_0^2}{\sigma}\frac{\partial}{\partial p}\left[\frac{R}{p}(-\mathbf{v}_g \cdot \nabla_p T)\right]. \quad (1.3.1)$$

Along a uniform, westerly, geostrophic current, there is no advection of geostrophic relative vorticity because there are no gradients in geostrophic vorticity.[3] Furthermore, there is no advection of Earth's vorticity because the geostrophic wind has no meridional component. Therefore there can be no local changes in height owing to vorticity advection, and hence vorticity advection alone cannot form an upper-level system in a uniform westerly current.

Now consider a uniform, westerly, geostrophic current, upon which there is superimposed a train of alternating cyclones and anticyclones; the result is a wavetrain having meridionally oriented troughs and ridges (Fig. 1.46). There is no advection of geostrophic vorticity along the trough and ridge axes, because the geostrophic vorticity there is a local maximum and minimum, respectively. Furthermore, there is no advection of Earth's vorticity along the trough and ridge axes, where the geostrophic component of the flow has no meridional

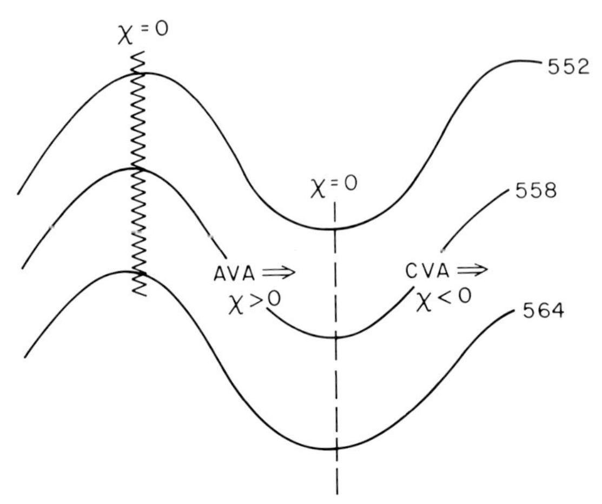

Figure 1.46 Illustration of the effect of vorticity advection on the geopotential-height tendency field. 500-mb height field in dam (solid lines) for a wavetrain in the westerlies. Areas of cyclonic vorticity advection (CVA) and anticyclonic vorticity advection (AVA) and height rises and falls are indicated.

component. Therefore heights cannot change locally along the ridge and trough axes owing to vorticity advection, and hence vorticity advection does not amplify the amplitude of the wavetrain.

However, if a geostrophic wavetrain is superimposed upon a westerly, geostrophic current that is strongest at some latitude, then troughs and ridges can intensify at the expense of a reduction in shear vorticity associated with the westerly current as a result of *barotropic instability*. This will be discussed briefly later in this chapter.

Differential temperature advection and differential diabatic heating can affect the height field so that upper-level systems form or intensify. Typically, an upper-level ridge forms or builds over a region of warm advection (i.e., temperature advection is decreasing with height and the height rises), and an

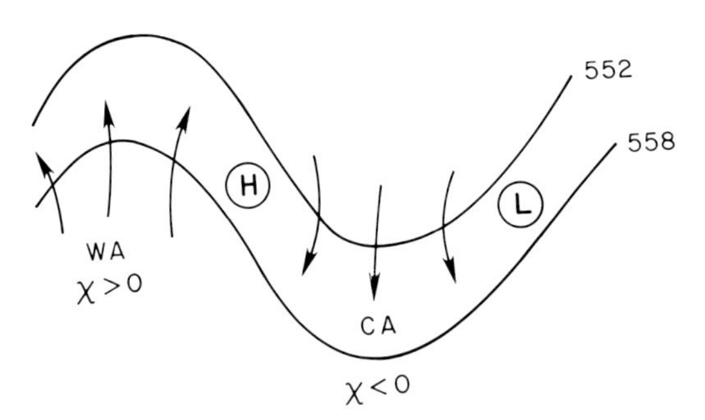

Figure 1.47 Illustration of the effect of low-level temperature advection on the geopotential-height tendency field. 500-mb height field in dam (solid lines) for a wavetrain in the westerlies. Low-level low- (*L*) and high- (*H*) pressure areas indicated. Low-level warm advection (WA) and cold advection (CA) are shown. The sense of geopotential-height tendency at 500 mb is indicated.

upper level trough forms or deepens over a region of cold advection (i.e., temperature advection is increasing with height and the height falls) (Fig. 1.47). Examples of a deepening trough over a region of cold advection and a building ridge over a region of warm advection are shown in Figs. 1.48 and 1.49, respectively.

It is also theoretically possible (but not observed in nature to the author's knowledge) for an upper-level trough to form or intensify in the presence of only warm advection above (i.e., temperature advection increasing with height) and for an upper-level ridge to form in the presence of only cold advection above (i.e., temperature advection decreasing with height). Also, upper-level troughs deepen or intensify if there is cold advection below *and* warm advection aloft; similarly, upper-level ridges build or intensity if there is warm advection below *and* cold advection aloft.

Strong diabatic heating at the surface can form or intensity an upper-level ridge or high, since diabatic heating decreases with height (Fig. 1.50, top). This

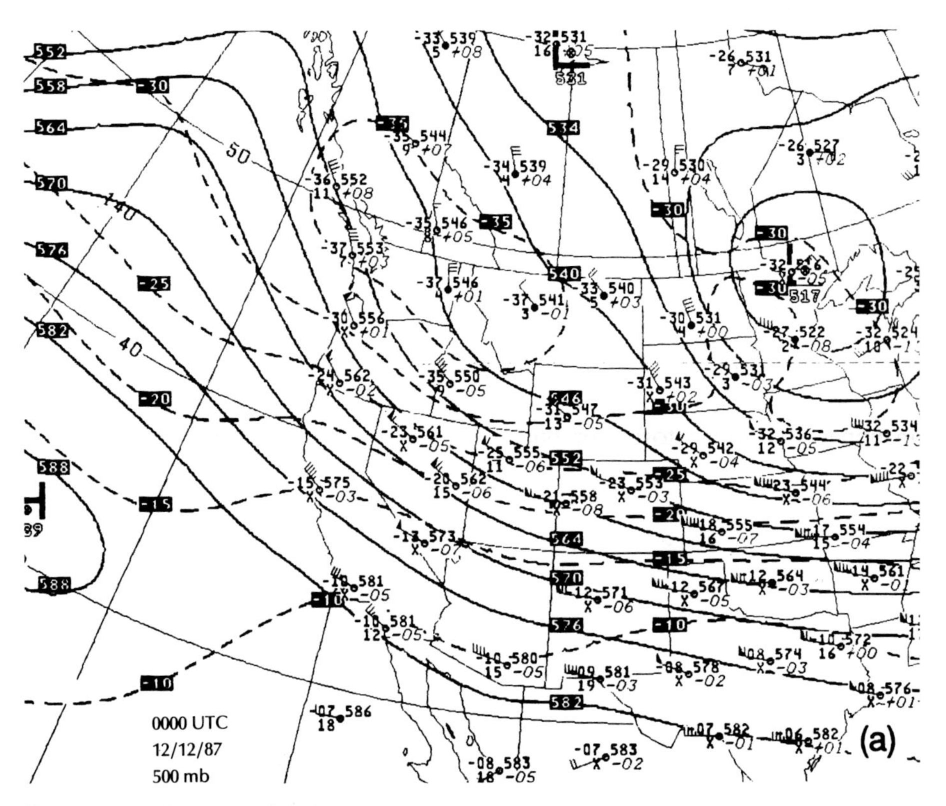

Figure 1.48 An example of a trough deepening over a region of cold advection. (a) NMC 500-mb analysis, 0000 UTC December 12, 1987; 12-h height tendencies plotted in dam under height; (b) as in (a), but for 850 mb; (c) as in (a), but at 0000 UTC December 13; (d) as in (b), but at 0000 UTC December 13; (e) as in (a), but at 0000 UTC December 14. A trough at 500 mb develops along the West Coast of the United States in response to strong cold advection at and below 500 mb.

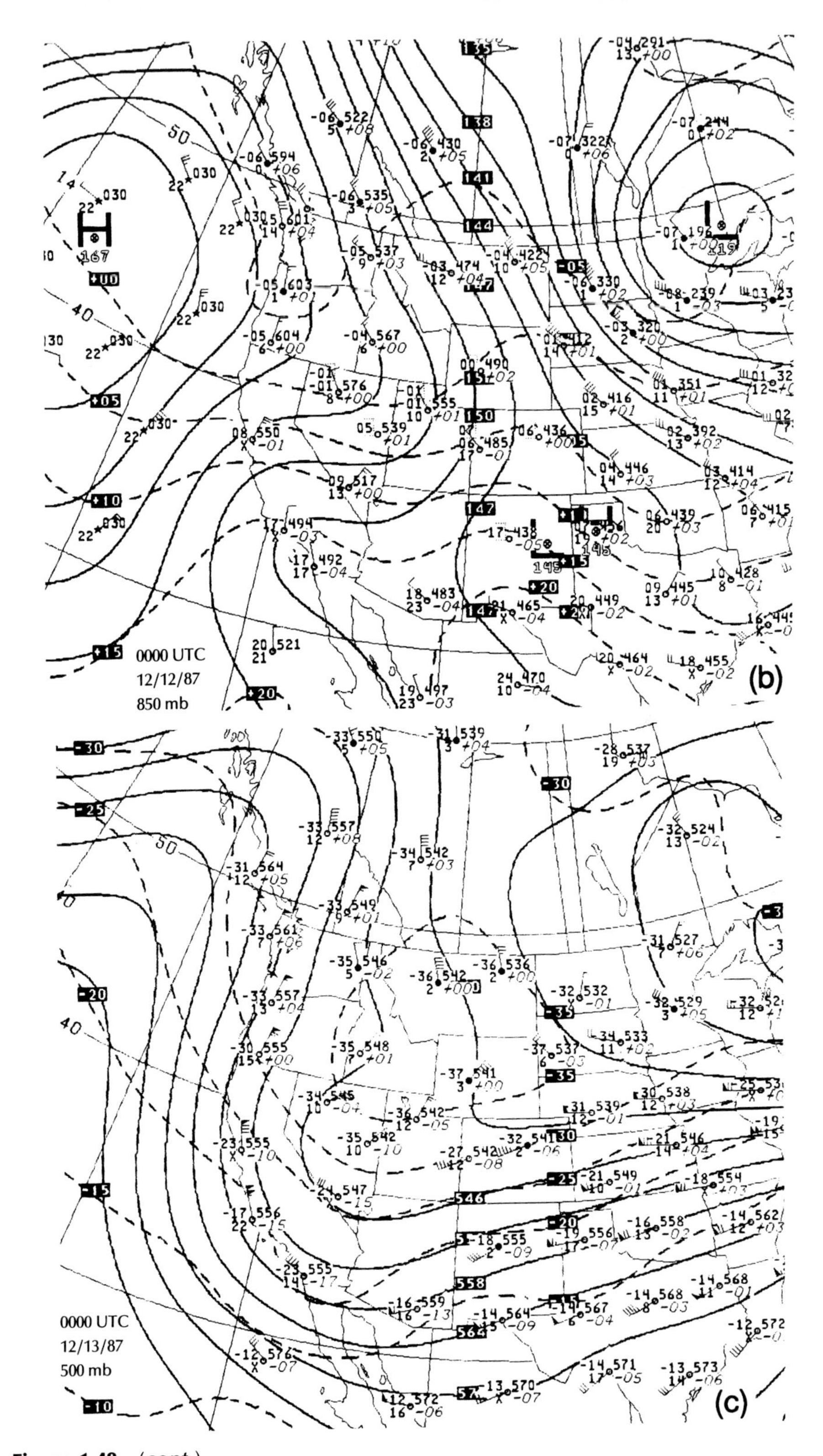

Figure 1.48 *(cont.)*

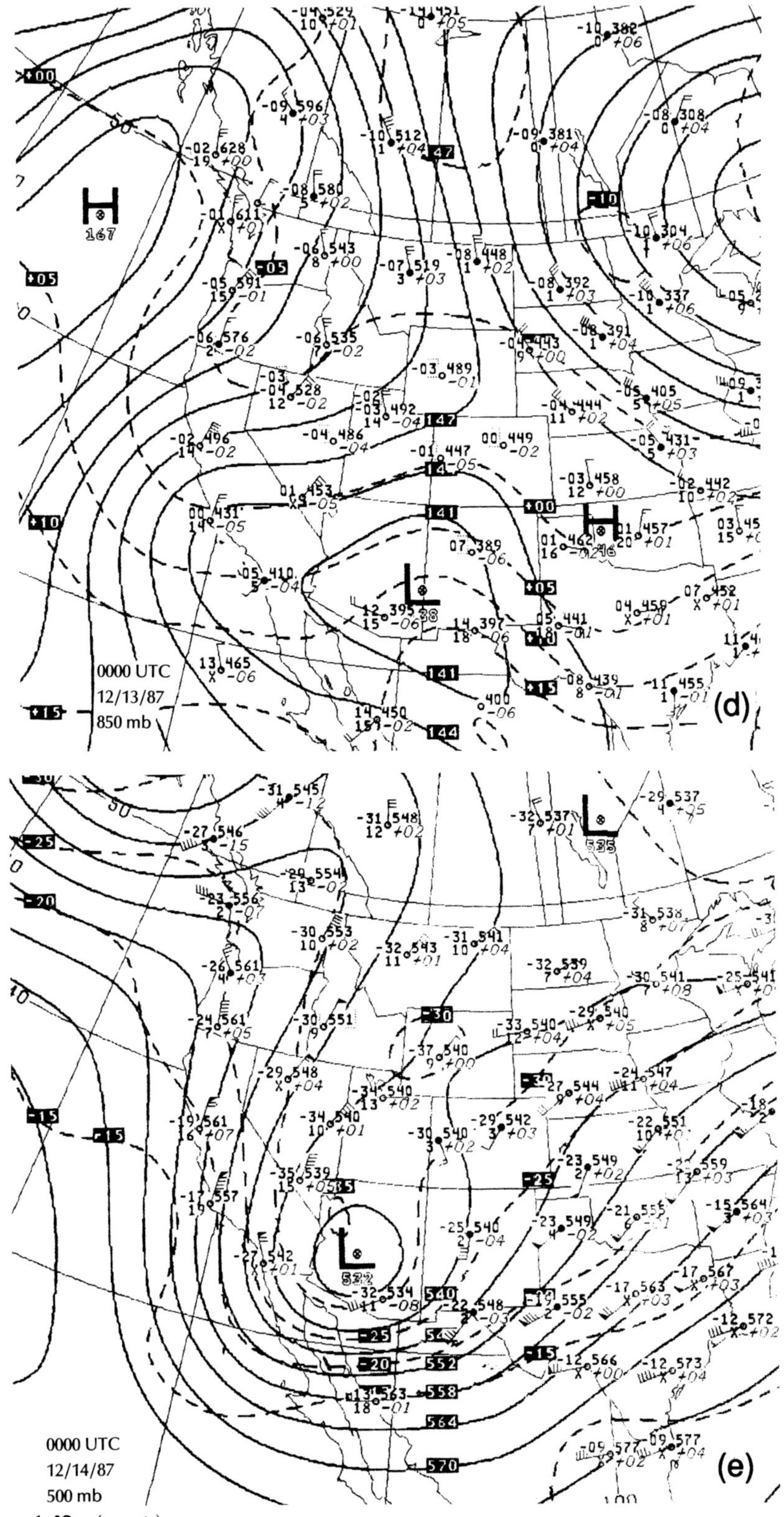

Figure 1.48 (*cont.*)

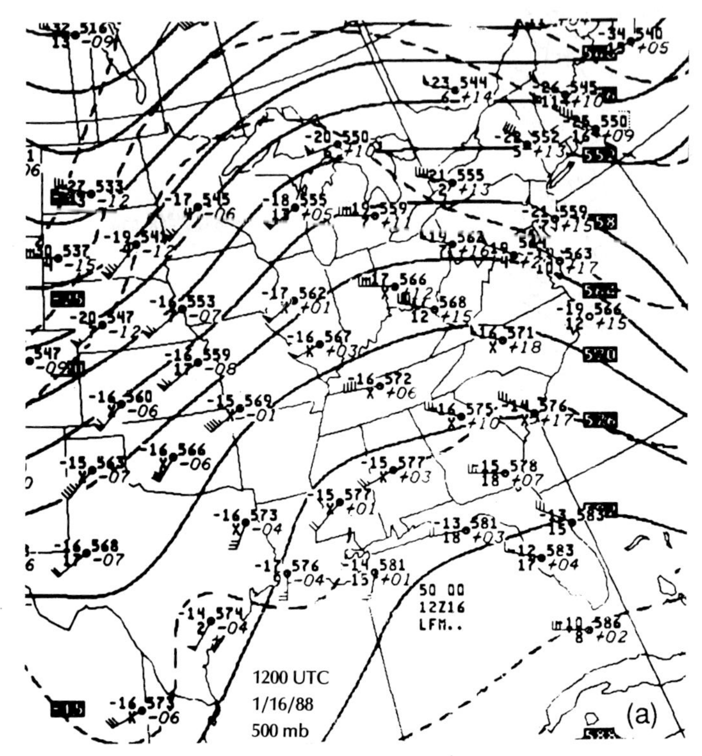

Figure 1.49 An example of a ridge building over a region of low-level warm advection. (a) NMC 500-mb analysis for 1200 UTC, January 16, 1988; 12-h height tendencies plotted in dam under height; (b) as in (a), but at 700 mb. In this example 500-mb heights have risen approximately 10–15 dam or greater during the previous 12 hours in the northeastern section of the United States. In this region there is a broad pattern of warm advection at 700 mb. The rapidly rising heights are due in part to the strong warm advection below 500 mb (the temperature advection at 500 mb is negligible), and in part to the further eastward progression of the trough off the east coast.

commonly occurs over desert areas and over regions of elevated terrain during the summer. At this time of the year, the solar angle is high, and the possible counteracting effects of vorticity advection and temperature advection are small because the "basic" westerly current is usually weak and displaced poleward from its wintertime position. Strong diabatic cooling at the surface can form or sustain an upper-level low or trough, since diabatic heating increases with height (Fig. 1.50, bottom). This happens over ice and snow fields that receive little sunlight. This effect is probably prominent in the winter poleward of the "basic" westerly current, when and where little solar insolation is received.

Suppose there is a localized region of latent heat release in the midtroposphere, so that diabatic heating increases with "height" in the lower half of the troposphere, and decreases with "height" in the upper half. Then heights will

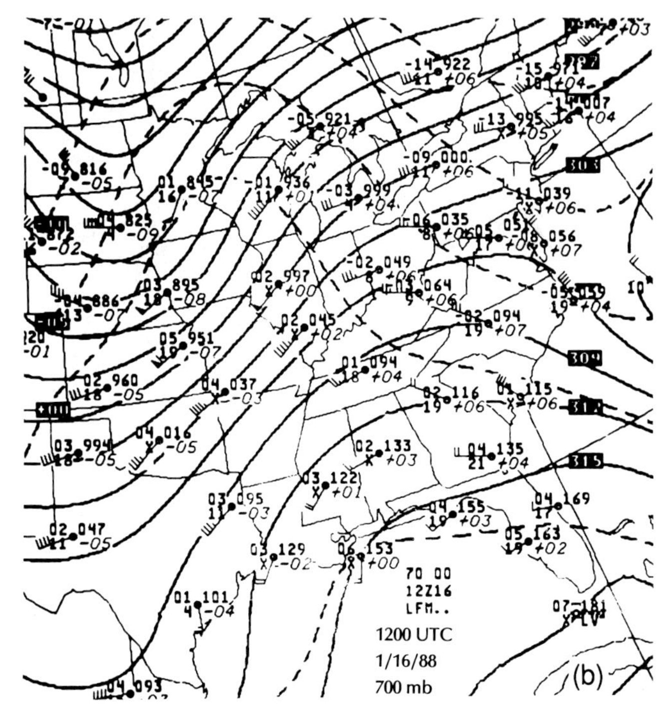

Figure 1.49 *(cont.)*

fall in the lower half of the troposphere, and rise in the upper half. The result is a warm-core cyclone at low levels, and anticyclone aloft (Fig. 1.51).

An alternative analysis of the formation of upper-level systems is the use of Eq. (1.1.10) evaluated at any pressure level:

$$\nabla_p^2 \chi = f_0[-\mathbf{v}_g \cdot \boldsymbol{\nabla}_p(\zeta_g + f)] + f_0^2 \frac{\partial \omega}{\partial p}. \tag{1.3.2}$$

We had used Eq. (1.1.10) previously to analyze the formation of surface (i.e., at p_0) systems. Although the analysis of the formation of upper-level systems using Eq. (1.3.1) has the virtue that vertical motion does not explicitly appear, one must use Eq. (1.1.10) to get the upper and lower boundary conditions, and it *is* necessary to know the vertical motion to use Eq. (1.1.10). From Eq. (1.3.2) we find that the formation of upper-level systems depends on the vertical gradient of vertical velocity. Above the level of nondivergence the height tendency is the opposite sign of what it is below. The vertical velocity is diagnosed from the quasigeostrophic ω equation.

For example, cold advection below 700 mb is accompanied by sinking motion. If at 500 mb the sinking motion decreases with height, then there is convergence, an increase in geostrophic vorticity, and a decrease in height. (At the surface there is divergence, a decrease in vorticity, and an increase in height. The thickness has decreased in the lower half of the troposphere owing to the cold advection.) We thus see another perspective on how upper-level trough intensification can be related to low-level cold advection.

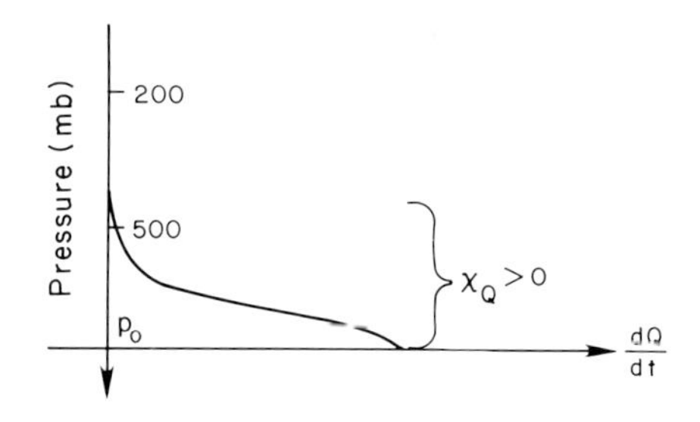

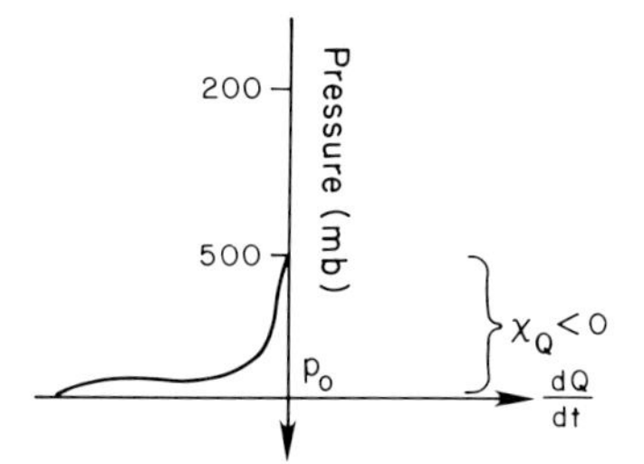

Figure 1.50 Illustration of (top) low-level diabatic heating (diabatic heating decreasing with height) associated with geopotential-height rises in the middle troposphere; (bottom) low-level diabatic cooling (diabatic heating increasing with height) associated with geopotential-height falls in the middle troposphere.

1.4. THE MOVEMENT OF UPPER-LEVEL TROUGHS AND RIDGES

The apparent movement of pressure and height systems is due to the horizontal gradient of the isallobaric field. Systems that move in the same direction as the basic current "progress," while systems that move in the opposite direction to the basic current "retrograde." For example, a westward-moving trough in the baroclinic westerlies is retrograding, while a westward-moving trough in the tropical easterlies is progressing.

1.4.1. Short Waves and Long Waves

Consider the following distribution of geopotential height on a surface of constant pressure for a traveling wave having a phase speed c and a zonal basic

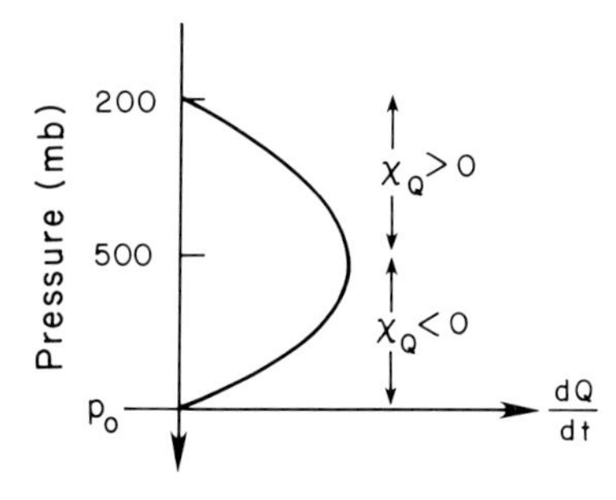

Figure 1.51 Illustration of diabatic heating at middle levels (diabatic heating increasing with height below middle levels, and decreasing with height above middle levels) associated with geopotential-height falls below, and rises above middle levels.

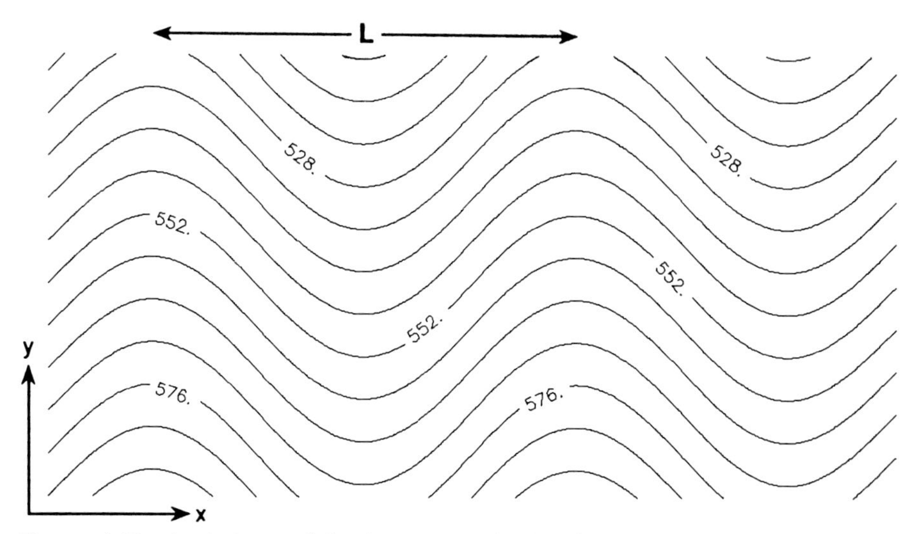

Figure 1.52 Analytic model of a wavetrain in the westerlies having a zonal wavelength *L*. Geopotential-height contours are labeled every 6 dam.

current of speed U (Fig. 1.52):

$$\Phi(x, y) = \hat{\Phi} \sin \frac{2\pi}{L}(x - ct) - f_0 U y + \Phi_0, \tag{1.4.1}$$

where $\hat{\Phi}$, Φ_0, L, c, f_0, and U are constants. This is a solution to the quasigeostrophic vorticity equation in which the "basic state" U is uniform and does not change with time. (In nature U actually varies as a function of latitude and is not time independent.) What are the relative contributions of advection of Earth's vorticity and advection of geostrophic vorticity? The horizontal components of the geostrophic wind are as follows:

$$u_g = -\frac{1}{f_0}\frac{\partial \Phi}{\partial y} = U \tag{1.4.2}$$

$$v_g = \frac{1}{f_0}\frac{\partial \Phi}{\partial x} = \frac{\hat{\Phi}}{f_0}\frac{2\pi}{L}\cos\frac{2\pi}{L}(x - ct). \tag{1.4.3}$$

The advection of Earth's vorticity is

$$-\mathbf{v}_g \cdot \nabla_p f = -v_g \beta = -\beta \frac{\hat{\Phi}}{f_0}\frac{2\pi}{L}\cos\frac{2\pi}{L}(x - ct), \tag{1.4.4}$$

and the advection of geostrophic vorticity is

$$-\mathbf{v}_g \cdot \nabla_p \zeta_g = U \frac{\hat{\Phi}}{f_0}\left(\frac{2\pi}{L}\right)^3 \cos\frac{2\pi}{L}(x - ct). \tag{1.4.5}$$

The ratio of the magnitude of geostrophic vorticity advection to Earth's vorticity advection is therefore

$$\frac{|-\mathbf{v}_g \cdot \nabla_p \zeta_g|}{|-\mathbf{v}_g \cdot \mathbf{v}_p f|} = \frac{U}{\beta}\left(\frac{2\pi}{L}\right)^2. \tag{1.4.6}$$

For a given basic current, the geostrophic advection of geostrophic (relative) vorticity is much greater than the geostrophic advection of Earth's vorticity if the wavelength is very short. Waves for which the advection of Earth's vorticity can be neglected in comparison to the advection of geostrophic vorticity are called *short waves*. The geostrophic advection of Earth's vorticity is much greater than the geostrophic advection of geostrophic (relative) vorticity if the wavelength is very long. Waves for which the

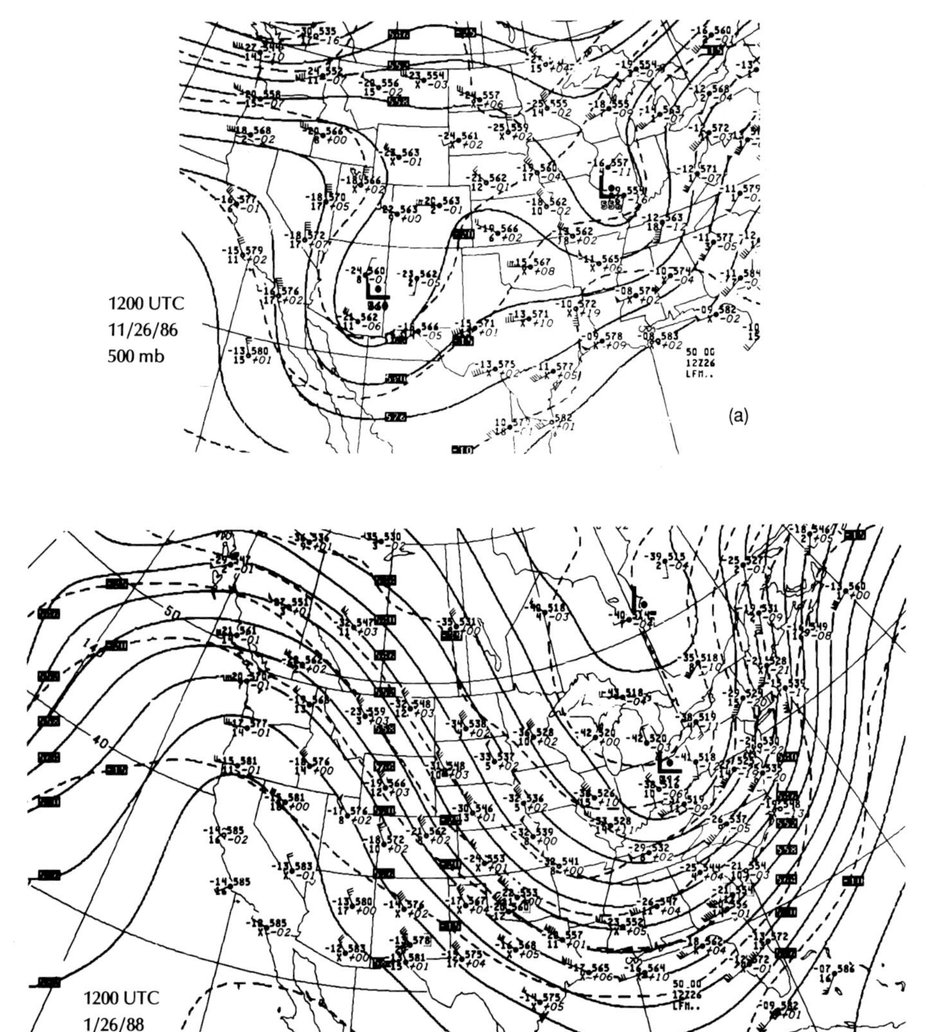

Figure 1.53 Example of a train of short waves and a long wave. (a) NMC 500-mb analysis of short-wave troughs at 1200 UTC, November 26, 1986, over Arizona and New Mexico, and over Illinois; (b) as in (a), but for a long-wave pattern at 1200 UTC, January 26, 1988.

advection of geostrophic vorticity can be neglected in comparison to the advection of Earth's vorticity are called *long waves*. In the atmosphere there is really a whole spectrum of waves, while our terminology implies that there is a dynamical separation of wavelengths. At some wavelength the advection of geostrophic vorticity is equal in magnitude to the advection of Earth's vorticity for a given basic current U and latitude. Examples of short waves and long waves are shown in Fig. 1.53.

However, weather forecasters often do not use the exact dynamical definition of the "short wave." Instead, they use the expression *short wave* in referring to the relative maximum in geostrophic vorticity associated with one trough portion of the wavetrain. "Short waves" are thus also referred to as vorticity maxima (or "vort maxes" or "shorts" by lingual conservatives). However, since vorticity may be associated with horizontal shear within the current in addition to curvature, the latter usage of the term *short wave* may not even refer to a wave in the wind field at all! For example, the cyclonic-shear side of a jet streak (see Chap. 2) in straight flow is also sometimes referred to as a short wave.

The movement of short waves and long waves. In a wavetrain of short waves there is cyclonic vorticity advection and height falls (according to the

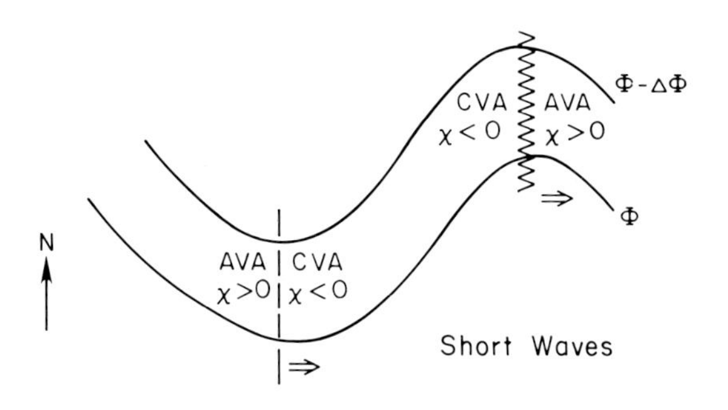

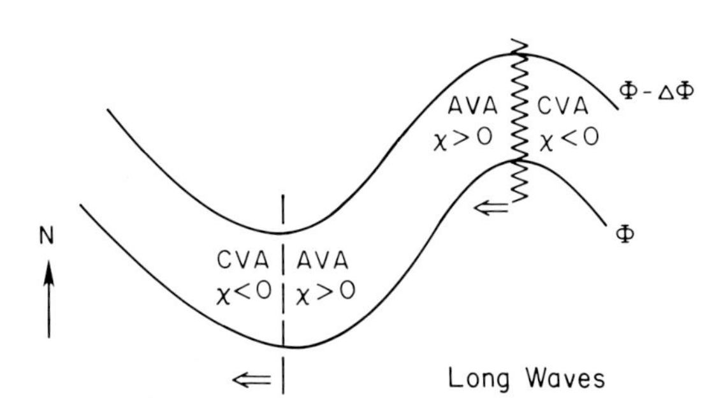

Figure 1.54 Vorticity advection in a wavetrain in the westerlies for (top) short waves and (bottom) long waves. Geopotential-height contours (solid lines). Motion of troughs and ridges is as indicated.

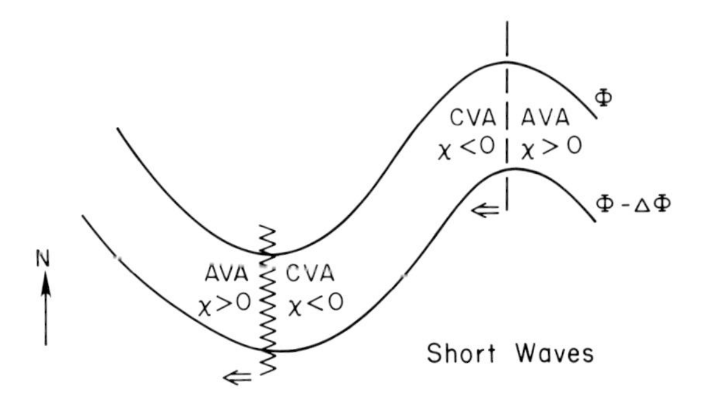

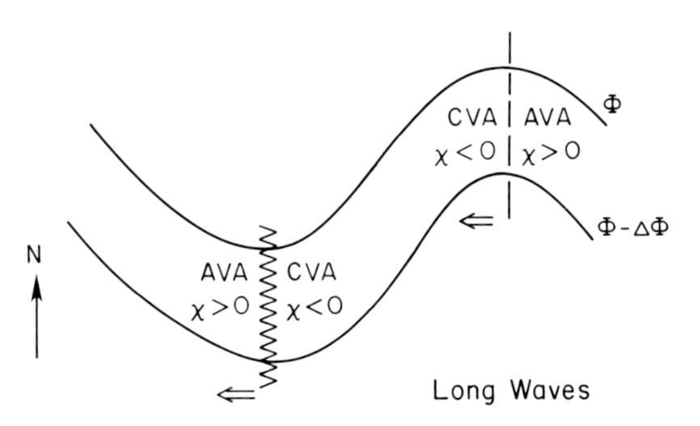

Figure 1.55 As in Fig. 1.54, but for a wavetrain in the easterlies.

height-tendency equation) downstream from troughs and upstream from ridges, and anticyclonic vorticity advection and height rises (according to the height-tendency equation) downstream from ridges and upstream from troughs (Figs. 1.54 and 1.55). In a wavetrain of long waves there is cyclonic vorticity advection and height falls downstream from troughs in the easterlies, upstream from ridges in the easterlies (Fig. 1.55), downstream from ridges in the westerlies, and upstream from troughs in the westerlies (Fig. 1.54). There is anticyclonic vorticity advection and height rises downstream from ridges in the easterlies, upstream from troughs in the easterlies (Fig. 1.55), downstream from troughs in the westerlies, and upstream from ridges in the westerlies (Fig. 1.54).

The effect of vorticity advection on a wavetrain of short waves is to make it progress along with the basic current (the current is thought of as the "carrier" of short-wave troughs),[4] while the effect of vorticity advection on a wavetrain of long waves is to make it retrograde if the basic current is westerly, or progress if the basic current is easterly. [The reader should note that although $\chi = 0$ along ridge and trough axes, where there is no vorticity advection, Φ *does* in fact change with time there; $\chi = \partial\Phi/\partial t = 0$ is valid only *at* the time that vorticity advection is zero; that is,

$$\lim_{\Delta t \to 0} \frac{\Phi(t + \Delta t) - \Phi(t - \Delta t)}{2\,\Delta t} = \left.\frac{\partial \Phi}{\partial t}\right|_{t=0} = 0. \tag{1.4.7}$$

The actual height at time $t = \Delta t$ is

$$\Phi(t = \Delta t) = \left(\lim_{\Delta t \to 0} \frac{\Phi(t = \Delta t) - \Phi(t = 0)}{\Delta t} \right) \Delta t + \Phi(t = 0)$$

$$= \left(\frac{\partial \Phi}{\partial t} \bigg|_{t = \Delta t/2} \right) \Delta t + \Phi(t = 0), \tag{1.4.8}$$

where $(\partial \Phi / \partial t)_{t = \Delta t/2}$ is *nonzero,* even though $(\partial \Phi / \partial t)_{t=0}$ is zero.]

In midlatitudes,

$$\beta \sim 10^{-11}\ \mathrm{s}^{-1}\ \mathrm{m}^{-1}, \tag{1.4.9}$$

and typical zonal wind speeds (U) are on the order of $10\ \mathrm{m\,s}^{-1}$. From Eq. (1.4.6) we see that in a wavetrain of short waves,

$$\frac{U}{\beta} \left(\frac{2\pi}{L} \right)^2 \gg 1, \tag{1.4.10}$$

and so the horizontal wavelength is much shorter than 6000 km. On the other hand, the horizontal wavelength for long waves is much longer than 6000 km. Since the circumference of the Earth in midlatitudes is on the order of 5 times its radius, which is on the order of 6000 km, long waves extend perhaps once or twice around the Earth (zonal wave number one or two). Waves that have wavelengths of several thousand kilometers or less are short waves. Wavetrains of approximately 3000 to 10,000 km wavelength are affected by both advection of geostrophic vorticity and Earth's vorticity.

It might be better to distinguish between "mobile" waves and "quasi-stationary" waves, rather than between short waves and long waves. From Eq. (1.4.6) we see, for example, that if the basic current U is very strong, then the progressive effects of geostrophic vorticity advection can be substantial even if L is relatively long. F. Sanders has found that in the Northern Hemisphere the "mobile" waves are evident most frequently for 5 days; more than half last for over 12 days. The few longest-lived waves have been observed to travel around the hemisphere at least twice!

If there were no temperature-advection effects or diabatic-heating effects, then a wavetrain would be stationary if

$$L = 2\pi (U/\beta)^{1/2}. \tag{1.4.11}$$

Since it is observed that not all systems of wavelength much longer than 6000 km actually retrograde, then according to quasigeostrophic theory temperature advection and/or diabatic heating must be important. If a trough in the westerlies tilts westward with height, then the region of maximum cold advection below may be found along and slightly upstream from the trough axis at some level, and the region of warm advection below may be found downstream from the trough axis at some level. Therefore in a qualitative sense the effects of differential temperature advection may be to retard the eastward movement of short-wave troughs in the westerlies that tilt with height toward the west (Fig. 1.56). However, according to Sanders' analytic model,

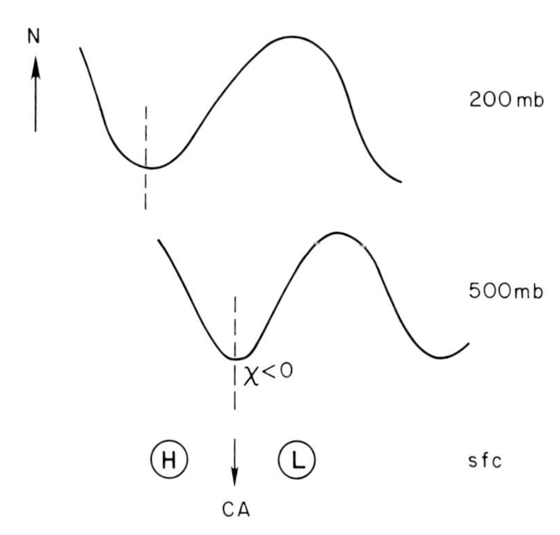

Figure 1.56 Effects of low-level cold advection in a system that tilts toward the west with height. 200- and 500-mb height-contour pattern as indicated, with heights decreasing with increased latitude; trough locations indicated by dashed lines. Surface low (L) and high (H) are indicated; there is cold advection (CA) at the surface west of L, east of H, underneath the 500-mb trough.

the effects of vorticity advection (see Fig. 1.38), for example, $-\mathbf{v}' \cdot \nabla_p f \ (\chi_{12})$ and $-\mathbf{v}_M \cdot \nabla_p \zeta_0 \ (\chi_{21})$, dominate the other forcing functions in effecting the solution to χ at middle to upper levels in the troposphere. The effects of temperature advection (e.g., χ_3^δ) at upper levels are relatively minor (this is not always true in nature, however), and it is therefore likely that differential diabatic heating acts as a "brake," which keeps some long-wave systems from retrograding. Of course we have simplified our discussion here by separating short waves from long waves, and by assuming that there is a train of waves. In reality, there is a whole spectrum of waves, and the resulting wavetrain does not necessarily have a regular period.

Effects of an upstream short wave on a long-wave trough. When a progressive short-wave trough in the westerlies approaches a stationary long-wave trough or a short-wave trough embedded equatorward in weaker westerlies, the effective horizontal wavelength decreases and the stationary trough may become progressive (Fig. 1.57) as geostrophic-vorticity advection becomes significant. It is said that the upstream trough has "kicked out" the stationary trough. The upstream trough is therefore sometimes referred to as the *kicker.* Often the mobile upstream trough takes the place of the former stationary trough. An example of this phenomenon is shown in Fig. 1.58. According to "Henry's rule," a stationary trough over the southwestern United States will be "kicked out" when the "kicker" gets to within 2200 km upstream from it.

Discontinuous retrogression. When a progressive short-wave trough moves downstream, while an upstream trough approaches and intensifies, the former trough may appear to retrograde suddenly as the latter trough intensifies

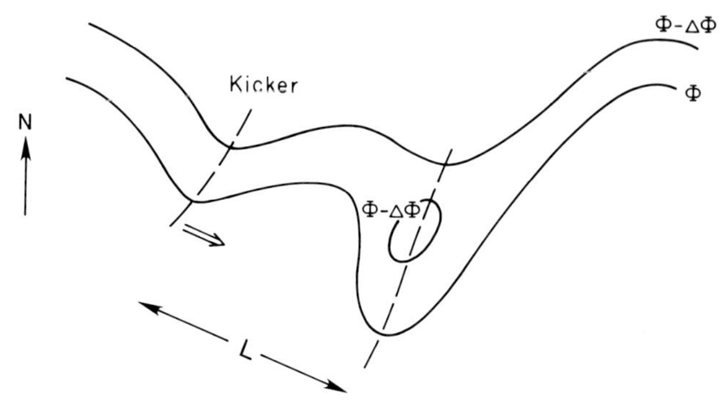

Figure 1.57 Illustration of a "kicker" advancing with respect to a stationary trough downstream, thereby reducing the wavelength (L) of the downstream trough. Geostrophic-vorticity advection therefore becomes more significant, and the downstream trough may progress eastward. Geopotential-height contours (solid lines).

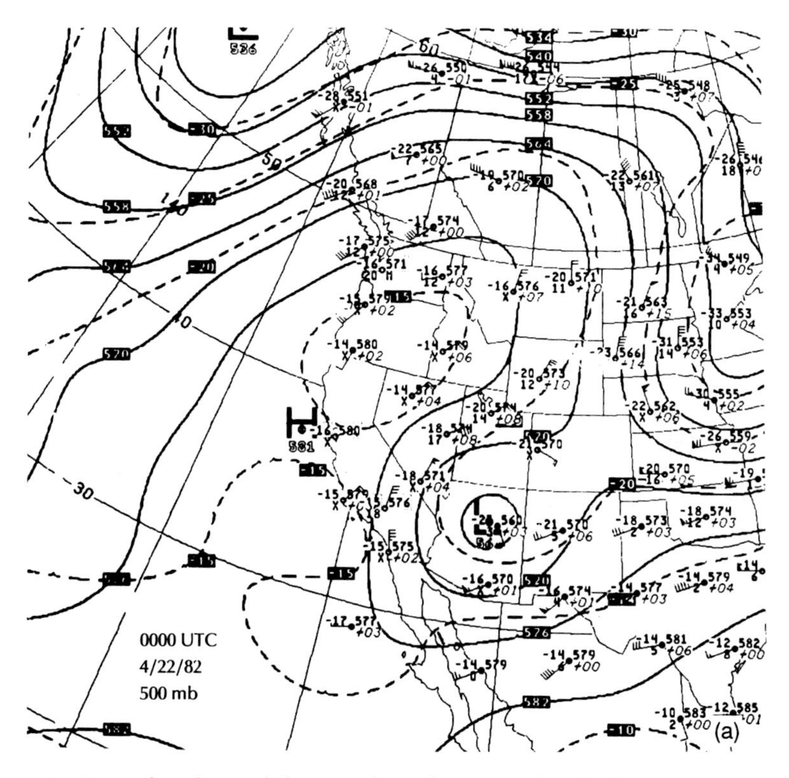

Figure 1.58 Example of a mobile trough in the westerlies kicking out a stationary downstream trough. (a) NMC 500-mb analysis at 0000 UTC, April 22, 1982; (b) as in (a), but for 0000 UTC, April 23; (c) as in (a), but for 0000 UTC, April 24; (d) as in (a), but for 0000 UTC, April 25. In (a) the cutoff low over Arizona has weakened over the past 12 hours, as evidenced by the height rises; there is a trough in the Gulf of Alaska around 145°W. In (b) we see that the trough has progressed eastward to almost 135°W; the cutoff low has continued to fill. As the trough reaches the coast (c) at about 125–130°W, about 2000 km from the longitude of the cutoff, the cutoff has begun to move east-northeastward. In (d) we find that the cutoff has continued to move east-northeastward, while the trough to the north has become tilted.

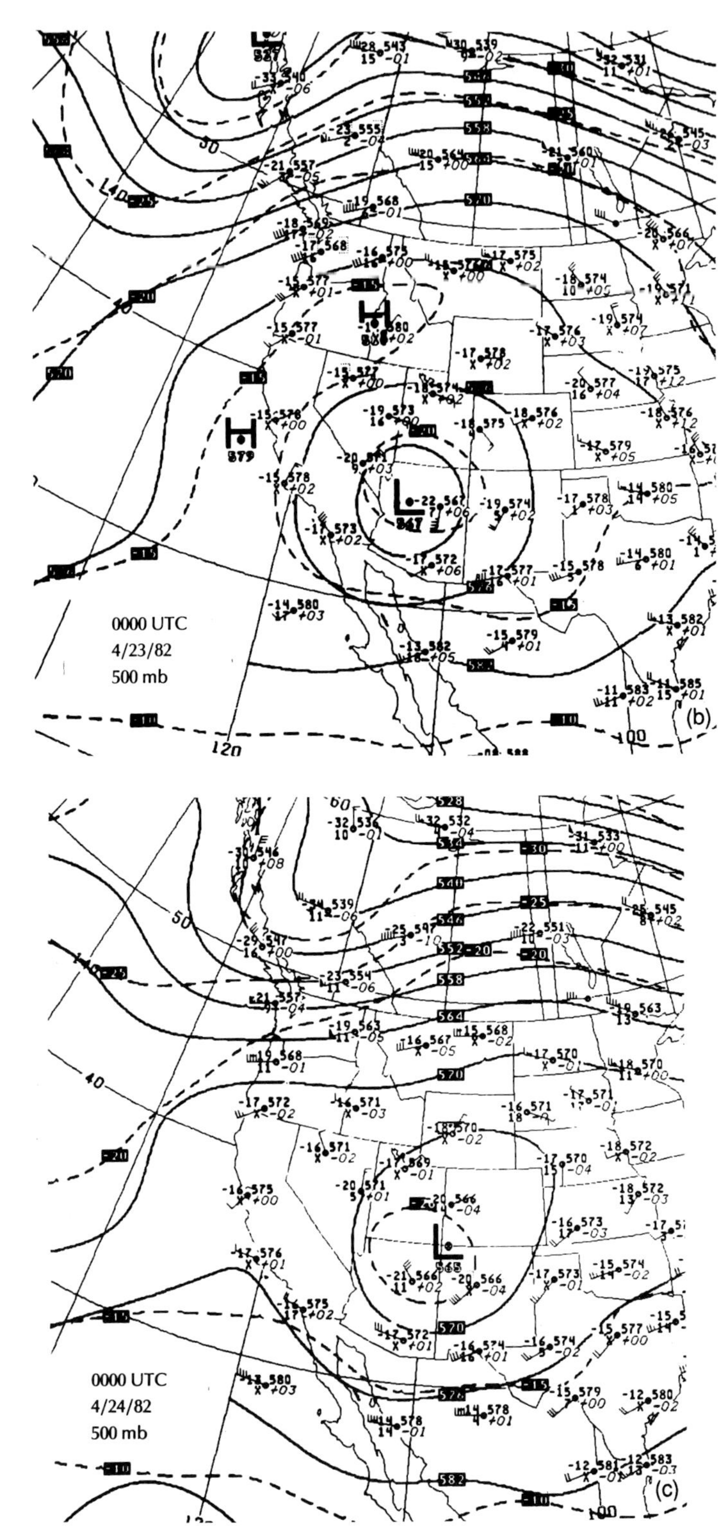

Figure 1.58 (*cont.*)

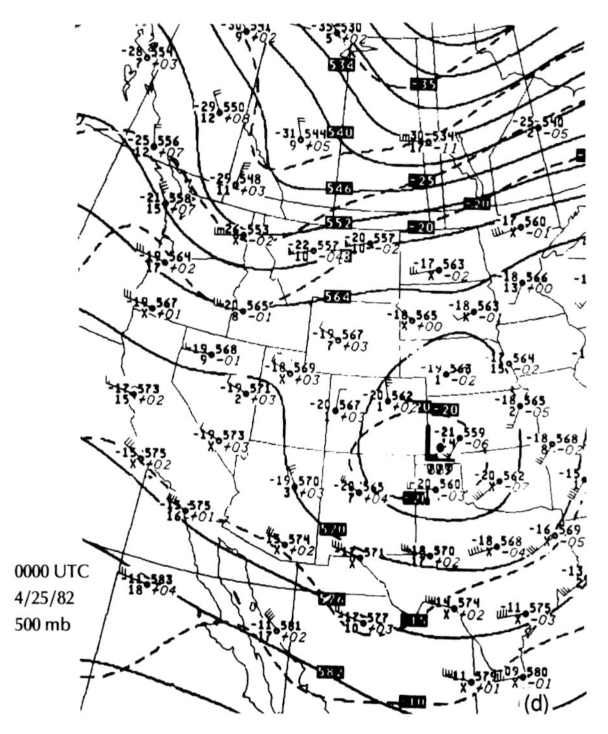

Figure 1.58 (*cont.*)

upstream from the old trough. This phenomenon is known as *discontinuous retrogression* and is illustrated in Fig. 1.59. It often occurs with other troughs around the hemisphere also, and is therefore not usually a local phenomenon. Typically, the wave number is 4 or 5. Although one would expect retrogression when the effective wavelength between two troughs becomes large, for example, when a progressive short-wave trough moves farther away from a stationary trough, this is not observed.

Effects of maxima in wind speed upstream and downstream from a trough on its movement. The wind speeds in the vicinity of a trough (of limited latitudinal extent) are often indicators of the motion of the trough:

> When the strongest winds are upstream from the trough, the trough tends to "dig" equatorward.
>
> When the strongest winds are downstream from the trough, the trough tends to "lift out" poleward.

These observations can be explained in terms of quasigeostrophic theory. Assume that the variations in wind speed in the vicinity of the trough are approximately geostrophic. If the region of maximum wind speed is upstream from a trough in the westerlies, then the geostrophic-vorticity maximum associated with the trough is also upstream from the trough: The relative maximum in geostrophic vorticity is located between the area of greatest

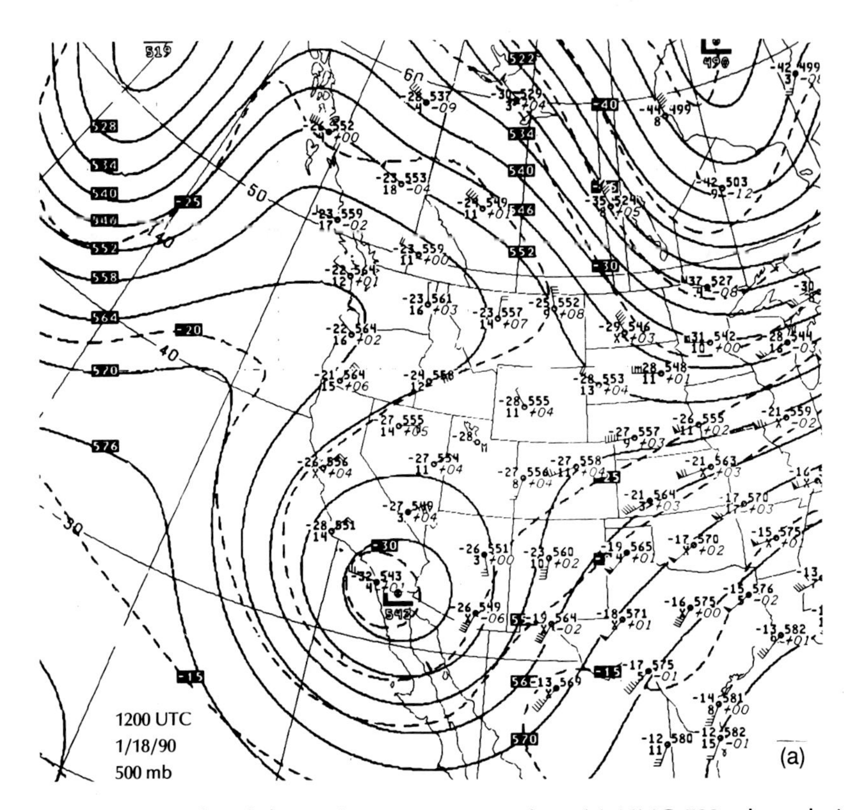

Figure 1.59 Example of discontinuous retrogression. (a) NMC 500-mb analysis at 1200 UTC, January 18, 1990; (b) as in (a), but for 1200 UTC, January 19; (c) as in (a), but for 1200 UTC, January 20. There is a trough approaching the west coast of the United States and Canada at 145–150°W, while there is a slowly moving cutoff low over Southern California (a). In (b) we see that the trough has progressed eastward to 130°–135°W, while the cutoff has moved to eastern Arizona. In (c) we note that the cutoff has progressed rapidly to north-central Kansas, while the upstream trough has "dropped" to a position off the California coast, replacing the old cutoff farther west than the cutoff's earlier position. This process looks like the "kicking out" process, except that the "kicker" replaces the trough that has gotten "kicked out," and the new cutoff is thereby established farther west, in the direction opposite to that of the flow.

cyclonic curvature, which is along the trough axis, and the area of greatest cyclonic shear, which is on the poleward side of the upstream wind maximum (Fig. 1.60). The region of maximum cyclonic vorticity advection and height falls is therefore near the "base" of the trough axis, not east of it. The trough therefore has an equatorward component of motion.

Similarly, when the region of maximum wind speed is downstream from a trough in the westerlies, the geostrophic-vorticity maximum associated with the trough is also downstream from the trough: The relative maximum in geostrophic vorticity is located between the area of greatest cyclonic curvature, along the trough axis, and the area of greatest cyclonic shear, on the poleward

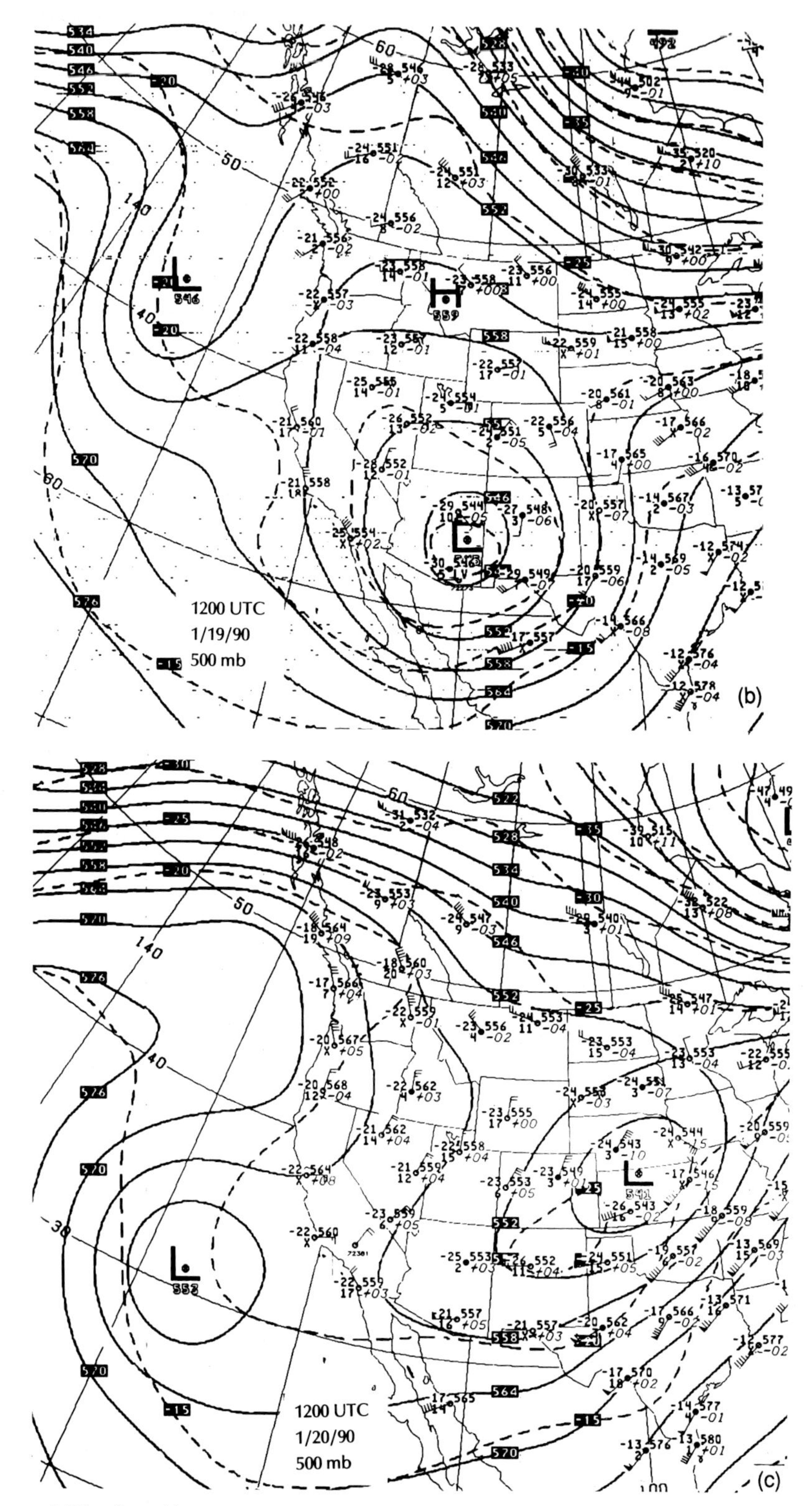

Figure 1.59 (*cont.*)

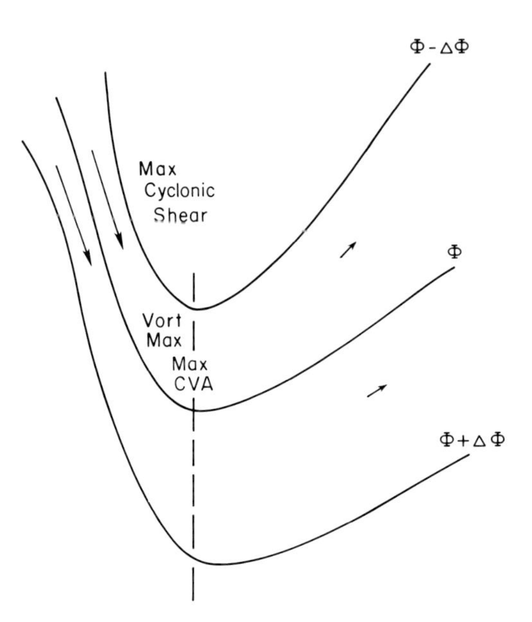

Figure 1.60 Vorticity advection in a short-wave trough whose maximum wind speeds are upstream from the trough axis. Geopotential height (solid lines). The greatest cyclonic *shear* is located upstream from the trough axis (dashed line); the trough axis is defined by the locus of relative maximum in *curvature* of the geopotential-height contours.

side of the downstream wind maximum (Fig. 1.61). The region of maximum cyclonic vorticity advection and height falls is therefore downstream and poleward from the "base" of the trough. The trough therefore has a poleward component of motion.

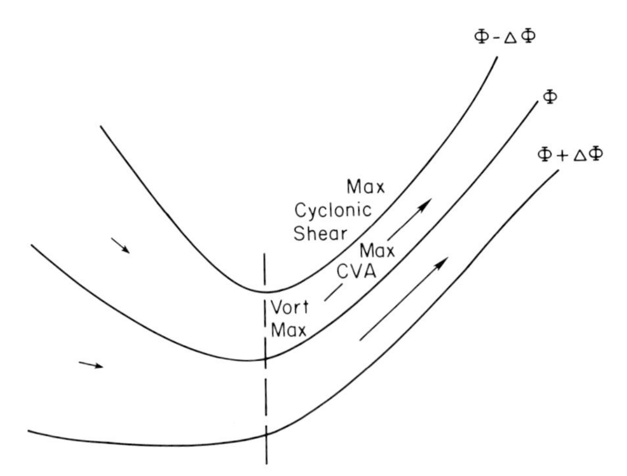

Figure 1.61 As in Fig. 1.60, but for a trough whose maximum wind speeds are downstream from the trough axis.

1.4.2. Group Velocity

Up to now we have considered the motion only of individual wave components, the *phase velocity*. Carl Gustav Rossby in 1945 considered the motion of groups of waves, the *group velocity,* marking the first time the concept of group velocity had been applied to meteorological phenomena.

Consider two sinusoidal geopotential-height fields, each having an amplitude $\hat{\Phi}$. The phase speed and wavelength of each are c_1, L_1, and c_2, L_2, respectively. Then the geopotential height of the sum of the two waves is

$$\Phi_{\text{total}} = \hat{\Phi}\left(\sin\frac{2\pi}{L_1}(x - c_1 t) + \sin\frac{2\pi}{L_2}(x - c_2 t)\right). \tag{1.4.12}$$

Using the trigonometric identity

$$\sin a + \sin b = 2\sin\tfrac{1}{2}(a+b)\cos\tfrac{1}{2}(a-b), \tag{1.4.13}$$

we find that

$$\begin{aligned}\Phi_{\text{total}} = 2\hat{\Phi}\sin&\left\{\frac{2\pi}{L_1 L_2}\left[\left(\frac{L_1+L_2}{2}\right)x - \left(\frac{c_1 L_2 + c_2 L_1}{2}\right)t\right]\right\}\\ \times\cos&\left\{\frac{2\pi}{L_1 L_2}\left[\left(\frac{L_2-L_1}{2}\right)x - \left(\frac{c_1 L_2 - c_2 L_1}{2}\right)t\right]\right\}.\end{aligned} \tag{1.4.14}$$

The total geopotential field is essentially the product of two traveling sinusoids. Suppose that

$$L_2 = L_1 + \delta L, \tag{1.4.15}$$

where $\delta L > 0$, and that the phase speeds are both eastward, so that $c_1 > 0$ and $c_2 > 0$; then the first part varies rapidly with x and t, while the second part varies slowly with x and t (Fig. 1.62). (We know from our experience with sound waves that when two sounds of nearly identical pitch are superimposed, we also hear the slow "beat frequency," a measure of the actual difference in frequency. When tuning a musical instrument against a standard tone, we know the instrument is in tune when the beat frequency goes to zero.) The speed of the "envelope" of the more rapidly varying wave is called the *group*

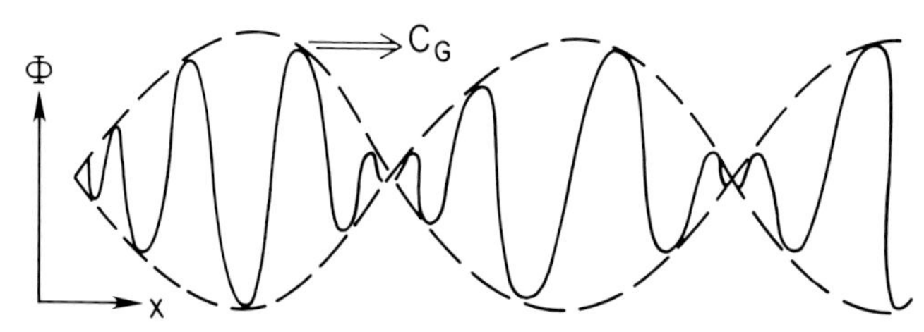

Figure 1.62 Illustration of group waves. The amplitude of the waves is plotted on the abscissa as a function of horizontal distance (x) at some instant. The envelope (dashed lines) of the shorter-wavelength (solid line) feature defines the group wave. The group wave moves with the speed c_G, while the individual waves move with the phase velocity.

velocity c_G, where

$$c_G = \frac{c_1 L_2 - c_2 L_1}{L_2 - L_1} \tag{1.4.16}$$

$$= \frac{c_1(L_2 - L_1) - L_1(c_2 - c_1)}{L_2 - L_1} \tag{1.4.17}$$

$$= c_1 - L_1 \frac{c_2 - c_1}{\delta L}. \tag{1.4.18}$$

In the limit as δL approaches zero

$$c_G = c - L\frac{dc}{dL}, \tag{1.4.19}$$

where c and L are the phase speed and horizontal wavelength of the rapidly varying wave.

Group velocity in an equivalent-barotropic quasigeostrophic atmosphere. Consider the simple geopotential height field again for a wavetrain in the westerlies (Fig. 1.63). If the atmosphere is baroclinic, but has no geostrophic temperature advection because the isotherms are parallel to the height contours, then the atmosphere is *equivalent barotropic.* We neglect for the sake of simplicity advection of temperature by the ageostrophic component of the wind, a process that is not quasigeostrophic, and friction. Thermodynamic processes will be assumed to be adiabatic. Suppose also that the horizontal wavelength is "short," so that progression of the troughs and ridges is governed mainly by advection of geostrophic vorticity. Then the wavetrain

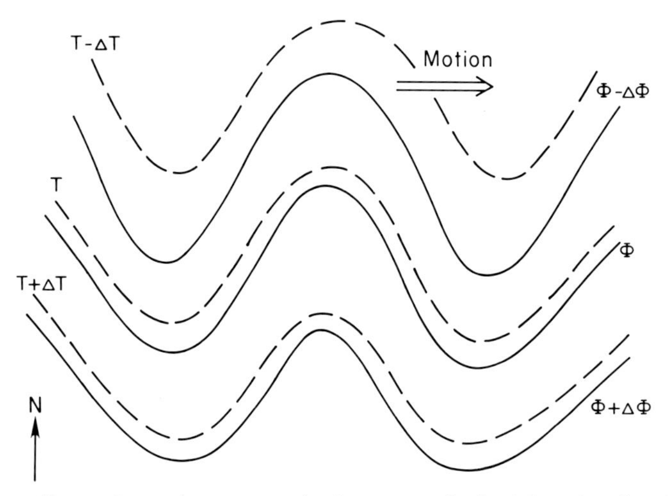

Figure 1.63 Illustration of a wavetrain in an equivalent-barotropic atmosphere. Geopotential-height contours (solid lines) for some pressure level, isotherms (dashed lines). The isotherms are parallel to the height contours.

progresses toward the east. As the wavetrain progresses toward the east, the wavetrain in the thickness field, which is coincident with the wavetrain in the height field, also progresses toward the east. An observer who is stationary with respect to the ground, but who is above the ground, would infer that local changes in temperature that occur as the equivalent barotropic wavetrain progresses are due to temperature changes resulting from vertical motion alone, since there is no temperature advection and no diabatic heating. However, an observer on the surface, which is assumed to have no slope, would find that $\omega = 0$ and therefore could not explain the local changes in temperature in terms of vertical motion, and hence would have to conclude that the atmosphere could not remain equivalent barotropic, and that temperature advection must be an important process at the surface. In fact, we know from observation that horizontal temperature advection is not only important at the surface, but is usually also strongest near the surface.

We substitute the height field (1.4.1) into the *barotropic vorticity equation,*

$$\frac{D_g}{Dt}(\zeta_g + f) = 0, \tag{1.4.20}$$

which is the quasigeostrophic vorticity equation valid at the level of nondivergence, and find that

$$c = U - \beta \Big/ \left(\frac{2\pi}{L}\right)^2. \tag{1.4.21}$$

[It should be recognized that Eq. (1.4.1) represents the solution to Eq. (1.4.20) linearized about the basic state $u_g = U$. In real life, u_g is neither time independent nor uniform; it changes owing to the *nonlinear* action of momentum advection: If u_g were not constant, then the advection term "$-(\mathbf{v}_g \cdot \nabla_p)\mathbf{v}_g$" would be a function of the *product* of terms containing $\mathbf{v}_g$ or its derivatives; if u_g were constant, then "$(-\mathbf{v}_g \cdot \nabla_p)\mathbf{v}_g$)" would be a function only of terms containing $\mathbf{v}_g$ or its derivatives, and hence would be linear.]

The "dispersion relation" (1.4.21) is known as the *Rossby-wave formula.* It shows how the phase speed of the wavetrain depends upon the horizontal wavelength L, the latitude (through β), and the basic zonal current U. At a given latitude, waves of different wavelengths are "dispersed" (i.e., are separated). Another example of dispersion is the separation of colors of visible light that occurs when sunlight passes through a glass prism. The boundary condition on the energy flux normal to the edge of the prism is essentially the electromagnetic analog of our atmospheric kinematic boundary condition. A consequence of this boundary condition on the energy flux is that, since each color of the spectrum propagates at a different phase speed, each travels out from the prism at a different angle.

The group velocity is different from the phase velocity only if the waves are dispersive. In a quasigeostrophic, equivalent-barotropic, frictionless atmosphere, we find from Eq. (1.4.21) and (1.4.19) that

$$c_G = U - \beta\bigg/\left(\frac{2\pi}{L}\right)^2 - L\frac{\partial}{\partial L}\left[U - \beta\bigg/\left(\frac{2\pi}{L}\right)^2\right]$$

$$= U + \beta\bigg/\left(\frac{2\pi}{L}\right)^2. \tag{1.4.22}$$

In this case the group velocity is faster than both the phase velocity and the zonal wind. The crests of the slowly varying envelope moving at the group velocity represent the high-amplitude sections of the more rapidly varying wave, and in a sense represent regions of amplification of the wave train, and perhaps cyclogenesis.

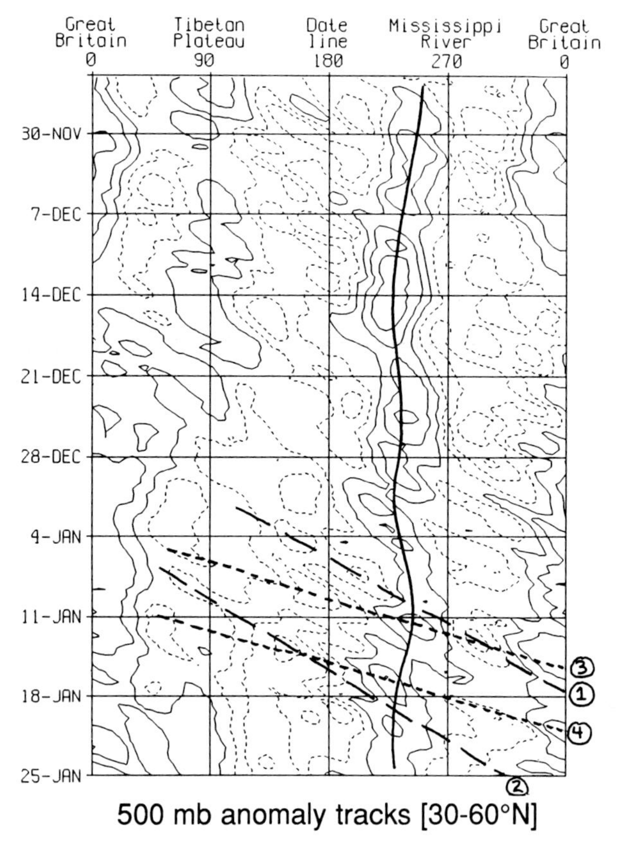

Figure 1.64 Hövmöller diagram from mid-November, 1989, to late January, 1990. The abscissa is degrees east of Great Britain; the ordinate is the day. Contours represent the 500-mb height anomaly averaged between 30 and 60°N. The contour interval is 60 m. Short dashed lines represent location of group waves; long dashed lines represent location of short-wave troughs; thick solid line represents location of quasistationary long-wave ridge. (Courtesy John McGinley, PROFS, Forecast Systems Laboratory, Boulder, Colorado)

Hövmöller diagrams. The relationship between the slowly moving, rapidly varying wavetrain, and the rapidly moving, slowly varying envelope is easily seen on a time-longitude plot of the geopotential height averaged over the latitude belt representing the westerlies (Fig. 1.64). This plot is called a *Hövmöller diagram* in honor of its inventor. Group velocity actually depends not only upon vorticity advection, but also temperature advection, and perhaps diabatic heating and, near the surface, friction. Using the Hövmöller diagram, we can identify individual, periodic short waves, and the group wave and make forecasts of wave amplification. For example, if the extrapolated position of a short-wave trough intersects a group wave, then we would forecast that the short-wave trough will intensify at the indicated longitude and on the given day.

1.4.3. Blocking

Having discussed certain aspects of the motion of systems, it is appropriate to consider a phenomenon characterized by the lack of motion. A "blocking" pattern exists when systems do not progress at all within the latitude belt of the baroclinic westerlies. The zonal movement of short waves is effectively halted, as the zonal current vanishes [Eq. (1.4.21)], and hence vorticity advection, the major mechanism by which upper level systems move, becomes negligible [Eq. (1.4.5)]. Blocking patterns make life miserable for many people affected by them, since those living near upper-level cyclones tend to experience a persistent combination of precipitation and relatively cool temperatures, while those living near upper-level anticyclones tend to experience drought conditions: Blocking situations are frequently accompanied by extreme meteorological events.

Much of the pioneering observational work on blocking was done around 1950 by D. Rex. In the Northern Hemisphere, blocking situations occur most frequently in April, and least frequently in August and September. Furthermore, the transition in blocking activity from late spring to midsummer is pronounced. Blocking activity often persists for 10 days or more.

The three most common types of blocking patterns are:

1. High-over-low block;
2. Omega block;
3. Stationary, high-amplitude ridge.

In the Northern Hemisphere the "high-over-low" block (Fig. 1.65) occurs most frequently over the west coasts of Europe and North America. This type of a block is often referred to as *split flow,* since the basic current "splits" around the block. (However, the expression *split flow* is used to describe such a flow pattern even if it does not persist.) Owing to the geographical preference for this type of block, horizontal variations in surface heating and/or topography probably play an important role in creating the block.

The flow pattern in the *omega block* has the shape of the capital Greek letter Ω (Fig. 1.66). It is a zonally oriented configuration of a high sandwiched

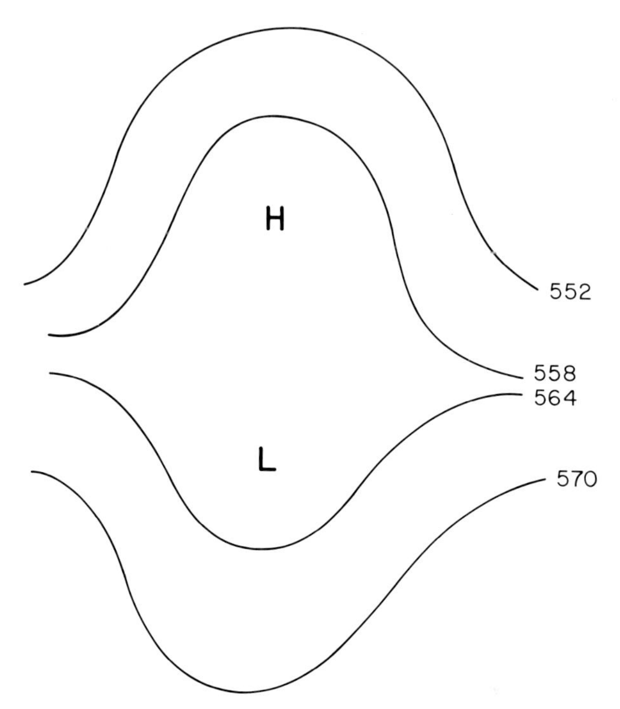

Figure 1.65 Idealized illustration of a high-over-low block in the Northern Hemisphere. 500-mb height contours in dam (solid lines). North is toward the top of the figure.

in between two lows. The third type of block, the stationary, high-amplitude ridge (Fig. 1.67), is associated mainly with hot, dry, dull weather.

The three types of blocking patterns we have just discussed are based upon subjective criteria. Objective classification schemes for blocking patterns may be based upon Hövmöller diagrams or zonal Fourier harmonic analysis. Blocking patterns are examples of the more general phenomenon of *persistent* (lasting longer than synoptic time scales) *flow anomalies.*

The dynamics of blocking patterns are not completely understood. It has

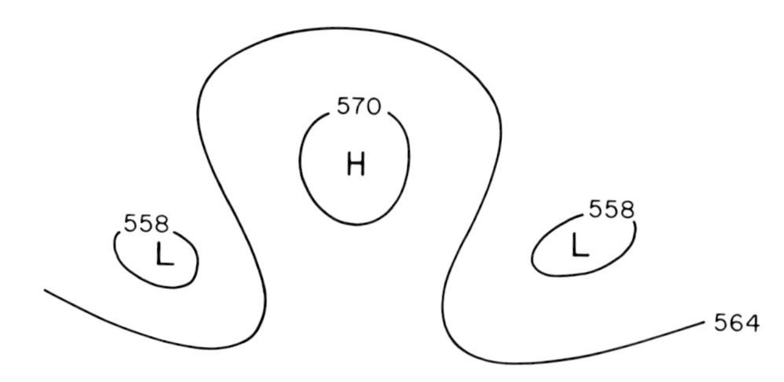

Figure 1.66 Idealized illustration of an omega block in the Northern Hemisphere. 500-mb height contours in dam (solid lines). North is toward the top of the figure.

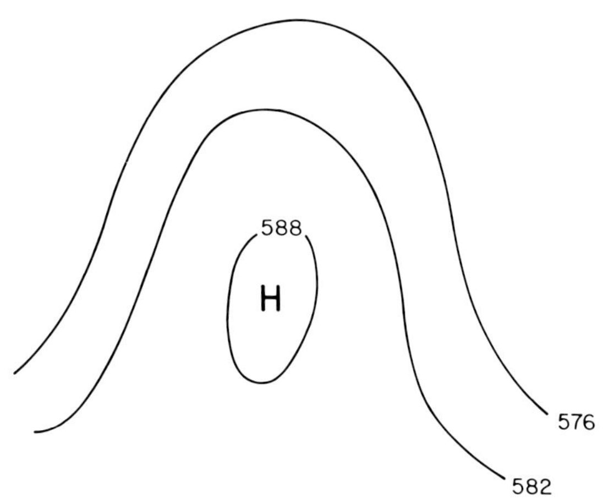

Figure 1.67 Idealized illustration of a stationary, high-amplitude ridge in the Northern Hemisphere. 500-mb height contours in dam (solid lines). North is toward the top of the figure.

been suggested that blocks might be the result of the resonant interaction between topographically forced (planetary-scale) waves and free (synoptic-scale) waves. It is also possible that blocking patterns might be *solitary waves*. Solitary waves in water are nonlinear, hump-shaped waves that are long lasting. They were first discovered in 1834 by J. Scott Russell, who observed one on the Edinburgh–to–Glasgow canal. In 1964 R. Long showed that wave motion mathematically similar to solitary-wave motion is possible in a westerly current in a rotating atmosphere. Such wave motion is stable because the dispersive effects of the time rate of change of, and mean advection of, perturbation vorticity terms are balanced by the "concentrating" effects of the nonlinear advection terms.

Strong surface cyclogenesis and anticyclogenesis are often associated with the creation of the pattern of closed highs and lows that occur in blocking situations. There is a conversion of available potential energy, which is associated with a strong meridional temperature gradient, into kinetic energy. The meridional temperature gradient is consequently reduced. As the meridional temperature gradient weakens, the zonal component of the thermal wind and the actual wind aloft weaken.

Hurd Willett in 1949 first suggested that blocking situations and periods of strong progression alternate over irregular periods of at least 2–3 weeks. Strong zonal flow, the *high-index pattern* (Fig. 1.68), weakens, moves equatorward, and becomes wavelike. Cyclogenesis and anticyclogenesis occur, and closed pressure centers appear. The latter pattern is referred to as a *low-index pattern* (Fig. 1.69). Eventually the westerlies increase as available potential energy increases in response to net radiational cooling at high latitudes, and warming at low latitudes. The alternation from high- to low-index flow has been reproduced in analog models of the atmosphere such as dishpans; it is called *vacillation*.

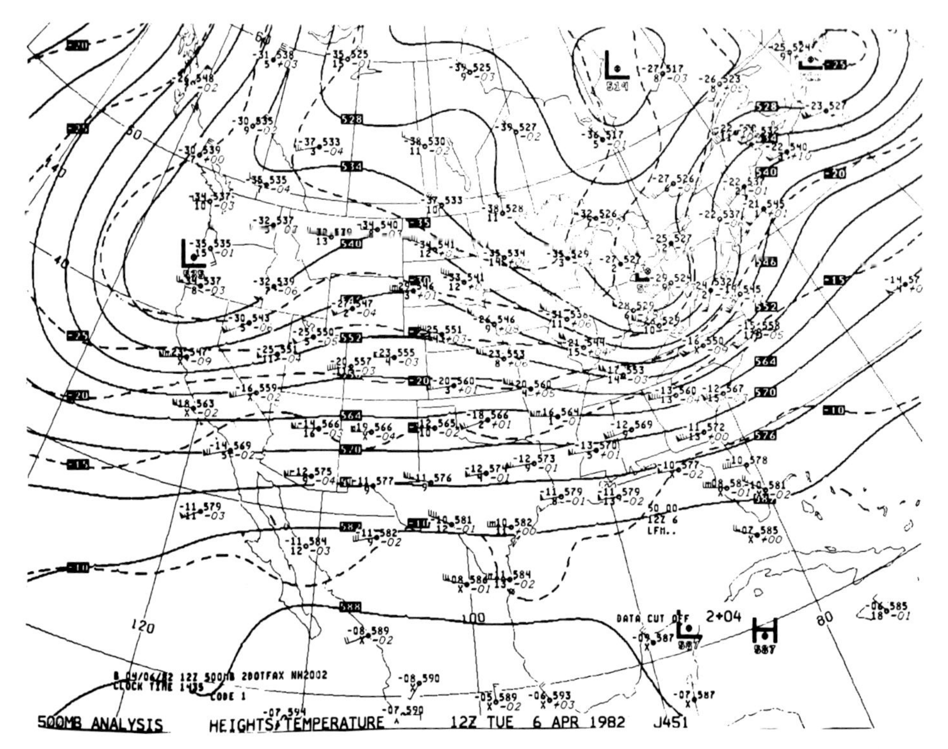

Figure 1.68 Example of a high-index pattern. NMC 500-mb analysis showing rapid low-latitude zonal flow, with intense troughs along the northern periphery of the current along the Pacific Northwest coast and over the Great Lakes, at 1200 UTC, April 6, 1982.

1.4.4. The Climatology of Height and Temperature at Standard Pressure Levels in the Troposphere

Climatological maps of height and temperature at selected pressure levels from 850 to 100 mb for the winter and summer seasons are shown in Figs. 1.70–1.93.

Winter climatology in the Northern Hemisphere. During the winter, areas of high height at 850 mb are found in the eastern Pacific and the Atlantic Oceans at 25° and 20°N, respectively, to the south of the eastern Pacific and Atlantic surface high-pressure areas and in the western Pacific near 20°N. Weak ridges of high height are located over Northwest Canada and Russia, coincident with a mean sea-level high-pressure area in the case of the former, and to the south in the latter. Centers of low height are seen to the west of Aleutians over the Bering Sea and over southern Greenland and to its west. The former extends to the west of the surface Aleutian low over to Siberia, and the latter extends to the west of the Icelandic low. A secondary trough is located over the southwest United States. Approximately three long-wave troughs are evident: One is over the eastern half of the United States; one is over the Central Pacific; one is over eastern Europe. Centers of cold air are located over the North Pole, northern Canada, and Siberia. Three major troughs in the

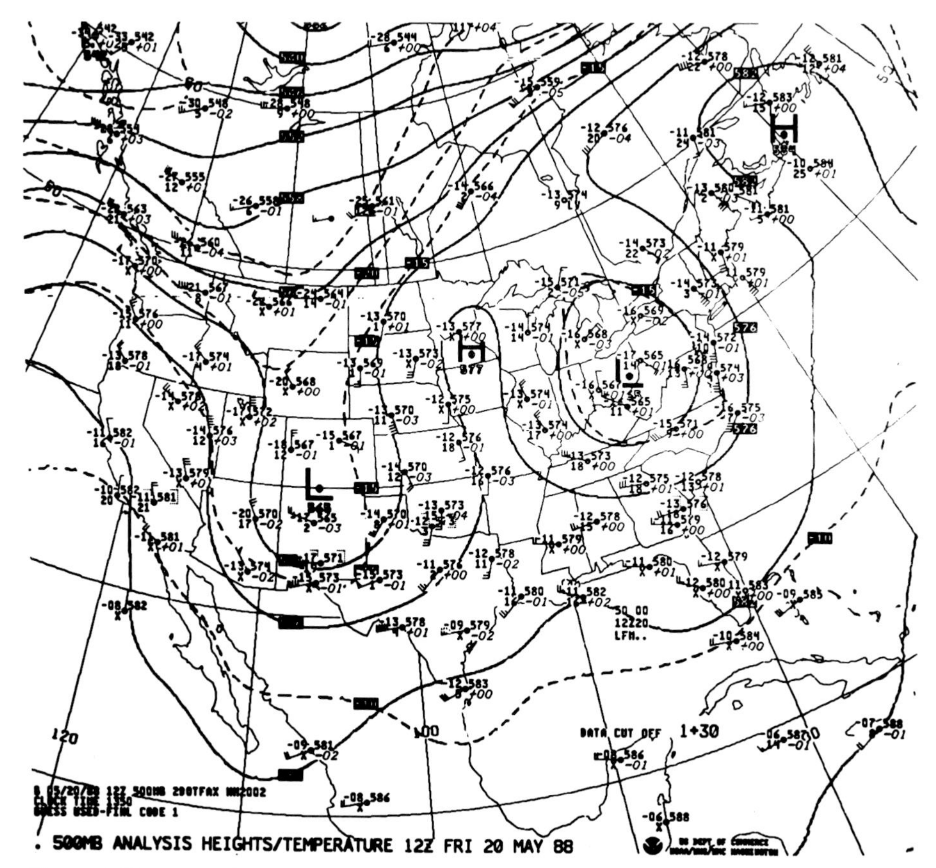

Figure 1.69 Example of a low-index pattern. NMC 500-mb analysis showing a train of troughs and high-amplitude ridges embedded in a very weak westerly current at 1200 UTC, May 20, 1988.

temperature field are also evident: One is over the east half of the United States; one is over the east coast of Asia; one weaker trough is over eastern Europe. The mean north–south temperature gradient is relatively intense [$\gtrsim$10 K (1000 km)$^{-1}$].

The 700-mb height pattern during the winter also shows three long-wave troughs, in locations similar to those of the 850-mb troughs. Ridges are found near 15–20°N in the Atlantic and western Pacific. A ridge also is found over North Africa and southeast Asia. The 700-mb temperature field exhibits two sharp troughs: one over eastern Asia and one over Hudson's Bay, and a broader one over eastern Europe.

The 500-mb winter height field also shows three long-wave troughs. A low is located near the North Pole, in between the east–Asian trough and the Hudson's Bay trough. The 500-mb isotherms are approximately parallel to the height contours: The atmosphere at 500 mb is therefore approximately equivalent barotropic in the mean. (The appearance of temperatures of −40°C

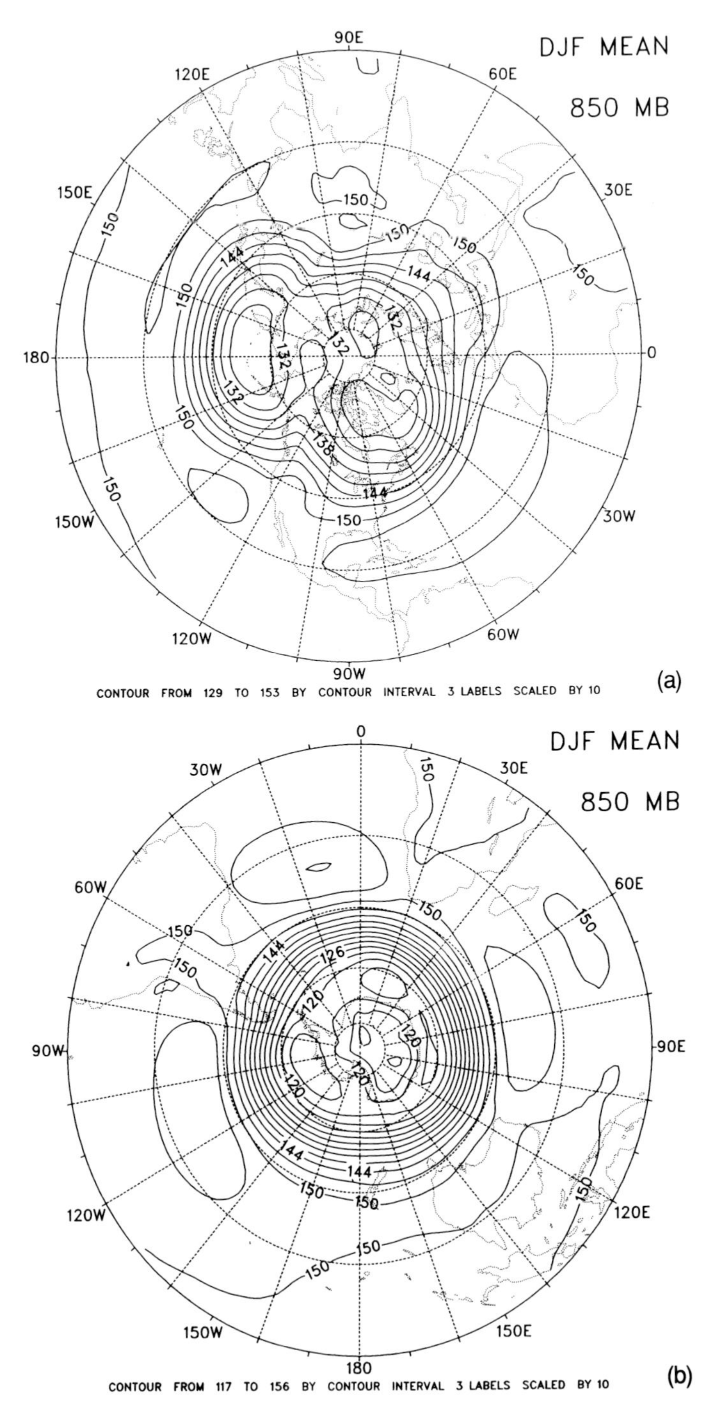

Figure 1.70 Average height (dam) at 850 mb for 1979–1988 in (a) the Northern Hemisphere for the winter and (b) the Southern Hemisphere for the summer (December, January, and February) (from ECMWF data, 1979–1988; courtesy Kevin Trenberth and Amy Solomon, NCAR).

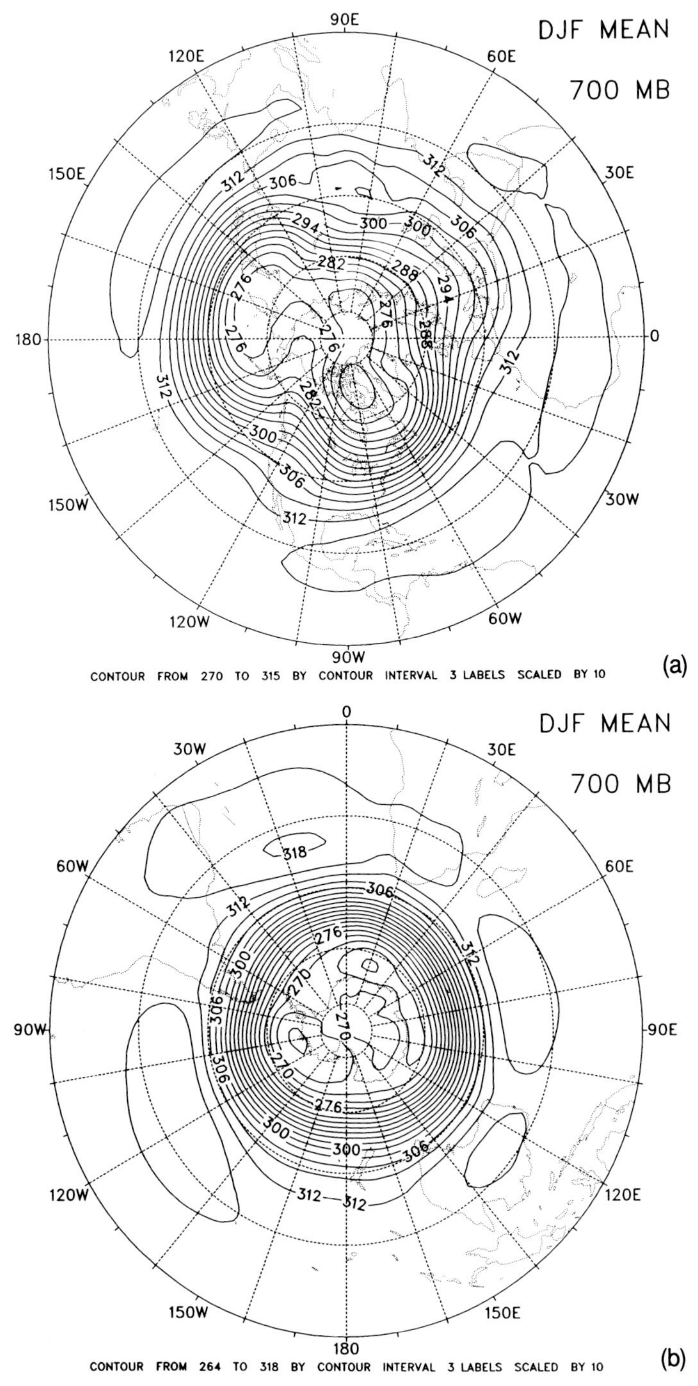

Figure 1.71 Average height (dam) at 700 mb for 1979–1988 in (a) the Northern Hemisphere for the winter and (b) the Southern Hemisphere for the summer (December, January, and February) (from ECMWF data, 1979–1988; courtesy Kevin Trenberth and Amy Solomon, NCAR).

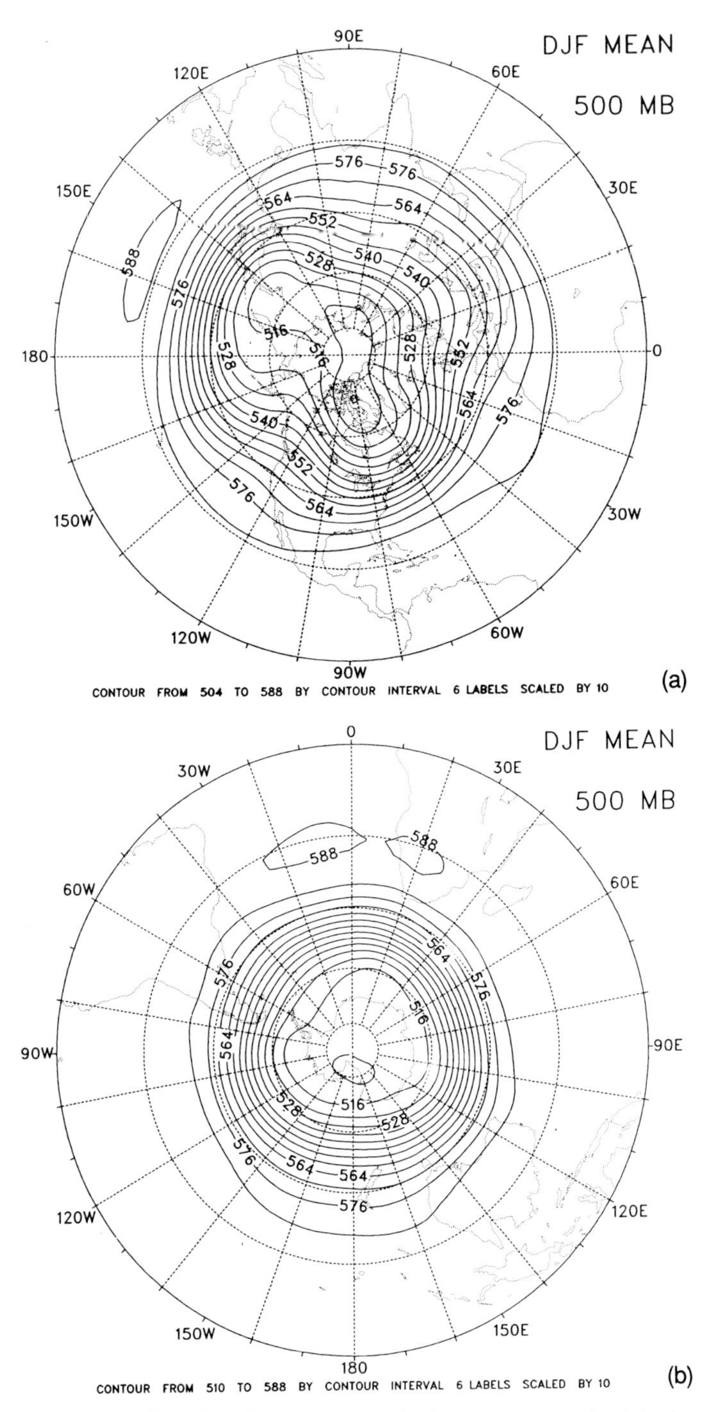

Figure 1.72 Average height (dam) at 500 mb for 1979–1988 in (a) the Northern Hemisphere for the winter and (b) the Southern Hemisphere for the summer (December, January, and February) (from ECMWF data, 1979–1988; courtesy Kevin Trenberth and Amy Solomon, NCAR).

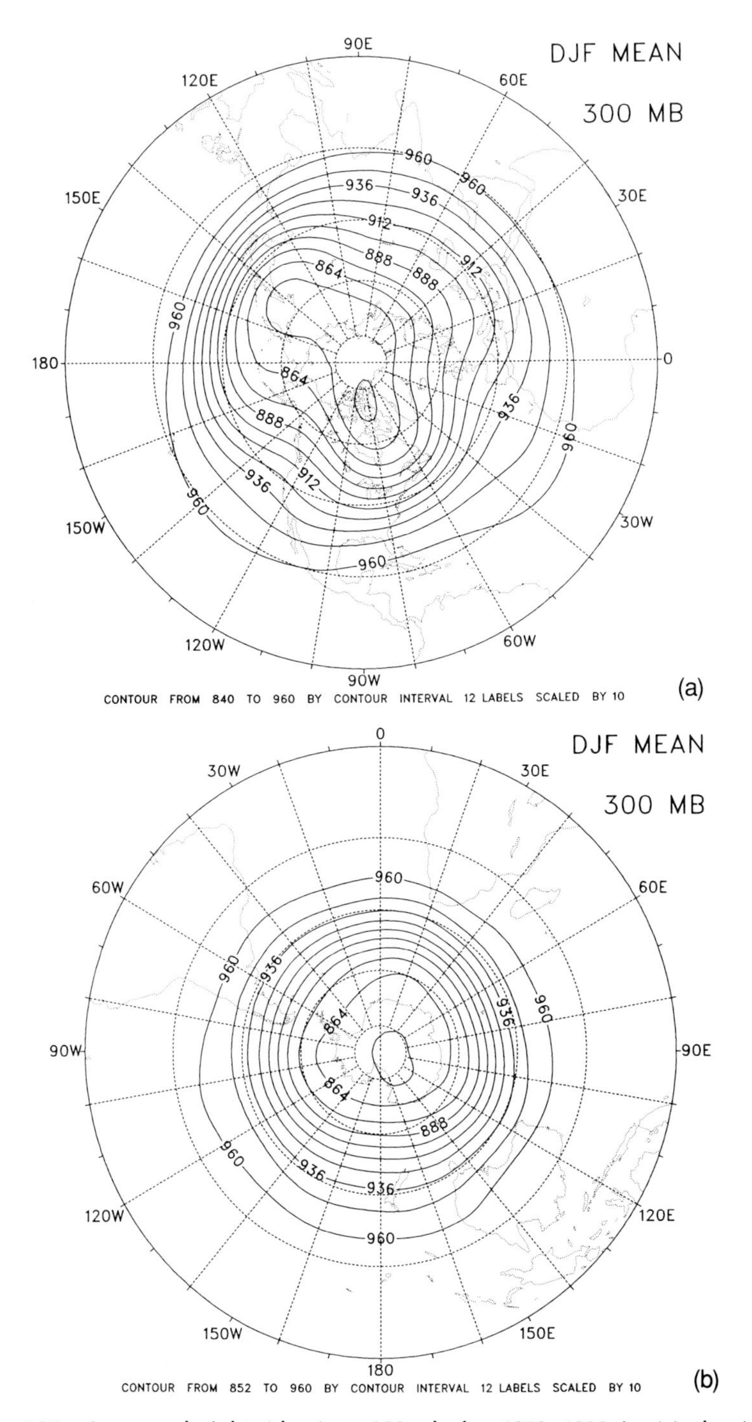

Figure 1.73 Average height (dam) at 300 mb for 1979–1988 in (a) the Northern Hemisphere for the winter and (b) the Southern Hemisphere for the summer (December, January, and February) (from ECMWF data, 1979–1988; courtesy Kevin Trenberth and Amy Solomon, NCAR).

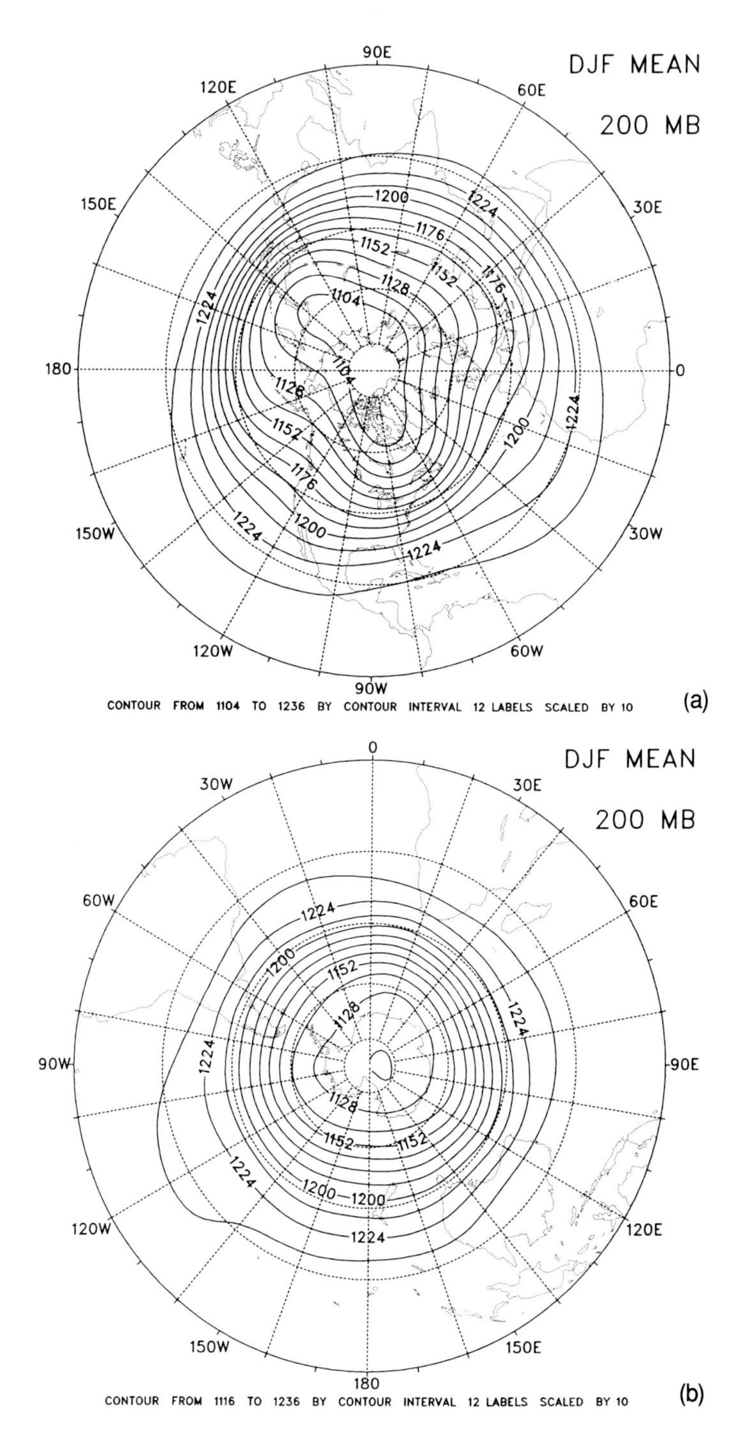

Figure 1.74 Average height (dam) at 200 mb for 1979–1988 in (a) the Northern Hemisphere for the winter and (b) the Southern Hemisphere for the summer (December, January, and February) (from ECMWF data, 1979–1988; courtesy Kevin Trenberth and Amy Solomon, NCAR).

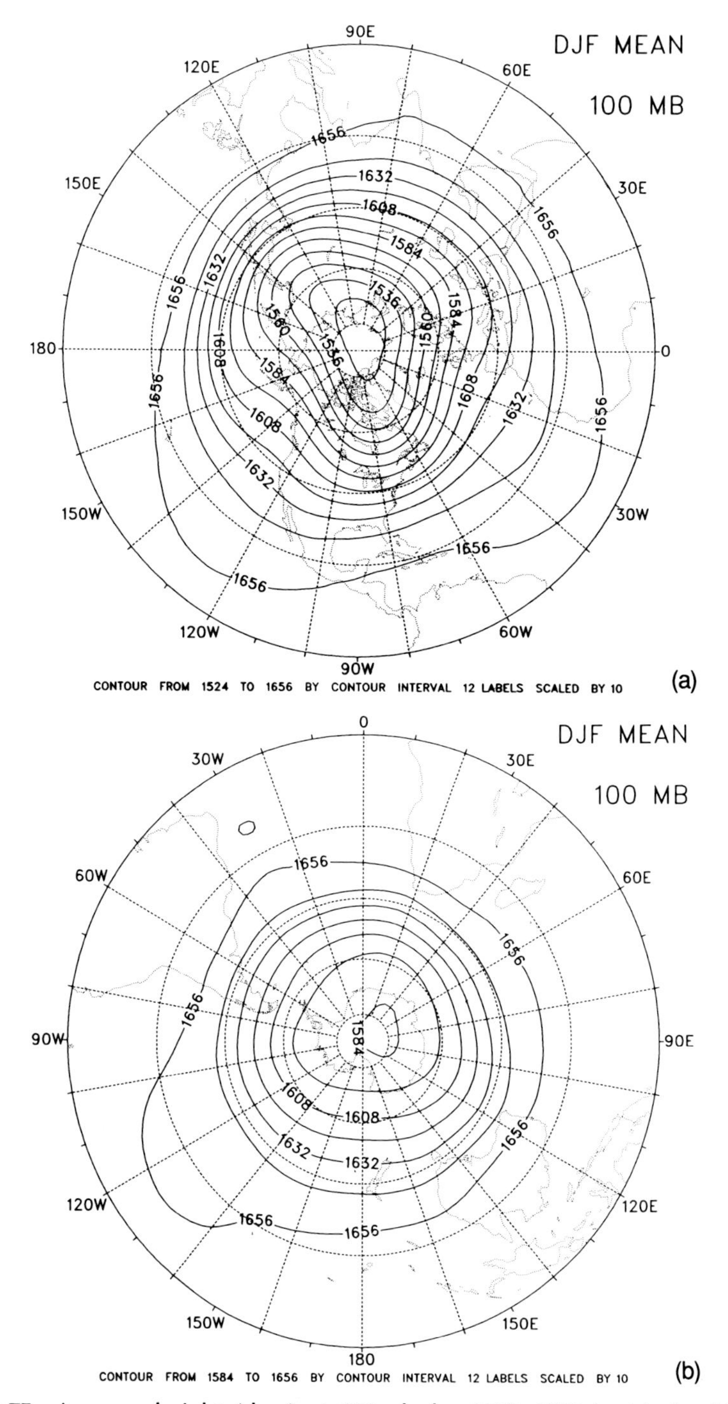

Figure 1.75 Average height (dam) at 100 mb for 1979–1988 in (a) the Northern Hemisphere for the winter and (b) the Southern Hemisphere for the summer (December, January, and February) (from ECMWF data, 1979–1988; courtesy Kevin Trenberth and Amy Solomon, NCAR).

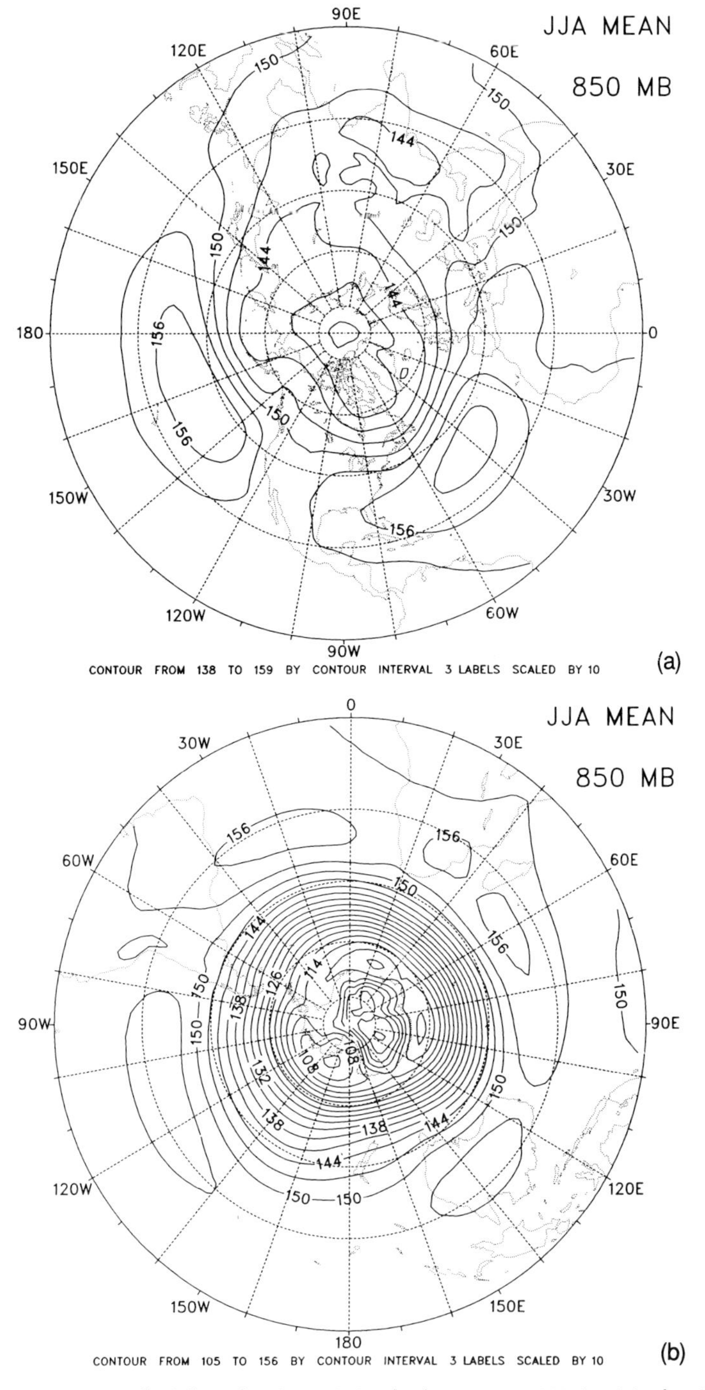

Figure 1.76 Average height (dam) at 850 mb for 1979–1988 in (a) the Northern Hemisphere for the summer and (b) the Southern Hemisphere for the winter (June, July, and August) (from ECMWF data, 1979–1988; courtesy Kevin Trenberth and Amy Solomon, NCAR).

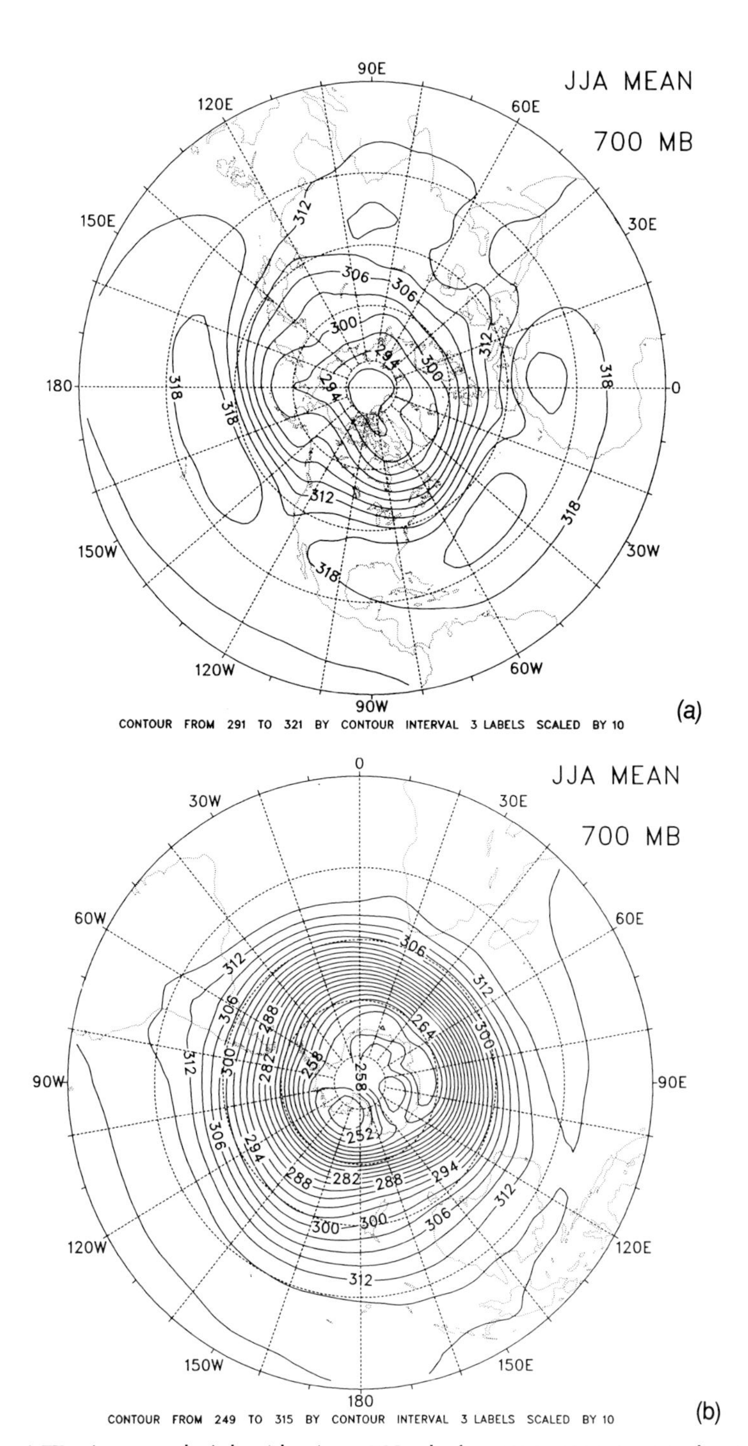

Figure 1.77 Average height (dam) at 700 mb for 1979–1988 in (a) the Northern Hemisphere for the summer and (b) the Southern Hemisphere for the winter (June, July, and August) (from ECMWF data, 1979–1988; courtesy Kevin Trenberth and Amy Solomon, NCAR).

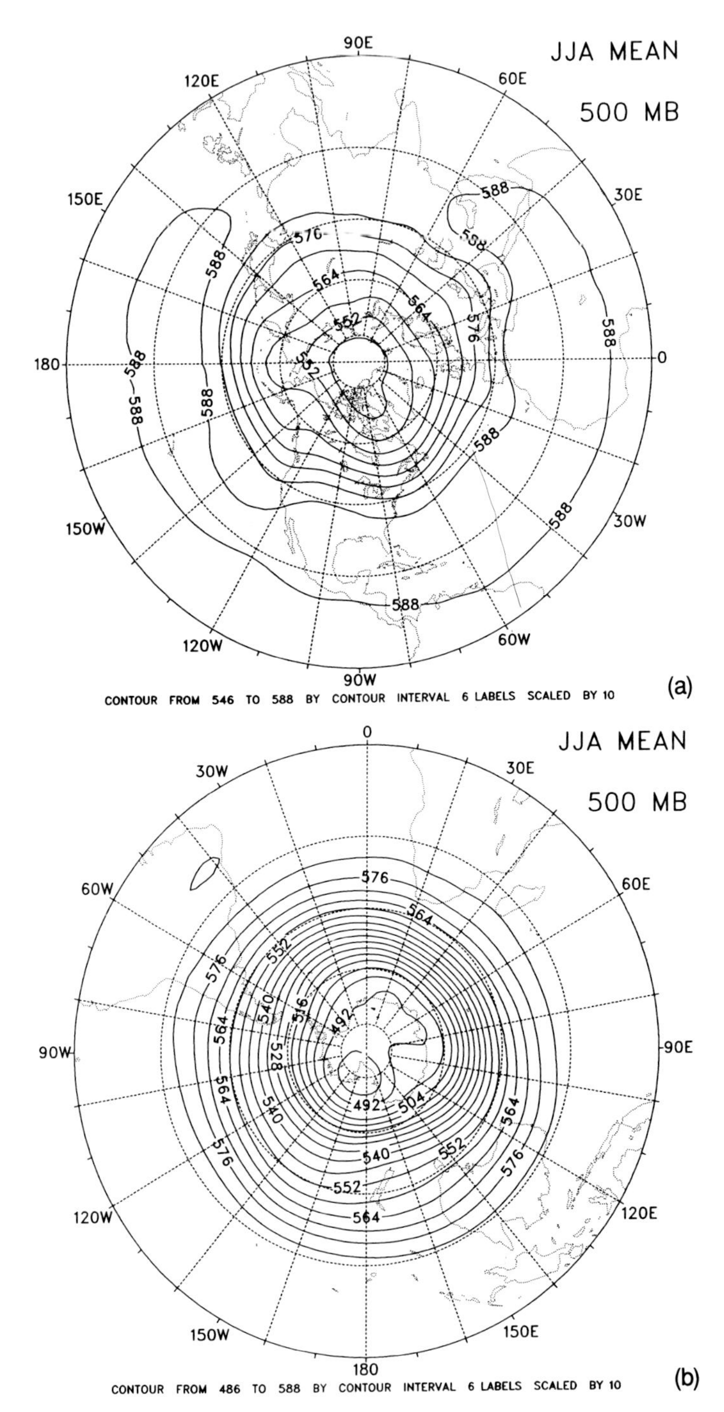

Figure 1.78 Average height (dam) at 500 mb for 1979–1988 in (a) the Northern Hemisphere for the summer and (b) the Southern Hemisphere for the winter (June, July, and August) (from ECMWF data, 1979–1988; courtesy Kevin Trenberth and Amy Solomon, NCAR).

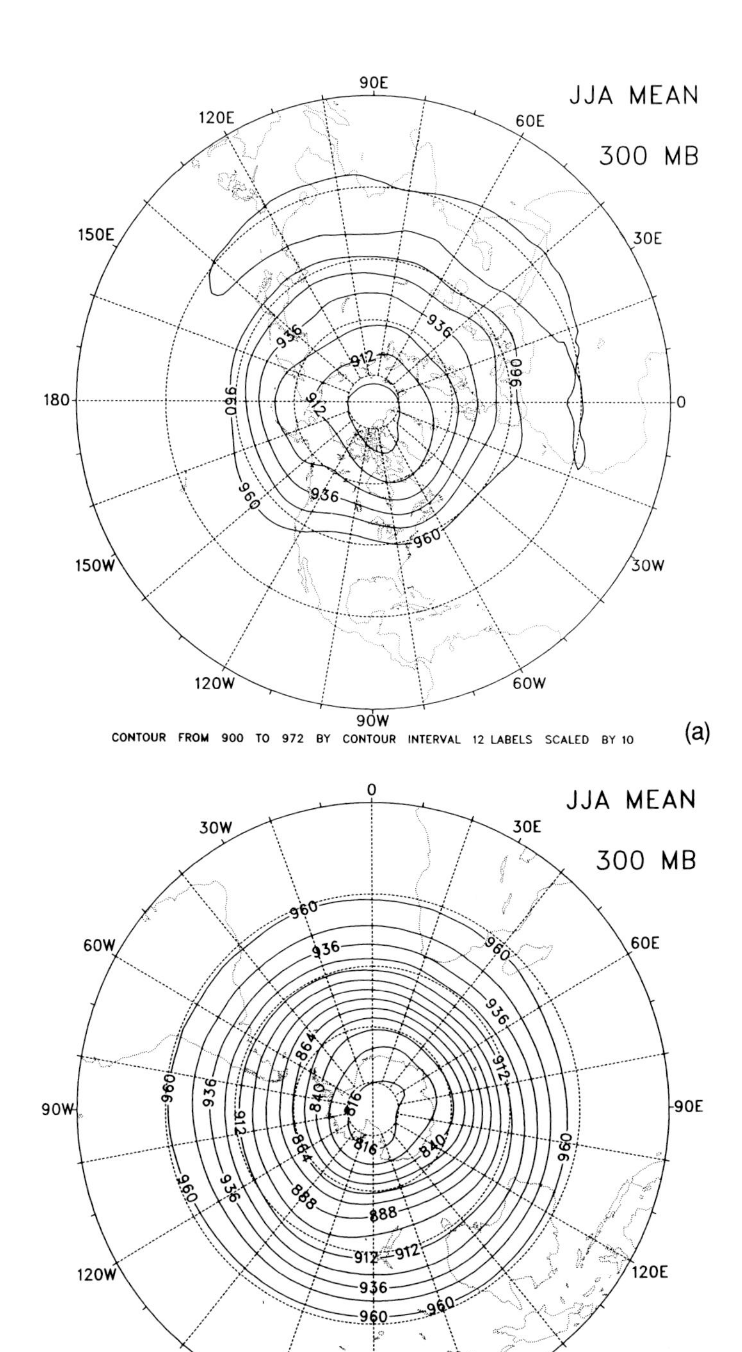

Figure 1.79 Average height (dam) at 300 mb for 1979–1988 in (a) the Northern Hemisphere for the summer and (b) the Southern Hemisphere for the winter (June, July, and August) (from ECMWF data, 1979–1988; courtesy Kevin Trenberth and Amy Solomon, NCAR).

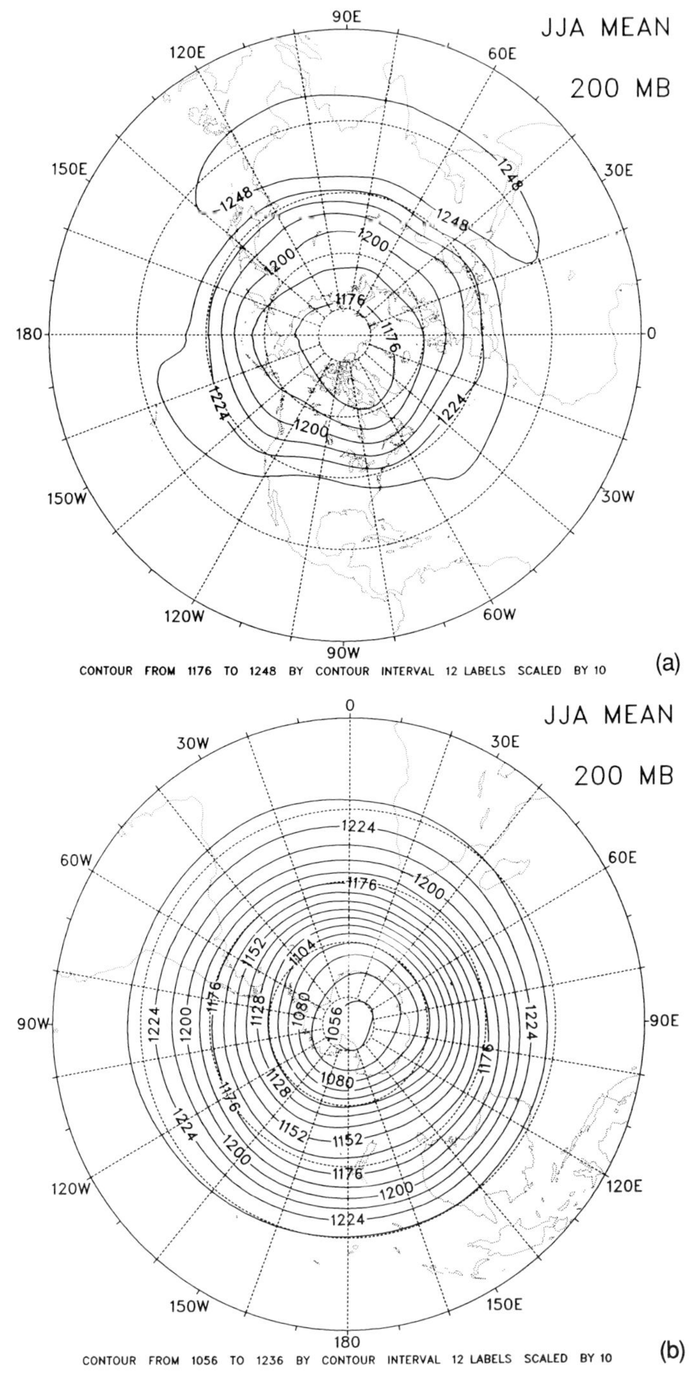

Figure 1.80 Average height (dam) at 200 mb for 1979–1988 in (a) the Northern Hemisphere for the summer and (b) the Southern Hemisphere for the winter (June, July, and August) (from ECMWF data, 1979–1988; courtesy Kevin Trenberth and Amy Solomon, NCAR).

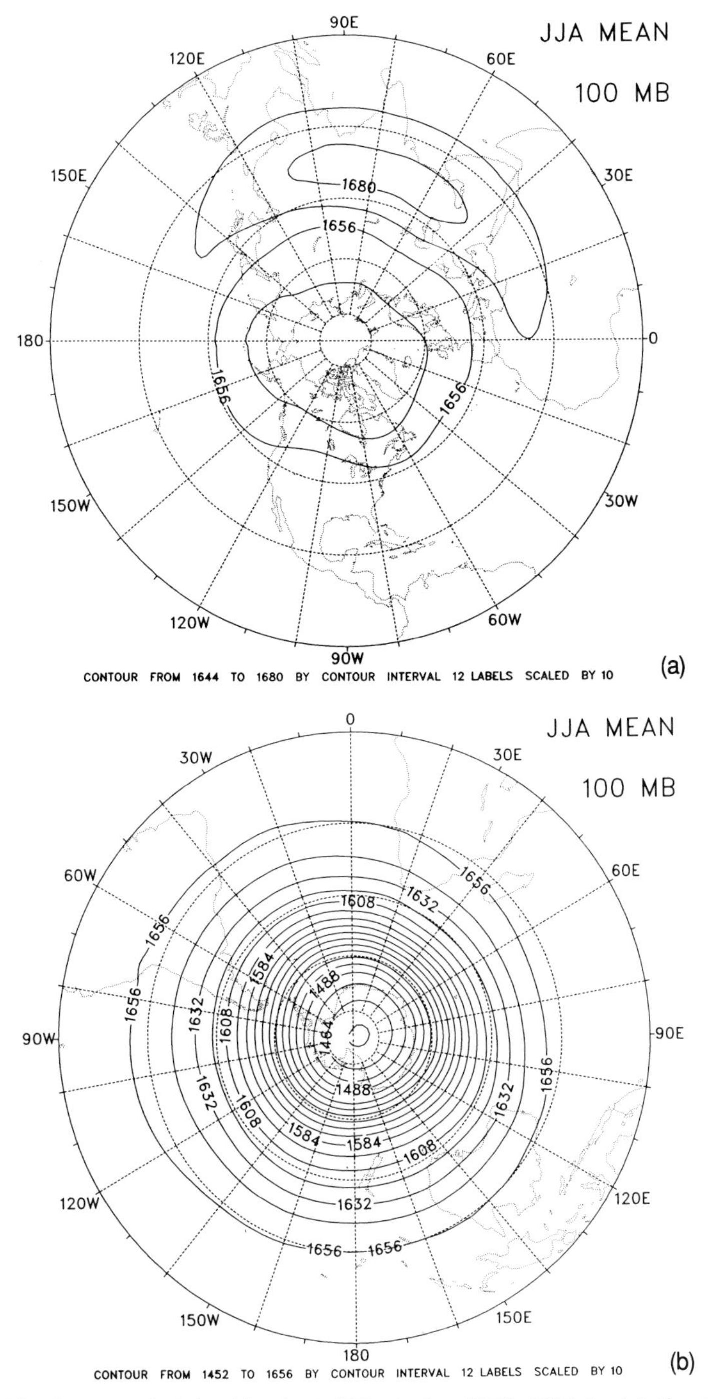

Figure 1.81 Average height (dam) at 100 mb for 1979–1988 in (a) the Northern Hemisphere for the summer and (b) the Southern Hemisphere for the winter (June, July, and August) (from ECMWF data, 1979–1988; courtesy Kevin Trenberth and Amy Solomon, NCAR).

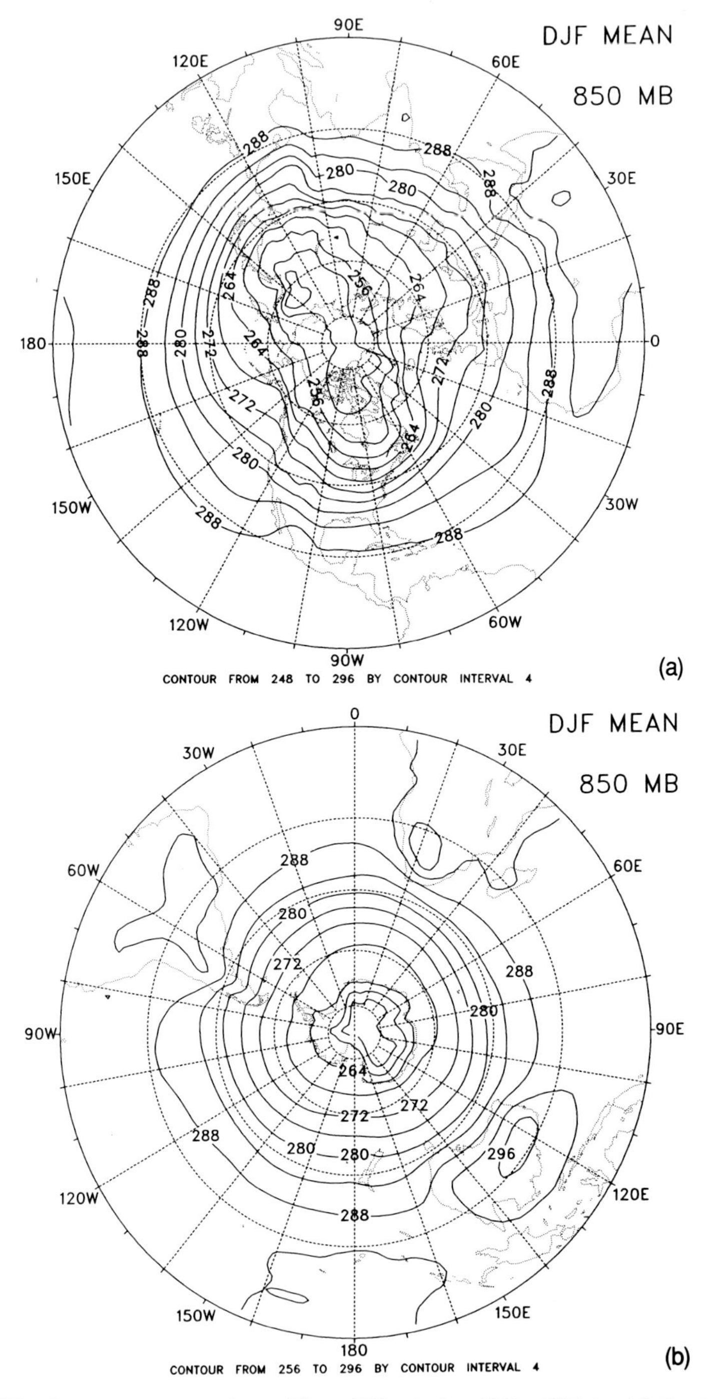

Figure 1.82 Average temperature (K) at 850 mb for 1979–1988 in (a) the Northern Hemisphere for the winter and (b) the Southern Hemisphere for the summer (December, January, and February) (from ECMWF data, 1979–1988; courtesy Kevin Trenberth and Amy Solomon, NCAR).

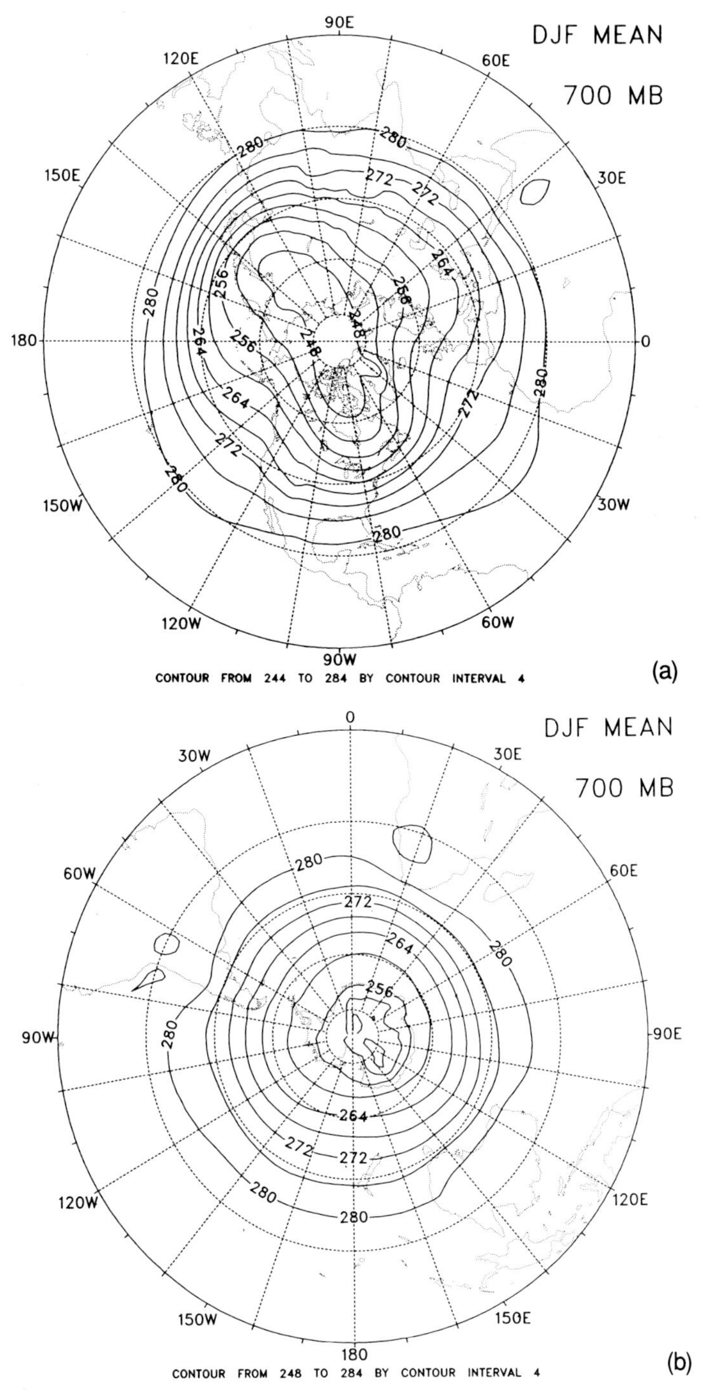

Figure 1.83 Average temperature (K) at 700 mb for 1979–1988 in (a) the Northern Hemisphere for the winter and (b) the Southern Hemisphere for the summer (December, January, and February) (from ECMWF data, 1979–1988; courtesy Kevin Trenberth and Amy Solomon, NCAR).

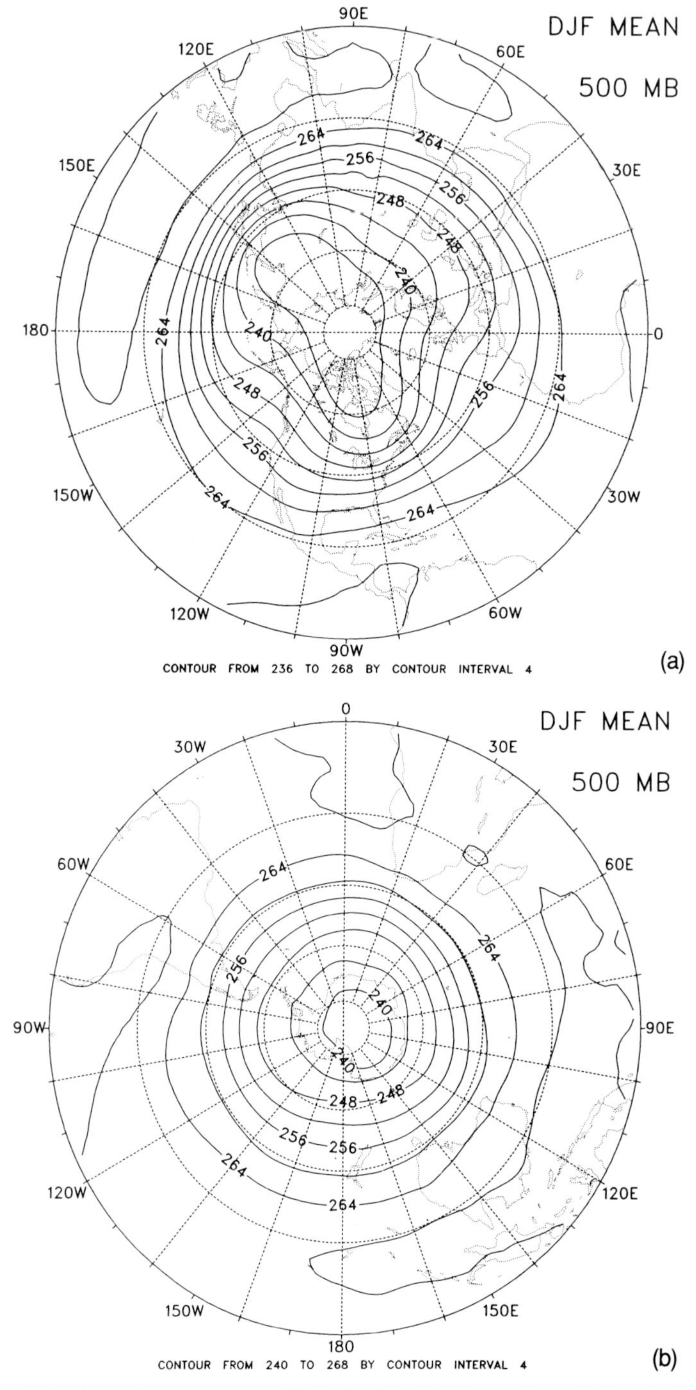

Figure 1.84 Average temperature (K) at 500 mb for 1979–1988 in (a) the Northern Hemisphere for the winter and (b) the Southern Hemisphere for the summer (December, January, and February) (from ECMWF data, 1979–1988; courtesy Kevin Trenberth and Amy Solomon, NCAR).

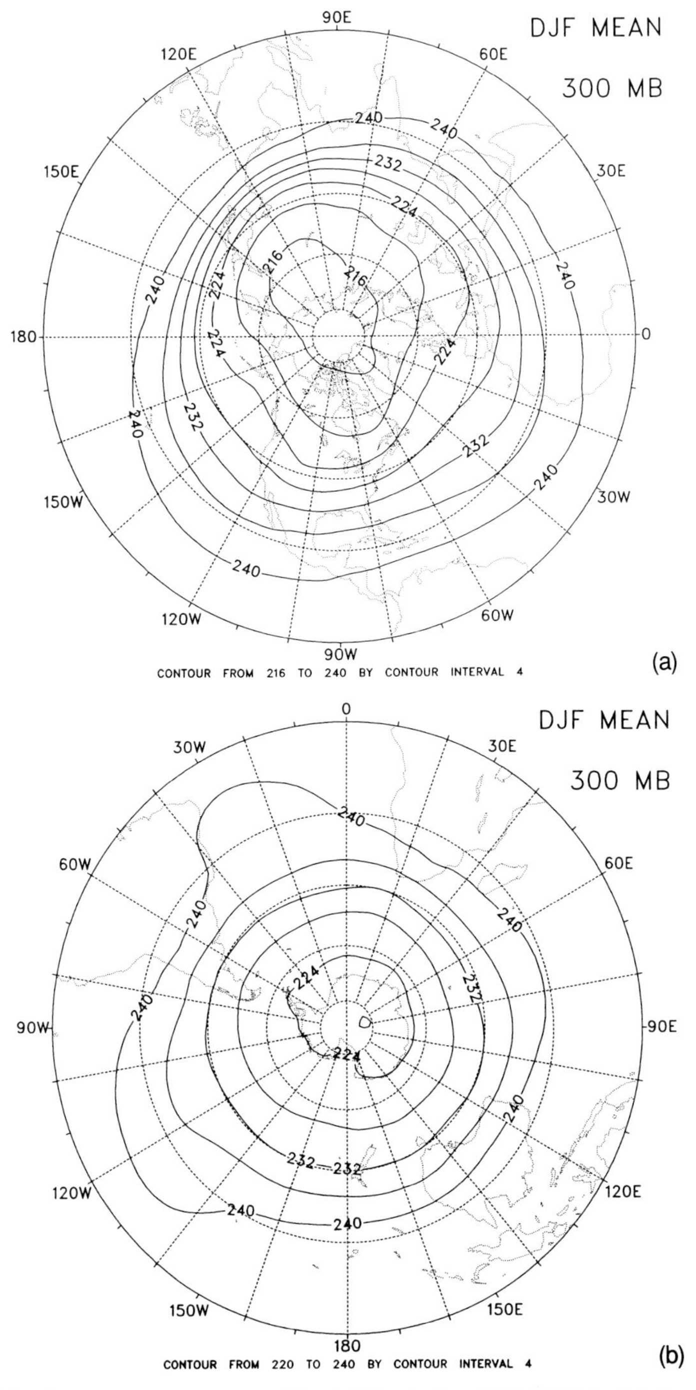

Figure 1.85 Average temperature (K) at 300 mb for 1979–1988 in (a) the Northern Hemisphere for the winter and (b) the Southern Hemisphere for the summer (December, January, and February) (from ECMWF data, 1979–1988; courtesy Kevin Trenberth and Amy Solomon, NCAR).

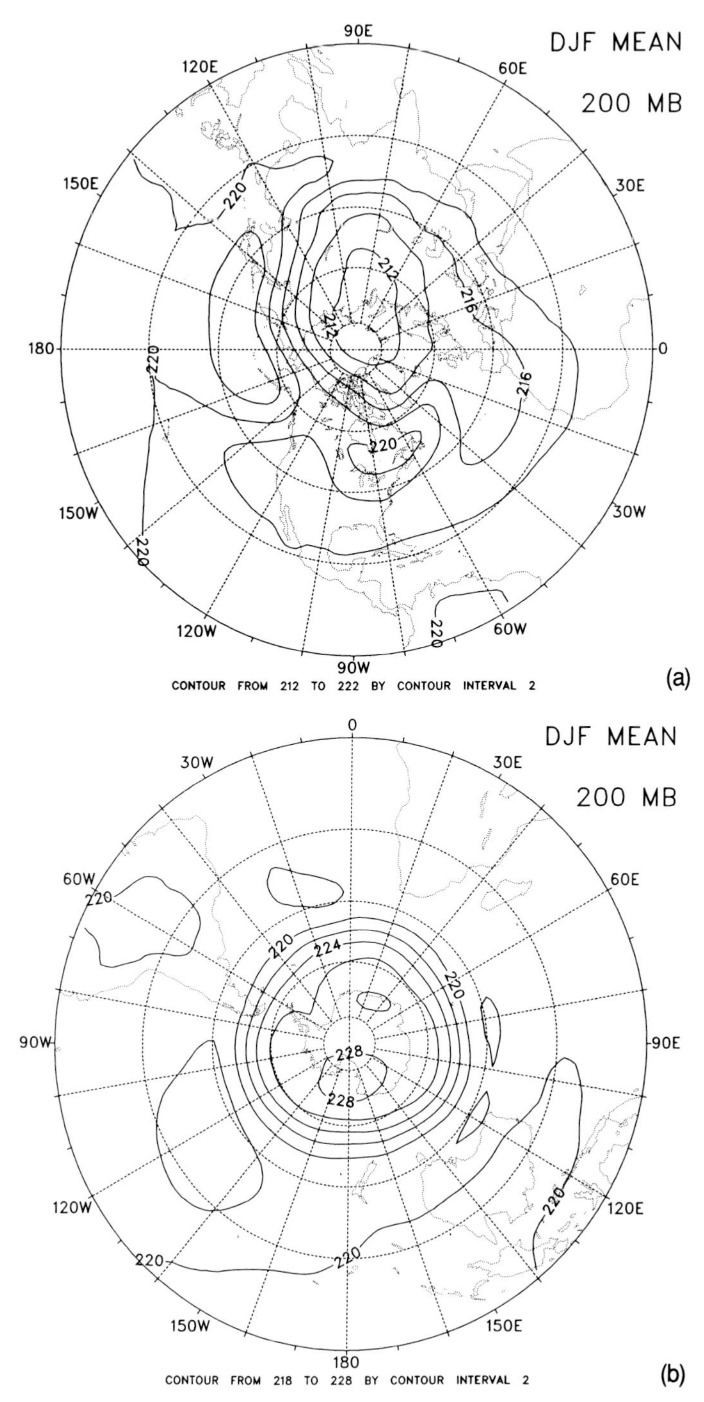

Figure 1.86 Average temperature (K) at 200 mb for 1979–1988 in (a) the Northern Hemisphere for the winter and (b) the Southern Hemisphere for the summer (December, January, and February) (from ECMWF data, 1979–1988; courtesy Kevin Trenberth and Amy Solomon, NCAR).

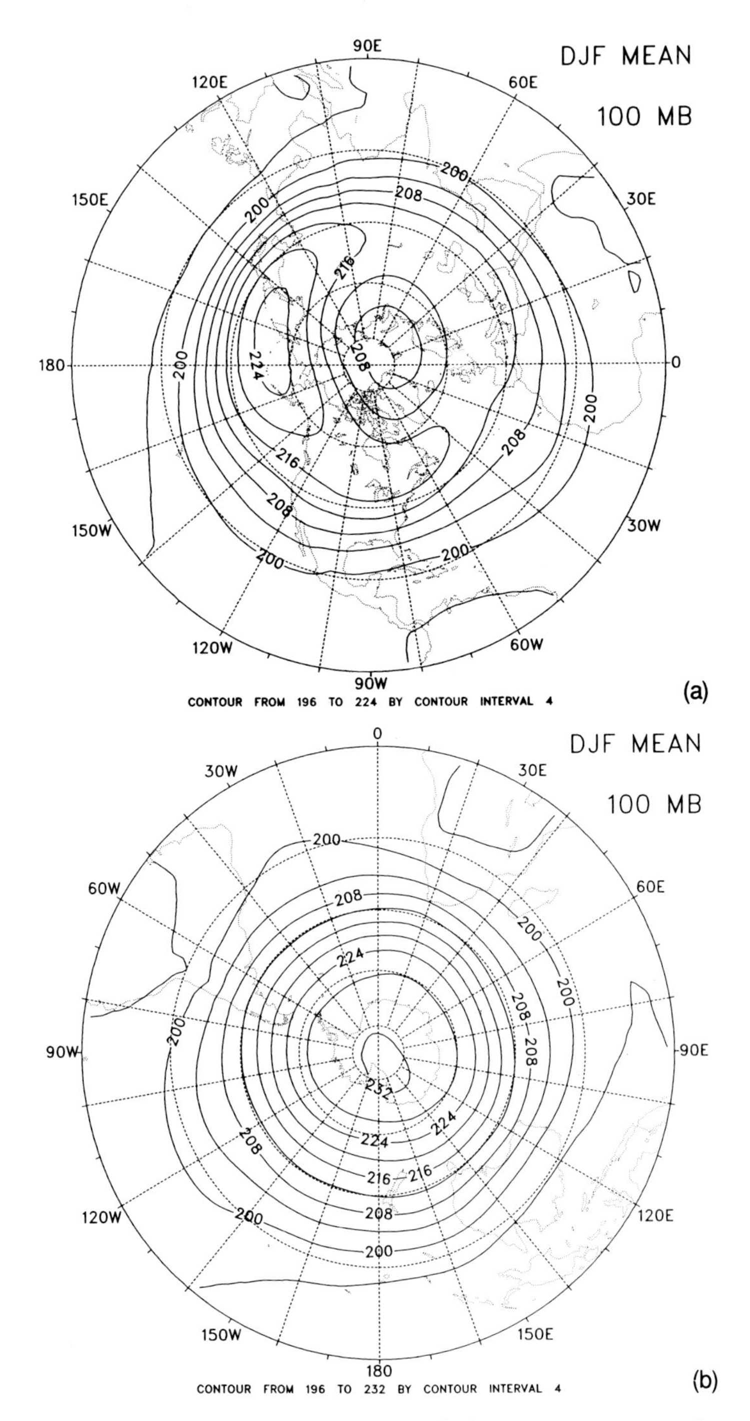

Figure 1.87 Average temperature (K) at 100 mb for 1979–1988 in (a) the Northern Hemisphere for the winter and (b) the Southern Hemisphere for the summer (December, January, and February) (from ECMWF data, 1979–1988; courtesy Kevin Trenberth and Amy Solomon, NCAR).

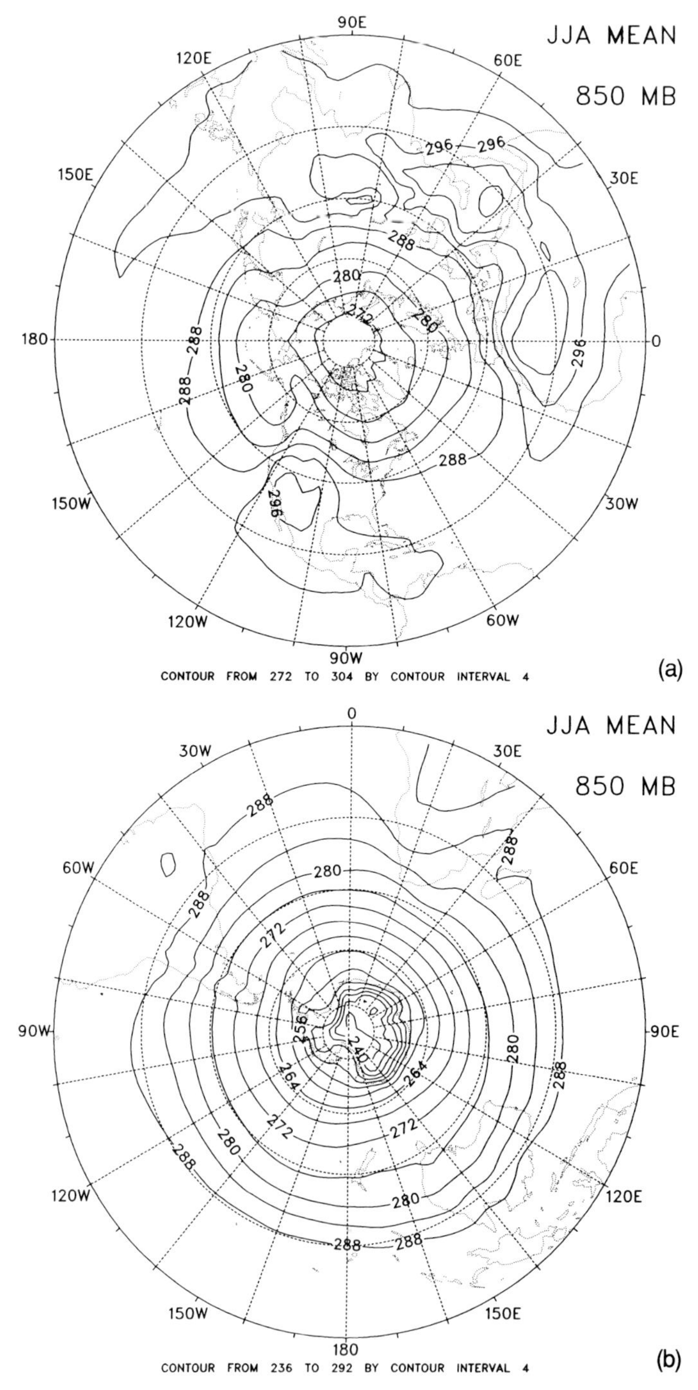

Figure 1.88 Average temperature (K) at 850 mb for 1979–1988 in (a) the Northern Hemisphere for the summer and (b) the Southern Hemisphere for the winter (June, July, and August) (from ECMWF data, 1979–1988; courtesy Kevin Trenberth and Amy Solomon, NCAR).

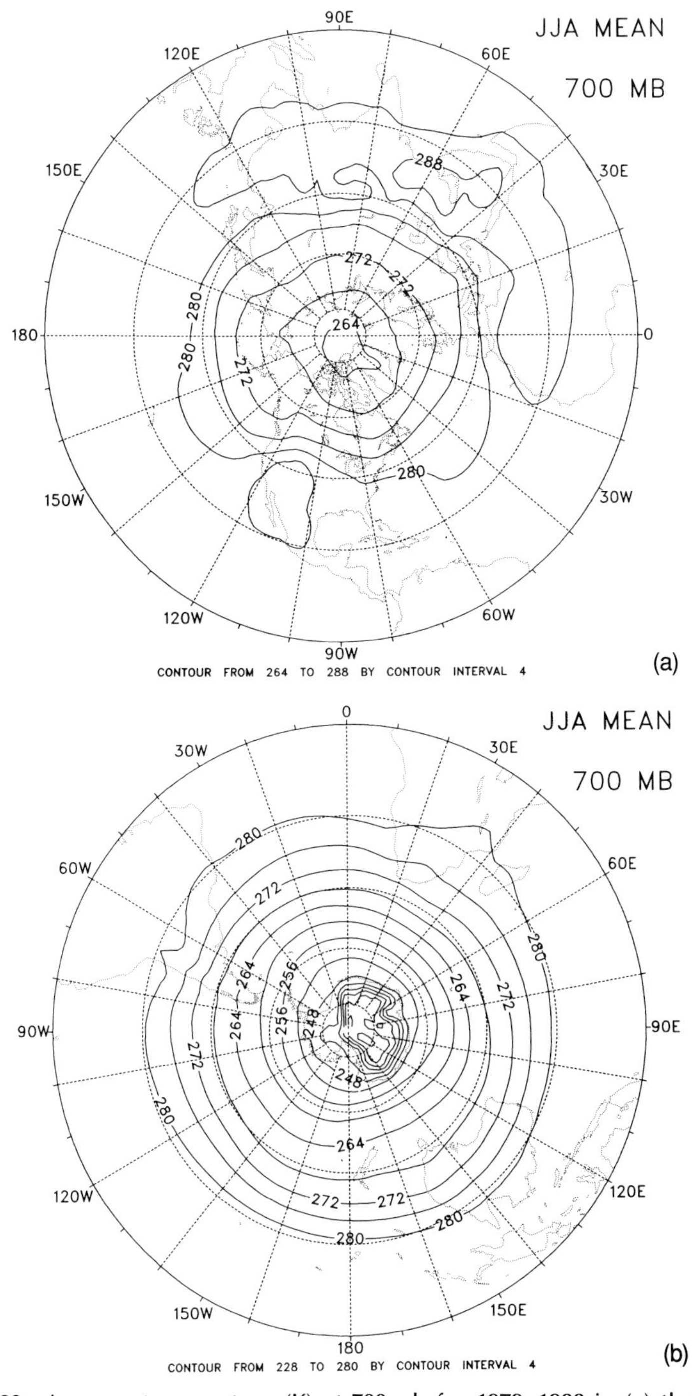

Figure 1.89 Average temperature (K) at 700 mb for 1979–1988 in (a) the Northern Hemisphere for the summer and (b) the Southern Hemisphere for the winter (June, July, and August) (from ECMWF data, 1979–1988; courtesy Kevin Trenberth and Amy Solomon, NCAR).

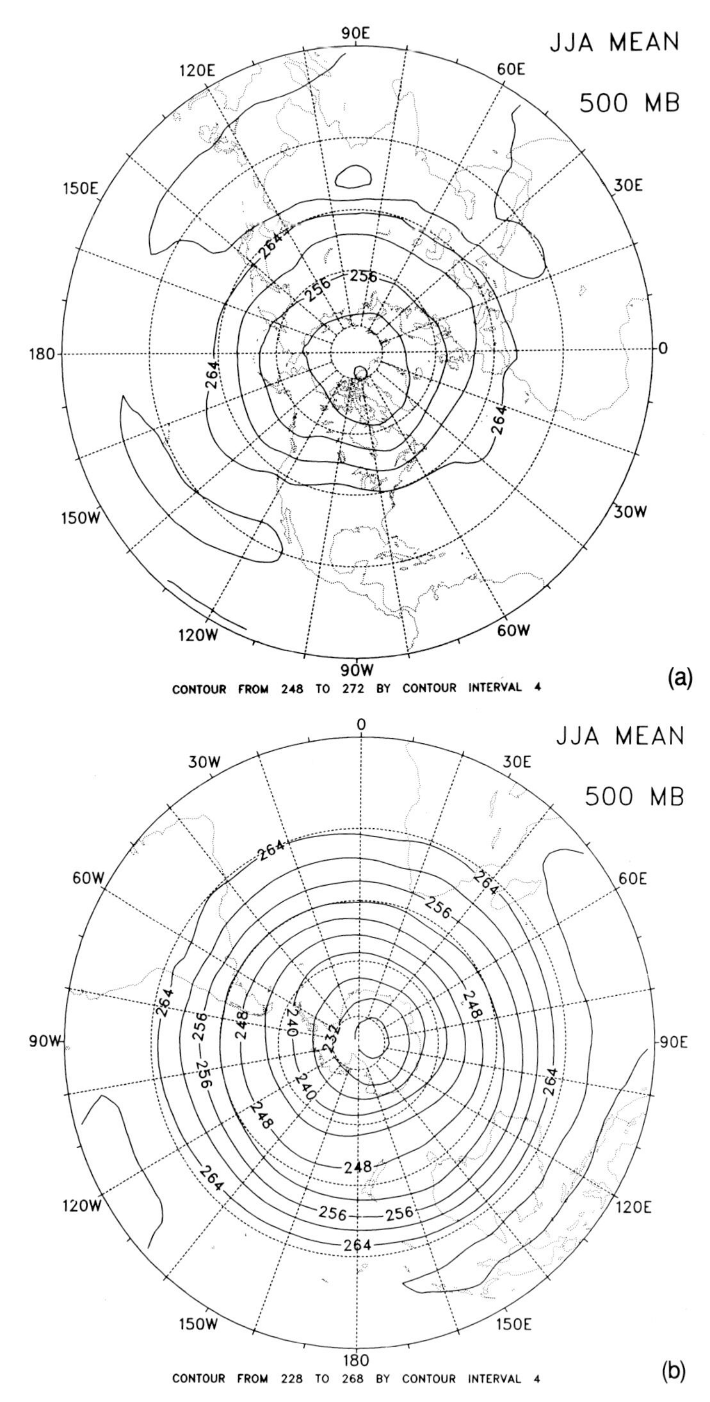

Figure 1.90 Average temperature (K) at 500 mb for 1979–1988 in (a) the Northern Hemisphere for the summer and (b) the Southern Hemisphere for the winter (June, July, and August) (from ECMWF data, 1979–1988; courtesy Kevin Trenberth and Amy Solomon, NCAR).

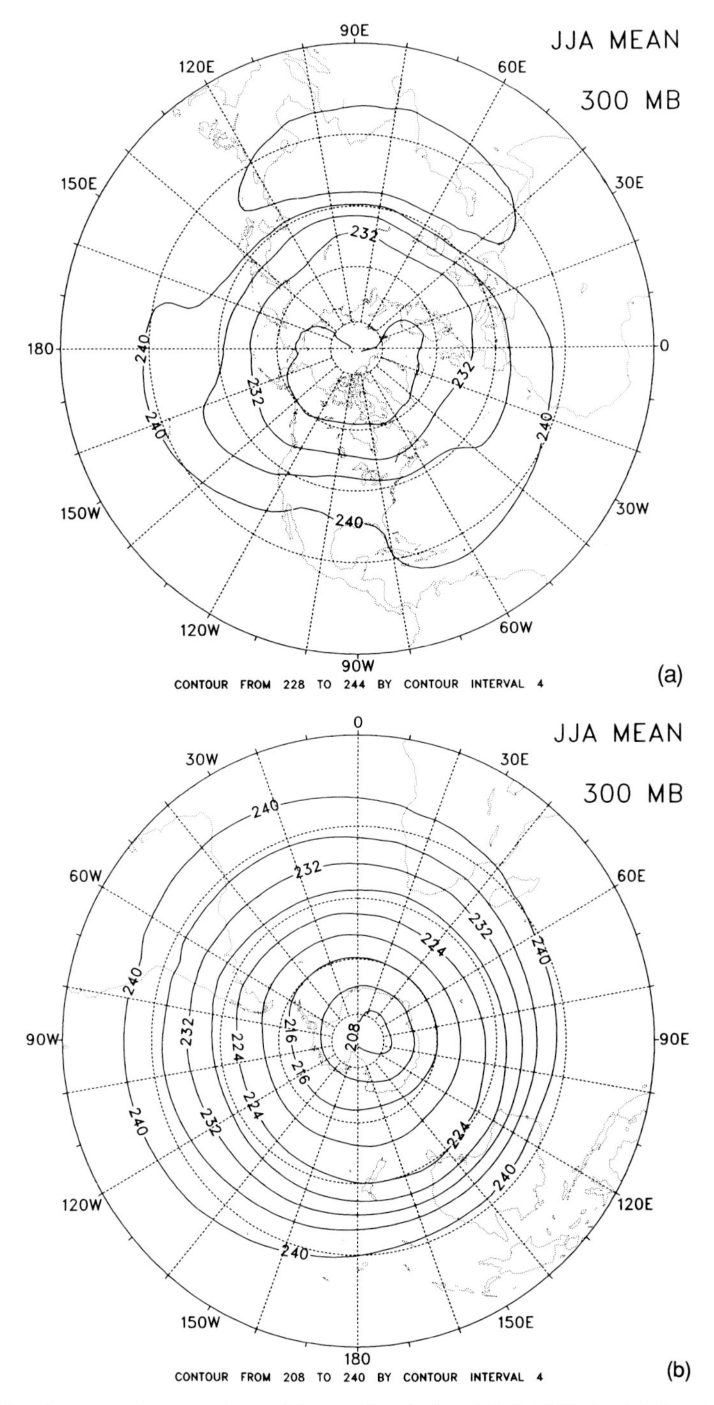

Figure 1.91 Average temperature (K) at 300 mb for 1979–1988 in (a) the Northern Hemisphere for the summer and (b) the Southern Hemisphere for the winter (June, July, and August) (from ECMWF data, 1979–1988; courtesy Kevin Trenberth and Amy Solomon, NCAR).

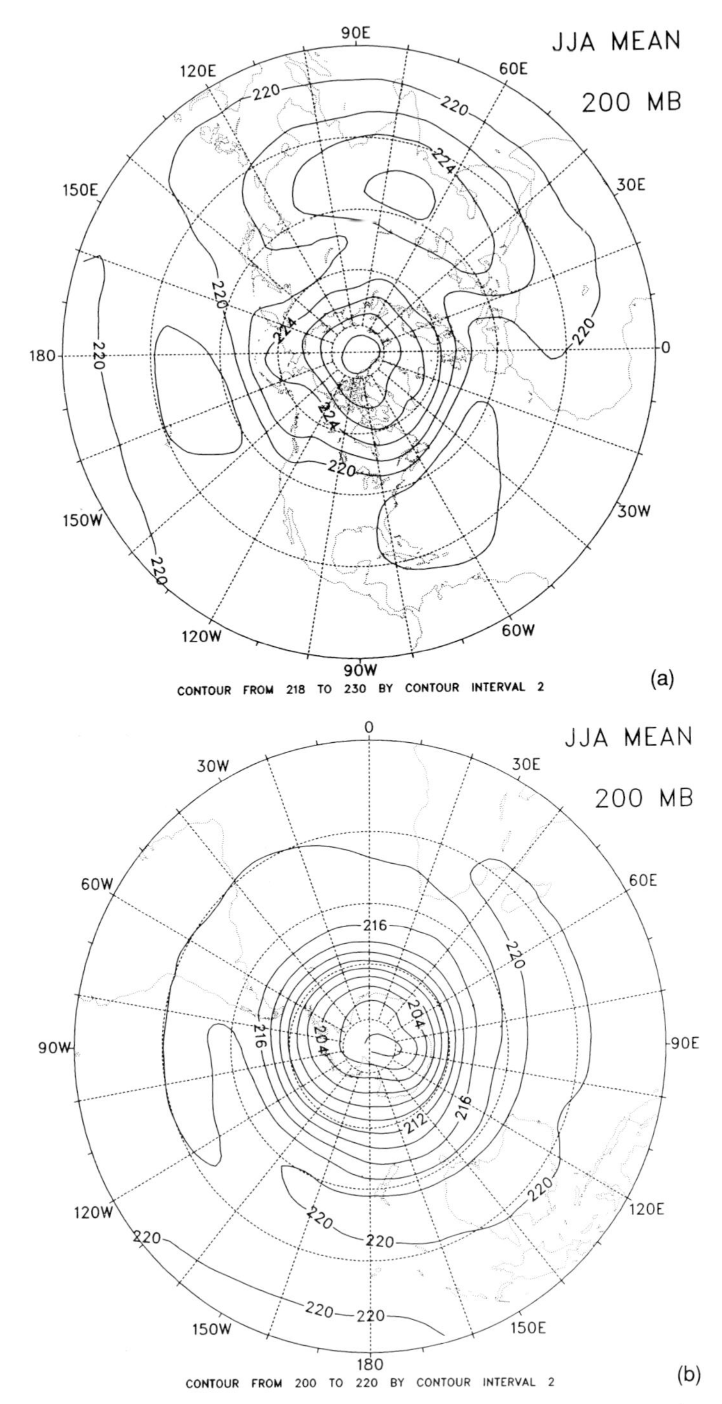

Figure 1.92 Average temperature (K) at 200 mb for 1979–1988 in (a) the Northern Hemisphere for the summer and (b) the Southern Hemisphere for the winter (June, July, and August) (from ECMWF data, 1979–1988; courtesy Kevin Trenberth and Amy Solomon, NCAR).

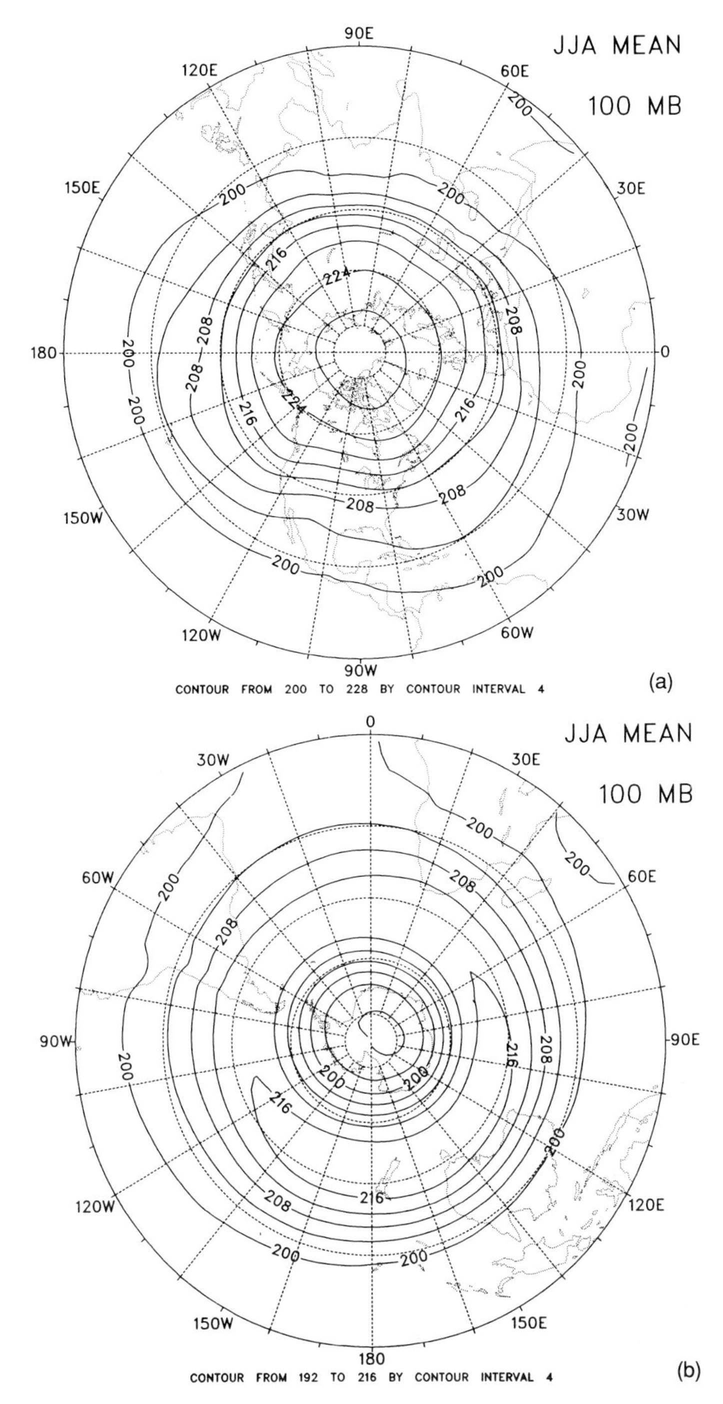

Figure 1.93 Average temperature (K) at 100 mb for 1979–1988 in (a) the Northern Hemisphere for the summer and (b) the Southern Hemisphere for the winter (June, July, and August) (from ECMWF data, 1979–1988; courtesy Kevin Trenberth and Amy Solomon, NCAR).

(233 K) or less is sometimes taken by some to be the unofficial beginning of winter!).

At 300 mb, during the winter, the three troughs are still evident and the atmosphere is still approximately equivalent barotropic. The strongest westerly geostrophic winds and meridional temperature gradients are found over east Asia.

The height field at 200 mb is similar to that at 300 mb. However, the temperature field is rather complex; gradients are weak, and the atmosphere is *not* equivalent barotropic. The tropopause usually intersects the 200-mb surface, and the location of this intersection varies from day to day. Hence, strong gradients may be found in some areas on a daily basis, but not in the mean. The coldest air (211 K, −62°C) is located over the Northern portion of Russia.

The wintertime 100-mb height field shows major long-wave troughs near the east coast of Asia, the east coast of North America, and parts of Russia and Eastern Europe. The north–south temperature gradient at 100 mb, which is in the stratosphere, is in general *reversed* from that below. The warmest temperatures are found near 55°N, where the tropopause is relatively low; a minor minimum is located near the North Pole; temperatures as cold as −77 to −82°C (196 to 191 K) are found in the tropics, where the tropopause is the highest.

It should be noted also that the low-level cold highs over Siberia and Northwest Canada do *not* appear aloft, while the ridges in the subtropical Atlantic and eastern Pacific *do* also appear at higher levels. The Aleutian and Icelandic lows, and perhaps the Italian lows, are also reflected aloft.

Spring climatology in the Northern Hemisphere.[5] The 850-mb height field during the spring is qualitatively similar to the 850-mb height field during the winter. However, the mean features are generally weaker. More pronounced warm regions are found over northern India and North Africa. In addition, a thermal ridge is found over the elevated terrain of western North America. The 700-mb mean height field in spring is also qualitatively similar to the 700-mb mean height field in winter. However, thermal maxima are now a bit better defined around 10–15°N. The mean spring 500-mb height field is qualitatively similar to the mean winter 500-mb height field. However, the Polar heights are about 16 dam higher, and hence the circumpolar westerly geostrophic vortex is weaker. The 500-mb isotherms are nearly parallel to the height contours. The 300-mb height and temperature fields during the spring are qualitatively similar to those during the winter. Weaker gradients are evident at high latitudes, however. The spring 200-mb height and temperature fields are similar to the winter height and temperature fields. However, the high-latitude temperatures are warmer by 10 K or more during the spring. The springtime 100-mb circumpolar geostrophic vortex is much weaker than it is during the winter. The warmest air during the spring is over Northern Canada and west of Alaska. The minor minimum found near the North Pole during the winter is not present during the spring.

Summer climatology in the Northern Hemisphere. During the summer the ridges at 850 mb are well defined at 30°N in the eastern Pacific and the Atlantic Oceans. Lows are found in the mean at the North Pole and north of India; a weak trough is found over the elevated terrain of the western United States. The temperature field at 850 mb exhibits a relatively weak mean pole to equator gradient. Hot areas are located over the desert southwest of the United States and the mountain regions of western North America, and the deserts of extreme North Africa and Saudi Arabia.

The Atlantic and Pacific ridges are also well defined at 700 mb during the summer; a high is found over Northern India. Four long-wave troughs are evident: one over eastern Canada, one west of the Aleutians (over the Bering Sea), one over Siberia, and another weaker one near Iceland. Thermal ridges are found in the mean over the southwest United States and North Africa, Saudi Arabia, and the Tibetan Plateau. These features correlate reasonably well with the pressure troughs and lows at lower levels.

The 500-mb mean summer height field exhibits weak circumpolar geostrophic westerlies poleward of about 40°N. Ridges are found around 20–30°N, especially over the oceans. A four-wave pattern is evident: One trough is over the east coast of the United States, one west of the Aleutians (over the Bering Sea), one over the northern part of Russia, and one west of Scandinavia. The mean summer temperature field at 500 mb also has a similar four-wave pattern. The mean meridional temperature gradient is relatively weak. The southern portion of Asia is anomalously warm.

The same (i.e., as at 500 mb) four-wave pattern is found in the summer 300-mb mean height field, which shows relatively weak circumpolar westerlies north of 40°N. A four-wave pattern is also seen in the 300-mb temperature field. A warm area is found over the Himalayas; the meridional temperature gradient is intense north of the Himalayas.

A weak four-wave pattern is found at 200 mb during the summer; zonally oriented ridges are located over the southern Asian continent and North Africa. The mean temperature gradient at 200 mb during the summer is reversed from that below: The warmest temperatures are found at the North Pole, where the tropopause is low, and the coldest temperatures are found over the subtropical Pacific and Atlantic Oceans, where the tropopause is high. Relatively warm air is also found north of the Himalayas.

The 100-mb mean summer height field shows a *very weak* circumpolar vortex. This is in accord with the poleward mean temperature gradient and the thermal wind relation. The 100-mb temperature field is truly extraordinary. The temperature at the North Pole is over 30 K *warmer* than the temperatures in the Tropics. There is a strong *easterly* thermal wind. A ridge at 100 mb is located over much of southern Asia, North Africa, and the United States.

Fall climatology in the Northern Hemisphere.[5a] The mean 850-mb height field during the fall shows the Pacific and Atlantic ridges between 20 and 30°N. A circumpolar westerly geostrophic vortex is found between 40 and 60°N. This vortex has three long-wave troughs embedded in it: One is near Iceland and

Greenland, one over the northern portion of Russia and one broad trough west of the Aleutians. The 850-mb temperature field shows a mean north–south temperature gradient, and thermal ridges over the southwest United States, and from North Africa, extending eastward to India.

At 700-mb during the fall the Atlantic and Pacific ridges persist. A ridge is also located over North Africa. The geostrophic circumpolar westerly vortex in the midlatitudes has increased in strength from its summer magnitude. Three long-wave troughs are evident: One is just west of Greenland, one over Russia and there is a broad trough west of the Aleutians. If one counts the sharper troughs (within the broad trough) over China and Japan, and over the Aleutians, then there are four troughs. The 700-mb mean fall temperature field shows major thermal troughs over northeast North Africa, northeastern Canada, and eastern Asia, and a minor trough over the northwestern portion of Russia and near the Aleutians.

At 500 mb during the fall there is a trough over northeast North America and a broad trough over the northeast portion of the Asian continent. The 500-mb mean temperature pattern (not shown) is similar to the mean height pattern.

At 300-mb during the fall the height and temperature fields are qualitatively similar to the 500-mb height and temperature fields.

Although the 200-mb mean fall height pattern is qualitatively similar to the 300- and 500-mb patterns, the 200-mb temperature field is radically different. Temperature gradients at 200 mb are relatively weak. The coldest air is found over the subtropical Atlantic and over the eastern Pacific and western United States. The warmest air is found north of Hudson's Bay, and over the Himalayas, and on to the east.

The fall 100-mb mean height field indicates weak geostrophic westerlies in midlatitudes, with long-wave troughs over eastern North America and eastern Asia. The temperature field at 100 mb during the fall shows a strong, south–north temperature gradient. The warmest air is found over the northwest portion of North America and extreme northeast Asia. The coldest temperatures are found in the tropics. Over the Pacific, temperatures at 60°N are 30 K warmer than they are at the equator!

Winter climatology in the Southern Hemisphere. The 850-mb height field shows an intense circumpolar cyclonic vortex between 40 and 60°S. The intensity of the vortex is greater than that of the corresponding Northern Hemisphere winter vortex. There are troughs located west of South Africa, about 140°W, and about 110°E. Low centers are found over Antarctica in the Eastern Hemisphere and along the edge of the Antarctican continent near 160°W, over the Ross Sea. Highs are found near 25°S over the Atlantic, off the west coast of South America, over Australia, and in the Indian Ocean and over South Africa. The coldest temperatures are found over Antarctica; an intense meridional temperature gradient is located along the edge of the continent.

There is an intense circumpolar cyclonic vortex at 700 mb, with long-wave mean troughs located near the longitude of South Africa, 140°W, and 110°E.

The pattern is not symmetrical with respect to the South Pole; moderate geostrophic westerlies are found down to 20°S between about 80°W and Australia; the geostrophic flow is weaker in the opposite hemisphere near 20°S. Lowest heights are found along the edge of the Antarctican continent around 160°W, over the Ross Sea. Ridges stretch westward off the coast of South America at 15°S to northern Australia; another ridge is found in the Indian Ocean at 20°S. The coldest temperatures are found over Antarctica in the Eastern Hemisphere.

At 500 mb there is a strong circumpolar cyclonic vortex, with a three-wave pattern. The lowest heights are located over the edge of the Antarctican continent around 160°W, over the Ross Sea. The geostrophic westerlies extend down to 20–25°S. The coldest air is found over Antarctica; the warmest air is found near 10°S from the Indian Ocean eastward to east of Australia.

The 300-mb height and temperature fields have patterns similar to those of the 500-mb height and temperature fields.

Strong geostrophic westerlies are found at 200 mb. The coldest air is located over Antarctica; the warmest air is found near 30°S from Madagascar east to about 160°W.

Strong geostrophic westerlies are found poleward of 25°S at 100 mb. The temperature gradient is reversed from what it is below equatorward of 40°S. The warmest air is found near 45–50°S, between 110°W and 60°E. The coldest air is located over Antarctica.

Summer climatology in the Southern Hemisphere. Highs are found at 850 mb near 25°S west of South America, in the Atlantic, over the Indian Ocean, and over Antarctica. Lows are located equatorward of the edge of the Antarctican continent. A trough extends across northern Australia into the Indian Ocean. The coldest air is found over Antarctica; the warmest air is found over northwest Australia, South Africa, and eastern South America.

The 700-mb height field is dominated by an intense cyclone centered over the South Pole; there are strong geostrophic westerlies between 40 and 60°S. Highs are found west of South America, in the Atlantic, in the Indian Ocean, and over western Australia. The coldest temperatures are found over Antarctica, with the mean isotherms mainly symmetric about the South Pole.

At 500 mb there is a band of strong geostrophic westerlies. The lowest heights are found over Antarctica near the dateline. At high latitudes there is a hint of a three-wave pattern, with troughs near the longitude of Africa, west of South America, and western Australia. The temperature field is relatively symmetric about the South Pole, as at 700 mb.

The 300-mb height-field pattern is similar to the 500-mb height-field pattern. Temperature gradients at 300 mb are relatively weak.

The 200-mb height-field pattern is similar to the 300-mb height-field pattern. However, the meridional temperature gradient is the reverse of the 300-mb temperature gradient; the warmest air is found over Antarctica near the dateline.

Weak geostrophic westerlies are found at 100 mb. The temperature gradient is relatively strong, and toward the South Pole; the warmest air is located over Antarctica.

Climatological summary. Belts of geostrophic westerlies are found in the midlatitudes, flanked by subtropical ridges equatorward and lows near the poles.

In the Northern Hemisphere the meridional temperature and height gradients in the troposphere are in general strongest in the winter, and weakest in the summer. In the Southern Hemisphere the meridional temperature and height gradients are also strong during the summer. Low-level temperature gradients are strongest around the rim of the Antarctican continent during the winter. The mean pole–equator directed temperature gradient in the troposphere reverses at the tropopause. The coldest temperatures are found at 100 mb during the Southern Hemisphere winter over Antarctica and during the Northern Hemisphere winter and Southern Hemisphere summer near the equator in the Pacific.

The difference in land mass between that in the Northern Hemisphere and that in the Southern Hemisphere, and the presence of a polar continent surrounded by an ocean in the Southern Hemisphere, must both play large roles in determining the different climatologies in each hemisphere. The mountains of Asia also play an important role in modulating the climatology in that section of the world.

1.5. INSTABILITY, CYCLOGENESIS, AND ANTICYCLOGENESIS

1.5.1. The Concept of Instability

When pressure (and height) systems intensify, the increased horizontal pressure gradient induces an increase in wind speed, and hence an increase in kinetic energy. The energetics of this process is the conversion of available potential energy associated with horizontal temperature gradients into kinetic energy through the lifting of "warm," relatively light air, and the sinking of "cold," relatively dense air. The spontaneous conversion of potential energy into kinetic energy is known and loved by every "bomb fan"[6] as *baroclinic instability*. The pioneering theoretical work on baroclinic instability was done by Charney and by Eady in the late 1940's.

Baroclinic instability stands in direct contrast to *barotropic instability*; the latter is the spontaneous conversion of kinetic energy associated with the mean flow into kinetic energy of a growing disturbance (and is still loved by bomb fans, but perhaps not as much). The pioneering theoretical work on barotropic instability was performed by Kuo about the same time that studies were first done on baroclinic instability. A necessary condition for barotropic instability in a zonal current is that the gradient of absolute vorticity associated with the basic current vanishes. This condition is often met near jets (Fig. 1.94). Under these circumstances, the basic current u_g decreases, vorticity associated with the basic current decreases, and vorticity associated with the wave grows.

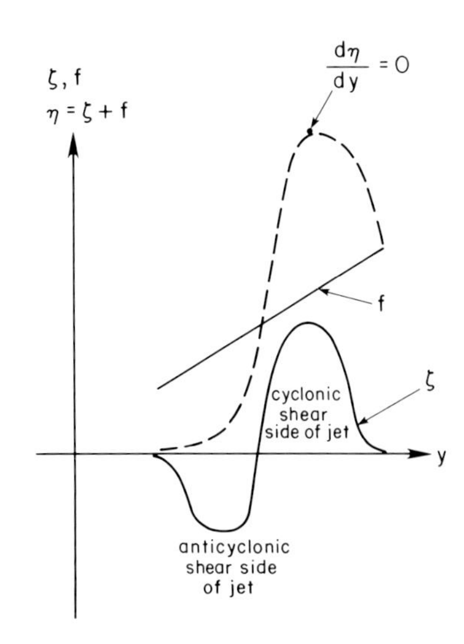

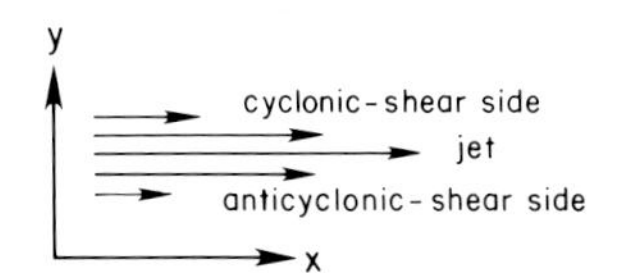

Figure 1.94 Example of how the barotropic instability criterion may be met near jets. (top) Idealized meridional profiles of relative vorticity (ζ), Coriolis parameter (f), and absolute vorticity (η; dashed line) for jet depicted (below).

The mathematical problem of determining the necessary and sufficient conditions for baroclinic and barotropic instability in an atmosphere that is continuously stratified (i.e., not a simple layered one), that is rotating (i.e., that has a Coriolis force), and that has a realistic basic current (i.e., not a zonal basic current, but one that is wavelike and is concentrated), is formidable. Our knowledge of the susceptibility of the atmosphere to baroclinic instability as a function of the form of the basic current is to a large extent based upon observation.

A mechanical analog to baroclinic instability in the atmosphere is the behavior of a coin resting on its edge (Fig. 1.95). Thus we see what money and atmospheric behavior have in common. If the coin is given a small push to the

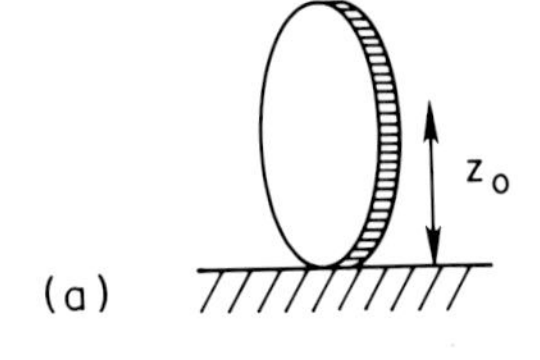

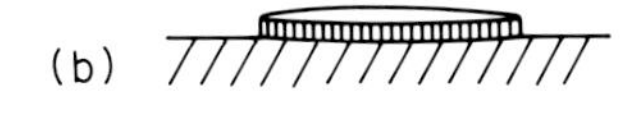

Figure 1.95 The potential energy of the coin in (a) is its mass times g times the height of its center of gravity z_0, which is at the center of the coin; in (b) its potential energy is its mass times g times the height of its center of gravity, which is only one-half the thickness of the coin. The available potential energy of the coin is the difference between its potential energy in (a) and its potential energy in (b).

side, it will fall over and come to a rest. When the coin tips over, its center of gravity is lowered, and thus its potential energy decreases. The motion of the coin represents kinetic energy that has been converted from potential energy. Not all the potential energy of the coin has been converted into kinetic energy. Since the coin has a finite thickness, it still has some potential energy left. At the risk of making a pun, we note that further "tipping" of the "coin" will not result in anymore conversion of potential energy into kinetic energy. The potential energy that can be converted into kinetic energy is called the *available potential energy*. The remainder is called the *unavailable potential energy*. The atmosphere is actually a bit more complicated because it is compressible and rotating about the Earth's axis.

1.5.2. Application of Sanders' Analytic Model to the Baroclinic Instability Process

The nature of baroclinic instability may be understood to some extent by considering the deepening rate of a cyclone and seeing how it depends upon its environment. In Sanders' quasigeostrophic, analytic model the geopotential height tendency (see Appendix 1) at the center of a surface cyclone is

$$\chi\left(\frac{L}{2} - \lambda,\ 0,\ 1000\ \text{mb}\right) = (K_{13}^{\chi} P_{13}^{\chi} - K_{14}^{\chi} P_{14}^{\chi}) \sin \frac{2\pi\lambda}{L}. \tag{1.5.1}$$

When the horizontally varying parts of the temperature and height fields are in phase, $\lambda = 0$, and there is no intensification. This configuration of the temperature field and height field is called an *occlusion*. Polar-front theorists viewed this configuration as one in which the warm-air mass has been "cut off," or occluded by the cold–air mass. The occlusion is the last part of the life history of midlatitude, synoptic-scale cyclones.

The maximum deepening rate occurs when the phase lag is one-quarter wavelength, that is, when $\lambda = L/4$. The physical effect associated with K_{13}^{χ}, which contributes to intensification, is divergence accompanying the vertical motion that is associated with the advection of perturbation thermal vorticity by the mean thermal wind. The parameter K_{13}^{χ} is related to the wavelength L, static stability γ, and meridional temperature gradient a is follows (see Appendix 1):

$$K_{13}^{\chi} = \frac{f_0 R a \hat{T} L}{4\pi T_0 \gamma}. \tag{1.5.2}$$

The divergence accompanying the vertical motion that is associated with the advection of Earth's vorticity by the perturbation thermal wind is the physical effect representing K_{14}^{χ}; it acts to counter intensification. The parameter K_{14}^{χ} is related to wavelength, static stability, and the beta effect (β) as follows (see Appendix 1):

$$K_{14}^{\chi} = \frac{f_0^2 \hat{T} \beta L^3}{4(2\pi)^3 T_0 \gamma}. \tag{1.5.3}$$

At long wavelengths, the advection of Earth's vorticity becomes more important than the advection of geostrophic vorticity, and hence there is no intensification. If there were no beta effect, waves of all wavelengths would deepen. The wavelength at which the deepening rate vanishes is called the *long-wave cutoff.* At short wavelengths, the advection of geostrophic vorticity is dominant. However, vorticity advection is small at very short wavelengths because K_{13}^{χ} approaches zero as the wavelength approaches zero [Eq. (1.5.2)]. [A linear stability analysis of a two-level quasigeostrophic model (Holton, 1979) predicts a short-wave cutoff also. The Sanders analytic model is continuously stratified and does not have a short-wave cutoff. The short-wave cutoff is therefore a spurious effect resulting from truncation error. However, the Sanders model does show that the deepening rate approaches zero as the wavelength is made vanishingly small. This point is academic, however, since

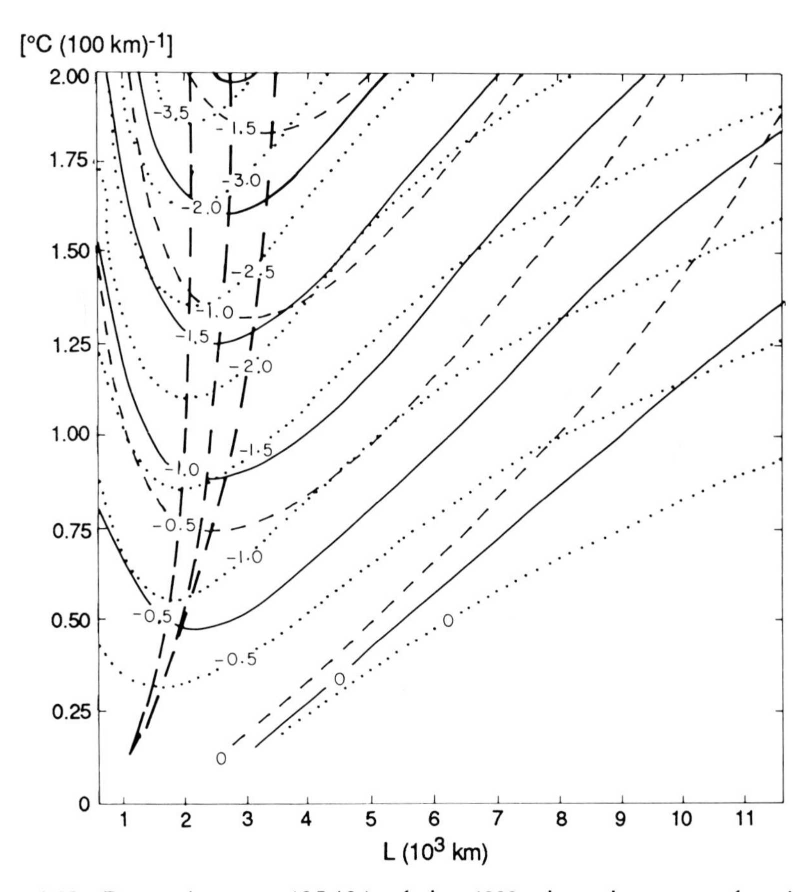

Figure 1.96 Deepening rate ($\partial\Phi/\partial t$) of the 1000-mb cyclone as a function of wavelength (L) and meridional temperature gradient (ordinate) for selected values of the vorticity-stability parameter and for $\hat{T} = 1°C$. The isopleths are labeled in units of $10^{-3}\,m^2\,s^{-3}$. The dashed, solid, and dotted lines are for values of the vorticity-stability parameter as in Fig. 1.39. The longer dashed lines are loci of the wavelength of maximum deepening rate (from Sanders, 1971). (Courtesy of the American Meteorological Society)

the model does not apply to very short wavelengths, for which the quasi-geostrophic assumptions are violated.]

The most rapid intensification of surface cyclones and anticyclones occurs when the wavelength is between 1500 and 3000 km (Fig. 1.96). We conclude that baroclinic instability is probably responsible for the formation of many midlatitude, synoptic-scale systems, because we observe that the horizontal scale of most cyclones and anticyclones is in this range: The evidence is circumstantial. A serious limitation of our analysis, however, is that it does not account for the nonlinear time evolution of the height field and temperature field: It only tells us what the height tendencies are at one time.

Several other aspects of our analysis are noteworthy. Since K_{13}^{χ} is proportional to a and inversely proportional to γ, the degree of baroclinic instability increases with meridional temperature gradient, and decreases with static stability. This makes physical sense, since the mean thermal wind is proportional to a, and the forcing function in the Trenberth formulation of the quasigeostrophic ω equation, which represents advection by the mean thermal wind, is inversely proportional to static stability. Thus the magnitude of the vertical circulation (ω and $\mathbf{v}_a$) that lifts "warm" air and "pushes down" "cold" air, and converts available potential energy into kinetic energy, is proportional to a and inversely proportional to γ. It should also be noted that small static stability is associated with smaller synoptic-scale disturbances. The wavelength of the mean 500-mb flow during the summer in the Northern Hemisphere [Fig. 1.78(a)], for example, when static stabilities are relatively low, is shorter than that during the winter [Fig. 1.72(a)]. It should be noted parenthetically that the flow is stronger during the winter when a is larger. The thermal wind shear is therefore larger, and the winds aloft are stronger. During the winter there is no heating at high latitudes at all, while during the summer there is some short-wave heating to offset the long-wave radiational cooling. In addition, the belt of strong westerly flow is displaced northward during the summer.

If the effect of friction in the boundary layer is large for short wavelengths, then there may actually be a short-wave cutoff, since geostrophic vorticity is inversely proportional to the horizontal wavelength [see Eqs. (1.4.2) and (1.4.3)], and the effect of friction is related to the vorticity. There is some suggestion that the effect of friction may be to lengthen the wavelength of maximum intensification.

1.5.3. Baroclinic Instability and the Flow Pattern

Inferences about baroclinic instability from the analytic model are limited in part because the temperature and height fields in the real atmosphere are not perfect sinusoids. If the flow pattern is not perfectly sinusoidal, then the results discussed earlier probably have to be modified before they can be applied. In particular, we expect that waves having certain shapes might be more or less baroclinically or barotropically unstable than others.

In addition, the reader should also note that the analytic solution is actually the solution to a linearized nonlinear equation. In other words, parameters that are given as constants, such as a, $\hat{T}$, etc., actually are modified by the wind field. We must therefore be very cautious in applying the Sanders model to an analysis of baroclinic instability.

"Diffluent" and "confluent" troughs. In 1937 Scherhag postulated that upper-level troughs, downstream from which the flow pattern is diffluent (Fig. 1.97), often intensify. Bjerknes in 1954 and Polster in 1960 presented evidence that a majority of all instances of conventional intensification of cyclones at the surface in the midlatitudes occur underneath *diffluent troughs*. The rest occur underneath parallel flow, while only a few cases occur underneath *confluent troughs* (Fig. 1.98). It is possible that diffluent troughs are associated with larger values of cyclonic vorticity advection than confluent troughs. This is not obvious, however: The geostrophic wind is relatively weak downstream from a diffluent trough. On the other hand, the isopleths of relative vorticity cut across height contours at a large angle, which is nearly perpendicular to the height contours. The latter argues for large vorticity advection, while the former argues for small vorticity advection.

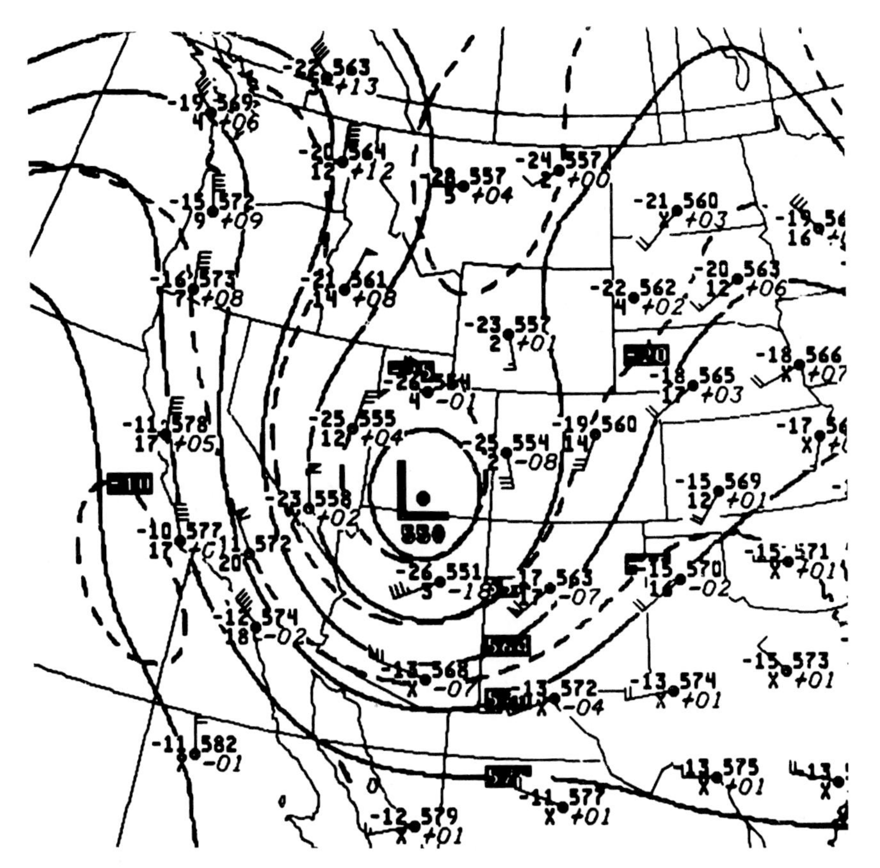

Figure 1.97 Example of a diffluent upper-level trough in the western United States. NMC 500-mb analysis at 0000 UTC, November 26, 1987.

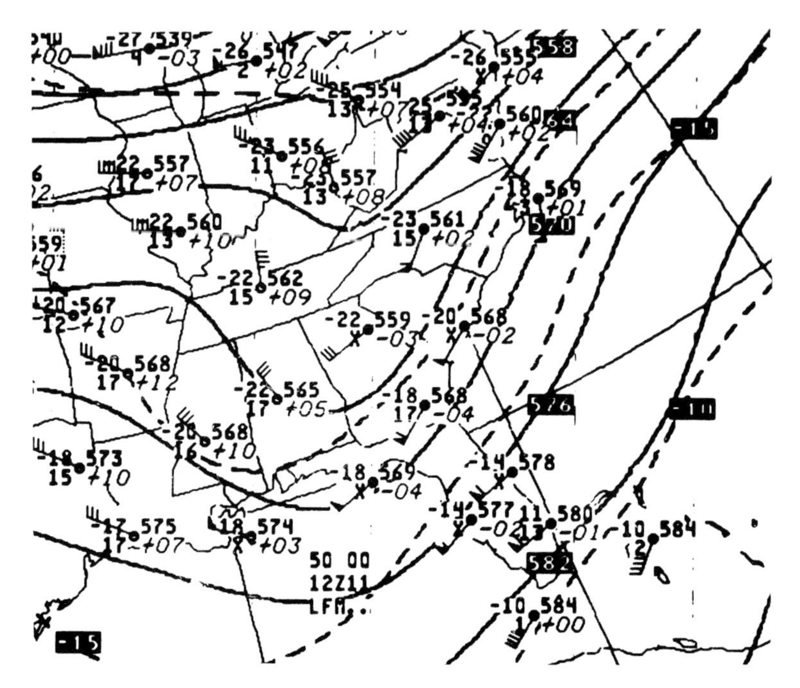

Figure 1.98 Example of a confluent upper-level trough in the southeastern United States. NMC 500-mb analysis at 1200 UTC, January 11, 1988.

Diffluent troughs have stronger winds upstream, while confluent troughs have stronger winds downstream. As shown earlier diffluent troughs should dig equatorward, while confluent troughs should lift poleward. The motion of the trough relative to the surface cyclone could modify the forcing functions in the ω equation, and hence affect the surface pressure tendencies.

It is also possible that the deceleration of air parcels entering the diffluent region downstream from a trough or the acceleration of air parcels entering the confluent region downstream from a trough is substantial, and hence quasi-geostrophic theory cannot explain the observed relationship between the degree of confluence or diffluence downstream from a trough and the amount of baroclinic instability.

"Tilted" waves. Another characteristic of atmospheric flow that may affect the degree of baroclinic instability is the "tilt" of the waves. Waves in the westerlies whose axes slope in the direction of the flow with latitude are said to be *positively tilted.* V. Starr found that positively tilted troughs are characterized by poleward eddy fluxes of angular momentum. The source of the angular momentum is in the tropics, where the tropical easterly flowing surface air acquires some of the Earth's angular momentum as it is slowed down by friction. This westerly momentum is then transported upward in the Hadley cell and poleward by the eddies associated with positively tilted troughs. *Negatively tilted* troughs in the westerlies slope in the direction opposite to the flow with latitude, and are associated with equatorward fluxes of angular momentum, that is, a removal of momentum from the westerly jet, possibly into a growing system owing to barotropic instability.

In 1977 Glickman et al. found observational evidence that negatively tilted upper-level troughs are more likely to be associated with convective activity than positively tilted troughs. This might be a consequence of a stronger vertical circulation and lower static stability.

Waves acquire a tilt when the zonal wind speed varies as a function of latitude. For example, according to the Rossby-wave formula [Eq. (1.4.21)], if the zonal basic current increases with latitude, then waves in the westerlies will become positively tilted. In fact, it is even possible that the wavetrain can become "fractured," as the rapidly moving poleward portion of the trough separates from the slowly moving equatorward portion (Fig. 1.99).

1.5.4. Special Types of Cyclogenesis and Anticyclogenesis

The "bomb" and "bombogenesis." Explosive cyclogenesis occurs most frequently over the ocean during the "cold" season (Fig. 1.100), downstream

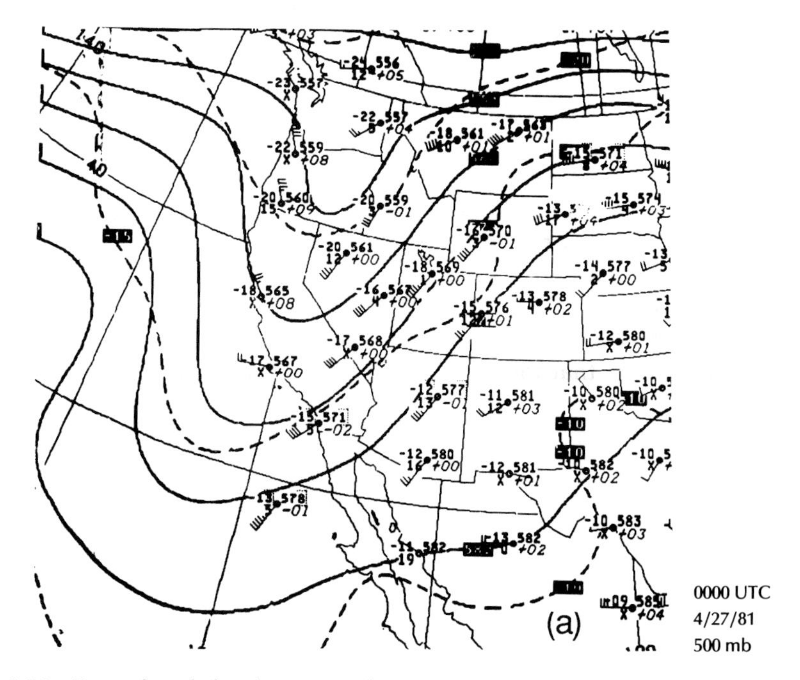

Figure 1.99 Example of the fracture of a trough into two pieces as a result of differential eastward motion in a sheared basic current. NMC 500-mb analyses for (a) 0000 UTC, April 27, 1981, (b) 1200 UTC, April, 27, (c) 0000 UTC, April, 28, (d) 1200 UTC, April 28. In (a) the trough extends from southern California up to Washington state. In (b) the trough extends from southern California up to Idaho and western Montana. The northern portion of the trough has progressed to eastern Montana and the Idaho–Wyoming border in (c). In (d) the southern California portion of the trough has become cut off, while the northern portion of the trough has progressed to the Dakotas.

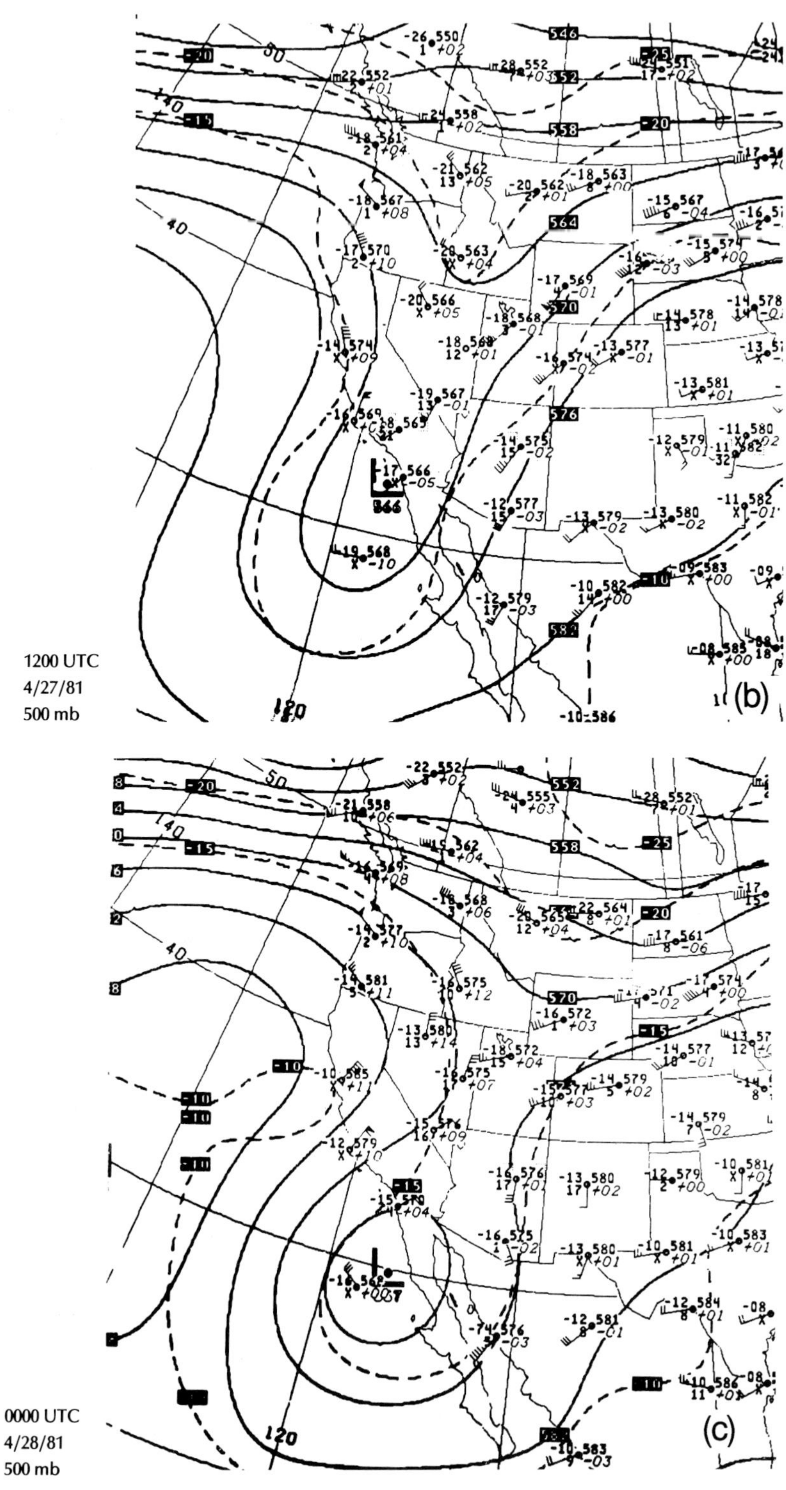

Figure 1.99 (*cont.*)

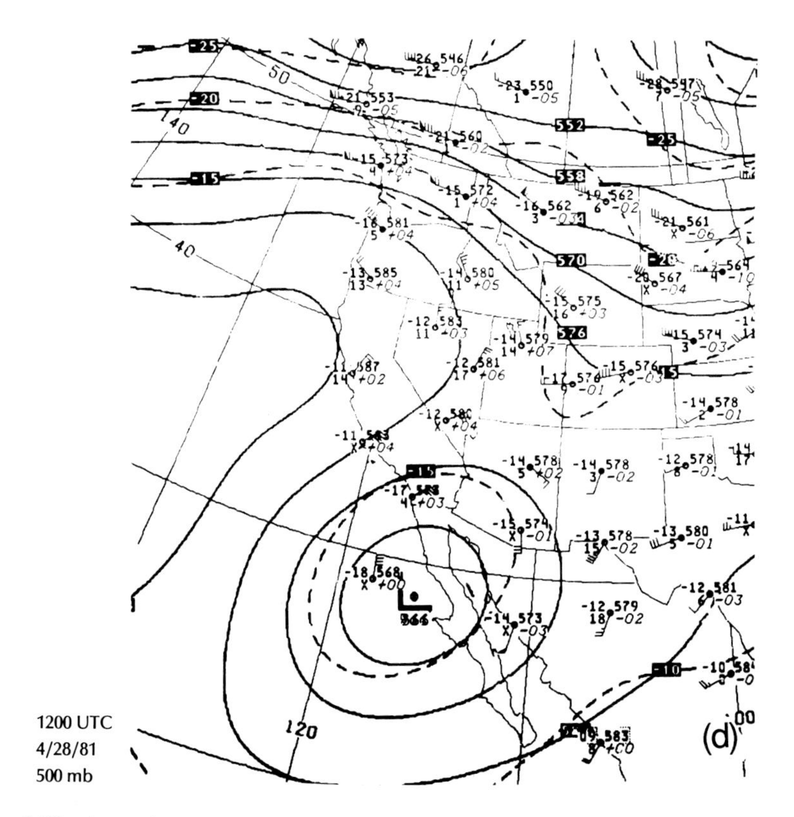

Figure 1.99 (*cont.*)

from mobile, diffluent, upper-level troughs, within or poleward of the maximum westerly current, and near the strongest sea-surface temperature gradients such as the northern edge of the Gulf Stream. Sanders and Gyakum in 1980 named cyclones that have deepening rates in excess of 1 mb h^{-1} for 24 h *bombs.* Bombs are especially dangerous to shipping. Are bombs ordinary cyclones with extraordinary deepening rates, the upper classes of cyclone society, or do they represent a fundamentally different phenomenon?

Deepening rates such as those found in bombs cannot be explained quantitatively on the basis of quasigeostrophic theory. A pressure tendency of -1 mb h^{-1} for 24 h or more is called a *bergeron,* in honor of the pioneer Norwegian meteorologist, Tor Bergeron. The following facts must be considered in a discussion of whether bombs are extreme events or unique events:

1. The turbulent transfer of sensible heat between the relatively warm ocean and cold continental air can create very low static stability near the surface. The forcing functions in the quasigeostrophic ω equation are therefore enhanced.
2. A temperature gradient is found at the sea surface along the edge of the Gulf Stream and the Kuroshio, which can act to enhance temperature advection at low levels.

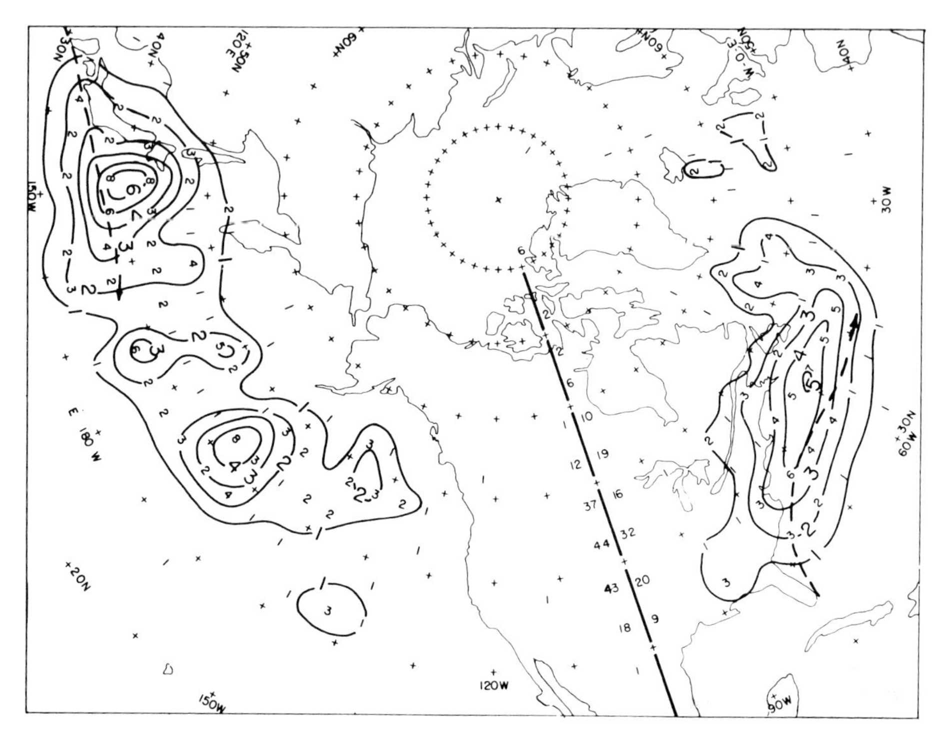

Figure 1.100 Distribution of bomb events during three cold seasons. Raw nonzero frequencies appear in each 5×5 quadrilateral of latitude and longitude. Isopleths represent smoothed frequencies, obtained as one-eighth the sum of four times the raw central frequency plus the sum of the surrounding raw frequencies. The column of numbers to the left and right of the heavy line along longitude 90°W represent, respectively, the normalized frequencies for each 5° latitude belt in the Pacific and Atlantic regions, using a normalization factor of cos 42.5°/cos ϕ, where ϕ is the latitude. Heavy dashed lines represent the mean winter position of the Kuroshio and the Gulf Stream (from Sanders and Gyakum, 1980). (Courtesy of the American Meteorological Society)

3. Stability analyses under conditions similar to those just described suggest that a shallow, short-wavelength cyclone may form, similar to the "bombs" observed.
4. The "air–sea interaction" instability proposed by Emanuel and Rotunno may also play an important role since the turbulent transfer of sensible heat and water vapor are enhanced as the surface wind speed increases. Thus, bombs may behave like tropical cyclones. In this case, a cyclone may develop through "conventional" quasigeostrophic processes, and then much further through the air–sea interaction process. Many bombs in fact show eyes in satellite photographs.
5. Boyle and Bosart have found unusually large values of temperature advection at high levels (Fig. 1.101) as a surface cyclone explosively deepened. Quasihorizontal temperature advection is ordinarily strongest near the ground, and weakest aloft. The unusually intense cold

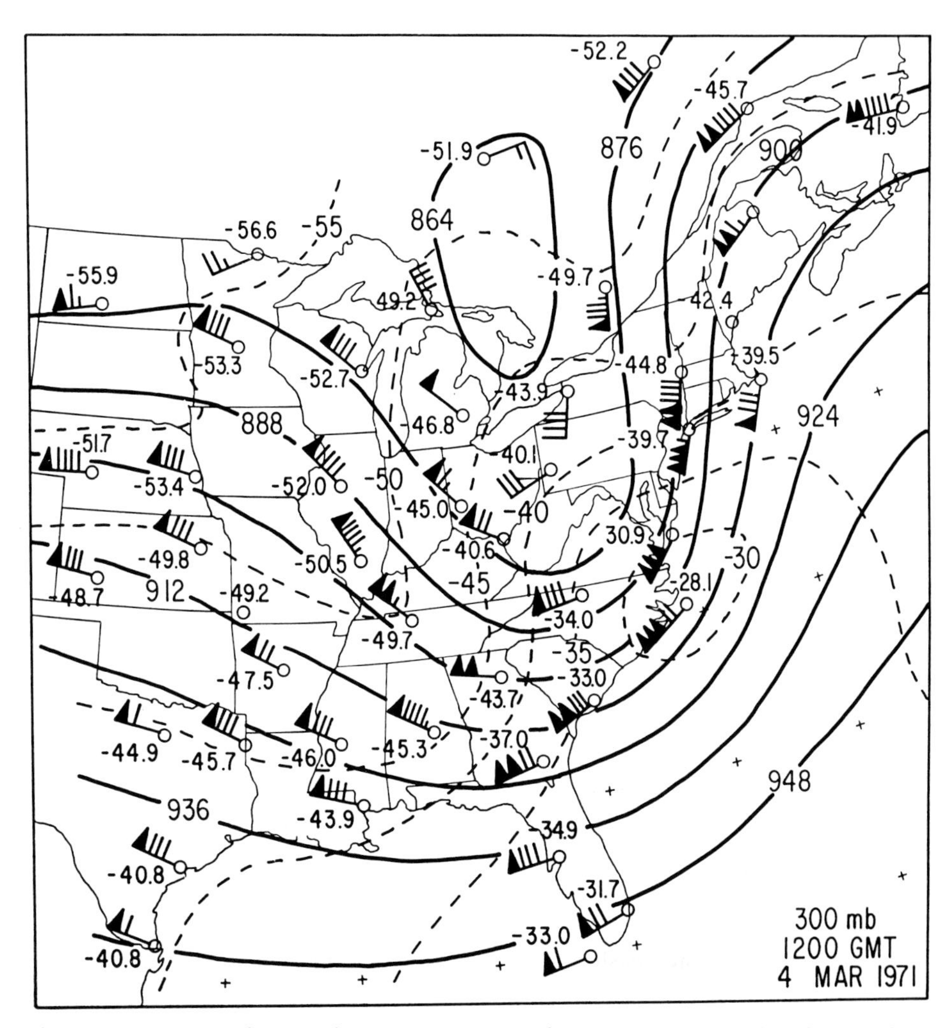

Figure 1.101 300-mb map for 1200 UTC, March 4, 1971. 300-mb heights in dam (solid lines); isotherms in °C (dashed lines). There is strong cold advection upstream from the trough, unusual warmth just downstream from the trough (the temperature at Hatteras is −28.1°C), and strong warm advection downstream from the trough (from Boyle and Bosart, 1986). (Courtesy of the American Meteorological Society)

advection and warm advection upstream and downstream, respectively, from an upper-level trough were due to the steep slope of the tropopause. The tropopause had been advected unusually far downward by strong subsidence upstream from the upper level trough and west of the surface cyclone. The tropopause had been advected down to 600 mb, as evidenced by high values of Ertel's potential vorticity, subsidence, and dry air. The use of potential vorticity as a tracer of air masses will be discussed later on in this chapter. Thus, relatively warm, stable air is located aloft, while relatively cool air is located near the

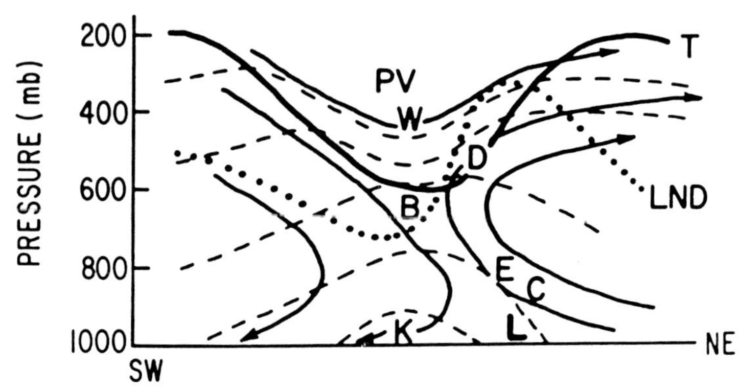

Figure 1.102 Vertical cross section of potential temperature (dashed lines) and airflow (solid lines) relative to the surface cyclone (L) in a plane aligned east–west across the upper-level trough and surface cyclone for the intense cyclogenesis phase. Stratospheric potential vorticity reservoir denoted by PV; sloping tropopause (heavy solid line) and sloping level of nondivergence (heavy dotted line) marked by T and LND, respectively. Lower (upper) tropospheric cold (warm) air reservoir denoted by K (W). Ascending (descending) air will have a component into (out of) the cross section. Cyclonic vorticity is produced in region B by vortex tube stretching accompanying subsidence, which is greatest at the LND. The vorticity so produced is advected downstream. Additional vorticity is generated in region D by horizontal convergence beneath the elevated LND, upstream of the surface cyclone. Air parcels exiting the trough aloft must lose cyclonic vorticity rapidly by horizontal divergence, which supports the development of a deep updraft and lower tropospheric warm air advection in locale C. Dynamically the sloping tropopause and lowered potential vorticity maximum allows a dome of cold air in the lower troposphere to remain vertically coupled with the potential vorticity maximum aloft. Surface cyclogenesis commences as the potential-vorticity maximum aloft begins to cross the boundary of the cold-air dome while moving toward warmer air. During the intense cyclogenesis phase the tropopause and potential vorticity maximum lower to the midtroposphere. The tropopause and LND are strongly inclined. The high LND over the cyclone center ensures convergence and cyclonic vorticity generation through a deep layer accompanying the development of a cutoff vortex. Air parcels that gained vorticity by vortex tube stretching during subsidence in region B gain further vorticity by convergence above region C as they are advected downstream across the steeply inclined LND. The steeply inclined tropopause over the surface cyclone center results in appreciable warm-air advection in the 400–200-mb layer, which further contributes to the intensity of development (from Boyle and Bosart, 1986). (Courtesy of the American Meteorological Society)

surface in the vicinity of the trough axis aloft (Fig. 1.102). The flow through the upper-level trough in the presence of the stratospheric air in the region of the low tropopause results in a strong temperature advection pattern. Cyclonic vorticity is generated aloft through convergence above the region of maximum sinking motion. This vorticity is subsequently advected downstream, and the differential vorticity advection forcing is enhanced. Furthermore, the warm advection downstream from the upper level trough aloft is superimposed over warm advection below, so that there is a *deep* column of rising air. The result is a deep layer of rising motion, convergence, and increase in vorticity. The column of convergence aloft associated with the region of subsidence is not as deep because the tropopause, which acts as a lid to vertical motion, is very low. It therefore appears as if the region of subsidence upstream from an upper level trough can, in some instances, enhance surface cyclogenesis.

6. The physical processes mentioned in items 1.–5. may all act to reinforce each other and synergistically to produce much more rapid surface pressure falls than would be otherwise associated with each alone.

Recent small-scale observations in oceanic bombs and their numerical simulations have revealed that a warm-core structure develops at low levels, which is not formed as advancing "cold" air cuts off (occludes) the retreating

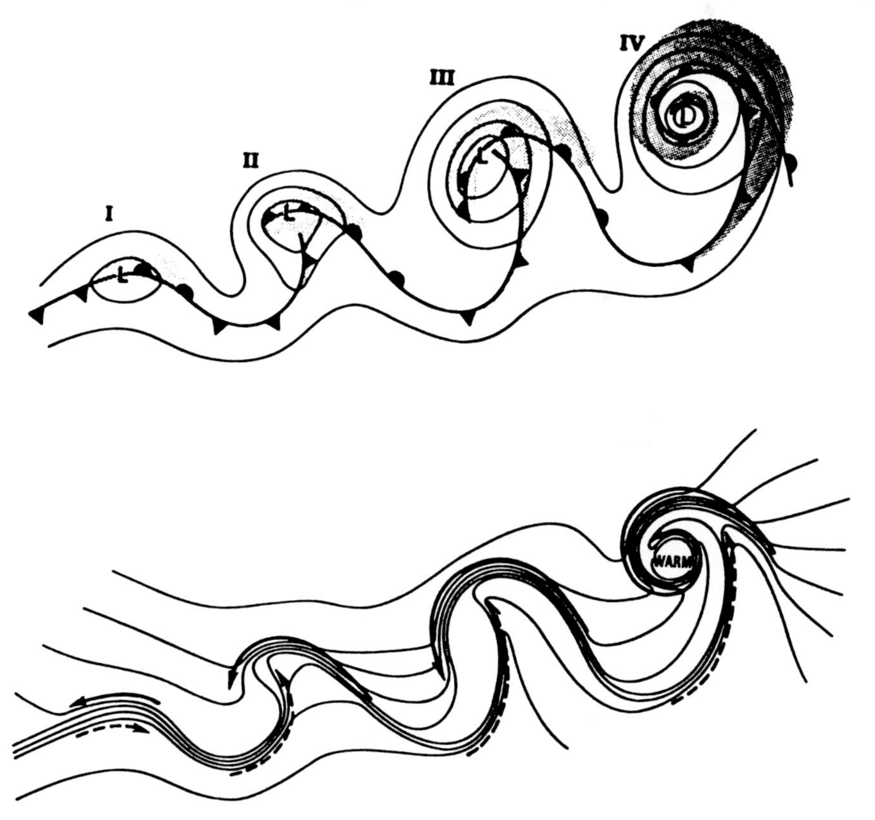

Figure 1.103 Idealized example of a seclusion and its antecedent evolution: (I) incipient frontal cyclone; (II) frontal fracture; (III) bent-back warm front and frontal T bone; (IV) warm-core frontal seclusion. Upper: sea-level pressure, solid lines; fronts, bold lines; and cloud signature, shaded. Lower: temperature, solid lines; cold and warm air currents, solid and dashed arrows, respectively (from Shapiro and Keyser, 1990).

"warm" air, as described in Polar–Front theory. Instead, it appears as if the warm core, which is warmer than the advancing cold air mass and colder than the retreating warm air mass, forms as relatively warm air is left behind near the center of the circulation. Mel Shapiro has named this feature (Fig. 1.103) a "seclusion."

The "polar" low, or "instant occlusion." The "polar" low, or "instant occlusion," first appears in "polar" air streams poleward of major frontal bands during the "cold" season (between October and April) (Fig. 1.104). In other words, it can form in the absence of strong surface warm advection, in the presence of reduced cold advection, neutral advection, or weak warm advection, and at the surface looks like the occluded phase of a cyclone from the very beginning. Some polar lows are therefore similar to the Type B cyclone described by Petterssen and Smyebe. However, some polar lows are associated with the margin between ice fields or cold land and relatively warm water, where there is shallow baroclinicity, and hence the potential for significant temperature advection. Other polar lows have been observed under cold-core upper-level lows. These have had intense cumulus convection and eyes, like hurricanes, and suggestions of a warm-core structure. The polar low may appear on satellite imagery as a comma-shaped cloud mass, which is in large part made up of convective cells.

The polar low sometimes has a horizontal wavelength as short as several hundred kilometers, and is associated with strong sensible heating from the ocean surface and conditional instability. Latent-heat release from cumulus convection, air–sea interaction instability, baroclinic instability in the face of low static stability, and barotropic instability may play roles in polar cyclogenesis.

However, the polar low can also appear over land (Fig. 1.105). Higher static stability over land probably precludes the possibility of rapid intensifica-

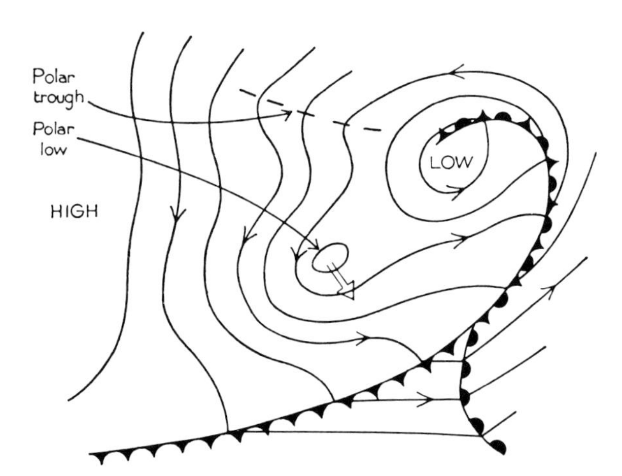

Figure 1.104 Schematic diagram of a polar low and polar trough to the rear of an occluding depression (from Reed, 1979; after Meteorological Office, 1962).

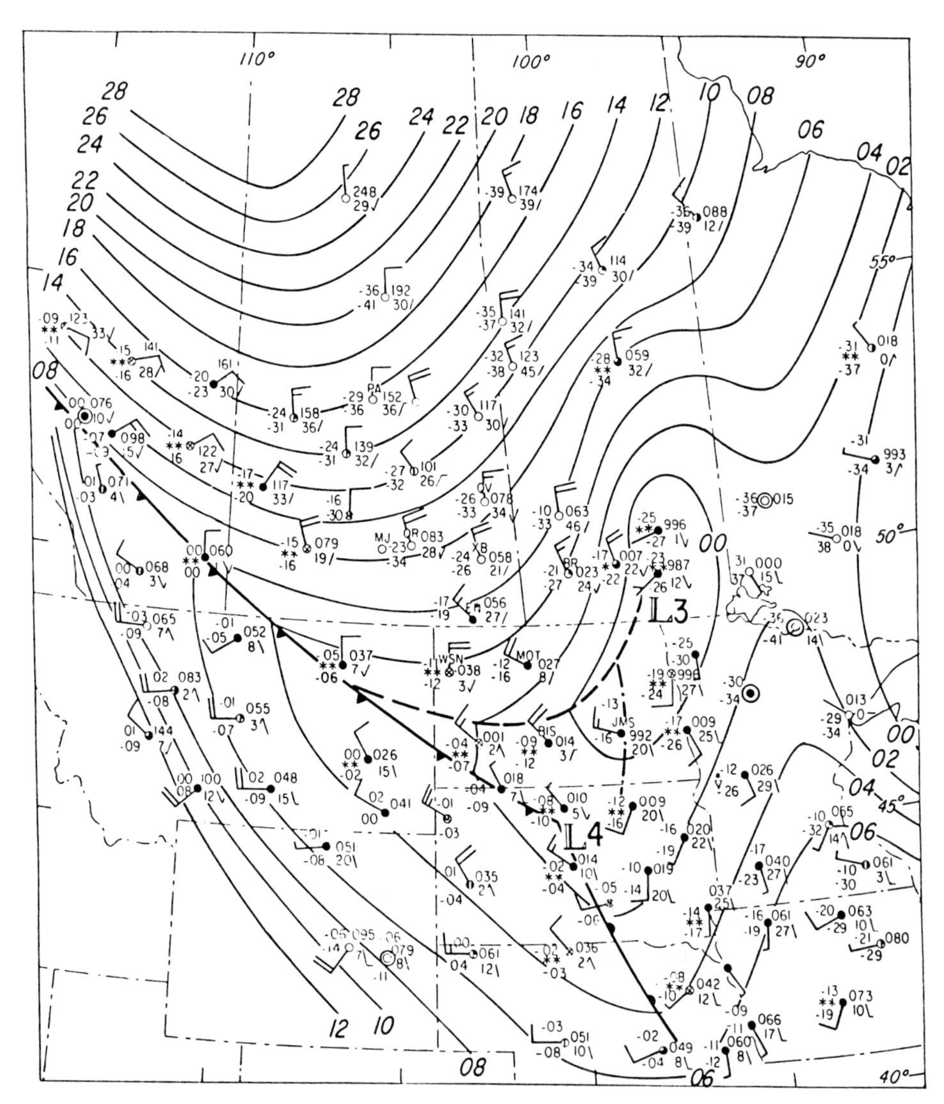

Figure 1.105 Example of a polar trough over land. Surface analysis for 1200 UTC, January 27, 1977. Full wind barb = 5 m s^{-1}, half wind barb = 2.5 m s^{-1}; temperatures and dew points in °C; sea-level pressure in mb × 10 without the leading "10" or "9"; isobars in mb without the leading "10." The heavy dashed line represents the polar trough and the heavy dashed-dotted line represents the wind-shift line (from Mullen, 1982). (Courtesy of the American Meteorological Society)

tion. Polar lows have been documented over the ocean near Greenland, north of Scandinavia, in the eastern North Pacific, and in the Mediterranean. Some polar lows over the ocean also are "bombs."

The "zipper" low. The surface temperature field within and around weak cyclones that move along the baroclinic zone associated with the coastal front (see Chap. 2) sometimes looks like the opening and closing of a zipper. The baroclinic zone is tightened ahead of the cyclone as a result of geostrophic deformation, while the baroclinic zone to the rear of the cyclone remains unchanged. The winds on either side of the frontal zone are approximately parallel to the frontal zone. There is no cold anticyclone north of the front. Consequently, there is no cold-air damming north of the front. Keshishian and Bosart have argued that zipper lows may "ripen" the atmosphere for further cyclogenesis by increasing environmental vorticity, convergence, and by reducing static stability.

The dryline–front intersection low. Shallow cyclones are often found at the intersection of the dryline (see Chap. 2) and a frontal zone (or outflow boundary) in the Southern Plains of the United States (Fig. 1.106). These lows are sometimes called *subsynoptic-scale lows* (SSLs). They often form in the

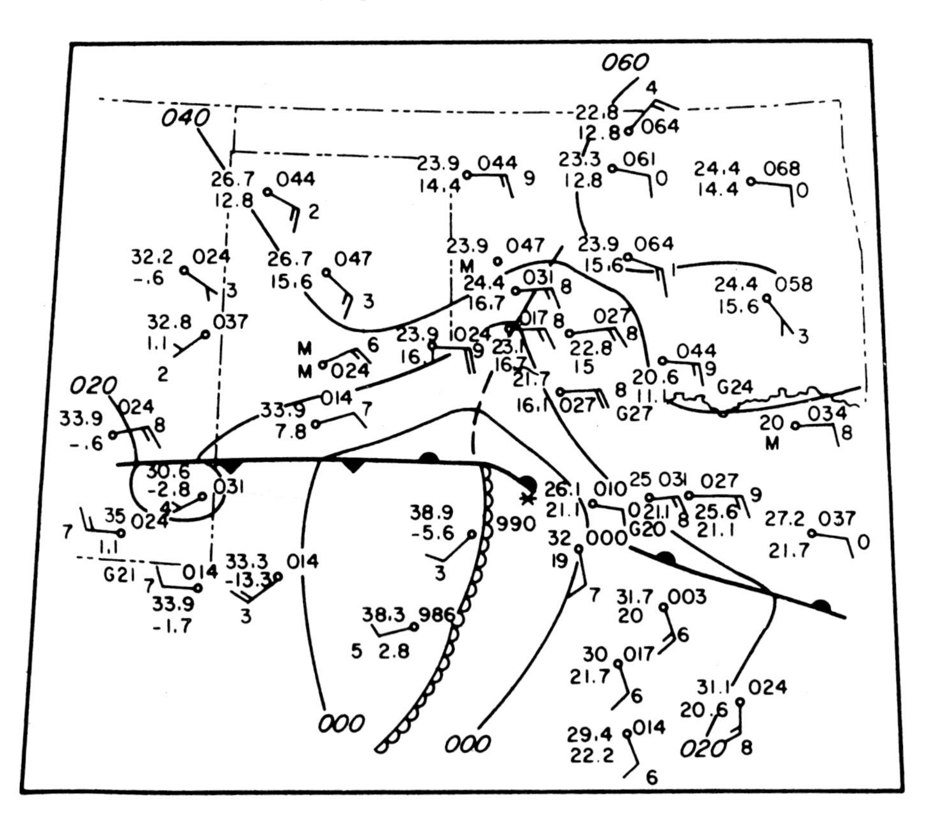

Figure 1.106 Example of a dryline–front intersection low (also referred to as a *sub-synoptic-scale low* (SSL) or *triple point*). Surface analysis for 0000 UTC, May 17, 1978. Pressures plotted are altimeter settings converted to tens of mb, with the 100's and 1000's digits omitted; wind direction in tens of degrees, with the 100's digits omitted, is plotted at the end of each wind barb; temperature and dew point in °C; whole wind barb = 5 m s^{-1}, half wind barb = 2.5 m s^{-1}; dryline noted by scalloped line; asterisk denotes location of a severe thunderstorm (from Bluestein and Parks, 1983). (Courtesy of the American Meteorological Society)

absence of strong synoptic-scale forcing, and are quasistationary. It is likely that diabatic heating in the boundary layer and topography play important roles in their formation. The cool air north of the front is usually moister than the hot air west of the dryline, but drier than the warm air east of the dryline.

The thermal low. Surface cyclones that form in response to intense diabatic heating at low levels over land during the warm season are called *thermal lows.* These cyclones are shallow, warm-core systems. Figure 1.107 illustrates the structure of a thermal low in the Southern Plains of the United States during the great summer heat wave of 1980. The cyclone at the surface does not show up at 700 mb.

The subtropical high. Deep anticyclones (Fig. 1.108), manifest at the surface as the "subtropical highs", are present nearly all the time over the ocean near

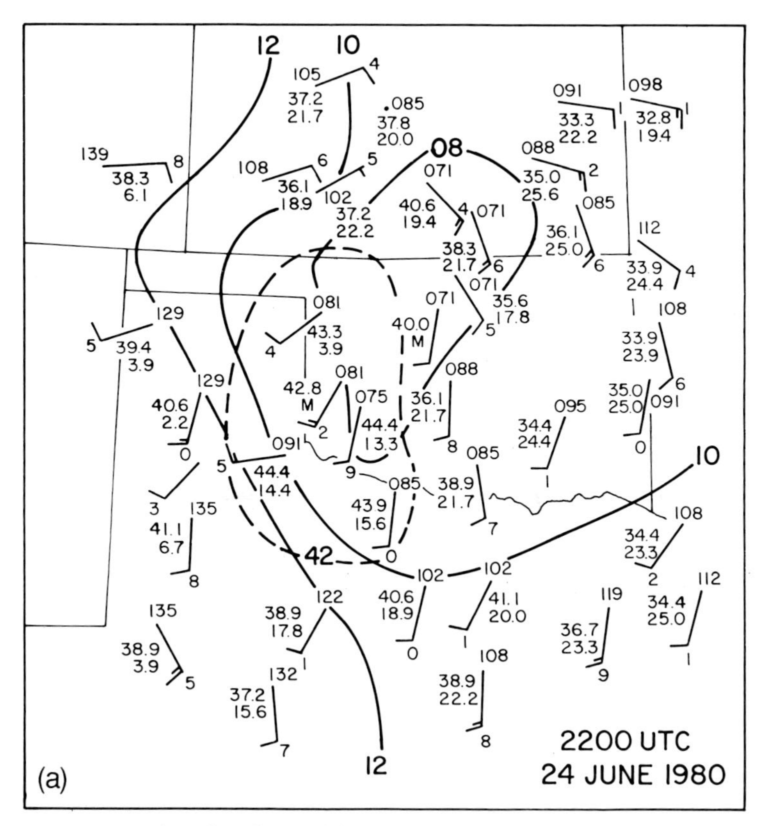

Figure 1.107 Example of a thermal low. (a) surface map for 2200 UTC, June 24, 1980, as in Fig. 1.106. 42°C isotherm (dashed line). (b) 700-mb map for 0000 UTC, June 25, 1980; 700-mb height plotted in m with the leading "3" omitted. Note that there is a surface cyclone centered near the northwest Oklahoma–Kansas border; the warmest air is displaced to the southwest slightly. At 700 mb there is an anticyclone centered in southwest Oklahoma–northwest Texas, near the warm core at the surface. This example occurred during the great heat wave of 1980 in the Southern Plains.

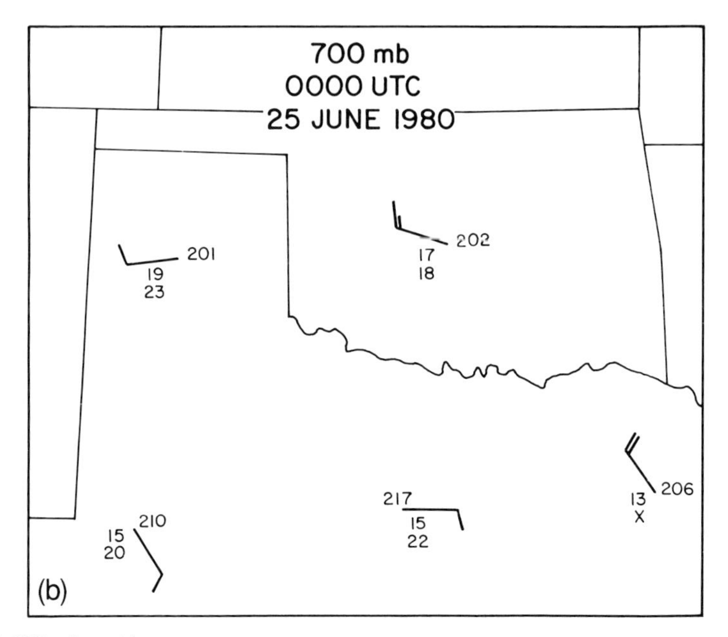

Figure 1.107 (*cont.*)

30° latitude. They are probably due to a combination of surface diabatic cooling and mobile anticyclones "beaching" themselves in the same place. The subtropical highs are located in the subsiding branch of the Hadley circulation, where one would expect to find anticyclogenesis in the mean. This region is sometimes called the *horse latitudes,* because in the early days of shipping, ships were trapped for days in the weak flow, and were forced to rid themselves of heavy cargo. Horses were evidently the first to go overboard. Not surprisingly, we note that the air in the horse latitudes is very "stable." The subtropical highs are not mirror images of the thermal lows: The thermal lows are shallow, while the subtropical highs are deep.

1.6. THE CLASSICAL MIDLATITUDE CYCLONE

1.6.1. A Quasigeostrophic Analysis of the Typical Life Cycle of a Midlatitude System

Idealized example of a developing system. On the basis of *surface* observations in Europe, J. Bjerknes, T. Bergeron, and H. Solberg of the "Norwegian School" in Bergen just after World War I formulated the "Polar–Front theory," which postulated that cyclones form along surface fronts. It is now generally accepted that a surface cyclone often, but not always, intensifies when an eastward-progressing area of cyclonic vorticity advection downstream from an upper level short-wave trough becomes

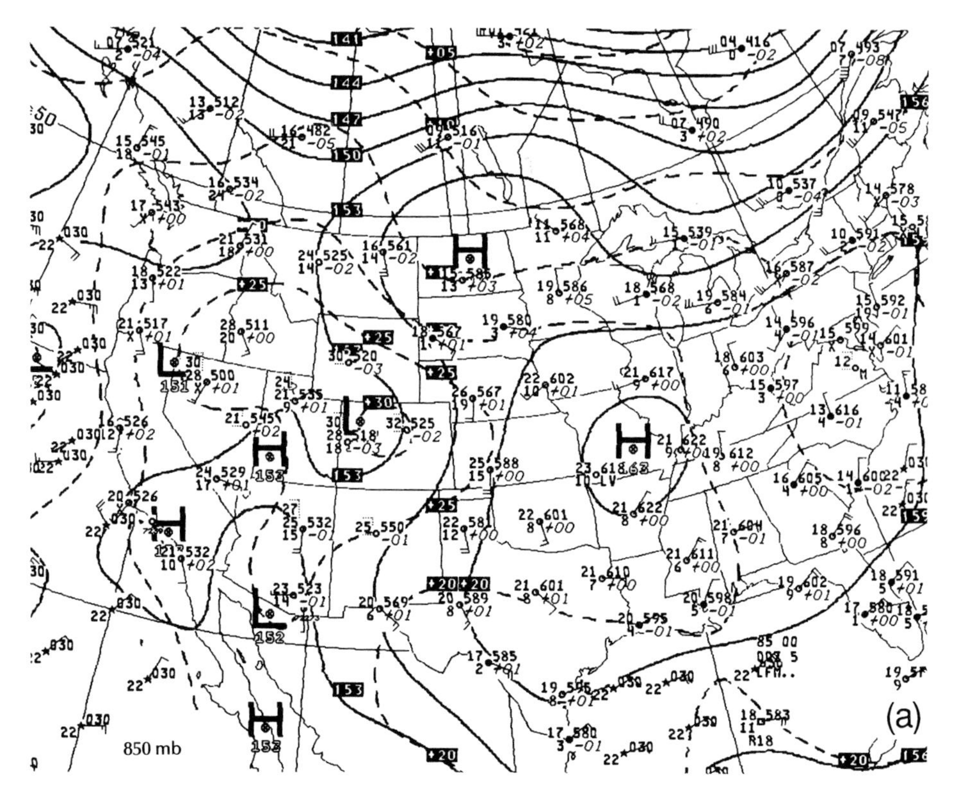

Figure 1.108 Example of the structure of the subtropical high. NMC analyses at 0000 UTC, September 5, 1990 for (a) 850, (b) 500, and (c) 200 mb. The example shown is over the land, where data are available, rather than over the data-sparse ocean. The anticyclone extends throughout the depth of the troposphere.

superimposed over a region of strong horizontal temperature gradient associated with a frontal zone at the surface. If the surface cyclone "opens up" to a trough aloft, a consequence of the thermal-wind relation, and if it leans toward the west with height, then there is both warm advection and vorticity advection becoming more cyclonic with height above the surface cyclone and eastward (Fig. 1.109) (This is a region characterized by cyclonic advection of geostrophic vorticity by the thermal wind.) In Fig. 1.109 we see that the thickness field (thermal wind) lags the wind field at 500 mb by a quarter of a wavelength. If the scale of the cyclone is short enough so that advection of Earth's vorticity is relatively small, and if the static stability is not too large, the cyclone may intensify as a result of the upward motion, and the concomitant low-level convergence. (Before the area of cyclonic vorticity advection aloft reached the baroclinic zone east of the surface cyclone, the differential vorticity-advection and temperature-advection forcing functions counteracted each other.) Rising motion in the relatively warm regions east and northeast of the surface low and sinking motions in the relatively cold regions west of the surface low result in a conversion of potential energy into kinetic energy, and the winds doth blow harder!

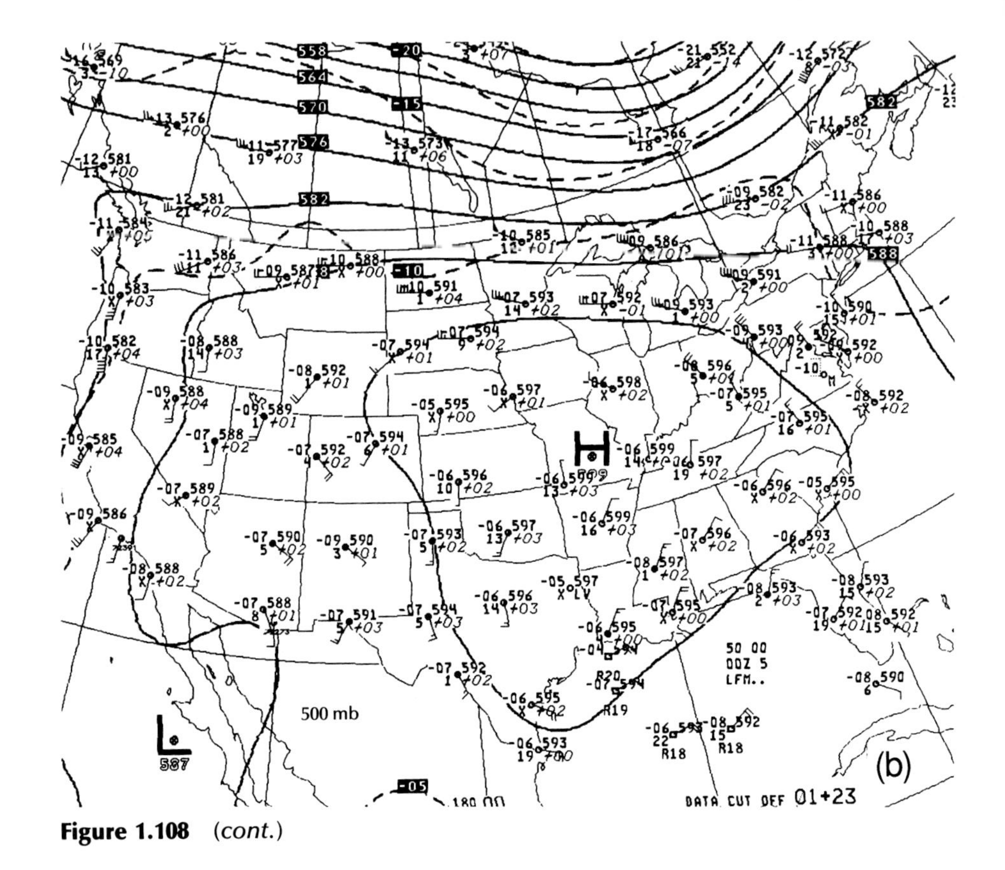

Figure 1.108 *(cont.)*

Warm advection at low levels downstream (relative to the flow in the middle troposphere; usually to the east or northeast in the Northern Hemisphere) from the cyclone and cold advection upstream (usually to the west or southwest in the Northern Hemisphere) result in the movement of the surface cyclone approximately in the direction of the winds in the middle troposphere (i.e., toward the east or northeast). Anticyclogenesis occurs at the surface in response to the surface cold advection and vorticity advection becoming more anticyclonic with height upstream from the upper level trough. (This is a region characterized by anticyclonic advection of geostrophic vorticity by the thermal wind.)

As warm, low-level air is advected poleward to the east and northeast of the surface low [Fig. 1.110(a)], and as cold, low-level air is advected equatorward, to the west and southwest of the low, the isotherms become more meridionally oriented [Fig. 1.110(b)], provided that the local temperature changes owing to vertical motions are smaller than the temperature changes owing to temperature advection. This is certainly the case on a level surface where $\omega = 0$. Vorticity rotates the isotherms. (In rapidly deepening cyclones, however, the vertical motions above the ground may be so intense that local temperature changes owing to vertical motions are larger than the temperature changes owing to advection.) The cold advection underneath the

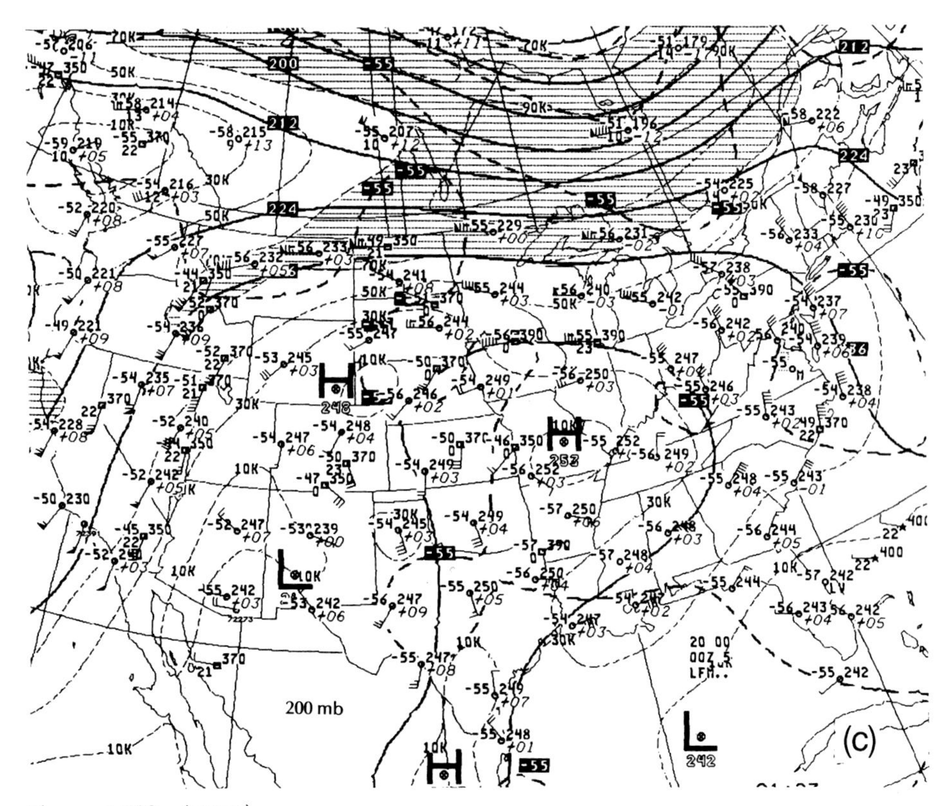

Figure 1.108 (*cont.*)

base of the upper-level trough results in a deepening of the trough. Warm advection to the east and northeast of the surface cyclone at low levels results in a building of the downstream ridge. The sharpening of the downstream ridge could also result in the sharpening of the downstream trough, and there may be "sympathetic development" as waves around the globe are influenced by local development.[7]

Meanwhile, back at the original developing system, the region of warm advection becomes situated more and more poleward of the cyclone, and the region of cold advection becomes situated more and more equatorward of the cyclone. Since surface cyclones tend to move to a large extent from areas of cold advection toward regions of warm advection, the classical midlatitude cyclone will move less rapidly in the zonal direction, and more rapidly in the meridional direction. A cold front trailing to the southwest and south of the surface cyclone progresses eastward more rapidly than the rate of eastward motion of the cyclone, so that the gap between the cold front and a warm front extending to the northeast and east of the cyclone, the "warm sector," diminishes in area (Fig. 1.110(c)).

The eastward movement of the upper-level trough is now much more rapid than the eastward component of movement of the surface cyclone, and eventually the upper-level trough catches up with and becomes superimposed

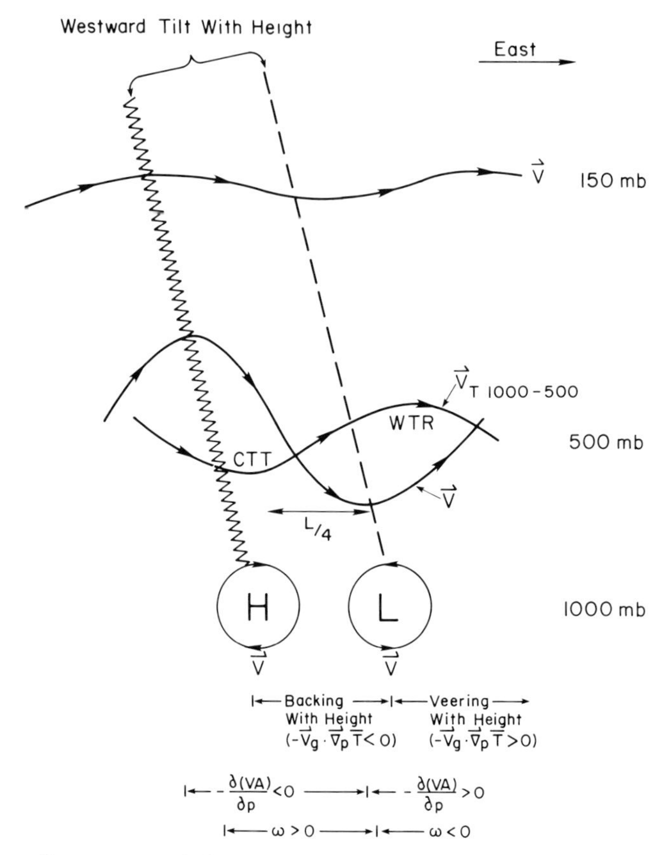

Figure 1.109 Illustration of the idealized temperature and wind structure of a developing baroclinic wave in the midlatitudes. The wind field at 1000, 500, and 150 mb are depicted as streamlines. The locations of surface low- and high-pressure areas are at L and H, respectively. The 1000–500 mb thermal wind ($\mathbf{v}_T$) is also depicted by a streamline. The wavetrain in the wind field tilts toward the west with height; the ridge and trough axes are marked by a sawtooth line and a dashed line, respectively. A warm thickness ridge in the 1000–500 mb layer is found at WTR; a cold thickness trough is found at CTT. CTT lags the trough in the wind field at 500 mb by a quarter of a wavelength ($L/4$). Regions of backing with height of the geostrophic wind (cold advection) at low levels and veering with height of the geostrophic wind (warm advection) are as indicated. Regions of vorticity advection becoming more cyclonic with height $[-\partial(\text{VA})/\partial p > 0]$ and more anticyclonic with height $[-\partial(\text{VA})/\partial p < 0]$ are as indicated. Regions of quasigeostrophic rising motion ($\omega < 0$) and sinking motion ($\omega > 0$) are located where shown.

on top of the surface cyclone. The cold front at the surface *appears* to catch up to the warm front northeast of the surface low and the warm sector at the surface is cut off, or "occluded." This may be due really to large surface-pressure falls in the cold air, resulting in the formation of a surface trough *in* the cold air (called an *occluded front*). The warm ridge near the cyclone corresponds to the area of occlusion described in Polar–Front theory (Fig.

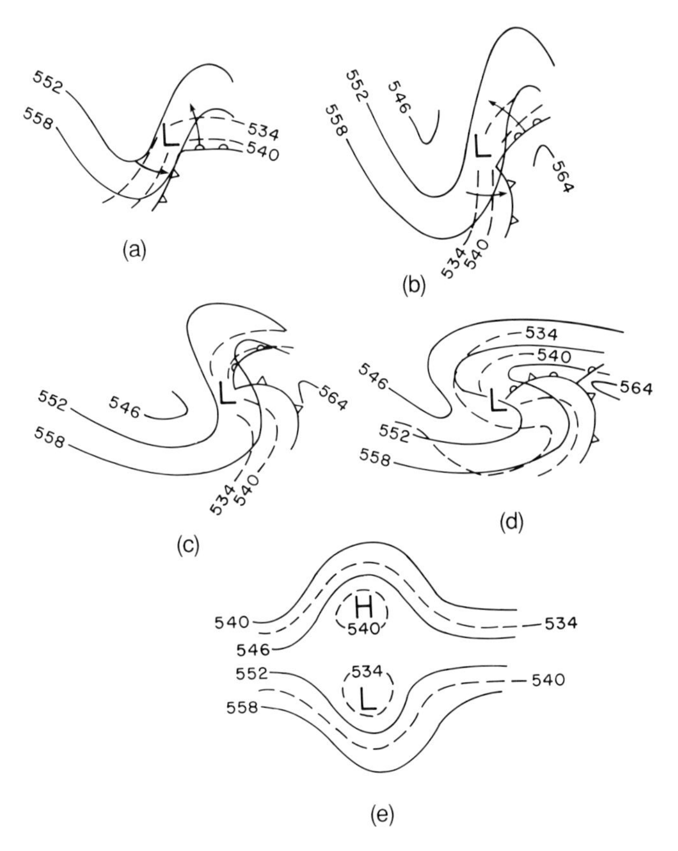

Figure 1.110 Illustration of the life cycle [(a)–(e) as a function of time; the time interval between successive panels of the figure is on the order of 1 day] of an idealized surface cyclone (L) and upper-level wave in the baroclinic westerlies. 500-mb height contours in dam (solid lines); 1000–500-mb thickness contours in dam (dashed lines). Locations of surface fronts as shown. Arrows in (a) and (b) depict surface wind field. Upper-level features in (e) are those of a high-over-low block. Description of relevant physical processes occurring in each panel given in the text. (This example is shown for illustrative purposes; it is not intended to serve as a model of all instances of the life cycle of cyclones).

1.110(d)). At this time, the upper level trough tends to "close off," that is, become a closed center of low pressure, while the "warm air" appears to have been pulled completely around and up over the surface cyclone (Fig. 1.110(e)). Surface development ceases because advection of upper level vorticity by the thermal wind vanishes. Furthermore, the movement of the upper-level system comes to a halt, because vorticity advection is decreased substantially owing to the near circular symmetry of the upper-level low. The surface cyclone also becomes nearly stationary because temperature advection has weakened. The cyclone *manqué* above the friction layer expires owing to the dissipative effects of friction, as divergence decreases the cyclone's vorticity. The upper-level low may become "cut off" from the basic westerly current and not feel the effects of geostrophic vorticity advection. Cut-off lows sometimes end up on the equatorward side of the basic westerly current. However, they do sometimes end up on the poleward side of the current. When they end up on the equatorward side, a ridge is sometimes built up on the poleward side, and the resulting configuration is a high-over-low block. The cyclone in the friction layer may persist as it is sustained by frictional convergence; however, cyclonic vorticity above the friction layer is destroyed by divergence associated with the rising motion below.

Case study of a developing system. An actual case of strong cyclogenesis is depicted in Fig. 1.111. The corresponding 300-mb pattern is shown in Fig. 1.112. Diagnostic computations of vertical velocity at 500 mb are shown in Fig. 1.113 for time period of most rapid development. Note the similarity of Fig. 1.110(b) to Fig. 1.111 and Fig. 1.112 at 1200 GMT (now called UTC), March 3. Figure 1.110(c) is similar to Fig. 1.111 and Fig. 1.112 at 1200, March 4, in that the 1000–500-mb thickness contours have a "pinched" appearance. By 1200, March 5, a cut-off low at 300 mb and center of cold 1000–500-mb thickness have appeared on the poleward side of the basic westerly current. A surface trough that extends eastward from this center corresponds to the "occlusion" in Polar–Front theory. However, unlike Fig. 1.110(c), Fig. 1.111 does not show a closed center of warm thickness north of the surface low.

Rising motion at 500 mb (Fig. 1.113) is due mostly to differential vorticity advection in the vicinity of the surface cyclone. However, strong sinking motion induced by cold advection is found just upstream from the upper-level trough. Convergence related to rising motion associated with both differential vorticity advection and temperature advection is largely responsible for the cyclone development at 900 mb (Fig. 1.114). On the other hand, vorticity advection is predominantly responsible for height tendencies at 300 mb (Fig. 1.115).

The patterns of **Q** vectors and isotherms at 800 mb are shown in Fig. 1.116. A strong divergence of **Q** is found north of the cold front in Texas and Oklahoma at 0000, March 3. A strong convergence in **Q** is found along the cold front and north of the surface low at 0000, March 4. At 0000, March 5 intense **Q** convergence is located along the trough extending eastward from the surface low. At 0000, March 4, strong **Q** convergence is found at 500 mb southwest of the surface low (Fig. 1.117).

Case study of the transformation of a warm-core, tropical cyclone into a cold-core system. An interesting case of the transformation of a warm-core, tropical system into a typical cold-core, extratropical system is illustrated in Fig. 1.118. At 1200 on June 20, 1972, there was an intense, closed circulation at 850 mb. The intensity of the circulation *decreased* with height above 500 mb; only a weak trough in the westerlies was evident at 200 mb (Fig. 1.118(a)). The surface circulation, which was the remains of Hurricane Agnes, was weak, presumably owing to friction. Temperature gradients at all levels were relatively weak. A surface cold front approached from the northwest (Fig. 1.118(b)), and by 0000, June 22, had caught up to the weak circulation at the

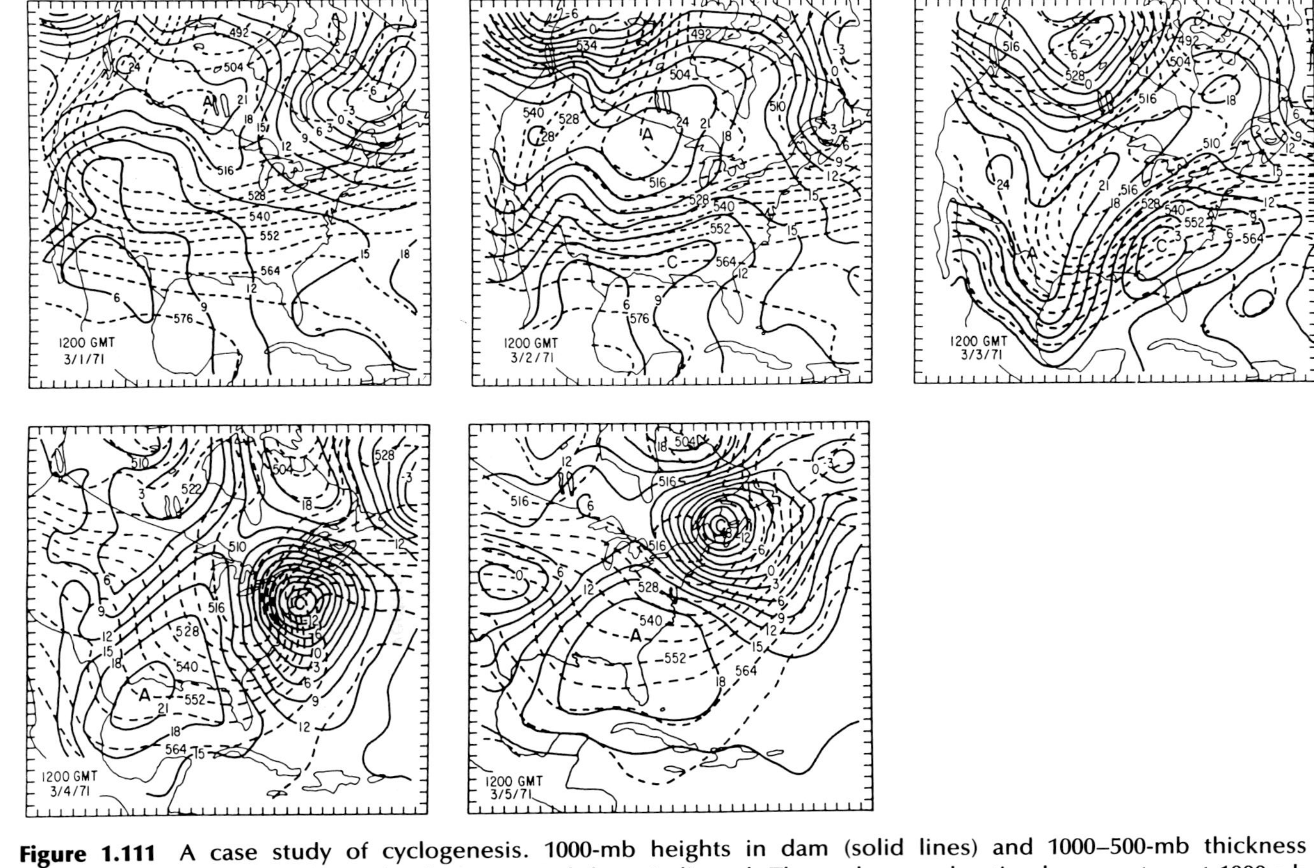

Figure 1.111 A case study of cyclogenesis. 1000-mb heights in dam (solid lines) and 1000–500-mb thickness contours in dam (dashed lines) for the times and dates indicated. The cyclone and anticyclone centers at 1000 mb are indicated by C and A, respectively (from Boyle and Bosart, 1986). (Courtesy of the American Meteorological Society)

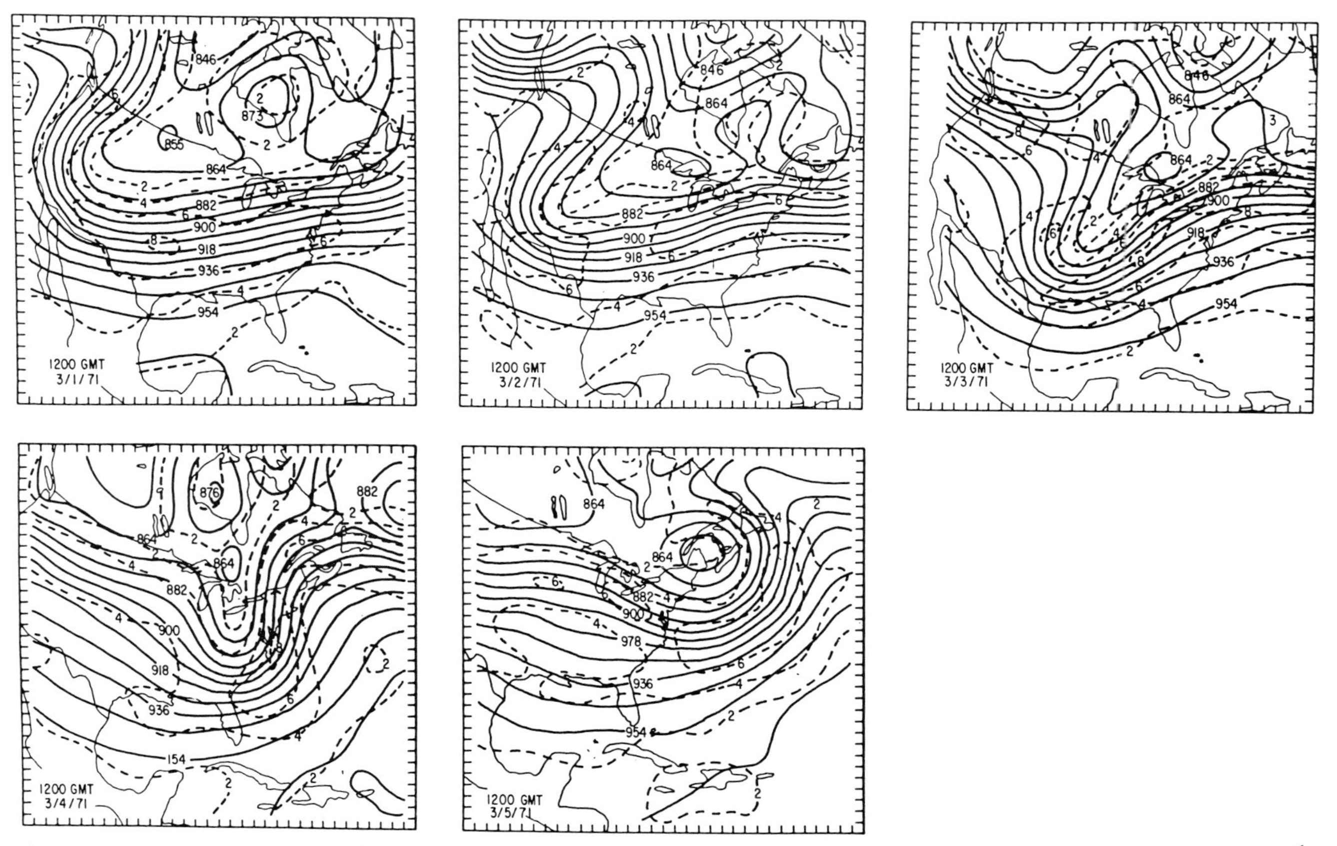

Figure 1.112 As in Fig. 1.111, but for 300-mb height contours in dam (solid lines) and 300-mb isotachs in dam s^{-1} (dashed lines) (from Boyle and Bosart, 1986). (Courtesy of the American Meteorological Society)

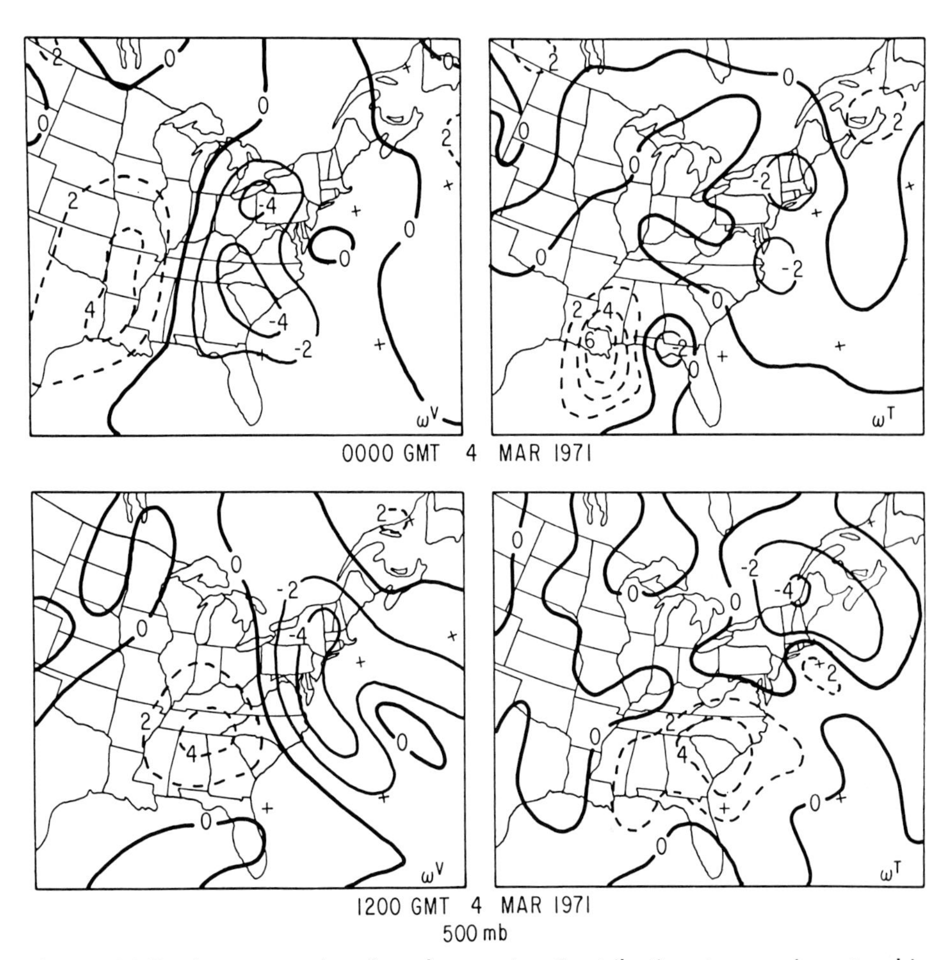

Figure 1.113 A case study of cyclogenesis. Contribution to quasigeostrophic vertical motion from differential vorticity advection (ω_V, left) and the Laplacian of temperature advection (ω_T, right) at 500 mb for 0000 UTC, March 4, 1971 (top) and 1200 UTC, March 4, 1971 (bottom) in 10^{-3} mb s^{-1}. Ascent denoted by solid lines; descent by dashed lines (from Boyle and Bosart, 1986). (Courtesy of the American Meteorological Society)

surface (not shown). A cold advection pattern developed to the west and southwest of the surface low, while a weaker warm advection pattern appeared to the north-northeast of the low. By 1200, June 22 (Fig. 1.118(c)), the surface cyclone had deepened, and the patterns of cold and warm advection had intensified. (The author was, at the time, in Washington, D.C., getting soaked with heavy rain.) By 1200 on June 23 (Fig. 1.118(d)), the system was "occluded" and cold core. The intensity of the circulation now *increased* with height, and relatively cold air was present near the cyclone, to the west and southwest.

From June 20–21 diabatic heating was most responsible for rising motion in the lower and middle troposphere. By 0000, June 22, differential vorticity

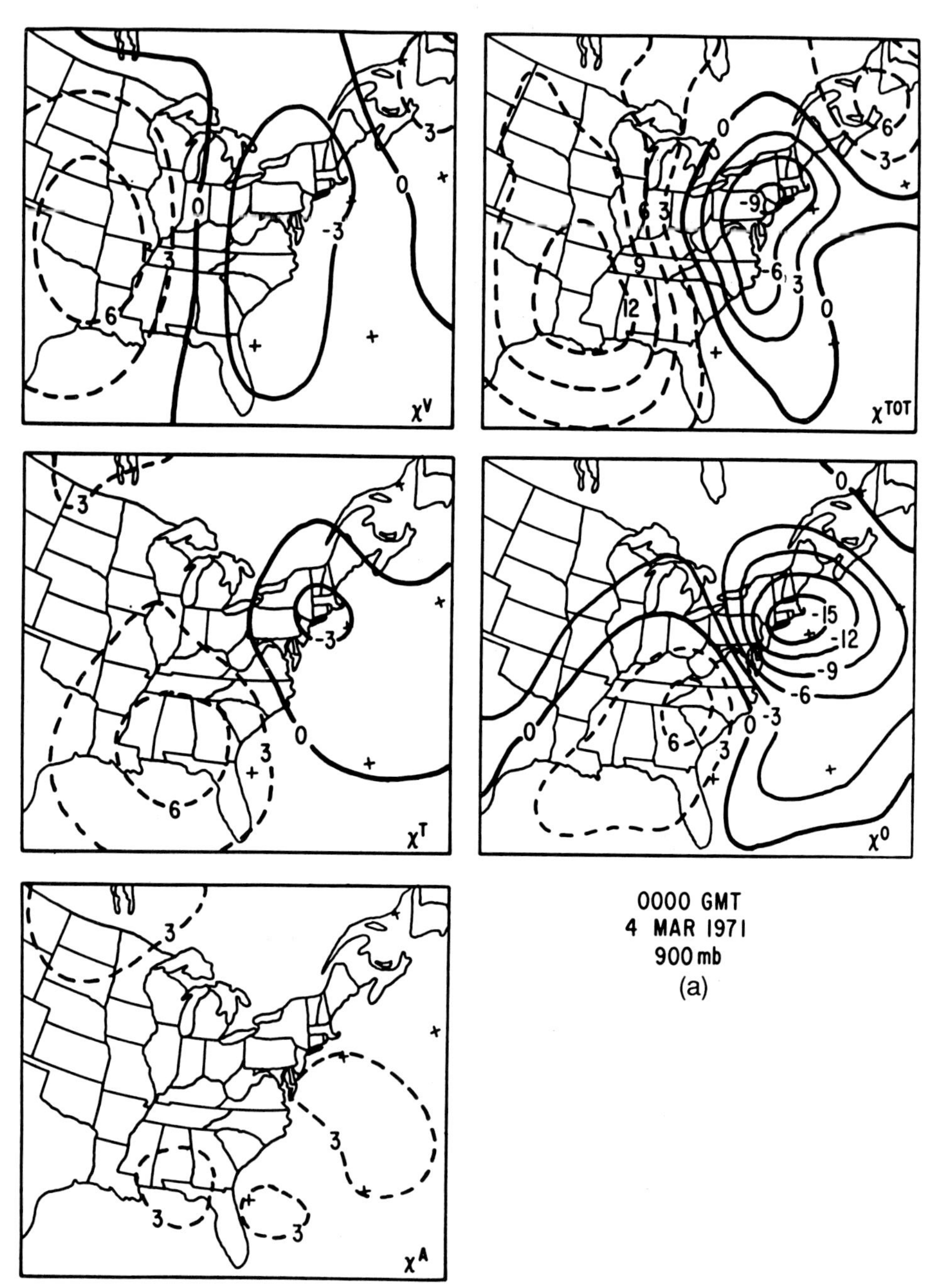

Figure 1.114 A case study of cyclogenesis. Partitioned height tendencies at 900 mb in dam $(12\text{ h})^{-1}$ for (a) 0000 UTC, March 4, 1971, and (b) 1200 UTC, March 4, 1971, as follows: χ^{V} (upper left), χ^{T} (middle left), χ^{tot} (upper right), χ^{o} (middle right), and χ^{A} (lower left). Rises and falls given by dashed and solid contours, respectively. The quasigeostrophic height-tendency equation may be written as follows: $\nabla^2\chi^{tot} = -f_0\mathbf{v}_g \cdot \nabla_p(\zeta_g + f) + f_0^2(\partial\omega/\partial p)$, where $\chi^{tot} = \chi^{A} + \chi^{V} + \chi^{T}$; χ^{A} is associated with vorticity advection, χ^{V} is associated with divergence and vertical motion induced by

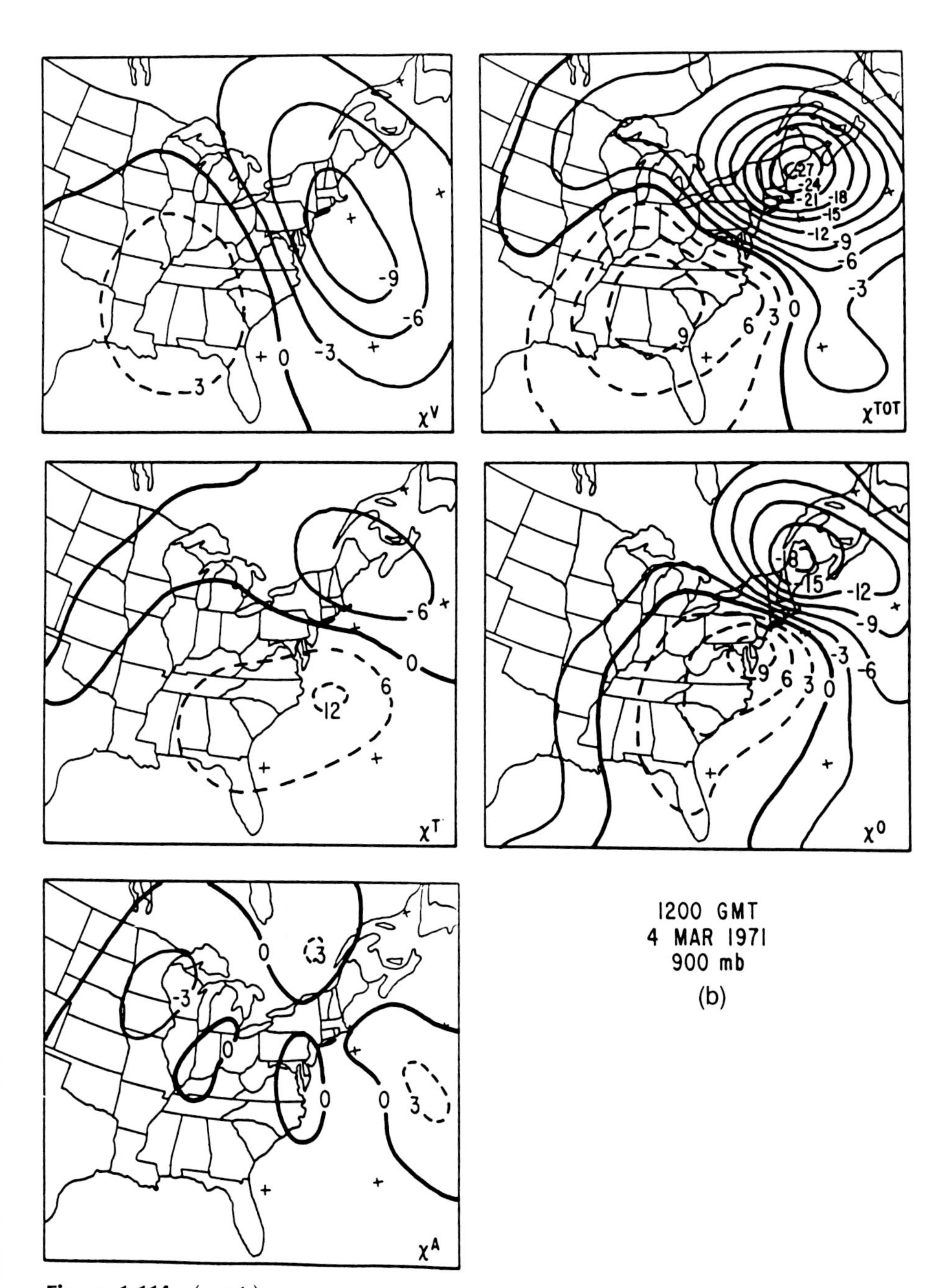

Figure 1.114 *(cont.)*
differential vorticity advection, and χ^T is associated with divergence and vertical motion induced by temperature advection. χ^o is the observed height tendency (from Boyle and Bosart, 1986). [Note that the contours for -12 and -18 dam $(12\text{ h})^{-1}$ are unintentionally omitted for χ^T near Maine in (b).] (Courtesy of the American Meteorological Society)

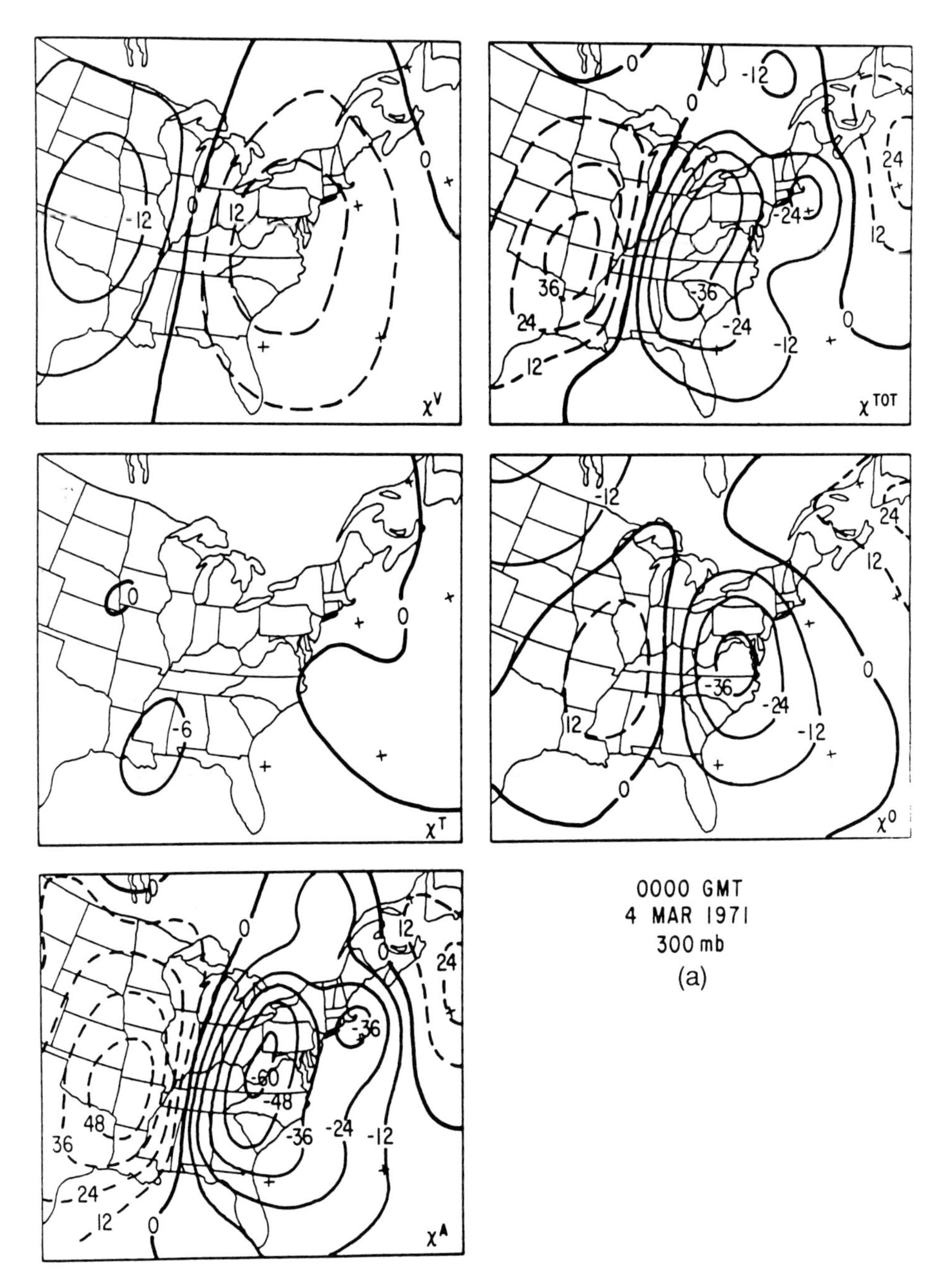

Figure 1.115 As in Fig. 1.114, but for height tendencies at 300 mb (from Boyle and Bosart, 1986). (Courtesy of the American Meteorological Society)

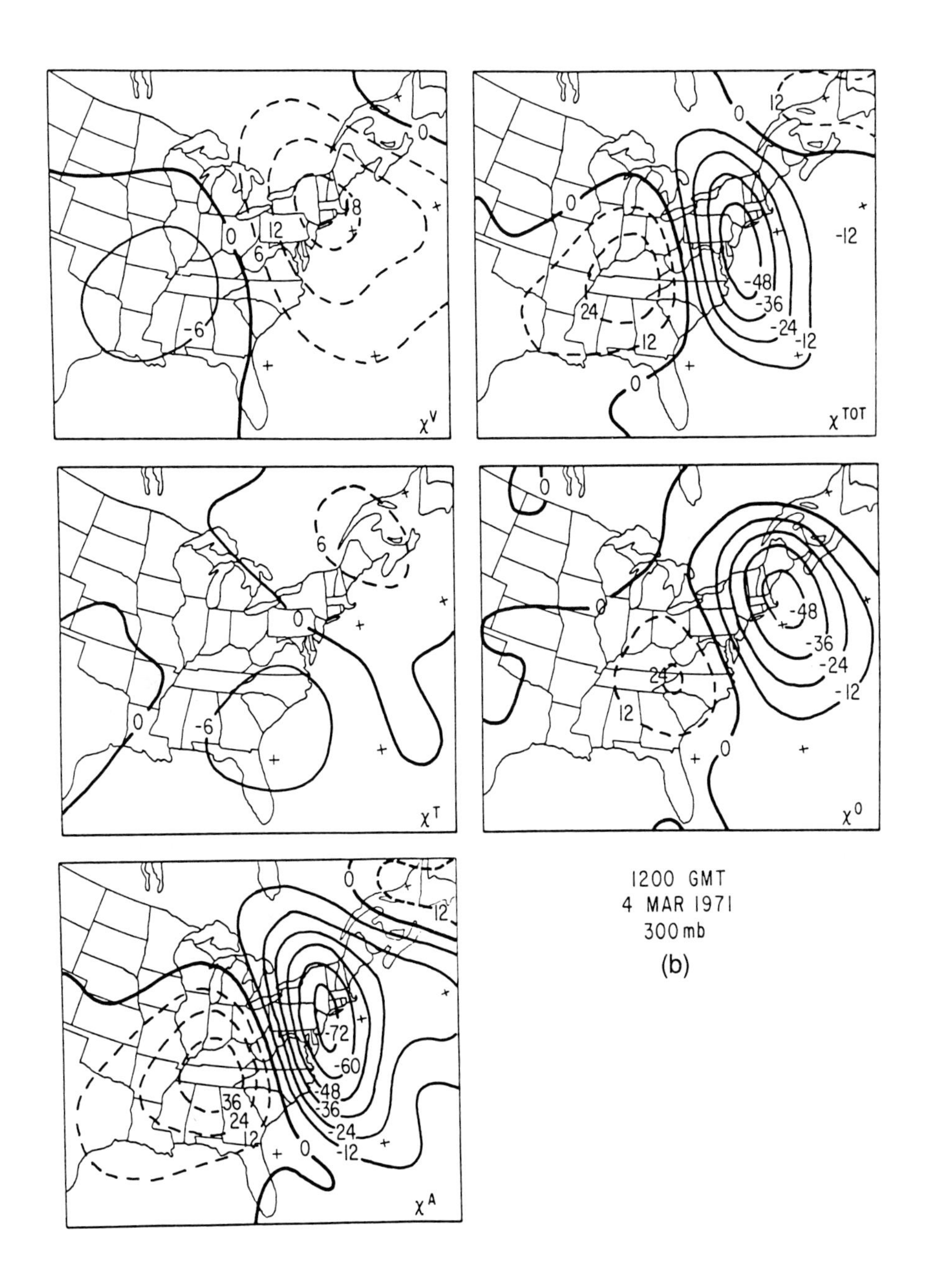

Figure 1.115 (*cont.*)

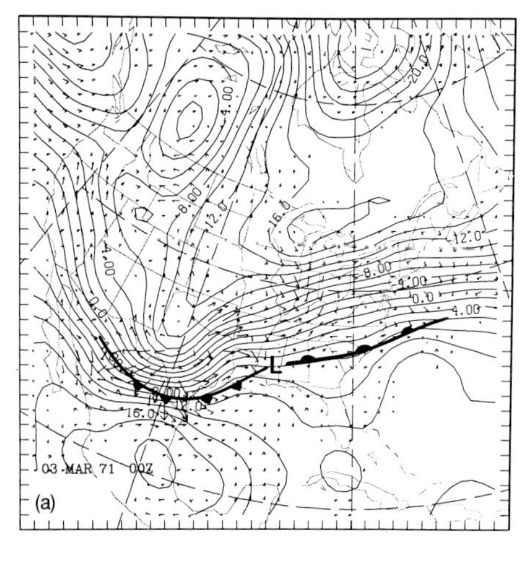

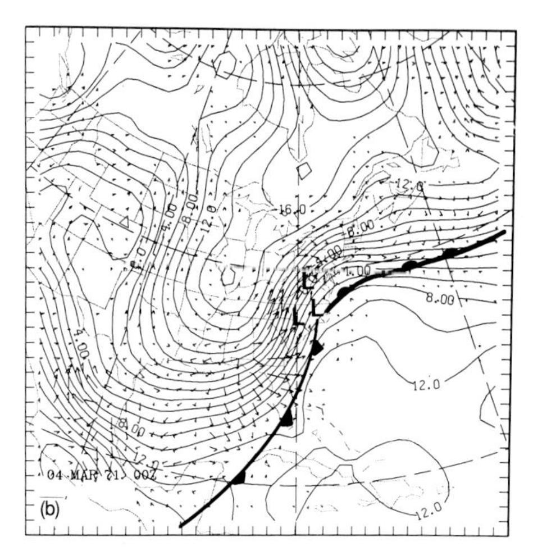

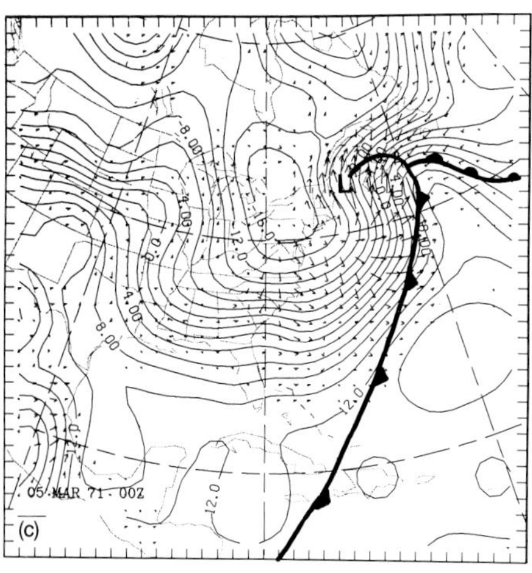

Figure 1.116 A case study of cyclogenesis. 800-mb **Q** vectors and isotherms (°C) for (a) 0000 UTC, March 3, 1971, (b) 0000 UTC, March 4, 1971, and (c) 0000 UTC, March 5, 1971. The longest vector represents a magnitude of 1.5×10^{-9} K s^{-1} m^{-1}; all others are scaled linearly to this value. Positions of the surface cyclone (L) and frontal systems are indicated (from Boyle and Bosart, 1986). (Courtesy of the American Meteorological Society)

advection and thermal advection at low levels had become more important. Very strong subsidence associated with cold advection was present at 1200, June 22, southwest of the low. Subsidence associated with cold advection had increased even more as of 0000, June 23. This had weakened by 1200, June 23. Rising motion in the upper troposphere was found west of the low at all times in association with differential vorticity advection.

Idealized example of a nondeveloping system. A nondeveloping baroclinic wave is shown in Fig. 1.119. Waves tilt toward the east with height, so that low-level warm advection east of the surface low is associated with height rises at the 500-mb trough; low-level cold advection west of the surface low is associated with height falls at the 500-mb ridge. Note that the thickness field (see the thermal wind) leads the height field at 500 mb. Sinking motion, low-level divergence, and pressure rises are found near the surface low.

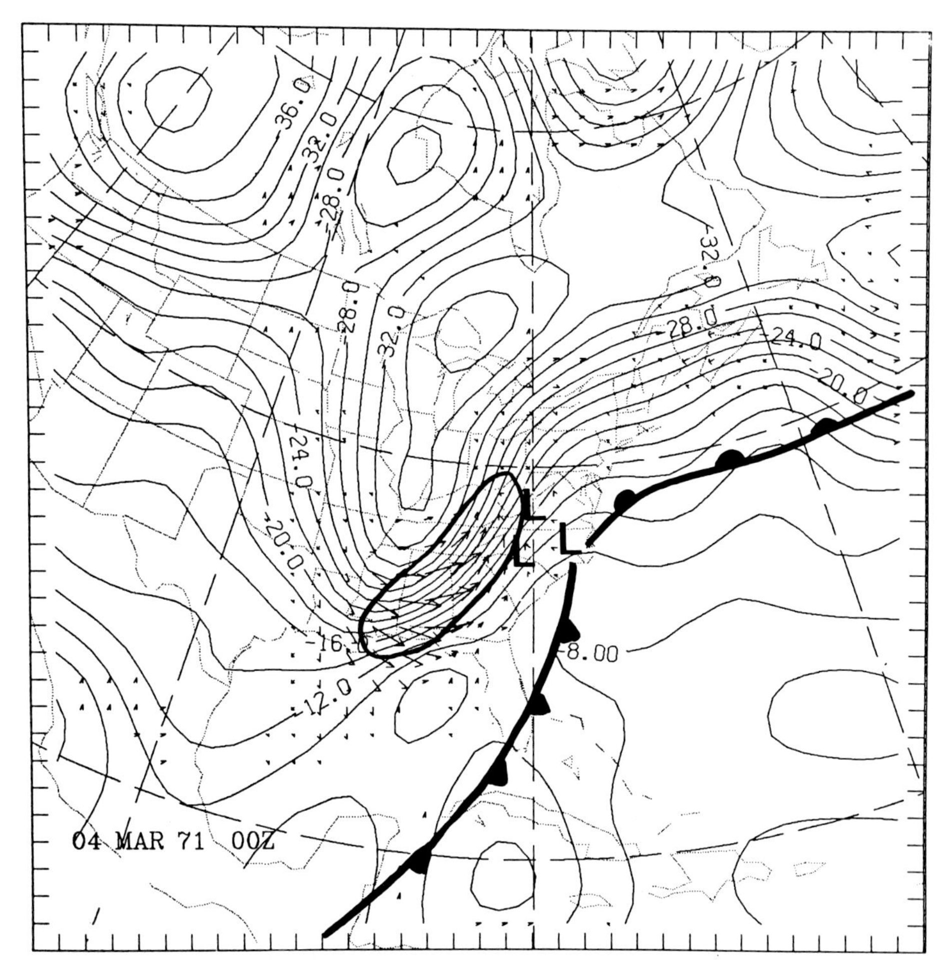

Figure 1.117 A case study of cyclogenesis. 500-mb **Q** vectors and isotherms (°C) for 0000 UTC, March 4, 1971. The heavy solid line encircles the region where $|\nabla_p T| > 3$ K $(100\text{ km})^{-1}$. The scaling factor for the **Q** vectors is 8×10^{-9} K s^{-1} m^{-1} (from Boyle and Bosart, 1986). (Courtesy of the American Meteorological Society)

1.6.2. The Cloud Pattern in a Classical Midlatitude Cyclone.

The cloud and precipitation pattern around an idealized, steady-state, mature cyclone can be inferred from an application of quasigeostrophic theory and air-trajectory analysis. The component of motion of a surface cyclone from west to east is slower than the westerly component of the winds at high levels in the troposphere, but more rapid than the average westerly component at the surface. In the reference frame of the surface cyclone, air enters at low levels from the east on its poleward side, and at high levels from the west (Fig. 1.120). Air does not enter the surface cyclone from the east at low levels, equatorward of the cyclone center (Fig. 1.121), because the surface wind has a

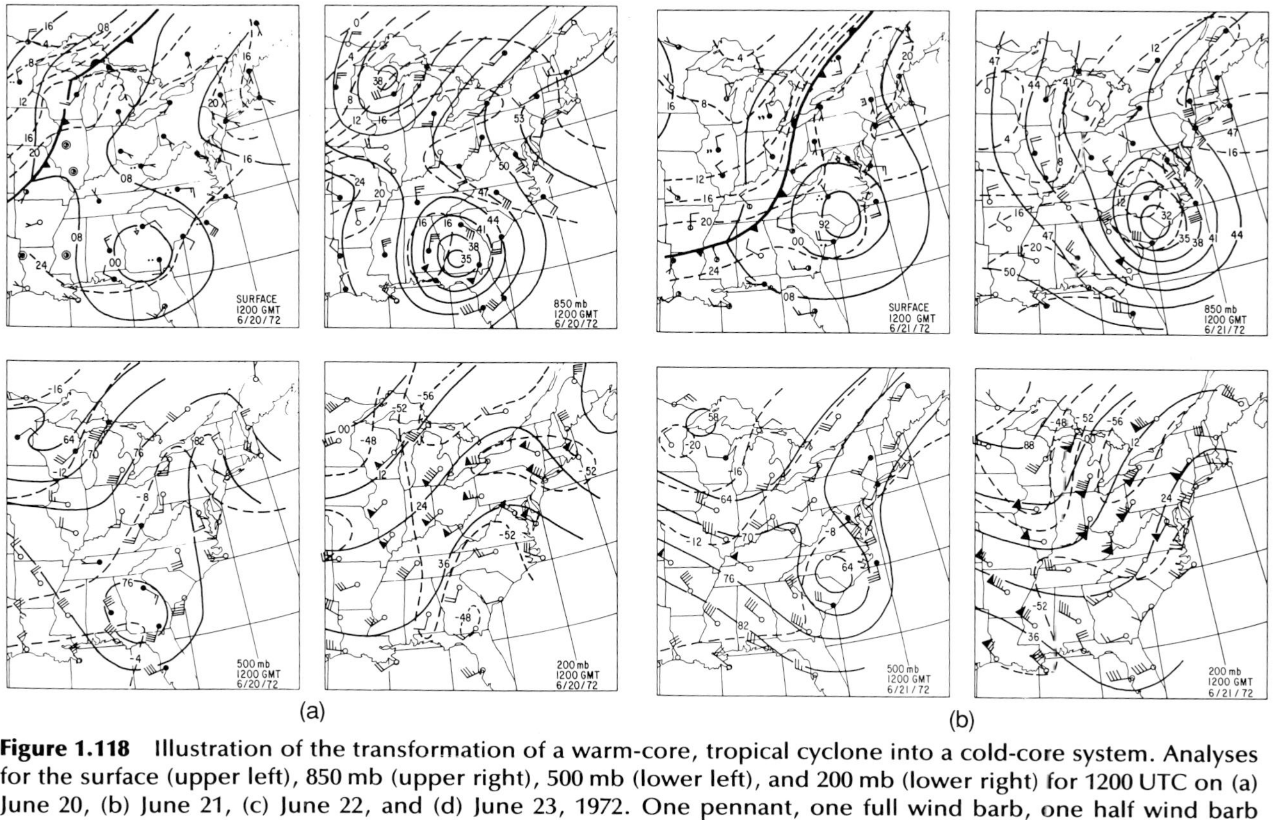

Figure 1.118 Illustration of the transformation of a warm-core, tropical cyclone into a cold-core system. Analyses for the surface (upper left), 850 mb (upper right), 500 mb (lower left), and 200 mb (lower right) for 1200 UTC on (a) June 20, (b) June 21, (c) June 22, and (d) June 23, 1972. One pennant, one full wind barb, one half wind barb represents 25, 5 and 2.5 m s^{-1}, respectively. Temperature in °C (dashed lines), heights in dam (solid lines) at 850, 500, and 200 mb. Sea-level isobars in upper left in mb without the leading "10." Weather and sky conditions as indicated. Solid circles at 850 and 500 mb denote relative humidities in excess of 70% (from DiMego and Bosart, 1982). (Courtesy of the American Meteorological Society)

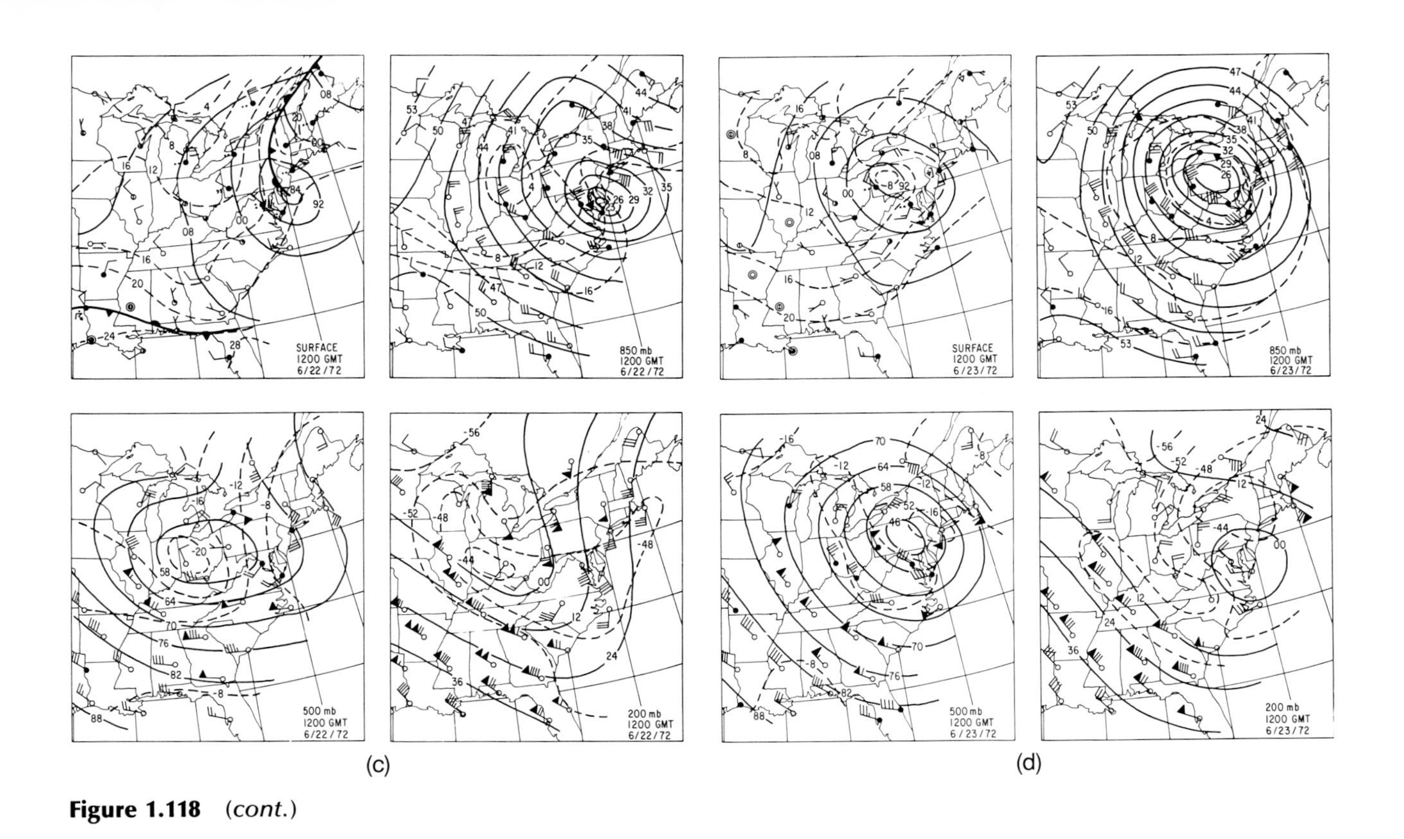

Figure 1.118 *(cont.)*

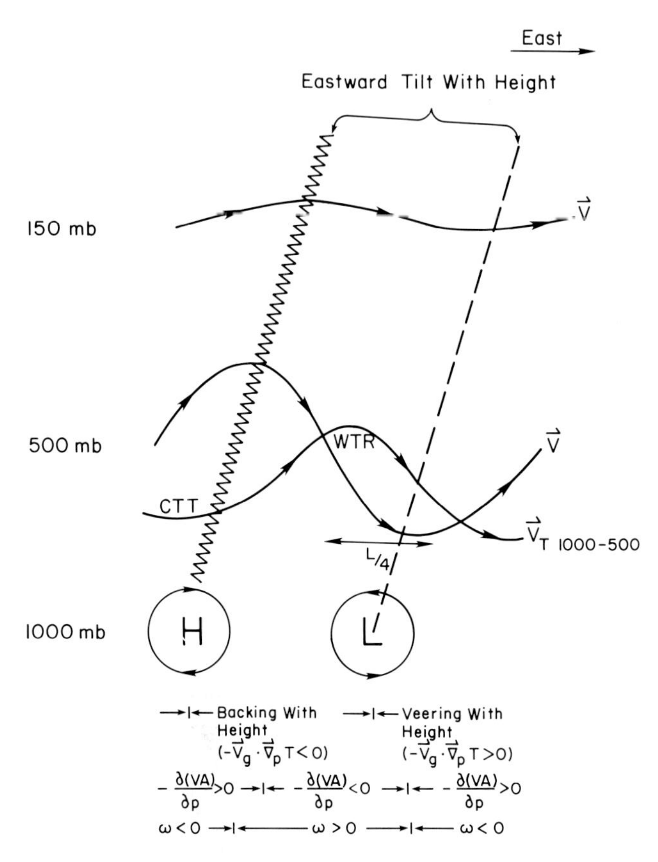

Figure 1.119 As in Fig. 1.109, but for a nondeveloping baroclinic wave. In this case the wavetrain tilts towards the east with height.

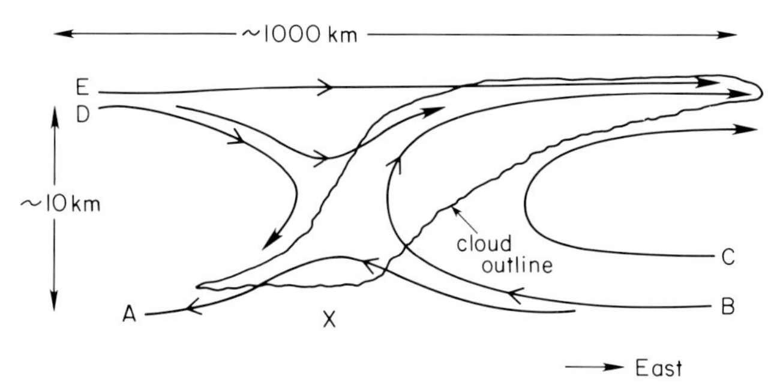

Figure 1.120 Illustration of idealized trajectories in the east–west vertical plane that cuts across the region just *poleward* of a midlatitude cyclone (X). Trajectory A ascends east of the cyclone and descends west of the cyclone. Trajectories B and C rise east of the cyclone and reverse direction aloft; condensation occurs at a lower level in B than in C because B begins at a lower level and is moister. Trajectory D reverses direction as it descends. Trajectory E passes through the system from west to east.

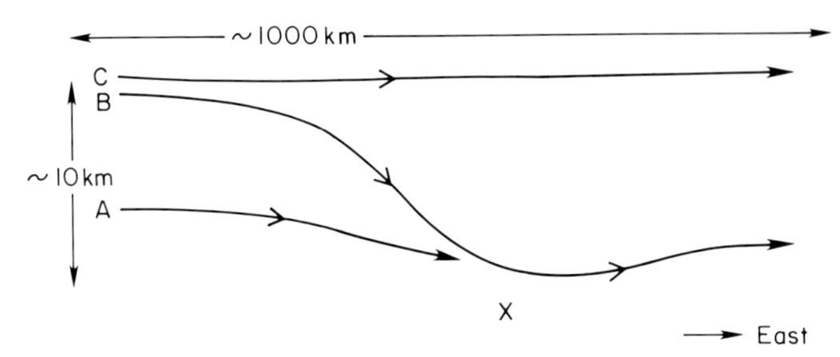

Figure 1.121 As in Fig. 1.120, but for an east–west vertical plane that cuts across the region just *equatorward* of a midlatitude cyclone. Trajectories A and B descend and pass by the cyclone from west to east. Trajectory C passes through the system from west to east.

moderate westerly component, which is as fast as or faster than the cyclone's rate of movement eastward. There is rising motion at and just northeast of the cyclone in association with warm advection and vorticity advection becoming more cyclonic with height (cyclonic advection of geostrophic vorticity by the thermal wind). The region well to the west and northwest of the cyclone is characterized by cold advection and vorticity advection becoming more anticyclonic with height (anticyclonic advection of geostrophic vorticity by the thermal wind), and hence there is sinking motion there.

Air that has had a history of ascent is likely to be relatively humid, while air that has had a history of descent is likely to be dry. Low-level air approaches the cyclone on its poleward side from the east, rises, and is cooled to saturation. Clouds and precipitation particles form and are carried back toward the east by the strong winds aloft. Material from the clouds is carried far downstream and often curves anticyclonically around the downstream ridge. This cloud material is usually composed of ice crystals, owing to the very cold temperatures aloft. Air that has originated just east of the cyclone at *low levels* rises, produces a relatively thick cloud, exits to the west of the cyclone, sinks, and dries out as it warms. Air that has originated at low levels farther to the east of the cyclone rises, produces a relatively high-based cloud, reverses direction relative to the cyclone, and flows downstream aloft. Air that has originated far to the east at higher levels may never reach saturation, since its level of origin is high where the air is relatively dry, and also because its vertical excursion is small. Thus, the eastern portion of the cloud system looks like a large-scale thunderstorm anvil, since the cloud base increases with height to the east. An observer poleward of an approaching Northern Hemisphere midlatitude cyclone first notices thin, cirrus clouds off to the southwest. Soon the cirrus thicken into cirrostratus, and middle-level altocumulus and altostratus appear. (This description is somewhat overly simplified because clouds often assume a layered appearance and also a banded appearance. The mesoscale aspects of the midlatitude cyclone are discussed in Chap. 3). Middle–to–upper-level air poleward of the cyclone descends as it approaches the cyclone from the west and reverses direction and flows back toward the west away from the cyclone.

Air enters the equatorward side of the cyclone at all levels from the west (Fig. 1.121). This air has had a history of subsidence and is therefore dry and clear. Some of this air flows southward of the cyclone's center and is then diverted poleward around to the east of the cyclone. This tongue of air is known as the *dry slot* or *dry tongue* or *dry intrusion.* It is lifted as it curves to the east and northeast of the surface cyclone. However, middle and upper-level clouds are usually not present because the air, having had a long history of subsidence, is so dry. When the dry slot rides over a mass of moist air near the surface, potential instability is increased and a nearly dry adiabatic layer is superimposed upon the moist air; the threat of severe cumulus convection is thus also increased (see Chap. 3).

Low-level air that has had a history of ascent from the east enters the poleward side of the cyclone and is diverted equatorward west of the cyclone center. This tongue of air is known as the *wrap-around moisture* and is

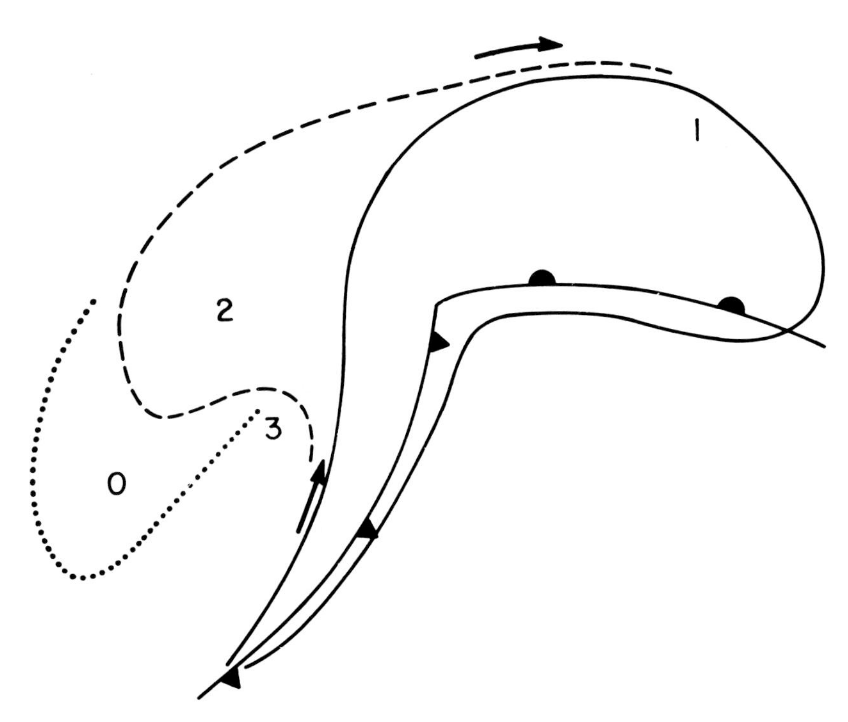

Figure 1.122 Schematic illustration of the comma-shaped cloud shield associated with a midlatitude wave–cyclone system. Early stage of cyclone development is associated with an area of dense middle- and upper-layer cloud contained within the solid boundary and labeled 1. The dashed line encloses area 2, which corresponds to the westward extension of the cloud canopy. The area enclosed by the dotted border labeled 0 denotes a layer of thin stratus within the boundary layer, the wrap-around moisture. The dry slot is represented by the narrow zone labeled 3. Arrows indicate location of major upper-tropospheric wind maxima; the winds, however, are not necessarily parallel to the cloud borders as shown (from Carlson, 1980). (Courtesy of the American Meteorological Society)

responsible for relatively shallow clouds just to the west and perhaps equatorward of the surface cyclone center. The entire cloud system takes on the shape of a comma: The wrap-around moisture forms the back of the comma head, while the dry intrusion creates a bow-shaped appearance to the hook of the comma (Figs. 1.122–1.125). Sometimes the wrap-around moisture and dry intrusion complete more than one-half a circuit about the cyclone and a spiral-shaped cloud band is formed (Fig. 1.123). Very intense oceanic cyclones sometimes have eye-like features, and thus look similar to tropical

Figure 1.123 Spiral-shaped comma cloud off the northwest coast of the U.S. at 1845 UTC, February 27, 1980. A second, much smaller comma cloud is also evident off the west coast of California, west of the "tail" associated with the former comma cloud; the tail of the main comma cloud appears to have been deformed by the small comma cloud. Another comma cloud is seen farther west in the Pacific. Cirrus clouds flowing anticyclonically about a ridge that is in between the cyclonic disturbances associated with the two main comma clouds are also visible. (The zonally oriented band of clouds at low latitudes is the intertropical convergence zone, which is not discussed in this text.)

Figure 1.124 Comma cloud with an eye (the President's Day snow storm) off the mid-Atlantic coast of the United States (from a GOES visible satellite photograph 1830 UTC, February 19, 1979). The photograph is misgridded; the geography should be shifted southeastward about 50 km relative to the picture (from Bosart, 1981). (Courtesy of the American Meteorological Society)

cyclones (Fig 1.124). The eyes may be the result of the occlusion of the clear, dry air.

The southern part of the "comma cloud," the "tail," is often composed of thunderstorms owing to its proximity to moist, low-level air that is potentially unstable. The relationship between the 500-mb flow and a comma-shaped water-vapor field is shown in Fig. 1.126 for a continental storm. The relationship among the mean wind, temperature, and moisture fields and a composite comma cloud are shown in Fig. 1.127. A cyclonic circulation is usually centered near the comma head at 850 and 700 mb, while a trough axis is positioned near the comma head at 500 mb. The core of strongest winds aloft blows from the southwest across the comma tail from the dry slot. Often a band of cirrus having small-scale transverse bands is located here also (Fig.

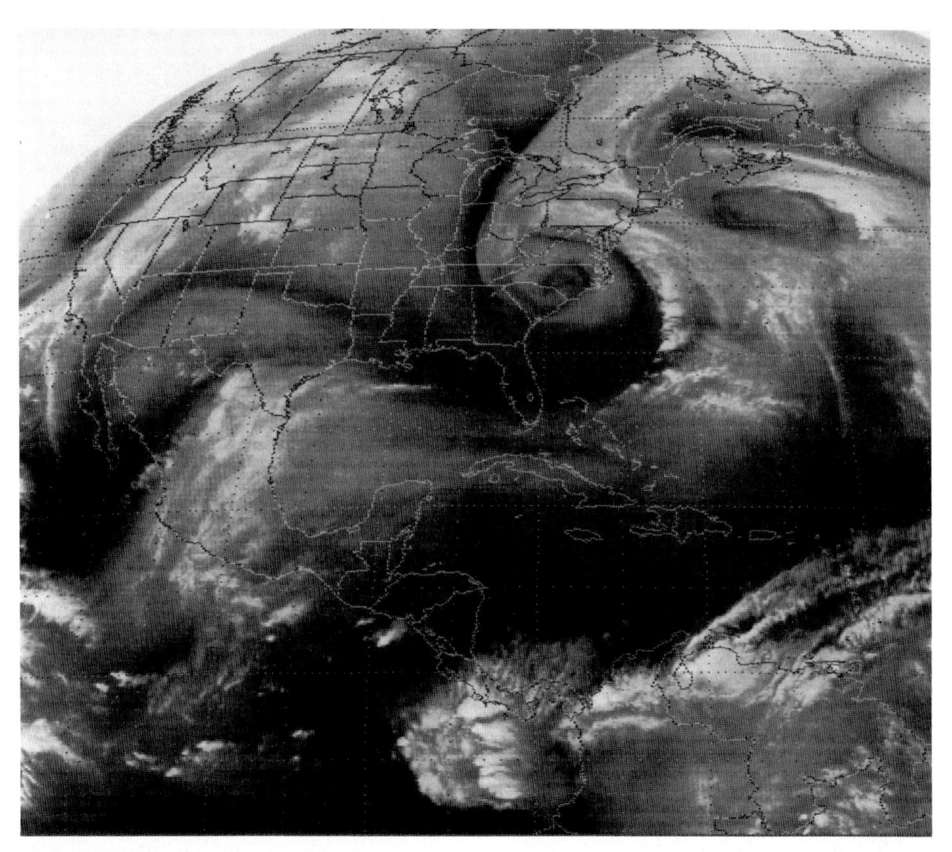

Figure 1.125 Comma cloud with a mesoscale dry (dark) region in the head (over North Carolina) (from a GOES water-vapor channel photograph 1501 UTC, April 7, 1988). The water-vapor channel is most sensitive to water vapor in the upper troposphere. Thunderstorm anvils are visible in the tail region of the comma cloud, to the east of the dry slot.

1.128). This band is often associated with an upper-level jet and will be discussed in Chap. 2.

The "English School" has contributed additional terminology (Fig. 1.129) that is sometimes associated with trajectories through midlatitude systems: The relatively warm air, whose origin is equatorward and eastward of the cyclone in the "warm sector" at low levels, and which rises into and becomes part of the comma cloud system, is called the *warm conveyor belt.* Relatively cold air whose origin is at low levels to the east of the cyclone and to the north of the warm front, and which rises into and becomes part of the comma cloud system, is called the *cold conveyor belt.*

The development of comma-cloud patterns is depicted in Fig. 1.130. In one instance the upper-level short-wave trough has caught up with the surface frontal zone. The development of a comma cloud in a "polar low" or "instant occlusion" is also seen in Fig. 1.130. The comma cloud links up with the wavy

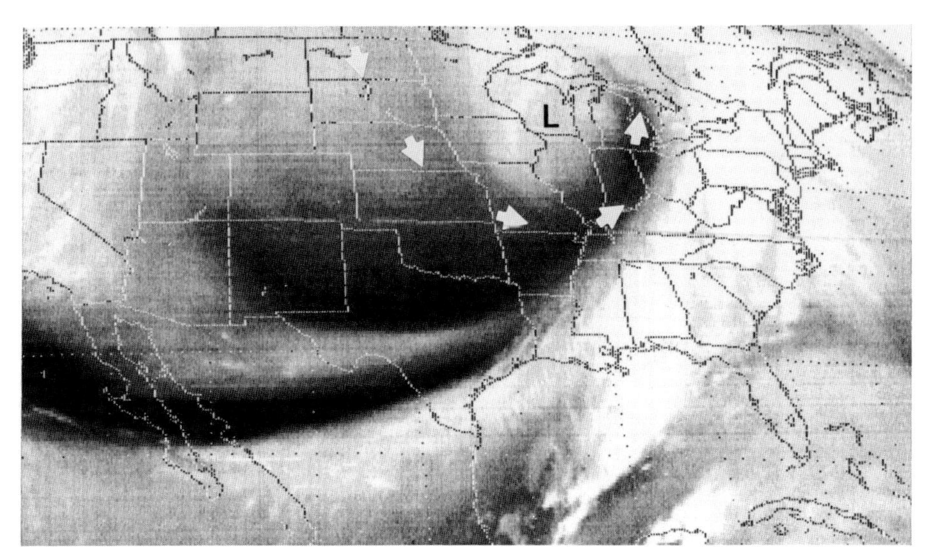

Figure 1.126 Comma cloud as seen in the infrared water-vapor channel of the GOES satellite at 0000 UTC, December 4, 1990. This photograph is particularly instructive because the comma cloud occurred over land, where rawinsonde data were available. The arrows depict the sense of the strongest flow at 500 mb, and the L denotes the center of the 500-mb low. The head of the comma is located near the low center, and the dry slot is aligned along the 500-mb flow. To the extent that moisture can be viewed as a conservative quantity, we can interpret the comma-shaped pattern in terms of the rotation of dry and moist air streams. Thus, we can look at comma-shaped upper-tropospheric patterns over data-void regions of the ocean and make a crude inference about the location of midtropospheric vorticity maxima. The example shown here was associated with a big snowstorm in Wisconsin and the surrounding region.

frontal band to the east. Comma clouds are also visible in satellite photographs in association with upper-level short-wave troughs that are well to the west of the surface frontal zone. One must therefore be very cautions in using a satellite-observed comma alone to infer the three-dimensional wind, temperature, and moisture structure of the atmosphere. The latter type of comma clouds are sometimes the location of subsequent thunderstorm formation.

1.6.3. Cyclogenesis along the East Coast of the United States

Some cyclones ("Type I") form along a surface frontal zone in and near the Gulf of Mexico. Most of them move through the southeastern states and northeastward along the Atlantic coast. Others ("Type II") form as "secondary" low-pressure areas at the surface, to the southeast of an occluding "primary" low-pressure area. The latter is usually located over the Ohio valley. The "secondary" low often develops along a shallow frontal zone separating cold air that has been "dammed up" east of the Appalachians and is

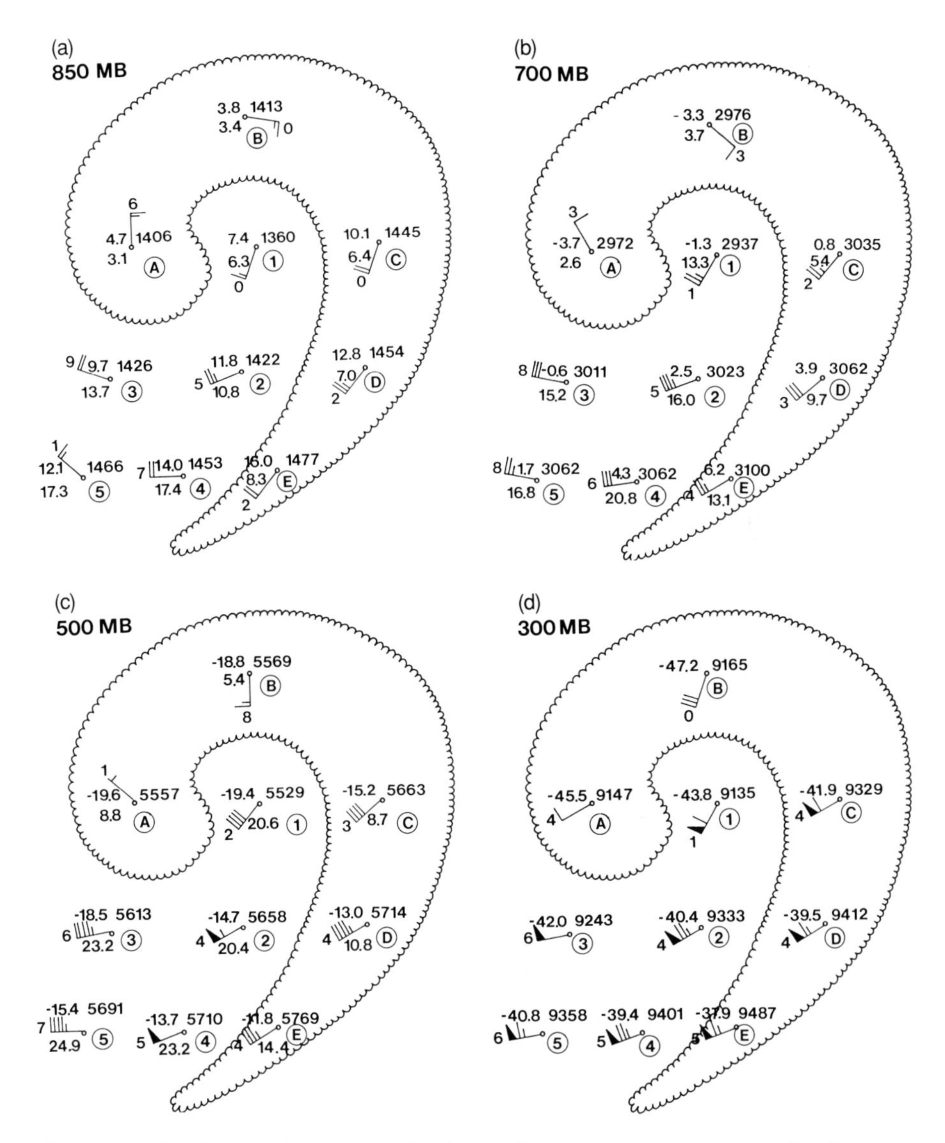

Figure 1.127 Composite upper-air data with respect to comma-cloud systems (outlined by scalloped line) in the Great Plains region of the United States during the spring at (a) 850, (b) 700, (c) 500, and (d) 300 mb. Temperature and dew-point depression in °C, height in m, full wind barb = 5 m s^{-1}, half wind barb = 2.5 m s^{-1}, pennant = 25 m s^{-1}. The 10's digit of the wind direction is plotted at the end of each wind staff (from Carr and Millard, 1985). (Disregard circled numbers and letters.) (Courtesy of the American Meteorological Society)

(a)

(b)

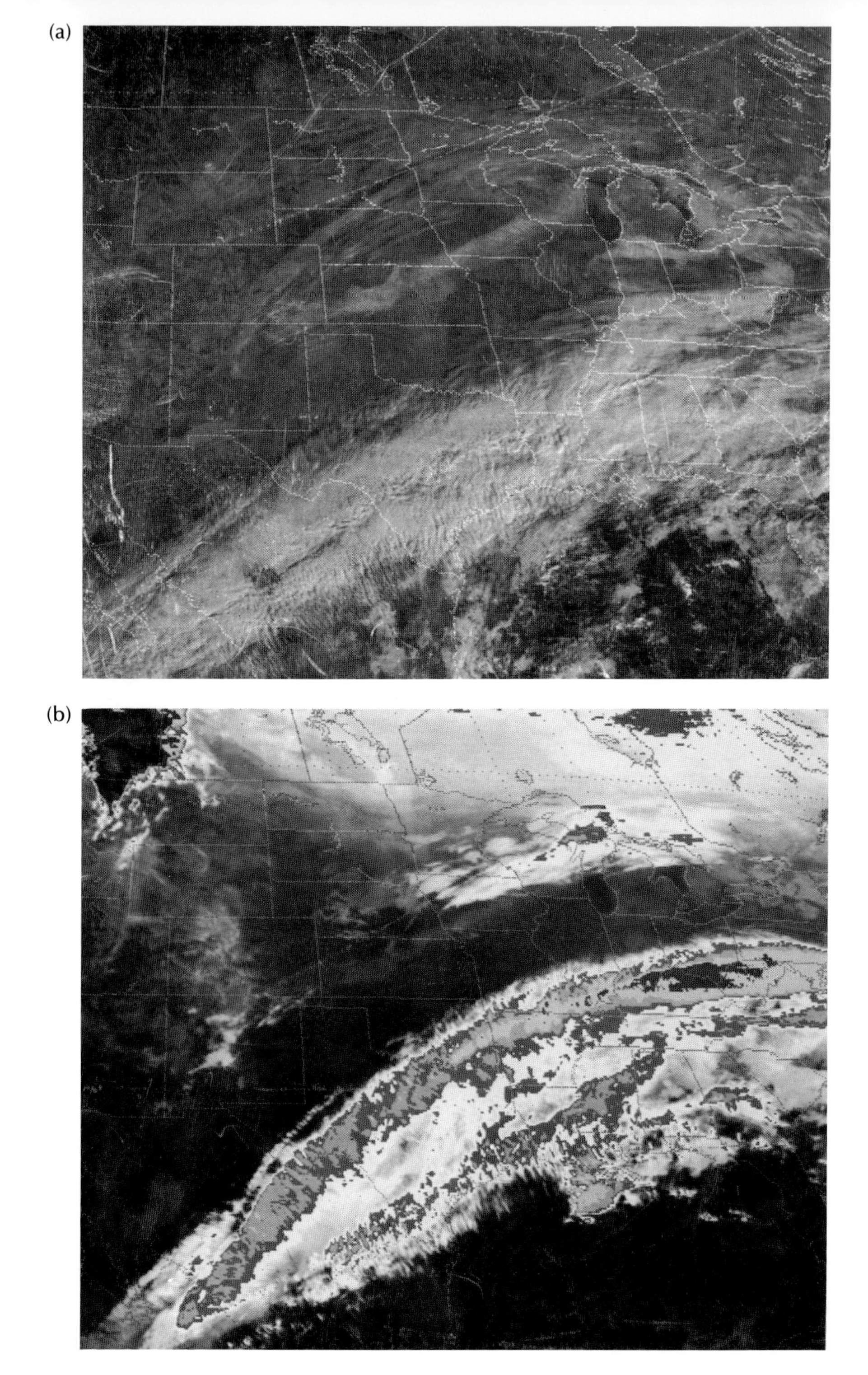

Figure 1.128 Small-scale transverse bands within a cirrus band over (a) Mexico and (b) the Texas Gulf Coast at (a) 1531 UTC and (b) 2231 UTC on January 5, 1990. The satellite photograph in (a) is visible, in (b) is enhanced infrared.

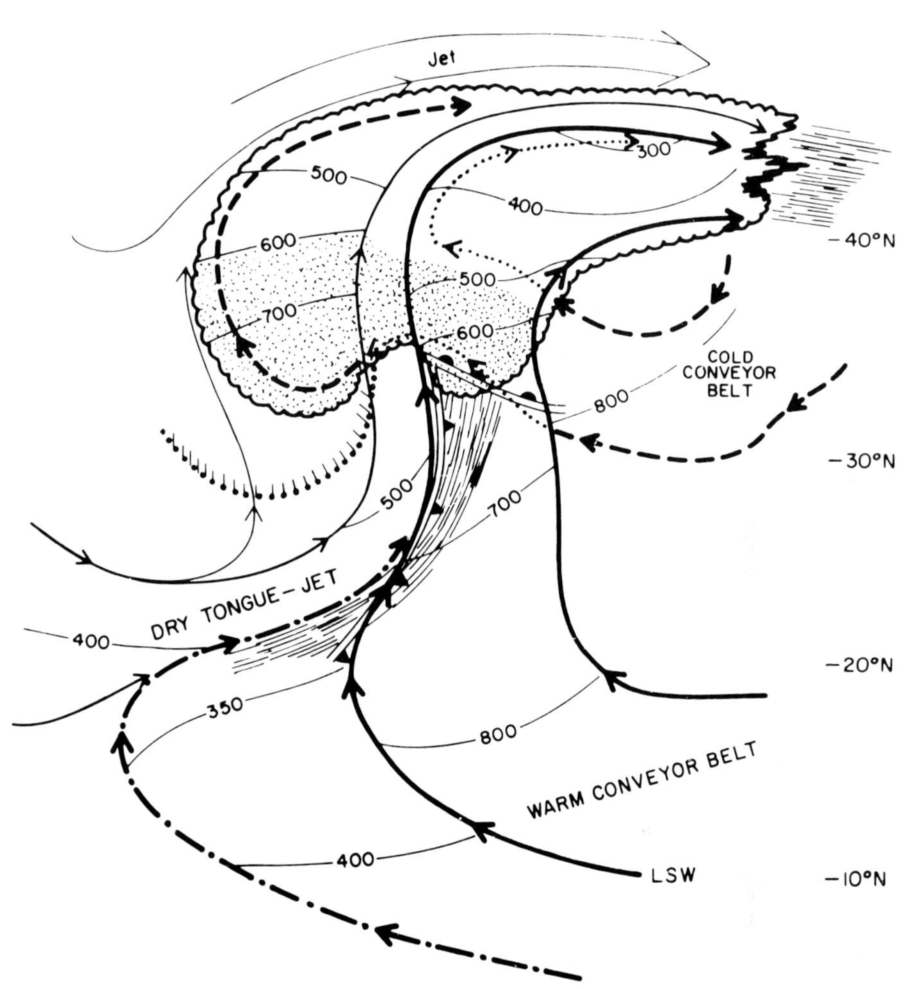

Figure 1.129 Relative airflow through a midlatitude wave cyclone: schematic composite of airflow through an idealized midlatitude cyclone. Heavy solid streamlines depict airflow at top of the warm conveyor belt. Dashed flow represents cold conveyor belt (drawn dotted when it lies beneath the warm conveyor belt or dry airstream.) Dot-dashed flow represents air originating at middle levels in the tropics. Thin solid streamlines pertain to dry air that originates at upper levels west of the trough. Thin solid lines denote the heights of the airstreams (mb) and are approximately normal to the direction of the respective air motion (isobars are omitted for the cold conveyor belt where it lies beneath the warm conveyor belt or beneath the jet stream flow). The region of dense upper- and middle-level layer cloud is represented by scalloping, and sustained precipitation by stippling. Streaks denote thin cirrus. The edge of the low-level stratus is shown by the curved border of small dots with tails. The major upper-tropospheric jet streams are labelled JET. The limiting streamline for the warm conveyor belt is labelled LSW (from Carlson, 1980). (Courtesy of the American Meteorological Society)

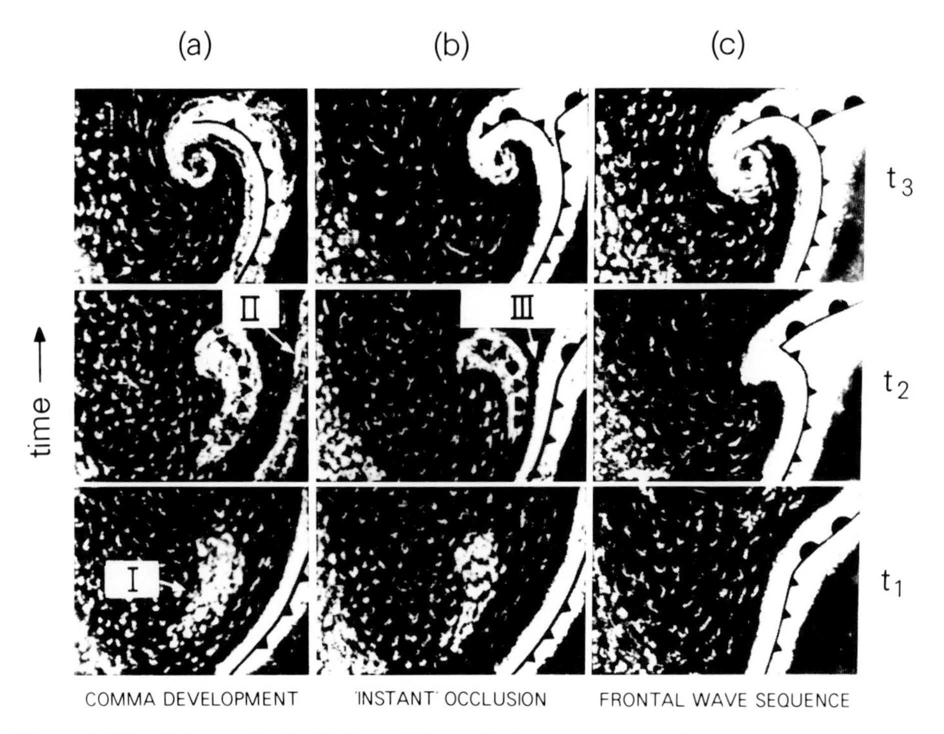

Figure 1.130 Schematic depiction of three basic sequences of vortex development evident in satellite imagery: (a) development of a comma cloud entirely within the cold air mass (i.e., a polar low), (b) development of an instant occlusion (intermediate situation between polar low and frontal wave), (c) development of a frontal wave. The figure (adapted from Zillman and Price, 1972) was derived from observations over the Southern Ocean, but it is printed vertically so as to apply to the Northern Hemisphere. Frontal symbols indicate one scheme for representing the various evolution sequences using the tools of conventional frontal analysis. The labels I, II, and III indicate a region of enhanced convection, a decaying cloud band, and a convective band merging with the frontal cloud band, respectively (from Browning, 1986). (Courtesy of the American Meteorological Society)

associated with the "Baker ridge," and warm, maritime, Atlantic air. This frontal zone is called a *coastal front,* and will be discussed in Chap. 2. Examples of Type I and Type II cyclogenesis are shown in Fig. 1.131. (The Type I and Type II cyclones were originally called "Type A" and "Type B" cyclones. We have changed their names to avoid confusion with those defined by Petterssen and Smebye.)

1.6.4. Cyclogenesis in the Lee of the Colorado Rockies

The drop in surface pressure along the lee slopes of the Rockies in response to strong westerly flow is not by itself sufficient to form a lee cyclone that intensifies and moves away from the mountains. An upper-level short-wave

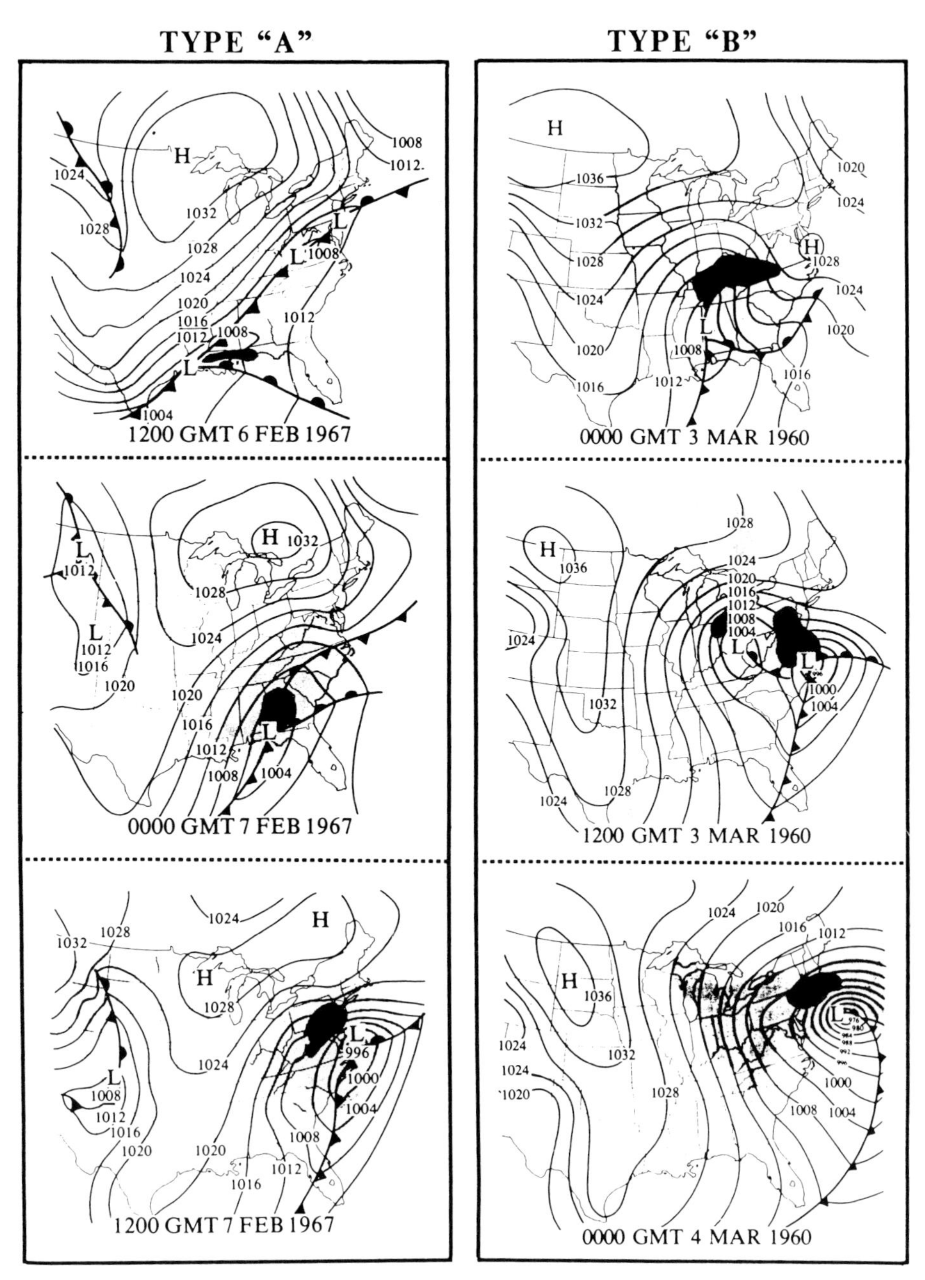

Figure 1.131 Examples of Type I (A) and Type II (B) cyclogenesis along the east coast of the United States. Analyses show frontal positions, sea-level isobars in mb (solid lines), high- (H) and low- (L) pressure centers, and precipitation (shaded; heavy shading denotes moderate to heavy precipitation) (adapted from Miller, 1946, by Paul Kocin).

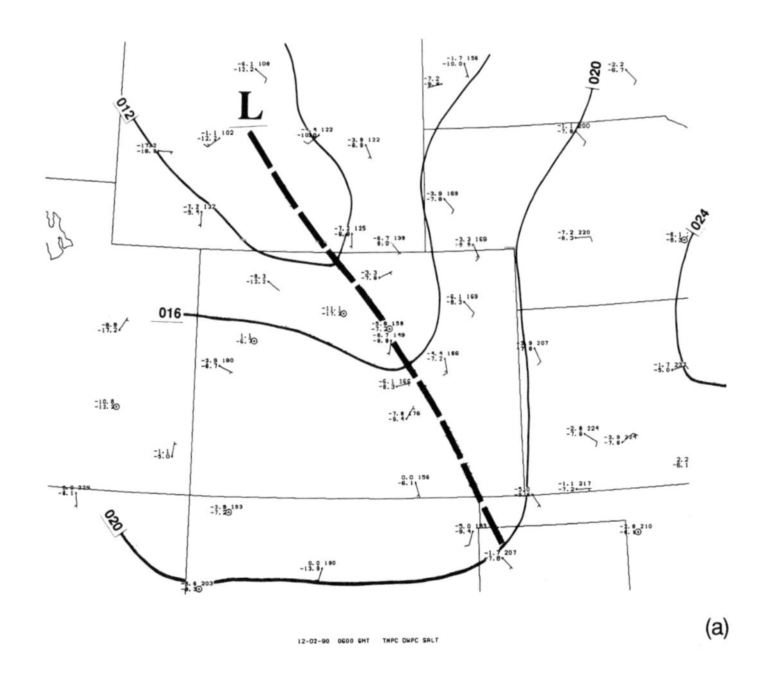

Figure 1.132 Example of the life history of a Colorado cyclone at the surface. Analyses of altimeter setting (solid lines) in mb without the leading "1" in a reference frame following the storm. (a) 0600 UTC, December 2, 1990; (b) 1200, December 2; (c) 1800, December 2; (d) 0000, December 3; (e) 0600, December 3; (f) 1200, December 3; (g) 1800, December 3; (h) 0000, December 4; (i) 0600, December 4; (j) 1200, December 4. Altimeter setting plotted in tens of mb without the leading "10." Temperature and dewpoint plotted in °C; whole wind barb = 5 m s^{-1}; half wind barb = 2.5 m s^{-1}. A low is located in northwestern Wyoming in (a), with a trough extending just east of or near the Continental Divide into southeastern Colorado. In (b) the weak low has shifted into southeastern Wyoming. The real action begins in (c), with a well-defined cyclone showing up over the Colorado–Oklahoma Panhandle–Kansas border. A dryline (see Chapter 2) extends southward from the low; a warm or stationary front appears to the east of the low; a trough extends northward from the low into the Nebraska Panhandle. In (d) we see that the pattern has shifted to the east-southeast, with the low centered over north-central Oklahoma. A cold front extends to the southwest of the low. The low then deepens, and the pattern shifts to the northeast (e). The trough extending from the low center in southeastern Missouri tilts toward the west with latitude. The low deepens further and tracks over central Illinois (f). In (g) we see that the low center has moved to northeastern Illinois, and the cold front has surged eastward ahead of the low; an occluded front appears, and the trough has rotated to a position now located to the southwest of the low. The low no longer deepens, and continues to move slowly northeastward, while the cold front continues to surge eastward through Ohio (h). The low has tracked to northeastern Michigan, while the cold front has pushed through much of Pennsylvania in (i); the occluded front that extends from the intersection of the cold front and warm front to the low center has increased in length. The low has broken away from the cold front and warm front in New Jersey and New York, while a new low appears over southeastern New York (j).

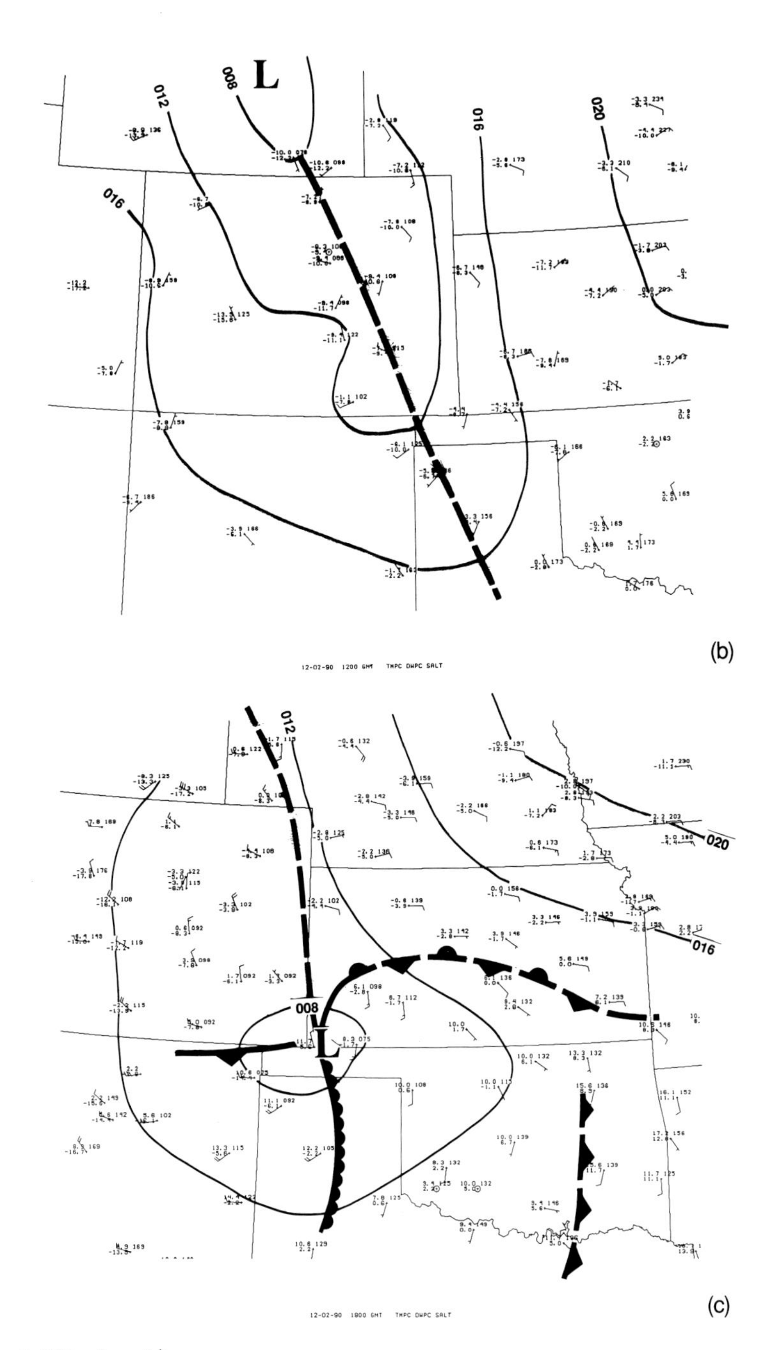

Figure 1.132 *(cont.)*

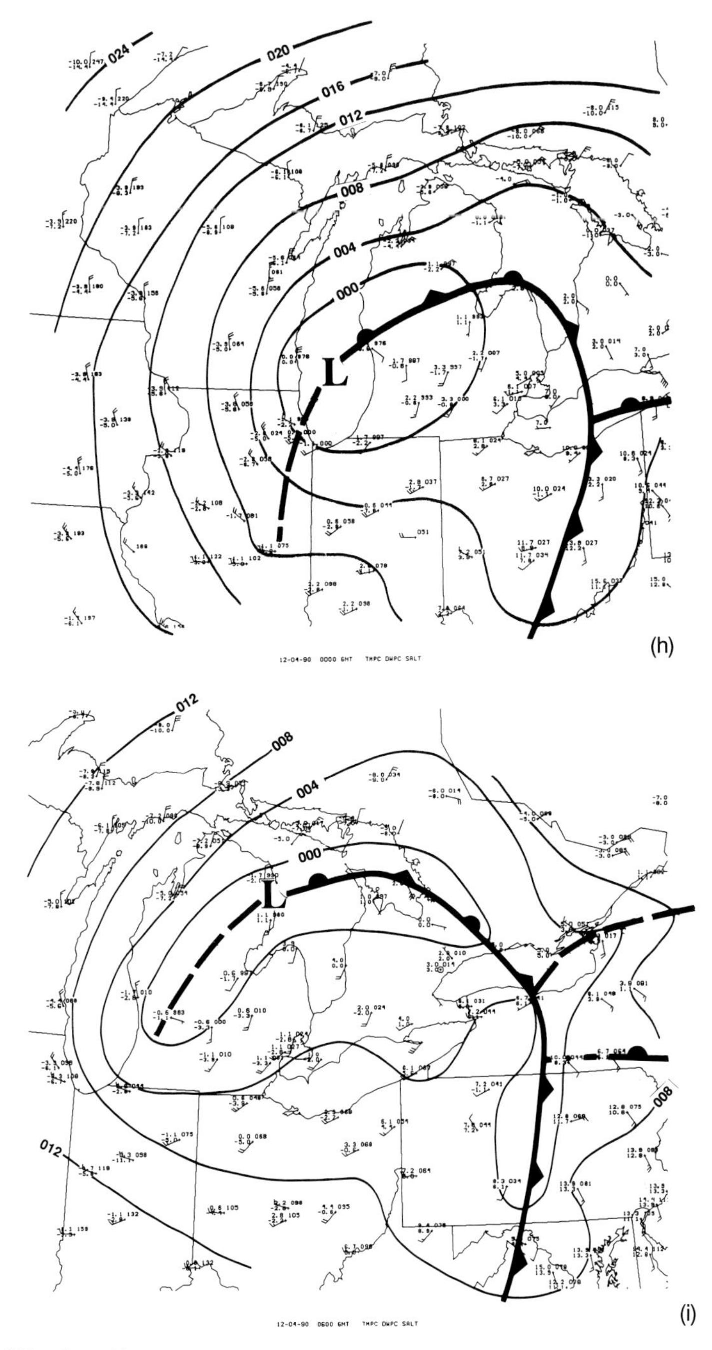

Figure 1.132 *(cont.)*

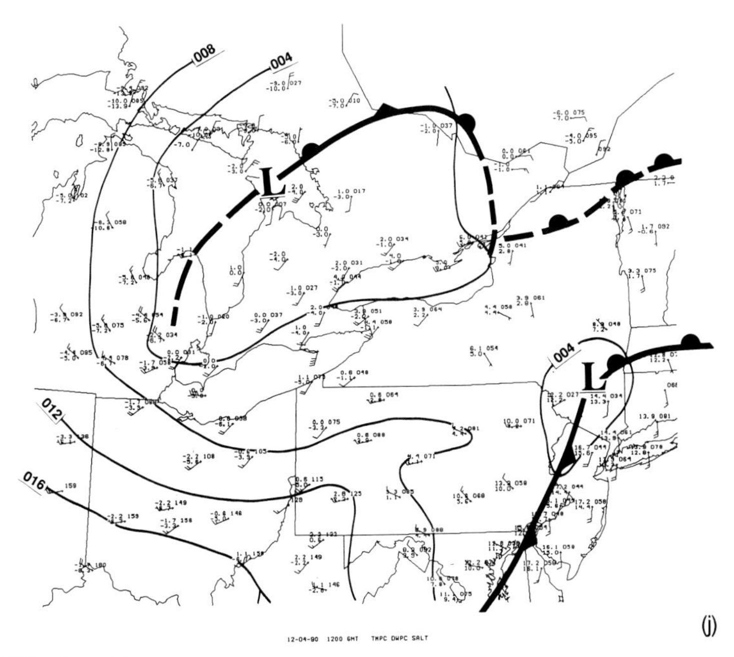

Figure 1.132 (*cont.*)

trough is usually necessary for initiating a lee cyclone that develops and goes through its life cycle. The lee cyclone initially moves southward, with the higher terrain lying to the right of the cyclone. Warm advection to the east of the cyclone is associated with pressure falls to the east. In addition, as the region of strong cyclonic vorticity advection aloft associated with the short-wave trough moves eastward away from the mountains, the pressure falls to the east add to the warm-advection pressure falls. The net result is that the surface cyclone moves toward the southeast, away from the mountains. As the cyclone moves away from the mountains, its southward component of motion vanishes, and thus it moves toward the east, and finally curves towards the northeast, owing to strong cold advection west and southwest of the center, and strong warm advection east and northeast of the center. Thus, "Colorado lows" usually end up near Chicago.[8] It is noteworthy that these lows do not usually form along frontal zones, but rather in a fixed location with respect to the mountains. The life cycle of a Colorado lee cyclone and the baroclinic wave aloft is depicted in Figs. 1.132–1.134. This example is especially instructive because it occurred over land under the watchful eyes of rawinsonde data, unlike many coastal lows in the eastern United States, which occur in part over the data-sparse ocean.

1.6.5. Cyclogenesis in the Lee of the Alps

Cyclones that form in the lee of the Alps are called *Genoa cyclones*. They occur in an environment of northwesterly flow aloft. The Alps do not present a

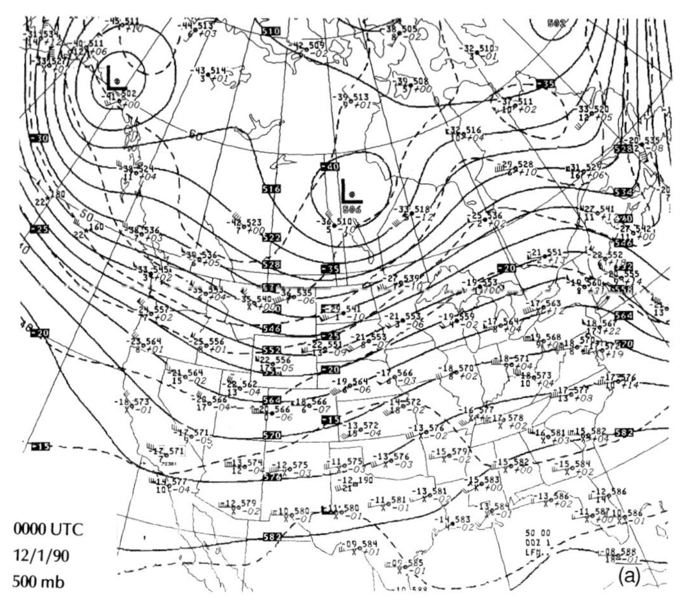

Figure 1.133 Example of the life history of a baroclinic wave associated with a Colorado cyclone. NMC 500-mb analyses for (a) 0000 UTC, December 1, 1990; (b) 1200, December 1; (c) 0000, December 2; (d) 1200, December 2; (e) 0000, December 3; (f) 1200, December 3; (g) 0000, December 4; (h) 1200, December 4. A 90-kt wind southwest of the cyclone over southeastern Alaska is a clue that a strong vorticity maximum is heading out over the Gulf of Alaska (a). The vorticity maximum is probably located over the data-void region of the Gulf of Alaska (b). The temperature is a frigid −48°C near the center of the cyclone. There is marked diffluence downstream from the vorticity maximum. Heights have fallen along the Pacific Northwest coast. The vorticity maximum has finally hit the data network in (c)! A 100-kt wind is observed over Vancouver Island. The vorticity maximum is associated with the cyclonic shear northeast of this strong northwesterly flow, and the cyclonic curvature associated with the diffluent, negatively tilted trough that has come onshore. The jet streak (maximum in wind speed within the strong current of wind; see Chapter 2) has moved southeastward in (d); the wind at Boise, Idaho is a whopping 110 kts. The trough downstream from the jet streak is amplifying over Utah and the Idaho–Wyoming border. At the same time [Fig. 1.132(b)] there is a lee trough in eastern Colorado and a weak lee cyclone in southeastern Wyoming. The 500-mb trough has amplified further in (e) over the Colorado–Kansas–Oklahoma border. The jet streak has moved even farther southeastward and is located over Amarillo, Texas. The 12-h height fall at Dodge city, Kansas, 15 dam, is relatively large. The surface cyclone [Fig. 1.132(d)] is located to the east of the 500-mb vorticity maximum. The trough that extends at the surface to the northwest of the surface cyclone is near the location of the 500-mb trough axis. In (f) the 500-mb trough has progressed eastward to Missouri; the jet streak is probably located in Missouri, downstream from Monnett, Missouri, where an 85-kt wind is reported, but upstream from Peoria, Illinois, which did not report a wind (the balloon was probably moving so quickly that the tracker lost the balloon). The surface low at this time [Fig. 1.132(f)] is located to the northeast of the trough at 500 mb, in central Illinois. In (g) we find that the trough has deepened even more. The jet streak is probably located over Ohio; the rawinsonde observation at Dayton, Ohio is missing. The center of the 500-mb trough is now located over south-central Wisconsin, while the surface cyclone is located just to the east over

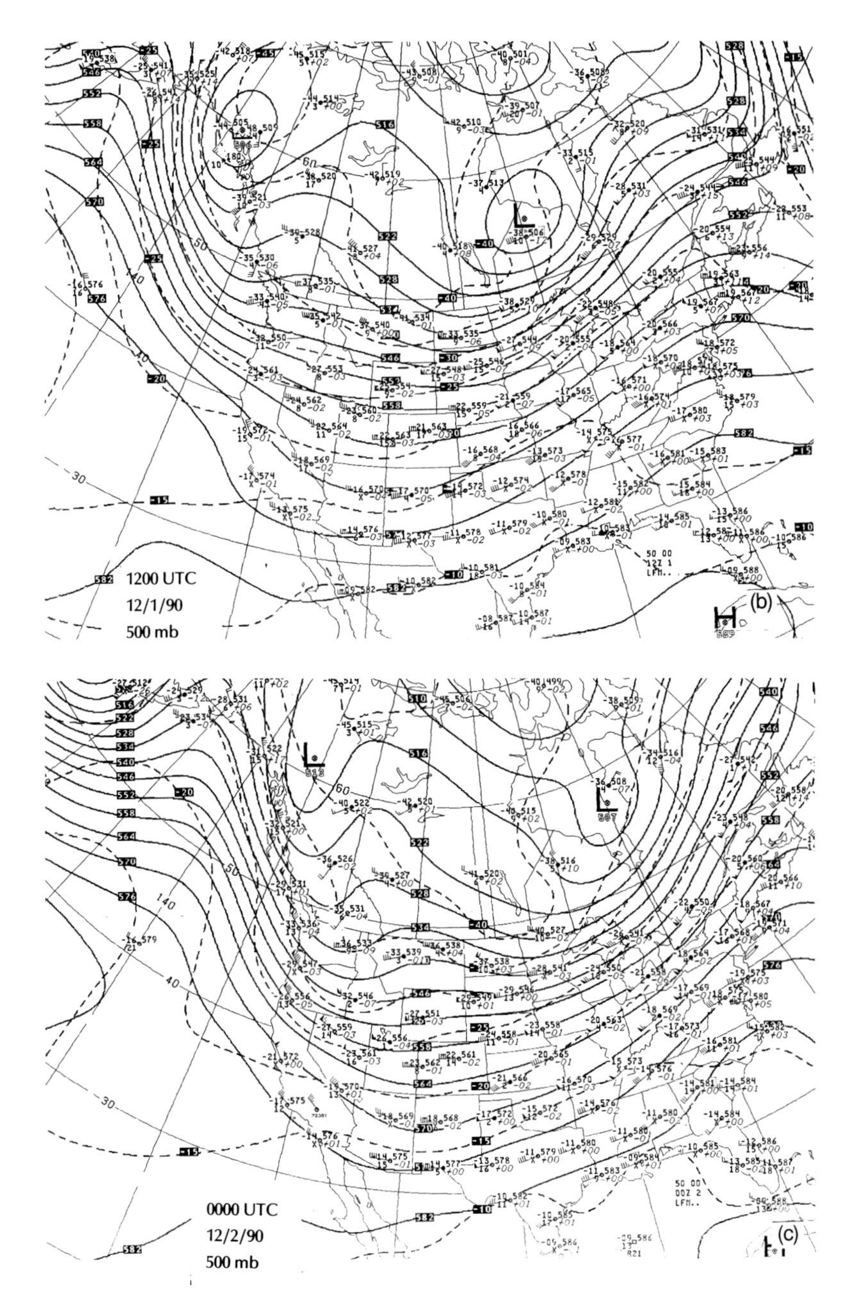

Figure 1.133 *(cont.)*
Lake Michigan [Fig. 1.132(h)]. In (h) the trough has moved northeastward to eastern Michigan and is near the surface low, which is no longer well defined; it now looks like a northeast–southwest-oriented trough [Fig. 1.132(j)]. The jet streak is located near Buffalo, New York, which has a 105-kt wind. Note that the jet streak was located upstream from the 500-mb trough before the trough amplified, and that it rotated around the base of the trough, and ended up downstream from the trough.

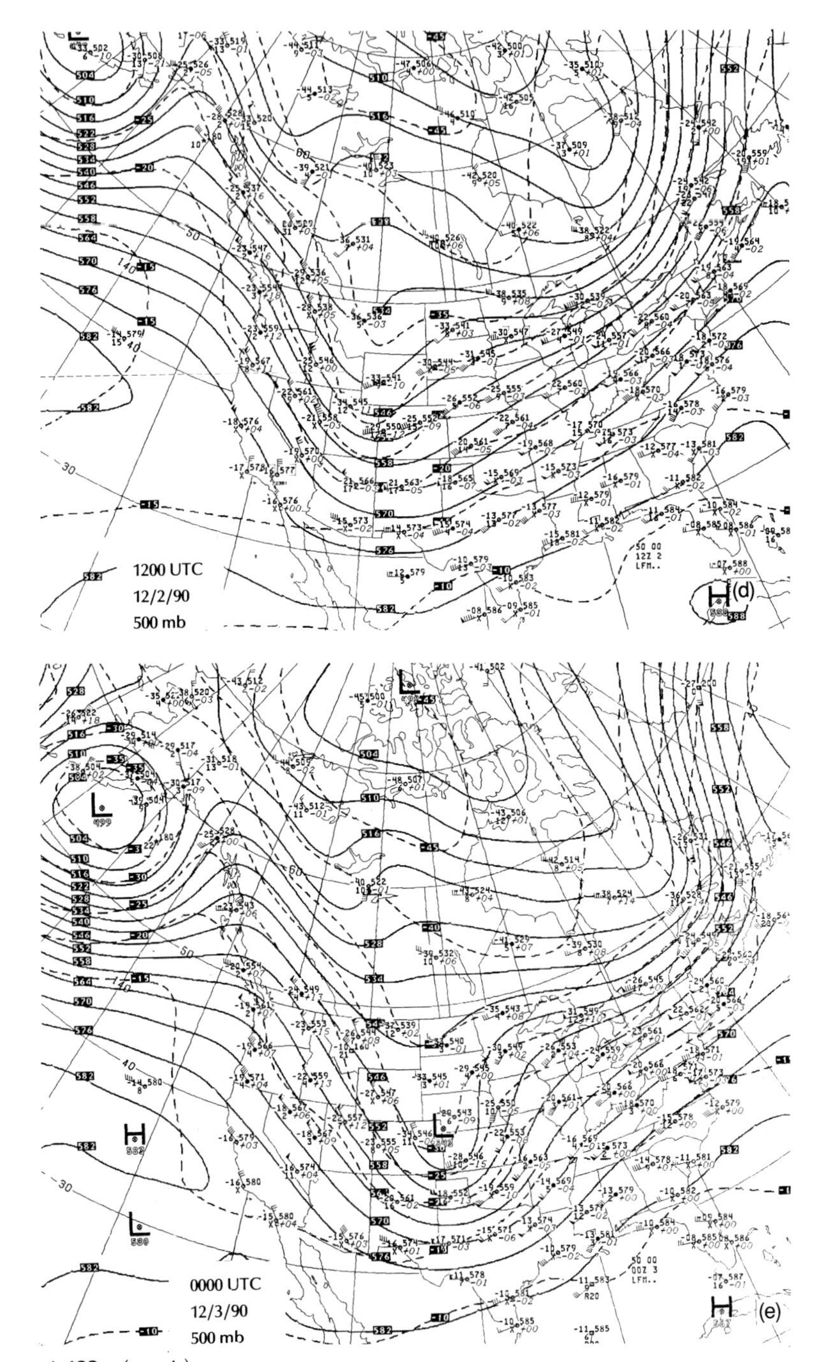

Figure 1.133 *(cont.)*

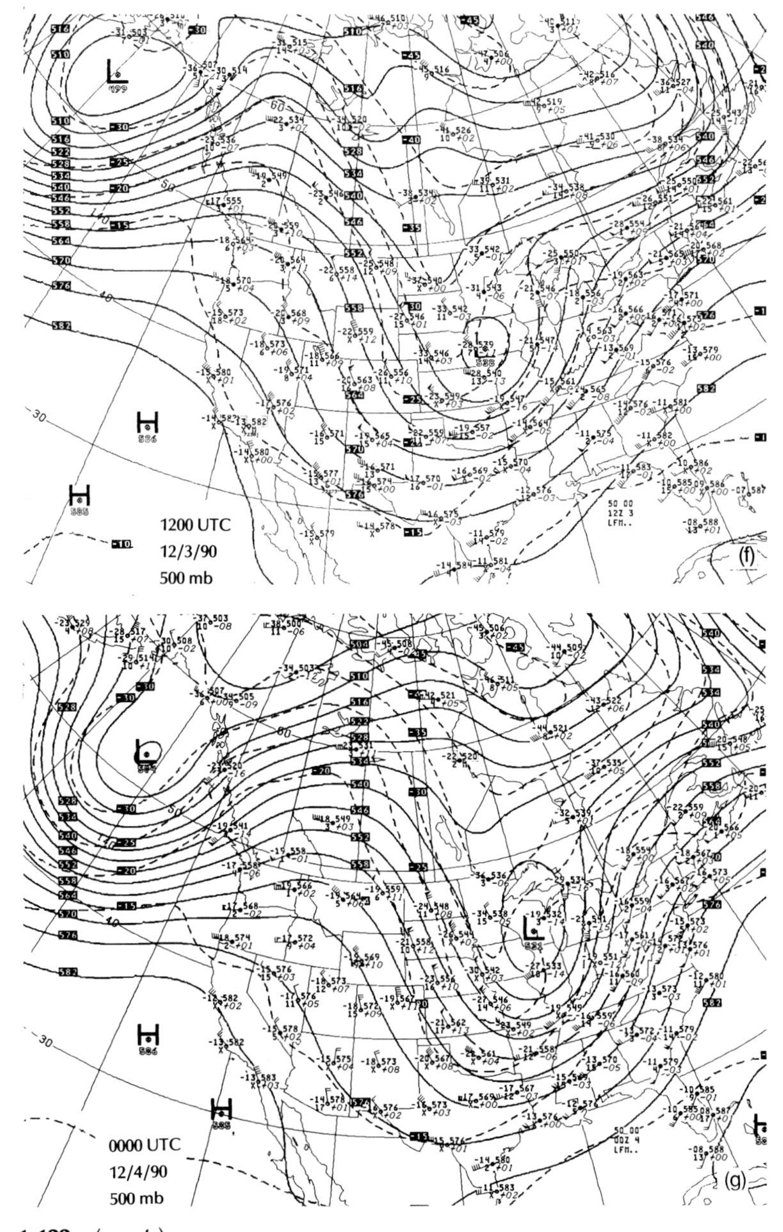

Figure 1.133 (*cont.*)

long obstacle to the flow aloft as do the Rockies. Some of the flow is forced around the mountains, while some is forced over the top, as in the Rockies. Cyclogenesis occurs as a low-level cold front crosses the Alps, and the temperature field becomes distorted. An upper-level trough moves southeastward in the northwesterly current. Before the cold front crosses the Alps, low-level cold advection opposes the differential vorticity advection associated

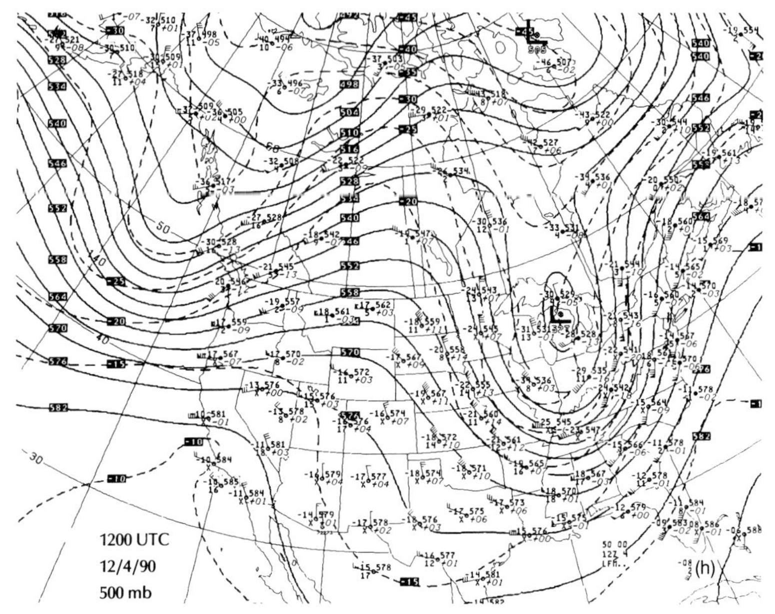

Figure 1.133 (*cont.*)

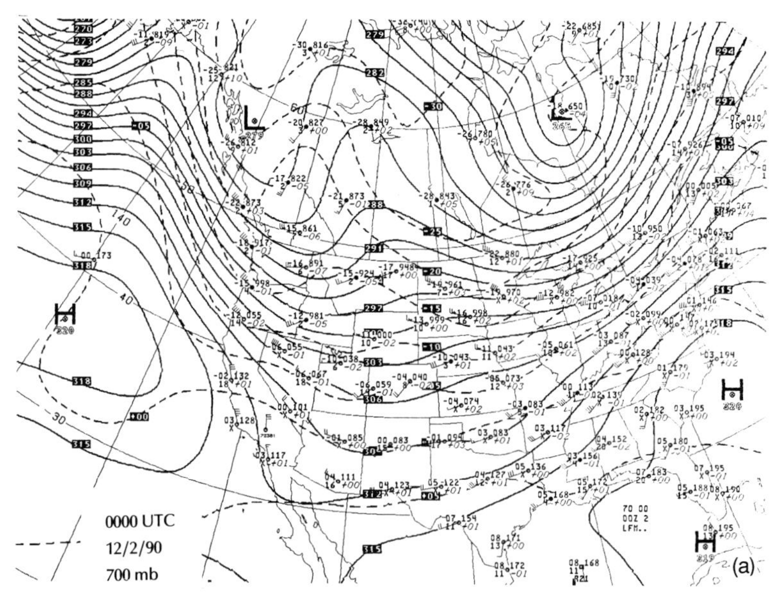

Figure 1.134 Example of the life history of a baroclinic wave associated with a Colorado cyclone. NMC 700-mb analyses for (a) 0000, December 2; (b) 1200, December 2; (c) 0000, December 3; (d) 1200, December 3; (e) 0000, December 4; (f) 1200, December 4. Note the pattern of cold advection into the base of the trough in (a)–(e); in (f) the cold-advection pattern has rotated around the base of the trough and is located southeast and east of the trough.

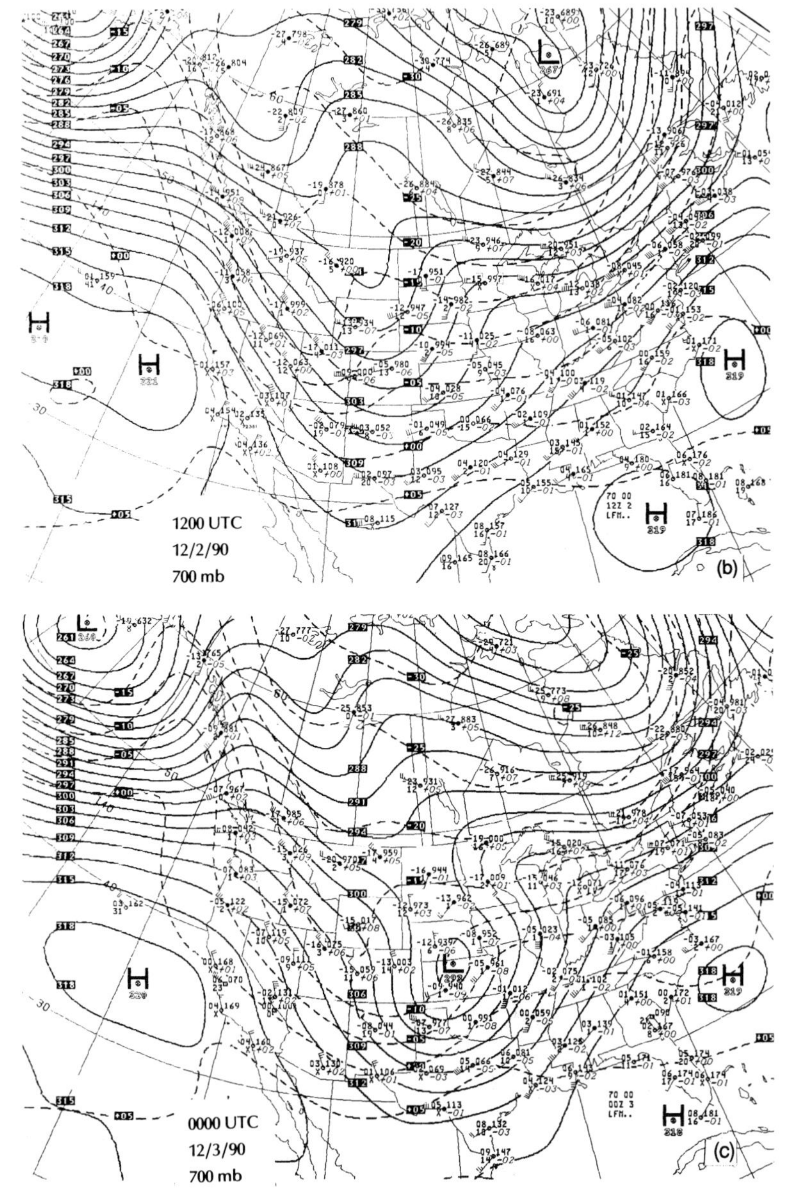

Figure 1.134 *(cont.)*

with the trough aloft, and hence cyclogenesis is inhibited (Fig. 1.135). The Alps "block" the cold air at the surface on the windward side, and hence low-level cold advection is inhibited on the lee side of the mountains. Hence, differential vorticity advection can act unopposed by temperature advection, and a cyclone can form on the lee side. The Genoa cyclone is in a sense a "secondary" cyclone, in that often a "primary" low exists to the north. The life of a Genoa cyclone is depicted in Fig. 1.136.

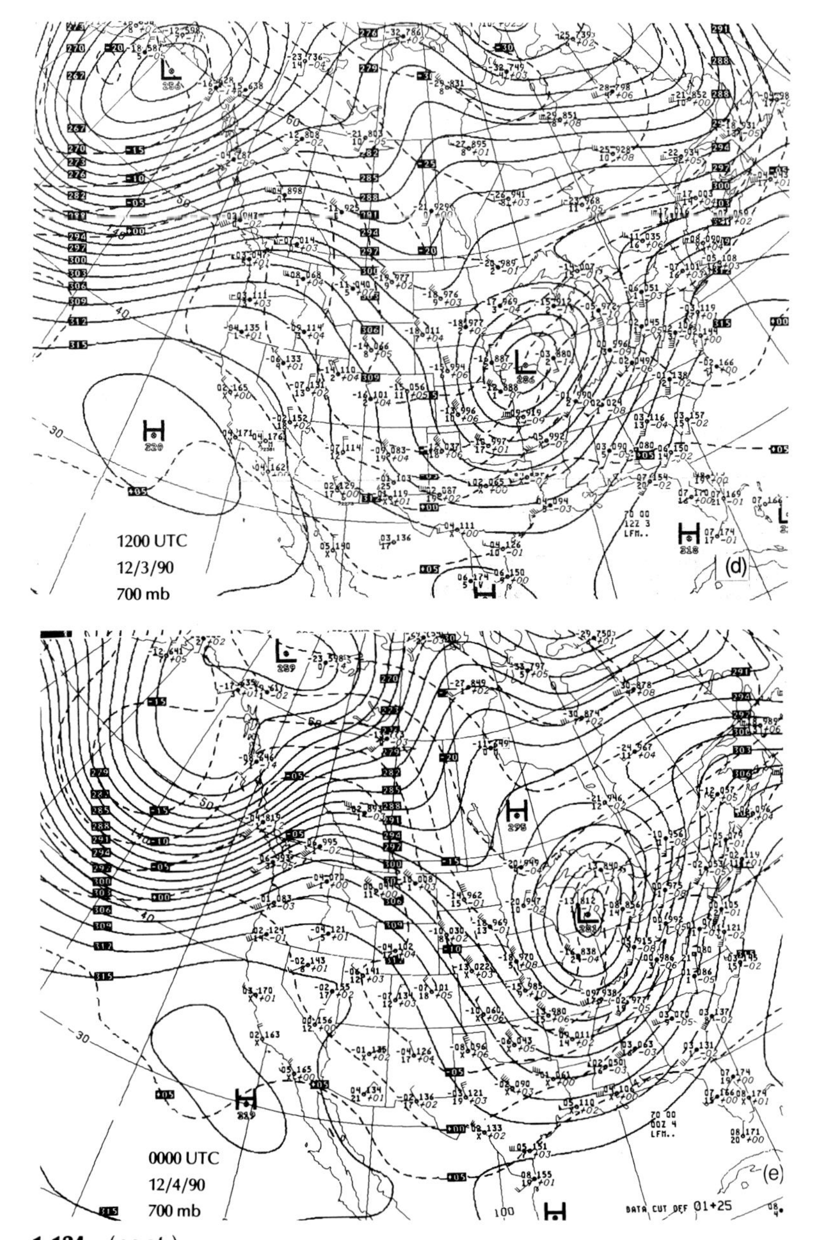

Figure 1.134 *(cont.)*

1.7. ANALYSIS OF MIDLATITUDE, SYNOPTIC-SCALE SYSTEMS USING THE BALANCE EQUATIONS

J. Charney in 1962 developed a system of equations based on scale analysis, using a formal power-series expansion in terms of the Rossby number, which is more general than the quasigeostrophic system of equations. For example, advection by the ageostrophic component of the wind field is, to some extent,

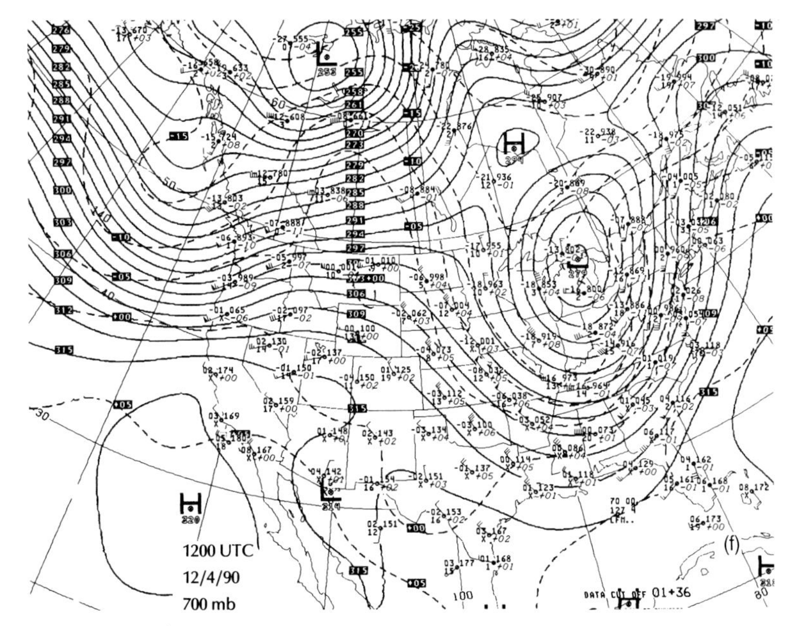

Figure 1.134 *(cont.)*

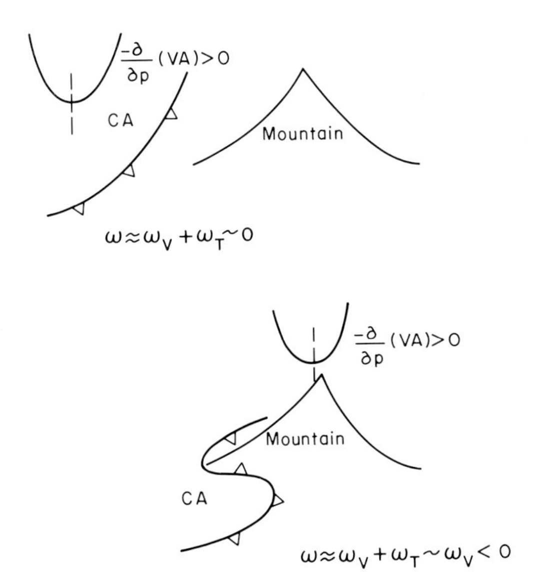

Figure 1.135 Effects of blocking by the Alps on surface cyclogenesis. (top) Vorticity advection becoming more cyclonic with height $[(-\partial(VA)/\partial p) > 0]$ associated with rising motion ($\omega_V < 0$); cold advection (CA) associated with sinking motion ($\omega_T > 0$). (bottom) Low-level cold air blocked by the mountains, and forced to flow around the sides. Vorticity advection becoming more cyclonic with height unopposed by cold advection; rising motion and low-level convergence induce surface cyclogenesis.

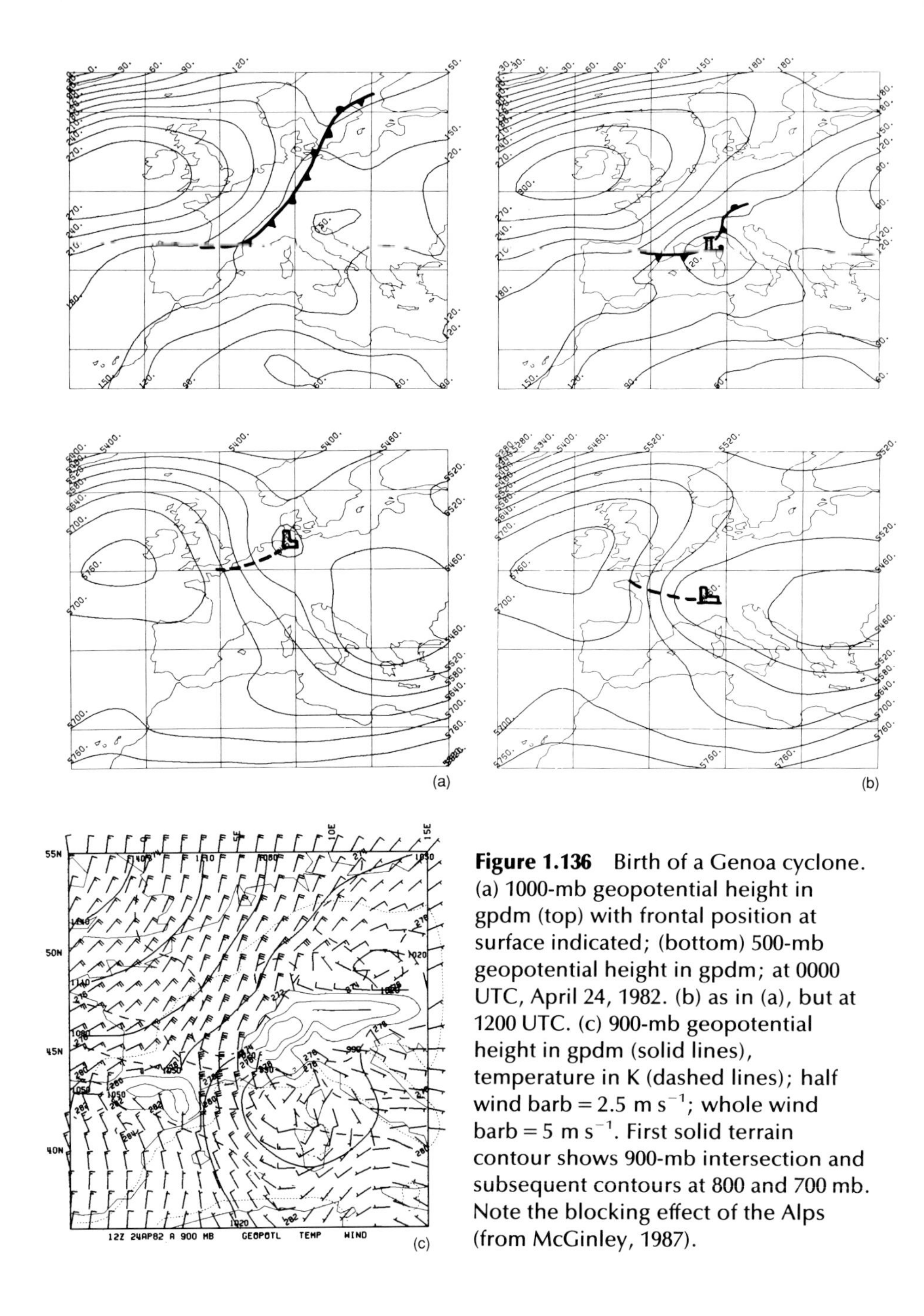

Figure 1.136 Birth of a Genoa cyclone. (a) 1000-mb geopotential height in gpdm (top) with frontal position at surface indicated; (bottom) 500-mb geopotential height in gpdm; at 0000 UTC, April 24, 1982. (b) as in (a), but at 1200 UTC. (c) 900-mb geopotential height in gpdm (solid lines), temperature in K (dashed lines); half wind barb = 2.5 m s^{-1}; whole wind barb = 5 m s^{-1}. First solid terrain contour shows 900-mb intersection and subsequent contours at 800 and 700 mb. Note the blocking effect of the Alps (from McGinley, 1987).

retained. However, the equations are more restrictive than the primitive equations.

Recall that the vorticity equation in a frictionless atmosphere is as follows:

$$\frac{\partial \zeta}{\partial t} = -\mathbf{v} \cdot \nabla_p(\zeta + f) - \omega \frac{\partial \zeta}{\partial p} - \delta(\zeta + f) - \hat{\mathbf{k}} \cdot \left(\nabla_p \omega \times \frac{\partial \mathbf{v}}{\partial p}\right). \qquad (1.7.1)$$

The divergence equation,

$$\frac{\partial \delta}{\partial t} = -\mathbf{v} \cdot \nabla_p(\zeta + f) \times \hat{\mathbf{k}} - \omega \frac{\partial \delta}{\delta p} + \zeta(\zeta + f) - \nabla_p \omega \cdot \frac{\partial \mathbf{v}}{\partial p}$$
$$- \nabla_p^2\left(\Phi + \frac{\mathbf{v} \cdot \mathbf{v}}{2}\right), \qquad (1.7.2)$$

is obtained by differentiating the inviscid u equation of motion with respect to x, and adding it to the inviscid v equation of motion differentiated with respect to y.

The horizontal wind field can be expressed in terms of a nondivergent, rotational part ($\mathbf{v}_r$) and a divergent, irrotational part ($\mathbf{v}_d$), such that

$$\mathbf{v} = \mathbf{v}_r + \mathbf{v}_d. \qquad (1.7.3)$$

This method of partitioning the wind field is an alternative to decomposing the wind field into geostrophic and ageostrophic components. It can be confusing, however: Both $\mathbf{v}_r$ and $\mathbf{v}_d$ can contain parts that have deformation and uniform translation; for example, the nondivergent part of the wind simply has no divergence—it can have deformation and uniform translation.

The nondivergent part of the wind field can be expressed in terms of a streamfunction ψ, such that

$$\mathbf{v}_r = \hat{\mathbf{k}} \times \nabla_p \psi, \qquad (1.7.4)$$

where

$$\nabla_p^2 \psi = \zeta. \qquad (1.7.5)$$

The divergent part of the wind field can be expressed in terms of a velocity potential χ (not to be confused with $\chi = \partial\Phi/\partial t$ in earlier sections), such that

$$\mathbf{v}_d = -\nabla_p \chi, \qquad (1.7.6)$$

where

$$-\nabla_p^2 \chi = \delta. \qquad (1.7.7)$$

Then

$$\mathbf{v} = \hat{\mathbf{k}} \times \nabla_p \psi - \nabla_p \chi. \qquad (1.7.8)$$

On the basis of scale analysis, it can be shown that twisting by the divergent part of the wind field can be neglected; hence, from Eqs. (1.7.1) and (1.7.8) we see that

$$\frac{\partial}{\partial t}\nabla_p^2\psi = -\mathbf{v}_r\cdot\nabla_p(\zeta+f)+\nabla_p\chi\cdot\nabla_p(\zeta+f)-\omega\frac{\partial}{\partial p}\nabla_p^2\psi$$

$$+(\nabla_p^2\chi)(\zeta+f)-\hat{\mathbf{k}}\cdot\nabla_p\omega\times\frac{\partial}{\partial p}(\hat{\mathbf{k}}\times\nabla_p\psi). \tag{1.7.9}$$

Therefore

$$\frac{\partial}{\partial t}\nabla_p^2\psi = -\mathbf{v}_r\cdot\nabla_p(\zeta+f)+\nabla_p\cdot[(\zeta+f)\nabla_p\chi]-\nabla_p\cdot\left(\omega\,\nabla_p\frac{\partial\psi}{\partial p}\right). \tag{1.7.10}$$

Terms involving the divergent part of the wind, and the $\nabla_p\omega\cdot\partial\mathbf{v}/\partial p$ term are neglected in the divergence equation; Eq. (1.7.2) may then be written as follows:

$$0 = -\hat{\mathbf{k}}\times\nabla_p\psi\cdot\nabla_p(\zeta+f)\times\hat{\mathbf{k}}+\zeta(\zeta+f)-\nabla_p^2\left(\Phi+\frac{\mathbf{v}_r\cdot\mathbf{v}_r}{2}\right). \tag{1.7.11}$$

Using Eq. (1.7.5) we find that

$$0=\nabla_p\cdot[(\zeta+f)\nabla_p\psi]-\nabla_p^2\Phi-\nabla_p^2\left(\frac{\mathbf{v}_r\cdot\mathbf{v}_r}{2}\right). \tag{1.7.12}$$

This equation is called the *balance equation.* It can be used to find ψ if Φ is known, or Φ if ψ is known. It is useful, for example, if we wish to compute the forcing functions in the quasigeostrophic ω and height-tendency equations, but do not have height data. The vorticity field can be computed from the wind field, if it is available. The streamfunction ψ can then be determined from Eq. (1.7.5), given proper boundary conditions. Then the geopotential height field Φ can be computed from Eq. (1.7.12), again given proper boundary conditions. Thus, since $\Phi(x,y,p)$ can be inferred from $\mathbf{v}(x,y,p)$ alone, and since temperature is related to the vertical gradient of Φ through the hydrostatic equation, we can actually obtain temperature information from the wind field. The accuracy of this method depends upon the validity of our simplifying assumptions used to derive the divergence equation. The balance equation is also used to initialize numerical models.

The adiabatic form of the thermodynamic equation may be written, using Eq. (1.7.8), as

$$\frac{\partial T}{\partial t} = -\mathbf{v}_r\cdot\nabla_p T+\nabla\chi\cdot\nabla_p T+\omega\sigma\frac{p}{R}, \tag{1.7.13}$$

where

$$\sigma = -\frac{RT}{p}\frac{\partial\ln\theta}{\partial p}.$$

The simplified vorticity equation, Eq. (1.7.10), the thermodynamic equation, Eq. (1.7.13), the balance equation, Eq. (1.7.12), and the continuity equation may be combined to form the following system of three simultaneous partial differential equations for ω, χ, and $\partial\psi/\partial t$:

$$\begin{aligned}\nabla_p^2(\sigma\omega)+f^2\frac{\partial^2\omega}{\partial p^2} &= -f\frac{\partial}{\partial p}[-\mathbf{v}_r\cdot\nabla_p(\zeta+f)] \\ &\quad -\frac{R}{p}\nabla_p^2(-\mathbf{v}_r\cdot\nabla_p T) \\ &\quad +\frac{\partial}{\partial t}\frac{\partial}{\partial p}\nabla_p^2\left(\frac{\mathbf{v}_r\cdot\mathbf{v}_r}{2}\right) \\ &\quad -f\frac{\partial}{\partial p}(\zeta\nabla_p^2\chi) \\ &\quad +f\frac{\partial}{\partial p}\left(\omega\frac{\partial}{\partial p}\nabla_p^2\psi\right) \\ &\quad +f\frac{\partial}{\partial p}\left(\nabla_p\omega\cdot\nabla_p\frac{\partial\psi}{\partial p}\right) \\ &\quad -f\frac{\partial}{\partial p}[\nabla_p\chi\cdot\nabla_p(\zeta+f)] \\ &\quad -\frac{R}{p}\nabla_p^2(\nabla_p\chi\cdot\nabla_p T) \\ &\quad -\beta\frac{\partial}{\partial p}\frac{\partial}{\partial y}\left(\frac{\partial\psi}{\partial t}\right) \qquad (1.7.14)\end{aligned}$$

$$\nabla_p^2\chi=\frac{\partial\omega}{\partial p} \qquad (1.7.15)$$

$$\nabla_p^2\frac{\partial\psi}{\partial t}=-\mathbf{v}_r\cdot\nabla_p(\zeta+f)+\nabla_p\cdot[(\zeta+f)\nabla_p\chi]-\nabla_p\cdot\left(\omega\nabla_p\frac{\partial\psi}{\partial p}\right). \qquad (1.7.16)$$

The first term in the ω equation, Eq. (1.7.14), represents differential absolute vorticity advection by the nondivergent part of the wind. The second represents the Laplacian of temperature advection by the nondivergent part of the wind. The third represents differential deformation. The remaining terms, in their respective order represent:

The differential divergence effect;
The differential vertical advection of relative vorticity;
Differential tilting;
Differential advection of absolute vorticity by the divergent part of the wind;
The Laplacian of temperature advection by the divergent part of the wind; and
The beta term of the divergence equation.

The variable $\partial\psi/\partial t$ is analogous to $\partial\Phi/\partial t$ in the quasigeostrophic system. The vertical velocity ω may be determined, using appropriate boundary conditions. Note that, unlike the quasigeostrophic ω equation, the streamfunction tendency (height tendency) has not been eliminated from the right-hand side of the equation. In addition, ω also appears explicitly on the right-hand side.

T. Krishnamurti in 1968 first used this omega equation [Eq. (1.7.14) to determine the typical values of each "forcing function" on the right-hand side of Eq. (1.7.14) for a typical midlatitude system. The first two forcing functions are dominant. They are similar to the two forcing functions in the quasi-geostrophic ω equation, except that Φ is replaced by ψ, and f_0 is replaced by f. The differential deformation term is also important. It diminishes the amount of rising motion downstream from an upper-level trough. The other terms are relatively small.

The streamfunction-tendency equation [Eq. (1.7.16)] is analogous to the quasigeostrophic height-tendency equation. However, ω still appears explicitly on the right-hand side; it has not been eliminated as it has been in the quasigeostrophic height-tendency equation.

1.8. ANALYSIS OF MIDLATITUDE, SYNOPTIC-SCALE SYSTEMS USING A GENERALIZED HEIGHT-TENDENCY EQUATION

A general diagnostic equation for geopotential-height tendency may be derived from the following thermodynamic equation, vorticity equation, and equation of continuity:

$$\frac{\partial T}{\partial t} = -\mathbf{v} \cdot \nabla_p T + \omega\sigma\frac{p}{R} + \frac{1}{C_p}\frac{dQ}{dt} \tag{1.8.1}$$

$$\frac{\partial \zeta}{\partial t} = -\mathbf{v} \cdot \nabla_p(\zeta + f) - \omega\frac{\partial \zeta}{\partial p} - \delta(\zeta + f) - \hat{\mathbf{k}} \cdot \nabla_p \omega \times \frac{\partial \mathbf{v}}{\partial p} + \hat{\mathbf{k}} \cdot \nabla \times \mathbf{F}_{\text{fric}} \tag{1.8.2}$$

$$\frac{\partial \omega}{\partial p} = -\delta. \tag{1.8.3}$$

We express the total wind in terms of geostrophic ($\mathbf{v}_{\text{g}}$) and ageostrophic ($\mathbf{v}_{\text{a}}$) components, so that

$$\mathbf{v} = \mathbf{v}_{\text{g}} + \mathbf{v}_{\text{a}}. \tag{1.8.4}$$

Then the vorticity of the total wind may be expressed in terms of the geopotential height field as follows:

$$\zeta = \zeta_{\text{g}} + \zeta_{\text{a}} = \frac{\nabla_p^2 \Phi}{f_0} + \zeta_{\text{a}}, \tag{1.8.5}$$

where ζ_{a} is the vorticity of the ageostrophic component of the wind. Since part of $\mathbf{v}_{\text{a}}$ is irrotational, ζ_{a} represents the vorticity of the nondivergent part of the ageostrophic wind. The hydrostatic equation can be expressed as

$$\frac{\partial \Phi}{\partial p} = -\frac{R}{p} T. \tag{1.8.6}$$

We replace f by f_0 in the $-\delta(\zeta + f)$ term in Eq. (1.8.2) because $|\beta y|$ is small compared to $|f_0|$. Given that

$$\chi = \frac{\partial \Phi}{\partial t},$$

we see from Eqs. (1.8.1), (1.8.2), (1.8.3), (1.8.5), and (1.8.6) that

$$\frac{\partial \chi}{\partial p} = -\frac{R}{p}(-\mathbf{v} \cdot \nabla_p T) - \omega\sigma - \frac{R}{pC_p}\frac{dQ}{dt} \tag{1.8.7}$$

$$\begin{aligned} \nabla_p^2 \chi = {} & f_0[-\mathbf{v} \cdot \nabla_p(\zeta + f)] - f_0 \omega \frac{\partial \zeta}{\partial p} + f_0 \frac{\partial \omega}{\partial p}(\zeta + f_0) \\ & - f_0 \hat{\mathbf{k}} \cdot \nabla_p \omega \times \frac{\partial \mathbf{v}}{\partial p} \\ & + f_0(\hat{\mathbf{k}} \cdot \nabla \times \mathbf{F}_{\text{fric}}) - f_0 \frac{\partial \zeta_a}{\partial t}. \end{aligned} \tag{1.8.8}$$

Differentiating Eq. (1.8.7) with respect to pressure, multiplying it by f_0^2/σ, adding the result to Eq. (1.8.8), and using Eq. (1.8.4), we get the following equation for χ:

$$\begin{aligned} \left(\nabla_p^2 + \frac{f_0^2}{\sigma}\frac{\partial^2}{\partial p^2}\right)\chi = {} & f_0[-\mathbf{v}_g \cdot \nabla_p(\zeta_g + f)] - \frac{f_0^2}{\sigma}\frac{\partial}{\partial p}\left(\frac{R}{p}(-\mathbf{v}_g \cdot \nabla_p T)\right) \\ & + f_0(\hat{\mathbf{k}} \cdot \nabla \times \mathbf{F}_{\text{fric}}) - \frac{f_0^2}{\sigma}\frac{\partial}{\partial p}\left(\frac{R}{p}\frac{1}{C_p}\frac{dQ}{dt}\right) \\ & - f_0^2 \omega \frac{\partial \ln \sigma}{\partial p} \\ & + f_0[-\mathbf{v}_a \cdot \nabla_p(\zeta + f)] + f_0(-\mathbf{v}_g \cdot \nabla_p \zeta_a) \\ & - \frac{f_0^2}{\sigma}\frac{\partial}{\partial p}\left(\frac{R}{p}(-\mathbf{v}_a \cdot \nabla_p T)\right) \\ & - f_0 \omega \frac{\partial \zeta}{\partial p} + f_0 \zeta \frac{\partial \omega}{\partial p} \\ & - f_0 \hat{\mathbf{k}} \cdot \nabla_p \omega \times \frac{\partial \mathbf{v}}{\partial p} - f_0 \frac{\partial \zeta_a}{\partial t}. \end{aligned} \tag{1.8.9}$$

The first five forcing functions in Eq. (1.8.9) are equivalent to those in the "complete" quasigeostrophic height-tendency equation, including the term involving the vertical variations in σ. The remainder represent the following ageostrophic effects, respectively:

Ageostrophic advection of absolute vorticity;
Geostrophic advection of ageostrophic vorticity;
Differential ageostrophic temperature advection;
Vertical advection of relative vorticity;
The divergence effect acting on relative vorticity;
Tilting;
Local changes in ageostrophic vorticity.

The latter term is especially difficult to calculate.

Equation (1.8.9) explicitly contains terms involving ω and $\mathbf{v}_a$ and may be used to assess the relative contribution of the nonquasigeostrophic terms to the quasigeostrophic terms if ω and $\mathbf{v}_a$ have been estimated satisfactorily. (Colucci and Walker first used a similar equation in 1986 to find χ at 500 mb using $\partial\Phi/\partial t = 0$ as the upper and lower boundary conditions.)

A corresponding generalized ω equation is not easily obtained from Eqs. (1.8.1), (1.8.2), and (1.8.3) because ω appears in both the vertical advection and tilting terms in Eq. (1.8.2).

1.9. ANALYSIS OF MIDLATITUDE, SYNOPTIC-SCALE SYSTEMS USING ISENTROPIC POTENTIAL VORTICITY THINKING

1.9.1. Historical Review

We will now provide an historical review of the use of potential temperature as a vertical coordinate. When upper-air observations became available during the 1930s, as a result of the invention of radiosondes, there was a debate in the meteorological community on what surfaces upper-air data should be plotted. Sir Napier Shaw in 1933 advocated the use of isentropic coordinates (i.e., potential temperature as the vertical coordinate). However, pressure became the operational choice in Germany, while height became the choice in the United States and Britain. In the late 1930s, as a result of research by Rossby, Namias, and their co-workers, the U.S. Weather Bureau started to transmit data in a form that could be used for isentropic analysis. Probably because computers were not available to reduce the labor in making computations, and probably because there was a flaw in the computation of the Montgomery streamfunction and its use to compute the geostrophic wind, isentropic coordinates were not widely accepted by operational meteorologists. Around 1945, operational isentropic analysis ceased, as did height analysis, and pressure coordinates became the operational standard. In retrospect, Danielsen in 1959 found that the Montgomery streamfunction M had been computed using an inaccurate method, which had nearly been discovered by Brooks in 1942. The terms Φ and C_pT were interpolated separately, without recognition that individual errors can lead to an inconsistency in the hydrostatic equation as expressed in isentropic coordinates. The more accurate method of computing M is to integrate the hydrostatic equation in isentropic coordinates directly (since Φ is computed from measurements of both T and p).

Potential vorticity evaluated on isentropic surfaces was shown to be conserved for frictionless, adiabatic flow by Rossby in 1940. In the early 1940s its use as a tracer of air masses was investigated by Starr and others. Kleinschmidt in the early 1950s attempted to explain cyclogenesis on the basis of Rossby's potential vorticity and its advection. He failed, however, to explain why fronts are often the site of cyclogenesis, an observation noted by the early Polar-Front theorists and accepted as canon ever since. Furthermore, he overemphasized the role of latent-heat release in creating potential vorticity (and hence, effecting cyclogenesis). It is interesting, in maintaining perspective, to note that quasigeostrophic theory had just been developed using pressure as a vertical coordinate, and that baroclinic instability theory had also been formulated in isobaric coordinates. In the mid-1950s Reed and Sanders used potential vorticity as a tracer, and established that air in middle and upper tropospheric fronts originated in the stratosphere (detailed in Chap. 2). Haynes and McIntyre (1987) showed that potential vorticity can be diluted or concentrated only by flow across isentropes; it cannot be created or destroyed within a layer bounded by isentropic surfaces.

The use of potential vorticity, evaluated on isentropic surfaces, for diagnosing the behavior of synoptic-scale systems, has recently enjoyed a resurgence in popularity. Hoskins, McIntyre, and Robertson, in a paper published in 1985, refer to the method of dynamical analysis using potential vorticity on isentropic surfaces as *IPV thinking*, where IPV stands for *isentropic potential vorticity*.

1.9.2. The Observed Distribution of IPV in the Atmosphere

Before we discuss the dynamical analysis of isentropic potential vorticity, it is useful to see what typical vertical distributions and fields of it on isentropic surfaces actually are in the atmosphere. We use the following form of Rossby's potential vorticity P:

$$P = -g(\zeta_\theta + f)\frac{\partial \theta}{\partial p}, \tag{1.9.1}$$

where

$$\zeta_\theta = \left(\frac{\partial v}{\partial x} - \frac{\partial u}{\partial y}\right)_\theta. \tag{1.9.2}$$

Henceforth we will refer to P simply as "potential vorticity." We remind ourselves that Eq. (1.9.1) is similar in form to Ertel's potential vorticity $[(C_p/\rho)(\nabla \times \mathbf{v} \cdot \nabla \ln \theta)]$ for a hydrostatic atmosphere, with potential temperature used as a vertical coordinate.

For typical midlatitude, synoptic-scale flow

$$P \simeq -gf\frac{\partial \theta}{\partial p}. \tag{1.9.3}$$

Typically

$$\frac{\partial \theta}{\partial p} \simeq -\frac{10\ \mathrm{K}}{100\ \mathrm{mb}}. \tag{1.9.4}$$

Thus, isentropic potential vorticity is on the order of

$$\begin{aligned} P &\simeq -(10\ \mathrm{m\,s^{-2}})(10^{-4}\ \mathrm{s^{-1}})\left(-\frac{10\ \mathrm{K}}{10\ \mathrm{kPa}}\right)\frac{1\ \mathrm{kPa}}{10^3\ \mathrm{kg\,m\,s^{-2}\,m^{-2}}} \\ &= 10^{-6}\ \mathrm{m^2\,s^{-1}\,K\,kg^{-1}} \equiv 1\ \mathrm{PVU}, \end{aligned} \tag{1.9.5}$$

where a PVU is a *potential vorticity unit.* This terminology[9] is introduced to simplify the units of IPV, which are cumbersome. Values of IPV less than approximately 1.5 PVU are usually associated with tropospheric air, while larger IPV values usually are associated with stratospheric air. Figure 1.137 shows the average vertical and meridional variation of IPV in the troposphere and lower stratosphere. Although the 350 K isentropic surface varies little in height (~200 mb) with latitude, it is in the stratosphere in middle and high latitudes, and in the troposphere in the tropics. On the other hand, the 300 K isentropic surface varies significantly in height (1000–300 mb) with latitude, but is almost always in the troposphere.

In the United States one is used to thinking in terms of analyses at standard pressure levels because operational analyses are drawn on constant pressure surfaces. We therefore will discuss briefly an example of Northern Hemisphere analyses of potential vorticity on the 325-K isentropic surface (Fig. 1.138), and compare them to isobaric analyses at 500 mb (Fig. 1.139).

At 1200 UTC, May 16, 1989, the 325-K tropopause [Fig. 1.138(a)], which is marked by IPV values of around 1.5 PVU, undulates across the continent from as far north as Canada along the west coast of North America and north of the Great Lakes to as far south as Baja and South Carolina. The region of stratospheric air generally found at very high latitudes (Fig. 1.37) is referred to as the *stratospheric reservoir.* Relative maxima in IPV form tongues of stratospheric air extruding equatorward; relative minima in IPV extrude poleward. At 500 mb [Fig. 1.139(a)] the IPV maxima are associated with cyclonic curvature and shear in troughs and relatively cool air; the minima are associated with anticyclonic curvature and shear in ridges and, to some extent, relatively warm air.

We consider the evolution of the IPV field by looking at a sequence of daily IPV analyses at 325 K (Fig. 1.138), and the corresponding geopotential-height analyses at 500 mb (Fig. 1.139). Between 1200 UTC on May 16, 1989, and 1200 UTC on May 18, the IPV maximum over Baja and the intermountain area of the western United States moved eastward, became narrower, and split off from the stratospheric reservoir to the north. It appeared as if the "umbilical" cord connecting the IPV maximum to the reservoir became

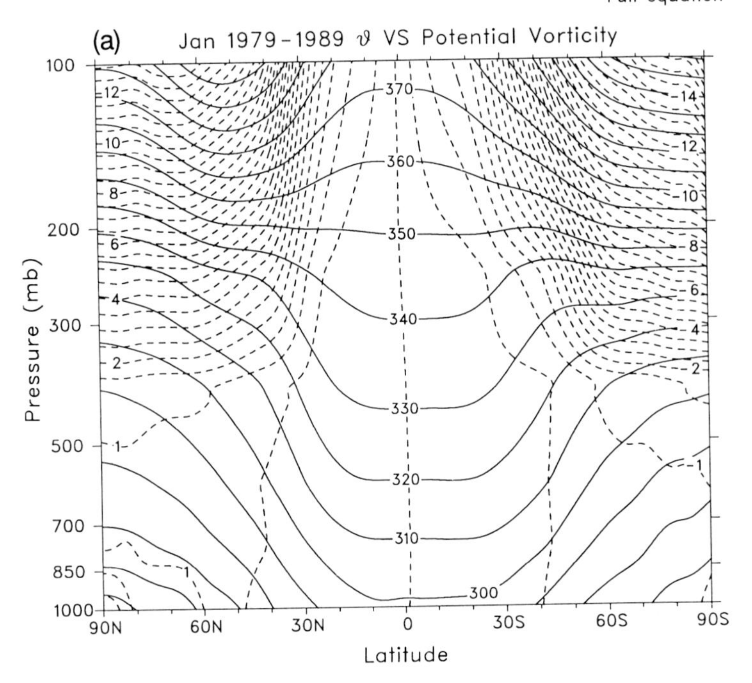

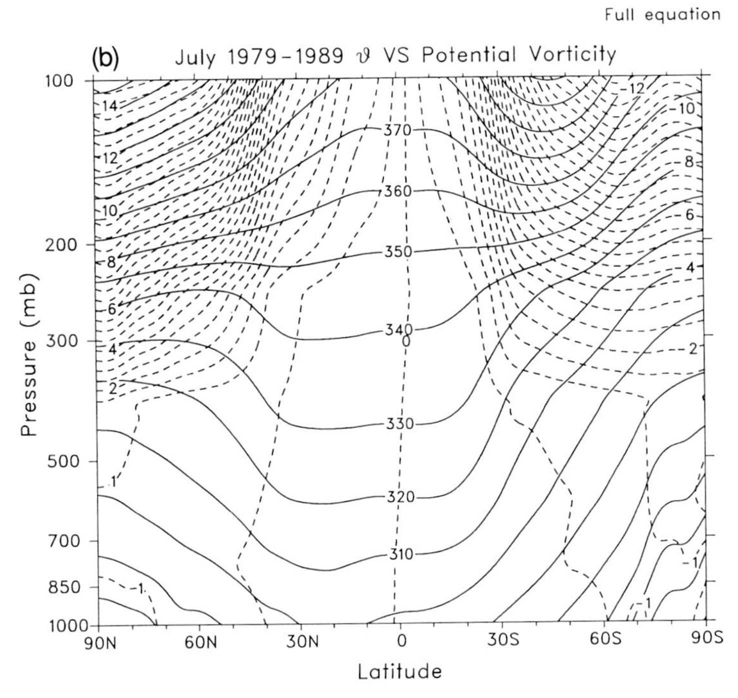

Figure 1.137 Vertical cross section of zonal mean of Ertel's potential vorticity in PVU (dashed lines) and potential temperature in K (solid lines) in (a) January and (b) July; from 1979–1989 initialized ECMWF data at seven levels, twice daily. (Courtesy Kevin Trenberth and Amy Solomon, NCAR)

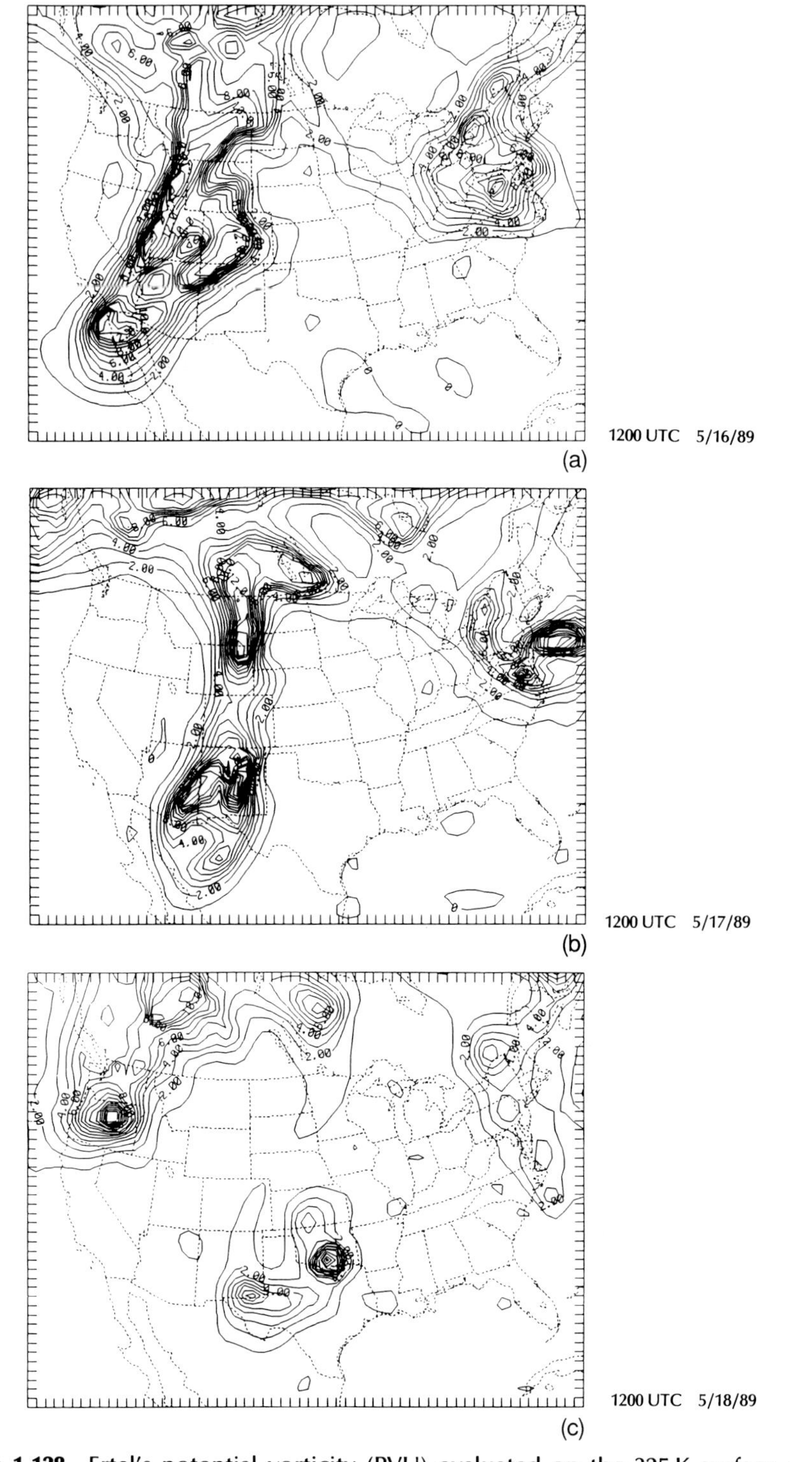

Figure 1.138 Ertel's potential vorticity (PVU) evaluated on the 325 K surface over the United States. (a) 1200 UTC, May 16; (b) 1200 UTC, May 17, and (c) 1200 UTC, May 18, 1989. (Courtesy Stan Benjamin, Tracy Smith, and Tom Schlatter, Forecast Systems, Laboratory, PROFS, Boulder, Colorado)

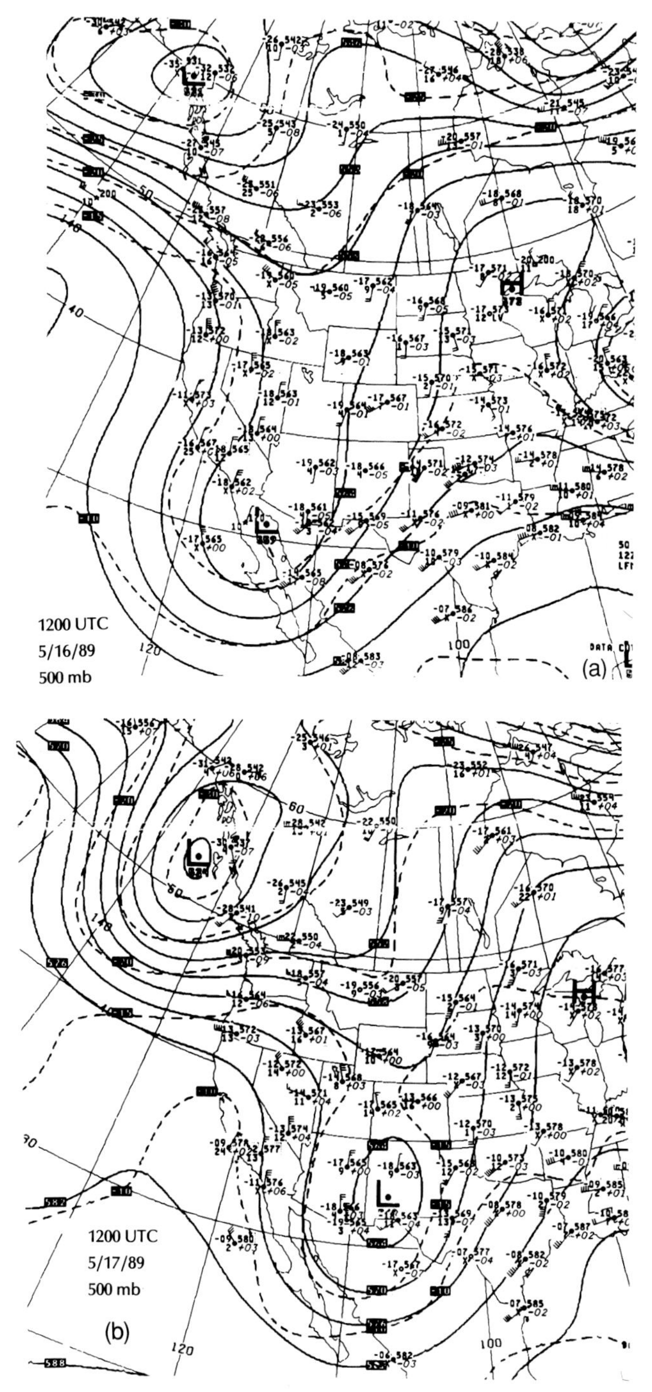

Figure 1.139 500-mb height contours in dam (solid lines), temperature and dew-point depression (°C), and winds from NMC's analysis for (a) 1200 UTC, May 16, (b) 1200 UTC, May 17, and (c) 1200 UTC, May 18, 1989.

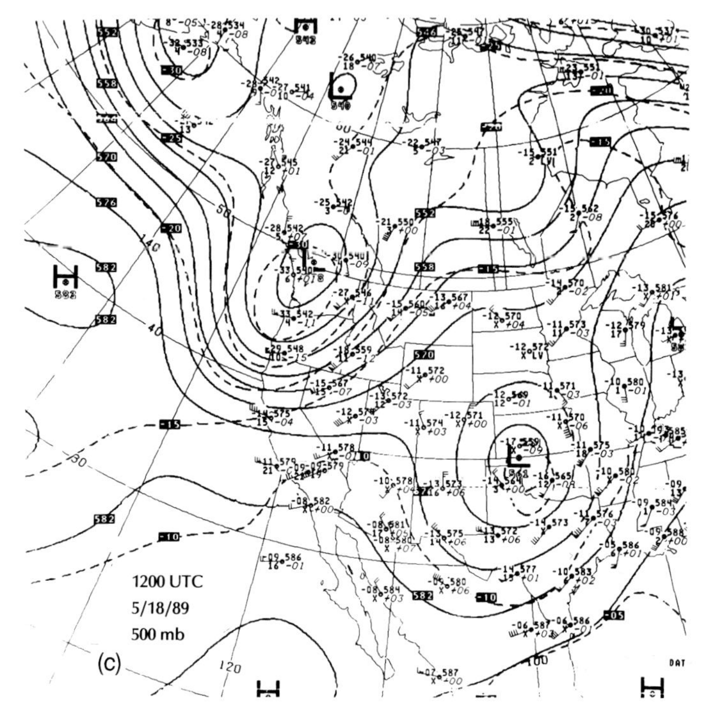

Figure 1.139 *(cont.)*

severed! (In a continuous fluid, material lines cannot be cut. Severance of the IPV contours implies that IPV is not being conserved. Small-scale turbulent mixing allows IPV contours to break apart and rejoin again.) At 500 mb, the trough moved eastward, closed off over the Oklahoma–Kansas border, and became cut off from the westerlies, which lay to the north in association with a new trough (the IPV maximum in the Pacific Northwest) entering the west coast of the United States. The cut-off low at 500 mb was associated with a "pool" of relatively cool air.

Sometimes IPV minima become pinched off (a non-conservative process) within the stratospheric reservoir, just as stratospheric IPV maxima become closed off within the lower-latitude tropospheric air mass. The closed-off maxima tend to be associated with blocking ridges. (IPV maxima and minima escape from the lair of the stratosphere and move about the troposphere. They sometimes are snatched back into the stratospheric reservoir!)

It therefore appears as if tongues of high-IPV, stratospheric air that extend equatorward from the high-IPV reservoir are associated with troughs in the height field and associated cyclonic flow, while tongues of low-IPV, tropospheric air that extend poleward are associated with ridges in the height field and associated anticyclonic flow. Furthermore, isolated regions of stratospheric IPV that are situated equatorward from the reservoir tend to be associated with troughs or closed lows in the height field and associated cyclonic flow; isolated regions of tropospheric IPV that are found embedded within the reservoir tend to be associated with ridges or closed highs in the height field and associated anticyclonic flow. The features seem to evolve through translation, rotation, and deformation by the wind field. This is to be expected if friction and diabatic heating are negligible, and potential vorticity is in fact to a large degree conserved, as was verified by Starr and others around 1940. We will henceforth refer to the regions of relatively high IPV (relative to other values of IPV on the same isentropic surface at the same latitude) as *positive IPV anomalies,* and to the regions of relatively low IPV as *negative IPV anomalies.* The latter will be associated with anticyclonic relative vorticity on isentropic surfaces, while the former will be associated with cyclonic relative vorticity on isentropic surfaces.[10]

The vertical structure of an upper-level, positive IPV anomaly is illustrated in Fig. 1.140. The tropopause is relatively low, the upper troposphere is relatively cold, and the lower stratosphere is relatively warm. On isentropic surfaces that pass through the upper troposphere under the positive IPV anomaly, the static stability is relatively low (compared to the stability elsewhere on the same isentropic surface); on isentropic surfaces that pass through the lower stratosphere in the positive IPV anomaly, the static stability is relatively high. For example, the 300 K surface is in a region of relatively closely packed isentropic surfaces away from the (cyclone associated with the) low-tropopause area, and in a region of relatively less densely packed isentropic surfaces (in the cyclone), just under the low-tropopause area. On the other hand, the 320-K surface is in a region of relatively closely packed isentropic surfaces (in the cyclone) just above the low-tropopause area, and in a region of less densely packed isentropic surfaces away from the (cyclone associated with the) low-tropopause area.

The vertical structure of an upper-level, negative IPV anomaly is illustrated in Fig. 1.141. The tropopause here is relatively high (near 200 mb), the upper troposphere is relatively warm, and, although not pronounced in this case, the lower stratosphere is relatively cold (see the 345-K surface). On isentropic surfaces that pass under the negative IPV anomaly (e.g., at 310 K), the static stability is relatively high; on isentropic surfaces that pass through the negative IPV anomaly (e.g., at 330 K), the static stability is relatively low.

It seems, then, that the atmospheric structures observed in association with positive and negative IPV anomalies at upper levels are opposite to each other, and therefore that there must be a "conceptual duality" between strong upper-level cyclones and anticyclones.

1.9.3. Upper-Level IPV Anomalies

The invertibility principle. We will now demonstrate that, to a first approximation, atmospheric structure may be regarded as a superposition of positive and negative IPV anomalies, and that the observed wind field is approximately equal to the *sum* of the wind fields associated with each IPV anomaly separately.

Let us assume that the *reference state* for the atmosphere is one of constant Coriolis parameter and zero relative vorticity; furthermore, let us assume that

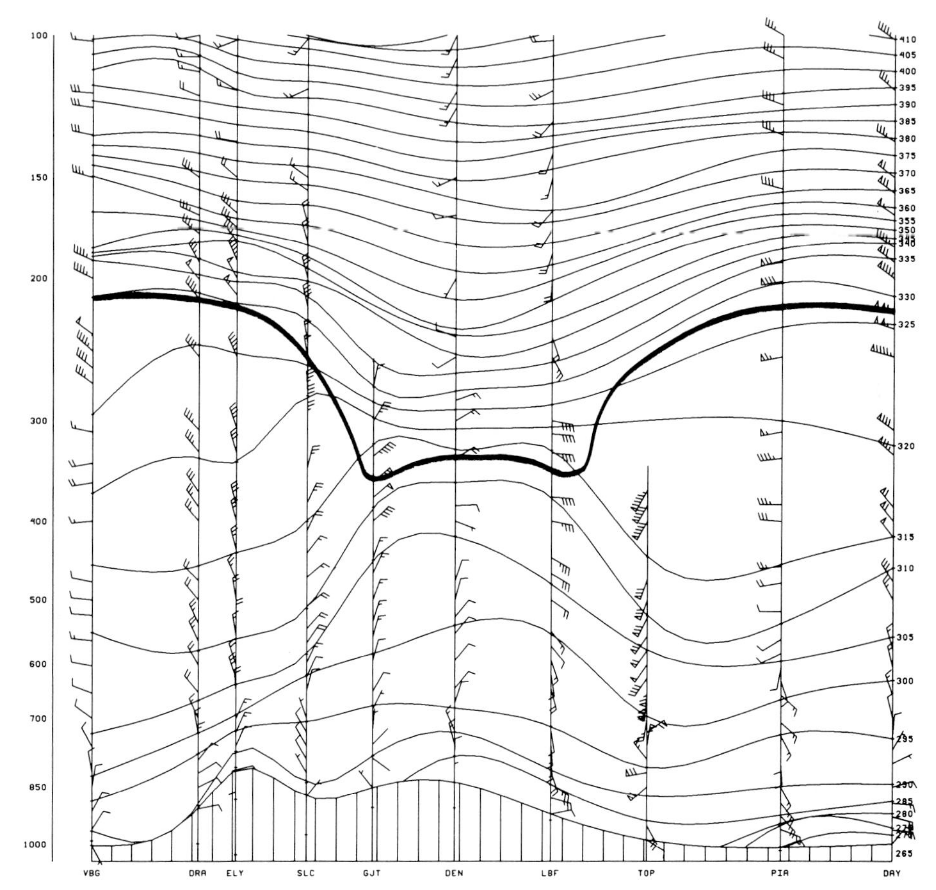

Figure 1.140 Vertical cross section through an upper-level, positive (cyclonic) IPV anomaly at 1200 UTC, March 7, 1990. The ordinate is the pressure in mb; the abscissa is the location from Vandenburg AFB, California (VBG), through Denver, Colorado (DEN), through Dayton, Ohio (DAY), on an approximate west-southwest to east-northeast line across the United States. Potential temperature in K (solid lines); winds plotted according to convention, half barb = 2.5 m s^{-1}, full barb = 5 m s^{-1}, pennant = 25 m s^{-1}. Tropopause is indicated by the thick, solid line.

reference pressure is a function only of potential temperature, and consequently that static stability is uniform on each isentropic surface. It follows that potential vorticity in the reference atmosphere is a function only of potential temperature. Since potential vorticity is conserved for frictionless, adiabatic flow, IPV anomalies can be created only by nonuniform (but mass-conserving) vertical displacements of isentropic surfaces. The actual IPV field is the sum of the reference state P_{ref} and the anomalies P', that is,

$$P(x,y,\theta) = P_{\text{ref}}(\theta) + P'(x,y,\theta). \tag{1.9.6}$$

Let us now address the following problem: If we know the distribution of IPV (i.e., the distribution of the product of absolute vorticity and static

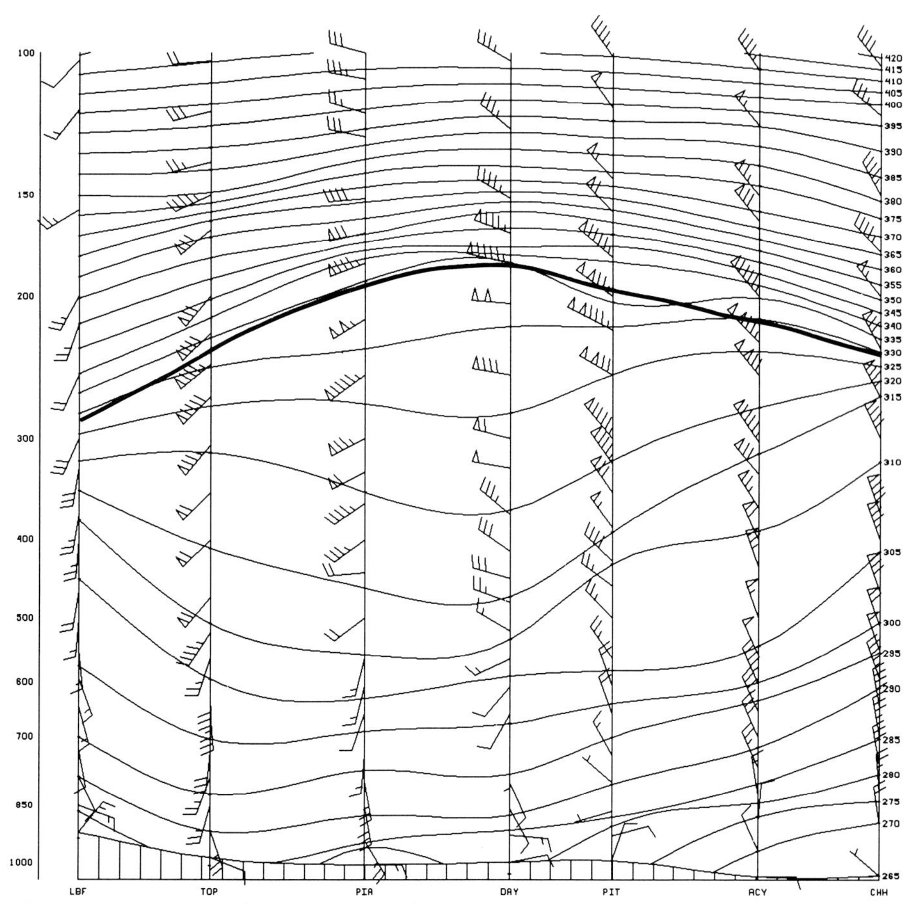

Figure 1.141 Vertical cross section through an upper-level, negative (anticyclonic) IPV anomaly at 0000 UTC, March 8, 1990; through North Platte, Nebraska (LBF), Dayton, Ohio (DAY), and Chatham, Massachusetts (CHH), on an approximate east–west line across the United States. Otherwise as in Fig. 1.140.

stability), how can we find out what the vorticity and static-stability fields are *individually*? Suppose that there is a *balance condition* imposed on the wind field. Then there is a *diagnostic* relationship between the wind field and a scalar field that represents mass such as Φ, M, or p. Geostrophic balance, gradient-wind balance, and (1.7.12) are familiar examples of "balance." We will demonstrate that it is possible to determine, uniquely, the distribution of both vorticity *and* static stability associated with an IPV field, given the proper boundary conditions. The balance condition, assumption of a reference state, and boundary conditions will allow us to do this: These constraints remove any arbitrariness, and only one set of vorticity and static stability values fit the puzzle of the "global" distribution of IPV; if we knew IPV locally only, we could not uniquely determine the local absolute vorticity and static stability. We must also be sure to remember that the balance condition we use must be compatible with the space and time scales of the observed air motion.

Suppose for simplicity that the atmosphere is in *gradient-wind balance* and *hydrostatic balance,* and that the wind and temperature fields are circularly symmetric. On an isentropic surface in polar coordinates (r, ϕ), gradient-wind balance is given by the following:

$$\frac{V^2}{r} + fV - \frac{\partial M}{\partial r} = 0, \tag{1.9.7}$$

where V is the wind speed. Hydrostatic balance is expressed as

$$\frac{\partial M}{\partial \theta} = C_p \left(\frac{p}{p_0} \right)^{R/C_p}. \tag{1.9.8}$$

Differentiating Eq. (1.9.7) with respect to θ and Eq. (1.9.8) with respect to r, we find that

$$\frac{\partial^2 M}{\partial \theta \, \partial r} = \frac{\partial V}{\partial \theta} f_{\text{loc}} \tag{1.9.9}$$

$$\frac{\partial^2 M}{\partial r \, \partial \theta} = K(p) \frac{\partial p}{\partial r}, \tag{1.9.10}$$

where

$$f_{\text{loc}} = f + \frac{2V}{r} \tag{1.9.11}$$

$$K(p) = \frac{\partial}{\partial p} \left[C_p \left(\frac{p}{p_0} \right)^{R/C_p} \right]. \tag{1.9.12}$$

The parameter f_{loc} represents the absolute rotation rate about the local vertical (f from Earth's rotation, and $2V/r$ from relative rotation about the origin). Subtracting Eq. (1.9.9) from Eq. (1.9.10), and rearranging terms, we obtain the following *thermal-wind* relation for the gradient wind (we usually define the thermal wind in terms of the geostrophic wind):

$$\frac{\partial V}{\partial \theta} = \frac{K(p)}{f_{\text{loc}}} \frac{\partial p}{\partial r}. \tag{1.9.13}$$

It relates the vertical wind shear in isentropic coordinates $(\partial V / \partial \theta)$ to the radial pressure gradient on an isentropic surface $(\partial p / \partial r)$, the latter of which is like the radial temperature gradient on an isobaric surface.

Isentropic potential vorticity Eq. (1.9.1) may be expressed as

$$P = \frac{\zeta_{a\theta}}{\sigma^*}, \tag{1.9.14}$$

where

$$\zeta_{a\theta} = \zeta_\theta + f \tag{1.9.15}$$

$$\sigma^* = -\frac{1}{g} \frac{\partial p}{\partial \theta}. \tag{1.9.16}$$

The former $(\zeta_{a\theta})$ is the absolute vorticity evaluated on an isentropic surface; the latter (σ^*) is a measure of static stability. Recall that in pressure coordinates static stability is given by the variation of potential temperature with pressure; since pressure plays the role of temperature in isentropic coordinates, static stability is given by the variation of pressure with potential temperature. One should be careful in comparing measures of static stability in pressure coordinates to those in isentropic coordinates because the static stability parameters are *reciprocals* of each other.

In circularly symmetric flow, relative vorticity on an isentropic surface expressed in polar coordinates is

$$\zeta_\theta = \frac{1}{r}\frac{\partial (rV)}{\partial r}. \tag{1.9.17}$$

From Eqs. (1.9.14) (1.9.15), and (1.9.17) we see that

$$\sigma^* P = \left(\frac{1}{r}\frac{\partial (rV)}{\partial r} + f\right). \tag{1.9.18}$$

Then the derivative of Eq. (1.9.18) with respect to r may be expressed, with the aid of Eqs. (1.9.14) and (1.9.15), as

$$\sigma^* \frac{\partial P}{\partial r} = \frac{\partial}{\partial r}\left(\frac{1}{r}\frac{\partial (rV)}{\partial r}\right) + \frac{\partial f}{\partial r} - \frac{\partial \sigma^*}{\partial r}\frac{\zeta_{a\theta}}{\sigma^*}. \tag{1.9.19}$$

From Eq. (1.9.16) it follows that

$$\frac{\partial \sigma^*}{\partial r} = -\frac{1}{g}\frac{\partial^2 p}{\partial r\,\partial \theta} = -\frac{1}{g}\frac{\partial}{\partial \theta}\left(\frac{\partial p}{\partial r}\right). \tag{1.9.20}$$

Substituting for $\partial p/\partial r$ in Eq. (1.9.20) from the thermal-wind relation, Eq. (1.9.13), we see that

$$\frac{\partial \sigma^*}{\partial r} = -\frac{1}{g}\frac{\partial}{\partial \theta}\left(\frac{f_{\text{loc}}}{K(p)}\frac{\partial V}{\partial \theta}\right). \tag{1.9.21}$$

Substituting Eq. (1.9.21) into Eq. (1.9.19), and rearranging terms, we obtain the following equation:

$$\frac{\partial}{\partial r}\left(\frac{1}{r}\frac{\partial (rV)}{\partial r}\right) + \frac{\zeta_{a\theta}}{\sigma^*}\frac{1}{g}\frac{\partial}{\partial \theta}\left(\frac{f_{\text{loc}}}{K(p)}\frac{\partial V}{\partial \theta}\right) + \frac{\partial f}{\partial r} = \sigma^* \frac{\partial P}{\partial r}. \tag{1.9.22}$$

This is a second-order, partial-differential equation for $V(r, \theta)$ forced by the radial gradient of IPV. It is highly nonlinear when $\zeta_{a\theta}$, σ^*, and f_{loc} are functions of r or θ. Given the proper boundary conditions for $V(r,\theta)$, Eq. (1.9.22) could, in principle, be solved. In other words, *if we know the distribution of IPV, then we also know the wind field*: This is the *invertibility principle.* We say that the wind field $V(r,\theta)$ is "induced" by the IPV-anomaly field. (The balance condition was necessary to link the wind field to the temperature field, and obtain a diagnostic equation for V.) One can think of the IPV-anomaly field inducing the wind field like a charge distribution inducing an electrical field.

For the sake of simplicity, we will treat f as a constant on the left-hand side of Eq. (1.9.22). Equation (1.9.22) is reminiscent of the quasigeostrophic ω equation, in which the vertical-velocity field is known if the distribution of geostrophic wind is known; it is also reminiscent of the Sawyer–Eliassen equation (see Chap. 2), in which the vertical ageostrophic circulation is known if the distribution of geostrophic wind is known. The equation is elliptic if

$$-4\frac{\zeta_{a\theta}}{\sigma^*}\frac{1}{g}\frac{f_{\text{loc}}}{K(p)} = -\frac{4Pf_{\text{loc}}}{gK(p)} < 0, \tag{1.9.23}$$

that is, if

$$Pf_{\text{loc}} > 0 \tag{1.9.24}$$

since $g > 0$ and $K(p) > 0$. Equation (1.9.24) implies that if the atmosphere is statically stable ($\sigma^* > 0$), inertially stable ($\zeta_{a\theta} > 0, f_{\text{loc}} > 0$) (see Chap. 2), and symmetrically stable ($P > 0$) (see Chap. 2), then Eq. (1.9.22) is elliptic.

If we replace $\zeta_{a\theta}$, σ^*, $K(p)$, and f_{loc} where they appear on the left-hand side of Eq. (1.9.22) by their values in the reference atmosphere, then

$$\zeta_{a\theta} = f_0 \tag{1.9.25}$$

$$\sigma^* = \sigma^*_{\text{ref}}(\theta) \tag{1.9.26}$$

$$K(p) = K_{\text{ref}}(\theta) \tag{1.9.27}$$

$$f_{\text{loc}} = f_0. \tag{1.9.28}$$

We also apply Eq. (1.9.26) to the right-hand side of Eq. (1.9.22). Equation (1.9.22) is then written as follows:

$$\frac{\partial}{\partial r}\left(\frac{1}{r}\frac{\partial(rV)}{\partial r}\right) + \frac{f_0^2}{\sigma^*_{\text{ref}}(\theta)}\frac{1}{g}\frac{\partial}{\partial\theta}\left(\frac{1}{K_{\text{ref}}(\theta)}\frac{\partial V}{\partial\theta}\right) = \sigma^*_{\text{ref}}(\theta)\frac{\partial P}{\partial r}. \tag{1.9.29}$$

The operator on the left-hand side of Eq. (1.9.29) is *linear,* so that the total response $[V(r,\theta)]$ to n IPV anomalies $[\sigma^*_{\text{ref}}(\partial P_1/\partial r + \partial P_2/\partial r + \cdots + \partial P_n/\partial r)]$ is the sum of the n individual responses $(V_1 + V_2 + \cdots + V_n)$, where V_i is the response to $\sigma^*_{\text{ref}}(\partial P_i/\partial r)$. (This situation is analogous to that of the quasigeostrophic ω equation, in which f is treated as a constant and quasihorizontal variations in σ are neglected in the operator: The total ω is then the sum of individual ω's forced by each forcing function.) It is important to recognize that ζ_θ and variations in f *are* accounted for in the forcing function on the right-hand side of Eq. (1.9.29), even though they are neglected in the operator on the left-hand side.

The ellipticity condition for Eq. (1.9.29) is simply that

$$\sigma^*_{\text{ref}} > 0. \tag{1.9.30}$$

The approximations expressed by Eqs. (1.9.25)–(1.9.28) are consistent with the assumption that IPV anomalies are relatively *weak*. Thus, Eq. (1.9.29) may be regarded as a "weak-anomaly" approximation: The response to weak anomalies is linear, while the response to strong anomalies is nonlinear. It is easier to understand the linear aspects of a dynamical process before tackling the nonlinear aspects. We therefore will work with the linear form [Eq. (1.9.29)] in much of our subsequent analyses.

The effects of scale. Suppose that an isolated IPV anomaly in the upper troposphere has a horizontal scale L. To what extent will it make itself felt above and below? This is like the opposite of the Rossby radius-of-deformation problem, in which one seeks the horizontal extent of the influence of a disturbance having a specified vertical scale. If the vertical scale in isentropic coordinates is $\Delta\theta$, and the azimuthal component of the wind is of order U, then

$$\frac{\partial}{\partial r}\left(\frac{1}{r}\frac{\partial(rV)}{\partial r}\right) \sim \frac{U}{L^2} \tag{1.9.31}$$

$$\frac{f_0^2}{\sigma^*_{\text{ref}}(\theta)\,g}\frac{1}{g}\frac{\partial}{\partial\theta}\left(\frac{1}{K_{\text{ref}}(\theta)}\frac{\partial V}{\partial\theta}\right) \sim \frac{f_0^2}{\sigma^*_{\text{ref}}g}\frac{1}{K_{\text{ref}}(\theta)}\frac{U}{(\Delta\theta)^2}. \tag{1.9.32}$$

As in the quasigeostrophic ω equation, we assume that the vertical and horizontal gradient terms in the elliptic operator are the same order of magnitude, so that, equating Eqs. (1.9.31) and (1.9.32), we find that

$$\Delta\theta \sim \frac{f_0 L}{\sqrt{\sigma^*_{\text{ref}}K_{\text{ref}}g}}. \tag{1.9.33}$$

Thus, the vertical scale of an IPV anomaly is proportional to the horizontal scale and inversely proportional to the square root of the static stability parameter σ^*. (This relationship is analogous to one obtained in quasi-geostrophic theory when the adjustment by the wind field is comparable to that by the temperature field, i.e. $L \sim NH/f$, where N is the Brunt-Väisälä frequency, and H is the vertical scale. It follows that $H \sim fL/N$.) A typical value of σ^*_{ref} is

$$\sigma^*_{\text{ref}} = -\frac{1}{g}\left(\frac{\partial p}{\partial\theta}\right)_{\text{ref}} \sim \frac{1}{10\ \text{m s}^{-2}}\,\frac{100\ \text{mb}}{10\ \text{K}}\,\frac{1\ \text{kPa}}{10\ \text{mb}} = 0.1\ \text{kPa m}^{-1}\,\text{s}^2\,\text{K}^{-1}, \tag{1.9.34}$$

and a typical value of K_{ref} is

$$K_{\text{ref}} = \frac{\partial}{\partial p}\left[C_p\left(\frac{p}{p_0}\right)^{R/C_p}\right] = \frac{R}{p}\left(\frac{p}{p_0}\right)^{R/C_p} = \frac{RT}{p\theta} = \frac{1}{\rho\theta}$$
$$\sim \frac{1}{(1\ \text{kg m}^{-3})(300\ \text{K})}. \tag{1.9.35}$$

Then substituting Eqs. (1.9.35) and (1.9.34) into Eq. (1.9.33), we find that in the midlatitudes, for IPV anomalies on the order of 1000 km across,

$$\Delta\theta \sim (10^{-4}\ \text{s}^{-1})(10^6\ \text{m}) \Big/ \left[\left(10^{-1}\ \text{kPa m}^{-1}\,\text{s}^2\,\text{K}^{-1}\,\frac{10^3\ \text{kg m s}^{-2}\,\text{m}^{-2}}{1\ \text{kPa}}\right.\right.$$
$$\left.\times\left(\frac{1}{300\ \text{kg m}^{-3}\,\text{K}}\right)10\ \text{m s}^{-2}\right]^{1/2}$$
$$\sim 60\ \text{K}.$$

This is as least as deep as the average depth of the troposphere (see Fig. 1.137). Therefore a synoptic-scale IPV anomaly at upper levels can induce a wind field all the way down to the ground.

Consider Eqs. (1.9.31), (1.9.32), and the right-hand side of Eq. (1.9.29). It follows that

$$\frac{U}{L^2} \sim \frac{f_0^2}{\sigma_{\text{ref}}^*} \frac{1}{g} \frac{1}{K_{\text{ref}}} \frac{U}{(\Delta\theta)^2} \sim \sigma_{\text{ref}}^* \frac{\tilde{P}}{L}, \tag{1.9.37}$$

where $\tilde{P}$ is the order of magnitude of the isentropic potential vorticity. Then for a given value of $\tilde{P}$, for f_0, σ_{ref}^*, and K_{ref} held fixed, small L must be associated with small U and vice versa, since

$$U \sim (\sigma_{\text{ref}}^* \tilde{P})L. \tag{1.9.38}$$

(In addition, for a given value of $\tilde{P}$ and L, U is enhanced by large σ_{ref}^*, i.e. by low static stability.) According to Eq. (1.9.33), small L must be associated with small $\Delta\theta$ and vice versa. Therefore small-scale IPV anomalies of a given strength induce weak wind fields, whose vertical influence is only in a shallow layer, while large-scale IPV anomalies of the same strength induce strong wind fields, whose vertical influence is deep. Thus, the response of the atmosphere to IPV anomalies is dependent upon their scale. (Since small-scale IPV anomalies induce weak wind fields, small-scale noise in the IPV field does not adversely affect the accuracy of the "inverted" wind field.)

Analyses of IPV (Fig. 1.138) appear to have more structure than isobaric analyses of height (Fig. 1.139). The reason for this is a mathematical consequence of dynamical principles. Equation (1.9.29) relates the horizontal gradient of IPV to second derivatives of the wind field with respect to both the horizontal and the vertical-coordinate variables. The wind field is thus the inverse Laplacian of the derivative of the IPV field. Since the inverse Laplacian acts as a smoother, the wind field must be smoother than the IPV gradient field (and the IPV field). But the horizontal gradient of the height field is related to the wind field through the gradient-wind formula. Therefore the IPV field is related to second-order derivatives of the height field, and hence must show a lot more structure than the height field: Second derivatives of a function usually enhance even the smallest variations in the function. For example, vorticity, which is given by the Laplacian of the streamfunction (or height field), shows much more structure than the streamfunction (or height field). To demonstrate this point another way, consider a continuous spectrum of the height field as a function of frequency (f), $W(f)$ (Fig. 1.142). The spectrum of the IPV field is proportional to the spectrum of the second derivative of the height field. Therefore the spectrum of the IPV field is a

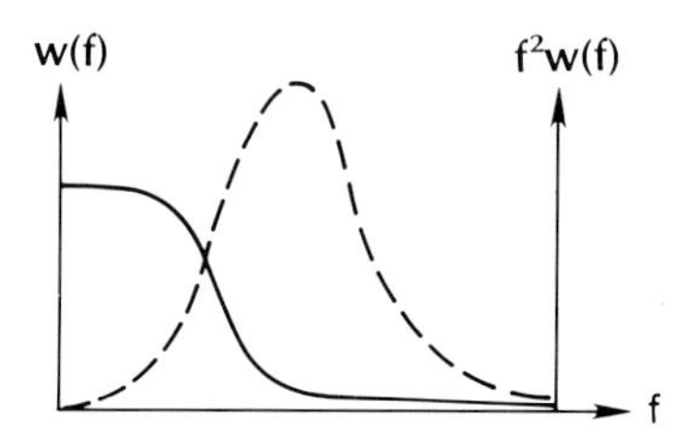

Figure 1.142 The peak in the spectrum of $f^2W(f)$ (dashed line) is at higher frequency than that of $W(f)$ (solid line), where f is the frequency.

function of $f^2W(f)$: The spectrum is depressed at low frequencies and enhanced at high frequencies. (If the spectrum is discrete, then there are no shifts in the spectrum to higher frequencies. For example, if the height field is a pure sine wave, the spectrum is a pair of impulse functions, and the IPV field is also a pure sine wave of the same frequency.) We therefore conclude that "IPV thinking" allows us to visualize finer-scale features more easily than traditional isobaric height and temperature analysis. (We note that vorticity analysis on constant-pressure surfaces also allows us to visualize finer-scale features more easily.)

The structure of upper-level IPV anomalies. Solutions of the fully nonlinear Eq. (1.9.22) to idealized, circularly symmetric, positive and negative IPV anomalies in the upper troposphere and their accompanying θ fields are given in Thorpe (1985). A schematic representation of the solutions is shown in Fig. 1.143. The reference state consists of a layer of high, constant static stability

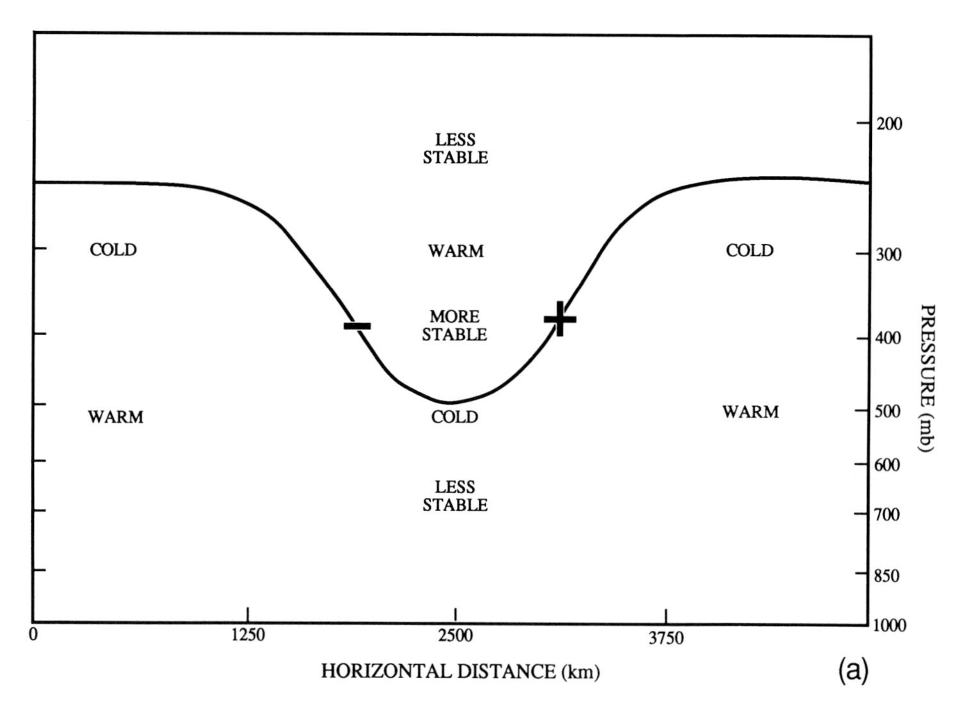

Figure 1.143 Schematic vertical cross section through an isolated, circularly symmetric (a) cyclonic IPV anomaly and (b) anticyclonic IPV anomaly. Tropopause (thick solid line); center of the anomaly is located near the tropopause at 2500 km; + and − indicate the location of maximum azimuthal flow into and out of the plane of the cross section, respectively; that is, in (a) there is a cyclonic vortex centered at 2500 km near 400 mb, which decreases in intensity above and below in accord with the thermal-wind relation; it is relatively cold below and warm above; in (b) there is an anticyclonic vortex centered at 2500 km just above 250 mb, which decreases in intensity above and below in accord with the thermal-wind relation; it is relatively warm below and cold above.

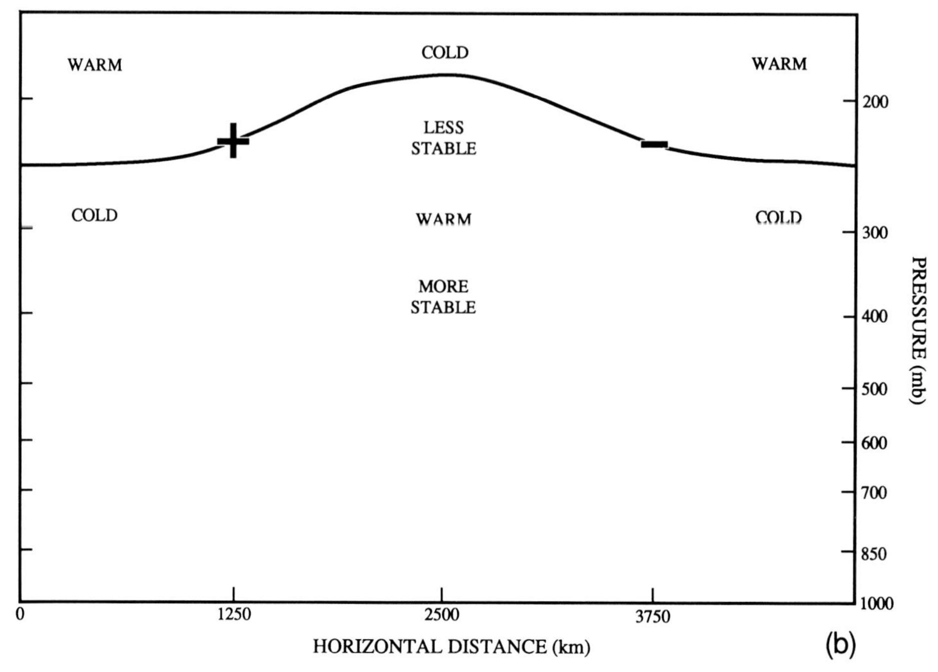

Figure 1.143 (*cont.*)

(the stratosphere) overlying a layer of relatively low, constant static stability (the troposphere). The boundary conditions are as follows:

$$V \rightarrow 0, \quad \text{as} \quad r \rightarrow \infty \tag{1.9.39}$$

$$\theta = \text{constant, at 60 and 1000 mb.} \tag{1.9.40}$$

The positive IPV anomaly is accompanied by a cyclonic vortex and high static stability (relative to the static stability outside the anomaly on the same isentropic surface); the negative IPV anomaly is accompanied by an anti-cyclonic vortex and low static stability. Thus, the relative vorticity and static-stability anomaly on isentropic surfaces are positively correlated. The static stability anomalies above and below the IPV anomaly are in the opposite sense of those in the IPV anomaly: Static stability is relatively low above and below the positive IPV anomaly, and relatively high above and below the negative IPV anomaly. The idealized structures of the upper-level, positive and negative IPV anomalies are similar to those of the observed examples (Figs. 1.140 and 1.141).

Both observations and theory indicate that an IPV anomaly is composed of individual anomalies of the same sense in *both* the absolute vorticity and the static stability fields. In fact, this partitioning is a consequence of thermal-wind balance. For example, if a positive IPV anomaly showed up entirely as a positive anomaly in relative vorticity (and hence in absolute vorticity) in some layer, then the isentropic surfaces must be squeezed together around the center of the vorticity anomaly so that there is a radial gradient of pressure on isentropic surfaces to accompany the vertical wind shear above and below the vorticity anomaly. Thus, static stability is relatively high. On the other hand, if the IPV anomaly showed up entirely as a positive anomaly in static stability, then there would be radial pressure gradients on isentropic surfaces, which would have to be accompanied by local vertical shear, and a resulting pattern of vorticity.

Cut-off cyclones and anticyclones are ordinarily defined in terms of closed isobars or height contours. According to this definition, cutoffs depend upon the reference frame from which they are viewed. If we add a strong, uniform current to a cut-off, the cutoff "opens up"! Since the dynamics of a phenomenon should not depend upon the reference frame from which we view the phenomenon, the conventional definition is somewhat lacking because it really makes no *dynamical* distinction between cutoffs and open waves. However, IPV anomalies, and the closed IPV contours associated with them, are independent of reference frame.

1.9.4. Surface Potential-Temperature Anomalies

Suppose that instead of upper-level IPV anomalies, we have anomalies of potential temperature at the surface, which we will assume is at 1000 mb. The warm core of a positive θ anomaly must be accompanied by anticyclonic thermal-wind shear. If the wind field is undisturbed high up, then there must be a cyclone at the surface. On the other hand, the cold core of a negative θ anomaly must be accompanied by cyclonic thermal-wind shear. Then there must be an anticyclone at the surface.

Solutions of Eq. (1.9.22) to idealized, circularly symmetric, positive and negative surface-temperature anomalies, with no potential-temperature variation along the tropopause, are given in Thorpe (1985). A schematic representation of the solutions is depicted in Fig. 1.144. Since pressure in the reference state is a function only of potential temperature, the ground is level, and mass is conserved, the potential temperature anomalies can be created only if isentropic surfaces just above the ground are brought down or up. In the case of a positive θ anomaly, static stability is relatively high just above the ground. In the case of a negative θ anomaly, static stability is relatively low near the anomaly. Therefore, like the upper-level IPV anomalies, positive θ anomalies at the surface are associated with vorticity and static stability anomalies that are positive, while negative θ anomalies are associated with vorticity and static stability anomalies that are negative. The positive surface-θ anomalies are therefore accompanied by a low-level positive, IPV anomaly, and the negative surface-θ anomalies are accompanied by a negative, low-level IPV anomaly. The positive low-level IPV anomaly is confined to a thin layer near the ground because isentropic surfaces are bunched in an infinitesimal layer near the ground. On the other hand, the negative low-level IPV anomaly extends upward in the manner of a "cold dome," because isentropic surfaces bulge upward near the anomaly.

The wind fields induced by synoptic-scale, low-level IPV anomalies typically extend all the way up to the tropopause, just as the wind fields

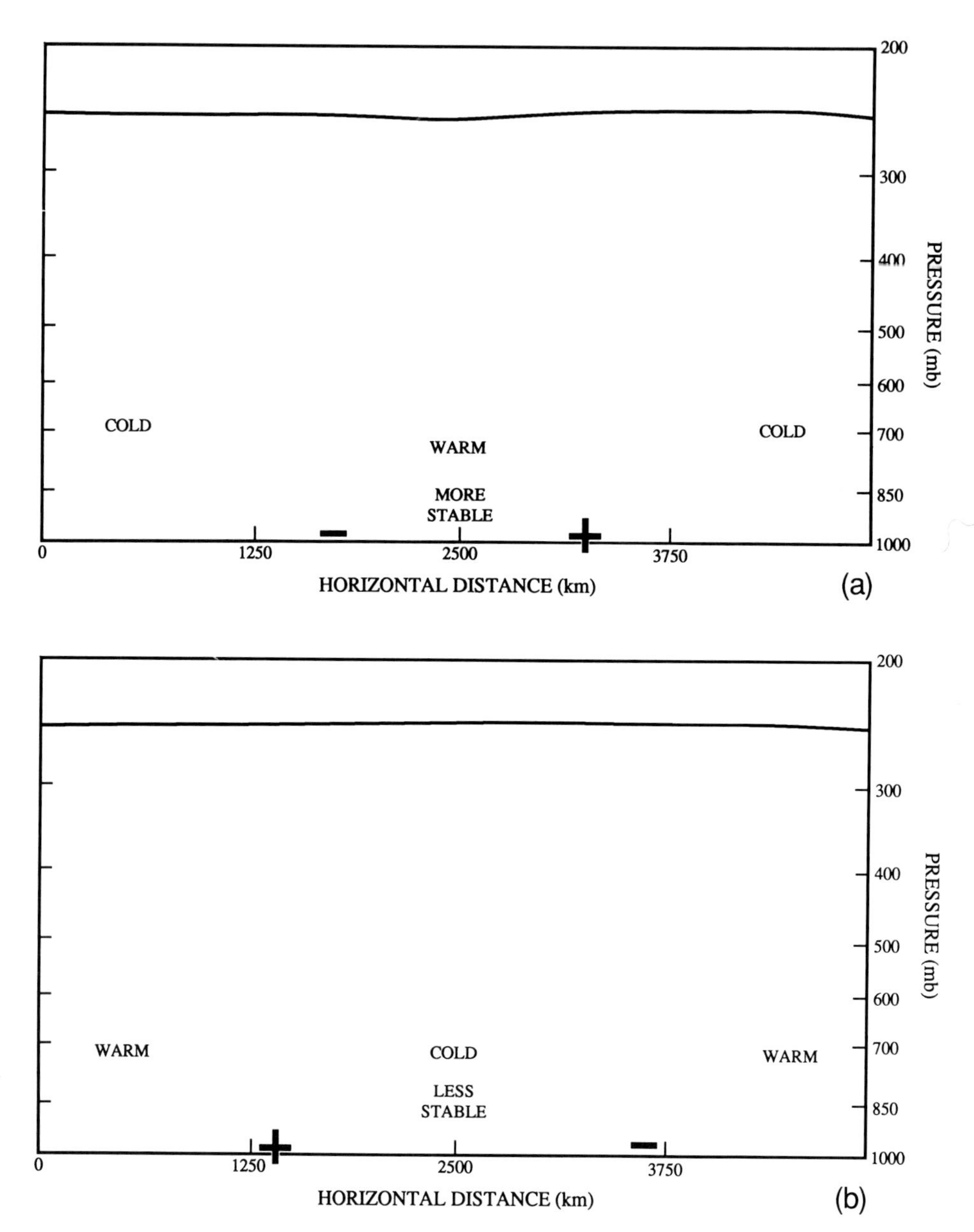

Figure 1.144 As in Fig. 1.143, but for (a) a warm potential temperature anomaly near the ground and (b) a cold potential temperature anomaly near the ground. The vortices depicted in (a) and (b) are strongest at the ground.

induced by upper-level anomalies typically extend all the way down to the ground.

1.9.5. Diagnosis of Vertical Motion Associated with IPV Anomalies Embedded in a Baroclinic Zone

Upper-level IPV anomalies. Suppose that a synoptic-scale, positive, *upper-level* IPV anomaly is embedded in a current of uniform, westerly vertical shear (i.e., in a baroclinic zone) in the midlatitudes in the Northern Hemisphere (Fig. 1.145). This essentially describes a short-wave trough in the baroclinic westerlies: The superposition of the strong westerly current aloft and the cyclonic vortex induced by the positive IPV anomaly results in an open-wave configuration. The westerly current transports the anomaly downstream. In the reference frame of the IPV anomaly the anomaly is stationary, by definition. However, a cyclonic vortex is induced below, where, owing to the vertical shear, there is a uniform easterly current relative to it. Even though the relative easterly current tries to transport the induced cyclonic vortex westward, the induced wind field must remain with its parent anomaly.

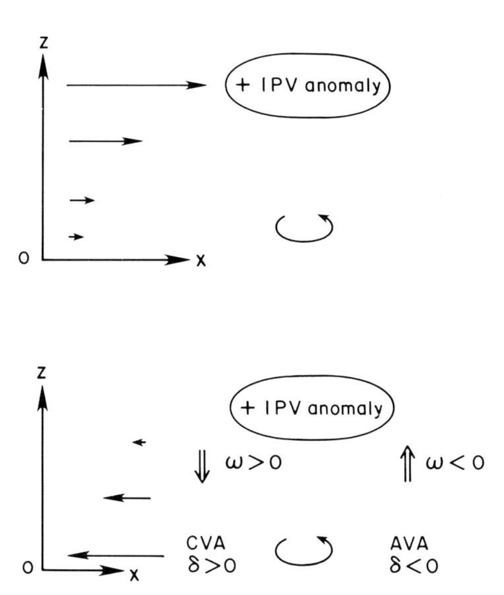

Figure 1.145 IPV anomaly, and vertical motion diagnosed from the vorticity equation: (top) cyclonic IPV anomaly near the tropopause is advected by the basic current, which is one of increasing westerly wind with height; cyclonic vortex is induced below the anomaly. (bottom) In the reference frame of the anomaly, below the anomaly, cyclonic vorticity associated with the induced vortex (which is steady in the reference frame of the anomaly) is advected (CVA) to the west of the anomaly, and anticyclonic vorticity is advected (AVA) to the east of the anomaly. According to the vorticity equation, there must be divergence to the west, and convergence to the east of the induced vortex. From continuity considerations, there must be rising motion to the east and sinking motion to the west of the anomaly.

Consider the quasigeostrophic vorticity equation in pressure coordinates. Suppose that in the reference frame of the positive IPV anomaly, geostrophic absolute vorticity does not change, even below where the induced wind field extends. Then

$$0 = -\mathbf{v}_g \cdot \nabla_p(\zeta_g + f) - \delta f_0. \tag{1.9.41}$$

The region *west* of the IPV anomaly is located downstream from the induced cyclonic vortex in the moving reference frame. The vorticity advection is cyclonic there at low levels $[-\mathbf{v}_g \cdot \nabla_p(\zeta_g + f) > 0]$, and hence from Eq. (1.9.41) we conclude that there must be horizontal divergence ($\delta > 0$); similarly, upstream from the induced cyclonic vortex there is anticyclonic vorticity advection, and convergence. Assuming that $\omega = 0$ at the level ground, we can use the continuity equation to deduce that there must be rising motion east of the IPV anomaly, and sinking motion west of the IPV anomaly. The wheel has just been reinvented!

We can do a similar analysis of the vertical-motion field using the adiabatic form of the thermodynamic equation. In the reference frame of the IPV anomaly, $\partial T/\partial t = 0$, so that

$$0 = -\mathbf{v}_g \cdot \nabla_p T + \omega\sigma \frac{p}{R}. \tag{1.9.42}$$

Below the positive anomaly the air is relatively cool (Fig. 1.146). Therefore in the reference frame of the eastward moving anomaly, below the anomaly, there is warm advection ($-\mathbf{v}_g \cdot \nabla_p T > 0$) to the east and cold advection to the west. From Eq. (1.9.42) we deduce that if the atmosphere is statically stable there must be rising motion ($\omega < 0$) to the east and sinking motion ($\omega > 0$) to the west.

Surface potential-temperature anomalies. Suppose now that there is a synoptic-scale, warm, *surface* potential-temperature anomaly embedded in a current of uniform westerly vertical shear in the midlatitudes (Fig. 1.147). This essentially describes a warm-core surface cyclone embedded in a north–south-

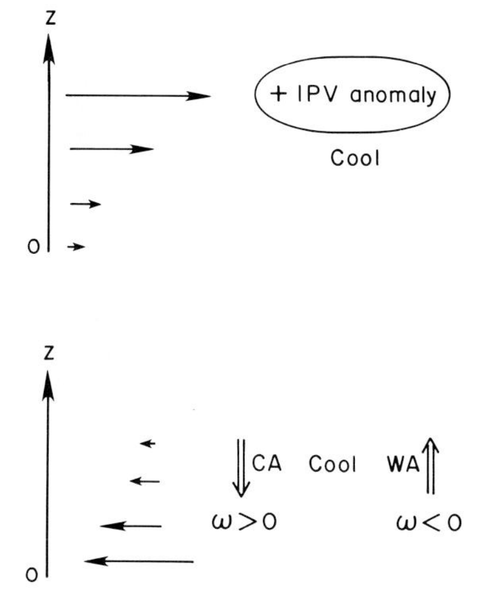

Figure 1.146 IPV anomaly, and vertical motion diagnosed from the thermodynamic equation: (top) cyclonic IPV anomaly near the tropopause is advected by the basic current, which is one of increasing westerly wind with height, as in Fig. 1.145; cool anomaly below the IPV anomaly is associated with the increased static stability in the anomaly. (bottom) In the reference frame of the anomaly, below the anomaly, cold air is advected (CA) to the west of the anomaly, and warm air is advected (WA) to the east of the anomaly. According to the thermodynamic equation, there must be sinking motion to the west, and rising motion to the east.

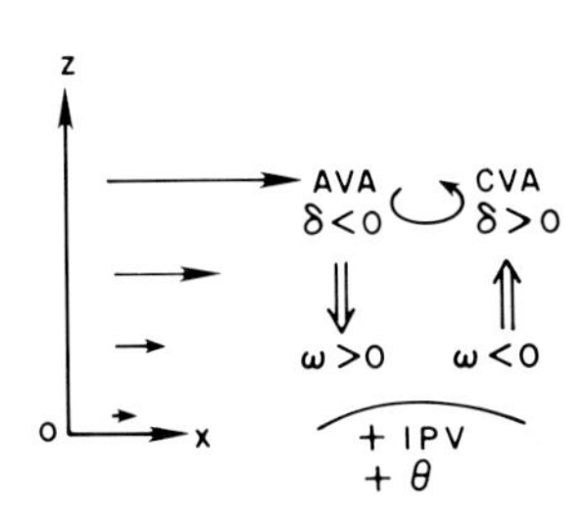

Figure 1.147 Surface positive (cyclonic) IPV anomaly, and vertical motion diagnosed from the vorticity equation: Basic current is as in Fig. 1.145. Cyclonic vortex is induced aloft by surface IPV anomaly, which is stationary, owing to the lack of wind at the surface. There is AVA upstream from the vortex aloft, and CVA downstream from the vortex aloft. According to the vorticity equation, there is convergence to the west of the vortex, and divergence to the east of the vortex. From continuity considerations, there must be sinking motion upstream from the vortex, and rising motion downstream from the vortex.

oriented temperature gradient. If there is no relative flow at the surface, then the westerly current aloft tries to advect the cyclonic vortex induced aloft downstream. However, the induced vortex remains tied to the stationary positive IPV anomaly at the surface. We can use Eq. (1.9.41) to deduce the divergence field. In the upper troposphere there is cyclonic vorticity advection and divergence to the east of the positive surface-IPV anomaly, and anticyclonic vorticity advection and convergence to the west. From continuity consideration there must be rising motion to the east and sinking motion to the west of the surface anomaly.

We now look at the same problem from the perspective of the thermodynamic equation (Fig. 1.148). Aloft, east of the surface anomaly, there is an induced southerly wind component that acts to advect warm air northward. Similarly, aloft, west of the surface anomaly, there is an induced northerly wind component, which acts to advect cold air southward. From Eq. (1.9.42) we deduce that there must be rising motion to the east and sinking motion to the west.

We now see that we can use either the quasigeostrophic vorticity equation or thermodynamic equation *alone* plus "IPV thinking" to diagnose the vertical-motion field. Our results are consistent with those using the quasigeostrophic ω equation: Vertical motion is upward downstream from upper-

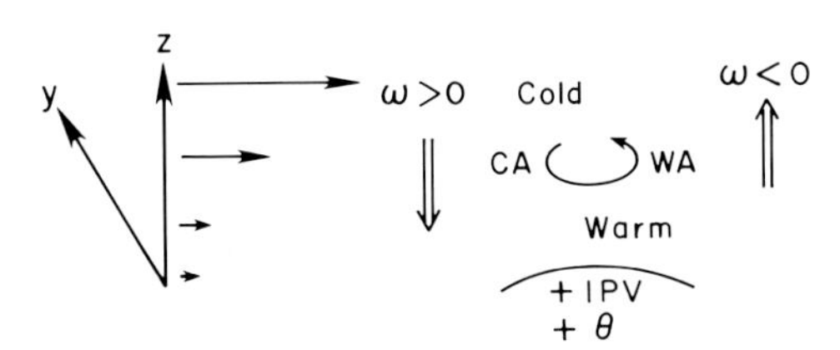

Figure 1.148 Surface positive (cyclonic) IPV anomaly, and vertical motion diagnosed from the thermodynamic equation: Cyclonic vortex is induced aloft by surface IPV anomaly, which is stationary, owing to the lack of wind at the surface. The temperature gradient vector points toward the south, owing to the westerly thermal wind. There is therefore warm advection (WA) east of the vortex, and cold advection (CA) west of the vortex. According to the thermodynamic equation, there must be rising motion to the east of the vortex, and sinking motion to the west of the vortex.

level vorticity maxima, where vorticity advection becomes more cyclonic with height (IPV advection is positive), and downward upstream from upper-level vorticity maxima, where vorticity advection becomes less cyclonic with height (IPV advection is negative). Vertical motion is upward east of surface cyclones embedded in a north–south-oriented temperature gradient, where there is warm advection, and downward to the west where there is cold advection.

1.9.6. Applications of "IPV Thinking" for Time-dependent Dynamical Processes

We will now use "IPV thinking" to explain the motion and development of synoptic-scale systems. This discussion is an alternative to the one that appeared earlier in this chapter. This approach is neither more nor less accurate than the quasigeostrophic diagnosis in pressure coordinates: "IPV thinking" merely provides another perspective. (If so, why do we even bother to discuss it? We do so because at the time of this writing it is in vogue! If *new* insights into the behavior of synoptic-scale systems are found using "IPV thinking," then it will become very important. On the other hand, if no new insights are found, then it may fade away, just as analysis using the pressure-tendency equation faded after the development of quasigeostrophic theory in the late 1940s. Quasigeostrophic theory, however, when it first appeared, explained phenomena not yet understood; it was not an alternative to another successful theory.)[11]

The basic ideas we will employ are as follows:

1. Atmospheric structure is composed of a superposition of upper-level positive and negative IPV anomalies, positive and negative surface potential-temperature anomalies, and a basic current. In conventional parlance, atmospheric structure is a combination of mobile upper-level troughs and ridges, and surface cyclones and anticyclones.
2. Gradient-wind balance holds to a first approximation. We assume that the magnitudes of the anomalies are weak enough so that quasi-geostrophic theory is valid: The diagnostic equation that relates the IPV field to the wind field has a linear operator. Furthermore, the atmosphere is statically stable so that the equation is elliptic. Then the total wind field induced by all the anomalies is the sum of the wind fields induced by each anomaly separately. For typical anomalies, we assume that the induced wind fields extend throughout the depth of the troposphere.
3. Diabatic heating and friction are ignored, so that potential vorticity is conserved. Hence, IPV anomalies are advected on isentropic surfaces and account for local changes in potential vorticity.
4. Each anomaly's induced wind field therefore *changes* the distribution of IPV.
5. The resulting new distribution of IPV is associated with *new* induced wind fields, and so on.

The continued interaction of (4) and (5) is the essence of "IPV thinking" applied to time-dependent dynamical processes.

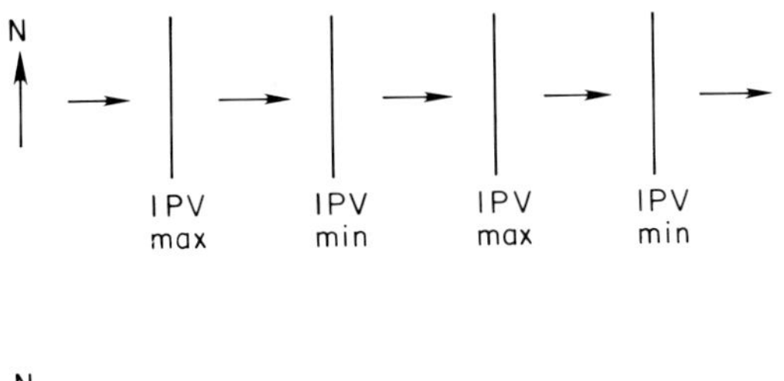

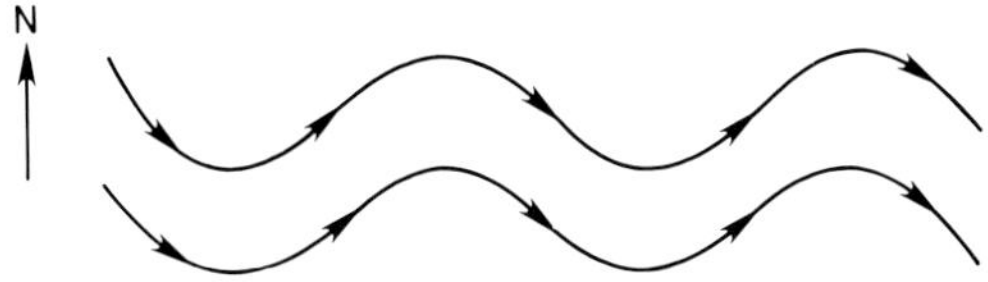

Figure 1.149 A wavetrain in the westerlies (top) from the perspective of IPV thinking and (bottom) depicted as streamlines.

1.9.7. The Motion of Upper-level Troughs and Ridges in the Baroclinic Westerlies

Consider a series of alternating positive and negative upper-level IPV anomalies, in the Northern Hemisphere, which are aligned in the east–west direction, and embedded in a uniform westerly current. Then IPV maxima are co-located with relative vorticity maxima (troughs), and IPV minima are co-located with relative-vorticity minima (ridges), and the wavetrain is simply advected downstream by the basic current (Fig. 1.149).

Now, consider a reference state in which the basic IPV gradient is due to the north–south variation in the Coriolis parameter (Fig. 1.150):

$$\nabla_\theta P = -g\frac{\partial\theta}{\partial p}\beta\hat{\mathbf{j}}. \tag{1.9.43}$$

We now superimpose a series of alternating positive and negative IPV anomalies that are confined within a latitude belt (Fig. 1.151). However, we will not consider any basic current for the moment. The series of IPV anomalies were produced by simply displacing air parcels in an east–west channel alternately toward the pole and equator (compare the IPV distribution in Fig. 1.150 to that in Fig. 1.151.)

Each positive IPV anomaly induces a cyclonic vortex, while each negative IPV anomaly induces an anticyclonic vortex. We in effect have an alternating

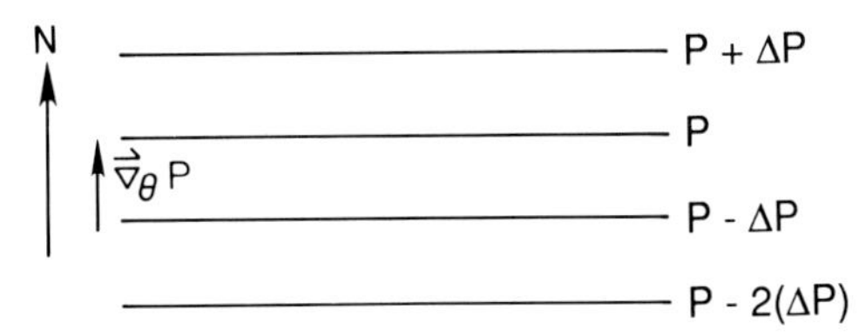

Figure 1.150 Reference state in which the variation in isentropic potential vorticity P (solid lines) is due to the meridional variation in the Coriolis parameter.

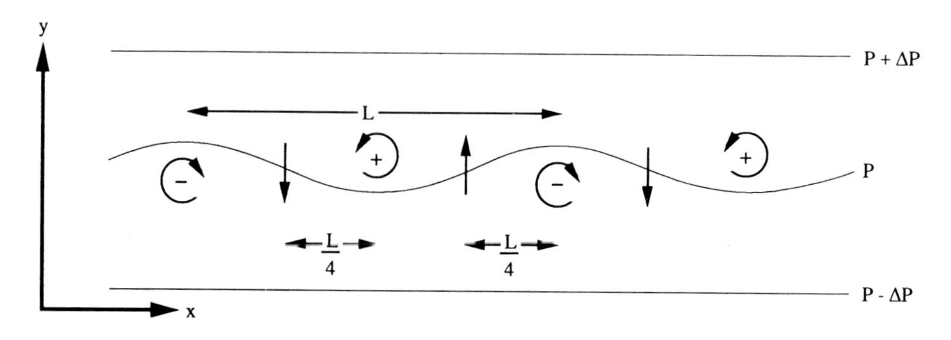

Figure 1.151 Wavetrain of potential vorticity anomalies confined within a latitude belt on a surface of constant potential temperature. IPV (solid lines); centers of positive and negative (cyclonic and anticyclonic, respectively, in the Northern Hemisphere) anomalies located at + and −, respectively. The x and y axes point toward the east and north, respectively. Sense of the induced wind field indicated by curved and straight arrows. One wavelength indicated by L (adapted from Hoskins *et al.*, 1985).

series of cyclones and anticyclones aligned in an east–west direction. The locations of the maximum southerly component of the induced velocity field are one-quarter of a wavelength to the west of the most poleward parcel displacements; the latter are located at the sites of negative IPV anomalies. The locations of the maximum northerly component of the induced velocity field are one-quarter of a wavelength to the west of the most equatorward parcel displacements; the latter are located at the sites of positive IPV anomalies. The induced wind field therefore advects low IPV northward just to the east of IPV maxima, and high IPV southward just to the west of IPV maxima. The net result is that the wave pattern in the IPV field, along with its induced velocity field, *propagates westward.*

Recall that large-scale IPV anomalies induce relatively strong wind fields, while small-scale IPV anomalies induce relatively weak wind fields. Therefore the effect of westward propagation is greatest for long waves, and smallest for short waves. We now add a basic westerly current, which acts to advect the wave pattern *eastward.* We conclude that in short waves the effect of eastward advection is dominant, while in long waves the effect of westward propagation is dominant. Thus, as was found earlier, Rossby waves embedded in the westerlies that have short wavelengths, progress eastward, while Rossby waves that have long wavelengths retrograde.

1.9.8. The Motion of Surface Cyclones and Anticyclones on Level Terrain

To consider the motion of surface cyclones and anticyclones, we use the same arguments used to explain the motion of troughs and ridges, but simply substitute surface-θ anomalies for upper-level IPV anomalies. Consider the reference state for θ at the surface depicted in Fig. 1.152. If we alternately displace air parcels poleward and equatorward within an east–west-oriented

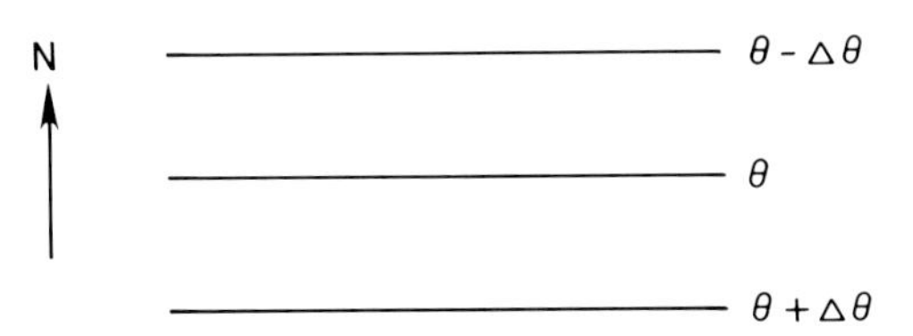

Figure 1.152 Reference state for potential temperature θ (solid lines) at the surface due to the pole-to-equator temperature gradient.

channel (remember, θ is conserved for adiabatic processes), then the configuration shown in Fig. 1.153 is produced. The warm-θ anomalies at the surface induce cyclones, while the cold-θ anomalies induce anticyclones. Since mass continuity requires that $\omega = 0$ at the surface, which is level, potential temperature changes locally only due to advection.

Maximum cold advection occurs one-quarter of a wavelength east of cold-θ anomalies, while maximum warm advection occurs one-quarter of a wavelength east of warm-θ anomalies. Therefore the wave in the potential temperature field, along with the induced surface cyclones and anticyclones, propagate eastward. Thus, as with isobaric, quasigeostrophic analysis, surface cyclones moves from regions of cold advection to regions of warm advection, while anticyclones move from regions of warm advection to regions of cold advection.

1.9.9. The Effects of Orography on the Motion of Surface Cyclones and Anticyclones

Suppose that the statically stable reference state, in which pressure is a function only of potential temperature and $\zeta_\theta = 0$, is situated over terrain that slopes upward to the west as in Fig. 1.154. Suppose air parcels near the surface are brought alternately upslope from the east and downslope from the west along north–south-oriented lines of constant elevation. In doing so, we are careful not to generate any static instability in the regions of downslope motion. The θ field that is thus produced is sketched in Fig. 1.155. It looks like the θ-field in Fig. 1.153 rotated by 90°.

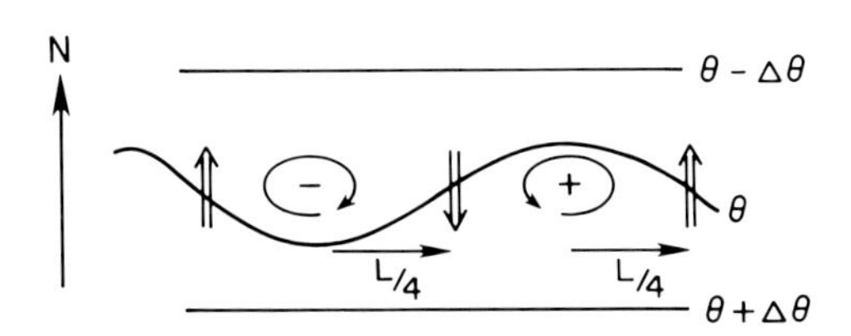

Figure 1.153 The reference state depicted in Fig. 1.152, disturbed by alternating northward and southward displacements. Cyclonic (+) and anticyclonic (−) vortices are induced by the positive and negative potential temperature anomalies, respectively. Maximum temperature advection occurs one-quarter of a wavelength from the centers of the temperature extrema.

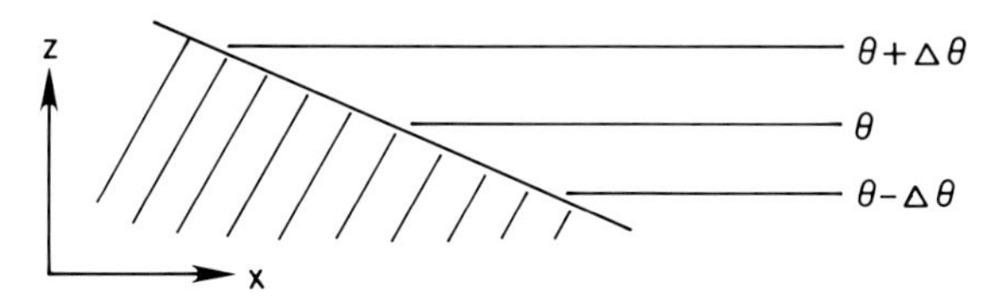

Figure 1.154 Vertical cross section of a reference state for the potential temperature θ (solid lines) in which there is sloping topography.

Cold anomalies, brought about by upslope motion (movement of lower-valued isentropes from below up to high elevation), induce anticyclonic vortices, while warm anomalies, brought about by downslope motion (movement of high-valued isentropes from above to lower elevation), induce cyclonic vortices. It is easy to see that the induced wind field causes the cyclones and anticyclones to propagate southward. Thus, as we showed earlier in this chapter using isobaric, quasigeostrophic theory, surface cyclones and anticyclones tend to move with higher terrain on their right (in the absence of other effects) in the Northern Hemisphere.

If, however, the terrain has a component of slope in the north–south direction, then advection of f must be considered also. In other words, for example, if upslope or downslope motions are induced that have a meridional component, then one must take into account the advection of IPV associated with f. For short-wavelength disturbances, this effect is small. However, if the wavelength is long, it must also be considered.

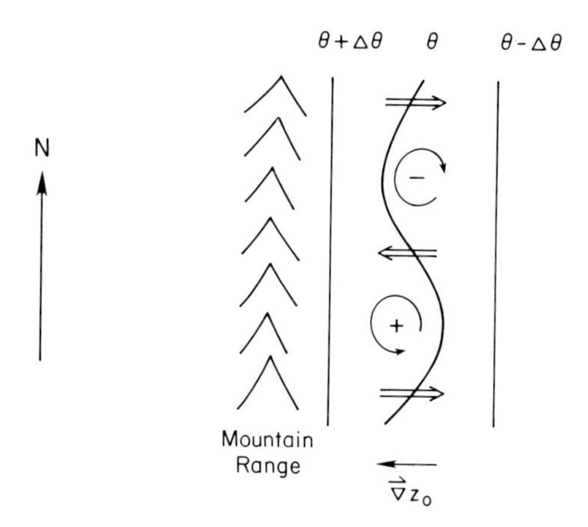

Figure 1.155 Reference state at the surface for the potential temperature along a sloping surface, which is disturbed by alternating upslope and downslope motions. Cyclonic (+) and anticyclonic (−) vortices are induced by the positive and negative potential-temperature anomalies. z_0 is the elevation above sea level. Arrows indicate sense of maximum upslope and downslope motions induced by the potential-temperature anomalies.

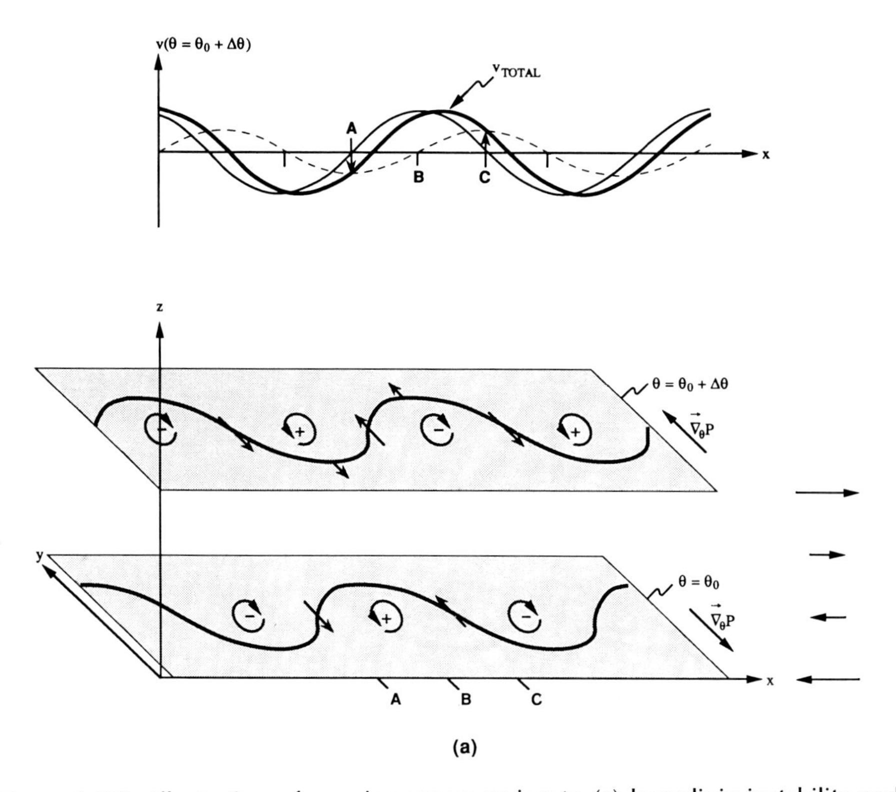

Figure 1.156 Illustration of growing waves owing to (a) baroclinic instability and (b) barotropic instability using IPV thinking. In (a) contours of potential vorticity on an upper and a lower surface of constant potential temperature are shown by a thick solid line. The sign of the potential vorticity anomalies are indicated, along with the sense of the induced wind fields. The wind field induced by the anomalies at key locations is indicated schematically by the vectors; the wind field in the upper level induced by the anomalies in the lower level is indicated at the locations of the farthest northward and southward displacements of the potential vorticity contour. The vertical wind profile relative to the resulting wave pattern is indicated qualitatively at the right. The direction of the potential vorticity gradients on both the upper and lower potential temperature surfaces is also indicated at the right. The top section shows the meridional component of the wind at the upper level induced by the potential vorticity anomalies at the upper level (thin solid line), the meridional component of the wind at the upper level induced by the anomalies at the lower level (dashed line), and the total induced meridional wind at the upper level, v_{TOTAL} (thick solid line). Locations A, B, and C represent respectively longitudes at which the potential vorticity contour on the upper surface is farthest south, the potential vorticity contour is undisturbed from its mean position, and the potential vorticity contour is farthest north. In (b) contours of potential vorticity on an isentropic surface are as shown. The sign of potential vorticity anomalies and their induced wind fields are shown as in (a), except that the anomalies are laid out at different latitudes rather than at different levels. In both (a) and (b) the potential vorticity contours at their locations of farthest northward and southward displacements are further displaced to the north and south.

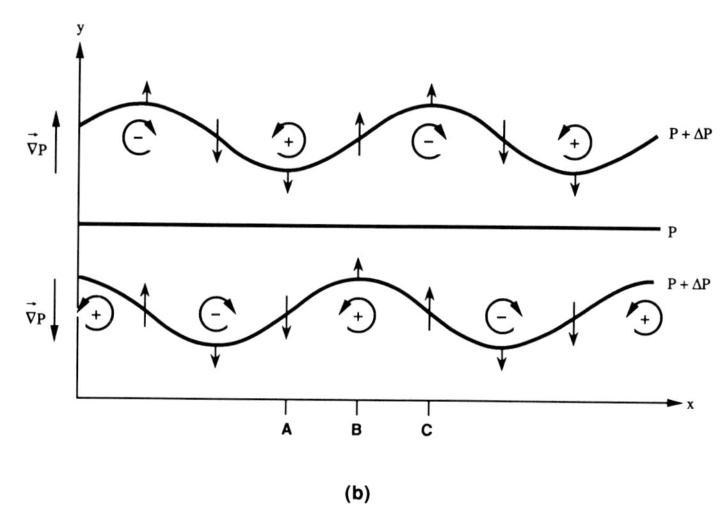

Figure 1.156 *(cont.)*

1.9.10. The Formation of Upper-Level Systems

Baroclinic instability. How do upper-level systems form from the perspective of "IPV thinking?" We mentioned earlier in this chapter that they can form as a result of baroclinic or barotropic instability.

Suppose that in each layer of a two-layer atmosphere there is an alternating train of positive and negative IPV anomalies confined within a channel. Furthermore, suppose that the upper wavetrain is shifted slightly to the west of the lower wavetrain: This is essentially a wavetrain in the baroclinic westerlies that tilts westward with height. In Fig. 1.156(a) we view the IPV and flow patterns in a reference frame moving along with the wavetrain, which moves at a speed intermediate between that of the mean flow in the upper layer, and the mean flow in the lower layer. In the top layer, IPV increases toward the north; in the bottom layer, IPV increases toward the south. How is this realized in the real atmosphere? The IPV gradient in the top layer is due in part to the meridional gradient of f. An isentrope that at high levels at low latitudes is in the troposphere, may be in the stratosphere at high latitudes, so that static stability increases toward the north on that isentrope. Finally, suppose that in both layers warm potential temperatures are found at low latitudes, while cold potential temperatures are found at high latitudes. If the temperature gradient is concentrated in the center of the channel, then according to the thermal-wind relation there must be westerly thermal-wind shear, with a westerly jet in the upper layer, at the center of the channel, flanked by cyclonic shear on the poleward side of the channel and anticyclonic shear on the equatorward side. Thus, there is a poleward gradient of Earth's vorticity, relative vorticity, and static stability, all of which contribute to a poleward IPV gradient in the upper layer. In the bottom layer there is an equatorward gradient in IPV as a result of the equatorward temperature gradient: According to the thermal-wind relation, there must be a weakness in westerly (or even easterly) winds at the center of the channel in the lower layer. There is therefore anticyclonic shear on the poleward side of the channel, flanked by cyclonic shear on the equatorward side. The meridional gradient in f, however, acts in the opposite direction. [It is not as easy to visualize the contribution of the meridional variation in static stability. If the potential-temperature surfaces in the reference state are quasihorizontal, then the "warm" ("cold") side is characterized by high (low) static stability.]

Suppose that at the interface between the two layers, there is no current. In order that the vertical shear be westerly, there must be a westerly current in the upper layer and an easterly current in the lower layer. The upper-level wavetrain's motion owing to propagation is toward the west. Because the IPV gradient in the lower layer is reversed, the lower-level wavetrain's motion owing to propagation is toward the east. Advection by the basic current, however, moves the wavetrain toward the east in the upper layer, and toward the west in the lower layer. If the wavelength is relatively short, the effects of advection overwhelm the effects of propagation, and the upper-level and lower-level wavetrains move in opposite directions.

However, the upper-level IPV anomalies induce vortices in the lower layer, which affect the distribution of IPV in the lower layer, while the lower-level IPV anomalies induce vortices in the upper layer, which affect the distribution of IPV in the upper layer. The sum of the wind fields in the top layer induced by the IPV anomalies in the top layer *and* in the bottom layer results in a *greater* poleward component of motion just west of the IPV minima and a *greater* equatorward component of motion west of the IPV maxima than would occur in the absence of the wind field induced by the lower layer. Thus, the rate of westward propagation is increased, so that the net rate of eastward motion is reduced. In a similar manner, the sum of the wind fields in the bottom layer induced by the IPV anomalies in the bottom layer *and* in the top layer results in a *greater* poleward component of motion just east of the IPV maxima and a greater equatorward component of motion east of IPV minima than would occur in the absence of the wind field induced by the upper layer. Thus, the rate of eastward propagation is increased below, so that the net rate of westward motion of the lower wavetrain is reduced. Therefore the wavetrains try to "lock" onto (phase with) each other: Each prevents the other from racing off in the opposite direction. (The reader should show, as an exercise, that if the IPV gradient in the bottom layer were in the same direction as the IPV gradient in the top layer, then the wavetrains do not lock onto each other.)

Now suppose that each wavetrain were shifted more downstream, so that there was less tilt in the vertical; that is, suppose the wavetrains were more in phase. Then the effect of the wind field induced by the lower wavetrain on the upper wavetrain, and the effect of the wind field induced by the upper wavetrain on the lower wavetrain would act to increase the individual propagation speeds. Propagation could then be strong enough to overwhelm the effects of advection, and the wavetrains would move to a configuration in which they were again tilted toward the west with height. A similar argument can be used to show that if the wavetrains were shifted more upstream, so that they tilted even more toward the west with height, then the effects of propagation would decrease, and advection by the basic current would restore the wavetrains to their original phase. Therefore there is an optimal phase difference for which the two wavetrains can be in phase with each other: As in the fairy tale "Goldilocks and the Three Bears," one is too much, one too little, and one just right. For very short wavelengths, however, propagation could never be significant if the basic current were strong, and therefore the wavetrains could not lock onto each other at all. On the other hand, for very long wavelengths, propagation could always overwhelm the effects of advection, and therefore the wavetrains could still not lock onto each other. Thus, for a given vertical shear, the two wavetrains can be in phase with each other only for a certain range of wavelengths.

If the wavelength is within the range for which the wavetrains can lock onto each other, then the total induced-velocity pattern is *less* than one-quarter of a wavelength out of phase with the displacement pattern. Thus, at the locations at which the IPV contours are displaced farthest north, the IPV contours are subjected to even more northward displacement; at the locations at which the IPV contours are displaced farthest south, the IPV contours are subjected to even more southward displacement. In other words, the waves grow in amplitude. We conclude then that for a certain range of wavelengths, which depends upon the vertical shear, troughs and ridges will grow in amplitude if they lean westward with height. This is what we found with our analysis of baroclinic instability using Sanders' quasigeostrophic analytic model (and Holton's two-layer quasigeostrophic model). (On the other hand, if the IPV gradient in the bottom layer were in the same direction as the IPV gradient in the top layer, there would not be any growth in amplitude.)

The reader should now, as an exercise, convince himself (or herself), using "IPV thinking, " that if a wavetrain leans toward the east with height, then for a certain range of wavelengths, the two wavetrains can lock onto each other and each will *decay* in amplitude. However, if each wavetrain were shifted slightly upstream or downstream, then the waves would continue to shift and eventually lean toward the west with height.

How does static stability affect baroclinic instability from the "IPV-thinking" perspective? From Eq. (1.9.33) we see that for a given wavelength, the depth of a layer in isentropic coordinates ($\Delta\theta$) affected by an IPV anomaly *increases* as the static stability parameter decreases. Since the static stability parameter in isentropic coordinates (see Eq. (1.9.34)) is inversely proportional to the static stability parameter in pressure coordinates, $\Delta\theta$ *decreases* as "static stability" decreases. However, since θ surfaces are farther apart when the static stability is low, the *actual* depth *increases* as the static stability decreases. (More simply put, since $H \sim (fL/N)$, H increases as N decreases.) Therefore the effect of propagation is enhanced at low static stabilities, because the wind field induced by a wavetrain at one level in the other level is enhanced. This implies that even though the induced wind field may be weak for typical static stabilities and short wavelengths, it may be relatively strong if the static stability is low enough; therefore it might actually be possible for short-wave wavetrains (which could not be in phase with each other at typical static stabilities) to lock onto each other. Furthermore, for very long wavelengths, the induced wind field is also stronger if the static stability is lower. The induced wind field may become so strong, in fact, that long-wave wavetrains that can still lock onto each other at typical static stability cannot do so at low static stability, because the effect of propagation is so strong. We conclude, then, that the effect of low static stability is to reduce the scales at which baroclinic instability occurs. Therefore we would expect to find shorter-wavelength disturbances in an environment of low static stability. (It was noted earlier that over the ocean during the winter in the presence of low static stability, surface cyclones, which form in response to upper-level troughs, are often very small scale.)

What are the effects of vertical shear (i.e., baroclinicity)? If vertical shear were increased, then the upper and lower waves would move in opposite directions, because the effects of propagation would be overwhelmed by advection. The waves could lock on and amplify, however, if the wavelength were longer, because propagation would be stronger for longer wavelengths.

Barotropic instability. To consider the possibility of barotropic instability, that is, growth of a disturbance with energy coming from the mean horizontal shear (rather than the mean vertical shear as in baroclinic instability), we can use Fig. 1.156(b). In this case we have two wavetrains side by side, with a reversal in the horizontal IPV gradient occurring at the interface between the two wavetrains. Thus, for positive values of IPV, there is a relative minimum in IPV in between the two wavetrains.

Since the meridional variation in f contributes to an increase in f with latitude, the decrease in IPV with latitude in the equatorward region may be realized if there is a zone of strong anticyclonic shear between the two regions. In nature, the winds may be westerly at all latitudes, with a jet north of the interface between the two regions (the Polar-Front jet?), and an increase in westerlies farther to the south (the subtropical jet?).

If the wavetrains tilt westward with latitude, then we can employ the identical analysis we used in the baroclinic instability case of westward tilt with height to find that this configuration of negatively tilted troughs and ridges is barotropically unstable for a range of wavelengths. In a similar manner, we can see that if the wavetrains tilt eastward with height, then the positively tilted troughs and ridges are barotropically stable for a range of wavelengths.

Nonlinear effects. When the troughs and ridges grow beyond a certain amplitude, nonlinear effects become important. In addition, the presence of poleward and equatorward boundaries limit the extent to which the wavetrains can grow. (In the real atmosphere, however, solid boundaries of vertical walls do not exist. However, we can regard regions into which the waves may not propagate laterally as barriers to wave growth.) The wave's shape becomes distorted as IPV surfaces hit the side "walls" of the lateral boundaries (Fig. 1.157), just as ocean waves move toward shore and break when they encounter the ocean bottom. As the wave grows further in amplitude, the IPV contours may wrap up, and deformation in the wind field reduces the scale of the IPV

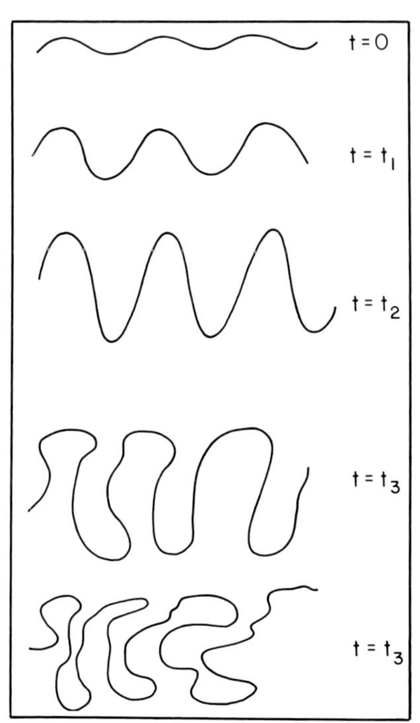

Figure 1.157 Idealized schematic of one IPV contour in a growing baroclinic wave as a function of time t. The contour grows in latitudinal extent between $t = 0$ and t_2; after t_2 its latitudinal extent is limited by "walls" to the north and the south. Its shape then becomes distorted, and the curve tends to fill more and more of the space, so that the distance between adjacent contours decreases.

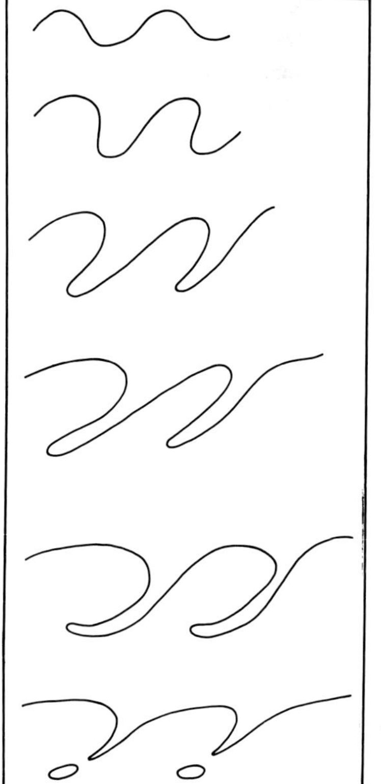

Figure 1.158 Illustration of the time evolution (from top to bottom) of an idealized baroclinic wave in a westerly current that increases in strength towards the north. Line of constant IPV (solid line).

anomalies. This represents a "downward cascade in scale." But small-scale anomalies induce weaker wind fields, and hence the growth process slows down and is self-limiting. This can also occur when an IPV anomaly is advected into a diffluent region, so that the anomaly is stretched in the meridional direction, and foreshortened in the zonal direction.

Another example, which is interesting, is that of a wavetrain that extends equatorward into a region of anticyclonic horizontal shear, as is found just south of the baroclinic westerlies. The wave takes on a northeast–southwest-oriented tilt, owing to differential advection associated with shearing deformation, and the gradient in IPV can become very large (Fig. 1.158). Eventually, the southern part of the wave may cut off, owing to turbulent mixing.

Finally, we note that during surface cyclogenesis the isotherms may wrap up so that the scale of potential-temperature anomalies becomes small.

1.9.11. The Formation of Surface Systems

Suppose that an upper-level, positive IPV anomaly is advected over a zone of strong, low-level, equatorward temperature gradient. This essentially describes an upper-level trough moving over a zonally oriented surface cold front. The anomaly induces a cyclonic vortex down to the frontal zone. Warm advection occurs east of the induced cyclone, and cold advection occurs west of the induced cyclone (Fig. 1.159). Thus, a positive potential-temperature anomaly forms just east of the surface cyclone, while a negative potential-temperature anomaly forms just west of the surface cyclone. The former is associated with its own induced cyclonic vortex; the latter is associated with its own induced anticyclonic vortex.

The combined effect of these two induced vortices is a strong equatorward wind component under the upper-level IPV anomaly. This induced equatorward wind component is felt aloft, and the IPV contours are advected farther equatorward, just where they are displaced farthest equatorward anyway. Thus the wave aloft grows in amplitude; furthermore, its eastward movement is slowed down, since the IPV contours just west of the anomaly are advected equatorward and the IPV contours just east of the anomaly are advected poleward. This leads to an even stronger surface cyclone and anticyclone, and so on, much to the delight of bomb fans!

We note that isobaric, quasigeostrophic theory also describes how surface cyclogenesis occurs when an upper-level trough moves over a low-level baroclinic zone. However, in the "IPV-thinking" analysis, we never needed to consider vertical motions explicitly. In the isobaric, quasigeostrophic explanation, it *was* necessary to consider explicitly the effects of vertical motions, which thermodynamically tend to counteract the effects of advection.

If the static stability is very low, for example, if the air is saturated and the lapse rate is nearly moist-adiabatic, or if the air is unsaturated and cold air is advected over warm water, then the effects of the induced wind fields extend through a deeper layer. Hence there is stronger temperature advection at the surface, and a stronger feedback to the IPV anomaly at upper levels: Cyclogenesis proceeds more rapidly, perhaps even "explosively." Earlier in

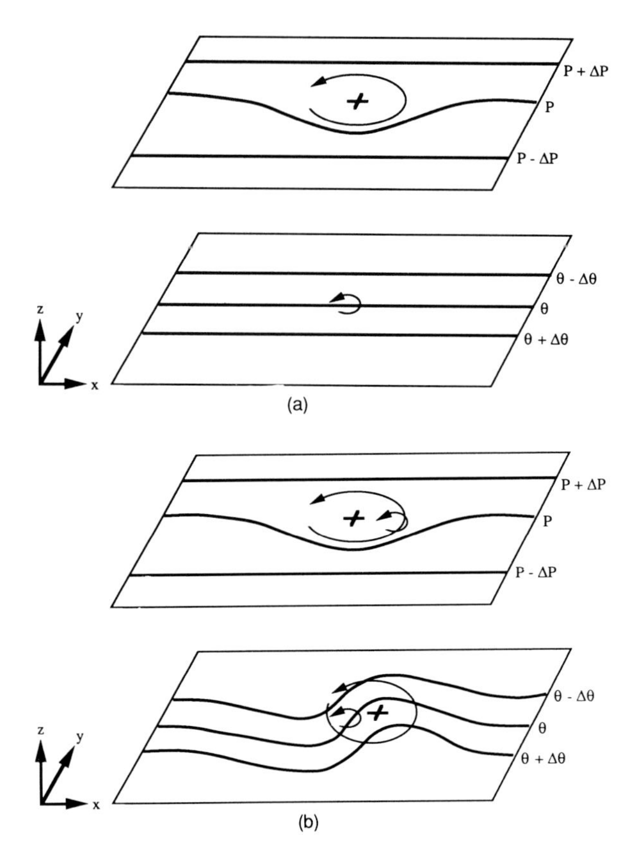

Figure 1.159 Illustration of the growth of an idealized wave-cyclone as an upper-level cyclonic IPV anomaly becomes situated over a low-level baroclinic zone. The top plane is a surface of constant potential temperature characteristic of the upper troposphere, with lines of constant IPV given by thick solid lines. The bottom plane is the ground surface with lines of constant potential temperature (θ) given by thick solid lines. The x, y, and z directions point toward the east, north, and upward, respectively. In (a) the upper-level cyclonic IPV anomaly, indicated by +, has just arrived over a region of significant low-level baroclinicity. The circulation induced by the anomaly is indicated by thin solid arrows, with the radius of the circulation proportional to the strength of the circulation. In (b) the warm anomaly at the surface east of the circulation (indicated by +), created by the warm advection pattern associated with the circulation, induces its own cyclonic circulation, which makes itself felt just to east of the upper-level IPV anomaly (adapted from Hoskins *et al.*, 1985).

this chapter it was also found, using isobaric, quasigeostrophic theory, that low static stability enhances the atmosphere's response to the forcing functions.

1.9.12. Lateral and Vertical Propagation of Waves at Upper Levels

Consider a wavetrain confined to a channel (area of limited latitudinal extent) in the baroclinic westerlies at upper levels (Fig. 1.160). This may be considered an alternating series of positive and negative IPV anomalies in a reference state in which IPV increases northward, owing to the increase of f with latitude. Suppose that the wavelength is short enough and the basic westerly current strong enough so that the wave moves eastward.

The cyclonic and anticyclonic vortices induced by the IPV anomalies disturb the field of IPV both to the north and south to some extent (the elliptic operator in the diagnostic equation requires "global" knowledge of the wind field). For example, low values of IPV are advected poleward east of positive IPV anomalies and west of negative IPV anomalies; high values of IPV are advected equatorward east of negative IPV anomalies and west of positive IPV

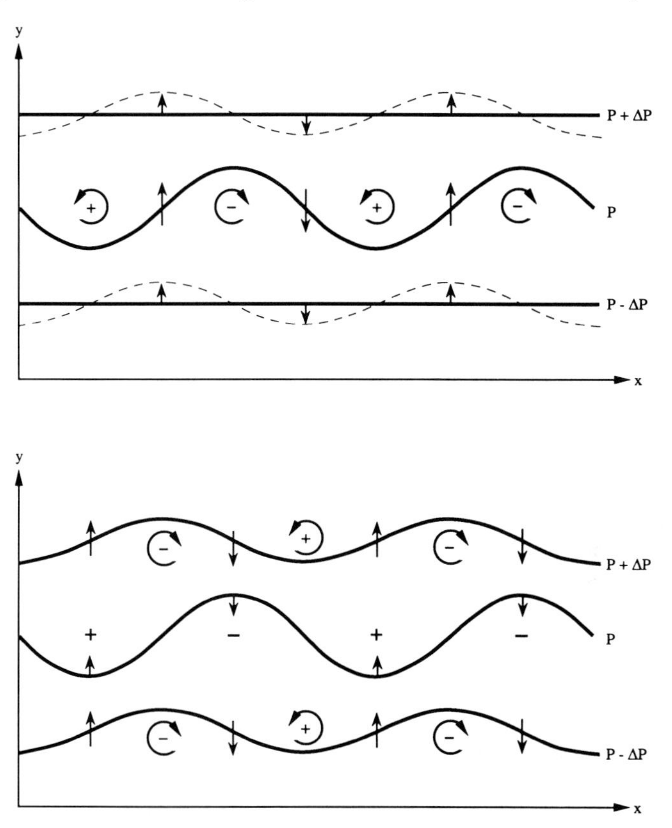

Figure 1.160 Illustration of horizontal Rossby-wave propagation. Pattern of potential vorticity anomalies and the sense of the induced circulations on an isentropic surface shown as in Fig. 1.156. At the top, the contours $P+\Delta P$ and $P-\Delta P$ are perturbed by the induced wind field as shown by the dashed contours. The new anomalies associated with the dashed potential vorticity contours are indicated in the bottom section, along with the wind field induced by these anomalies. These new anomalies induce a wind field that acts to suppress the amplitude of the original wave shown at the top. To consider vertical Rossby-wave propagation, simply change the y axis to the z axis, and let the upper, middle, and lower contours represent respectively the horizontal distribution of potential vorticity at an upper, intermediate, and a lower potential temperature surface.

anomalies. The net result is that wavetrains are induced both to the south and north, which lag (to the west) the original wavetrain.

The induced wavetrains to the north and south induce their own wind fields, which has the effect of dampening the original wavetrain. For example, in Fig. 1.160, the northerly winds induced west of positive IPV anomalies and east of negative IPV anomalies, in the northern wavetrain, move regions of large poleward displacement of the IPV contours equatorward; poleward-flowing winds induced west of negative IPV anomalies and east of positive IPV anomalies move regions of large equatorward displacement of the IPV contours poleward. In other words, the induced wavetrains to the north and south of the original wavetrain leave little behind. The new wavetrains in turn induce new wavetrains to the north and south, and so on. The net results of all this are northward- and southward-propagating wavetrains. Thus, "IPV thinking" can also explain the meridional motions of trough and ridges that are due to propagation.

The ability of a wavetrain to propagate laterally is a function of the horizontal IPV gradient and the wavelength: Long waves should propagate laterally more rapidly than short waves because they induce stronger wind fields. For a given wavelength, lateral propagation is more rapid in regions of stronger horizontal IPV gradients. Since IPV gradients are usually strongest near the tropopause, lateral propagation should be most efficient there, especially for long waves. Thus the tropopause can act as a "waveguide" for the longer waves.

If the ordinate in Fig. 1.160 is replaced by z, then a wavetrain will induce new wavetrains above and below that lag (to the west) the original wavetrain. Again, the original wavetrain will dampen, while new wavetrains are induced above and below. This vertical propagation mechanism works as long as there is an IPV gradient. If a level is reached at which the IPV gradient reverses (i.e., where there is no horizontal IPV gradient), then the waves become "trapped." The vertical-propagation mechanism can be used to describe how waves excited in the stratosphere might in some cases have an effect on tropospheric weather systems.

1.9.13. The Effects of Diabatic Heating and Friction on Upper-level IPV Anomalies

How do upper-level IPV anomalies decay? In order to answer this question, we need to account for the effects of diabatic heating and of friction. Advection can only move the anomalies from one place to another. Advection, for example, may, under some unusual circumstances, slowly move a cut-off, positive IPV anomaly poleward toward the stratospheric reservoir whence it came. Similarly, a cut-off negative IPV anomaly may be advected equatorward away from the stratospheric reservoir. However, the anomalies themselves have not changed in value.

The cyclonic vortex induced by an upper-level, positive IPV anomaly is destroyed *near the ground* by boundary-layer friction. Thus, cyclonic thermal shear is produced in the boundary layer, since the induced cyclone increases in strength with height there. This thermal shear coincides with a cold potential-temperature anomaly and a negative IPV anomaly. The negative IPV anomaly induces an anticyclonic vortex whose effects extend up to upper levels. However, the induced anticyclone there is not strong enough to cancel out the upper-level cyclone induced by the upper-level, positive IPV anomaly. Diabatic heating must play a vital role in ultimately destroying the upper-level IPV anomaly because it is the only other mechanism left that can.

The anticyclonic vortex induced by an upper-level, negative IPV anomaly is also destroyed by friction in the boundary layer. It is left as an exercise for the reader to show that a cyclonic vortex is induced in the boundary layer, as a result of friction, and is not strong enough to cancel out the anticyclone aloft. Again, we are led to the conclusion that diabatic heating must play a vital role in ultimately destroying the upper-level IPV anomaly.

To make a rough, quantitative estimate of the effects of diabatic heating, consider the equations for absolute vorticity and continuity in isentropic coordinates, which are expressed as follows:

$$\frac{\partial \eta}{\partial t} + \mathbf{v} \cdot \nabla_\theta \eta + \frac{D\theta}{Dt}\frac{\partial \eta}{\partial \theta} + \eta\, \nabla_\theta \cdot \mathbf{v} = \mathbf{k} \cdot \frac{\partial \mathbf{v}}{\partial \theta} \times \nabla\left(\frac{D\theta}{Dt}\right) + \mathbf{k} \cdot \nabla \times \mathbf{F}_{\text{fric}} \qquad (1.9.44)$$

$$\frac{\partial}{\partial t}\left(\frac{\partial \theta}{\partial p}\right) + \mathbf{v} \cdot \nabla_\theta\left(\frac{\partial \theta}{\partial p}\right) - \frac{\partial \theta}{\partial p}\nabla_\theta \cdot \mathbf{v} = \frac{\partial \theta}{\partial p}\frac{\partial}{\partial \theta}\left(\frac{D\theta}{Dt}\right) - \frac{D\theta}{Dt}\frac{\partial}{\partial \theta}\left(\frac{\partial \theta}{\partial p}\right), \qquad (1.9.45)$$

where $\eta = \zeta_\theta + f = \zeta_{a\theta}$. Note that two terms involving diabatic heating appear in Eq. (1.9.45), the *diabatic* form of the continuity equation. Multiplying Eq. (1.9.44) by $-g\,\partial\theta/\partial p$, and adding Eq. (1.9.45) multiplied by $-g\eta$ to it, we find that

$$\frac{D}{Dt}\left(-g(\zeta_\theta + f)\frac{\partial \theta}{\partial p}\right) = -g\frac{\partial \theta}{\partial p}(\zeta_\theta + f)\frac{\partial}{\partial \theta}\left(\frac{D\theta}{Dt}\right) - g\frac{\partial \theta}{\partial p}\mathbf{k} \cdot \frac{\partial \mathbf{v}}{\partial \theta} \times \nabla\left(\frac{D\theta}{Dt}\right)$$

$$- g\frac{\partial \theta}{\partial p}\mathbf{k} \cdot \nabla \times \mathbf{F}_{\text{fric}}. \qquad (1.9.46)$$

Making the substitutions $P = -g(\zeta_\theta + f)\,\partial\theta/\partial p$ and $\sigma^* = -(1/g)\,\partial p/\partial\theta$ into Eq. (1.9.46), and combining the first two terms on the right-hand side, we find that Eq. (1.9.46) can be written as

$$\frac{DP}{Dt} = \frac{1}{\sigma^*}(\nabla \times \mathbf{v} + f\mathbf{k}) \cdot \nabla\left(\frac{D\theta}{Dt}\right) + \frac{1}{\sigma^*}\mathbf{k} \cdot \nabla \times \mathbf{F}_{\text{fric}}. \qquad (1.9.47)$$

Neglecting the horizontal component of absolute vorticity in Eq. (1.9.47) for the sake of simplicity, and using Eq. (1.9.14), we can express the potential-vorticity equation as

$$\frac{DP}{Dt} = P\frac{\partial}{\partial \theta}\left(\frac{D\theta}{Dt}\right) + \frac{1}{\sigma^*}\mathbf{k} \cdot \nabla \times \mathbf{F}_{\text{fric}}. \qquad (1.9.48)$$

In the absence of friction, (1.9.48) may be written as

$$\left(\frac{\partial}{\partial t} + \mathbf{v} \cdot \nabla_{\theta}\right) P = P^2 \frac{\partial}{\partial \theta}\left(\frac{D\theta}{Dt}\frac{1}{P}\right) \tag{1.9.49}$$

where

$$\mathbf{v} \cdot \nabla_{\theta} = u\left(\frac{\partial}{\partial x}\right)_{\theta} + v\left(\frac{\partial}{\partial y}\right)_{\theta}. \tag{1.9.50}$$

We now estimate the right-hand side of (1.9.49).

Since P increases with height (and θ) near the tropopause, and the contribution to $D\theta/Dt$ from latent-heat release decreases with height in the upper troposphere, we infer that the right-hand side of Eq. (1.9.49) is negative: In other words, the diabatic heating term can in fact act as a destroyer of IPV. Refer to the cold, upper-level low (positive IPV anomaly) shown in Fig. 1.140. At 320 K, in the stratosphere, $(D\theta/Dt)(1/P) \approx 0$ because $D\theta/Dt$ is small in magnitude (and negative in sign) and P (not shown in Fig. 1.140) is very large; at 310 K, $P \sim 1$ PVU (not shown in Fig. 1.140). Suppose that $D\theta/Dt = 2.5\ \mathrm{K\,d^{-1}}$ at 310 K owing to latent-heat release. Then at 315 K, where $P \sim 2$ PVU,

$$P^2 \frac{\partial}{\partial \theta}\left(\frac{D\theta}{Dt}\frac{1}{P}\right) \sim 4\ \mathrm{PVU}^2 \frac{(-2.5\ \mathrm{K\,d^{-1}})(1/1\ \mathrm{PVU})}{10\ \mathrm{K}} \sim -1\ \mathrm{PVU\,d^{-1}}. \tag{1.9.51}$$

Therefore the upward decrease of latent-heat release near the tropopause (perhaps augmented by radiational cooling at cloud top) can destroy or substantially reduce the strength of an upper-level, positive IPV anomaly. However, well inland, during the winter, when the atmosphere is relatively dry, diabatic heating rates owing to latent-heat release are much less, and upper-level, positive IPV anomalies (cut-off cyclones) can last for a relatively long time.

Since diabatic heating is usually greatest just below the middle troposphere, where the combined effect of upward motion and water-vapor content is greatest, IPV is *created* below, where P does not vary too much vertically, and $D\theta/Dt$ *increases* with height. While an upper-level, positive IPV anomaly weakens, an IPV anomaly below therefore increases in strength. The net effect is that the IPV anomaly propagates downward; presumably, the anomaly eventually reaches the boundary layer, as the layer of precipitation associated with the cutoff low becomes progressively shallower, and friction finally destroys the IPV anomaly.

On the other hand, in an upper-level, negative IPV anomaly there is no latent-heat release; only radiative cooling acts. Strong static stability under the anomaly acts to suppress convection. Refer to the upper-level, negative IPV anomaly shown in Fig. 1.141. In the stratosphere, at 335 K, $(D\theta/Dt)(1/P)$ is nearly zero because P (not shown in Fig. 1.141) is very large. At 325 K, $D\theta/Dt \sim -1\ \mathrm{K\,d^{-1}}$ from radiative cooling, and $P \sim 0.5$ PVU (not shown in Fig. 1.141). Therefore at 330 K, where $P \sim 1$ PVU (not shown in Fig. 1.141)

$$P^2 \frac{\partial}{\partial \theta}\left(\frac{D\theta}{Dt}\frac{1}{P}\right) \sim 1\ \mathrm{PVU}^2 \frac{(-1\ \mathrm{K\,d^{-1}})(1/0.5\ \mathrm{PVU})}{10\ \mathrm{K}} = 0.2\ \mathrm{PVU\,d^{-1}}. \tag{1.9.52}$$

Thus, it takes a relatively long time to destroy an upper-level, negative IPV anomaly. Blocking highs may be more persistent than cut-off lows for this reason. This explanation, however, does not take nonlinear processes into account. Dole and Gordon (1983) have found that in the Northern Hemisphere long-duration, large-magnitude positive (anticyclonic) height anomalies are in fact more numerous than long-duration, large-magnitude negative (cyclonic) height anomalies during the winter.

NOTES

1. One also recalls the story of Fermat, who "had a very short and simple proof to his 'Last Theorem,' but kept it to himself so that others could have the pleasure of discovering it in their turn," much to the torment of future generations of mathematicians. I had also heard the story that he had run out of space on his paper!
2. Note that k, q, and r each are affected by L (see Appendix 1).
3. Actually we usually find a westerly current flanked poleward and equatorward by weaker flow. The point here is that there is no along-the-flow gradient in vorticity.
4. It is truly exciting to look upstream over the data-sparse Pacific for new short waves to reach the west coast of the United States. What will they do? Will they grow and produce a big surface cyclone? What will happen to short waves coming over the Rocky Mountains? Will there be lee cyclogenesis? Sometimes the basic current is composed of northern and southern branches: Although short waves then move along in two separate "lanes," they sometimes interact with each other.
5. Material discussed in this section is not illustrated.

5a. Ibid.

6. Aficianado of rapidly developing cyclones.
7. This process reminds one of the slogan "Think globally, act locally;" here the atmosphere acts locally and reacts globally.
8. In the 1950s R. Harms, the meteorologist in charge of the Milwaukee forecast office, referred to the path taken by the classic Colorado cyclone as a "Panhandle (i.e., the Oklahoma Panhandle) hook."
9. An alternative to using the PVU is to drop the factor of g in Eq. (1.9.3), and use mb as a unit of pressure, so that potential vorticity is on the order of 10^{-5} K mb^{-1} s^{-1}.
10. In the Southern Hemisphere synoptic-scale IPV is always negative; positive IPV anomalies are associated with anticyclonic vorticity, and negative IPV anomalies with cyclonic vorticity. Perhaps we should refer to IPV anomalies as cyclonic or anticyclonic. To avoid confusion, we henceforth will refer to Northern Hemisphere examples only.
11. The author acknowledges Fred Sanders for this perspective.

REFERENCES

General

Bosart, L. F., 1985: Weather forecasting (Chap 4). *Handbook of Applied Meteorology* (D. Houghton, ed.), Wiley, New York, 205–79.

Holton, J. R., 1979: *An Introduction to Dynamic Meteorology.* Academic Press, New York.

Newton, C. W. and E. O. Holopainen, eds., 1990: *Extratropical Cyclones. The Erik Palmén Memorial Volume.* Amer. Meteor. Soc., Boston.

Palmén, E., and C. W. Newton, 1969: *Atmospheric Circulation Systems.* Academic Press, New York.

Petterssen, S., 1956: *Weather Analysis and Forecasting (Vol. I). Motion and Motion Systems.* McGraw–Hill, New York.

Section 1.1

Bjerknes, J., 1951: Extratropical cyclones. *Compendium of Meteorology* (T. F. Malone, ed.), Amer. Meteor. Soc., Boston, 577–98.

Bosart, L. F., 1981: The Presidents' Day snowstorm of 18–19 February 1979: A subsynoptic scale event. *Mon. Wea. Rev.* **109,** 1542–66.

—, and S. C. Lin, 1984: A diagnostic analysis of the Presidents' Day storm of February 1979. *Mon. Wea. Rev.* **112,** 2148–77.

Boyle, J. S., and L. F. Bosart, 1983: A cyclone/anticyclone couplet over North America: An example of anticyclone evolution. *Mon. Wea. Rev.* **111,** 1025–45.

Boyle, J. S., and L. F. Bosart, 1986: Cyclone–anticyclone couplets over North America. Part II: Analysis of a major cyclone event over the eastern United States. *Mon. Wea. Rev.* **114,** 2432–65.

Chen, T.-G., and L. F. Bosart, 1979: A quasi-Lagrangian vorticity budget of composite cyclone–anticyclone couplets accompanying North American polar air outbreaks. *J. Atmos. Sci.* **36,** 185–94.

Pearce, R. P., 1974: The design and interpretation of diagnostic studies of synoptic-scale atmospheric systems. *Quart. J. Roy. Meteor. Soc.* **100,** 265–85.

Petterssen, S., 1955: A general survey of factors influencing development at sea level. *J. Meteor.* **12,** 36–42.

Petterssen, S., and S. J. Smebye, 1971: On the development of extratropical storms. *Quart. J. Roy. Meteor. Soc.* **97,** 457–82.

Section 1.1.4

Bosart, L. F., and F. Sanders, 1981: The Johnstown Flood of July 1977: A long-lived convective system. *J. Atmos. Sci.* **38,** 1616–42.

Danard, M. B., 1964: On the influence of released latent heat on cyclone development. *J. Appl. Meteor.* **3,** 27–37.

Manabe, S., 1956: On the contribution of heat released by condensation to the change in pressure pattern. *J. Meteor. Soc. Japan* **34,** 308–20.

Petterssen, S., D. L. Bradbury, and K. Pedersen, 1962: The Norwegian cyclone models in relation to heat and cold sources. *Geofys. Publ.* **24,** 243–80.

Rotunno, R., and K. Emanuel, 1987: An air–sea interaction theory for tropical cyclones. Part II: Evolutionary study using a nonhydrostatic axisymmetric numerical model. *J. Atmos. Sci.* **44,** 542–61.

Tracton, M. S., 1973: The role of cumulus convection in the development of extratropical cyclones. *Mon. Wea. Rev.* **109,** 573–93.

Section 1.1.7

Charney, J. G. and A. Eliassen, 1964: On the growth of the hurricane depression. *J. Atmos. Sci.* **21,** 68–75.

Emanuel, K. A., 1986: An air-sea interaction theory for tropical cyclones. Part I. *J. Atmos. Sci.* **43,** 585–604.

Emanuel, K. A. and R. Rotunno, 1989: Polar lows as Arctic hurricanes. *Tellus* **41A,** 1–17.

Rotunno, R. and K. Emanuel, 1987: op. cit.

Sutcliffe, R. C., 1947: A contribution to the problem of development. *Quart. J. Roy. Meteor. Soc.* **73,** 370–83.

Section 1.1.8

Bosart, L. F., 1984: The Texas coastal rainstorm of 17–21 September 1979: An example of synoptic-mesoscale interaction. *Mon. Wea. Rev.* **112,** 1108–33.

Colucci, S. J., 1976: Winter cyclone frequencies over the eastern United States and adjacent western Atlantic, 1964–1973. *Bull. Amer. Meteor. Soc.* **57,** 548–53.

Reitan, C. H., 1974: Frequencies of cyclones and cyclogenesis for North America, 1951–1970. *Mon. Wea. Rev.* **102,** 861–68.

Taylor, K. E., 1986: An analysis of the biases in traditional cyclone frequency maps. *Mon. Wea. Rev.* **114,** 1481–90.

Wexler, H., 1951: Anticyclones. *Compendium of Meteorology* (T. F. Malone, ed.), Amer. Meteor. Soc., Boston, 621–29.

Whittaker, L. M., and L. H. Horn, 1981a: Geographical and seasonal distribution of North American cyclogenesis, 1958–1977. *Mon. Wea. Rev.* **109,** 2312–22.

—, and —, 1981b: Northern Hemisphere extratropical cyclone activity for four midseason months. *J. Climo.* **4,** 297–310.

Zishka, K. M., and P. J. Smith, 1980: The climatology of cyclones and anticyclones over North America and surrounding ocean environs for January and July 1950–77. *Mon. Wea. Rev.* **108,** 387–401.

Section 1.2

Zishka, K. M., and P. J. Smith, 1980: The climatology of cyclones and anticyclones over North America and surrounding ocean environs for January and July 1950–77. *Mon. Wea. Rev.* **108,** 387–401.

Section 1.2.4

Austin, J. M., 1947: An empirical study of certain rules of forecasting the motion and intensity of cyclones. *J. Meteor.* **4,** 16–20.

Sanders, F., 1971: Analytic solutions of the non-linear omega and vorticity equation for a structurally simple model of disturbances in the baroclinic westerlies. *Mon. Wea. Rev.* **99,** 393–407.

—, and J. R. Gyakum, 1980: Synoptic-dynamic climatology of the "bomb." *Mon. Wea. Rev.* **108,** 1589–606.

Section 1.3

Boyle, J. S., and L. F. Bosart, 1986: Cyclone–anticyclone couplets over North America. Part II: Analysis of a major cyclone event over the eastern United States. *Mon. Wea. Rev.* **114,** 2432–65.

Sanders, F., 1988: Life history of mobile troughs in the upper westerlies. *Mon. Wea. Rev.* **116,** 2629–48.

Section 1.4.1

Blackmon, M. L., 1976: A climatological spectral study of the 500 mb geopotential height of the Northern Hemisphere. *J. Atmos. Sci.* **33,** 1607–23.

Cressman, G. P., 1949: Some effects of wave-length variations of the long waves in the upper westerlies. *J. Meteor.* **6,** 56–60.

Henry, W. K., 1978: The Southwest low and "Henry's Rule." *Natl. Wea. Dig.* **3,** 6–12.

Petterssen, S., 1952: On the propagation and growth of jet-stream waves. *Quart. J. Roy. Meteor. Soc.* **78,** 337–53.

Section 1.4.2

Cressman, G. P., 1949: Some effects of wave-length variations of the long waves in the upper westerlies. *J. Meteor.* **6,** 56–60.

Hövmöller, E., 1949: The trough-and-ridge diagram. *Tellus* **1,** 62–66.

Riehl, H., 1952: Forecasting in Middle Latitudes. *Meteor. Mono.* **1,** No. 5, Amer. Meteor. Soc., Boston.

Rossy, C.-G., 1945: On the propagation of frequencies and energy in certain types of ocean and atmospheric waves. *J. Meteor.* **2,** 187–203.

Section 1.4.3

Berggren, R., B. Bolin, and C. G. Rossby, 1949: An aerological study of zonal motion, its perturbation and breakdown. *Tellus* **1,** 14–37.

Charney, J. G., and J. G. DeVore, 1979: Multiple flow equilibria in the atmosphere and blocking. *J. Atmos. Sci.* **36,** 1205–16.

—, J. Shukla, and K. C. Mo, 1981: Comparison of a barotropic blocking theory with observation. *J. Atmos. Sci.* **38,** 762–79.

—, and D. M. Straus, 1980: Form-drag instability, multiple equilibria and propagating planetary waves in the baroclinic, orographically forced, planetary wave systems. *J. Atmos. Sci.* **37,** 1157–76.

Colucci, S. J., A. Z. Loesch, and L. F. Bosart, 1981: Spectral evolution of a blocking episode and comparison with wave interaction theory. *J. Atmos. Sci.* **38,** 2092–111.

Dole, R. M., and N. D. Gordon, 1983: Persistent anomalies of the extratropical Northern Hemisphere wintertime circulation: Geographical distribution and regional persistence characteristics. *Mon. Wea. Rev.* **111,** 1567–86.

Egger, J., 1978: Dynamics of blocking highs. *J. Atmos. Sci.* **35,** 1788–801.

Elliott, R. D., and T. B. Smith, 1949: A study of the effects of large blocking highs on the general circulation of the Northern Hemisphere westerlies. *J. Meteor.* **6,** 67–85.

Frederiksen, J. S., 1983: A unified three-dimensional instability theory of the onset of blocking and cyclogenesis. II: Teleconnection patterns. *J. Atmos. Sci.* **40,** 2593–609.

Hartmann, D., and S. Ghan, 1980: A statistical study of the dynamics of blocking. *Mon. Wea. Rev.* **108,** 1144–59.

Hide, R., 1953: Some experiments on thermal convection in a rotating liquid. *Quart. J. Roy. Meteor. Soc.* **79,** 161.

Kalnay-Rivas, E., and L. Merkine, 1981: A simple mechanism for blocking. *J. Atmos. Sci.* **38,** 2077–91.

Loesch, A. Z., 1974a: Resonant interactions between unstable and neutral baroclinic waves, Part I. *J. Atmos. Sci.* **31,** 1177–201.

—, 1974b: Resonant interactions between unstable and neutral baroclinic waves, Part II. *J. Atmos. Sci.* **31,** 1202–17.
Long, R. R., 1964: Solitary waves in the westerlies. *J. Atmos. Sci.* **21,** 197–200.
Lorenz, E. N., 1963: Mechanics of vacillation. *J. Atmos. Sci.* **20,** 449–64.
Mullen, S. L., 1987: Transient eddy forcing of blocking flows. *J. Atmos. Sci.* **44,** 3–22.
Namias, J., 1950: The index cycle and its role in the general circulation. *J. Meteor.* **7,** 130–39.
—, 1978: Multiple causes of the North American abnormal winter 1976/77. *Mon. Wea. Rev.* **106,** 279–95.
—, and P. F. Clapp, 1951: Observational studies of general circulation patterns. *Compendium of Meteorology* (T. F. Malone, ed.), Amer. Meteor. Sci., Boston, 551–67.
Reinhold, B. B., and R. T. Pierrehumbert, 1982: Dynamics of weather regimes: Quasi-stationary waves and blocking. *Mon. Wea. Rev.* **110,** 1105–45.
Rex, D. F., 1950a: Blocking action in the middle troposphere and its effects on regional climate. I: An aerological study of blocking. *Tellus* **2,** 196–211.
—, 1950b: Blocking action in the middle troposphere and its effects on regional climate. II: The climatology of blocking action. *Tellus* **2,** 275–301.
—, 1951: The effect of Atlantic blocking upon European climate. *Tellus* **3,** 1–16.
Sumner, E. J., 1954: A study of blocking in the Atlantic–European sector of the Northern Hemisphere. *Quart. J. Roy. Meteor. Soc.* **80,** 402–16.
Taubensee, R. E., 1977: Weather and circulation of December 1976: Extremes of dryness in the west and midwest. *Mon. Wea. Rev.* **105,** 368–73.
Tung, K. K., and R. S. Lindzen, 1979: A theory of stationary long waves, Part I: A simple theory of blocking. *Mon. Wea. Rev.* **107,** 714–34.
Wagner, A. J., 1977: Weather and circulation of January 1977: The coldest month on record in the Ohio Valley. *Mon. Wea. Rev.* **105,** 553–60.
White, W. B., and N. E. Clark, 1975: On the development of blocking ridge activity over the central North Pacific. *J. Atmos. Sci.* **32,** 489–502.
Willett, H. C., 1949: Long-period fluctuations of the general circulation of the atmosphere. *J. Meteor.* **6,** 34–50.

Section 1.4.4

Trenberth, K. E., and J. G. Olson, 1988: ECMWF Global Analyses 1979–1986: Circulation statistics and data evaluation. *NCAR Tech. Note* TN 300 + STR.

Section 1.5.1

Charney, J. G., 1947: The dynamics of long waves in a baroclinic westerly current. *J. Meteor.* **4,** 135–162.
Eady, E. T., 1949: Long waves and cyclone waves. *Tellus* **1,** 33–52.
Kuo, H. L., 1949: Dynamic instability of a two-dimensional nondivergent flow in a barotropic atmosphere. *J. Meteor.* **6,** 105–22.

Section 1.5.3

Bjerknes, J., 1954: The diffluent upper trough. *Arch. Meteor. Geophys. Bioklimatol.* **A7,** 41–46.
Glickman, T. S., N. J. MacDonald, and F. Sanders, 1977: New findings on the apparent relationship between convective activity and the shape of 500 mb troughs. *Mon. Wea. Rev.* **105,** 1060–61.

MacDonald, N. J., 1976: On the apparent relationship between convective activity and the shape of 500 mb troughs. *Mon. Wea. Rev.* **12,** 1618–22.

Polster, G., 1960: Über die Bildung und Vertiefung von Zykonen und Frontwellenentwicklugen am konfluenten Hohentrog. *Meteor. Abhandl., Inst. Meteor. Geophys. Freien Univ., Berlin* **14,** 1–70.

Scherhag, R., 1937: Bermekungen über die Bedeutung der Konvergenzen und Divergenzen des Geschwindigkeitsfeldes für die Drückänderungen. *Beitr. Physik Atmosphare* **24,** 122–29.

Starr, V. P., 1968: *The Physics of Negative Viscosity Phenomena.* McGraw–Hill, New York.

—, and R. D. Rosen, 1972: A variance analysis of angular momentum in stellar and planetary atmospheres. *Tellus* **24,** 73–87.

Section 1.5.4

Anthes, R. A., Y.-H. Kuo, and J. R. Gyakum, 1983: Numerical simulations of a case of explosive marine cyclogenesis. *Mon. Wea. Rev.* **111,** 1174–88.

Berry, E., and H. Bluestein, 1982: The formation of severe thunderstorms at the intersection of a dryline and a front: The role of frontogenesis. *Preprints, 12th Conf. on Severe Local Storms,* San Antonio, 597–602.

Bluestein, H., and C. R. Parks, 1983: A synoptic and photographic climatology of low-precipitation severe thunderstorms in the Southern Plains. *Mon. Wea. Rev.* **111,** 2034–46.

Boyle, J. S., and L. F. Bosart, 1986: Cyclone–anticyclone couplets over North America. Part II: Analysis of a major cyclone event over the Eastern United States. *Mon. Wea. Rev.* **114,** 2432–65.

Businger, S., and R. J. Reed, 1989: Cyclogenesis in cold air masses. *Wea. and Forecasting,* **4,** 133–56.

Duncan, C. N., 1977: A numerical investigation of polar lows. *Quart. J. Roy. Meteor. Soc.* **103,** 255–67.

Emanuel, K. A., 1986: op. cit.

Emanuel, K. A. and R. Rotunno, 1989: op. cit.

Gyakum, J. R., 1983a: On the evolution of the QE II storm. I: Synoptic aspects. *Mon. Wea. Rev.* **111,** 1137–55.

—, 1983b: On the evolution of the QE II storm. II: Dynamic and thermodynamic structure. *Mon. Wea. Rev.* **111,** 1156–73.

Harrold, T. W., and K. A. Browning, 1969: The polar low as a baroclinic disturbance. *Quart. J. Roy. Meteor. Soc.* **95,** 719–30.

Keshishian, L. G., and L. F. Bosart, 1987: A case study of extended East Coast frontogenesis. *Mon. Wea. Rev.* **115,** 100–17.

Mansfield, D. A., 1974: Polar lows: The development of baroclinic disturbances in cold air outbreaks. *Quart. J. Roy. Meteor. Soc.* **100,** 541–54.

Mullen, S. L., 1979: An investigation of small synoptic-scale cyclones in polar air streams. *Mon. Wea. Rev.* **107,** 1636–47.

—, 1982: Cyclone development in polar air streams over the wintertime continent. *Mon. Wea. Rev.* **110,** 1664–76.

—, 1983: Explosive cyclogenesis associated with cyclones in polar air streams. *Mon. Wea. Rev.* **111,** 1537–53.

Petterssen, S. and Smebye, 1971, op. cit.

Rasmussen, E., 1979: The polar low as an extratropical CISK disturbance. *Quart. J. Roy. Meteor. Soc.* **105,** 531–49

—, 1985: A case study of a polar low development over the Barents Sea. *Tellus* **37A,** 407–18.

Reed, R. J., 1979: Cyclogenesis in polar air streams. *Mon. Wea. Rev.* **107,** 38–52.

—, and M. D. Albright, 1986: A case study of explosive cyclogenesis in the Eastern Pacific. *Mon. Wea. Rev.* **114,** 2297–319.

Sanders, F., and J. R. Gyakum, 1980: Synoptic–dynamic climatology of the "bomb." *Mon. Wea. Rev.* **108,** 1589–606.

Sardie, J. M., and T. T. Warner, 1983: On the mechanism for development of polar lows. *J. Atmos. Sci.* **40,** 869-81.

Shapiro, M. A., L. S. Fedor, and T. Hampel, 1987: Research aircraft measurements of a polar low over the Norwegian Sea. *Tellus* **39A,** 272–306.

—, and D. Keyser, 1990: Fronts, jet streams, and the tropopause. *Extratropical Cyclones* (Chap. 10), Palmén Memorial Volume (C. W. Newton and E. O. Holopainen, eds.), Amer. Meteor. Soc., 167–91.

Tegtmeier, S., 1974: The role of the surface, sub-synoptic low pressure system in severe weather forecasting. M.S. Thesis, School of Meteorology, Univ. of Okla., Norman.

Winston, J. S., 1955: Physical aspects of rapid cyclogenesis in the Gulf of Alaska. *Tellus* **7,** 481–500.

Section 1.6.1

Bergeron, T., 1928: Über die dreidimensional verknupfende Wetteranalyse. I. *Geofys. Publ., Norske Videnskaps-Akad. Oslo* **5**(6), 1–111.

Bjerknes, J., 1919: On the structure of moving cyclones. *Geofys. Publ., Norske Videnskaps-Akad. Oslo* **1**(1), 1–8.

—, 1935: Investigations of selected European cyclones by means of serial ascents. *Geofys. Publ., Norske Videnskaps-Akad. Oslo* **11**(4), 1–18.

—, and E. Palmén, 1937: Investigations of selected European cyclones by means of serial ascents. *Geofys. Publ., Norske Videnskaps-Akad. Oslo* **12**(2), 1–62.

—, and H. Solberg, 1921: Meteorological conditions for the formation of rain. *Geofys. Publ., Norkse Videnskaps-Akad. Oslo* **2**(3), 1–60.

—, and H. Solberg, 1922: Life cycle of cyclones and the polar front theory of atmospheric circulation. *Geofys. Publ., Norske Videnskaps-Akad. Oslo* **3**(1) 1–18.

DiMego, G. J., and L. F. Bosart, 1982: The transformation of Tropical Storm Agnes into an extratropical cyclone. Part I: The observed fields and vertical motion computations. *Mon. Wea. Rev.* **110,** 385–411.

Palmén, E., 1951: The aerology of extratropical disturbances. *Compendium of Meteorology* (T. F. Malone, ed.), Amer. Meteor. Soc., 599–620.

Section 1.6.2

Bosart, L. F., 1981: The President's Day snowstorm of 18–19 February 1979: A subsynoptic-scale event. *Mon. Wea. Rev.* **109,** 1542–66.

Browning, K. A., 1986: Conceptual models of precipitation systems. *Weather and Forecasting* **1,** 23–41.

—, and F. F. Hill, 1985: Mesoscale analysis of a polar trough interacting with a polar front. *Quart. J. Roy. Meteor. Soc.* **111,** 445–62.

Carlson, T. N., 1980: Airflow through midlatitude cyclones and the comma cloud pattern. *Mon. Wea. Rev.* **108,** 1498–509.

Carr, F. H., and J. P. Millard, 1985: A composite study of comma clouds and their association with severe weather over the Great Plains. *Mon. Wea. Rev.* **113,** 370–87.

Weldon, R., 1979: *Part IV: Cloud Patterns and the Upper Air Wind Field.* Satellite Training Course Notes, National Environmental Satellite Service.

Zillman, J. W. and P. G. Price, 1972: On the thermal structure of mature Southern Ocean Cyclones. *Aust. Meteor. Mag.* **20,** 34–48.

Section 1.6.3

Baker, D. G., 1970: A study of high pressure ridges to the east of the Appalachian Mountains. Ph.D. Thesis, Massachusetts Institute of Technology.

Kocin, P. J., and L. W. Uccellini, 1984: A review of major East Coast snowstorms. *Preprints, 10th Conf. on Weather Forecasting and Analysis,* Clearwater Beach, Fla., Amer. Meteor. Soc., Boston, 189–198.

Mather, J. R., H. Adams, and G. A. Yoshioka, 1964: Coastal storms of the eastern United States. *J. Appl. Meteor.* **3,** 693–706.

Miller, J. E., 1946: Cyclogenesis in the Atlantic coastal region of the United States. *J. Meteor.* **3,** 31–44.

Pagnotti, V., and L. F. Bosart, 1984: Comparative diagnostic case study of East Coast secondary cyclogenesis under weak versus strong synoptic-scale forcing. *Mon. Wea. Rev.* **112,** 5–30.

Petterssen, S., 1941: Cyclogenesis over southeastern United States and the Atlantic coast. *Bull. Amer. Meteor. Soc.* **22,** 269–270.

Section 1.6.4

McClain, E. P., 1960: Some effects of the Western Cordillera of North America on cyclonic activity. *J. Meteor* **17,** 104–115.

Newton, C. W., 1956: Mechanisms of circulation change during a lee cyclogenesis. *J. Meteor.* **13,** 528–39.

Section 1.6.5

Buzzi, A., and S. Tibaldi, 1978: Cyclogenesis in the lee of the Alps: A case study. *Quart. J. Roy. Meteor. Soc.* **104,** 271–87.

McGinley, J., 1982: Diagnosis of Alpine lee cyclogenesis. *Mon. Wea. Rev.* **110,** 1271–87.

—, 1987: A variational objective analysis scheme for analysis of the ALPEX data set. *Meteor. Atmos. Phys.* **36,** 5–23.

Section 1.7

Charney, J. G., 1955: The use of the primitive equations of motion in numerical prediction. *Tellus* **7,** 22–26.

—, 1962: Integration of the primitive and balance equations. *Proc. Intern. Symp. Numerical Weather Prediction,* Tokyo, 131–52.

DiMego, G. J., and L. F. Bosart, 1982: The transformation of Tropical Storm Agnes into an extratropical cyclone. Part I: The observed fields and vertical motion computations. *Mon. Wea. Rev.* **110,** 385–411.

Haltiner, G. J., and R. T. Williams, 1980: *Numerical Prediction and Dynamic Meteorology*. Wiley, New York.

Krishnamurti, T. N., 1968a: A diagnostic balance model for studies of weather systems of low and high latitudes, Rossby number less than 1. *Mon. Wea. Rev.* **96,** 197–207.

—, 1968b: A study of a developing wave cyclone. *Mon. Wea. Rev.* **96,** 208–17.

Pagnotti, V., and L. F. Bosart, 1984: Comparative diagnostic case study of East Coast cyclogenesis under weak versus strong synoptic-scale forcing. *Mon. Wea. Rev.* **112,** 5–30.

Section 1.8

Colucci, S. J., and D. R. Walker, 1986: Large-scale circulation changes during the Presidents' Day storm: Implications for predicting the breakdown of 500 mb blocking cyclones. *Preprints, 11th Conf. on Weather Forecasting and Analysis,* Kansas City, MO, Amer. Meteor. Soc., Boston, Mass, 89–91.

Section 1.9

Bleck, R., 1973: Numerical forecasting experiments based on the conservation of potential vorticity on isentropic surfaces. *J. Appl. Meteor.* **12,** 737–52.

Bretherton, F. P., 1966: Baroclinic instability and the short wavelength cut-off in terms of potential vorticity. *Quart. J. Roy. Meteor. Soc.* **92,** 335–45.

Brooks, E. M., 1942: Simplification of the acceleration potential in an isentropic surface. *Bull. Amer. Meteor. Soc.* **23,** 195–203.

Danielsen, E. F., 1959: The laminar structure of the atmosphere and its relation to the concept of a tropopause. *Arch. Meteor. Geophys. Bioklim.* **A11,** 293–332.

Davis, C. A., and K. A. Emanuel, 1991: Potential vorticity diagnostics of cyclogenesis. *Mon. Wea. Rev.* **119,** 1929–53.

Haynes, P. H., and M. E. McIntyre, 1987: On the evolution of vorticity and potential vorticity in the presence of diabatic heating and frictional or other forces. *J. Atmos. Sci.* **44,** 828–41.

Hoskins, B. J., M. E. McIntyre, and A. W. Roberston, 1985: On the use and significance of isentropic potential vorticity maps. *Quart. J. Roy. Meteor. Soc.* **111,** 877–946.

Kleinschmidt, E., 1950a: Über Aufbau und Entstehung von Zyklonen (1. Teil). *Meteor. Runds.* **3,** 1–6.

—, 1950b: Über Aufbau und Entstehung von Zyklonen (2. Teil). *Meteor. Runds.* **3,** 54–61.

—, 1951: Über Aufbau und Entstehung von Zyklonen (3. Teil). *Meteor. Runds.* **4,** 89–96.

Namias, J., 1939: The use of isentropic analysis in short-term forecasting. *J. Aeronaut. Sci.* **5,** 295–98.

Reed, R. J., and F. Sanders, 1953: An investigation of the development of a mid-tropospheric frontal zone and its associated vorticity field. *J. Meteor.* **10,** 338–49.

Rossby, C.-G., and collaborators, 1937: Isentropic analysis. *Bull. Amer. Meteor. Soc.* **18,** 201–9.

—, 1940: Planetary flow patterns in the atmosphere. *Quart. J. Roy. Meteor. Soc.* **66,** Suppl., 68–87.

Shaw, Sir N., 1933: *Manual of Meteorology (Vol. 3): The Physical Processes of Weather,* Cambridge Univ. Press.

Starr, V. P., and M. Neiburger, 1940: Potential vorticity as a conservative property. *J. Marine Res.* **3,** 202–10.

Thorpe, A. J., 1985: Diagnosis of balanced vortex structure using potential vorticity. *J. Atmos. Sci.* **42,** 397–406.

Section 1.9.13

Dole, R. M., and N. D. Gordon, 1983: Persistent anomalies of the extratropical Northern Hemisphere wintertime circulation: Geographical distribution and regional persistence characteristics. *Mon. Wea. Rev.* **111,** 1567–86.

PROBLEMS

1.1. Consider the figure. In what direction will the surface low-pressure center move? Use quasigeostrophic theory. Temperature T (solid lines).

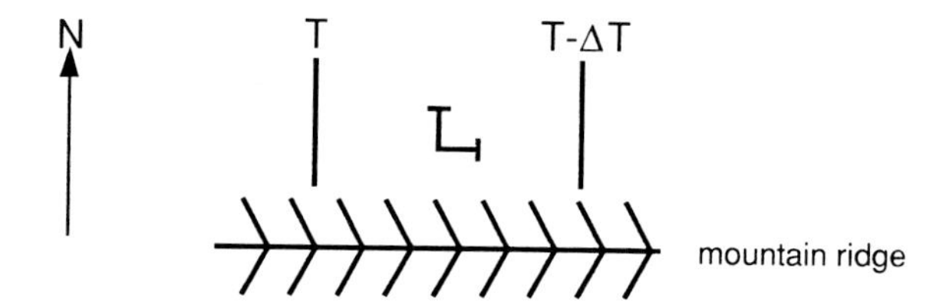

1.2. What must the direction of the geostrophic wind along lines A and B (bottom) in the figure be in order that the troughs and ridges at

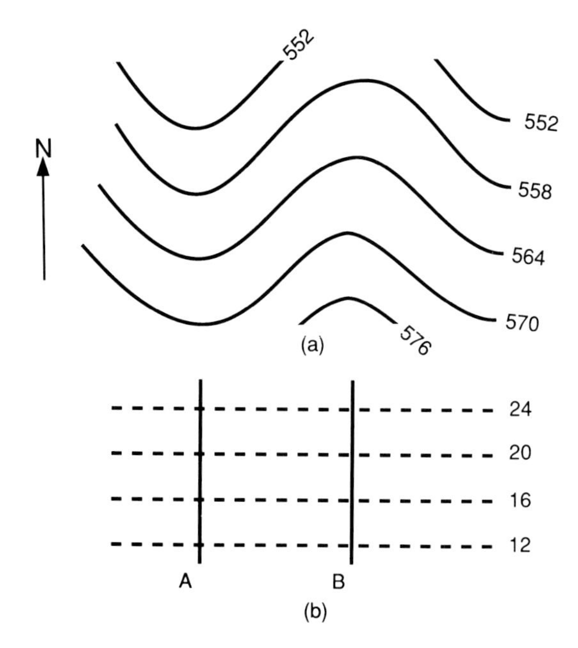

500 mb (top) intensify? Why? Use quasigeostrophic theory. (a) 500-mb height in dam (solid lines); (b) temperature in °C (dashed lines) below 500 mb.

1.3. What are the important physical mechanisms responsible for vertical motion in a quasigeostrophic, frictionless, adiabatic atmosphere well above the ground? How are each related to rising and sinking motion? Please be succinct!

1.4. What physical mechanism is responsible for *intensification* of upper-level troughs and ridges in an adiabatic, quasigeostrophic atmosphere?

1.5. Consider the 500-mb height field (solid lines; dam) shown in the figure. (a) If L is large, where is there CVA? AVA? (b) If L is small, where is there CVA? AVA?

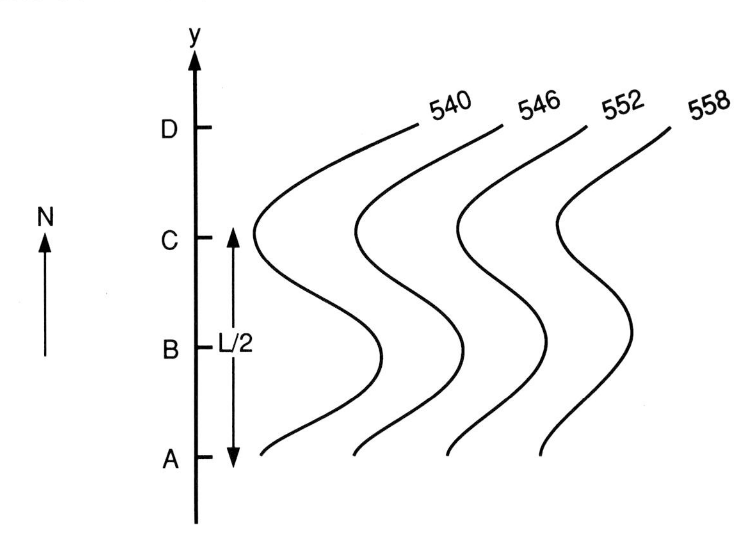

1.6. Suppose there is a circular 1000-mb low downstream from a 500-mb ridge in the westerlies, and upstream from a 500-mb trough. Ignore friction and diabatic heating. Use quasigeostrophic theory to obtain a qualitative estimate of the vertical motion field at 700 mb from each forcing function.

1.7. Define thermal vorticity. Show an example of anticyclonic thermal vorticity.

1.8. Suppose a *circular* surface low-pressure area lies between two topographically identical north–south-oriented mountain ranges. Quasigeostrophically, how will the shape of the surface low change owing to orographic effects alone? Why?

1.9. Consider a wavetrain in the easterlies. Where is there CVA and AVA for (a) long waves and (b) short waves? Explain.

1.10. Suppose there is a surface high-pressure area downstream from an upper-level trough in the westerlies, but upstream from an upper-level ridge. Use quasigeostrophic theory to obtain a qualitative estimate of the vertical motion field owing to temperature advection (ω_T) between the trough and ridge.

1.11. What direction will an upper-level, cut-off, barotropic, circular low-pressure area move? Why? Use quasigeostrophic theory. Neglect diabatic heating.

1.12. Under what condition(s) will there be sinking motion at 500 mb if there is CVA at 500 mb? Explain your answer in the context of quasi-geostrophic theory. Neglect diabatic heating and temperature advection.

1.13. In what direction will a surface windward ridge move in the easterlies, east of a north–south mountain range, if "warm" air lies to the south and "cold" air lies to north? Explain using quasigeostrophic theory.

1.14. Suppose there is a circular, barotropic high-pressure area at the surface. How does the system try to get back into quasigeostrophic balance if there is surface friction? Neglect diabatic processes and advection of the Earth's vorticity.

1.15. What are the possible air trajectories through an extratropical high-pressure area in an east–west vertical cross section south of the center? Assume that there is an upper-level ridge to the west (upstream) and an upper-level trough to the east (downstream). Assume, also, that the ridge is moving eastward at a *speed* less that that of the strong westerly winds aloft, and more than that of the winds around the surface high.

1.16. Explain why AVA at 500 mb associated with a "short" wave in an equivalent barotropic atmosphere implies sinking motion. Neglect diabatic heating and, of course, friction.

1.17. At 35°N, what is the horizontal wavelength of stationary barotropic Rossby waves in a mean westerly current of 15 m s^{-1}?

1.18. Consider the long-wave geopotential-height (solid lines) field shown in the figure. (a) If the atmosphere is equivalent barotropic, which way will the wavetrain move? Why? (b) If the atmosphere is equivalent barotropic, where will there be sinking motion?

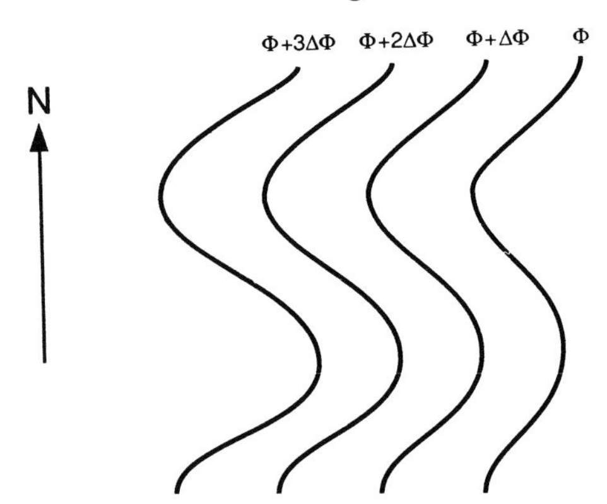

1.19. At what lettered points in the figure is there rising motion (a) if the wavelength of the feature is "long?" Why? (b) if the wavelength of the feature is "short?" Why? (Use quasigeostrophic reasoning, and neglect diabatic heating and friction.) Assume that there is no geostrophic wind at the ground, which is level. Geopotential height (solid lines).

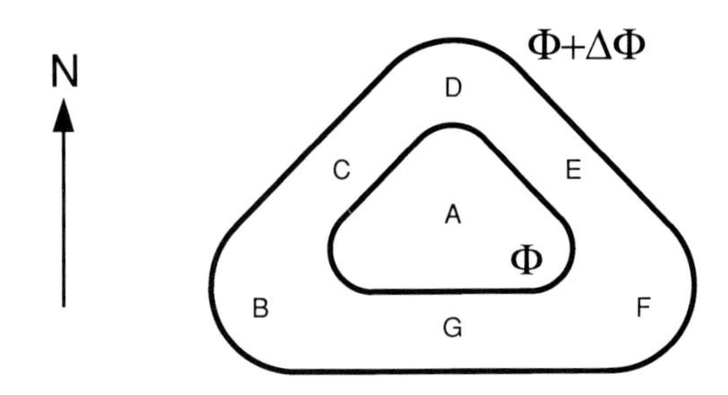

1.20. What is the sense (up or down) of the vertical motion above points A, B, C, and D in the figure. Neglect diabatic heating, friction, and advection of Earth's vorticity. Use quasigeostrophic theory. Surface isobars (solid lines); surface potential temperature (dashed lines).

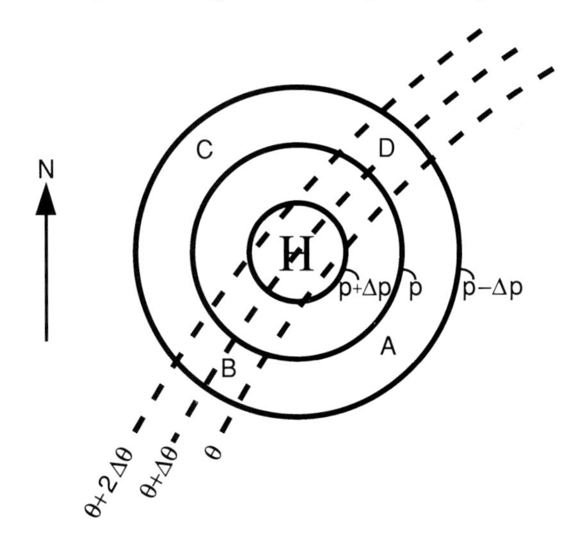

1.21. In a quasigeostrophic, equivalent barotropic atmosphere, the phase velocity C is related to the basic current U and the horizontal wavelength L by the relation

$$C = U - \frac{\beta}{(2\pi/L)^2},$$

where $\beta = 2\Omega \cos\phi/a$, ϕ is the latitude, and Ω is the rotation rate of the Earth. For $a = 6371$ km and $L = 4000$ km, at what basic current will the phase velocity be zero at 40°N?

1.22. Explain why Colorado surface low-pressure areas often track southeastward to Oklahoma, recurve northeastward, and end up near Chicago. Use quasigeostrophic theory. Be as complete (but as precise) as you can!

1.23. (a) What is the sign of χ (associated with each forcing function) at each lettered point indicated in the figure. Assume that the characteristic wavelength is short, there is no temperature advection below 500 mb,

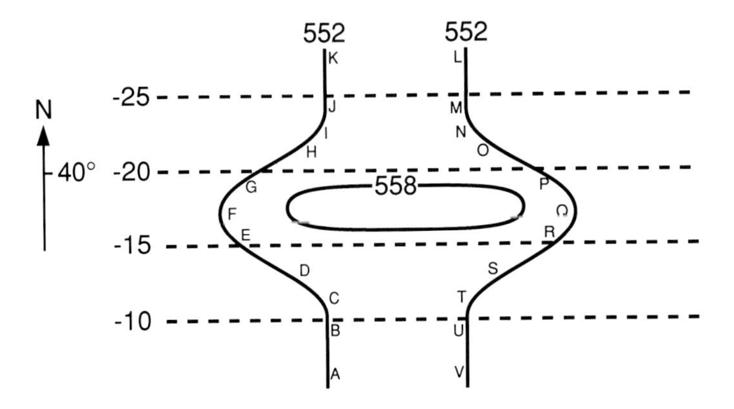

and temperature (thickness) advection above 500 mb is the same as it is at 500 mb. Neglect diabatic heating and friction. Use quasigeostrophic theory. 500-mb height in dam (solid lines); isotherms (dashed lines) in °C. (b) Repeat (a), except assume the characteristic wavelength is long.

1.24. Consider point P on the figure in Problem 1.23. If $w = +3.5$ cm s^{-1}, what is ω in μb s^{-1}? Use quasigeostrophic theory.

1.25. Suppose that the feature depicted in the figure moves southward. (a) Use quasigeostrophic theory to find the sign of δ at the tropopause level at each lettered point. Show your reasoning as completely as you can. (b) If there is no vorticity advection or diabatic heating or friction, in what direction must the temperature gradient point? Why? Assume $\omega = 0$ at the tropopause. Elevation in m MSL (solid lines); height in m for pressure surface near 800 m MSL (dashed lines).

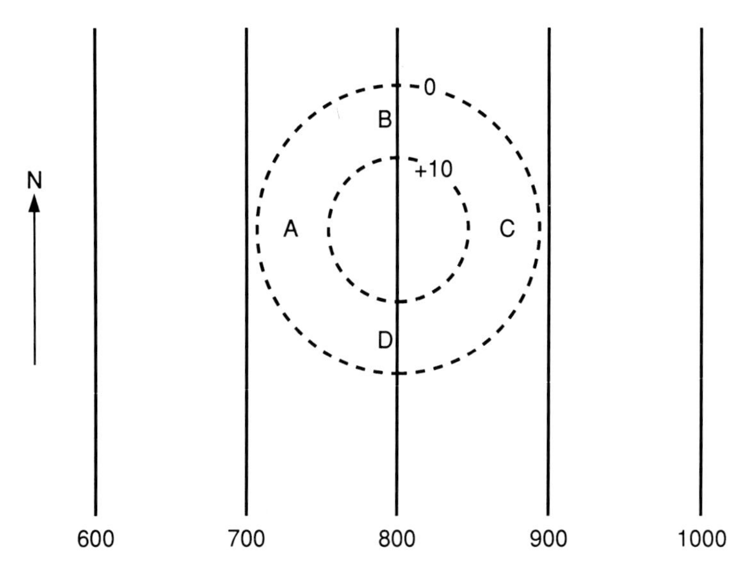

1.26. (a) At each numbered point marked in the figure, will the geopotential height rise or fall? Neglect advection of Earth's vorticity, temperature advection, and diabatic heating. (b) Repeat (a), but neglect advection

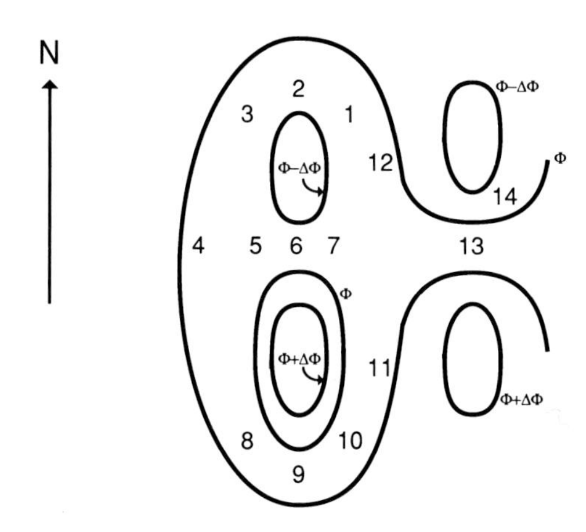

of geostrophic vorticity rather than advection of Earth's vorticity. Geopotential height (solid lines).

1.27. Suppose that the temperature increases locally. According to quasi-geostrophic theory, explain: (a) What physical processes could be responsible for the temperature increase? (b) What happens to the thermal vorticity? (c) What is the atmosphere's response? That is, what is the secondary circulation?

1.28. Where (in relation to wavelike upper-level features) would one expect to find the formation of a surface high-pressure area in an equivalent barotropic, frictionless, adiabatic, Northern Hemisphere atmosphere having a level surface and in which the basic current is *easterly* and the thermal wind is *easterly*? Why?

1.29. Demonstrate mathematically that an increase in thermal vorticity must be accompanied by a decrease in thickness.

1.30. Explain how upslope motion in a frictionless, adiabatic, quasi-geostrophic atmosphere is accompanied by height rises near the surface.

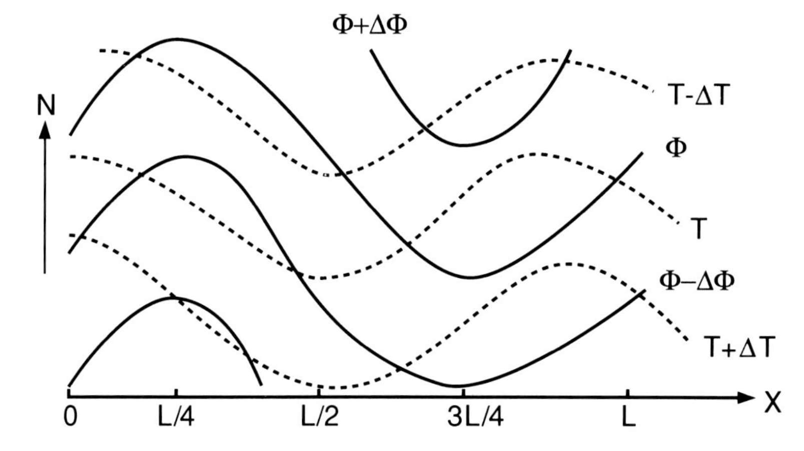

1.31. For what values of x in the figure is there rising motion? Sinking motion? Indeterminate? (That is, forcing functions act in opposition to each other—a quantitative calculation is needed.) Explain your answers! Use quasigeostrophic reasoning. Geopotential height (solid lines); isotherms (dashed lines).

1.32. What are typical order-of-magnitude values for each of the following for synoptic-scale, midlatitude motions? Include the correct units for each. (a) Differential geostrophic vorticity advection at 750 mb downstream from a trough at 500 mb, and coincident with the center of a cyclone at 1000 mb; (b) geostrophic advection of Earth's vorticity; (c) local rate of change of geostrophic relative vorticity.

1.33. Suppose that there is no diabatic heating and no local change in temperature, and the temperature (dashed lines; °C) and 500-mb height (solid lines; dam) fields are as given in the figure. What must the static stability parameter be at point P if ω is $-1\ \mu b\, s^{-1}$? Use quasi-geostrophic theory.

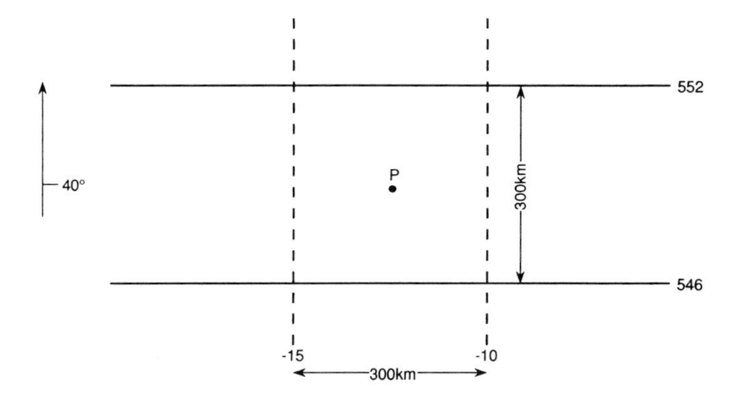

1.34. For what values of x in the figure will the surface pressure rise most rapidly? Why? Use quasigeostrophic theory. Elevation in m MSL (solid lines); surface temperature in °C (dashed lines); wind field (vectors).

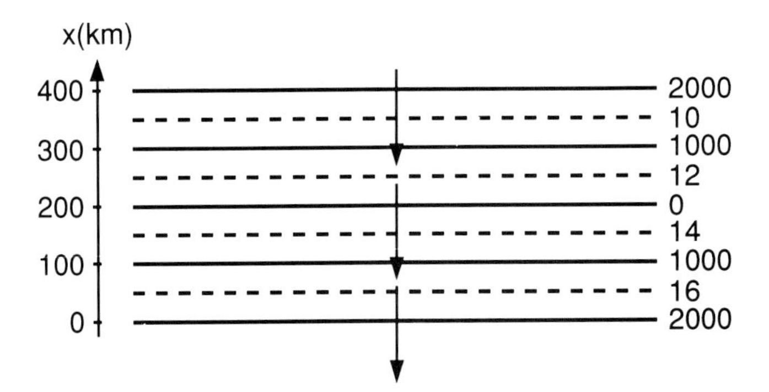

1.35. Neglect diabatic heating. (a) Which way will the high shown in the figure move if the wavelength is short? (b) long? Why? Use quasi-geostrophic theory. 500-mb height in dam (solid lines); 500-mb temperature in °C (dashed lines).

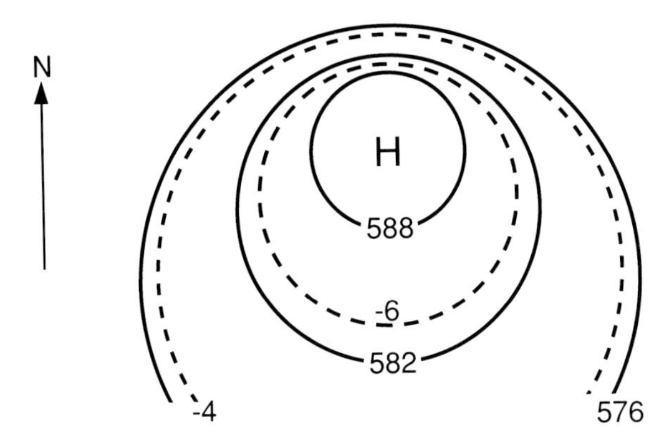

1.36. Suppose that

$$\Phi(x,y,p = 1000 \text{ mb}) = \hat{\Phi}_0 \cos[(2\pi/L)(x+\lambda)] \cos[(2\pi/L)y]$$

and

$$T(x,y,p) = T_m(p) - [1 - \alpha \ln(1000/p)]\{ay + \hat{T} \cos[(2\pi/L)x] \cos[(2\pi/L)y]\}.$$

(a) Show that

$$\begin{aligned}\Phi(x,y,p) = \Phi_m(p) &+ \hat{\Phi}_0 \cos[(2\pi/L)(x+\lambda)] \cos[(2\pi/L)y] \\ &+ R\{ay + \hat{T} \cos[(2\pi/L)x] \cos[(2\pi/L)y]\}\{\ln(p/1000) \\ &+ (\alpha/2)[\ln(p/1000)]^2\}.\end{aligned}$$

(b) Sketch for $|y| \leq L/4$ the 1000-mb isotherms at intervals of 5°C, labeled as departure from map average. Let $a = 10°\text{C}/1000$ km, $\hat{T} = 10°C$, and $L = 4000$ km. (c) Sketch the isopleths of 1000-mb geopotential at intervals of 600 $\text{m}^2\,\text{s}^{-2}$ for $\hat{\Phi}_0 = 1000\ \text{m}^2\,\text{s}^{-2}$ and $\lambda = L/8$. (d) Sketch the isopleths of 500-mb geopotential at intervals of 600 $\text{m}^2\,\text{s}^{-2}$, labeled as departures from the map-average value. Let $\alpha = 0.721$. (e) Repeat (d) for 300 mb. (f) Draw a graph of the vertical profile of the mean zonal geostrophic wind $[u_g(p)]$ at 40°N. Use a and α from (a)–(e).

1.37. Derive, for the *fun* of it (!), the differential vorticity-advection forcing function given $\Phi(x,y,p)$ in (a) of Problem 1.36. (Be thankful you are not asked to *solve* the ω equation!)

1.38. Show that the following are zero: (a) $-(\partial/\partial p)(-\mathbf{v}_0 \cdot \nabla_p \zeta_0)$; (b) $-(\partial/\partial p)(-\mathbf{v}_0 \cdot \nabla_p f)$; (c) $-\mathbf{v}' \cdot \nabla_p \zeta'$; (d) $-\mathbf{v}_M \cdot \nabla_p f$; (e) $-\mathbf{v}_M \cdot \nabla_p T_M$; (f) $-\mathbf{v}' \cdot \nabla_p T'$.

1.39. Show that (a) $-\mathbf{v}_0 \cdot \nabla_p \zeta' = -(-\mathbf{v}' \cdot \nabla_p \zeta_0)$; (b) $-\mathbf{v}_M \cdot \nabla_p T' = -(-\mathbf{v}' \cdot \nabla_p T_M)$.

1.40. Write the **Q**-vector form of the quasigeostrophic ω equation in terms of $\mathbf{v} = \mathbf{v}_0 + \mathbf{v}_M + \mathbf{v}'$ and $T = T_m + T_M + T'$.

1.41. Using Sanders' analytic model, compute the following at $x = 3L/8$, $y = L/8$ (i.e., northeast of the surface low) at 850 mb: (a) $-\mathbf{v}_g \cdot \nabla_p T$ (the local change in temperature in °C d^{-1} owing to geostrophic temperature advection); (b) $\omega\sigma p/R$ (the local change in temperature owing to vertical motion in °C d^{-1}). Is the effect of the secondary circulation on temperature more or less than that of the geostrophic "forcing?" (c) Repeat (a) at 500 mb; (d) Repeat (b) at 500 mb. Let $\alpha = 0.722$, $a = 10°\mathrm{C}/1000\ \mathrm{km}$, $\hat{\Phi}_0 = 1000\ \mathrm{m^2\ s^{-2}}$, $\hat{T} = 10°\mathrm{C}$, $\lambda = L/4$, latitude = 35°N, and $L = 3000$ km.

1.42. Derive a differential equation in isentropic coordinates, which relates the tangential wind field $V(r, \theta)$ to the distribution of potential vorticity $P(r, \theta)$. Assume circular symmetry and cyclostrophic balance. Neglect diabatic heating and friction. What is the ellipticity condition?

1.43. Consider a negative (anticyclonic) IPV anomaly in the middle troposphere. Assume gradient-wind balance and adiabatic, frictionless flow. (a) Make a qualitative sketch of a vertical cross section of θ and V. (b) Suppose that the anomaly is stationary, and that a uniform wind current blows from east to west at the tropopause, above the IPV anomaly. (There is no basic current at the level of the IPV anomaly). Diagnose the vertical motion field qualitatively, using both the vorticity and thermodynamic equations in pressure coordinates.

1.44. Suppose that the ground slopes upward toward the north (as it does, for example, in northern India) and that there is a negative IPV anomaly in the upper troposphere. (a) Where will there be surface cyclogenesis? surface anticyclogenesis? Justify your answers using "IPV thinking." (b) Repeat (a), but suppose the ground slopes upward toward the *south* (as it does, for example, north of the Himalayas). In both (a) and (b) the horizontal scale of the IPV anomaly is large enough that its effect extends to the ground. Also, the effect of variations in f on the basic state of isentropic potential vorticity is significant.

1.45. Consider a positive potential-temperature anomaly at the surface over North Africa, where warm, desert air lies to the north and cooler, moist, marine air lies to the south. Suppose the vertical wind profile is as given in the figure. There is no meridional wind component. Assume the effect of the surface IPV anomaly extends up to 12 km. Use "IPV

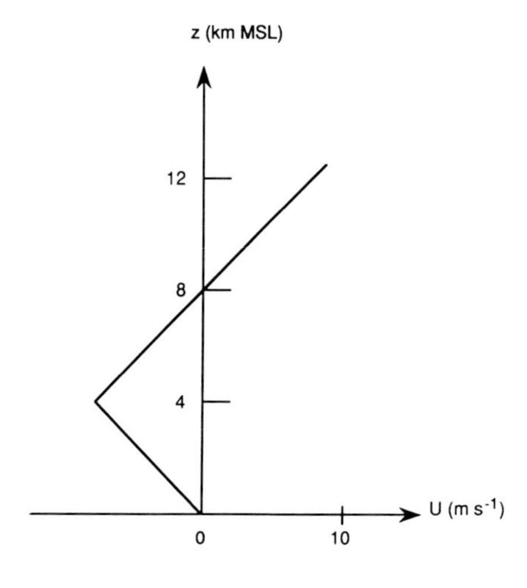

thinking" to answer the following: (a) What type of synoptic feature is present at the surface? Cyclone? Anticyclone? Neither? (b) Diagnose the vertical velocity field aloft using the vorticity equation on an isobaric surface. Assume $\omega = 0$ at the ground and at 12 km.

1.46. Write down the expression for isentropic potential vorticity. Find an order-of-magnitude estimate of it along a sea-breeze front. Assume a sea-breeze front is on the order of 10 km across. Be sure to express your answer in the correct units.

1.47. (a) Explain why a jet streak (isotach maximum in a jet) in the upper westerlies may be thought of as a positive–negative IPV couplet. (b) Assume the basic IPV state is a result of the meridional gradient in f. Which way will the jet streak *move*? Why? Use IPV thinking.

1.48. Suppose a surface low-pressure area in the Northern Hemisphere is located downstream from an upper-level ridge, and upstream from an upper-level trough in the baroclinic westerlies. Neglect friction, diabatic heating, and topography; assume the atmosphere is nearly equivalent barotropic at middle and upper levels. (a) If the temperature-gradient vector at low levels points toward the south, will the amplitude of the wavetrain in the middle troposphere increase, decrease, or remain the same? (b) If the temperature-gradient vector at low levels points toward the south, which way will the surface low move? (c) Repeat (a), but with the temperature-gradient vector at low levels pointing toward the north. (d) Repeat (b), but with the temperature-gradient vector at low levels pointing toward the north. Justify all your answers using quasigeostrophic theory.

1.49. Use "IPV thinking" to discuss under what conditions (if any) a vertically stacked wavetrain in the baroclinic westerlies could become baroclinically unstable.

2

Fronts and Jets

> The Sky is low—the Clouds are mean.
> A Travelling Flake of Snow
> Across a Barn or through a Rut
> Debates if it will go—
>
> A Narrow Wind complains all Day
> How some one treated him
> Nature, like Us is sometimes caught
> Without her Diadem.
>
> EMILY DICKINSON

2.1. THE RELATIONSHIP BETWEEN FRONTS AND JETS

The observed concentration of intense temperature gradients and strong winds into narrow zones in the troposphere is extremely intriguing. The reason for this is not obvious! Visitors from another planet might be struck by the beauty of the long, narrow bands of clouds associated with fronts on Earth. Recall how impressed we were upon viewing the first close-up photographs of Saturn's rings. In this chapter we will describe the observational characteristics of these zones, explain why they exist, and thereby explain why temperature and momentum are not more uniformly distributed.

We first acquired some motivation to develop a dynamical theory for synoptic-scale systems in middle latitudes and consequently discussed quasi-geostrophic theory (Chap. 5, Vol. I). This was *followed* by its application (Chap. 1, Vol. II). In this chapter we will reverse our approach: First we will describe the kinematics and the observed formation and movement of fronts and jets, and then we will discuss dynamical theories for their behavior. In some instances (for example, for "bombs" and polar lows) quasigeostrophic theory was pushed to and beyond its limits in Chap. 1. Many significant features in the atmosphere cannot be satisfactorily explained by quasigeostrophic theory. Among them are fronts, across which the horizontal scale is only 100 km or less, and some jets, whose wind speeds are 50 m s^{-1} or greater, and across which the horizontal scale is only hundreds of kilometers. If the length of a front or jet is on the order of 1000 km, then the Rossby number is usually reasonally small for flow along the front or jet so that geostrophic balance is approximately maintained across, but not along, the front or jet (Fig. 2.1). *Both* fronts and jets are usually marked by a concentration of isotherms and strong vertical shear, in accord with the thermal-wind relation.

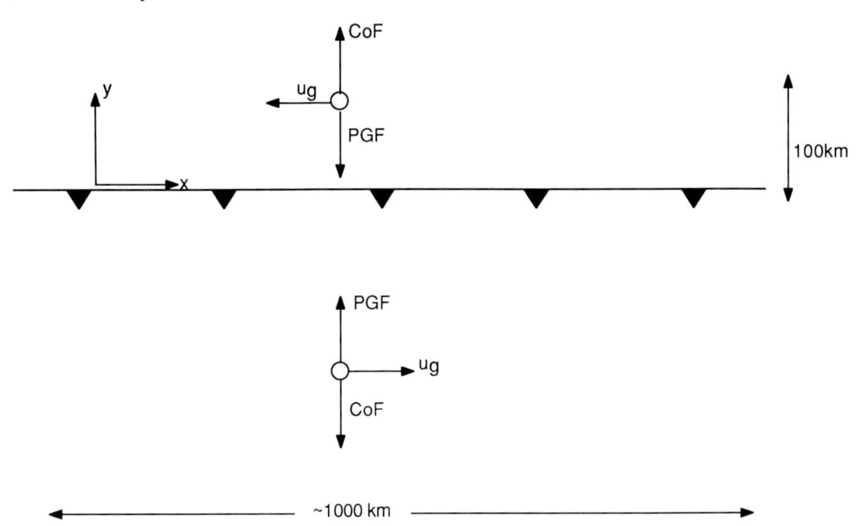

Figure 2.1 Horizontal scales of a surface front and balance-of-force diagrams on both sides of the front. Pressure-gradient force (PGF), Coriolis force (CoF), and along-the-front component of the geostrophic wind (u_g) (after Bluestein, 1986). (Courtesy of the American Meteorological Society)

Therefore, a theory that explains fronts may also explain jets. Upper-level fronts and jets in fact are often discussed together as "jet–front systems."

A front is most frequently defined as an "elongated" zone of "strong" temperature gradient and relatively high static stability. It has also been defined in terms of density or moisture only, or derived variables like potential temperature and equivalent potential temperature. In the broadest sense, the front is a *boundary between two air masses,* in analogy to the advancing line between opposing sides in warfare. As during World War I, often all is not quiet along the western front. By "strong" we usually mean an order of magnitude or more greater than the typical synoptic-scale strength of 10 K $(1000 \text{ km})^{-1}$ (or 10 g kg^{-1} water vapor mixing ratio per 1000 km). Such a gradient this intense is sometimes called a *hypergradient.* An "elongated" zone is one whose length is roughly a half an order of magnitude or more longer than its width. The width of the boundary can vary from an infinitesimal distance up to hundreds of kilometers.

A jet is an "intense," "narrow," quasihorizontal or horizontal current of wind that is associated with "strong" vertical shear. The adjective *intense* usually means wind speeds of at least 30 m s^{-1} for the upper portions of the troposphere, and 15 m s^{-1} for the lower portions of the troposphere. A "narrow" current is one whose width is at least a half of to an order of magnitude less than its length. Vertical shear is "strong" if it is at least 5–10 m s^{-1} km^{-1}, that is, at least a half of to an order to magnitude larger than typical synoptic-scale shear. An isotach maximum embedded within a jet is called a *jet streak.*

Fronts and jets are hybrid phenomena because each is characterized by two different horizontal scales that differ by as much as an order of magnitude. Therefore it would at first glance be expected that a proper explanation of the formation and behavior of fronts and jets whose length scales are 1000 km and whose width scales are 100 km requires use of both synoptic-scale, quasi-geostrophic dynamical principles and mesoscale dynamical principles. Fronts whose length scales are 100 km and whose width scales are 10 km are usually related to topographical features, and require reference to mesoscale and in some cases even smaller-scale dynamics for proper explanations. Nonhydrostatic effects, however, will not be considered here.

The importance of fronts and jets in the atmosphere is immense because large variations in meteorological conditions exist across them. Hence, the ability to forecast the weather often involves detailed knowledge of the motion of fronts and their structure on the mesoscale. Furthermore, the location and structure of jets is especially important to the aviation industry. Clear–air turbulence in the strongly sheared region of a jet and upper-level front is a hazard to aircraft. Upper-level fronts are sometimes responsible for the transport of ozone and other materials from the stratosphere into the troposphere. Also, the occurrence of severe thunderstorms is often related to the position of fronts and jets.

2.2. CONSEQUENCES OF THE CROSS-FRONT SCALE

2.2.1. The Front as a Discontinuity in Temperature

Suppose that a front is described as a boundary between two different air masses as in Polar-Front theory. If an air mass is characterized by its density, then the density varies discontinuously across the frontal boundary. In order that the acceleration induced by the pressure-gradient force be finite across the front, the pressure must be continuous across the front. According to the ideal gas law, then, temperature in addition to density must be discontinuous (Fig. 2.2).

If variations in the x direction (along the front) are negligible, and pressure is in a steady state, then the differential of p can be written as

$$dp = \frac{\partial p}{\partial y} dy + \frac{\partial p}{\partial z} dz. \tag{2.2.1}$$

Along the frontal boundary (Fig. 2.3), it follows then that in a hydrostatic atmosphere

$$\frac{dz}{dy} = \left[\left(\frac{\partial p}{\partial y} \right)_{\mathrm{c}} - \left(\frac{\partial p}{\partial y} \right)_{\mathrm{w}} \right] \Big/ g(\rho_{\mathrm{c}} - \rho_{\mathrm{w}}), \tag{2.2.2}$$

where the subscripts w and c refer to the warm and cold sides of the front, and p_{c} and p_{w} are equal at the frontal boundary (the "dynamic" boundary

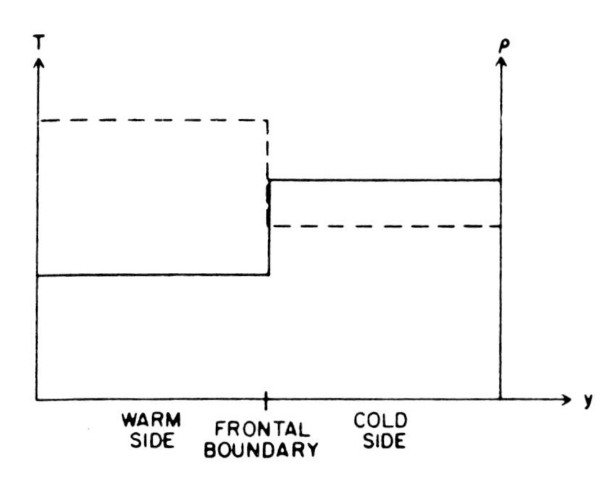

Figure 2.2 The front as a discontinuity in density ρ (solid line) and temperature T (dashed line) (from Bluestein, 1986). (Courtesy of the American Meteorological Society)

condition). Now dz/dy must be nonzero, or else there would be no front. Hence there must be a discontinuity in the cross-front pressure gradient across the frontal boundary. In particular, the cross-front pressure gradient is larger *in value* (but *not* greater in *magnitude* necessarily) on the cold side. If the along-the-front pressure gradient is nonzero, the isobars may be "kinked" or bent at the front (Fig. 2.4).

Using the along-the-front component of the geostrophic-wind relation in height coordinates

$$u_g = -\frac{1}{f\rho}\frac{\partial p}{\partial y}, \tag{2.2.3}$$

where f is the Coriolis parameter, it is seen from Eq. (2.2.2) that

$$\frac{dz}{dy} = \frac{f(\rho_w u_{gw} - \rho_c u_{gc})}{g(\rho_c - \rho_w)}. \tag{2.2.4}$$

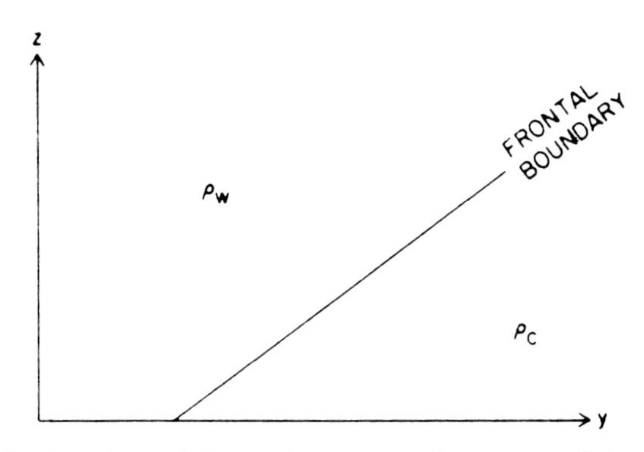

Figure 2.3 A sloping frontal boundary in an incompressible atmosphere. The subscripts w and c indicate warm and cold sides of the front (from Bluestein, 1986). (Courtesy of the American Meteorological Society)

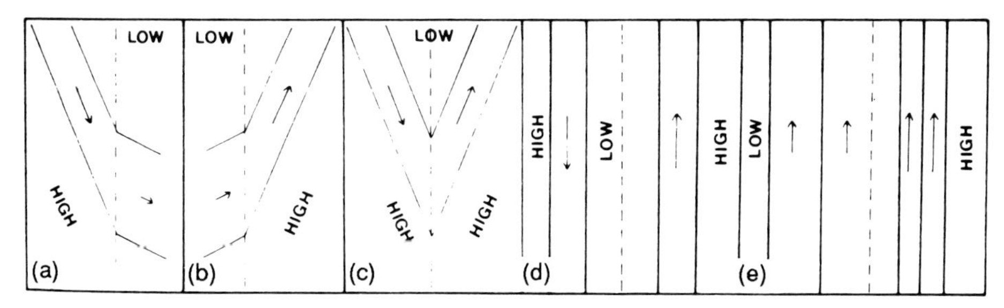

Figure 2.4 Examples ((a)–(e)) of surface pressure fields (isobars, solid lines) associated with frontal boundaries (dashed lines). Vectors indicate geostrophic wind (from Bluestein, 1986; adapted from Petterssen, 1956). (Courtesy of the American Meteorological Society)

Equation (2.2.4) may be written approximately as

$$\frac{dz}{dy} \approx \frac{f\bar{\rho}(u_{gw} - u_{gc})}{g(\rho_c - \rho_w)}, \tag{2.2.5}$$

where

$$\bar{\rho} = \frac{\rho_w + \rho_c}{2}. \tag{2.2.6}$$

This is essentially a statement of thermal-wind balance, since hydrostatic and geostrophic balance are assumed. In order that "dense" air not lie on top of "lighter" air, which is an unstable configuration, the front must slope toward the "heavy," cold air. It follows from the last statement and from Eq. (2.2.5) that

$$u_{gw} - u_{gc} > 0, \tag{2.2.7}$$

so that there must be cyclonic geostrophic shear vorticity across the front (Fig. 2.4). Unfortunately, the geostrophic vorticity is infinite at the front, an obviously unrealistic consequence of this model (except perhaps for the situation in which there is an elongated tornado embedded in the front!).

If there is also an along-the-front pressure gradient [see, e.g., Fig. 2.4(c)], then the geostrophic wind also has a cross-front component (v_g). It follows that, since mass must be conserved at the frontal boundary,

$$\rho_w v_{gw} = \rho_c v_{gc}, \tag{2.2.8}$$

where v_{gw} and v_{gc} are the cross-front geostrophic wind components on the warm and cold sides, respectively. However, according to the kinematic boundary condition,

$$v_{gw} = v_{gc}. \tag{2.2.9}$$

But *both* Eqs. (2.2.8) and (2.2.9) cannot be satisfied because $\rho_w \neq \rho_c$. This is another unrealistic consequence of the model.

The model does, however, reasonably specify the slope of a front (dz/dy). From Eq. (2.2.4) and the ideal gas law, the following relation, "Margules'" formula, is obtained:

$$\frac{dz}{dy} = \frac{f\bar{T}}{g}\frac{(u_{gw} - u_{gc})}{T_w - T_c}, \tag{2.2.10}$$

where $\bar{T}$ is some representative temperature. It can be shown that strictly speaking $\bar{T}$ is not necessarily the average of T_w and T_c: In fact,

$$\bar{T} < T_c, \quad \text{if } u_{gc} > 0 \quad \text{and} \quad u_{gw} > 0,$$
$$\bar{T} > T_w, \quad \text{if } u_{gc} < 0 \quad \text{and} \quad u_{gw} < 0,$$
$$T_c < \bar{T} < T_w, \quad \text{if } u_{gc} < 0 \quad \text{and} \quad u_{gw} > 0.$$

Since the fractional change in T is small compared to the fractional change in u_g, we may, for computational purposes, use Eq. (2.2.10), where $\bar{T}$ is the average of T_w and T_c. For typical values of wind and temperature in midlatitude fronts,

$$\frac{dz}{dy} \sim \frac{(10^{-4}\ \text{s}^{-1})(300\ \text{K})}{(10\ \text{m s}^{-2})}\frac{10\ \text{m s}^{-1}}{10\ \text{K}} \sim \frac{1}{300}. \tag{2.2.11}$$

One cannot infer naively that strong fronts (large $T_w - T_c$) have smaller slopes than weak fronts, because strong fronts, it will soon be seen, are likely to have larger values of vorticity (i.e., larger values of $u_{gw} - u_{gc}$ across the front) for dynamical reasons. Also, for given values of temperature contrast and vorticity, the slope of a front is shallower at low latitudes than it is at midlatitudes.

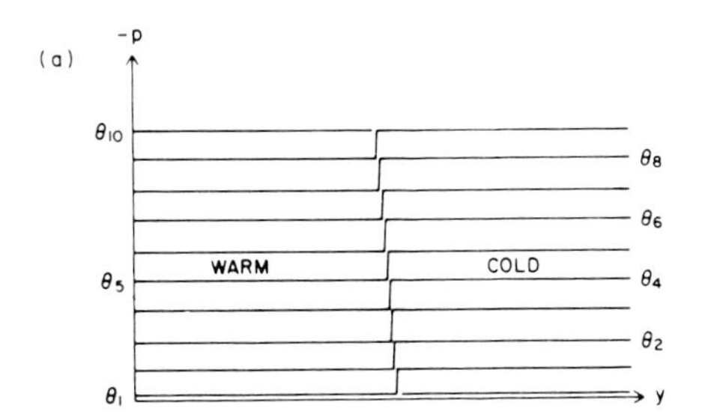

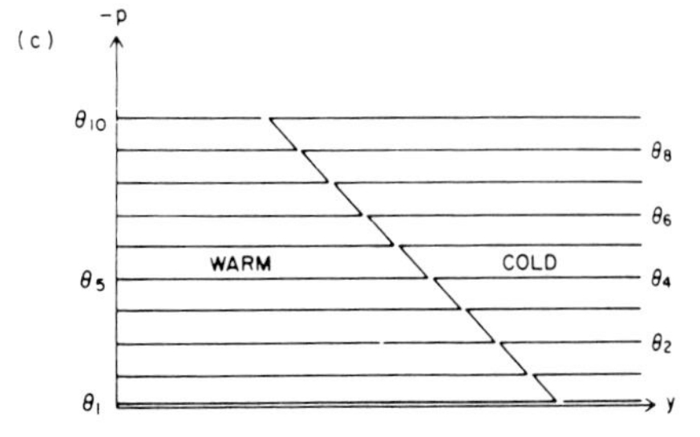

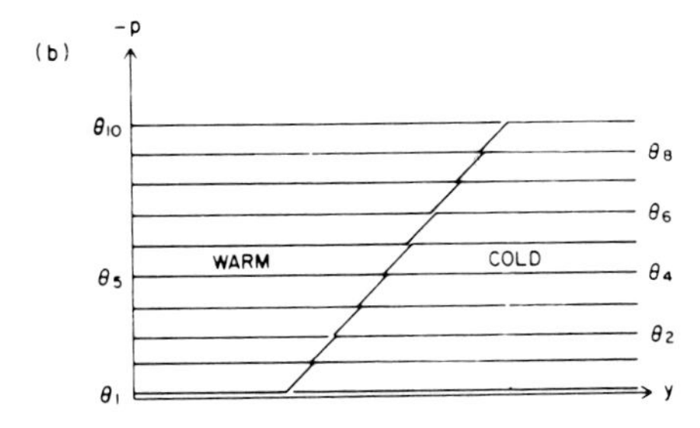

Figure 2.5 The vertical distribution of potential temperature θ (solid lines) across a frontal boundary that (a) is vertical, (b) slopes toward the "cold" air, and (c) slopes toward the "warm" air (from Bluestein, 1986). (Courtesy of the American Meteorological Society)

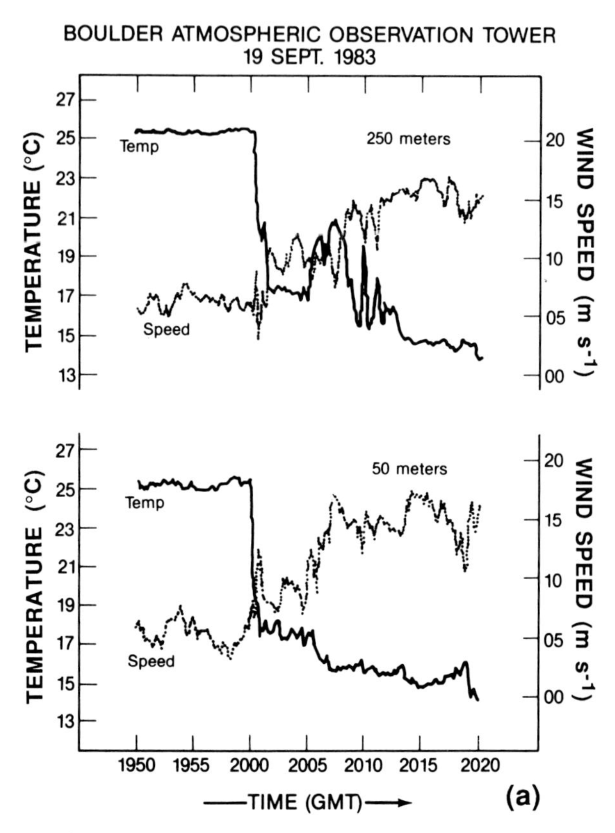

Figure 2.6 Example of the passage of a surface front just east of the Rocky Mountains. (a) Temperature in °C (solid lines) and wind speed in m s^{-1} (dotted lines) for 1950–2020 UTC (GMT), September 19, 1983, measured at the 250-m level (top) and 50-m level (bottom) of the Boulder Atmospheric Observation (BAO) tower in Boulder, Colorado. (b) Analysis of potential temperature in K (solid lines) as a function of height above the ground and time as in (a) up to 600 m AGL. Tower wind vectors preceding and following frontal passage are plotted at the indicated times. The cross-frontal scale is only a few kilometers or less. Such a scale is much less than the spacing between adjacent conventional surface observing stations (from Shapiro et al., 1985). (Courtesy of the American Meteorological Society)

Figure 2.5 shows vertical cross sections ("frontal" lobotomies) through quasihorizontal discontinuities in temperature in a continuously stratified (i.e., ρ is a monotonic, continuous function of height; in Fig. 2.3, ρ varies discontinuously with height, and takes on only one of two values) atmosphere. In Fig. 2.5(c) we see that if "cold" air "overlies" warm air, there is a local region of static instability: θ decreases with height in the frontal zone. The atmosphere as depicted in Fig. 2.5(a) is not realistic, because in the frontal zone the thermal wind is infinite. The situation shown in Fig. 2.5(b) is analogous to that in Fig. 2.3.

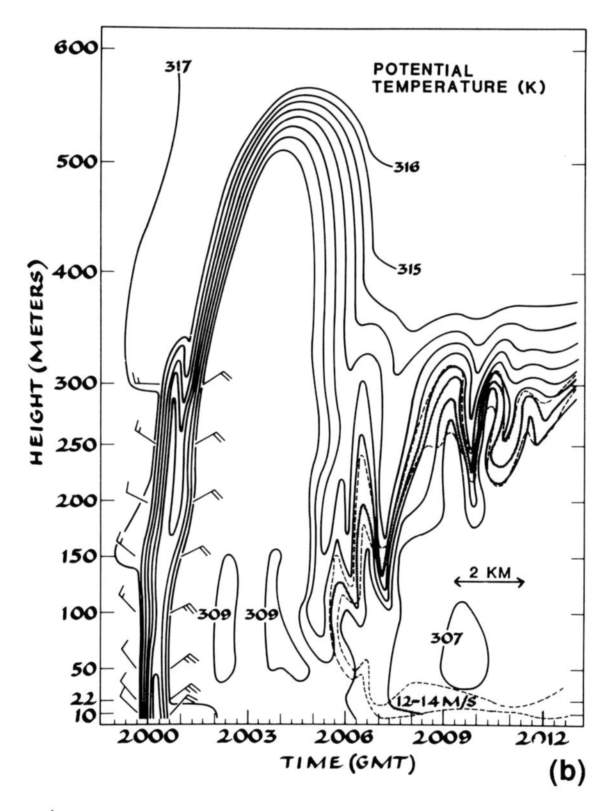

Figure 2.6 *(cont.)*

2.2.2. The Front as a Discontinuity in Temperature Gradient

Fronts are usually modeled as a discontinuity in temperature *gradient,* not in temperature. Temperature, however, is continuous across the front. Thus, a front is regarded as a finite zone, and not as an infinitesimal boundary, as usually depicted on weather maps. The zone may be as narrow as 1 km or less (Fig. 2.6), or as wide as 10–100 km (Fig. 2.7). When it gets as narrow as hundreds of meters, it may appear or behave like a "density" or "gravity" current, that is like evaporatively cooled outflow from convective storms (see Chap. 3). A density current is the flow of a relatively dense fluid that is embedded within a lighter fluid. The driving force is the hydrostatic pressure gradient across the boundary between the dense and light fluids (Fig. 2.8): When the horizontal scale across the front is very small, the horizontal pressure gradient may be very large and unbalanced, or balanced by friction.

A zone of strong horizontal temperature gradient may tilt with height toward either the warm or cold air mass (Fig. 2.9). When the zone of strong

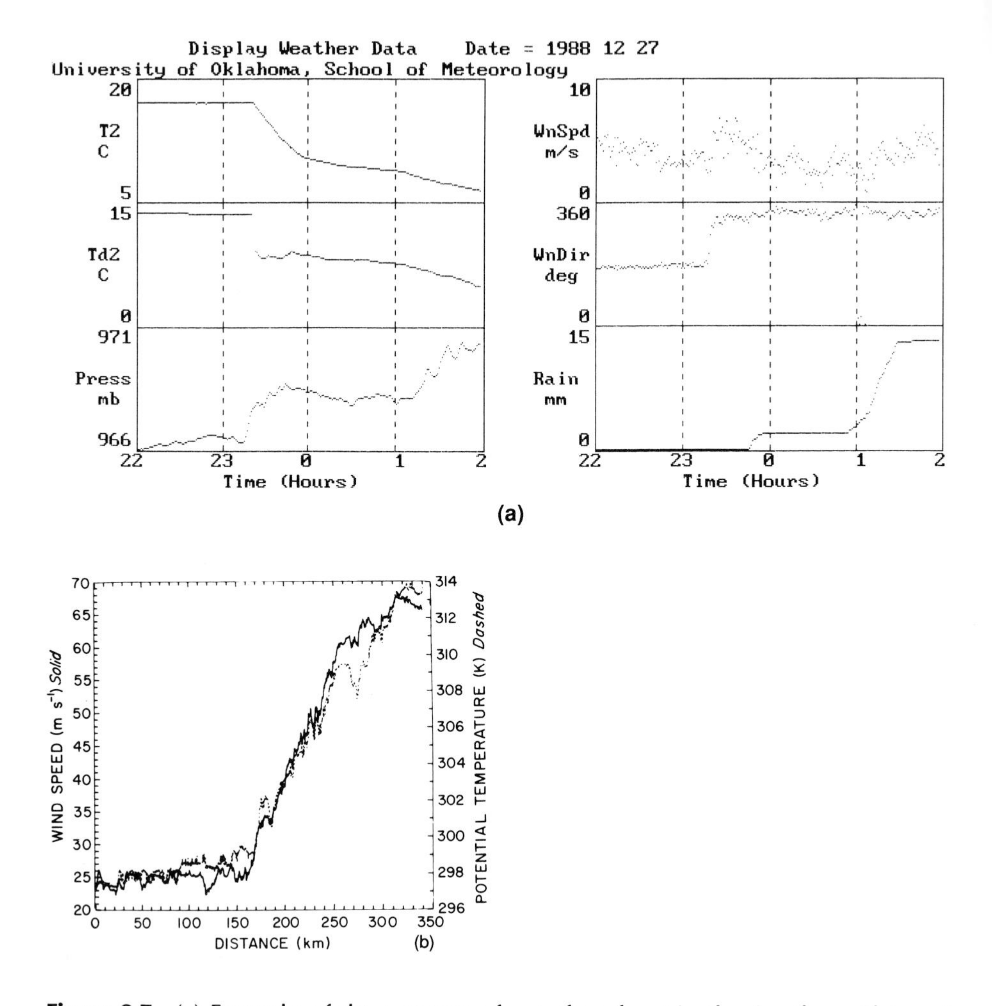

Figure 2.7 (a) Example of the passage of a surface front in the Southern Plains of the United States. Temperature $T2$ (°C), dew point $Td2$ (°C), station pressure (mb), wind speed ($\mathrm{m\,s^{-1}}$), wind direction (deg), and cumulative rainfall (mm) as a function of time (local standard time) at Norman, Oklahoma, on December 27, 1988. Frontal passage occurs at about 23 h 20 min; the temperature falls rapidly, the dew point drops off even more rapidly, and the pressure rises rapidly. The wind picks up in speed and shifts from south to north when the front passes by. Rainfall does not begin until after the front has gone by. The sharpest drop in temperature occurs during the first 30 min after frontal passage. If the front moves at $20\,\mathrm{m\,s^{-1}}$ and maintains its intensity, the scale across the front is $30\,\mathrm{min} \times 20\,\mathrm{m\,s^{-1}}$ (i.e., approximately 30–40 km). Such a scale is much shorter than the spacing between adjacent conventional surface observing stations. (Data courtesy Fred Brock, School of Meteorology, University of Oklahoma.) (b) Example of the profile across a midtropospheric front. Wind speed in $\mathrm{m\,s^{-1}}$ (solid line) and potential temperature in K (dashed line) at 465 mb in the southwestern United States on April 16, 1976, as recorded by an instrumented aircraft. The scale across the front is approximately 125 km (from Shapiro, 1978). (Courtesy of the American Meteorological Society)

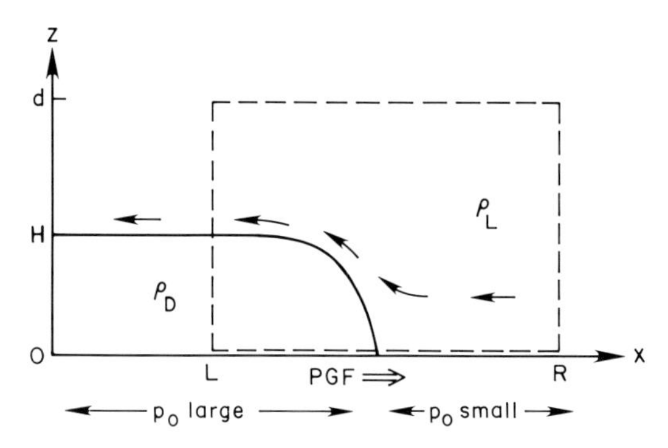

Figure 2.8 Schematic diagram of a stagnant density current. A dense fluid having density ρ_D undercuts a less dense fluid having density ρ_L. In this example the denser fluid is *H* units deep. The hydrostatic surface pressure p_o under the density current is higher than the hydrostatic surface pressure in the ambient fluid, owing to the heavier fluid in the density current. The sharp pressure gradient force (**PGF**), which is directed from the density current to the ambient fluid, accelerates the denser air in the direction opposite to the surface density gradient (from left to right in the figure). The flow relative to the density current is indicated by arrows. In order for a front to be considered a density current, the pressure gradient force must overwhelm all other forces. Letters *R, L,* and *d,* and the dashed line, will be discussed in the text.

temperature gradient slopes toward the warm air [Fig. 2.9(b)], it is characterized by low static stability rather than high static stability, and therefore does not exactly fit our definition of a front. A zone of strong temperature gradient that slopes toward the cold air, on the other hand, is characterized by high static stability, and does in fact fit our definition of a front [Fig. 2.9(a)]. The configuration shown in Fig. 2.9(b), owing to the low static stability, may be susceptible to cumulus convection if adequate moisture is available. When high-θ_w air at low levels ahead of a front is overrun by low-θ_w air aloft, then the front in terms of θ_w tilts toward the warm air, and cumulus convection may occur. This type of a front has been called a *split front,* since the θ_w cold front aloft is displaced from the surface θ_w cold front.

The strength of a front is the magnitude of the horizontal or quasihorizon-

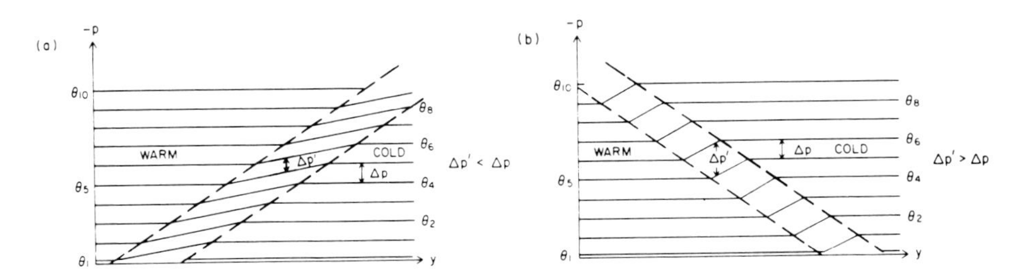

Figure 2.9 The vertical distribution of potential temperature θ (solid lines) across a frontal zone (dashed lines) that (a) slopes toward the "cold" air; (b) slopes toward the "warm" air (from Bluestein, 1986). (Courtesy of the American Meteorological Society)

tal temperature gradient. It is often convenient to define the strength of a front as the magnitude of the horizontal potential temperature gradient on a constant-pressure surface, since potential temperature is conserved for adiabatic processes.

2.3. KINEMATICS AND THERMODYNAMICS OF FRONTOGENESIS

2.3.1. Two-Dimensional Frontogenesis

The formation of a front is called *frontogenesis,* while the decay of a front is called *frontolysis.* These processes may be described quantitatively in terms of the *frontogenetical function* (also called the *frontogenesis function*):

$$F = \frac{D}{Dt} |\nabla_p \theta| . \tag{2.3.1}$$

The frontogenetical function quantifies the amount of change in potential temperature gradient following air-parcel motion. Although positive values of F do not necessarily indicate that a front will in fact form, and negative values of F do not necessarily indicate that an existing front will dissipate, a description of the atmosphere in terms of F has proved to be a useful concept for understanding frontal processes.

Strictly speaking, one should consider F in a reference frame moving along with the front, that is, in a "quasi-Lagrangian" reference frame. Although a front may neither intensify nor weaken, an air parcel experiences frontogenesis as it enters the frontal zone and frontolysis as it leaves it. In practice, it is difficult to work in a quasi-Lagrangian frame because a front's motion vector usually varies as a function of location along the front. In any event, the physical processes described by F are Galilean invariant and are therefore of fundamental importance.

Consider, for example, a frontal zone aligned along the x axis. Suppose that the isentropes (isotherms of potential temperature) are parallel to the front and there are no variations in the wind field along the front. (In the real atmosphere, there are usually along-the-front variations in wind. They are not considered in this discussion in order to simplify the problem, and to focus on the significant physical mechanisms responsible for frontogenesis. They are considered in the following, however.) Suppose also that the temperature on a constant-pressure surface decreases to the north, that is, with increasing y (Fig. 2.10). Let us use the thermodynamic equation expressed in the following form:

$$\frac{D\theta}{Dt} = \left(\frac{p_0}{p}\right)^{\kappa} \frac{1}{C_p} \frac{dQ}{dt}, \tag{2.3.2}$$

where $\kappa = R/C_p$, and dQ/dt includes latent heat exchange and radiation. From Eqs. (2.3.1) and (2.3.2) it is found that

$$F = \frac{D}{Dt}\left(-\frac{\partial \theta}{\partial y}\right)_p = \left(\frac{\partial v}{\partial y}\right)_p \left(\frac{\partial \theta}{\partial y}\right)_p + \left(\frac{\partial \omega}{\partial y}\right)_p \frac{\partial \theta}{\partial p} - \frac{1}{C_p}\left(\frac{p_0}{p}\right)^{\kappa}\left(\frac{\partial}{\partial y}\right)_p \left(\frac{dQ}{dt}\right). \tag{2.3.3}$$

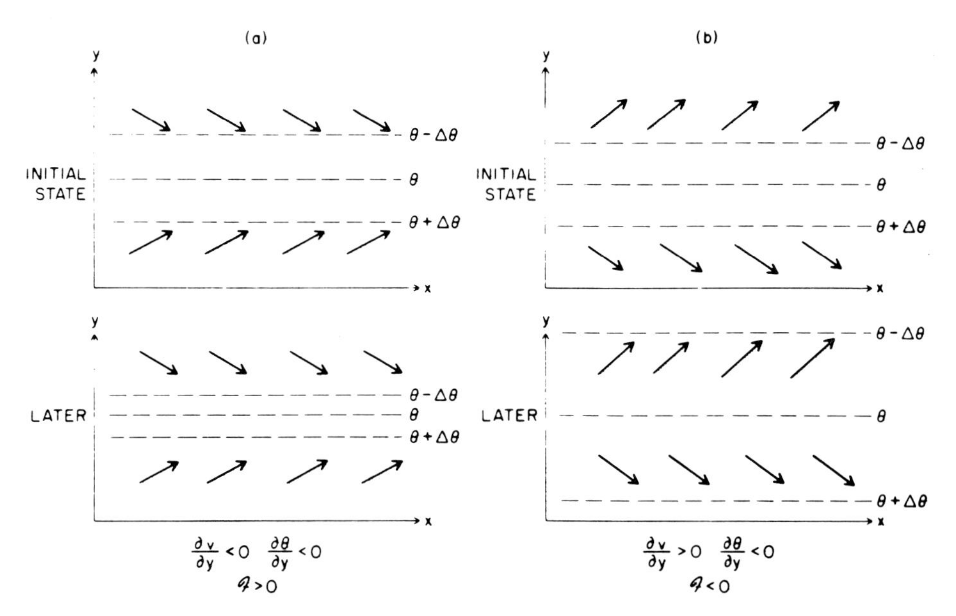

Figure 2.10 The effects of (a) confluence (frontogenesis) and (b) diffluence (frontolysis) acting on a quasihorizontal potential-temperature (dashed lines) gradient (from Bluestein, 1986). (Courtesy of the American Meteorological Society)

The first term, $(\partial v/\partial y)(\partial\theta/\partial y)$, represents the kinematic effect of confluence (diffluence) on the quasihorizontal temperature gradient. Negative $\partial v/\partial y$, confluence, acts to increase $|\nabla_p \theta| = -\partial\theta/\partial y$ [Fig. 2.10(a)], while positive $\partial v/\partial y$, diffluence, acts to decrease $|\nabla_p \theta|$ [Fig. 2.10(b)]. Confluence and diffluence contribute to both convergence (divergence) and quasihorizontal (nondivergent) deformation. The first term represents the thermodynamic effect of a quasihorizontal gradient in quasihorizontal temperature advection. Thus, cold advection on the cold side and warm advection on the warm side increase the temperature gradient.

The second term, $(\partial\omega/\partial y)(\partial\theta/\partial p)$, represents kinematically the tilting of the vertical potential temperature gradient $(\partial\theta/\partial p)$ onto the horizontal (Fig. 2.11). The second term also represents thermodynamically a quasihorizontal gradient of adiabatic temperature change owing to a quasihorizontal gradient in vertical motion. In a statically stable atmosphere, rising motion and its associated adiabatic cooling on the cold side, and sinking motion and its associated adiabatic warming on the warm side, increase the temperature gradient.

The third term, $(-1/C_p)(p_0/p)^\kappa(\partial/\partial y)(dQ/dt)$, represents a quasihorizontal variation in diabatic heating. For example, heating of the warm side of a front by the sun during the day, if it is clear there, without heating on the cold, cloudy side, if it is cloudy there, is a frontogenetical process. At night the effect of long-wave cooling is dominant, so that cooling on the clear, warm side, with restricted cooling on the cold, cloudy side, represents a frontolytical process. Differential heating along the boundary between snow cover and bare ground can act frontogenetically. If a surface cold front is very shallow, and

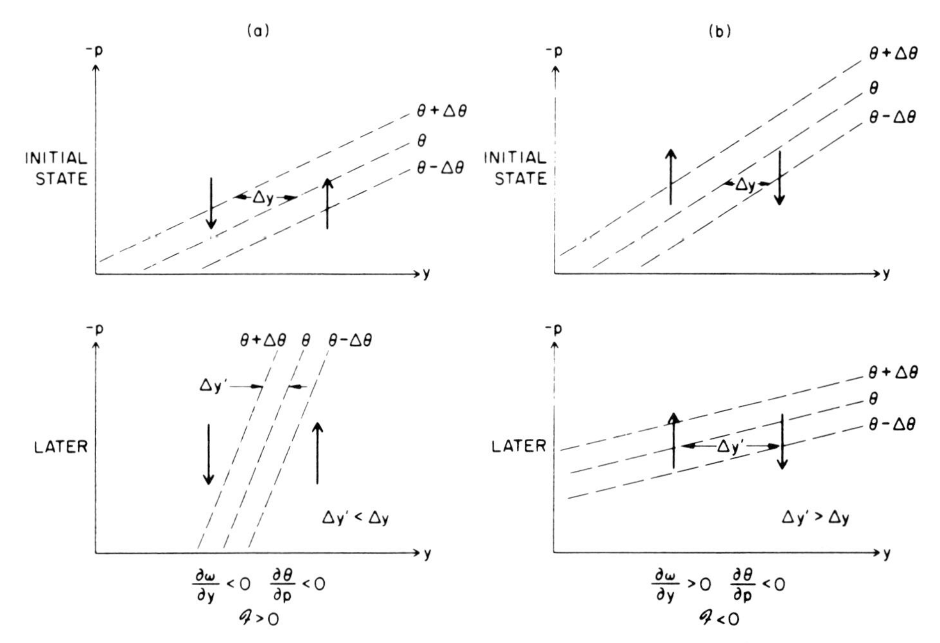

Figure 2.11 The effects of tilting on a vertical potential temperature (dashed lines) gradient: (a) frontogenesis; (b) frontolysis (from Bluestein, 1986). (Courtesy of the American Meteorological Society)

relatively warm air is present above the shallow cold air mass, then heating by the sun may result in the total destruction of the cold air mass, and hence also in the destruction of the front. In the High Plains area of the United States, a shallow cold front may be turned into a dryline by this mechanism. The efficiency of a horizontal variation in diabatic heating in contributing to F increases with "height" via the factor $(p_0/p)^\kappa$.

The kinematic effects of the horizontal component of the wind field can be examined in more detail if the effects of tilting and diabatic heating are neglected, while the restriction that the wind does not vary along the front is removed. Expanding Eq. (2.3.1) for quasihorizontal motion only, we find that

$$F = \frac{\partial}{\partial t}|\nabla_p \theta| + \mathbf{v} \cdot \nabla_p |\nabla_p \theta|$$
$$= \frac{\partial}{\partial t}\left[\left(\frac{\partial \theta}{\partial x}\right)^2 + \left(\frac{\partial \theta}{\partial y}\right)^2\right]^{1/2} + \mathbf{v} \cdot \nabla_p\left[\left(\frac{\partial \theta}{\partial x}\right)^2 + \left(\frac{\partial \theta}{\partial y}\right)^2\right]^{1/2}. \tag{2.3.4}$$

It follows that

$$F = \frac{1}{|\nabla_p \theta|}\left[\frac{\partial \theta}{\partial x}\frac{D_p}{Dt}\left(\frac{\partial \theta}{\partial x}\right) + \frac{\partial \theta}{\partial y}\frac{D_p}{Dt}\left(\frac{\partial \theta}{\partial y}\right)\right]. \tag{2.3.5}$$

Now

$$\frac{\partial}{\partial x}\left(\frac{D_p \theta}{Dt}\right) = \frac{D_p}{Dt}\left(\frac{\partial \theta}{\partial x}\right) + \frac{\partial u}{\partial x}\frac{\partial \theta}{\partial x} + \frac{\partial v}{\partial x}\frac{\partial \theta}{\partial y} = 0 \tag{2.3.6}$$

$$\frac{\partial}{\partial y}\left(\frac{D_p \theta}{Dt}\right) = \frac{D_p}{Dt}\left(\frac{\partial \theta}{\partial y}\right) + \frac{\partial u}{\partial y}\frac{\partial \theta}{\partial x} + \frac{\partial v}{\partial y}\frac{\partial \theta}{\partial y} = 0 \tag{2.3.7}$$

if the atmosphere is adiabatic. From the latter two equations and Eq. (2.3.5), the following is obtained:

$$F = \frac{1}{|\nabla_p \theta|}\left[-\left(\frac{\partial\theta}{\partial x}\right)^2 \frac{\partial u}{\partial x} - \frac{\partial\theta}{\partial y}\frac{\partial\theta}{\partial x}\frac{\partial v}{\partial x} - \frac{\partial\theta}{\partial x}\frac{\partial\theta}{\partial y}\frac{\partial u}{\partial y} - \left(\frac{\partial\theta}{\partial y}\right)^2 \frac{\partial v}{\partial y}\right]. \tag{2.3.8}$$

But

$$\frac{\partial u}{\partial x} = \tfrac{1}{2}(\delta + D_1) \tag{2.3.9}$$

$$\frac{\partial v}{\partial x} = \tfrac{1}{2}(\zeta + D_2) \tag{2.3.10}$$

$$\frac{\partial u}{\partial y} = -\tfrac{1}{2}(\zeta - D_2) \tag{2.3.11}$$

$$\frac{\partial v}{\partial y} = \tfrac{1}{2}(\delta - D_1), \tag{2.3.12}$$

where δ, ζ, D_1, and D_2 are horizontal divergence, vertical vorticity, and the components of deformation ($D_1 = \partial u/\partial x - \partial v/\partial y$, $D_2 = \partial v/\partial x + \partial u/\partial y$). Then

$$F = -\frac{1}{2\,|\nabla_p \theta|}\left[\left(\frac{\partial\theta}{\partial x}\right)^2(\delta + D_1) - \left(\frac{\partial\theta}{\partial x}\right)\left(\frac{\partial\theta}{\partial y}\right)(\zeta - D_2) + \left(\frac{\partial\theta}{\partial x}\right)\left(\frac{\partial\theta}{\partial y}\right)(\zeta + D_2) + \left(\frac{\partial\theta}{\partial y}\right)^2(\delta - D_1)\right]. \tag{2.3.13}$$

Rotating the coordinate system so that $D_2' = 0$, that is, so that either the axis of dilatation or axis of contraction lies along the x' axis, we find that

$$F = -\tfrac{1}{2}|\nabla_p \theta|\left\{\delta + D_1'\left[\left(\frac{\partial\theta}{\partial x}\right)^2 - \left(\frac{\partial\theta}{\partial y}\right)^2\right]\Big/|\nabla_p \theta|^2\right\}. \tag{2.3.14}$$

Let α be the angle $\nabla_p \theta$ makes with the x' axis (Fig. 2.12). Then

$$\cos\alpha = \left(\frac{\partial\theta}{\partial x'}\right)\Big/|\nabla_p \theta| \tag{2.3.15}$$

$$\sin\alpha = \left(\frac{\partial\theta}{\partial y'}\right)\Big/|\nabla_p \theta|. \tag{2.3.16}$$

The angle between the isotherms and the x' axis is b, where

$$b - \alpha = 90°. \tag{2.3.17}$$

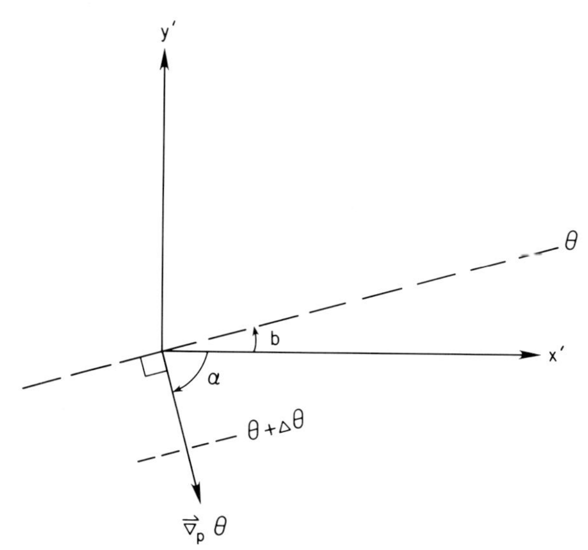

Figure 2.12 Geometrical relationship between the orientation of the axes of dilatation and contraction and the isotherms of potential temperature θ (dashed line), and the angles b and α, measured counterclockwise and clockwise, respectively, from the x' axis. When $D > 0$, x' is along the axis of dilatation, and y' is along the axis of contraction; when $D < 0$, x' is along the axis of contraction, and y' is along the axis of dilatation.

It follows that

$$F = -\tfrac{1}{2}|\nabla_p \theta||D_1'(\sin^2 b - \cos^2 b) + \delta] \qquad (2.3.18)$$

$$= \tfrac{1}{2}|\nabla_p \theta|(D \cos 2b - \delta), \qquad (2.3.19)$$

where D is the resultant deformation in the new coordinate system.

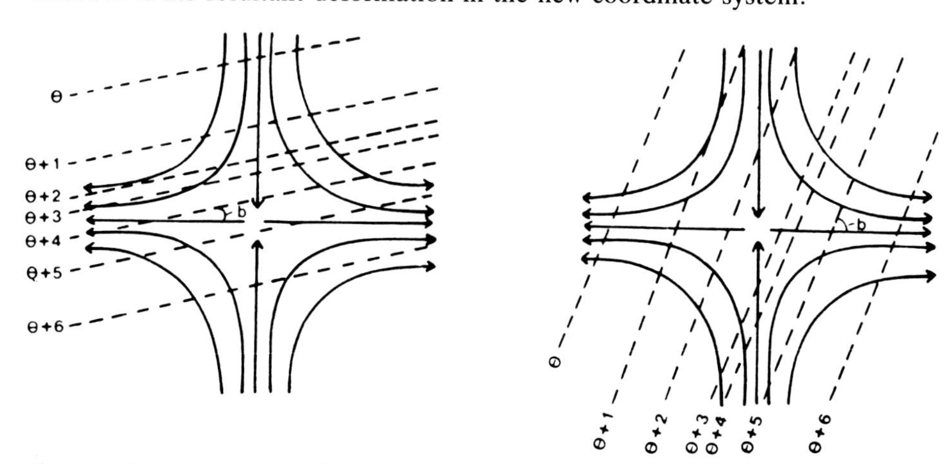

Figure 2.13 The relationship between the relative orientation of the axis of dilatation and the isotherms of potential temperature (dashed lines) as measured by angle b, and frontogenesis. (left) Frontogenesis ($0° < b < 45°$), $F > 0$. (right) Frontolysis ($45° < b < 90°$), $F < 0$ (from Bluestein, 1986; after Petterssen, 1956). (Courtesy of the American Meteorological Society)

When the x' axis is oriented along the axis of dilatation (axis of contraction), D is positive (negative). Thus, the effect of horizontal deformation alone is to promote frontogenesis when the axis of dilatation lies within 45° of the isotherms, and to promote frontolysis when the axis of dilatation lies between 45° and 90° of the isotherms (Fig. 2.13).

Convergence acts frontogenetically and divergence acts frontolytically independent of the orientation of $\nabla_p \theta$. Although deformation and divergence (convergence) appear in Eq. (2.3.19), vorticity does not. Vorticity can only rotate the isotherms; it cannot push them closer together or farther apart. However, vorticity may rotate the isotherms so that they become more aligned with the axis of dilatation. Hence, vorticity can act *indirectly* to affect frontogenesis. (This matter will be addressed in the following.)

2.3.2. Three-dimensional frontogenesis

With the advent of a network of upper-air observations in the late 1940s, the concept of a frontal zone was broadened to include vertical gradients of potential temperature. The complete three-dimensional frontogenetical function is therefore defined as

$$F = \frac{D}{Dt} |\nabla \theta|. \tag{2.3.20}$$

Height is used as the vertical coordinate so that the vertical component of the gradient of θ can be combined more readily with the horizontal gradient to obtain the three-dimensional gradient. The thermodynamic equation, Eq. (2.3.2), is used to eliminate $D\theta/Dt$, so that

$$\begin{aligned} F = \frac{1}{|\nabla \theta|} \Bigg(& \frac{\partial \theta}{\partial x} \left\{ \frac{1}{C_p} \left(\frac{p_0}{p} \right)^{\kappa} \left[\frac{\partial}{\partial x} \left(\frac{dQ}{dt} \right) \right]_1 - \left(\frac{\partial u}{\partial x} \frac{\partial \theta}{\partial x} \right)_2 - \left(\frac{\partial v}{\partial x} \frac{\partial \theta}{\partial y} \right)_3 - \left(\frac{\partial w}{\partial x} \frac{\partial \theta}{\partial z} \right)_4 \right\} \\ & + \frac{\partial \theta}{\partial y} \left\{ \frac{1}{C_p} \left(\frac{p_0}{p} \right)^{\kappa} \left[\frac{\partial}{\partial y} \left(\frac{dQ}{dt} \right) \right]_5 - \left(\frac{\partial u}{\partial y} \frac{\partial \theta}{\partial x} \right)_6 - \left(\frac{\partial v}{\partial y} \frac{\partial \theta}{\partial y} \right)_7 - \left(\frac{\partial w}{\partial y} \frac{\partial \theta}{\partial z} \right)_8 \right\} \\ & + \frac{\partial \theta}{\partial z} \left\{ \frac{p_0^{\kappa}}{C_p} \left[\frac{\partial}{\partial z} \left(p^{-\kappa} \frac{dQ}{dt} \right) \right]_9 - \left(\frac{\partial u}{\partial z} \frac{\partial \theta}{\partial x} \right)_{10} - \left(\frac{\partial v}{\partial z} \frac{\partial \theta}{\partial y} \right)_{11} - \left(\frac{\partial w}{\partial z} \frac{\partial \theta}{\partial z} \right)_{12} \right\} \Bigg) \end{aligned} \tag{2.3.21}$$

The reader should verify that Eqs. (2.3.3) and (2.3.19) are special cases of Eq. (2.3.21).

Terms 1, 5, and 9 are the "diabatic" terms; terms 2, 3, 6, and 7 are the "horizontal-deformation" terms; terms 10 and 11 are the "vertical-deformation" terms; terms 4 and 8 are the "tilting" terms; and term 12 is the "vertical divergence" term. The physical interpretation of each term is just a generalization of the physical interpretation given to each term in Eq. (2.3.3).

Some of the terms in the frontogenetical function have been computed from observations (Fig. 2.14). Surface fronts are influenced largely by horizontal deformation and horizontal gradients in diabatic heating. Temperature advection is usually strongest near the ground; sensible heat fluxes are strong when the surface is much warmer or colder than the overlying air mass.

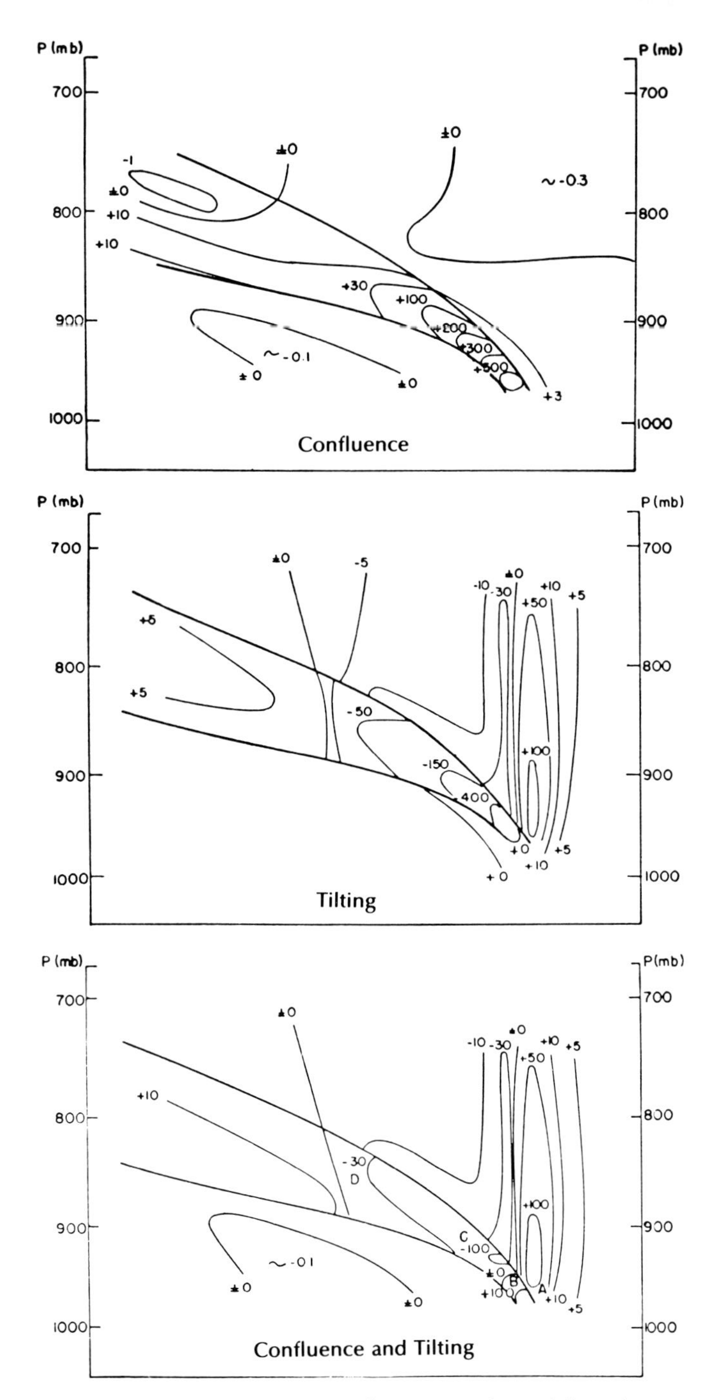

Figure 2.14 Calculation of the (top) confluence and (middle) tilting terms in the two-dimensional frontogenetical function as given by $[(\partial v/\partial y)(\partial \theta/\partial y)]$ and $[(\partial w/\partial y)(\partial \theta/\partial z)]$ (the vertical coordinate is height z, not p). Units expressed as 3-h changes in horizontal temperature gradient in °C $(100\,\text{km})^{-1}$. Positive values indicate frontogenesis in the temperature field. Net frontogenetical effect for an adiabatic atmosphere given in (bottom) (from Sanders, 1955). (Courtesy of the American Meteorological Society)

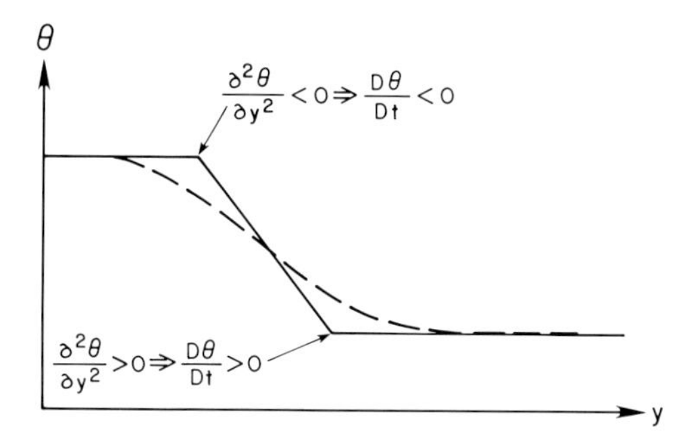

Figure 2.15 Effect of diffusion on a temperature gradient. Initial distribution of potential temperature θ as a function of y (solid line). Diffusion smooths out the kinks in the temperature field as indicated; final distribution of potential temperature (dashed line).

Near a level ground surface, where vertical motion is negligible, the tilting mechanism has no effect. On the other hand, tilting may be important when the slope of the ground with respect to a constant elevation above sea level changes significantly: The derivative of the slope is the significant feature, as the temperature gradient changes in response to surface variations in upslope or downslope motion. Another possibility is that the terrain slopes uniformly, but the upslope or downslope component of the wind is not uniform.

The tilting mechanism is very important in the middle and upper troposphere, where vertical motions are largest in magnitude. Since the static stability is strongest at and above the tropopause, the tilting mechanism is in general most significant in the upper troposphere. In the middle troposphere horizontal temperature advection is usually relatively small. As a result, the tilting mechanism is dominant. Horizontal and vertical deformation may be important near the tropopause if it slopes; thus horizontal temperature advection and its horizontal variation may be significant there.

Finally, suppose that the diffusion of heat is included in the thermodynamic equation, Eq. (2.3.2), so that

$$\frac{D\theta}{Dt} = K\,\nabla^2\theta, \qquad (2.3.22)$$

where K is an eddy diffusion coefficient (having units of $m^2\,s^{-1}$). If K is positive, then extrema in temperature are destroyed by diffusion. For example, the spatial variation in temperature depicted in Fig. 2.15 by the solid line will evolve into that depicted by the dashed line; that is, the magnitude of the temperature gradient will decrease.

2.4. OBSERVATIONAL ASPECTS OF FRONTS

In this section we will briefly consider some observational aspects of frontal zones. The kinematic structure of fronts will be emphasized. The dynamical aspects of fronts will be discussed in Section 5 of this chapter.

2.4.1. The Surface Front

A front whose intensity is strongest near the ground is called a *surface front* (or *low-level front* or *lower tropospheric front*). Surface fronts are usually found downstream from upper-level troughs and upstream from upper-level ridges, in a synoptic-scale environment characterized by rising motion; they may also be found, however, downstream from upper-level ridges and upstream from upper-level troughs. (Surface fronts are currently analyzed using Polar–Front theory symbols, even though many fronts do not fit the Polar–Front conceptual model.)

The cold front. When a "cold" air mass advances equatorward, equatorward and eastward, or eastward relative to the "warm" air mass, the front is called a *cold front.* (The cold front associated with a surface cyclone used to be called a *squall line.*) When the "cold" air advances westward or equatorward and

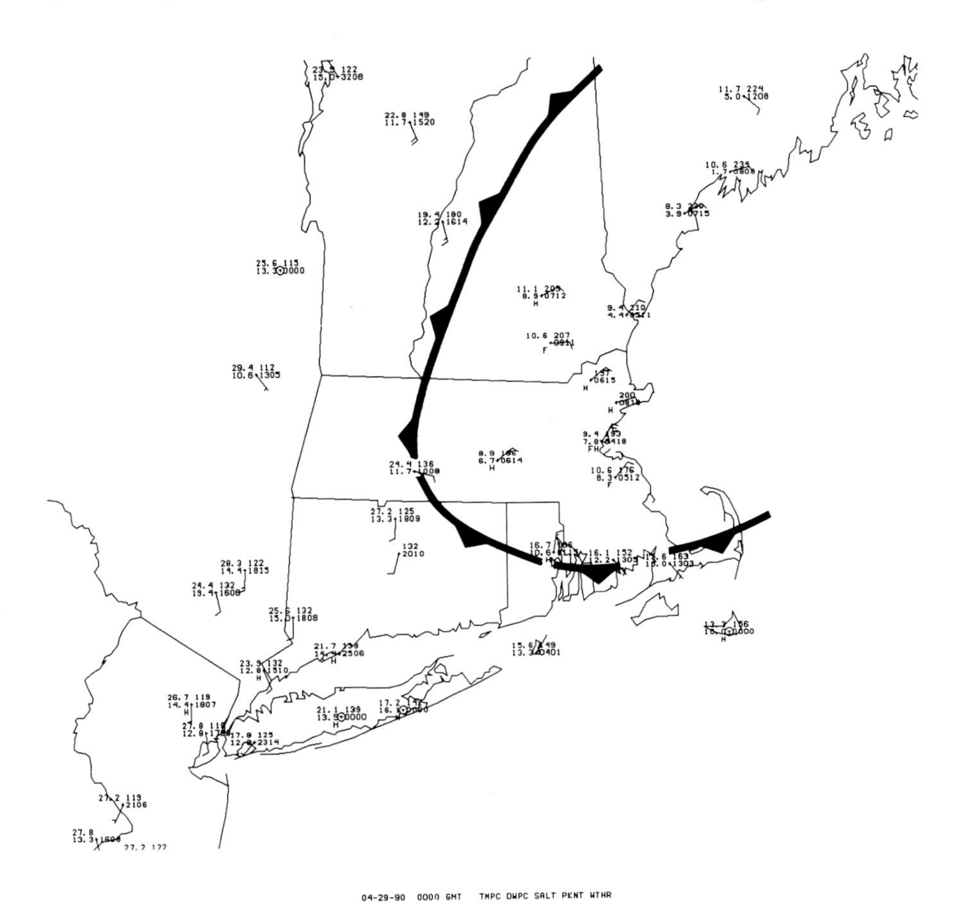

Figure 2.16 Example of a New England back-door cold front 0000 UTC, April 29, 1990. Winds east of the front are off the cold Atlantic; winds west of the front are off the warmer land. Temperature and dew point (°C) plotted according to convention. Pressure plotted is altimeter setting in tens of mb without the leading "10." Full wind barb = 5 m s^{-1}; half wind barb = 2.5 m s^{-1}.

westward, the front is called a *back-door cold front* because synoptic-scale weather systems in middle latitudes usually have an eastward component of motion. The former usually trail off in an equatorward direction from a surface cyclone, and are followed by a cold, surface anticyclone. Synoptic-scale vertical motion is usually upward, and an upper-level trough is found upstream. The latter are usually associated with a cold, surface anticyclone that expands toward the equator, equatorward and westward, or westward. Synoptic-scale vertical motion is usually downward, and an upper-level ridge is found upstream with respect to the flow aloft. Back-door cold fronts (Fig. 2.16) are common in the northeastern part of the United States during the spring and summer, east of the Rocky Mountains any time of year, and sometimes in the Great Plains during the period from fall through spring. The spring and summer New England back-door cold front is influenced strongly by the relatively cold ocean upstream.

The surface frontal zone (i.e., band of surface temperature gradient much greater in magnitude than the synoptic-scale quasihorizontal or horizontal temperature gradient) is usually found on the "cold" side of the wind shift associated with a surface pressure trough (Fig. 2.17). The wind veers (backs) along the wind-shift line equatorward (poleward) of the cyclone. It is common practice to identify the front symbol (line along which there are toothlike triangles pointing from the cold side toward the warm side, perhaps heralding the "biting" cold to come!) with the wind-shift line along which there is a maximum in cyclonic vorticity. Just as or shortly after the wind has shifted, the temperature drops and the pressure rises (Fig. 2.7(a)). One is in the frontal zone from the time the wind shifts direction and the temperature begins to drop until the time the temperature no longer falls rapidly.

A surface analysis, soundings, and a vertical cross section illustrating the structure of an intense cold front in the Plains of the United States are shown in Figs. 2.18, 2.19, and 2.20.

The surface front is most intense at the ground, and weakens with height (Fig. 2.20). It can intensify dramatically on time scales as short as 12 hours. The frontal zone is marked by strong static stability; it is associated with strong vertical wind shear, which is associated to some extent with the surface temperature gradient, in accord with the thermal wind relation. The frontal zone is steepest near the ground. A narrow jet of rising air is often located just

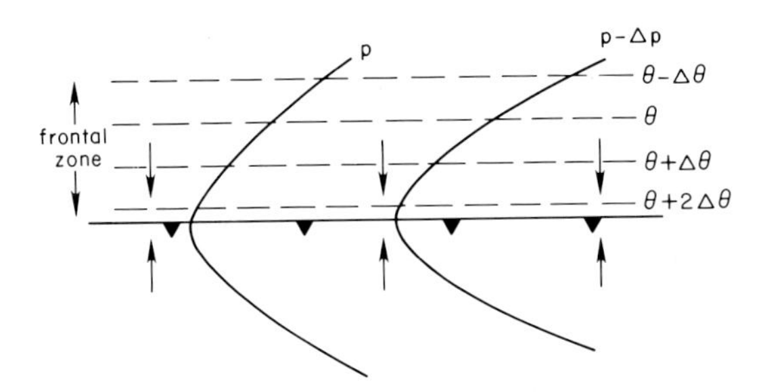

Figure 2.17 Idealized depiction of the distribution of potential temperature θ (dashed lines), pressure p (solid lines), and the divergent part of the wind field (vectors) (in the reference frame of the front) along a surface frontal zone.

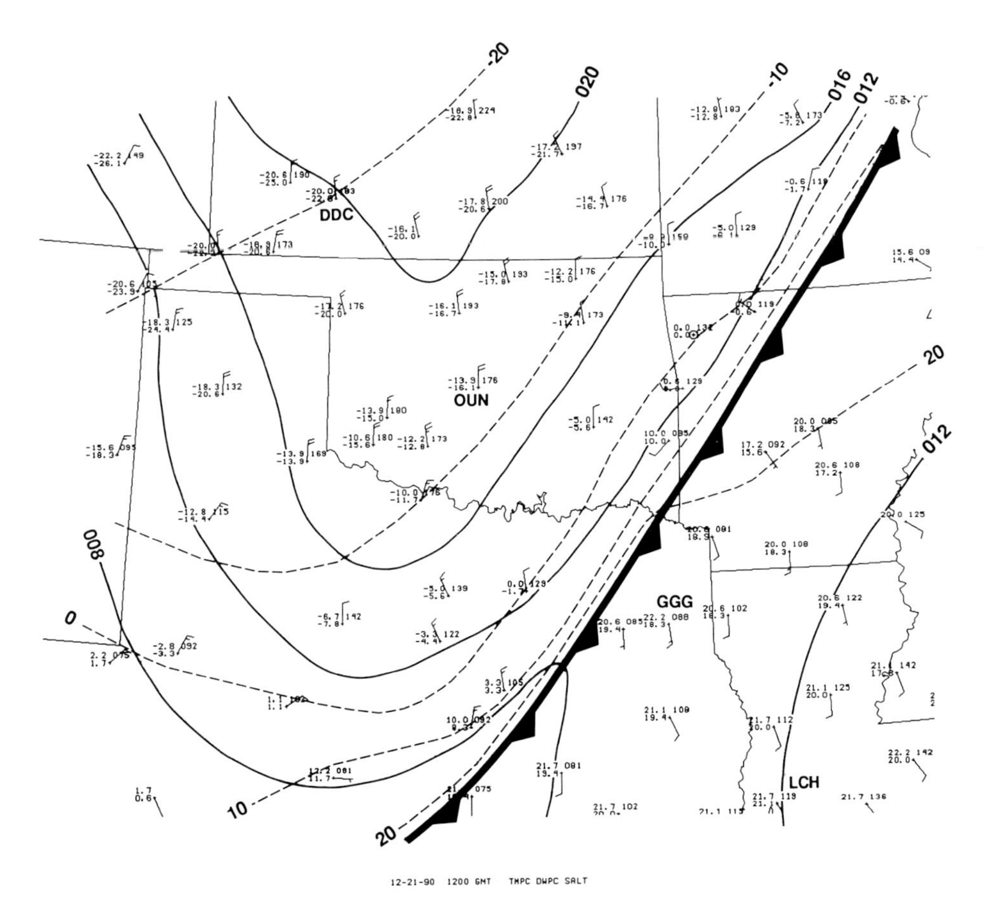

Figure 2.18 Surface analysis of an intense cold front in the Southern Plains of the United States at 1200 UTC (0600 LST), December 21, 1990. Winds south of the front have a southerly component; winds north of the front have a northerly component. Altimeter-setting analysis in mb without the leading 1 (solid lines); isotherms in °C (dashed lines); altimeter setting plotted in tens of mb, without the leading 10; temperature and dewpoint in °C; whole wind barb = 5 m s^{-1}; half wind barb = 2.5 s^{-1}. Sites of Lake Charles, Louisiana (LCH), Longview, Texas (GGG), Norman, Oklahoma (OUN), and Dodge City, Kansas (DDC) are marked; vertical cross section shown in Fig. 2.20 passes through these stations.

above the leading edge of the surface front. A narrow cloud line ("rope cloud") or band of precipitation may be located along the rising jet if there is sufficient moisture (Fig. 2.21). Below the frontal zone in the "cold" air at low levels, the lapse rate may be nearly dry adiabatic, and hence there is turbulent vertical mixing that creates gusty surface winds.

In the Southern Plains of the United States, the southward movement of cold anticyclones in the lee of the Rocky Mountains is occasionally accompanied by a boundary along which there is a rapid increase in wind speed and gustiness, but no wind shift or rapid change in temperature. These features are usually analyzed as fronts, and may become associated with stronger temperature gradients as they move southward (Fig. 2.22). Sometimes the wind gradually backs from southerly to northerly to the south of (i.e., in advance of) a cold front in the Southern Plains, and hence it is difficult to locate the front; sometimes the wind veers in advance of a front as a prefrontal trough (Fig. 2.23).

The movement of a cold front is not necessarily determined by the difference between the wind components normal to the front on either side of it. Movement is best correlated with the wind component normal to the front in the "cold" air. The movement of the surface cold front should be dependent upon the front-normal isallobaric gradient. The trough associated with a frontal zone moves from a region where surface pressures are rising to a region where surface pressures are falling (Fig. 2.24). Because the zone of temperature gradient is located on the "cold" side of the wind-shift line, we expect to find strong cold advection there, and according to quasigeostrophic theory, sinking motion and surface pressure rises. Thus, it is not surprising that frontal movement is associated with the front-normal wind speed in the "cold" air. However, if the front behaves like a density current, quasigeostrophic dynamics cannot be used to explain frontal movement because the air motions are nowhere geostrophic. Density-current motion will be discussed later in this chapter.

The distribution of clouds on either side of a front depends upon the vertical motion, the front-relative flow, and the availability of moisture upstream from the front (Fig. 2.25). The front-relative flow in eastward-moving fronts is likely to be rear to "front," owing to strong westerlies aloft; the front-relative flow in equatorward-moving fronts is more likely to be "front" to rear, especially if there is a poleward component of flow aloft downstream from a trough.

The classic Southern Plains analysis by Sanders in 1955 shows rising motion aloft on the cold side of the surface front. Thus, clouds may form and be advected toward the "cold" side of the front [Fig. 2.25(a)], and moisture may be mixed down to the surface in the presence of a near dry adiabatic lapse rate (as in Figs. 2.19 and 2.20). If the air "ahead" of the front, where vertical motion is weak, is relatively dry, it is clear there. Thus, it may become cloudy after surface frontal passage.

However, if the surface cold front extends equatorward from a developing cyclone, just downstream from a strong upper-level trough, the air "behind" the front aloft in the dry slot may have had a history of descent, and is therefore dry [Fig. 2.25(c)]. If the front-relative flow of this dry air is from the "cold" to the "warm" side of the front, then it will clear up just after frontal passage.

The downslope motion in the lee of the Appalachians can contribute to clearing behind East Coast cold fronts. On the other hand, if the Great Lakes are not frozen and the trajectory of cold air upstream from the front passes over them, then sufficient moisture may be evaporated into the air to produce cloudiness on the windward side of the Appalachians.

Back-door cold frontal passage in New England is frequently marked by fog and low clouds from the Atlantic. The "marine push" of western Washington is usually characterized by fog and low clouds from the Pacific. It

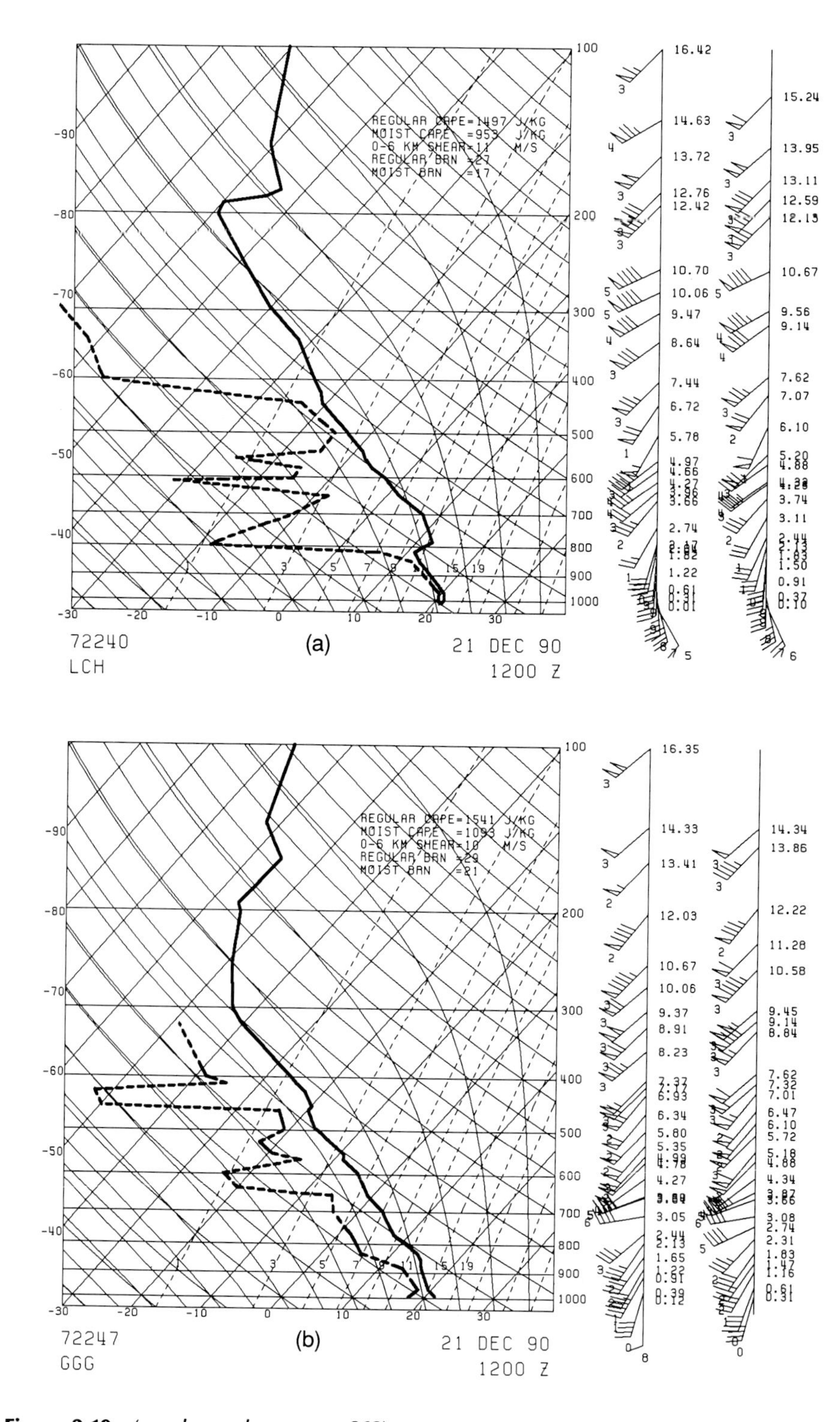

Figure 2.19 (*see legend on page 263*)

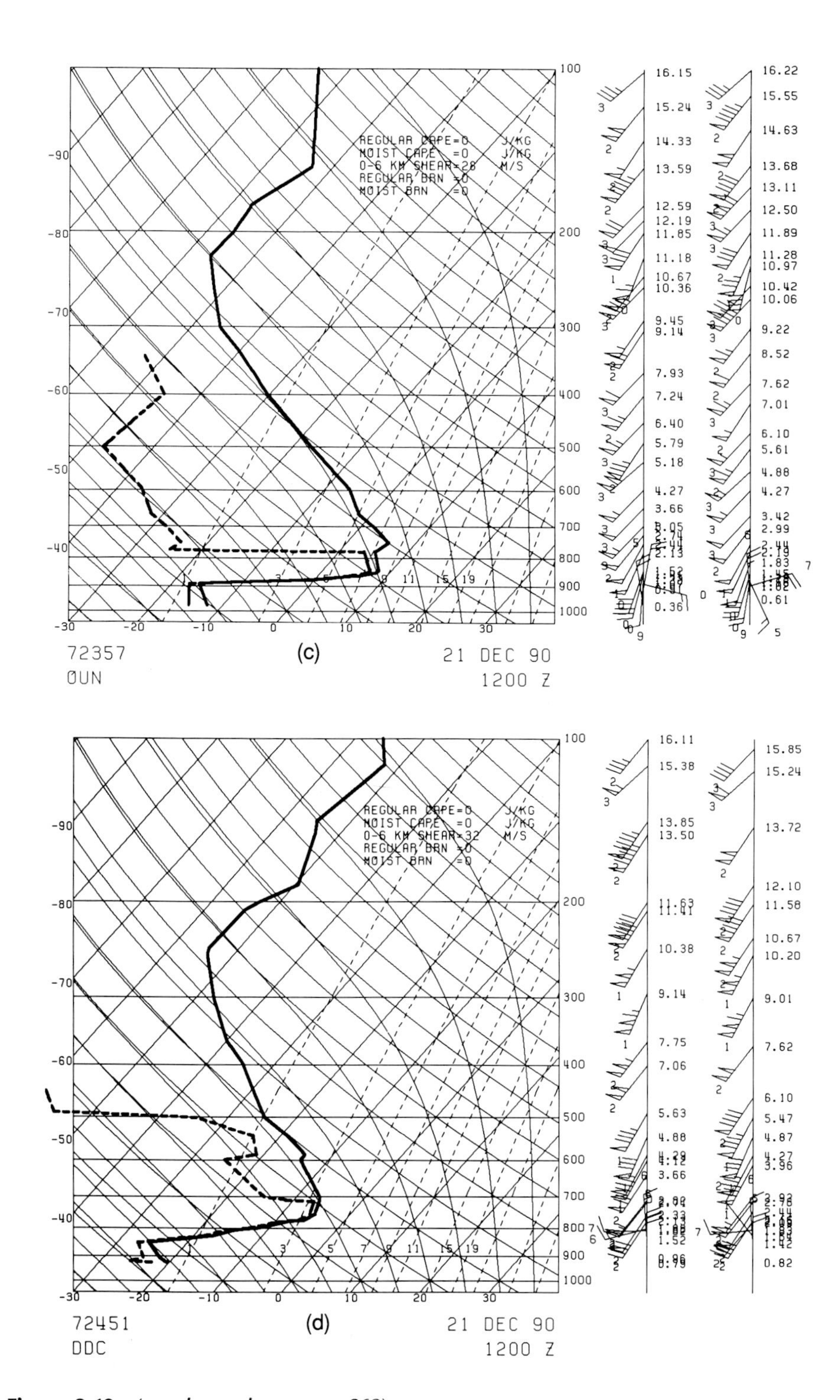

Figure 2.19 (*see legend on page 263*)

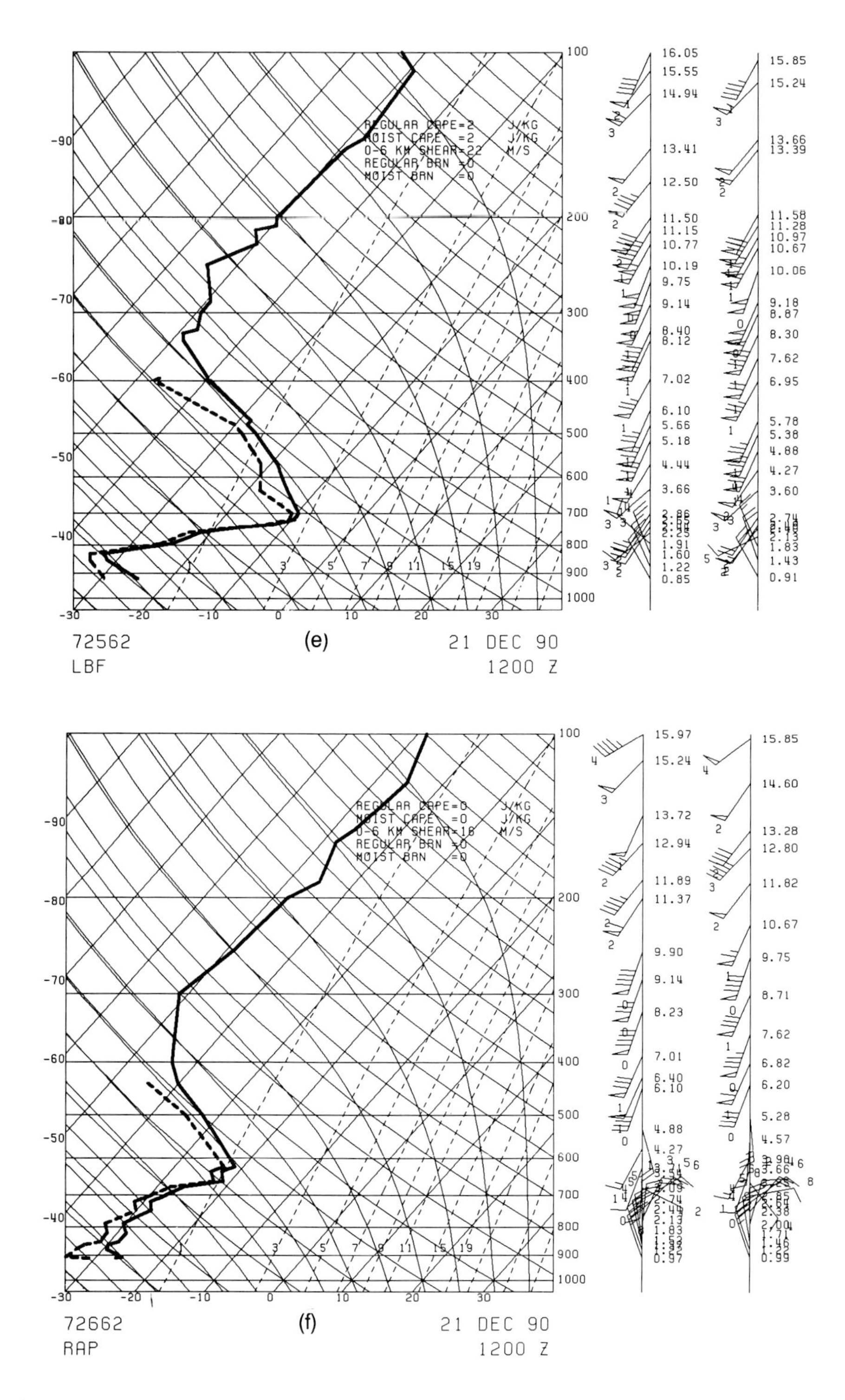

Figure 2.19 (*see legend on page 263*)

Figure 2.19 Soundings for 1200 UTC, December 21, 1990, at (a) Lake Charles, Louisiana (LCH), (b) Longview, Texas (GGG), (c) Norman, Oklahoma (OUN), (d) Dodge City, Kansas (DDC), (e) North Platte, Nebraska (LBF), and (f) Rapid City, South Dakota (RAP). (LBF is north of DDC; RAP is northwest of LBF) Skewed abscissa is temperature (°C); logarithmic ordinate is pressure (mb). Temperature plotted as solid line; dew point plotted as dashed line. Winds plotted to the right at heights above sea level in m; pennant = 25 m s^{-1}; whole barb = 5 m s^{-1}; half barb = 2.5 m s^{-1}. LCH and GGG are on the warm side of the front; the other stations are on the cold side of the front. It was snowing at OUN, despite the above-freezing temperatures between 885 and 690 mb! The frontal inversion between 900 and 850 mb, which is saturated, is pronounced, to say the least. The rapid veering with height of the winds in the frontal zone suggests warm advection above the shallow cold air mass; wind shear in the inversion layer is strong. The air above the frontal zone is relatively warm and dry. The DDC, LBF, and RAP soundings also show the strong, saturated, frontal stable layer (inversion); the height of the top of the cold air mass increases from 850 mb at OUN to 750 mb at DDC to 700 mb at LBF to about 650 mb at RAP. The backing of the winds with height in the inversion layer at DDC and LBF suggests cold advection; the veering of the winds with height at RAP suggests warm advection. There is an approximately 100-mb-deep layer next to the ground north of the frontal zone that has a nearly dryadiabatic lapse rate. The stratosphere above DDC, LBF, and RAP is marked by a deep, nearly isothermal layer. The lapse rate above the frontal zone, but below the tropopause, is relatively steep.

---

is related to ocean–land temperature contrast; the flow associated with the marine push is highly ageostrophic.

Frontal passage in the lee of the Rockies that is accompanied by the southward plunge of cold anticyclones to the east results frequently in cold, moist upslope flow from the northeast, east, or southeast. On the other hand, when a surface cold front is followed by an anticyclone that had its origin west of the Divide, then the flow is relatively dry, and has a northwesterly or westerly downslope component, and hence is warmer than that associated with the former case. The surface air is of Pacific origin, and hence these fronts are often referred to as *Pacific fronts.*

Pacific fronts that pass over the Rockies and reach the Southern Plains sometimes become dry lines, as the cool air mass is warmed and dried after its passage over the Continental Divide. The colder Arctic air masses that plunge southward east of the Rockies are called *Northers* or *blue Northers.* The flow behind these fronts is usually highly ageostrophic, especially near the mountains. The flow of cold Canadian air, behind a cold front, toward the Northeast United States, is sometimes referred to as the *Montreal Express.*

Cold fronts moving onto the Pacific coast of the United States are usually modified considerably by diabatic heating and evaporation at the ocean surface. After frontal passage, westerly or northwesterly flow may result in cloudiness owing to upslope along the coastal mountains, unless there is a strong dry slot upstream. Cold air masses associated with continental Canadian air to the east of the mountains may filter into the Pacific Northwest over

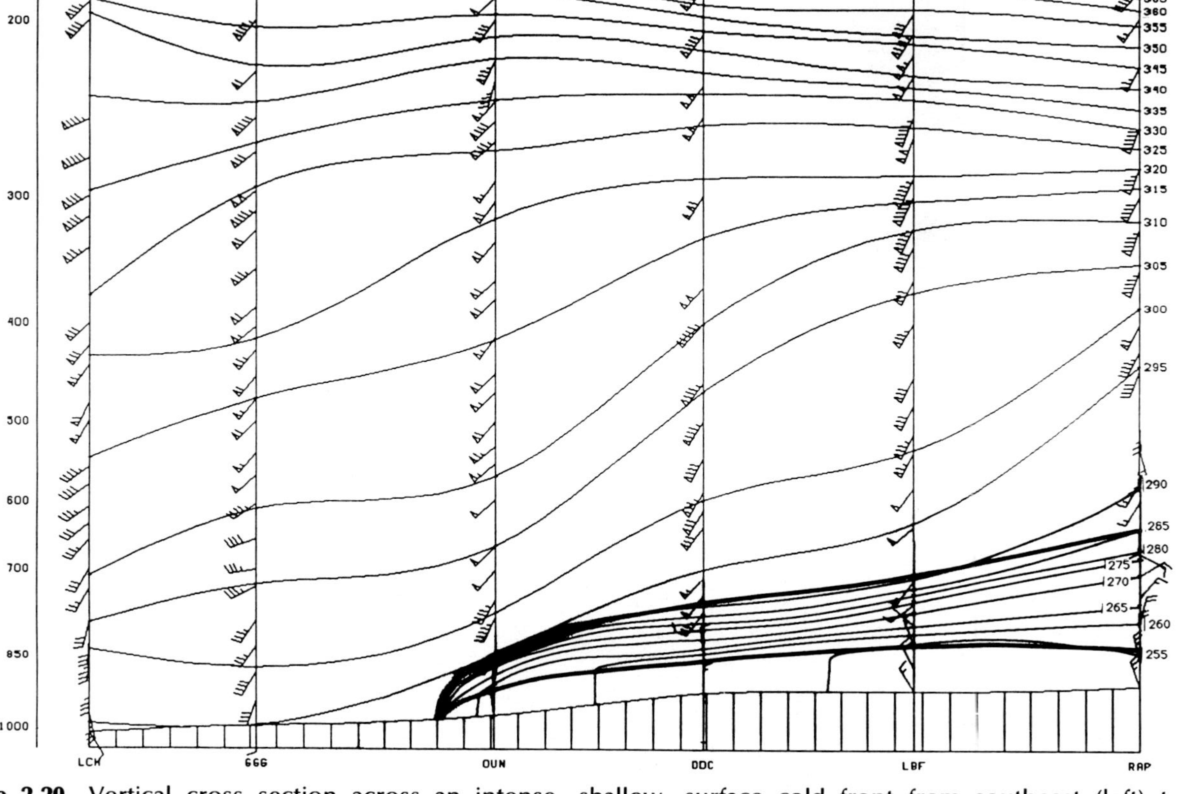

Figure 2.20 Vertical cross section across an intense, shallow, surface cold front from southeast (left) to northwest (right) from Lake Charles, Louisiana (LCH) to Rapid City, South Dakota (RAP) at 1200 UTC, December, 21 1990 (cf. Fig. 2.19). Potential-temperature isotherms in K (thin solid lines); pennants $= 25\ \mathrm{m\ s^{-1}}$; whole barbs $= 5\ \mathrm{m\ s^{-1}}$; half barbs $= 2.5\ \mathrm{m\ s^{-1}}$. Pressure in mb is plotted to the left. Boundaries of the frontal zone (thick solid lines). The distance from OUN to GGG is approximately 400 km.

Figure 2.21 "Rope cloud" along the leading edge of a cold front in the central United States at 2231 UTC, March 11, 1988, as seen by the visible channel of the GOES satellite. The rope cloud extending from central Texas through southeastern Oklahoma and up into Missouri erupted into a line of thunderstorms.

passes and through river basins. Downslope on the west side of the mountains may be responsible for clear skies.

In southern Australia, during the winter, an equatorward flow of Antarctic "cold" air from the south, behind a cold front, is modified considerably by heating from the relatively warm ocean. On the other hand, during the summer, the ocean is cool compared to the continent, and hence the air behind a cold front is very cold relative to the air over the hot summer continent to the north. Thus, unlike Northern Hemisphere cold fronts, Australian cold fronts are more intense during the summer than they are during the winter. During the spring and summer along the southeast coast of Australia, these fronts are called *southerly bursters* or *southerly busters*. Since many bursters develop as a result of ocean–land temperature contrast, the burster may be similar in some respects to the New England back-door cold front, the marine push of the Pacific Northwest, and the sea-breeze front. The southerly bu(r)ster is accompanied by strong ageostrophic flow on the cold side of the

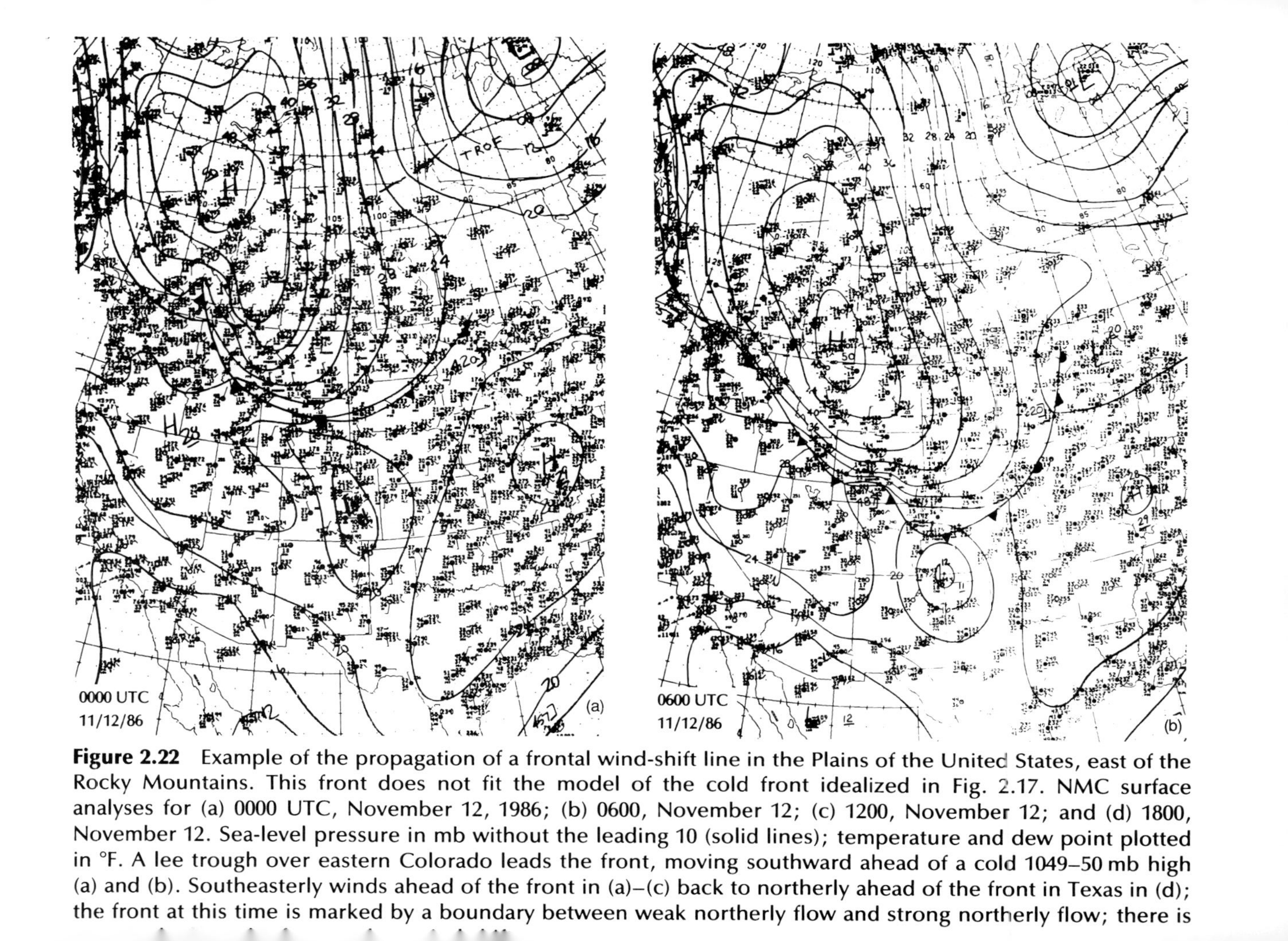

Figure 2.22 Example of the propagation of a frontal wind-shift line in the Plains of the United States, east of the Rocky Mountains. This front does not fit the model of the cold front idealized in Fig. 2.17. NMC surface analyses for (a) 0000 UTC, November 12, 1986; (b) 0600, November 12; (c) 1200, November 12; and (d) 1800, November 12. Sea-level pressure in mb without the leading 10 (solid lines); temperature and dew point plotted in °F. A lee trough over eastern Colorado leads the front, moving southward ahead of a cold 1049–50 mb high (a) and (b). Southeasterly winds ahead of the front in (a)–(c) back to northerly ahead of the front in Texas in (d); the front at this time is marked by a boundary between weak northerly flow and strong northerly flow; there is

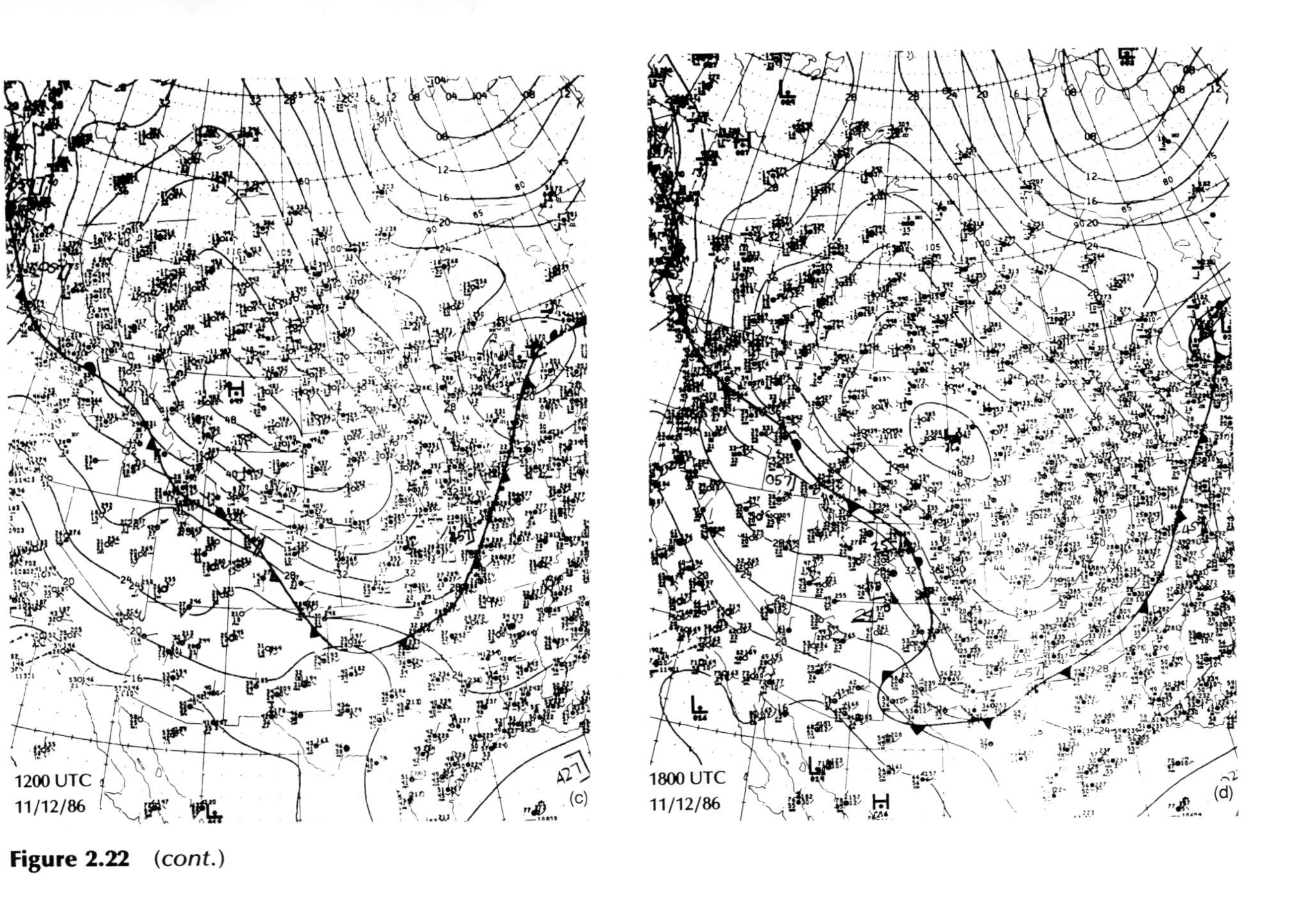

Figure 2.22 (*cont.*)

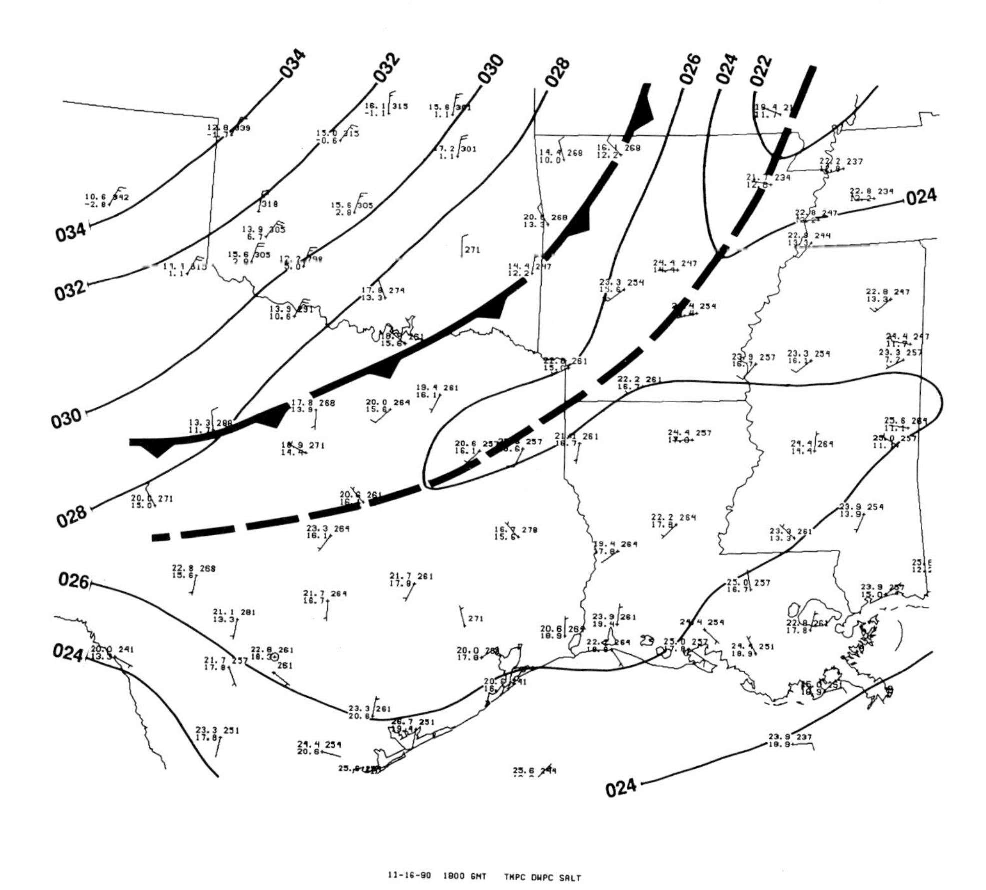

Figure 2.23 Example of a prefrontal trough in the United States at 1800 UTC (1200 LST), November 16, 1990. Altimeter setting in mb without the leading 1 (solid lines). Temperature and dew point plotted in °C; altimeter setting plotted in tens of mb, without the leading 10; whole wind barb = 5 m s^{-1}; half barb = 2.5 m s^{-1}.

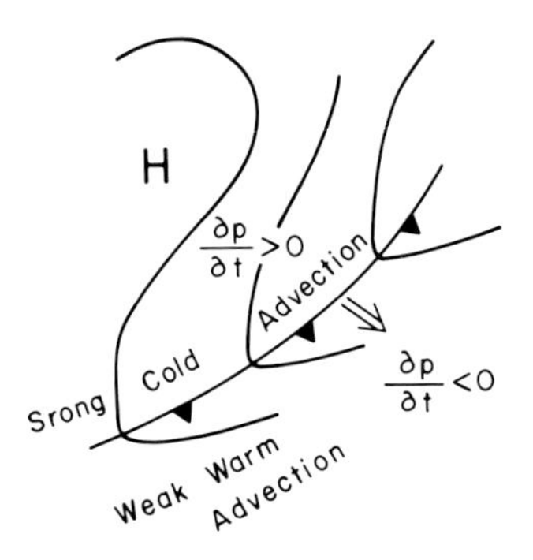

Figure 2.24 Motion of frontal surface trough and postfrontal surface ridge. Isobars (solid lines). Surface pressures fall ahead of the frontal trough in response to weak warm advection and rise behind the frontal trough in response to the strong cold advection. The frontal trough propagates against the isallobaric gradient, from the region of greatest pressure rises toward the region of greatest pressure falls.

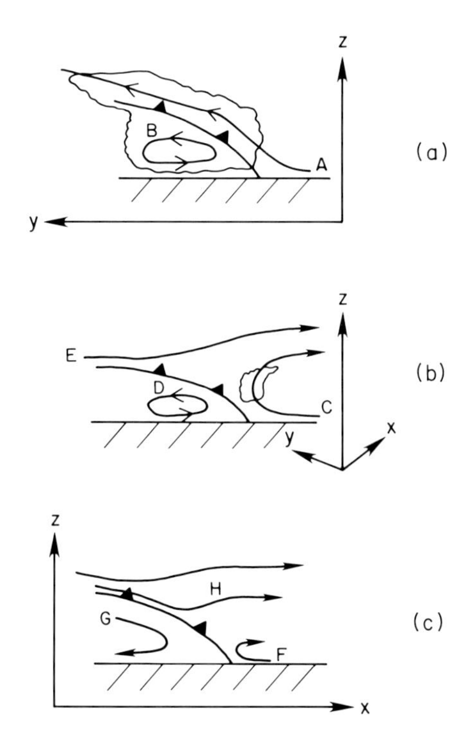

Figure 2.25 Examples of different possible types of front-relative air trajectories shown in a vertical cross section normal to the front. *x* points toward the east, and *y* points toward the north. (a) Cloudy north of the front. Air at A rises over the front and cools to saturation. Air in trajectory B beneath the frontal zone is recirculated. (b) Clear northwest of the front. Air in C approaches the front, rises to saturation, and reverses direction aloft. Air in D is recirculated behind and beneath the frontal zone. Air in E is dry and passes ahead of the frontal zone aloft. (c) Clear both to the west and east of the front. Air in F approaches the front, rises briefly, and reverses direction at low levels; since little vertical motion occurs, the air remains unsaturated. Air in G approaches the frontal zone, subsides, and reverses direction. Air in H subsides, and passes by the front aloft.

front. The "Pampero–Sucio" along the east side of the Andes during the summer may be a similar phenomenon.

The warm front. When the "warm" air mass advances poleward, poleward and eastward, and sometimes eastward relative to the "cold" air mass, the surface front is called a *warm front.* The structure of a warm front is illustrated in Figs. 2.26–2.28. (The warm front associated with a surface cyclone used to be called a *steering line.*) Warm frontal zones usually slope less in the vertical than cold frontal zones.

The poleward or poleward and eastward movement of a warm front is usually associated with strong low-level warm advection east or poleward and east of a developing or intensifying surface cyclone. However, not all surface cyclones are associated with warm fronts, or with warm fronts with well-defined wind-shift lines.

The wind veers when a warm front passes. Low clouds and fog often clear away, as drier air not having a history of ascent arrives. In the Southern Plains, however, warm frontal passage may mark the "return" of stratocumulus overcast from the Gulf of Mexico.

Relatively warm, Atlantic air sometimes circulates from the east around the north side of a cyclone near the Great Lakes and eventually advances westward, southwestward, southward, and even southeastward (Fig. 2.29). This feature could be called a *back-door warm front.*

In the High Plains of the United States, the passage of a lee trough usually marks the shift to dry, potentially warm air from the west (Figs 2.30 and 2.31).

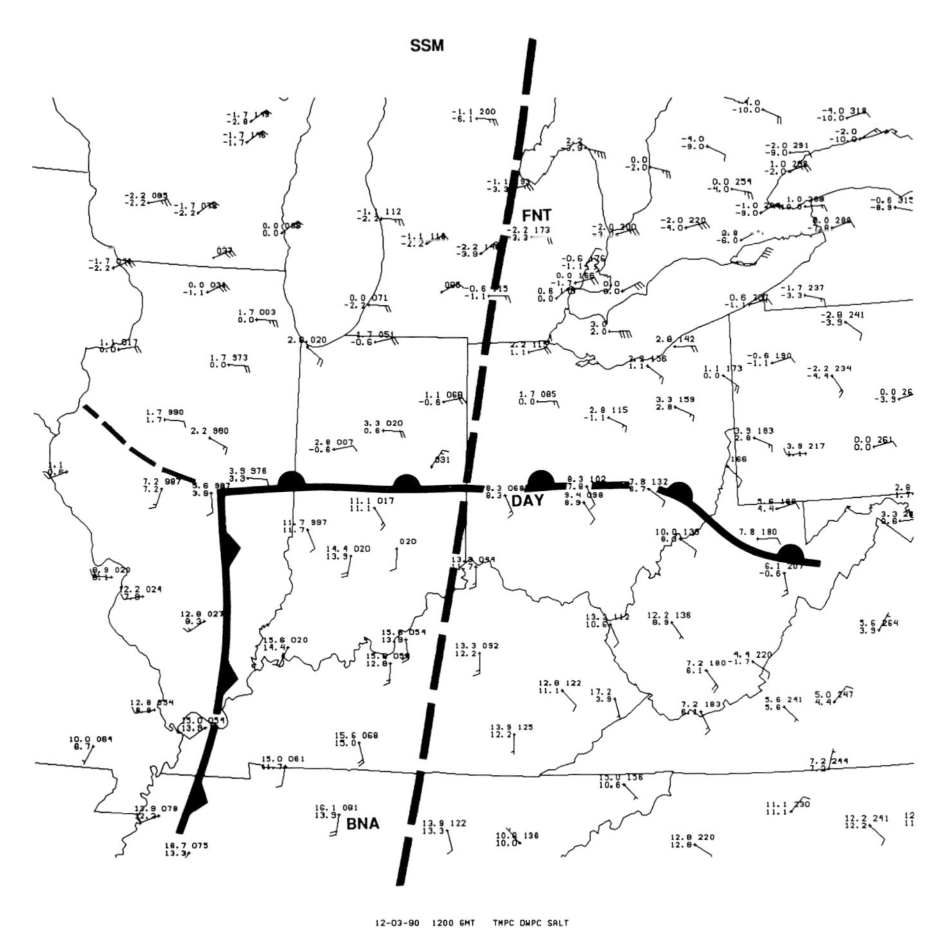

Figure 2.26 Surface analysis of a warm front in the Midwest and Great Lakes regions of the United States at 1200 UTC (0600 and 0700 LST), December 3, 1990. South of the warm front the winds are from the southeast or south; north of the warm front the winds are generally from the east. Altimeter setting plotted in tens of mb, without the leading 10 or 9; temperature and dew point in °C; whole barb = 5 m s^{-1}; half barb = 2.5 m s^{-1}. Dashed line indicates location of vertical cross section shown in Fig. 2.28.

This warming along the leading edge of the downslope air may be rather pronounced, especially when the air mass being displaced is cold, Arctic air that had arrived after the passage of a Norther. This type of a warm front is sometimes referred to as a *chinook front.* During the spring the "chinook" front may become a dryline. Strong downslope wind storms (Fig. 2.32) are common in the foothills of the Rockies as the lee trough/chinook front passes east of the Continental Divide. The potential for damaging winds is a function of the wind direction, vertical wind profile, and static stability upstream. The behavior of downslope wind storms depends upon the dynamics of gravity-wave motion in the presence of topographic obstacles.

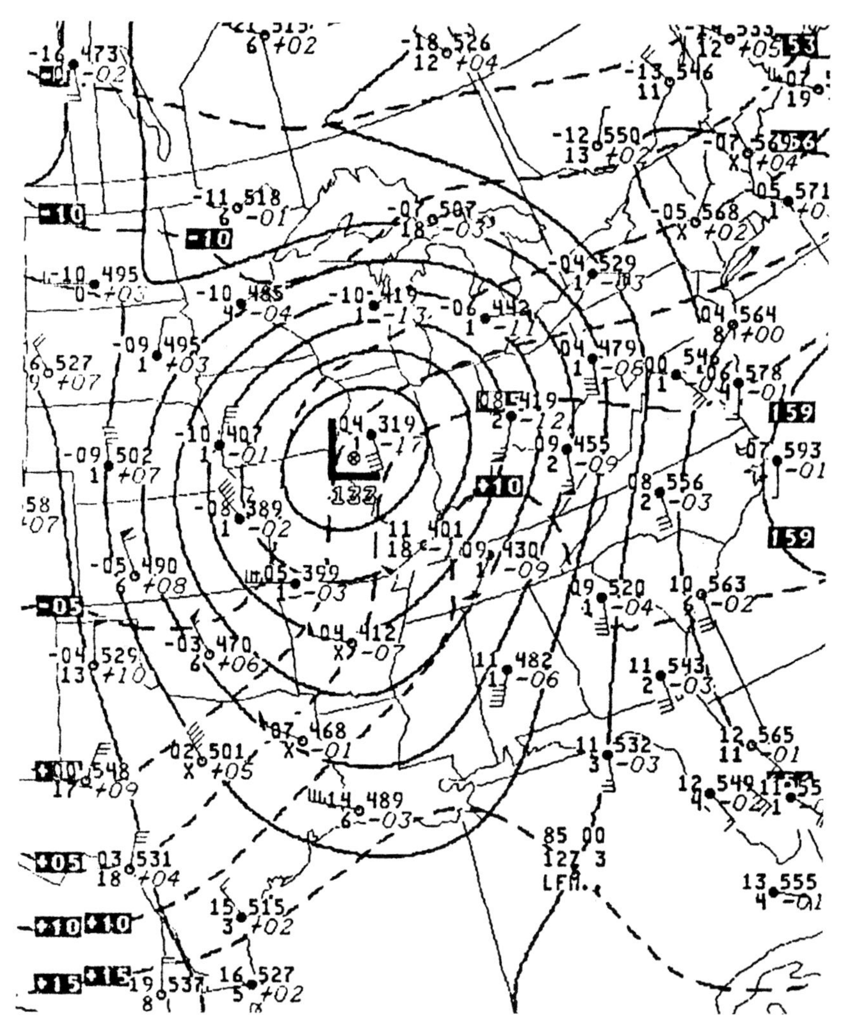

Figure 2.27 NMC analysis at 850 mb of the warm front shown at the surface in Fig. 2.26 at 1200 UTC, December 3, 1990. The warm front extends eastward from the cyclone over western Illinois. The wind shifts from southerly at 32.5 m s^{-1} south of the warm front to easterly at weak speeds north of the warm front.

The stationary front. When neither the "cold" air mass nor the "warm" air mass advances much relative to each other, the front that separates the two is referred to as *stationary* or *quasistationary*. Rising motion above the frontal zone is associated quasigeostrophically with warm advection. This brand of rising motion is often referred to as *overrunning*[1], since the "warm" air mass "overruns" the "cool" air mass. Precipitation associated with overrunning is usually stratiform and light owing to the high static stability in the frontal zone. (The expression *overrunning* can be misleading because it is often a kinematic, rather than a dynamic description of how air is being lifted. However, if the

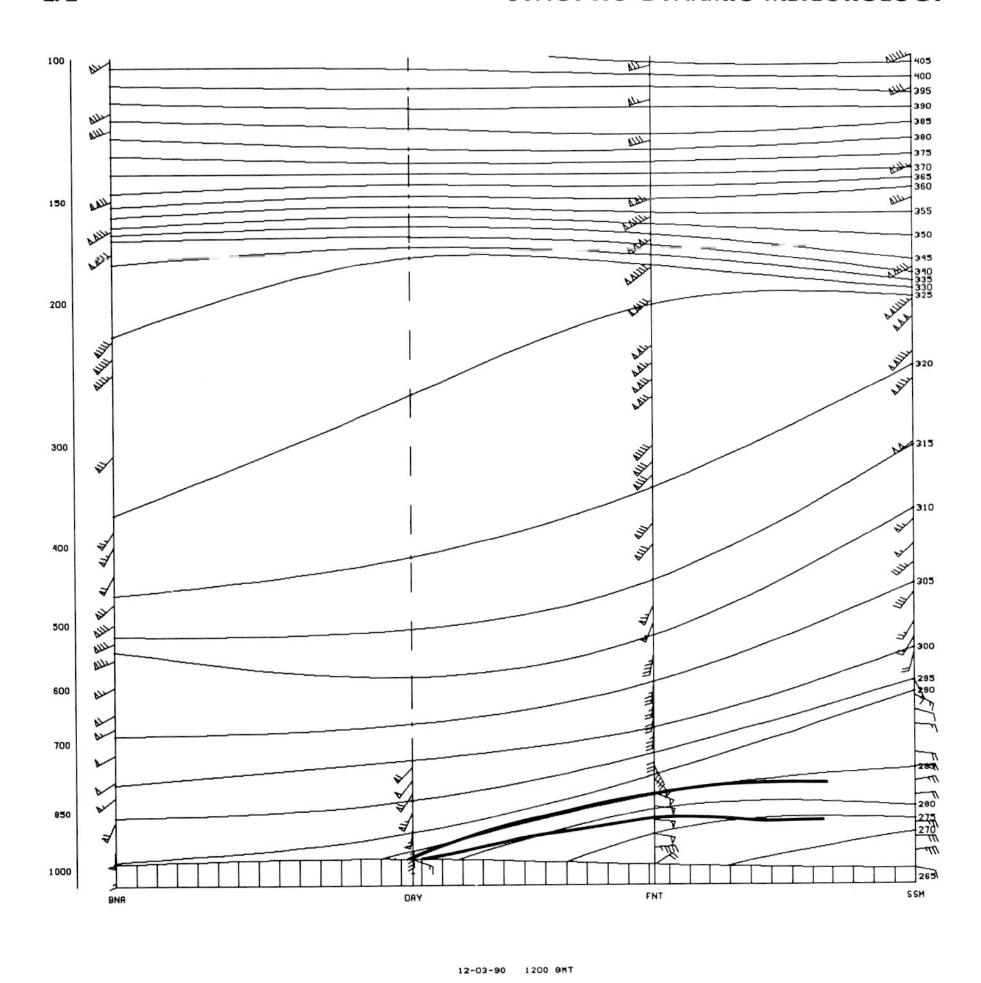

Figure 2.28 Vertical cross section across the warm front analyzed in Fig. 2.26 and shown in Fig. 2.27 at 1200 UTC, December 3, 1990. Cross section extends from the south (left) at Nashville, Tennessee (BNA) through Dayton, Ohio (DAY) and Flint, Michigan (FNT) up to the north (right) at Sault Ste. Marie, Michigan (SSM). Potential-temperature isotherms in K (thin solid lines); pennants = 25 m s^{-1}; whole barbs = 5 m s^{-1}; half barbs = 2.5 m s^{-1}. Pressure in mb is plotted to the left. Boundaries of the frontal zone (thick solid lines). The distance from DAY to FNT is approximately 400 km. The 1200 UTC rawinsonde observation made at DAY was taken while the warm front was passing by the station; the 1200 UTC surface observation was made later, after the warm front had already passed.

front behaves like a density current, then the warm air is in fact being lifted directly as a result of the frontal boundary.)

As cold fronts slow to a halt, often as a result of cyclogenesis to the west or equatorward and to the west, they become stationary. It is common for the stationary front to move back poleward as a warm front as the new cyclone deepens and moves poleward and eastward.

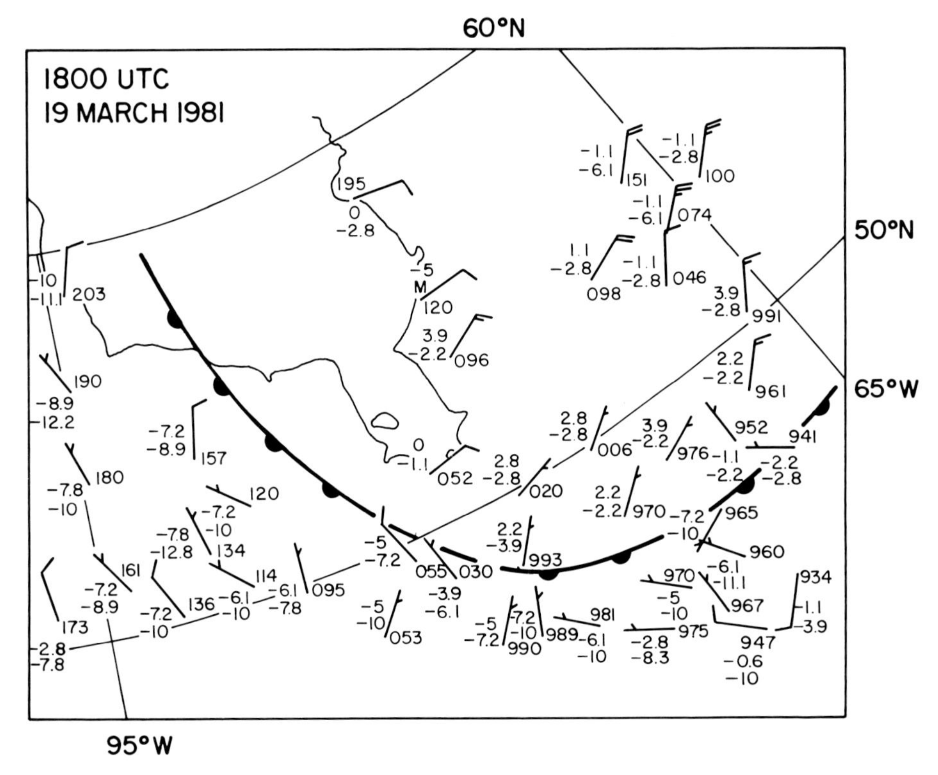

Figure 2.29 Surface analysis of a "back-door" warm front in the Great Lakes region of the United States and Canada at 1800 UTC, March 19, 1981. Sea-level pressure plotted in tens of mb, without the leading 10 or 9; temperature and dew point in °C; whole barb = 5 m s^{-1}; half barb = 2.5 m s^{-1}.

When precipitation falls along and north of a stationary front, flooding is sometimes possible, since the region experiencing precipitation may do so for a long time. For example, although the eastern section of the "Mei-Yu" (plum rain) front in China and "Baiu" front in Japan is a cold front, the western section is quasistationary; it is associated with weak temperature gradients and strong moisture gradients (like some Pacific fronts in the Plains of the United States). Rainfall associated with the Mei-Yu and Baiu front during late spring is the greatest for the year.

The occluded front. According to Polar–Front theory, if a cold front overtakes a warm front equatorward of a cyclone, the resulting wind and temperature field is referred to as an *occlusion,* since the "warm" air at the surface is blocked off from the ground. The surface boundary along which the cold front meets the warm front is called an *occluded front.* The vertical structure of an occlusion is described by quasigeostrophic theory during the final part of the life of an extratropical cyclone as a thickness ridge: The "warm" air mass is pinched in between the upstream and downstream "cold" air masses.

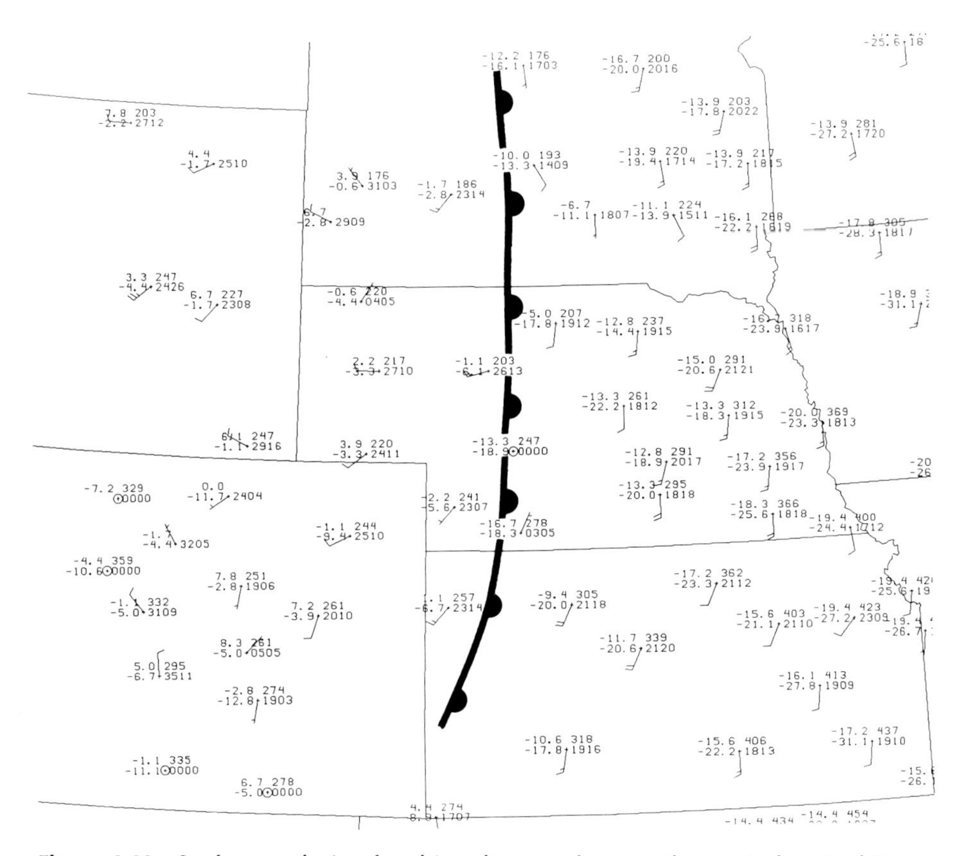

Figure 2.30 Surface analysis of a chinook warm front in the central United States at 1800 UTC, December 23, 1989. Temperature and dew point plotted in °C. Altimeter setting in tens of mb, with the leading 10 omitted. Whole and half wind barbs represent 5 and 2.5 m s^{-1}, respectively. Wind speed (kts) and direction (deg) in ddff format plotted under the altimeter setting. The front separates relatively warm, dry air from the southwest from cold air from the south (having wrapped around an anticyclone to the southeast).

Occluded fronts in which the advancing "cold" air is warmer than the retreating "cold" air are called *warm occlusions*; *cold occlusions* are occluded fronts in which the advancing "cold" air is colder than the retreating cold air.

Occluded fronts that move across the Pacific Northwest and then downslope east of the Rockies are usually warm occlusions. A relatively mild, maritime air mass is warmed through adiabatic compression in the lee of the mountains and advances relative to a retreating, cold, continental air mass. These occluded fronts are called *trowals* in Canada. In the lee of the mountains, these fronts may behave more like warm fronts.

Occluded fronts in the eastern United States are often cold occlusions, as relatively cold, continental air advances relative to retreating warm Atlantic air.

However, verification of the classic occlusion through analysis of observa-

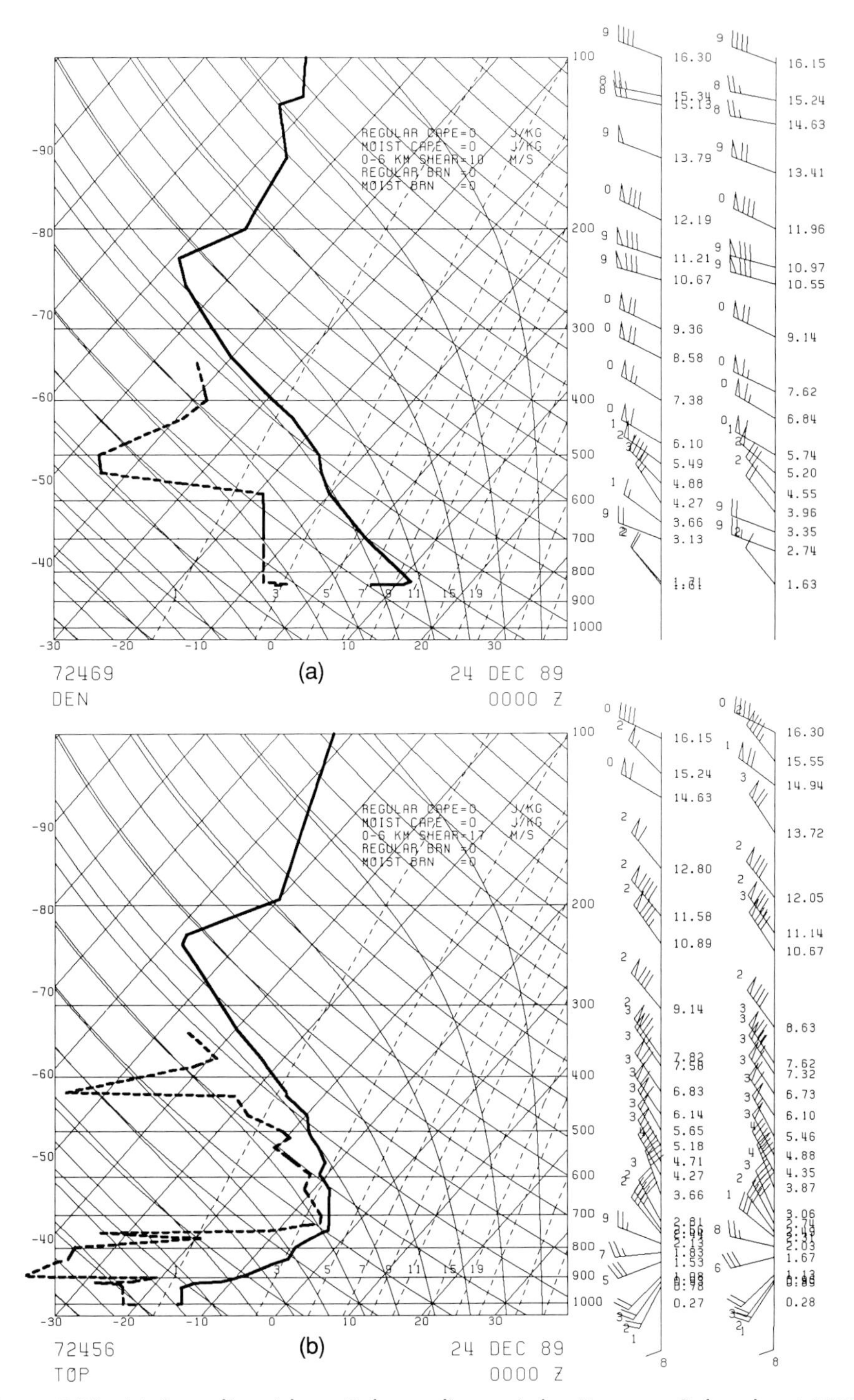

Figure 2.31 (a) Sounding (skew T–log p diagram) for Denver, Colorado at 0000 UTC December 24, 1989, which shows a deep, dry air mass subsiding in the lee (from the northwest, in this case) of the Rocky Mountains. The Denver sounding is characteristic of the air mass west of the chinook front. (b) Sounding for Topeka, Kansas, which shows a shallow, cold air mass surmounted by a frontal inversion, in which the winds veer with height, indicating warm advection. The Topeka sounding is characteristic of the air mass east of the chinook front.

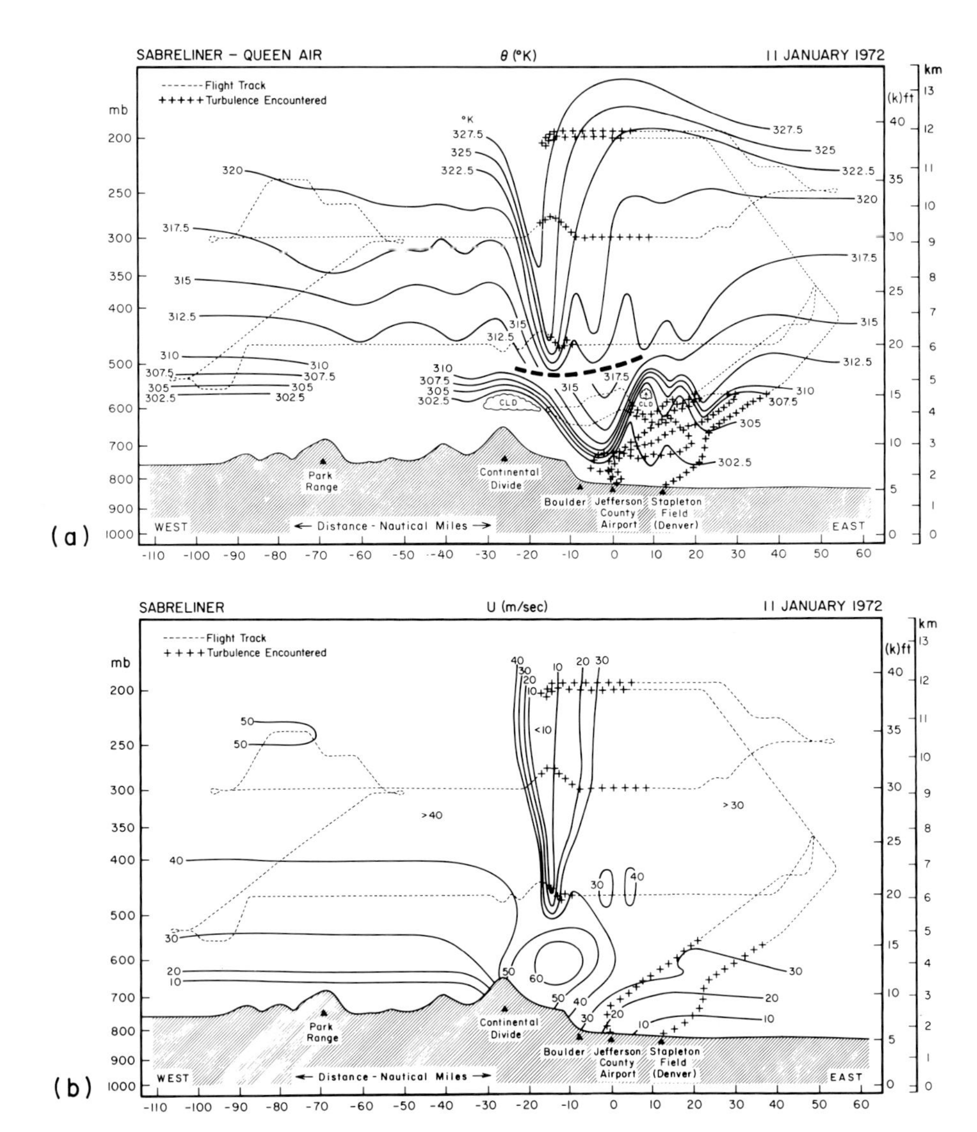

Figure 2.32 Downslope windstorm in the lee of the Rockies. (a) Cross section of the potential temperature field (K) along an east–west line through Boulder, Colorado, as obtained from analysis of aircraft data on January 11, 1972. For steady adiabatic flow, these isentropes are good indicators of the streamlines of the air motion. Data above the dashed line were collected 1700–2000 MST, while those below this line were collected 1330–1500 MST. Flight tracks are indicated by the dashed lines, except for crosses in turbulent portions. Note the wavelike character to the potential-temperature field, whose trough at low levels lies east of the mountains, and tilts with height upstream (i.e., to the west). (b) Contours of zonal wind speed ($\mathrm{m\,s^{-1}}$) along the same cross section as in (a). The analysis below 500 mb was partially obtained from vertical integration of the continuity equation, assuming two-dimensional steady-state flow. Note the strong (in excess of $50\ \mathrm{m\,s^{-1}}$) downslope flow between the Continental Divide and Boulder (from Klemp and Lilly, 1975). (Courtesy of the American Meteorological Society)

tional data has been difficult!. There are currently several other conceptuaal models for the formation of occluded fronts. For example, an occlusion can form in response to a deepening oceanic low on the "cold-air" side of a front even if the warm sector is not overrun by an advancing cold front. The core of air on the "cold side" that is warmer than its immediate surroundings, but cooler than air in the warm sector, is called a *warm seclusion* (Fig. 2.33). The warm seclusion may form when a cold front and warm front break apart from each other; this has been referred to as *frontal fracture.* Cold air wraps around the low to close off an area of relatively warm air. Occluded fronts can also form *in situ* in response to surface trough formation north of the intersection of a cold front and a warm front.

The coastal front. The *coastal front* is a shallow (i.e., less than 1 km deep) mesoscale zone of strong horizontal temperature gradient (on the order of 10°C/10 km) at the surface that separates relatively warm, maritime air from cold, continental air (Fig. 2.34). The coastal front has a time scale of 6–12 hours, which is somewhat less than the synoptic time scale of days. Coastal fronts are often found during the early and middle winter along the east coast of New England and the Carolina and Texas coasts, all of which are curved in a concave manner. During the early winter, there can be a substantial land–sea temperature contrast, because the ocean temperatures are still relatively warm compared to the air temperatures over land. When cold, continental air flows offshore along a concave coastline, there is a relative maximum in diabatic heating.

Coastal fronts usually behave like quasistationary fronts or warm fronts. They often move toward the "cold" air mass, and are marked by a convergence zone, along which there is a shift in wind direction. The maximum amount of accumulated precipitation is found along a band on the "cold" side of the front. Sometimes the coastal front separates frozen precipitation (on the "cold" side) from rain (along the "warm" side), or heavy precipitation from light precipitation.

Coastal frontogenesis in New England occurs when there is an anticyclone associated with a cold air mass to the north or northeast of New England, and when a trough approaches from the west at low to middle levels. The cold air flows southward and southwestward, and may become trapped by the Appalachian mountains. This phenomenon is known also as *cold-air damming,* since the relatively dense, cold air cannot be lifted easily to the west over the mountains: Less energy is required to deflect the airflow around the mountains. The quasistationary ridge that forms as a hydrostatic consequence of the trapped cold air mass is often referred to as the *Baker ridge* (Fig. 2.34).

Although a similar damming phenomenon occurs just east of the Rockies, it is obviously not associated with any coastal frontogenesis. Like the air flow behind "northers," the flow associated with the cold ridge is highly ageostrophic.

The combination of cold air over the land and relatively warm air over the ocean produces a land-to-ocean directed horizontal temperature gradient. Although differential friction between the land and ocean can account for a

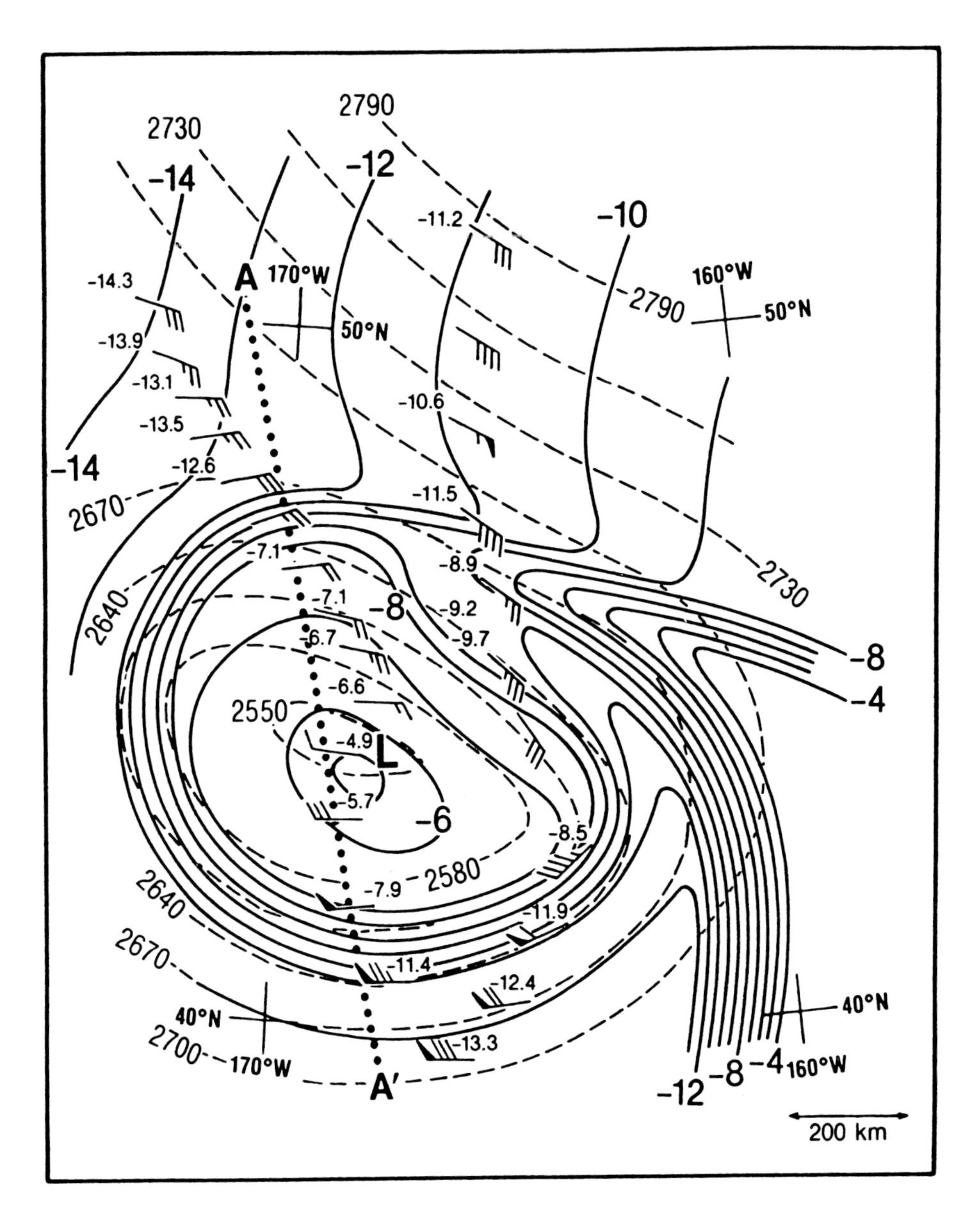

Figure 2.33 An example of a warm seclusion at its level of maximum baroclinicity: 700-mb analysis valid for 0000 UTC, March 10, 1987. Isotherms (solid lines) in °C; height (dashed lines) in m; whole wind barb = 5 m s^{-1}; half barb = 2.5 m s^{-1}; temperature plotted in °C. Data, which are from dropwindsondes, have been space-to-time adjusted. The temperature at the center of the warm seclusion is around −5°C; it is warmer than the surrounding air, but cooler than the air in the warm sector to the east (from Shapiro and Keyser, 1990). (Courtesy of the American Meteorological Society)

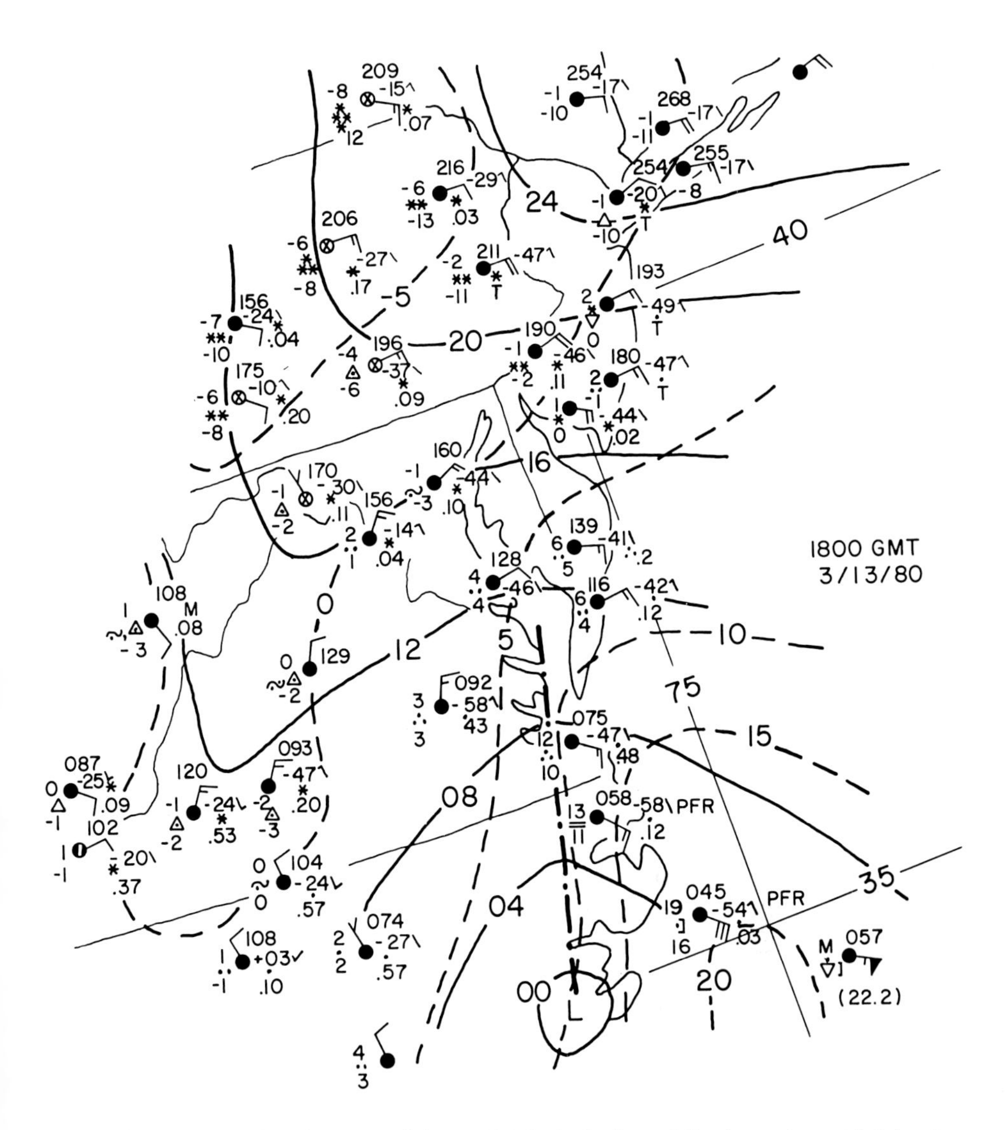

Figure 2.34 Example of a coastal front (thick dash-dotted line) on the mid-Atlantic coast of the United States 1800 UTC (GMT), March 13, 1980. Sea-level pressure in mb, without the leading 10 plotted (solid lines); isotherms in °C (dashed lines); temperature and dew point plotted according to convention in °C; pressure plotted in tens of mb, without the leading 10; weather as shown; whole wind barbs, half wind barbs, and pennants 5, 2.5, and 25 m s^{-1}, respectively. The "Baker ridge" is located in northwestern portion of the analysis, along and east of the Appalachians. (Courtesy Lance Bosart)

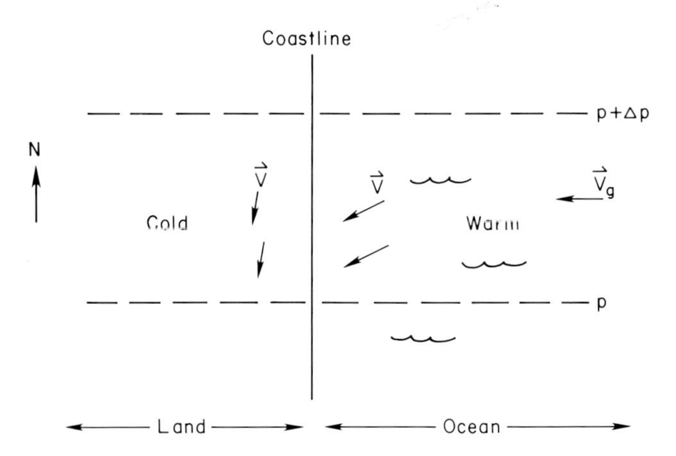

Figure 2.35 Illustration of the effect of differential friction along a coastline. Suppose the geostrophic wind is easterly as shown; sea-level isobars (dashed lines). Higher values of friction over the cold land create a larger cross-isobar flow component than there is over the warm ocean. Along the coastline convergence acts on the east–west temperature gradient between ocean and land to promote frontogenesis.

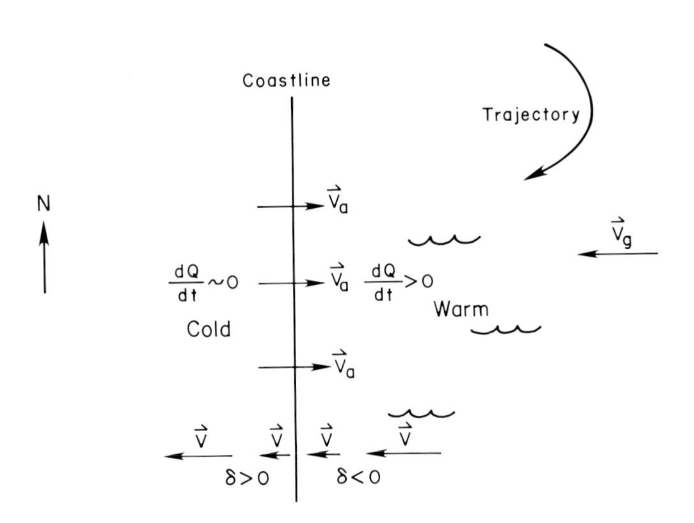

Figure 2.36 Illustration of the effects of differential diabatic heating on the formation of the coastal front. Cold continental air flows off shore, is heated from below, and curves anticyclonically back over to land ($\mathbf{v}_g$), where it is no longer heated from below. A solenoidal circulation, like the reverse of the sea breeze, forms, creating an ageostrophic flow of air ($\mathbf{v}_a$) from the land to the ocean. Convergence offshore tightens the land–ocean temperature gradient, and thus promotes frontogenesis.

band of convergence when there is onshore flow (Fig. 2.35), it is not generally considered to be significant.

In addition, heating of the cold, continental air from below by the relatively warm ocean results in a solenoidally induced *land breeze*, in which an offshore wind component along the shore contributes to a band of convergence along the coast (Fig. 2.36). This band of convergence acts to increase the extant temperature gradient, that is, to promote frontogenesis. Geostrophic deforma-

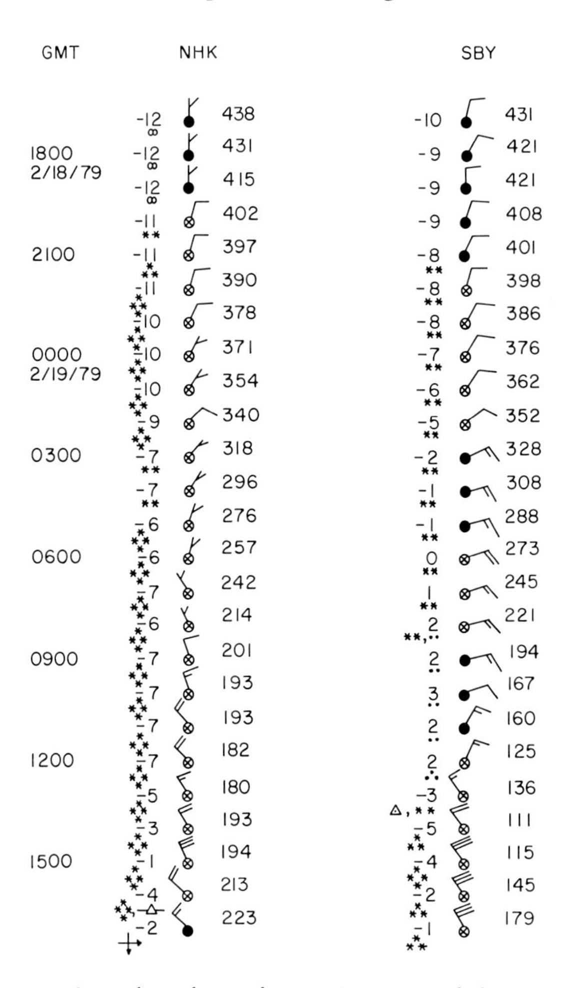

Figure 2.37 Time series of surface observations at Salisbury, Maryland (SBY), along the coast, and at Patuxent River, Maryland (NHK), inland from February 18, to February 19, 1979, during the Presidents' Day snowstorm. Temperature (°C), current weather, and sea-level pressure (tens of mb, with the leading 10 omitted); whole and half wind barbs indicate speeds of 5 and 2.5 m s^{-1}, respectively. At SBY the wind veers, and the temperature rises above freezing, as the coastal front passes to the west; it later moves back towards the east, as the wind backs and the temperature falls back down below freezing. The coastal front remains east of NHK, where it remains relatively cold and heavy snow is observed (from Bosart, 1981). (Courtesy of the American Meteorological Society)

tion is not responsible for significantly tightening the temperature gradient. The role of the approaching trough aloft is to increase the easterly wind component over the ocean as the surface pressure falls south of the cold anticyclone.

A spectacular case of coastal frontogenesis is depicted in Fig. 2.37. Note the passage of the coastal front at SBY between 0100 and 0300 UTC as a warm front: The wind subsequently veered, the temperature rose 5°C from 0100 to 0400 GMT (UTC), and light snow changed to light rain several hours later. Inland, at NHK, the winds backed and heavy snow fell. The cyclone that formed along this coastal front, turned into the infamous Presidents' Day storm of 1979.

The dryline. The *dryline* is a narrow zone of strong horizontal dew-point temperature gradient at the surface (Fig. 2.38). It is a boundary that separates moist, maritime air from dry, continental air. In the United States the dryline

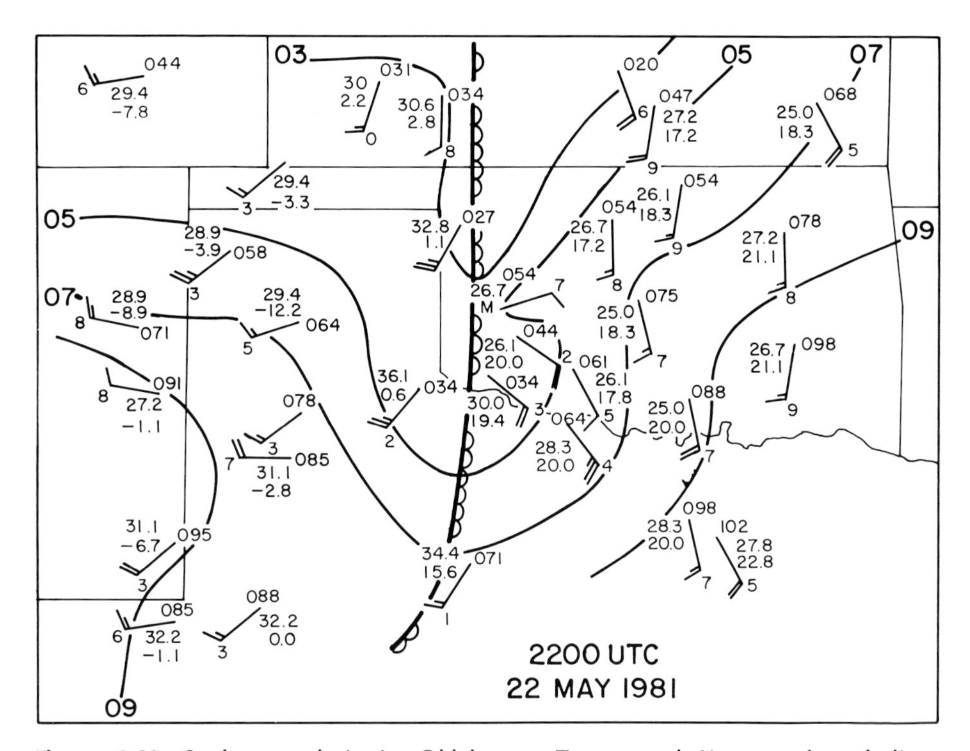

Figure 2.38 Surface analysis in Oklahoma, Texas, and Kansas of a dryline (scalloped line) under "quiescent" conditions at 2200 UTC, May 22, 1981. Altimeter setting (solid lines) in mb without the leading 10; temperature and dew point plotted in °C; altimeter setting plotted in tens of mb, without the leading 10; whole barb = 5 m s^{-1}; half barb = 2.5 m s^{-1}. At the time of this analysis tornadic storms were occurring just east of the dryline in western Oklahoma. Winds east of the dryline are generally from the south and southeast, while winds west of the dryline are from the southwest and west. Dew points east of the dryline are around 20°C; dew points west of the dryline are near and below 0°C.

is located frequently (about 40% of the time) over the Plains region during the spring and early summer. The dryline separates air that has had a history of contact with the Gulf of Mexico, from air that has had a history of contact with the hot, dry, elevated land mass of the southwestern United States and Mexico. The dryline also occurs over India before the onset of the Southwest monsoon, and over Central West Africa.

Thunderstorms frequently form along or just east of the dryline. During the daytime the area west of the dryline is often characterized by strong, gusty winds, blowing dust, and oppressive heat and dryness.

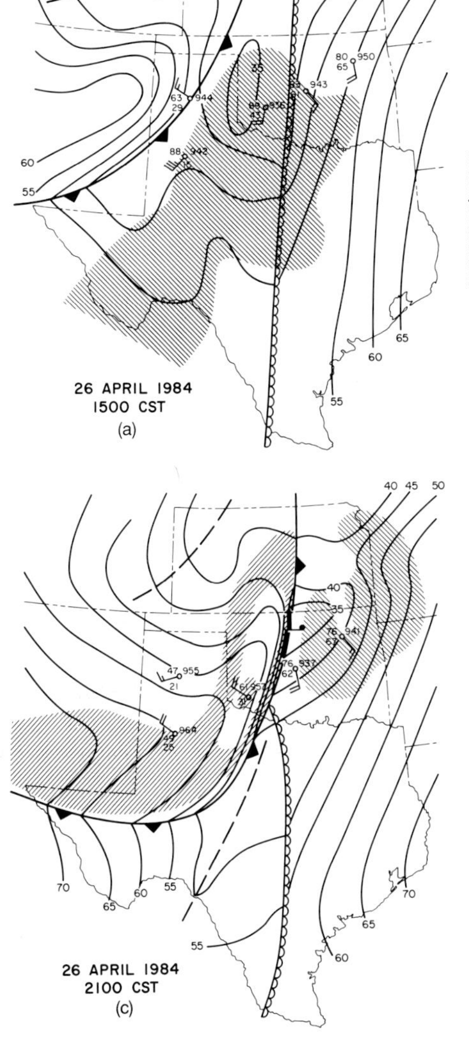

Figure 2.39 Analysis of a cold front catching up to a dryline (scalloped line) at (a) 1500 CST, (b) 1800, and (c) 2100, April 26, 1984, and transforming it into a cold front. Isobars (solid lines) in inches of mercury × 100 without the leading 29. Temperature and dew point plotted in °F; pressure plotted is altimeter setting in inches of mercury × 100, without the leading 2; whole wind barb = 5 m s^{-1}; half wind barb = 2.5 m s^{-1}. Severe thunderstorms formed along the dryline during the afternoon, and a severe squall line formed along the cold front when it caught up to the dryline (from Burgess and Curran, 1985). (Courtesy of the American Meteorological Society)

A wind-shift line is usually located near, but not necessarily along, the zone of sharpest dew-point gradient. Winds usually have a westerly component on the "dry" side, and frequently have a slight easterly component on the "moist" side. Just as we located surface fronts along the wind-shift line rather than along the zone of maximum horizontal temperature gradient, we locate the dryline along its wind-shift line, rather than along its zone of strongest dew-point gradient. The location of the dryline in the United States usually corresponds approximately to the 9 $g\,kg^{-1}$ mixing ratio isohume. The dryline sometimes appears on radar as a "fine-line," owing to a sharp gradient in the index of refraction or to insects. A pressure trough is often, but not always, located along the dryline.

Under "quiescent" (i.e., in the absence of strong synoptic-scale forcing) conditions, the dryline moves eastward during the day, and westward at night. The eastward movement during the day may be sporadic and not as clearly evident as the movement of an ordinary front. Sometimes a surface cold front catches up to a dryline and transforms it into a cold front (Fig. 2.39). On the other hand, a Pacific front, modified by its passage over the mountains of New Mexico, may turn into a dryline when it reaches West Texas, owing to the advection of moist air from the Gulf of Mexico, and downslope drying. The classical dryline, on the other hand, is not associated with any fronts; it is accompanied by a diurnal oscillation in the zonal component of the temperature gradient, which is easterly directed at night and westerly directed during the day. The modified Pacific front is usually associated with an easterly directed temperature gradient at all times of the day. The dryline in the United States is rarely found east of 96°W longitude. It can, however, be advected that far eastward and more by an intense synoptic-scale cyclone (Fig. 2.40).

The vertical structure of typical drylines is shown in Figs. 2.41 and 2.42 for special and conventional data, respectively. The dryline boundary is usually nearly vertical from the surface up to about 1–1.5 km AGL (above its surface location); it is nearly horizontal off to the east. Potentially cool, moist air is "capped" to the east of the dryline by a layer of strong static stability such as an inversion. Potentially warm, dry air, which overlies the inversion to the east of the dryline, is present at all levels to the west. The magnitude of the dew-point gradient is typically at least 15°C $(100\ km)^{-1}$. However, gradients as large as 9°C km^{-1} have been observed. The density (virtual temperature) gradient, however, is not as great as it could be in a dry atmosphere, since the warm (light) air to the west of the dryline is dry, while the cool (heavy) air to the east is moist; dry air is heavier than moist air at the same temperature. Diurnal temperature variations are less east of the dryline than west of it owing to the trapping of longwave radiation by clouds and water vapor on the east side.

The mechanisms responsible for the formation and movement of the dryline in the United States are as follows: Moisture from the Gulf of Mexico is advected northward in response to a lee trough and an anticyclone at the surface to its east. Suppose the "moist layer" extends from the surface up to some fixed pressure level. Then the moist layer is relatively deep to the east, where the surface elevation is relatively low, and relatively shallow to the west

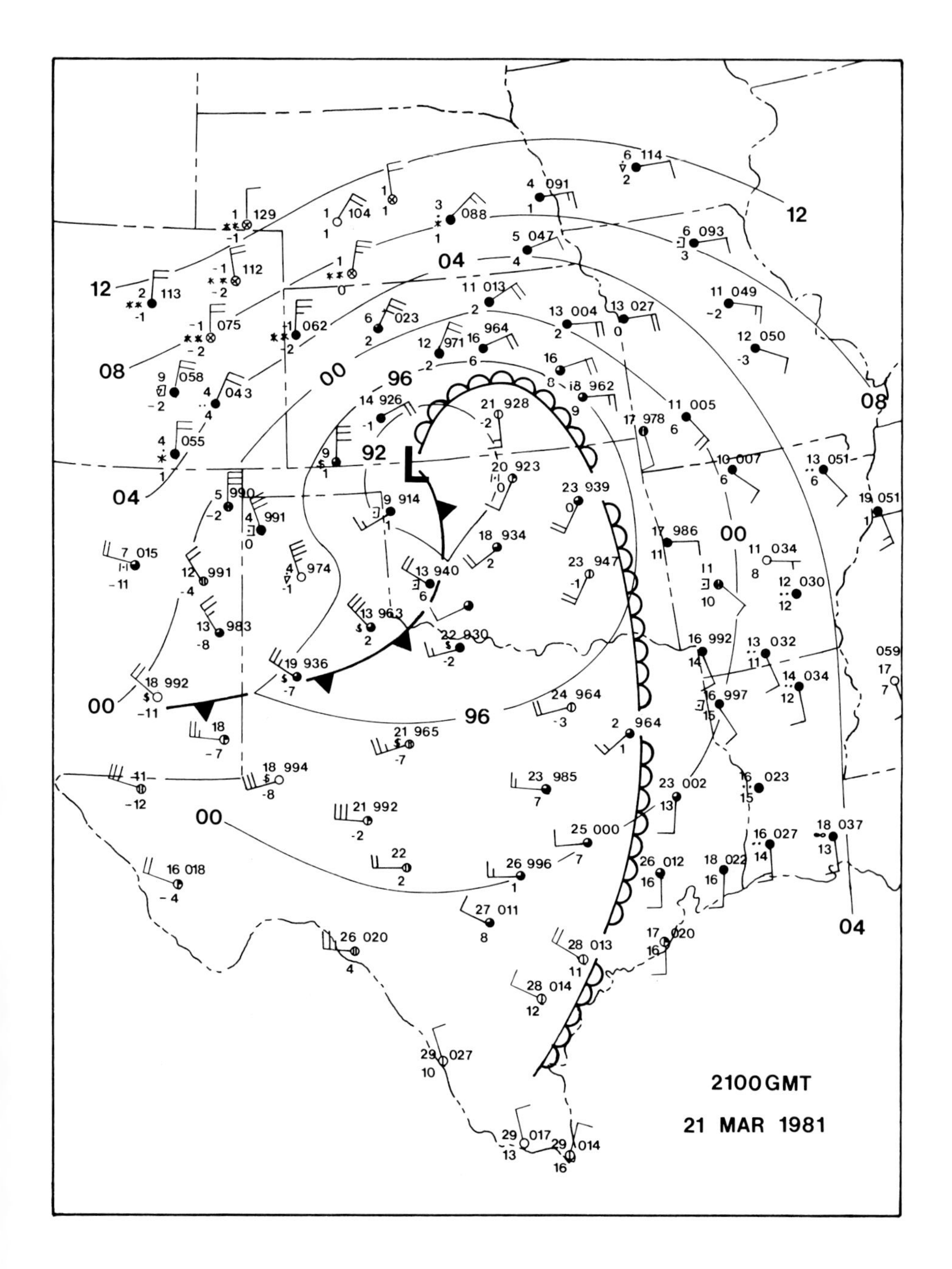

Figure 2.40 Example of a dryline (scalloped line) being advected far to the east by a strong synoptic-scale system 2100 UTC (GMT), March 21, 1981. Temperature and dewpoint in °C. Sea-level pressure in tens of mb, without the leading 9 or 10. Whole wind barb = 5 m s^{-1}; half wind barb = 2.5 m s^{-1}. Sea-level isobars in mb (solid lines), without the leading 9 or 10. Blowing dust is often observed when strong surface winds are found behind the dryline (from Carr and Millard, 1985). (Courtesy of the American Meteorological Society)

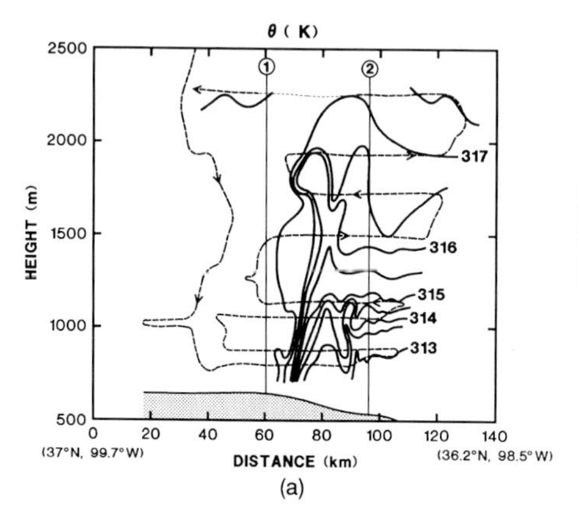

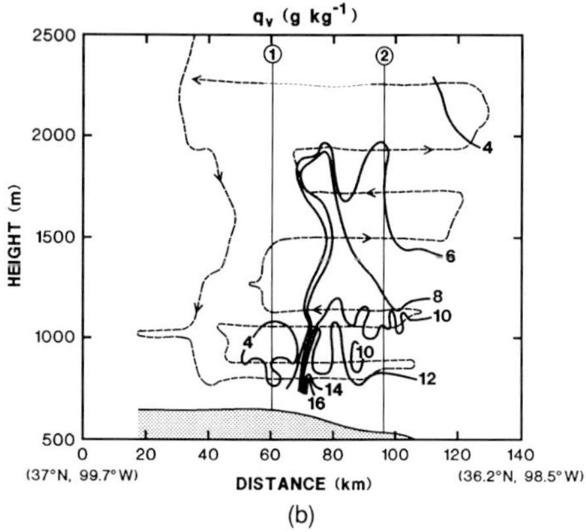

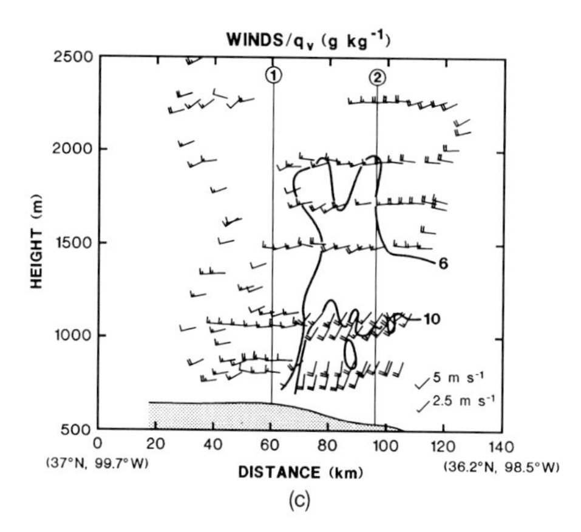

Figure 2.41 Example of the vertical structure across a quiescent dryline as determined from an aircraft 2019-2228 UTC, May 24, 1989. (a) Vertical cross section of potential temperature (K); (b) as in (a), but for water-vapor mixing ratio (g kg^{-1}); (c) winds and water-vapor mixing ratio. Flight track indicated in (a) and (b) by dashed line. (Courtesy C. Ziegler, National Severe Storms Laboratory)

where the surface elevation is relatively high (Fig. 2.43). At some longitude the top of the moist layer intersects the ground surface. A capping inversion (also called a *lid*) over the moist layer is produced as hot air from the relatively high Mexican Plateau and New Mexico flows over the cooler marine air. The Mexican and New Mexican air mass is characterized by a deep, dryadiabatic layer, owing to daytime solar heating. Since the heating rate depends upon soil moisture, we expect that the deep dryadiabatic layer is enhanced as a result of the arid soil over the high plateau. The "elevated mixed layer" is advected over the cool, moist layer, so that mixing between the two layers is inhibited. The top of the elevated mixed layer is itself usually capped by a stable layer (see, e.g., Fig. 2.42). This stable layer, which is usually found between 600 and 400 mb, represents the top of the deep, mixed layer over elevated terrain; it has been advected eastward, over the low-level moist layer. The flow of stable, moist air underneath the flow of neutral, dry air has been called *underrunning*. The "cap" underneath the elevated mixed layer may be maintained further by radiational cooling at the bottom of the stable layer, since the mixing ratio drops sharply with height.

During the day, and when not under the influence of a strong synoptic-scale system, the dryline's motion is due to the cross-dryline variation in vertical

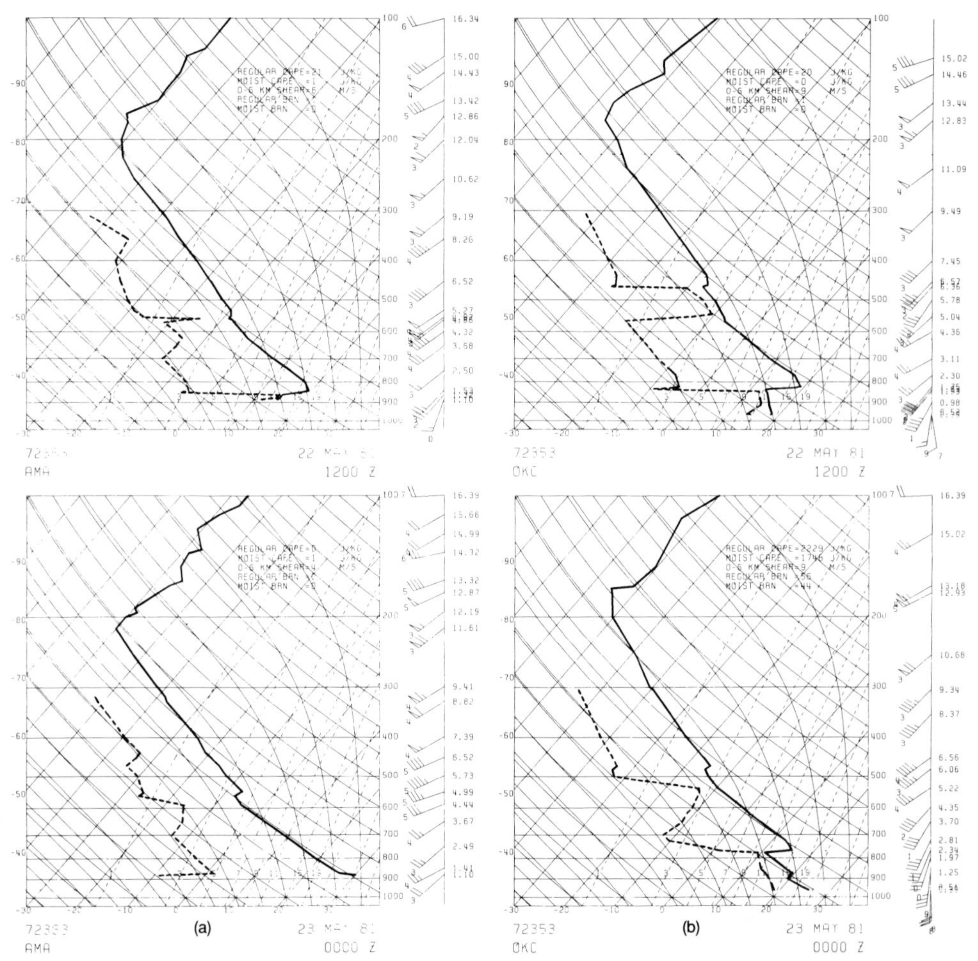

Figure 2.42 Soundings characteristic of the air masses (a) west and (b) east of the dryline on May 22, 1981. Soundings (skew T–log p diagrams) for (a) Amarillo, Texas at 1200 UTC (top) and at 0000 UTC, May 23 (bottom); (b) Oklahoma City, Oklahoma at 1200 UTC (top) and at 0000 UTC, May 23 (bottom). Skewed abscissa is temperature (°C); logarithmic ordinate is pressure (mb). Temperature and dew point plotted as solid line and dashed line, respectively. Winds plotted at the right at the plotted heights (km MSL); pennant = 25 m s^{-1}; whole barb = 5 m s^{-1}; half barb = 2.5 m s^{-1}. The morning Amarillo sounding has a very shallow moist layer near the ground, which mixes out completely with the dry air above when the surface temperature warms up to 24°C; at early evening, there is a deep dryadiabatic layer from the surface up to 560 mb; a shallow stable layer is found at the top of the dryadiabatic layer; a very shallow superadiabatic layer is found near the ground. The morning Oklahoma City sounding has a 130-mb deep moist layer near the ground, which is nearly well mixed, capped by a sharp inversion, and surmounted by a deep nearly dryadiabatic layer extending up to 560 mb; by early evening the moist layer has thickened somewhat, while the capping inversion remains. The dryadiabatic layer above the capping inversion has approximately the same potential temperature (313 K) as the deep dryadiabatic layer to the west at Amarillo.

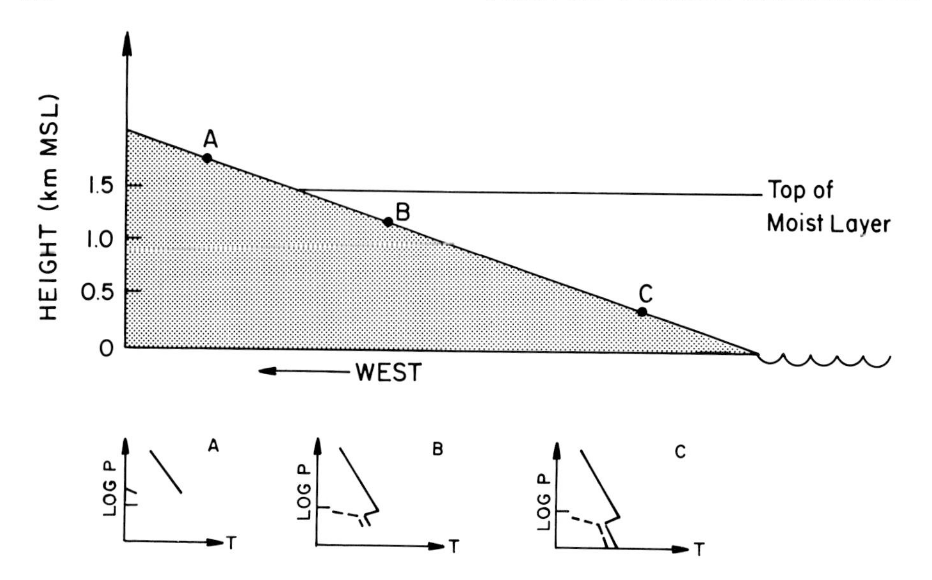

Figure 2.43 Schematic vertical cross section of the dryline and its relationship to topography. Idealized soundings (temperature, solid lines; dew point, dashed lines) at points A, B, C (bottom) represent the conditions west, just east, and far east of the dryline.

mixing. Since the height of the stable layer above the ground (i.e., the depth of the moist layer) increases toward the east, the net amount of surface heating by the sun needed to destabilize the stable moist layer increases toward the east. In addition, the increase in soil moisture in the east, where the total annual precipitation is greater, increases further the net amount of heating needed to destabilize the stable layer. The static stability is reduced until the lapse rate becomes dryadiabatic just to the east of the dryline; the dry air from above the stable layer is then mixed with the moist air from below, the surface dew point drops and the dryline *propagates* (not necessarily in a continuous fashion) eastward. The wind usually veers with height, owing to the decrease in friction with height and the concomitant low-level warm advection in the southerly air stream. The surface wind veers because higher westerly momentum aloft is mixed down to the surface.

After sunset, the clear, dry air west of the dryline cools off more rapidly than the cloudy, moist air to the east even in the face of windy conditions at the surface; a radiation inversion forms, which effectively ends the process of vertical mixing. The surface winds decrease in speed and back in direction in response to the pressure-gradient force associated with the lee trough. Low-level moisture that lies to the east is advected westward, and the dryline is advected westward. When radiational cooling acts to cool the dry air mass substantially, the retreating dryline may act like a warm front. On the other hand, it may act like a density current, especially early in the evening, when the air east of the dryline is relatively cool or if cool, thunderstorm outflow east of the dryline is advected westward.

Several theories have been proposed to explain the frequent occurrence of convective activity along the dryline. West of the dryline the winds at the surface usually veer with time, while east of the dryline the winds either back slightly or do not change direction. Hence, owing to the low-level convergence there may be upward vertical motion, which may become strong enough along the dryline to overturn the potentially unstable air mass just east of the dryline.

Sometimes sub-synoptic scale "bulges" or mesoscale waves (Figs. 2.44) along the dryline are associated with localized convergence. The nature of these waves is not well understood.

It has also been suggested that thermals or some types of waves originating to the west of the dryline may trigger convection as they travel eastward across the dryline. Or, dry thermals along the dryline may entrain low-level moist air from the east, and release conditionally instability. Outflow boundaries

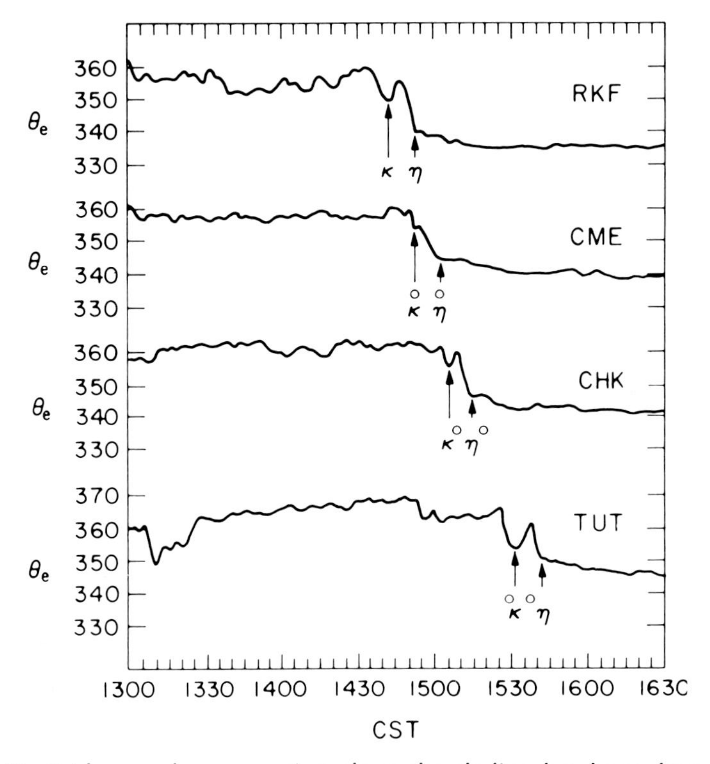

Figure 2.44 Evidence of wave motion along the dryline has been found in data from surface mesonetwork stations. This figure shows time series (time in CST) of measurements of equivalent potential temperature (θ_e) in K made at four stations spaced approximately 10–15 km apart along a north-northeast to south-southwest line. Station RKF is farthest south, and station TUT is farthest north, etc. Waves in the time series at each site are identified as κ and η. The waves were moving along the dryline with a phase speed of approximately 20 m s^{-1} (from McCarthy and Koch, 1982). (Courtesy of the American Meteorological Society)

traveling westward from convective storms may interact with the dryline to enhance convergence locally.

It had been proposed that a mechanism that is responsible for "thermohaline" convection in sea water could be responsible for sustained, buoyant, upward motion along the dryline. The nonlinear biconstituent diffusion of heat and water vapor in the atmosphere is analogous to that of heat and salt in sea water. It had been hypothesized that as heat diffuses toward the east and water vapor diffuses toward the west, a minimum in density, reflected in the virtual temperature, could be produced along the dryline. However, since the warm, moist air east or the dryline has a slightly higher heat capacity than the hot, dry air to the west, the temperature of the mixture at the dryline is actually slightly cooler than each alone.

Another hypothesis for a mechanism that produces vertical motion along the dryline is the *inland sea breeze* circulation (Fig. 2.45), proposed by Sun and Ogura. It may be set up along the dryline as a result of the varying rate of diabatic heating across the dryline: The diabatic heating rate is high on the clear, dry side of the dryline, and lower on the partly cloudy or cloudy, moist side of the dryline. If the soil on the dry side is dry, and the soil on the moist side is wet, then the differential diabatic heating is enhanced.

The interaction between synoptic-scale forcing and the dryline is not well understood. When a short-wave trough approaches the dryline in the United States, it is entering a region where the static stability in the lower troposphere is smaller. Thus, the atmosphere's quasigeostrophic response is enhanced. If the lapse rate is dryadiabatic, then there can be *no* response thermodynamically, and the atmosphere cannot remain in quasigeostrophic equilibrium; inertial-gravity waves must be generated.

2.4.2. The Middle–Upper Tropospheric Front

The *middle–upper tropospheric front,* which is also called an *upper-level front* or *hyperbaroclinic zone,* is a zone of strong, quasihorizontal or horizontal temperature gradient and high static stability in the middle and upper troposphere, which does not necessarily extend to the surface. It often forms downstream from an upper-level ridge and upstream from an amplifying upper-level trough, in a synoptic-scale environment characterized by sinking

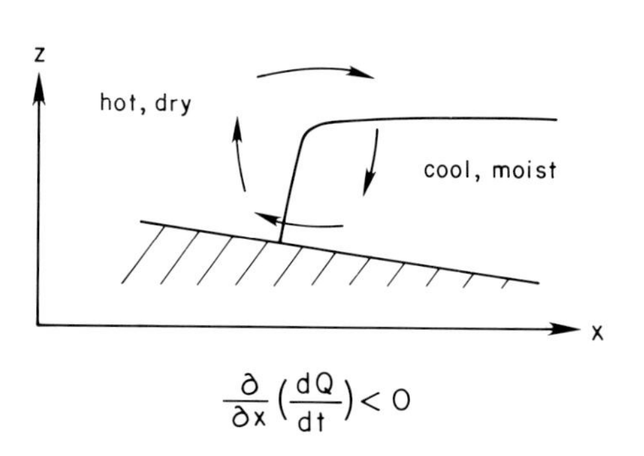

Figure 2.45 Illustration of the "inland sea breeze" mechanism for producing a vertical circulation across the dryline. As soon as the surface air on the dry side of the dryline becomes warmer than the air on the moist side, a hydrostatic pressure gradient develops that forces air from the cool (moist) side towards the warm (dry) side; the complete vertical circulation is as depicted by the arrows.

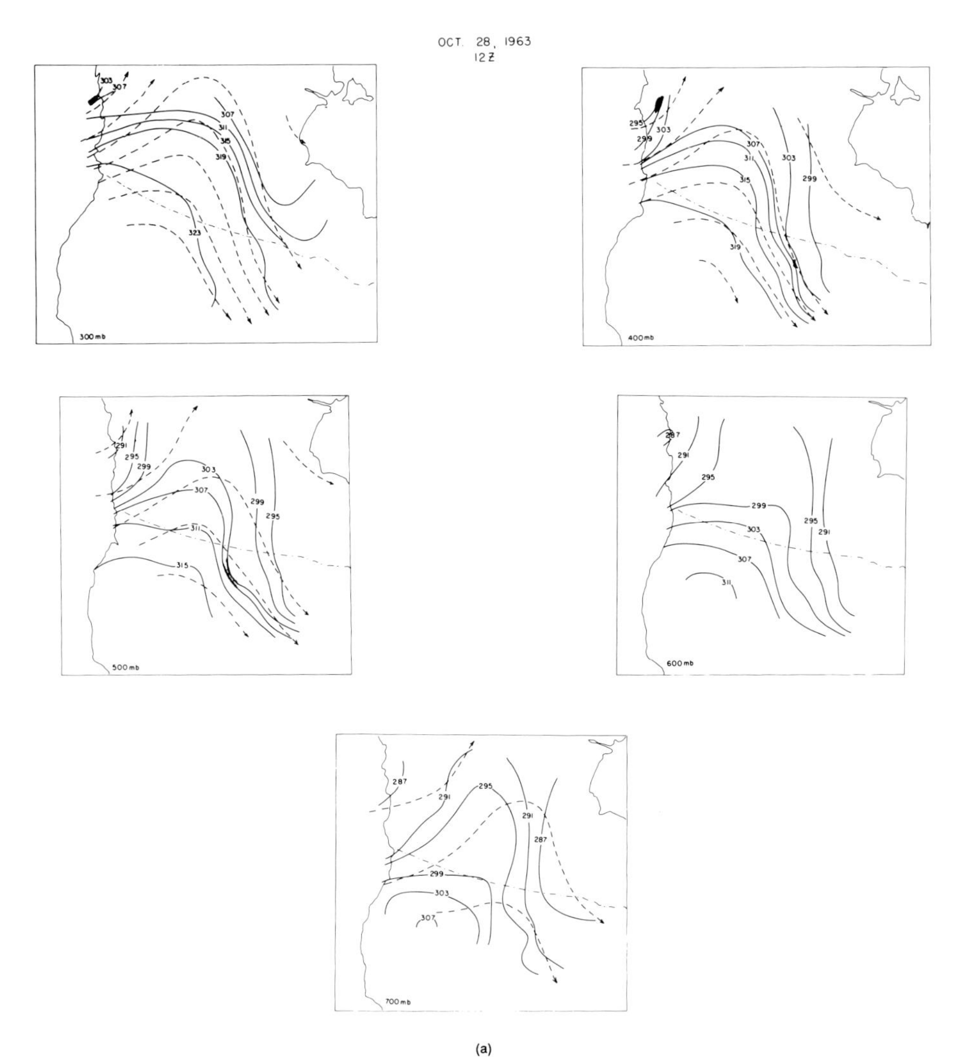

Figure 2.46 Illustration of the evolution of a middle–upper tropospheric front in the United States on October 28–30, 1963, at the indicated pressure levels and times in *Z* (UTC). Height contours at 12-dam intervals (dashed lines); isotherms of potential temperature in K (solid lines). The strongest quasihorizontal temperature gradients [shading denotes region of gradient in excess of 10 K (100 km)$^{-1}$] at 1200 UTC, October 28 (a) were located downstream from the ridge at 400 and 500 mb. By 0000 UTC, October 29 (b) the front extended down to 600 mb; it had propagated farther downstream to a position just upstream from the trough near the Great Lakes at a speed less than that of the wind. At 1200 UTC, October 29 (c) the region of strongest horizontal temperature gradients associated with the front was found wrapping around the trough, and extending all the way down to 800 mb southeast of the trough aloft. By 0000 UTC, October 30 (d) the front was apparent at and below 400 mb near the base of the trough. (Courtesy Fred Sanders)

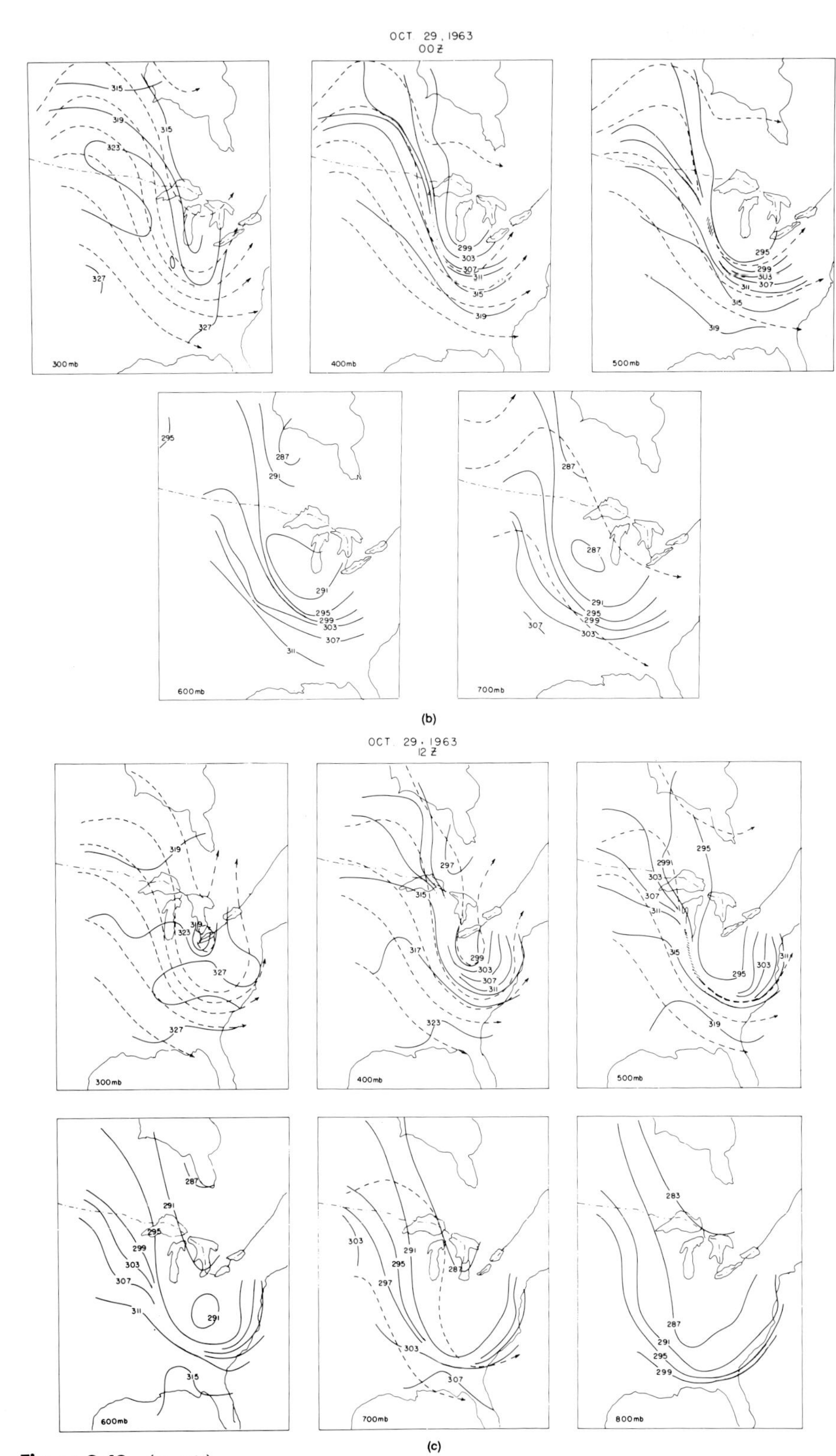

Figure 2.46 (*cont.*)

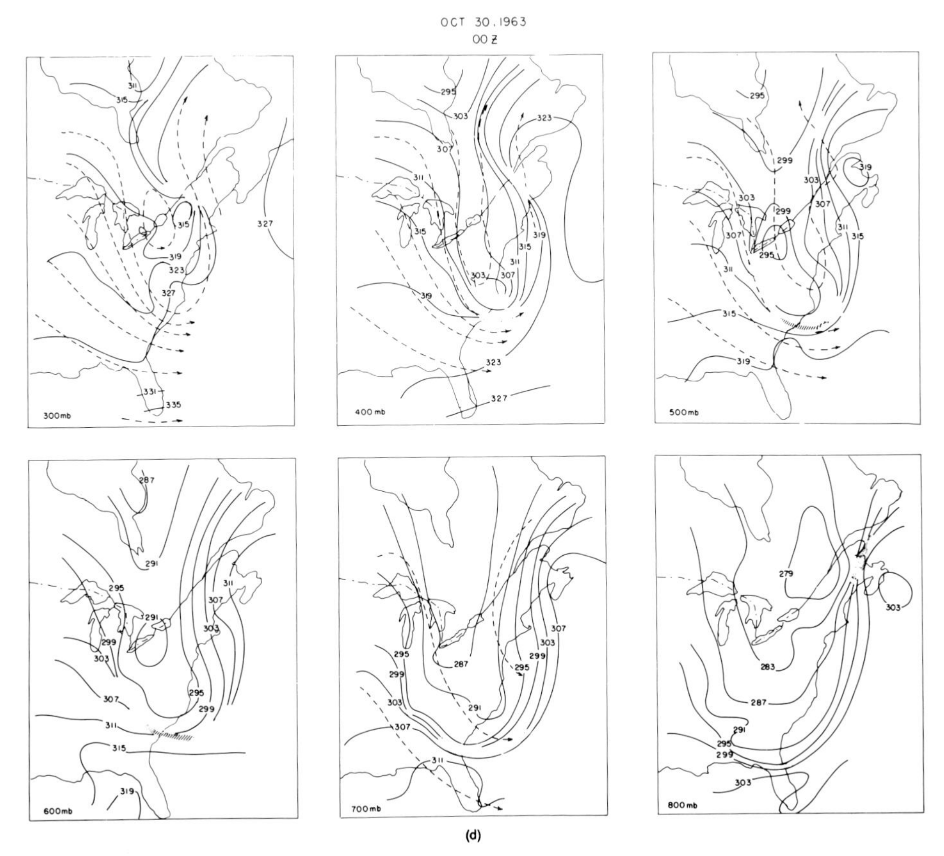

Figure 2.46 (*cont.*)

motion; it may propagate downstream and pass around the downstream trough (Fig. 2.46). Unlike surface fronts, the middle–upper tropospheric fronts are not described as "cold" or "warm," because the "cold" and "warm" air masses usually move approximately parallel to the winds. Thus, these fronts behave like segments of quasistationary fronts that move parallel to themselves. (At present, the upper-level front is not usually represented on maps by a symbol, unlike the surface front, which *is* represented by a symbol.)

Rawinsonde observations have been the main source of data that have been used to define the structure of these fronts. Unfortunately, the spacing between most rawinsonde stations in North America is 300–400 km, which is too large to resolve very well the horizontal frontal structure. However, excellent vertical resolution is available from each sounding. Aircraft have been used to probe the fine-scale horizontal structure. The middle–upper tropospheric front is somewhat mysterious in that it is not experienced by Earth-bound beings, and is generally in "clear" air, owing to subsidence: Only air travelers feel its effects directly. An example of a vertical cross section across a midlatitude upper-level front analyzed from conventional rawinsonde data is shown in Fig. 2.47. A sounding taken through the front is shown in Fig. 2.48.

In accord with the thermal wind relation, strong vertical wind shear accompanies the zone of strong quasihorizontal and horizontal temperature gradient. The Richardson number (Ri), which is a measure of the importance of buoyancy forces (as given by the static stability, or square of the Brunt–Väisälä frequency) to inertial accelerations, is defined as follows:

$$\mathrm{Ri} = \left(g\frac{\partial \ln \theta}{\partial z}\right) \Big/ \left(\frac{\partial |\mathbf{v}|}{\partial z}\right)^2. \tag{2.4.1}$$

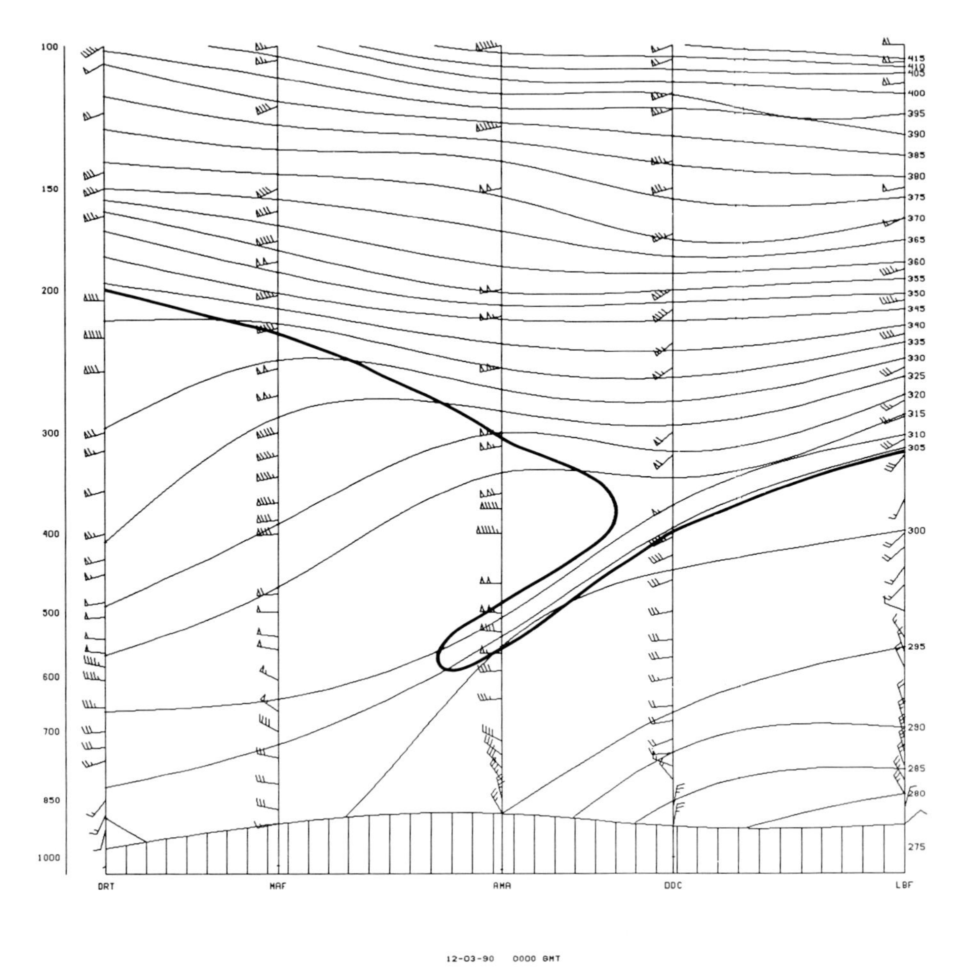

Figure 2.47 Vertical cross section across a middle–upper tropospheric front in the central United States at 0000 UTC, December 3, 1990 from the south (left) at Del Rio, Texas (DRT) through Midland, Texas (MAF), Amarillo, Texas (AMA), Dodge City, Kansas (DDC), and to the north (right) at North Platte, Nebraska (LBF). Isotherms of potential temperature (solid lines) in K; pennant = 25 m s^{-1}; whole wind barb = 5 m s^{-1}; half wind barb = 2.5 m s^{-1}. Tropopause indicated by thick solid line; note the fold around AMA. Distance from AMA to DDC is approximately 350 km.

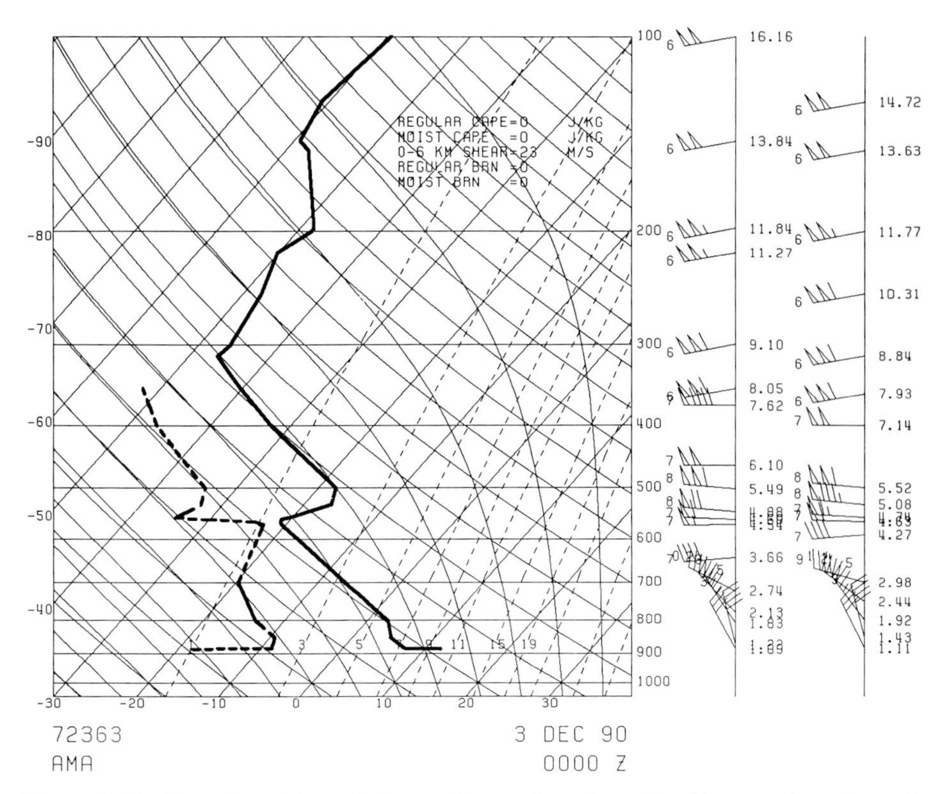

Figure 2.48 Sounding (skew T–log p diagram) at Amarillo, Texas taken through the middle–upper tropospheric front shown in Fig. 2.47 at 0000 UTC, December 3, 1990. The frontal zone is marked by the stable layer (in this case an inversion) between 560 and 500 mb. The wind increases in speed rapidly from about 25 m s^{-1} at the bottom of the stable layer, to 55 m s^{-1} at the top of the stable layer. The intense vertical wind shear of 30 m s^{-1} over 1 km implies a horizontal temperature gradient of nearly 8 K (100 km)$^{-1}$, if the wind were geostrophic (alas, data are not available on this scale to resolve the horizontal temperature gradient). The "first" tropopause is at the base of the inversion at 560 mb; the wind speed peaks at 55 m s^{-1} at the top of the inversion. The "second" tropopause is found at 310 mb; the wind speed peaks at 55 m s^{-1} again near this level. The air between 560 and 500 mb is in the stratosphere, as is the air above 310 mb. The stratospheric air in the 560–500 mb layer is part of a tropopause fold. The tropospheric layer between 500 and 310 mb has a lapse rate only slightly less than dryadiabatic.

When Ri is small, the flow becomes turbulent. Regions of clear-air turbulence are sometimes encountered in middle–upper tropospheric frontal zones, and hence the middle–upper tropospheric front can be dangeorus to aircraft. Like the "invisible man," the middle–upper tropospheric front can raise havoc and not be seen.

As a middle–upper tropospheric front propagates around and downstream from an upper-level trough, it descends, and the upper-level jet intensifies. What happens to the tropopause under these circumstances? The tropopause is located at the level at which relatively high tropospheric lapse rates are replaced by much lower stratospheric lapse rates (and where the winds are strongest, i.e., where the thermal wind shear reverses direction) that is, where the vertical gradient of static stability is very large.

Conservation of potential temperature following an air parcel in adiabatic flow implies that

$$\frac{\partial \theta}{\partial t} + \mathbf{v} \cdot \nabla_p \theta + \omega \frac{\partial \theta}{\partial p} = 0. \tag{2.4.2}$$

Differentiating Eq. (2.4.2) with respect to pressure, we see that

$$\frac{D}{Dt}\left(\frac{\partial \theta}{\partial p}\right) = -\frac{\partial \mathbf{v}_a}{\partial p} \cdot \nabla_p \theta + \delta \frac{\partial \theta}{\partial p}, \tag{2.4.3}$$

where $\mathbf{v}_a$ is the ageostrophic wind. In other words, static stability is not necessarily conserved for adiabatic motion, even though potential temperature is conserved.

However, Ertel's potential vorticity $[-(\zeta + f)_\theta(\partial\theta/\partial p)]$ which contains a measure of static stability $(\partial\theta/\partial p)$ as a factor, *is* conserved for adiabatic motion. Furthermore, Ertel's potential vorticity is relatively small in the troposphere because $-\partial\theta/\partial p$ is relatively small there, and because ζ_θ is small or even anticyclonic. The relative vorticity is generally small in part because some westerly shear appears as anticyclonic shear on the sloping isentropic surface (Fig. 2.49). However, in the stratosphere Ertel's potential vorticity is large because $-\partial\theta/\partial p$ is very large. Typical values of Ertel's potential vorticity and of potential temperature are shown in Fig. 1.137. In the stratosphere, values in excess of several PVU are found.

It is useful to define the tropopause in terms of Ertel's potential vorticity or its vertical gradient. As the frontal zone forms in the upper troposphere (in the absence of diabatic heating), the tropopause is advected downward at an angle such that it is found eventually at two levels: This process is known as

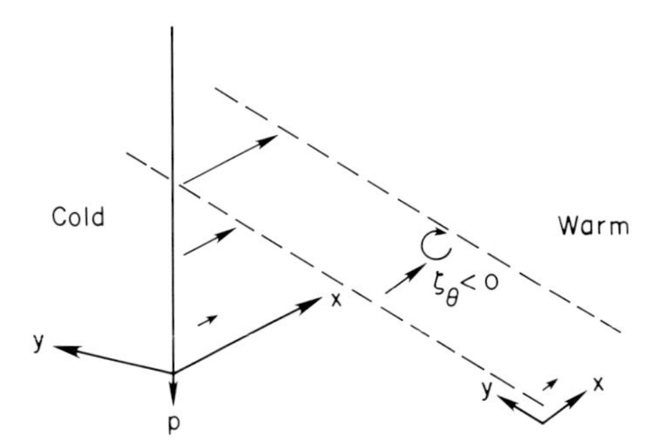

Figure 2.49 Illustration of how westerly vertical wind shear appears as anticyclonic shear on a surface of constant potential temperature (plane defined by dashed lines). Westerly vertical shear in the geostrophic wind implies a southerly directed temperature gradient, and a potential-temperature surface that tilts upward toward the north (with increasing *y*).

tropopause folding (Figs. 2.47 and 2.48). (Think of the tropopause as a blanket over the troposphere, which is being crimped at some point and pulled down.) When the tropopause becomes sloped, as it does in a fold, quasi-horizontal temperature advection can become substantial because the zone of strong temperature gradient is near the level of maximum wind speeds.

The influx of stratospheric air into the troposphere along tropopause folds in the middle–upper tropospheric frontal zone is important in that ozone and radioactive debris from nuclear weapons can thus enter our environment. In fact, much of the early research done on the middle–upper tropospheric front was motivated by a need to know how radioactive debris from above-the-ground nuclear explosions could get back into the troposphere.

Potential vorticity is sometimes anomalously large near the level of maximum wind above the tropopause fold; potential vorticity is therefore not necessarily conserved here, and hence turbulent mixing must be playing an important role. Just above and below the level of maximum wind, the Richardson number is small because the magnitude of the vertical shear is large; hence, turbulence is expected, and it can act to limit the strength of the frontal zone just below the level of maximum wind. It also follows that, owing to mixing, the folded tropopause is not a material surface, and atmospheric constituents such as ozone and chlorofluorocarbons can be exchanged through turbulent eddies back and forth between the stratosphere and troposphere.

2.5. DYNAMICS OF SURFACE FRONTOGENESIS

In Section 3 of this chapter we discussed the mechanisms by which a temperature gradient can be changed. Only the thermodynamic equation and the lexicon of the kinematics of the wind field were used. We now consider the dynamical consequences of a change in temperature gradient, that is, of frontogenesis and frontolysis.

2.5.1. Quasigeostrophic Frontogenesis

Consider Eq. (2.3.3) (the x axis is oriented along the isotherms as earlier) evaluated at the ground on a level surface; the kinematic lower boundary condition is therefore $\omega = 0$. Furthermore, suppose that diabatic heating and diffusion are neglected. The two-dimensional frontogenetical function is

$$F = \frac{D}{Dt}\left(-\frac{\partial \theta}{\partial y}\right) = \left(\frac{\partial v}{\partial y}\right)\left(\frac{\partial \theta}{\partial y}\right). \tag{2.5.1}$$

If

$$v = v_g$$

and the confluence represented by $-\partial v_g/\partial y$ is held fixed through time t, then it is seen after integrating Eq. (2.5.1) from time 0 to time t that

$$\left(-\frac{\partial\theta}{\partial y}\right)_t = \left(-\frac{\partial\theta}{\partial y}\right)_0 \exp\left[\left(-\frac{\partial v_g}{\partial y}\right)(t)\right]. \tag{2.5.2}$$

Thus, the initial horizontal gradient will increase exponentially with an e-folding time of $(-\partial v_g/\partial y)^{-1}$, which is on the order of 10^5 s, or one day. In other words, it would take on the order of 2.5 days for the temperature gradient to increase an order of magnitude. (This is just an order-of-magnitude estimate. With diffusion, this could be even longer, since diffusion acts to smooth out extrema. In addition, our analysis is linear in the sense that the geostrophic confluence is constrained to be a constant. Actually, the geostrophic confluence will change in response to the secondary circulation.) Observations indicate that fronts can form much more rapidly (the temperature gradient can increase an order of magnitude in one day), and therefore the kinematic action of geostrophic confluence on a temperature gradient is an incomplete explanation of frontogenesis.

Suppose that the atmosphere, which initially is in thermal wind balance, does not adjust to the increase in meridional temperature gradient induced by geostrophic deformation. Consider the simplified case in which

$$v = v_g(y), \tag{2.5.3}$$

$$\frac{\partial v_g}{\partial y} < 0, \tag{2.5.4}$$

$$u = u_g(x,p), \tag{2.5.5}$$

$$\theta = \theta(y,p), \tag{2.5.6}$$

$$\frac{\partial\theta}{\partial y} < 0. \tag{2.5.7}$$

The thermal wind relation in terms of potential temperature for Eq. (2.5.6) is

$$-\frac{\partial u_g}{\partial p} = \frac{R}{f_0 p}\left(\frac{p}{p_0}\right)^{\kappa}\left(-\frac{\partial\theta}{\partial y}\right). \tag{2.5.8}$$

It was shown in Vol. I that

$$\frac{D_g}{Dt}\left(-\frac{\partial u_g}{\partial p}\right) = -\frac{R}{f_0 p}\left(\frac{p}{p_0}\right)^{\kappa}\frac{D_g}{Dt}\left(-\frac{\partial\theta}{\partial y}\right). \tag{2.5.9}$$

The effect of geostrophic deformation acting to increase the horizontal temperature gradient is therefore to force the atmosphere *away* from thermal wind balance: The temperature gradient becomes too large for the vertical shear.

The adjustment process is depicted in Fig. 2.50 for the case in which the temperature gradient is changed impulsively. Consider a meridional temperature gradient such that it is cold to the north and warm to the south. If the temperature gradient is suddenly increased by deformation, then the thickness

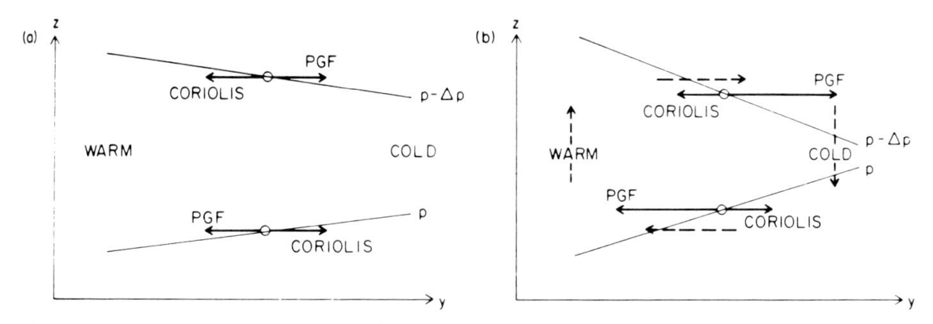

Figure 2.50 Illustration of the adjustment process that occurs when a horizontal temperature gradient is increased through the action of geostrophic deformation: The development of an ageostrophic circulation. Isobars (solid lines); forces (vectors). (a) Initial state of geostrophic balance; (b) imbalance of forces after horizontal potential temperature gradient has been increased. Ageostrophic circulation (dashed vectors) (from Bluestein, 1986). (Courtesy of the American Meteorological Society)

to the south increases hydrostatically, while the thickness to the north decreases, so that the isobars become more steeply sloped. Since the horizontal pressure gradient has been increased, there is now an imbalance of forces: A net force acts to the north aloft and to the south below. Thus, air accelerates in these directions and, owing to continuity, air rises and cools adiabatically on the warm side, and sinks and warms adiabatically on the cold side. The quasihorizontal temperature gradient is therefore decreased. When the air has been in motion long enough to be affected by the Earth's rotation, the northward-moving air aloft is deflected to the right and the wind acquires an additional westerly component, while the southward-moving air below is deflected to the right and the wind acquires an additional easterly component. Thus, the vertical shear is increased.

The adjustment process shown in Fig. 2.50 may also be explained as follows: If the quasihorizontal or horizontal temperature gradient is increased through the action of geostrophic deformation, then the thermal wind shear must increase in order to maintain near geostrophic and hydrostatic balance. The geostrophic wind accelerates aloft and decelerates below to effect the increase in thermal wind (Fig. 2.51). Since the ageostrophic wind is perpendicular and to the left of the parcel acceleration vector (in the Northern Hemisphere), there must be ageostrophic flow aloft from the "warm" side to the "cold" side, and below from the "cold" side to the "warm" side. From continuity considerations, there must be rising motion in the "warm" air and sinking motion in the "cold" air. Fronts in which the "warm" air rises are called *anafronts*. These fronts are likely being subject to frontogenetical, geostrophic forcing. (Fronts in which the "warm" air sinks are called *katafronts*. These fronts are likely to be subject to frontolytical, geostrophic forcing.)

The adjustment process depicted in Fig. 2.50 may also be explained using

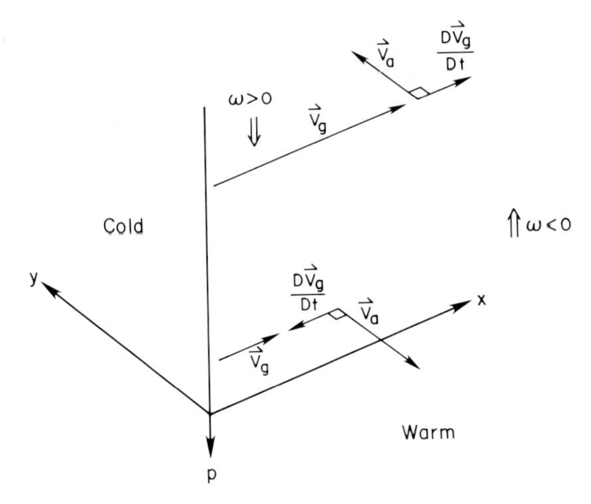

Figure 2.51 Another perspective on the adjustment process depicted in Fig. 2.50. An increase in westerly, geostrophic, vertical wind shear (x points toward the east, y toward the north) to accommodate the increase in north–south temperature gradient brought about by geostrophic deformation must be accompanied by accelerations in the geostrophic wind as shown. In the Northern Hemisphere the ageostrophic wind ($\mathbf{v}_a$) is directed perpendicular and to the left of the parcel acceleration vector as shown. We assume that, to the north and south of the zone along which the temperature gradient is being increased, there are no parcel accelerations and no ageostrophic wind. By continuity, then, there must be rising motion on the warm side, and sinking motion on the cold side of the zone along which frontogenesis is occurring.

Q vectors. The **Q** vector expressed in terms of potential temperature is

$$\mathbf{Q} = -\frac{R}{\sigma p}\left(\frac{p}{p_0}\right)^{\kappa}\begin{pmatrix} \dfrac{\partial v_g}{\partial x} & \dfrac{\partial \theta}{\partial y} \\ \dfrac{\partial v_g}{\partial y} & \dfrac{\partial \theta}{\partial y} \end{pmatrix} = \begin{pmatrix} Q_1 \\ Q_2 \end{pmatrix}, \tag{2.5.10}$$

where the coordinate system is oriented such that $\partial\theta/\partial x = 0$. Note that Eq. (2.5.10) is written with θ (instead of T) as a dependent variable. (Pressure is still an independent variable: The coordinate system is isobaric, not isentropic, even though potential temperature appears.) Now recall the following from Vol. I:

$$\nabla_p \omega - \frac{f_0^2}{\sigma}\frac{\partial \mathbf{v}_a}{\partial p} = -2\mathbf{Q}. \tag{2.5.11}$$

This equation was obtained from the quasigeostrophic equations of motion and the quasigeostrophic thermodynamic equation. The left-hand side of Eq. (2.5.11) represents the ageostrophic response to the geostrophic forcing given by the right-hand side. For Eqs. (2.5.3), (2.5.4), (2.5.6), and (2.5.7), Eq.

(2.5.11) is simply

$$\frac{\partial \omega}{\partial y}-\frac{f_0^2}{\sigma}\frac{\partial v_a}{\partial p}=-2Q_2>0, \tag{2.5.12}$$

with

$$Q_2=-\frac{R}{\sigma p}\left(\frac{p}{p_0}\right)^{\kappa}\frac{\partial v_g}{\partial y}\frac{\partial \theta}{\partial y}<0. \tag{2.5.13}$$

The ageostrophic adjustment to the temperature field is represented by $\partial\omega/\partial y$, and the adjustment to the wind field is represented by $(-f_0^2/\sigma)(\partial v_a/\partial p)$. For a length scale on the order of the Rossby radius of deformation NH/f_0, which is synoptic scale, the adjustments to the temperature and wind fields are of the same order of magnitude. [In fact, Hoskins and West's (1979) definition of a front includes a statement that the length scale along the front is on the order of the Rossby radius of deformation.] Therefore, $\partial\omega/\partial y>0$; since $f_0^2/\sigma>0$, it follows that $-\partial v_a/\partial p>0$. Taking into account continuity, it is seen that there is subsidence on the cool side and rising motion on the warm side. This acts to decrease the horizontal temperature gradient according to Eq. (2.3.3)

$$\frac{D_g}{Dt}\left(-\frac{\partial \theta}{\partial y}\right)=\frac{\partial\bar{\theta}(p)}{\partial p}\frac{\partial \omega}{\partial y}, \tag{2.5.14}$$

where $\partial\bar{\theta}(p)/\partial p$ is a measure of static stability. In other words, the atmosphere is brought back toward thermal wind balance by adjusting the temperature field back towards what it had been.

There is furthermore a southerly ageostrophic wind aloft, and a northerly ageostrophic wind below. This acts to increase the geostrophic shear according to the vertical gradient of the quasigeostrophic equation of motion in the x direction,

$$\frac{D_g}{Dt}\left(-\frac{\partial u_g}{\partial p}\right)=-f_0\frac{\partial v_a}{\partial p}. \tag{2.5.15}$$

In other words, the atmosphere is brought back toward thermal wind balance by increasing the vertical shear to accommodate the increased temperature gradient. Therefore ω and v_a comprise an ageostrophic, thermally direct circulation that obeys LeChatelier's principle, and acts to restore the atmosphere to geostrophic and hydrostatic balance.

From Vol. I (cf. Eq. (5.7.119)) we find that

$$\mathbf{Q}=\frac{R}{\sigma p}\left(\frac{p}{p_0}\right)^{\kappa}\frac{D_g}{Dt}(\boldsymbol{\nabla}_p\theta). \tag{2.5.16}$$

Thus, the $\mathbf{Q}$ vector is proportional to the rate of change of quasihorizontal potential temperature gradient vector following geostrophic motion. It is therefore similar to the frontogenetical function for geostrophic motion. From

Vol. I we also find that

$$F = \frac{\sigma p}{R} \frac{1}{|\nabla_p \theta|} (\nabla_p \theta \cdot \mathbf{Q}). \tag{2.5.17}$$

There is therefore frontogenesis when $\mathbf{Q}$, the vector geostrophic rate of change of $\nabla_p \theta$, is oriented within 90° of $\nabla_p \theta$ itself. In this case there is a thermally direct circulation. When the vector geostrophic rate of change of $\nabla_p \theta$, $\mathbf{Q}$, is oriented between 90° and 180° from $\nabla_p \theta$ itself, there is frontolysis. In this case there is a thermally indirect circulation.

Returning again to the case depicted in Fig. 2.50 [i.e., Eqs. (2.5.3)–(2.5.7)], we note that on a level ground $\omega = 0$, so that frontogenesis can proceed according to Eq. (2.5.1). That is,

$$F = \frac{\partial v_g}{\partial y} \frac{\partial \theta}{\partial y} > 0. \tag{2.5.18}$$

Above the ground $\omega \neq 0$, and the tilting of the vertical temperature gradient onto the horizontal by the thermally direct circulation $[(\partial \omega / \partial y > 0; (\partial \omega / \partial y)(\partial \theta / \partial p) < 0]$ acts frontolytically. Therefore, in the absence of diabatic heating and diffusion, the frontogenetical process occurs most rapidly at the ground, where the frontolytical effects of tilting vanish.

The strength of the surface temperature gradient is ultimately limited by diffusion. However, we saw in Fig. 2.6 that very sharp gradients can form in spite of diffusion. [Temperature gradients in fact can become so sharp that the hydrostatic pressure difference across the front is too large for geostrophy and the strong pressure-gradient force moves the frontal boundary. These surface fronts behave like density currents seen in thunderstorm outflow boundaries and gust fronts. In these features the role of Earth's rotation (f) is secondary, and hence quasigeostrophic equilibrium is not relevant.]

From the equation of continuity, we deduce that there is convergence at the surface under the rising branch of the circulation. Convergence acting on Earth's vorticity produces cyclonic geostrophic vorticity. It is the only source, according to quasigeostrophic dynamics, for cyclonic geostrophic vorticity at the surface. On a level surface there is no tilting because $\omega = 0$. Vorticity advection cannot *produce* vorticity; it only moves vorticity from one place to another. Furthermore, friction acting on a uniform, level surface cannot produce vorticity; it in fact destroys vorticity if the eddy coefficient of diffusion is constant.

Quasigeostrophic frontogenesis therefore requires that a convergent, cyclonic wind-shift line form at the surface on the warm side of a frontal zone where there is the rising branch of a vertical circulation. (In fact, it has been suggested that the fundamental process of frontogenesis is the concentration of vorticity.) The band of increasing cyclonic geostrophic vorticity is also associated with height falls on a pressure surface near the ground, and therefore pressure falls at the surface. Thus, a trough forms at the surface along the warm side of the frontal zone. Strong surface frontogenesis is accompanied by a strong vertical circulation, and hence an intense surface trough, and band of cyclonic geostrophic vorticity.

In summary, quasigeostrophic theory can account for the production of a sharp (but not infinite) temperature gradient at the surface over a finite duration of time. Quasigeostrophic theory also explains the decrease in strength of surface fronts with height (Figs. 2.20 and 2.52), and the formation of a convergent, cyclonic wind-shift line and pressure trough along the warm side of the frontal zone.

The weaknesses of quasigeostrophic frontogenesis are as follows:

1. The frontogenetical process at the ground is relatively slow.
2. The frontal zone does not tilt with height (Fig. 2.52), as is observed.
3. The field of relative vorticity contains regions of large anticyclonic as well as cyclonic vorticity; anticyclonic vorticity is produced at the surface under the sinking branch of the vertical circulation where $-\delta f_0 < 0$.
4. Regions of static instability may be produced (Fig. 2.52) at low levels on the "warm side."

Weaknesses (1), (2), and (4) are due to the neglect of advection of potential temperature by the ageostrophic component of the wind field. Weakness (3) is due to the neglect of geostrophic vorticity in the divergence term of the vorticity equation. These points will be discussed further in this chapter. Owing to their several unrealistic characteristics, quasigeostrophic fronts have been called *pseudofronts*.

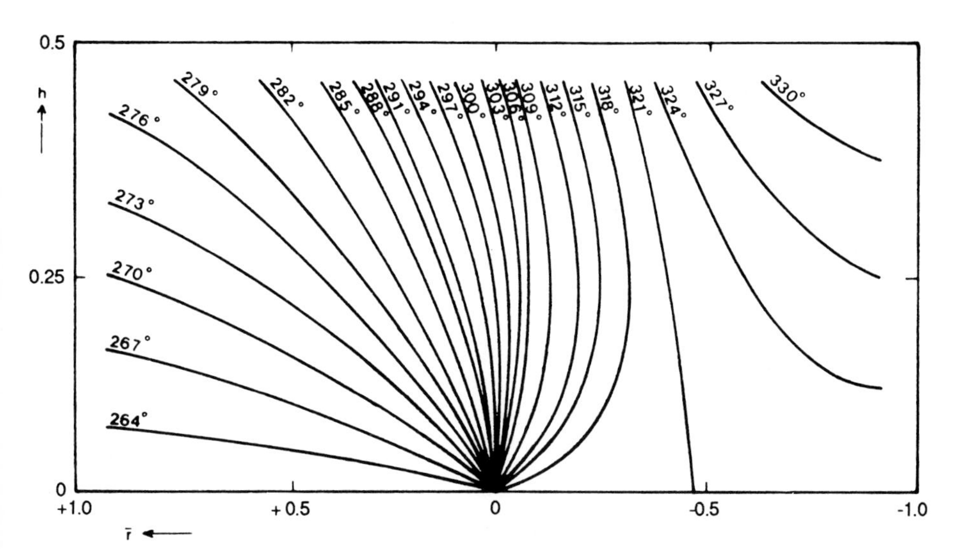

Figure 2.52 Vertical cross section through a front produced by quasigeostrophic processes only. Isotherms of potential temperature in K (solid lines); abscissa and ordinate represent scaled horizontal and vertical coordinates (from Stone, 1966). (Courtesy of the American Meteorological Society)

Quasigeostrophic principles can therefore account for some, but not all, of the features associated with surface fronts. It would seem that quasigeostrophic theory cannot be applied to the study of fronts because the Rossby number (U/fL, where U and L are wind and length scales) is too large for motions *across* the front (i.e., for small L). However, A. Eliassen has suggested that if local ($\partial \mathbf{v}/\partial t$) and convective $(\mathbf{v} \cdot \nabla)\mathbf{v}$ accelerations are of comparable magnitude and *opposite sign,* then the conventional definition of the Rossby number may not be appropriate. Air parcels may reside in the frontal zone for a long period of time, and hence parcel accelerations can be small and the narrow width of the front is not necessarily relevant. In other words, $|D\mathbf{v}/Dt|$ is not necessarily on the order of $U/(L/U)$, because L/U is really larger than that that follows from using an L characteristic of the width of the frontal zone. The conventional definition of Ro using characteristic scales of U and L is valid for an Eulerian framework. It is appropriate to use a Lagrangian definition of Ro, which in this case is not the same as the Eulerian definition.

The vector frontogenetical function and its relationship to the* Q *vector. In Section 3.1 we discussed the kinematics and thermodynamics of two-dimensional frontogenesis. The frontogenetical function, Eq. (2.3.1), was defined to quantify the physical processes that effect changes in the *magnitude* of the quasihorizontal potential temperature gradient. Let us now generalize the concept of a frontogenetical function to include also changes in the *direction* of the quasihorizontal potential temperature gradient.

Let

$$\mathbf{F}_p = \frac{D_p}{Dt} \nabla_p \theta, \tag{2.5.19}$$

where $D_p/Dt = \partial/\partial t + \mathbf{v} \cdot \nabla_p$, and $\mathbf{F}_p$ is the *vector* frontogenetical function for quasihorizontal motion: The vector frontogenetical function is the rate of change of the quasihorizontal potential temperature gradient following quasi-horizontal air-parcel motion. We now demonstrate that Petterssen's frontogenetical function, Eq. (2.3.1), is proportional to one component of the vector frontogenetical function in a natural coordinate system in which the x axis is oriented along the potential-temperature isotherms (i.e., the isentropes), and the y axis points in the direction of the colder air. (This natural coordinate system is like the one used to evaluate the **Q** vector.) The other component quantifies changes in the direction of $\nabla_p \theta$.

Equation (2.5.19) may be expressed in component form as follows:

$$\mathbf{F}_p = F_x \hat{\mathbf{i}} + F_y \hat{\mathbf{j}}, \tag{2.5.20}$$

where

$$\hat{\mathbf{j}} = -|\nabla_p \theta|^{-1} \nabla_p \theta, \tag{2.5.21}$$

and

$$\hat{\mathbf{i}} = \hat{\mathbf{j}} \times \hat{\mathbf{k}}. \tag{2.5.22}$$

The along-the-isotherm component of $\mathbf{F}_p$,

$$F_x = \hat{\mathbf{i}} \cdot \frac{D_p}{Dt} \nabla_p \theta. \tag{2.5.23}$$

If $\nabla_p \theta$ turns so that a positive (negative) $\hat{\mathbf{i}}$ component appears, then $\nabla_p \theta$ has been rotated in a counterclockwise (clockwise) direction. Substituting from Eq. (2.5.22), we write Eq. (2.5.23) as

$$F_x = \hat{\mathbf{j}} \times \hat{\mathbf{k}} \cdot \frac{D_p}{Dt} \nabla_p \theta. \tag{2.5.24}$$

With the aid of the vector identity

$$\mathbf{a} \times \mathbf{b} \cdot \mathbf{c} = \mathbf{a} \cdot (\mathbf{b} \times \mathbf{c}) \tag{2.5.25}$$

for arbitrary vectors **a**, **b**, and **c**, Eq. (2.5.24) can be written as

$$F_x = \hat{\mathbf{j}} \cdot \hat{\mathbf{k}} \times \frac{D_p}{Dt} \nabla_p \theta. \tag{2.5.26}$$

The along-the-countergradient of θ component of F_p,

$$F_y = \hat{\mathbf{j}} \cdot \frac{D_p}{Dt} \nabla_p \theta = -\frac{\nabla_p \theta}{|\nabla_p \theta|} \cdot \frac{D_p}{Dt} \nabla_p \theta = -\frac{D_p}{Dt} |\nabla_p \theta|, \tag{2.5.27}$$

which is the negative of Petterssen's frontogenetical function (for quasihorizontal air motion). The negative sign appears because $\hat{\mathbf{j}}$ points in the countergradient direction. Thus, F_y represents the change in magnitude of $\nabla_p \theta$, while F_x represents the change in direction of $\nabla_p \theta$.

It was shown earlier [Eq. (2.3.19)] how F_y is proportional to the magnitude of the quasihorizontal potential-temperature gradient, the resultant deformation, the relative orientation of the axis of dilatation with respect to the potential-temperature isotherms, and divergence. We now derive an analogous formula for F_x. Expanding Eq. (2.5.26), and using Eq. (2.5.21), we find that

$$F_x = |\nabla_p \theta|^{-1} \left[\frac{\partial \theta}{\partial x} \frac{D}{Dt} \left(\frac{\partial \theta}{\partial y} \right) - \frac{\partial \theta}{\partial y} \frac{D}{Dt} \left(\frac{\partial \theta}{\partial x} \right) \right]. \tag{2.5.28}$$

Substituting Eqs. (2.3.6) and (2.3.7) (the adiabatic form of the thermodynamic equation for quasihorizontal motion differentiated with respect to x and y) into Eqs (2.5.28), we obtain the following:

$$F_x = |\nabla_p \theta|^{-1} \left[-\left(\frac{\partial \theta}{\partial x} \right)^2 \frac{\partial u}{\partial y} - \frac{\partial \theta}{\partial x} \frac{\partial \theta}{\partial y} \frac{\partial v}{\partial y} + \frac{\partial \theta}{\partial x} \frac{\partial \theta}{\partial y} \frac{\partial u}{\partial x} + \left(\frac{\partial \theta}{\partial y} \right)^2 \frac{\partial v}{\partial x} \right]. \tag{2.5.29}$$

We now substitute Eqs. (2.3.9), (2.3.10), (2.3.11), and (2.3.12) into Eq. (2.5.29) and see that in terms of vorticity, divergence, and deformation

$$F_x = \tfrac{1}{2} |\nabla_p \theta|^{-1} \left[\left(\frac{\partial \theta}{\partial x} \right)^2 (\zeta - D_2) - \frac{\partial \theta}{\partial x} \frac{\partial \theta}{\partial y} (\delta - D_1) \right.$$
$$\left. + \frac{\partial \theta}{\partial x} \frac{\partial \theta}{\partial y} (\delta + D_1) + \left(\frac{\partial \theta}{\partial y} \right)^2 (\zeta + D_2) \right]. \tag{2.5.30}$$

The value of F_x is independent of the orientation of the coordinate system so that if $D_2' = 0$, that is, if either the axis of dilatation or axis of contraction lies along the x' axis, we find that

$$F_x = \tfrac{1}{2} |\nabla_p \theta|^{-1} \left[\zeta + \left(2D_1' \frac{\partial \theta}{\partial x'} \frac{\partial \theta}{\partial y'} \right) \Big/ |\nabla_p \theta|^2 \right]. \tag{2.5.31}$$

Substituting Eqs. (2.3.15), (2.3.16), and (2.3.17) into Eqs. (2.5.31), it follows that

$$F_x = \tfrac{1}{2} |\nabla_p \theta| \, (D \sin 2b + \zeta), \tag{2.5.32}$$

where b is the angle (measured in a counterclockwise direction) between the x' axis and the isotherms.

When the x' axis is oriented along the axis of dilatation (axis of contraction), D is positive (negative). Thus, the effect of horizontal deformation alone is to rotate the isotherms (and $\nabla_p \theta$) in a clockwise direction when the axis of dilatation lies within 90° of the isotherms, and to rotate the isotherms in a counterclockwise direction when the axis of dilatation lies between 90° and 180° of the isotherms (Fig. 2.53). The maximum rate of rotation occurs when the isotherms and axis of dilatation are 45° apart. In nearly all cases,[2] the isotherms are rotated toward the axis of dilatation.

Cyclonic vorticity acts to rotate the isotherms in a counterclockwise direction, and anticyclonic vorticity acts to rotate the isotherms in a clockwise direction, independent of the orientation of $\nabla_p \theta$.

In a natural coordinate system for θ defined by Eqs. (2.5.21) and (2.5.22),

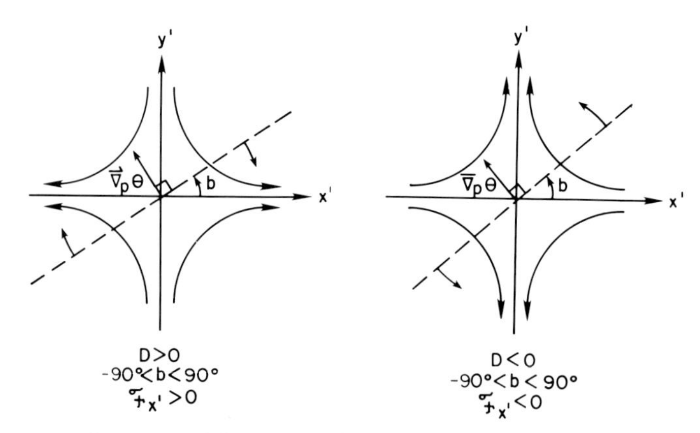

Figure 2.53 $F_{x'}$, that part of the vector frontogenetical function that represents the change in direction of the quasihorizontal potential temperature gradient $\nabla_p \theta$ for various values of D (resultant deformation) and b (the angle between the x' axis, the axis of dilatation for $D > 0$, axis of contraction for $D < 0$) and the potential-temperature isotherms (dashed line), measured in a counterclockwise direction from the x' axis. The isotherms (and temperature-gradient vector) are rotated by the quasihorizontal wind shear as indicated.

$\partial\theta/\partial x = 0$. It follows from Eqs. (2.5.27) and (2.3.13) that in this natural coordinate system.

$$F_y = \frac{1}{2}\left(-\frac{\partial\theta}{\partial y}\right)^{-1}\left(\frac{\partial\theta}{\partial y}\right)^2(\delta - D_1) = -\frac{\partial\theta}{\partial y}\frac{\partial v}{\partial y}. \tag{2.5.33}$$

From Eq. (2.5.30) we find that

$$F_x = \frac{1}{2}\left(-\frac{\partial\theta}{\partial y}\right)^{-1}\left(\frac{\partial\theta}{\partial y}\right)^2(\zeta + D_2) = -\frac{\partial\theta}{\partial y}\frac{\partial v}{\partial x}. \tag{2.5.34}$$

Suppose that the wind field is geostrophic; then from Eqs. (2.5.33) and (2.5.34) we see that the vector frontogenetical function can be expressed as

$$\mathbf{F}_p = -\begin{pmatrix} \dfrac{\partial v_g}{\partial x} & \dfrac{\partial\theta}{\partial y} \\ \dfrac{\partial v_g}{\partial y} & \dfrac{\partial\theta}{\partial y} \end{pmatrix}. \tag{2.5.35}$$

Comparing Eq. (2.5.35) with the expression for the **Q** vector in its natural coordinate system (cf. Eq. (2.5.16)), we see that

$$\mathbf{Q} = \frac{R}{\sigma p}\left(\frac{p}{p_0}\right)^{\kappa}\frac{D_g}{Dt}(\nabla_p\theta) = \frac{R}{\sigma p}\left(\frac{p}{p_0}\right)^{\kappa}\mathbf{F}_p, \tag{2.5.36),}$$

where

$$\frac{D_g}{Dt} = \frac{\partial}{\partial t} + \mathbf{v}_g \cdot \nabla.$$

Thus, the **Q** vector represents the vector frontogenetical function for the geostrophic wind.

The forcing function [in the (inviscid, adiabatic form of the) quasi-geostrophic ω equation less the β term] is $-2\nabla_p \cdot \mathbf{Q}$. From Eqs. (2.5.36), (2.5.19), and (2.5.20), it follows that the forcing function may be regarded as having one part that is due to the change in the magnitude of the temperature gradient (F_y), and another that is due to the change in the direction of the temperature gradient (F_x). Since the potential-temperature isotherms tend to get rotated toward the axis of dilatation, and the change in orientation of the temperature gradient determines the directional component of the vector frontogenetical function, and the relative orientation of the isotherms with respect to the orientation of the axis of dilatation determines the magnitude of the vector frontogenetical function, it is useful to plot deformation and the potential-temperature field together. (Plotting vorticity also is necessary to estimate F_x qualitatively.) The latter is depicted as a set of isotherms, and the former is depicted as a set of tick marks; the length of each tick mark is proportional to the magnitude of the resultant deformation, while the orientation of each depicts the axis of dilatation.

Although both F_x and F_y usually are of the same order of magnitude and therefore make a comparable contribution to the **Q** vector, the actual forcing

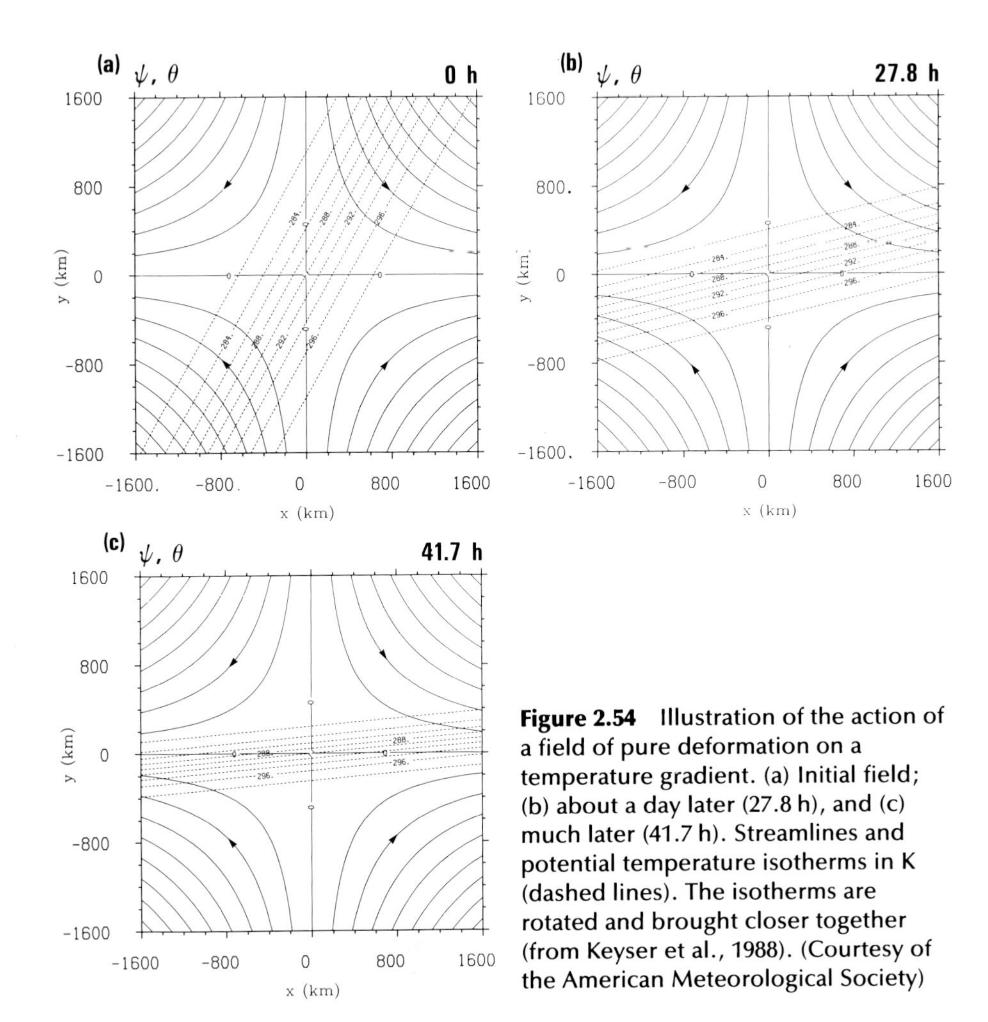

Figure 2.54 Illustration of the action of a field of pure deformation on a temperature gradient. (a) Initial field; (b) about a day later (27.8 h), and (c) much later (41.7 h). Streamlines and potential temperature isotherms in K (dashed lines). The isotherms are rotated and brought closer together (from Keyser et al., 1988). (Courtesy of the American Meteorological Society)

function $-2\,\nabla_p \cdot \mathbf{Q}$ depends mostly on F_y in the case of a pure deformation field acting on a temperature gradient (Figs. 2.54, and 2.55).

In the case of shear in the meridional wind component acting on a meridional temperature gradient, $-2\,\nabla_p \cdot \mathbf{Q}$ depends significantly on both F_x and F_y; however, it appears that F_y contributes toward a frontal-scale

→

Figure 2.55 Vector frontogenetical function (K m^{-1} s^{-1}; scale shown in lower right of a1 panel) and its horizontal divergence for the field of pure deformation [Fig. 2.54(b)]. Left column: frontogenetical function vectors and potential temperature (K). Right column: frontogenetical function vectors and their divergence (which represents the quasigeostrophic forcing function in the ω equation). Positive and zero contours (solid lines), negative contours (dashed lines). (a) Total vector frontogenetical function and its divergence; (b) magnitude component ($F_y = F_n$) and its divergence; (c) direction component ($F_x = F_s$) and its divergence (from Keyser et al., 1988). (Courtesy of the American Meteorological Society)

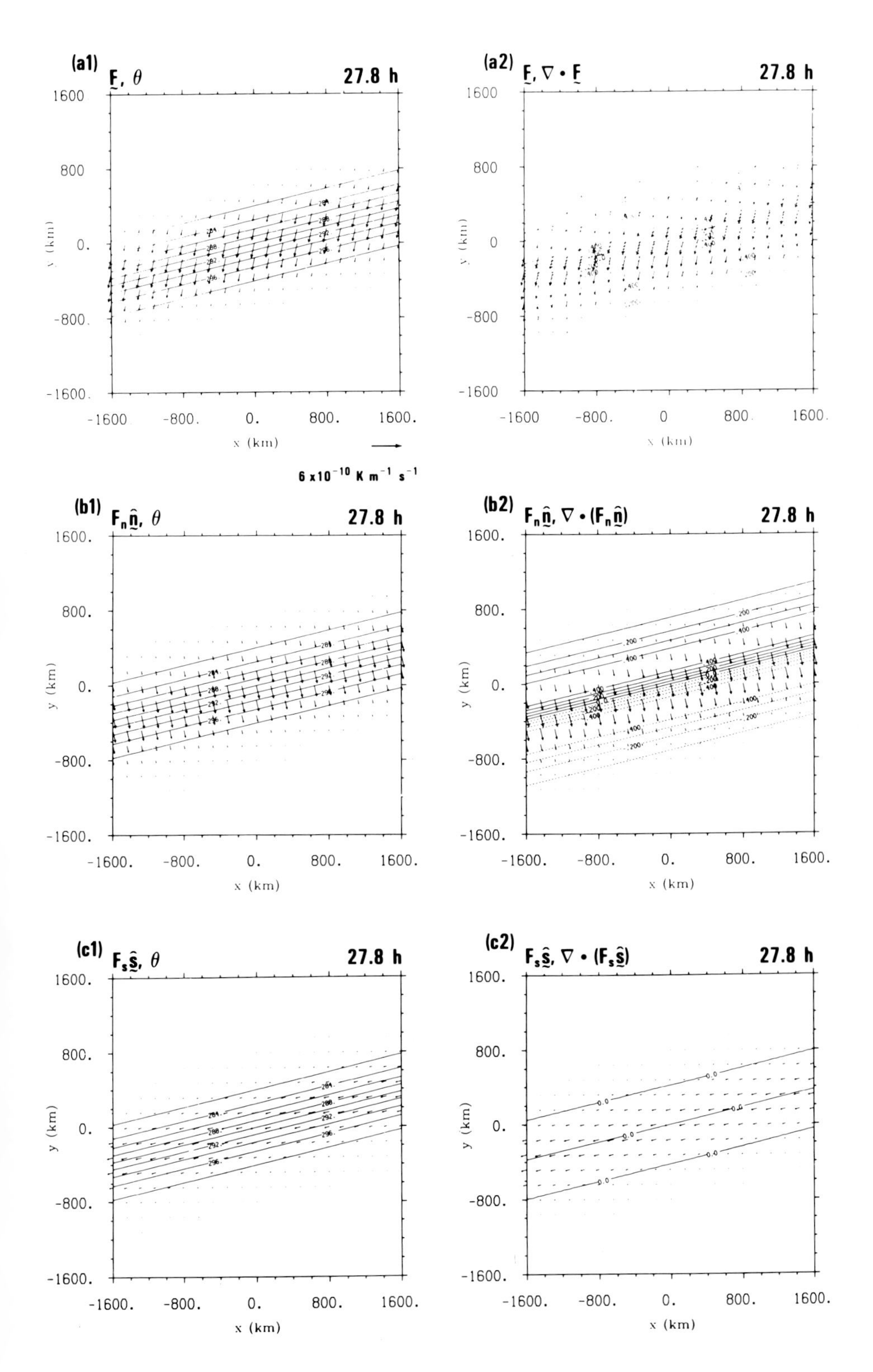
(a1)
F, θ
27.8 h
(a2)
F, ∇ • F
27.8 h
6 x10⁻¹⁰ K m⁻¹ s⁻¹
(b1)
Fₙn̂, θ
27.8 h
(b2)
Fₙn̂, ∇ • (Fₙn̂)
27.8 h
(c1)
Fₛŝ, θ
27.8 h
(c2)
Fₛŝ, ∇ • (Fₛŝ)
27.8 h
y (km)
x (km)

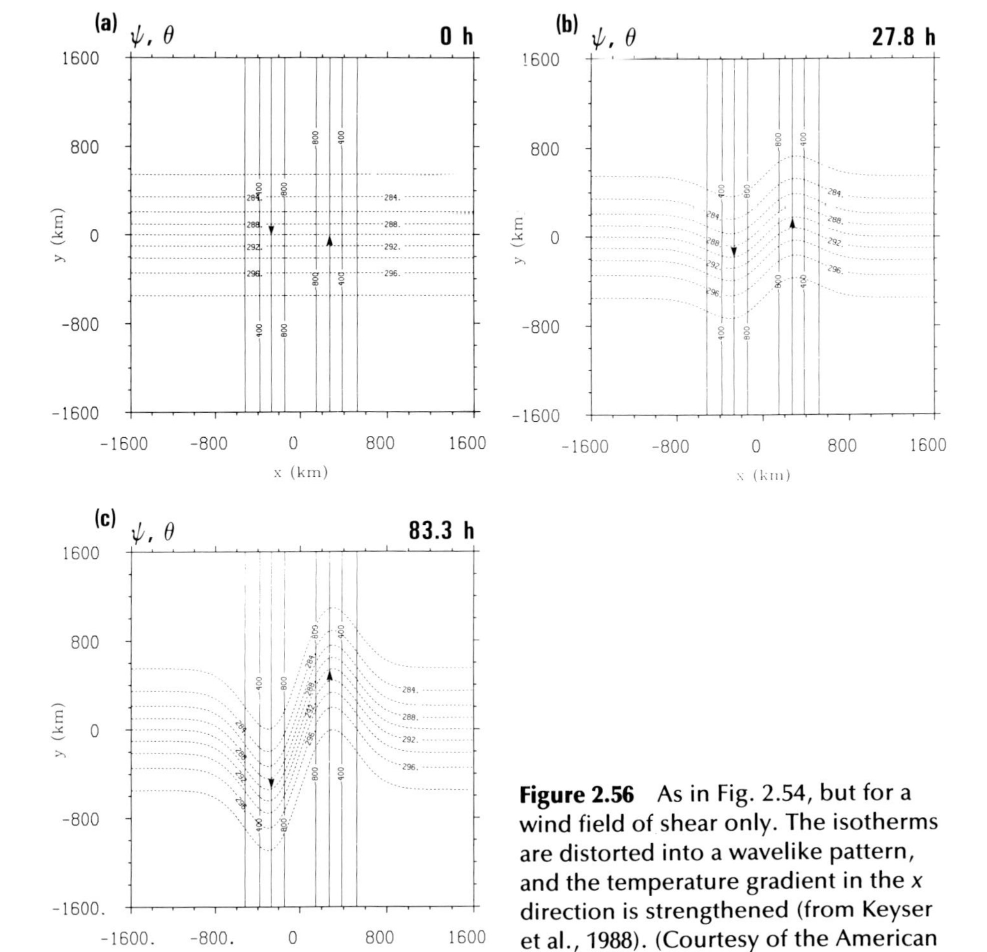

Figure 2.56 As in Fig. 2.54, but for a wind field of shear only. The isotherms are distorted into a wavelike pattern, and the temperature gradient in the x direction is strengthened (from Keyser et al., 1988). (Courtesy of the American Meteorological Society)

circulation, while F_x contributes toward a circulation on the scale of the wave disturbance produced as the isotherms are disturbed by the shear (Figs. 2.56 and 2.57).

In the case of a meridional temperature gradient being acted upon by a cyclonic vortex, $-2\,\nabla_p \cdot \mathbf{Q}$ also depends significantly on both F_x and F_y; however, as in the shear case, F_y seems to contribute toward a frontal-scale circulation, while F_x contributes more to a circulation on the scale of the vortex (Figs. 2.58, 2.59).

2.5.2. The Geostrophic-Momentum Approximation and Semigeostrophic Frontogenesis

The geostrophic-momentum approximation. The total wind may be expressed as the sum of the geostrophic and ageostrophic components:

$$\mathbf{v} = \mathbf{v}_g + \mathbf{v}_a. \tag{2.5.37}$$

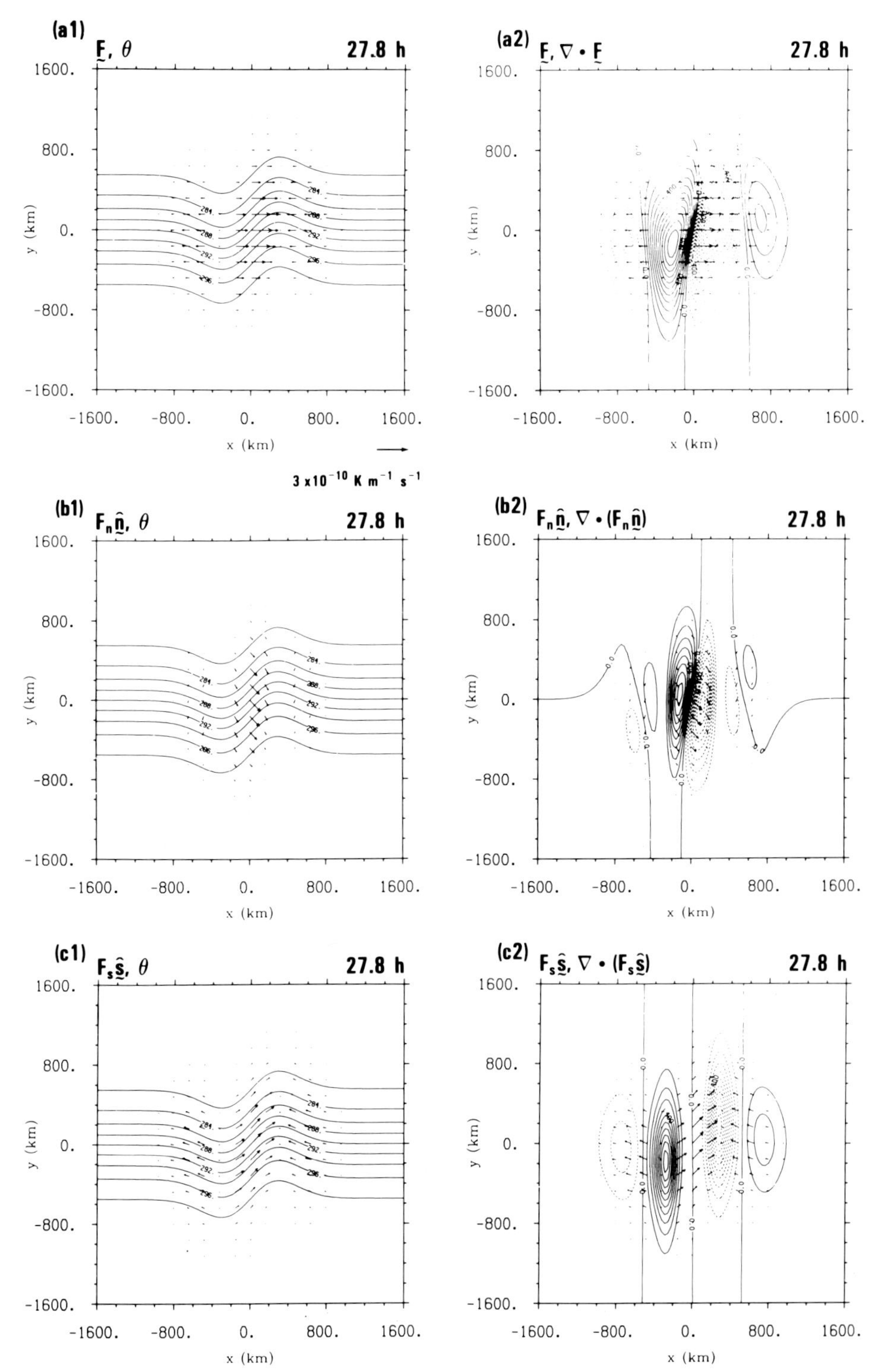

Figure 2.57 As in Fig. 2.55, but for the field of shear only [Fig. 2.56(b)]. Vector scale is halved with respect to that in Fig. 2.55 (from Keyser et al., 1988). (Courtesy of the American Meteorological Society)

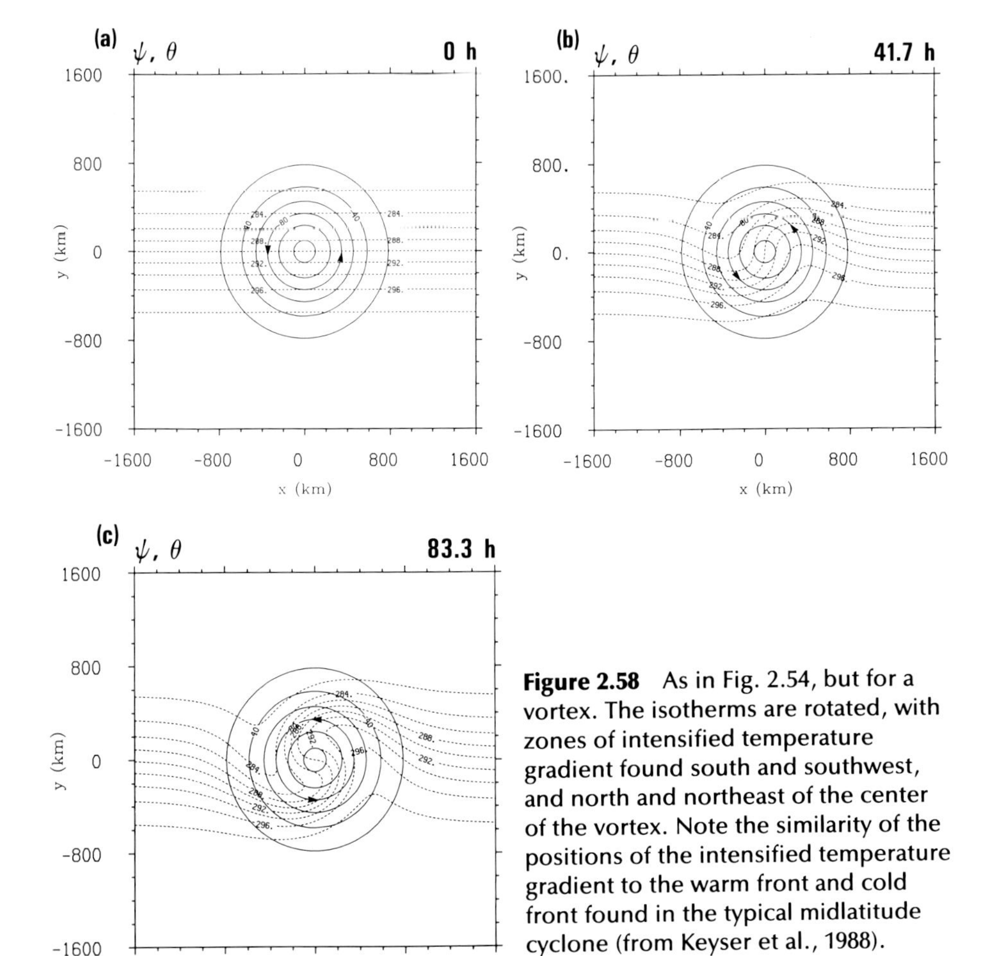

Figure 2.58 As in Fig. 2.54, but for a vortex. The isotherms are rotated, with zones of intensified temperature gradient found south and southwest, and north and northeast of the center of the vortex. Note the similarity of the positions of the intensified temperature gradient to the warm front and cold front found in the typical midlatitude cyclone (from Keyser et al., 1988). (Courtesy of the American Meteorological Society)

Recall that the frictionless form of the equation of motion can be written as:

$$\mathbf{v}_{\mathrm{a}} = \frac{1}{f}\hat{\mathbf{k}} \times \frac{D\mathbf{v}}{Dt}. \tag{2.5.38}$$

Then from Eqs. (2.5.37) and (2.5.38) it follows that

$$\mathbf{v} = \mathbf{v}_{\mathrm{g}} + \frac{1}{f}\hat{\mathbf{k}} \times \frac{D\mathbf{v}}{Dt}. \tag{2.5.39}$$

Substituting Eq. (2.5.37) and (2.5.38) into Eq. (2.5.39), we obtain:

$$\mathbf{v} = \mathbf{v}_{\mathrm{g}} + \frac{1}{f}\hat{\mathbf{k}} \times \left[\frac{D}{Dt}\left(\mathbf{v}_{\mathrm{g}} + \frac{1}{f}\hat{\mathbf{k}} \times \frac{D\mathbf{v}}{Dt}\right)\right]. \tag{2.5.40}$$

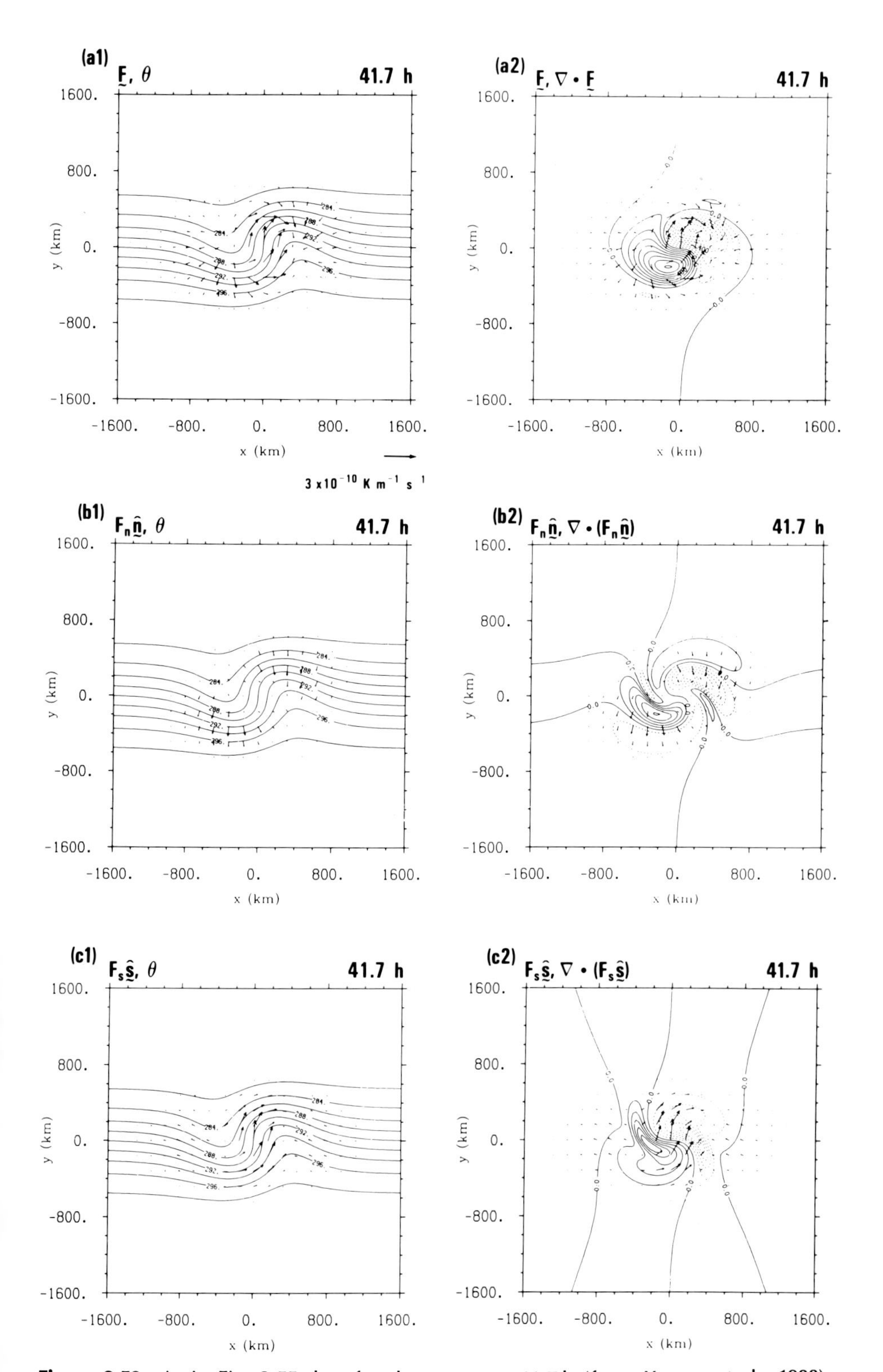

Figure 2.59 As in Fig. 2.55, but for the vortex at 41.7 h (from Keyser et al., 1988). (Courtesy of the American Meteorological Society)

Expanding Eq. (2.5.40) and solving for $\mathbf{v}_a$, we find that

$$\mathbf{v} - \mathbf{v}_g = \mathbf{v}_a = \frac{1}{f}\hat{\mathbf{k}} \times \frac{D\mathbf{v}_g}{Dt} - \frac{1}{f^2}\frac{D^2\mathbf{v}}{Dt^2}. \tag{2.5.41}$$

This recursive process could be continued until there is an infinite series [i.e., substitute Eqs. (2.5.37) and (2.5.38) for $\mathbf{v}$ into the right-hand side of Eq. (2.5.41) over and over again]. Suppose that for time scales longer than $1/f$, $|(1/f)D^2\mathbf{v}/Dt^2| \ll |D\mathbf{v}_g/Dt|$, and hence the $(1/f^2)D^2\mathbf{v}/Dt^2$ term may be neglected. The equation of motion subject to the "geostrophic momentum approximation" is

$$\mathbf{v}_a = \frac{1}{f}\hat{\mathbf{k}} \times \frac{D\mathbf{v}_g}{Dt} \qquad \text{or} \tag{2.5.42}$$

$$\frac{D\mathbf{v}_g}{Dt} = -f(\hat{\mathbf{k}} \times \mathbf{v}_a), \tag{2.5.43}$$

where

$$\frac{D}{Dt} = \frac{\partial}{\partial t} + (\mathbf{v}_g + \mathbf{v}_a) \cdot \boldsymbol{\nabla}_p + \omega\frac{\partial}{\partial p}. \tag{2.5.44}$$

Equation (2.5.43) is similar to the quasigeostrophic equation of motion, except that the ageostrophic and vertical advection of geostrophic momentum terms are retained (see Table 2.1). The geostrophic-momentum approximation, which was first introduced by Eliassen in 1948, is the substitution of the individual rate of change of geostrophic momentum for the individual rate of change of (total, i.e., geostrophic *and* ageostrophic) momentum.

It should be noted that the geostrophic advection of ageostrophic momentum is *not* retained in the equation of motion, even though the ageostrophic advection of geostrophic momentum is retained. The latter may be of the same order of magnitude as the former, especially in three–dimensional flow fields having substantial curvature. In this case the geostrophic-momentum approximation is not valid. [A more complicated set of equations known as the hypogeostrophic equations are valid (McWilliams and Gent, 1980) under these circumstances. They will not be discussed here.] It will be shown shortly,

Table 2.1 Various forms of the inviscid horizontal equations of motion in pressure coordinates

Geostrophic	$0 = -f_0\hat{\mathbf{k}} \times \mathbf{v}_g - \boldsymbol{\nabla}_p\Phi$
Quasigeostrophic	$\frac{\partial \mathbf{v}_g}{\partial t} + (\mathbf{v}_g \cdot \boldsymbol{\nabla}_p)\mathbf{v}_g = \frac{D_g\mathbf{v}_g}{Dt} = -f\hat{\mathbf{k}} \times \mathbf{v} - \boldsymbol{\nabla}_p\Phi$
Geostrophic-momentum approximation	$\frac{\partial \mathbf{v}_g}{\partial t} + (\mathbf{v} \cdot \boldsymbol{\nabla}_p)\mathbf{v}_g + \omega\frac{\partial \mathbf{v}_g}{\partial p} = \frac{D\mathbf{v}_g}{Dt} = -f\hat{\mathbf{k}} \times \mathbf{v} - \boldsymbol{\nabla}_p\Phi$
Full equation of motion	$\frac{\partial \mathbf{v}}{\partial t} + (\mathbf{v} \cdot \boldsymbol{\nabla}_p)\mathbf{v} + \omega\frac{\partial \mathbf{v}}{\partial p} = \frac{D\mathbf{v}}{Dt} = -f\hat{\mathbf{k}} \times \mathbf{v} - \boldsymbol{\nabla}_p\Phi$

however, that if the wind field is nearly straight (i.e., if parcel accelerations in the cross-stream direction can be neglected, and hence there is no along-the-stream component to the ageostrophic wind), then the geostrophic-momentum approximation is a good one. For this reason, the geostrophic-momentum approximation is well applied to frontal analysis, but not necessarily to the analysis of baroclinic waves, in which the wind field is curved.

The thermodynamic equation may be expressed as follows:

$$\frac{D_p\theta}{Dt} + \omega\frac{\partial\theta}{\partial p} = \frac{1}{C_p}\left(\frac{p_0}{p}\right)^{\kappa}\frac{dQ}{dt}, \tag{2.5.45}$$

where

$$\frac{D_p}{Dt} = \frac{\partial}{\partial t} + (\mathbf{v}_{\mathrm{g}} + \mathbf{v}_{\mathrm{a}}) \cdot \nabla_p. \tag{2.5.46}$$

In contrast to the quasigeostrophic thermodynamic equation, the term representing advection of temperature by the ageostrophic part of the wind is retained.

The geostrophic-momentum approximation is analogous in a sense to the hydrostatic approximation: According to the latter, the time rate of change of vertical momentum is neglected; however, advection by the vertical component of the wind is retained because vertical derivatives are large. According to the former, the time rate of change of ageostrophic momentum is neglected in comparison with the time rate of change of geostrophic momentum; however, advection by the ageostrophic part of the wind is retained because horizontal derivatives can be large, for example, near fronts and jets.

The Sawyer–Eliassen equation. We now will analyze the dynamics of a frontal zone using the geostrophic-momentum approximation. It will be assumed for simplicity that the front is oriented along the x axis. However, this does not necessarily imply that $\partial\theta/\partial x = 0$ (Fig. 2.60); there can be variations

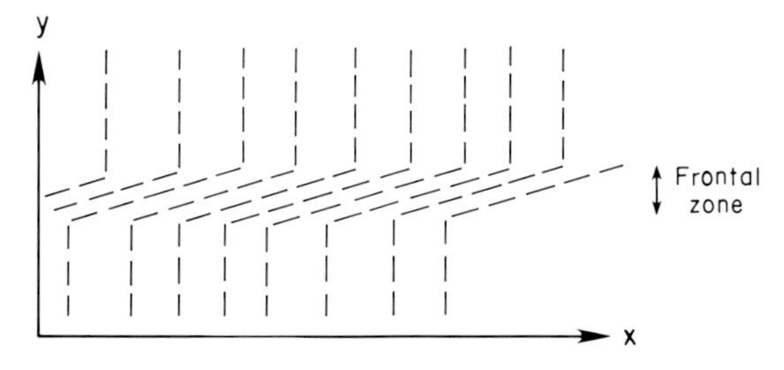

Figure 2.60 Illustration of an idealized frontal zone in which the temperature gradient is not normal to the frontal zone; that is, there is a component to the temperature gradient that lies along the front. Isotherms of potential temperature (dashed lines).

in temperature along the front. Furthermore, suppose there are accelerations along the front only; then across-the-front accelerations are given by

$$\frac{Dv_g}{Dt} = 0, \tag{2.5.47}$$

and hence from Eq. (2.5.42)

$$u_a = 0. \tag{2.5.48}$$

Since the flow is nearly straight, there is no curvature vorticity, so that

$$\frac{\partial v_g}{\partial x} = 0. \tag{2.5.49}$$

After combining the x components of Eqs. (2.5.43) and (2.5.45) and using the equation of continuity and the thermal wind relation, it is found that

$$\frac{R}{f_0 p}\left(\frac{p}{p_0}\right)^{\kappa} \frac{\partial}{\partial y}\left(-v_a \frac{\partial \theta}{\partial y} - \omega \frac{\partial \theta}{\partial p}\right) + \frac{\partial}{\partial p}\left[-v_a\left(f_0 - \frac{\partial u_g}{\partial y}\right) + \omega \frac{\partial u_g}{\partial p}\right]$$

$$= 2\frac{R}{f_0 p}\left(\frac{p}{p_0}\right)^{\kappa}\left(\frac{\partial \theta}{\partial y}\frac{\partial v_g}{\partial y} - \frac{\partial \theta}{\partial x}\frac{\partial u_g}{\partial y}\right) - \frac{R}{C_p f_0 p}\frac{\partial}{\partial y}\left(\frac{dQ}{dt}\right). \tag{2.5.50}$$

(If the flow were *not* straight, then both $u_a(\partial u_g/\partial x)$ and $u_g(\partial u_a/\partial x)$, which have the same magnitude, could not be neglected in the x-component of Eq. (2.5.43). Since the geostrophic-momentum approximation neglects the latter term, the approximation would not be valid.) This diagnostic equation relates the two dependent variables v_a and ω to the geostrophic wind field, the horizontal temperature-gradient field, diabatic heating, static stability, and absolute vorticity. The ageostrophic circulation lies in the y–p plane only.

Finding solutions to v_a and ω is simplified if a vertical streamfunction ψ is defined such that

$$v_a = -\frac{\partial \psi}{\partial p} \tag{2.5.51}$$

$$\omega = \frac{\partial \psi}{\partial y}. \tag{2.5.52}$$

Then from Eqs. (2.5.51), (2.5.52) and (2.5.8), Eq. (2.5.50) becomes

$$\frac{\partial^2 \psi}{\partial y^2}\left[-\frac{\partial \theta}{\partial p}\frac{R}{f_0 p}\left(\frac{p}{p_0}\right)^{\kappa}\right] + \frac{\partial^2 \psi}{\partial y\,\partial p}\left(2\frac{\partial u_g}{\partial p}\right) + \frac{\partial^2 \psi}{\partial p^2}\left(f_0 - \frac{\partial u_g}{\partial y}\right)$$

$$= 2\frac{R}{f_0 p}\left(\frac{p}{p_0}\right)^{\kappa}\left(\frac{\partial \theta}{\partial y}\frac{\partial v_g}{\partial y} + \frac{\partial \theta}{\partial x}\frac{\partial u_g}{\partial y}\right) - \frac{R}{C_p f_0 p}\frac{\partial}{\partial y}\left(\frac{dQ}{dt}\right). \tag{2.5.53}$$

This linear, second-order equation, is known as the *Sawyer–Eliassen equation.*

It relates the dependent variable ψ to other independent variables. Let

$$a = -\frac{R}{f_0 p}\left(\frac{p}{p_0}\right)^{\kappa}\frac{\partial\theta}{\partial p} \tag{2.5.54}$$

$$b = 2\frac{\partial u_g}{\partial p} = 2\frac{R}{f_0 p}\left(\frac{p}{p_0}\right)^{\kappa}\frac{\partial\theta}{\partial y} \tag{2.5.55}$$

$$c = f_0 - \frac{\partial u_g}{\partial y}, \tag{2.5.56}$$

where a, b, and c are assumed to be constant. Then we can express Eq. (2.5.53) as follows:

$$a\frac{\partial^2\psi}{\partial y^2} + b\frac{\partial^2\psi}{\partial y\,\partial p} + c\frac{\partial^2\psi}{\partial p^2} = \text{forcing}. \tag{2.5.57}$$

Equation (2.5.57) is elliptic and therefore always has unique solutions if $b^2 - 4ac < 0$, that is, if

$$\left(\frac{\partial u_g}{\partial p}\right)^2 + \left[\frac{R}{f_0 p}\left(\frac{p}{p_0}\right)^{\kappa}\right]\frac{\partial\theta}{\partial p}\left(f_0 - \frac{\partial u_g}{\partial y}\right) < 0. \tag{2.5.58}$$

(It will be shown later (2.5.261) that this ellipticity condition is equivalent to the necessary condition for symmetric stability.)

The right-hand side of Eq. (2.5.53) represents forcing owing to

1. Changes in the across-the-front temperature gradient due to geostrophic stretching deformation along the front (Fig. 2.61)

$$2\frac{R}{f_0 p}\left(\frac{p}{p_0}\right)^{\kappa}\left(\frac{\partial v_g}{\partial y}\frac{\partial\theta}{\partial y}\right);$$

2. Changes in the across-the-front temperature gradient as geostrophic shearing deformation tilts the along-the-front temperature gradient into the cross-front direction (Fig. 2.62)

$$2\frac{R}{f_0 p}\left(\frac{p}{p_0}\right)^{\kappa}\left(\frac{\partial u_g}{\partial y}\frac{\partial\theta}{\partial x}\right);\ \text{and}$$

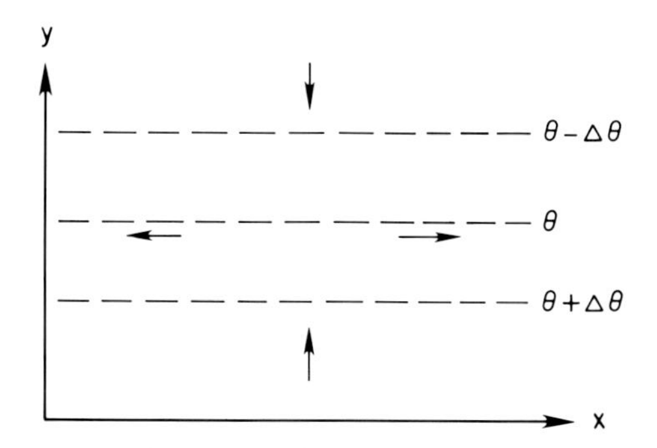

Figure 2.61 Illustration of geostrophic stretching deformation changing the across-the-front temperature gradient. Isotherms of potential temperature (dashed lines); geostrophic wind (vectors). $\partial u_g/\partial x - \partial v_g/\partial y > 0$, $\partial u_g/\partial x + \partial v_g/\partial y = 0$; $(\partial v_g/\partial y)(\partial\theta/\partial y) > 0$.

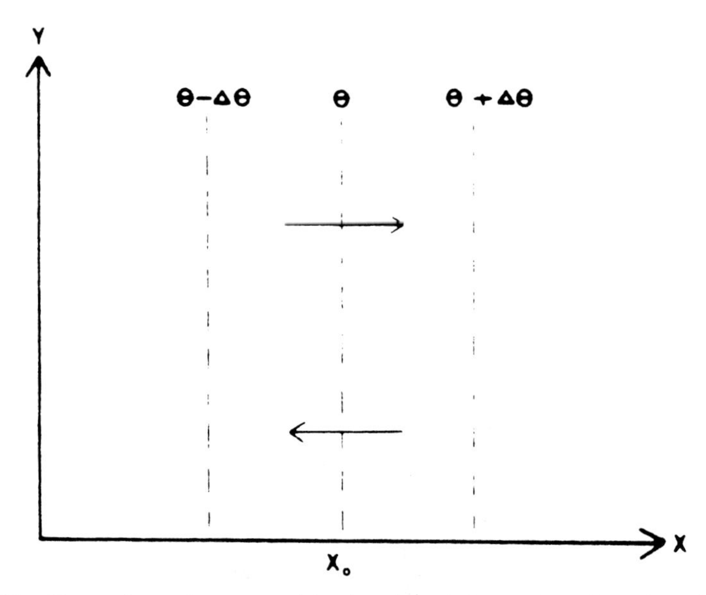

Figure 2.62 Illustration of geostrophic shearing deformation changing the across-the-front temperature gradient through tilting of the along-the-front component of the temperature gradient into the cross-front direction (from Bluestein, 1986). (Courtesy of the American Meteorological Society)

3. Differential diabatic heating

$$-\frac{R}{C_p f_0 p}\frac{\partial}{\partial y}\left(\frac{dQ}{dt}\right).$$

Note that forcing functions (1) and (2), i.e. $2(R/f_0 p)(p/p_0)^{\kappa}(\partial \mathbf{v}_g/\partial y)\cdot \boldsymbol{\nabla}_p\theta$, are proportional to the y component of the **Q** vector. (In fact, this is the origin of the **Q**-vector formulation!)

We use for simplicity the boundary condition

$$\psi = 0 \tag{2.5.59}$$

at the top, bottom, and at the lateral boundaries. It follows from Eq. (2.5.51) and (2.5.52) that no flow is allowed into or out from the area encompassed by the "vertical" (i.e., in the y–p plane) circulation. The solution to Eq. (2.5.57) in (y–p) space is a field of elliptically oriented streamlines (Fig. 2.63, right).

We can now discuss how the intensity, direction, and tilt of the vertical circulation depend upon the forcing, the static stability (a), the geostrophic vertical shear along the front, or equivalently, the quasihorizontal temperature gradient across the front (b), and the geostrophic absolute vorticity (c).

The intensity of the v_a–ω vertical circulation,

$$\frac{\partial^2\psi}{\partial y^2}+\frac{\partial^2\psi}{\partial p^2}=\nabla_x^2\psi, \tag{2.5.60}$$

is proportional to the magnitude of the forcing according to Eq. (2.5.57). Large changes in the across-the-front temperature gradient owing to geo-

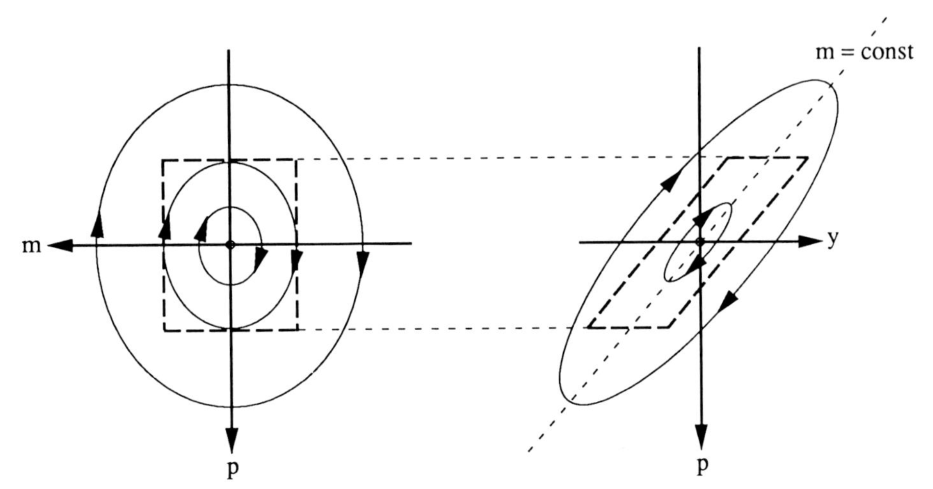

Figure 2.63 Streamlines around a positive point source in the *m*–*p* plane (left; to be discussed soon) and *y*–*p* plane (right) (after Eliassen, 1962).

strophic deformation or differential diabatic heating are accompanied by a strong vertical circulation. However, along warm fronts the effect of $(\partial u_g/\partial y)(\partial\theta/\partial x)$ often is frontolytical, while along cold fronts this effect tends to be frontogenetical (Fig. 2.64). Therefore vertical circulations along cold fronts tend to be stronger than those along warm fronts. Since $\partial^2\psi/\partial y^2$ and $\partial^2\psi/\partial p^2$ are inversely proportional to a and c, respectively [see Eq. (2.5.57)], the intensity of the vertical circulation is enhanced by low static stability or low geostrophic absolute vorticity.

When the vertical circulation [in the y–$(-p)$ plane] is in a clockwise direction,

$$-\frac{\partial v_a}{\partial p}=\frac{\partial^2\psi}{\partial p^2}>0 \tag{2.5.61}$$

$$\frac{\partial \omega}{\partial y}=\frac{\partial^2\psi}{\partial y^2}>0. \tag{2.5.62}$$

A counterclockwise circulation is thus associated with a negative $\partial^2\psi/\partial p^2$ and $\partial^2\psi/\partial y^2$. At the center of the circulation

$$\frac{\partial \omega}{\partial p}=-\frac{\partial v_a}{\partial y}=0=\frac{\partial^2\psi}{\partial y\,\partial p}. \tag{2.5.63}$$

Therefore from Eq. (2.5.57) we see that positive forcing (from a point source or a constant source) is accompanied by a clockwise circulation, and negative forcing by a counterclockwise circulation (Fig. 2.65), if the atmosphere is statically stable ($a>0$) and if the geostrophic absolute vorticity is cyclonic ($c>0$).

The eccentricity of the elliptical vertical circulation depends upon the relative magnitudes of absolute vorticity and static stability. If the absolute vorticity (c) is small compared to the static stability term (a), then $|\partial v_a/\partial p|$ is

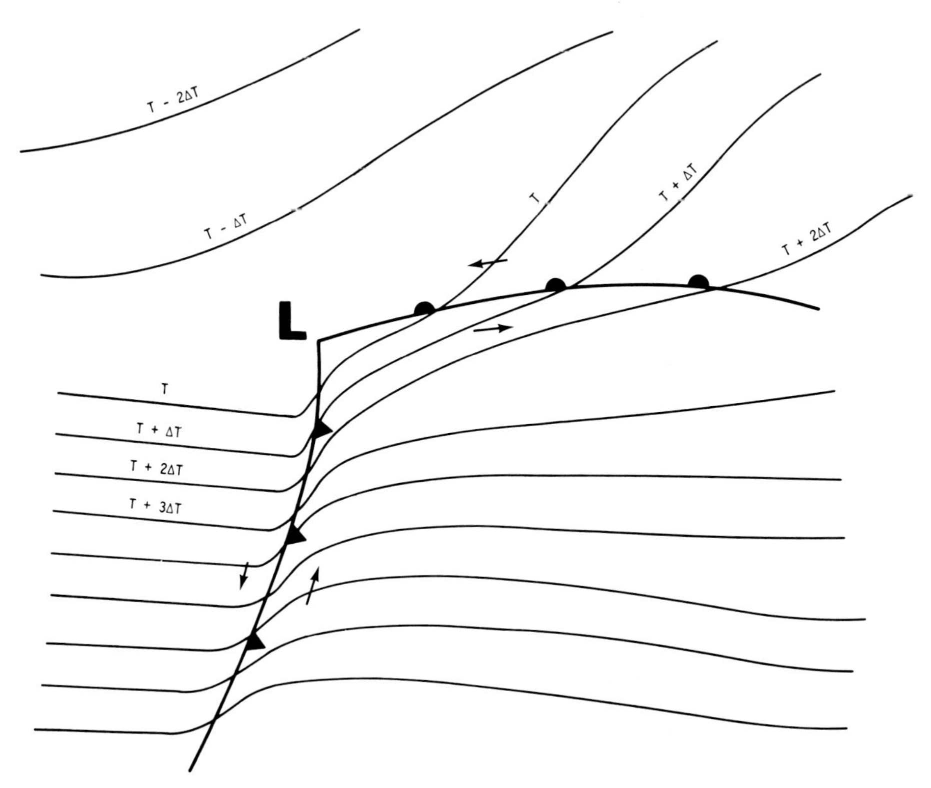

Figure 2.64 Illustration of the frontolytical effects of shear along the warm front, and the frontogenetical effects of shear along the cold front in an idealized midlatitude cyclone (after Gidel, 1978).

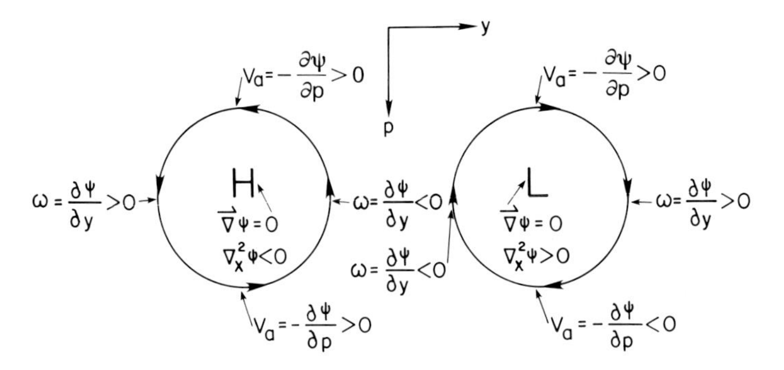

Figure 2.65 Sense of direction of streamlines around positive (H) and negative (L) point sources. Relationship between ageostrophic component of the wind normal to the front ($\mathbf{v}_a$), vertical velocity (ω), and the streamfunction (ψ) as shown. The streamfunction is the solution to the Sawyer–Eliassen equation.

large compared to $|\partial\psi/\partial y|$. Since v_a and ω are zero on the lateral boundaries and the top and bottom boundaries, respectively, then the circulation is composed mostly of the horizontal "branch." On the other hand, if the absolute vorticity (c) is large compared to the static stability term (a), the circulation is composed mostly of the vertical "branch." In other words, if the inertial stability is small compared to the static stability, horizontal motions dominate while vertical motions are inhibited. If the inertial stability is large compared to the static stability, vertical motions dominate while horizontal motions are inhibited.

The tilt of the elliptical, vertical circulation depends upon b, the vertical shear of the along-the-front component of the geostrophic winds, that is, the quasihorizontal across-the-front temperature gradient. We will soon see exactly what this dependence is.

The vorticity equation subject to the geostrophic-momentum approximation. Consider the u and v components of motion in Eq. (2.5.43) subject to the beta-plane approximation:

$$\frac{\partial u_g}{\partial t} + u_g\frac{\partial u_g}{\partial x} + u_a\frac{\partial u_g}{\partial x} + v_g\frac{\partial u_g}{\partial y} + v_a\frac{\partial u_g}{\partial y} + \omega\frac{\partial u_g}{\partial p} = f_0 v_a + \beta y v_g \qquad (2.5.64)$$

$$\frac{\partial v_g}{\partial t} + u_g\frac{\partial v_g}{\partial x} + u_a\frac{\partial v_g}{\partial x} + v_g\frac{\partial v_g}{\partial y} + v_a\frac{\partial v_g}{\partial y} + \omega\frac{\partial v_g}{\partial p} = -f_0 u_a - \beta y u_g. \qquad (2.5.65)$$

Differentiating Eq. (2.5.65) with respect to x and subtracting from the resulting equation Eq. (2.5.64) differentiated with respect to y, we get the following, after having rearranged the terms, neglected advection of ageostrophic vorticity, and used the definitions of δ and ζ:

$$\frac{\partial \zeta_g}{\partial t} + (u_g + u_a)\frac{\partial \zeta_g}{\partial x} + (v_g + v_a)\frac{\partial \zeta_g}{\partial y} + \omega\frac{\partial \zeta_g}{\partial p} + v_g\beta$$
$$= -\delta(\zeta_g + f) + \hat{\mathbf{k}}\cdot\frac{\partial \mathbf{v}_g}{\partial p}\times\nabla_p\omega. \qquad (2.5.66)$$

Equation (2.5.66) is the Eulerian form of the vorticity equation subject to the geostrophic-momentum approximation. The Lagrangian form of the vorticity equation subject to the geostrophic-momentum approximation is

$$\frac{D}{Dt}(\zeta_g + f) = -\delta(\zeta_g + f) + \hat{\mathbf{k}}\cdot\frac{\partial \mathbf{v}_g}{\partial p}\times\nabla_p\omega. \qquad (2.5.67)$$

Thus, if the geostrophic-momentum approximation is made, then divergence (and convergence) acts not only on f as it does in the quasigeostrophic vorticity equation, but also on ζ_g. The rate at which anticyclonic vorticity is produced in a region of divergence is less than the rate at which cyclonic vorticity is produced in a region of convergence, because $\zeta_g + f$ for $\zeta_g < 0$ is less than $\zeta_g + f$ for $\zeta_g > 0$ given that $f > 0$. Hence the strip of cyclonic vorticity along fronts is more prominent, and not equally as prominent as the strip of anticyclonic vorticity as predicted by quasigeostrophic theory.

The semigeostrophic equations. Let us now transform our coordinate system [i.e., (x,y,p,t)] into one in which air parcels move along with the geostrophic wind (i.e., on "Geostrophic Airlines"). This new coordinate system is called *geostrophic coordinates*. The purpose of the following discussion is to show that *in geostrophic coordinates* (X,Y,P,t^*), the dynamical equations subject to the geostrophic-momentum approximation are similar in form to the quasi-geostrophic equations in the old, untransformed coordinate system.

From the definition of geostrophic coordinates, we know that

$$\frac{DX}{Dt} = u_g \quad (2.5.68)$$

$$\frac{DY}{Dt} = v_g, \quad (2.5.69)$$

where X and Y are the quasihorizontal geostrophic coordinates. From Eq. (2.5.43) it follows that

$$\frac{Du_g}{Dt} = f(v - v_g) \quad (2.5.70)$$

$$\frac{Dv_g}{Dt} = -f(u - u_g). \quad (2.5.71)$$

It is seen from Eqs. (2.5.68) and (2.5.71) that

$$u_g = u + \frac{1}{f}\frac{Dv_g}{Dt} = \frac{DX}{Dt}. \quad (2.5.72)$$

Similarly, from Eqs. (2.5.69) and (2.5.70) it is seen that

$$v_g = v - \frac{1}{f}\frac{Du_g}{Dt} = \frac{DY}{Dt}. \quad (2.5.73)$$

If

$$X = x + \frac{v_g}{f}(x,y,p,t) \quad (2.5.74)$$

$$Y = y - \frac{u_g}{f}(x,y,p,t), \quad (2.5.75)$$

then Eqs. (2.5.72) and (2.5.73) are satisfied [i.e., Eqs. (2.5.74) and (2.5.75) are the indefinite integrals of Eqs. (2.5.72) and (2.5.73)]. In geostrophic coordinates

$$P = p \quad (2.5.76)$$

$$t^* = t. \quad (2.5.77)$$

We will now transform the geostrophic-momentum form of the equations of motion, Eqs. (2.5.64) and (2.5.65), and the thermodynamic equation, Eq. (2.5.45), into geostrophic coordinates. For simplicity,[3] the meridional variation in f will be neglected, so that we can let $f = f_0$. Since the geostrophic wind is a function of space and time, it follows from Eqs. (2.5.74), (2.5.75), (2.5.76), and (2.5.77) that

$$X = X(x,y,p,t) \tag{2.5.78}$$

$$Y = Y(x,y,p,t) \tag{2.5.79}$$

$$P = P(p) \tag{2.5.80}$$

$$t^* = t^*(t). \tag{2.5.81}$$

This transformation is not a simple one!

Using the chain rule and Eqs. (2.5.74), (2.5.75), (2.5.76), and (2.5.77), we find that

$$\frac{\partial}{\partial x} = \left(1 + \frac{1}{f_0}\frac{\partial v_g}{\partial x}\right)\frac{\partial}{\partial X} - \frac{1}{f_0}\frac{\partial u_g}{\partial x}\frac{\partial}{\partial Y} \tag{2.5.82}$$

$$\frac{\partial}{\partial y} = \frac{1}{f_0}\frac{\partial v_g}{\partial y}\frac{\partial}{\partial X} + \left(1 - \frac{1}{f_0}\frac{\partial u_g}{\partial y}\right)\frac{\partial}{\partial Y} \tag{2.5.83}$$

$$\frac{\partial}{\partial p} = \frac{1}{f_0}\frac{\partial v_g}{\partial p}\frac{\partial}{\partial X} - \frac{1}{f_0}\frac{\partial u_g}{\partial p}\frac{\partial}{\partial Y} + \frac{\partial}{\partial P} \tag{2.5.84}$$

$$\frac{\partial}{\partial t} = \frac{1}{f_0}\frac{\partial v_g}{\partial t}\frac{\partial}{\partial X} - \frac{1}{f_0}\frac{\partial u_g}{\partial t}\frac{\partial}{\partial Y} + \frac{\partial}{\partial t^*}. \tag{2.5.85}$$

The derivatives $\partial/\partial X$ and $\partial/\partial Y$ are *inversely* proportional to the Jacobian of the transformation

$$J = \begin{vmatrix} \dfrac{\partial X}{\partial x} & \dfrac{\partial X}{\partial y} \\ \dfrac{\partial Y}{\partial x} & \dfrac{\partial Y}{\partial y} \end{vmatrix} = \frac{\partial(X,Y)}{\partial(x,y)} = \frac{\partial X}{\partial x}\frac{\partial Y}{\partial y} - \frac{\partial Y}{\partial x}\frac{\partial X}{\partial y}. \tag{2.5.86}$$

For the set of equations, Eqs. (2.5.82) and (2.5.83),

$$J = 1 + \frac{\zeta_g}{f_0} + \frac{1}{f_0^2}\frac{\partial(u_g,v_g)}{\partial(x,y)}. \tag{2.5.87}$$

The transformation is possible only if

$$J \neq 0. \tag{2.5.88}$$

Thus, the transformation may be impossible if the geostrophic vorticity is comparable in magnitude to Earth's vorticity, and anticyclonic. In practice, one must sometimes "modify" observational data so that Eq. (2.5.88) is satisfied: J might be zero, for example, on the anticyclonic-shear side of a strong ridge; the data might be modified so that the anticyclonic shear is reduced.

From Eqs. (2.5.74), (2.5.75), (2.5.76), and (2.5.77) it also follows that

$$x = x(X, Y, P, t^*) \tag{2.5.89}$$

$$y = y(X, Y, P, t^*) \tag{2.5.90}$$

$$p = p(P) \tag{2.5.91}$$

$$t = t(t^*). \tag{2.5.92}$$

Using the chain rule and Eqs. (2.5.74), (2.5.75), (2.5.76), and (2.5.77), we find that

$$\frac{\partial}{\partial X} = \left(1 - \frac{1}{f_0}\frac{\partial v_g}{\partial X}\right)\frac{\partial}{\partial x} + \frac{1}{f_0}\frac{\partial u_g}{\partial X}\frac{\partial}{\partial y} \tag{2.5.93}$$

$$\frac{\partial}{\partial Y} = -\frac{1}{f_0}\frac{\partial v_g}{\partial Y}\frac{\partial}{\partial x} + \left(1 + \frac{1}{f_0}\frac{\partial u_g}{\partial Y}\right)\frac{\partial}{\partial y} \tag{2.5.94}$$

$$\frac{\partial}{\partial P} = -\frac{1}{f_0}\frac{\partial v_g}{\partial P}\frac{\partial}{\partial x} + \frac{1}{f_0}\frac{\partial u_g}{\partial P}\frac{\partial}{\partial y} + \frac{\partial}{\partial p}. \tag{2.5.95}$$

The derivatives $\partial/\partial x$ and $\partial/\partial y$ are *inversely* proportional to the inverse Jacobian of the transformation

$$J^{-1} = \begin{vmatrix} \dfrac{\partial x}{\partial X} & \dfrac{\partial x}{\partial Y} \\ \dfrac{\partial y}{\partial X} & \dfrac{\partial y}{\partial Y} \end{vmatrix} = \frac{\partial(x, y)}{\partial(X, Y)} = \frac{\partial x}{\partial X}\frac{\partial y}{\partial Y} - \frac{\partial y}{\partial X}\frac{\partial x}{\partial Y}. \tag{2.5.96}$$

For the set of equations (2.5.93) and (2.5.94)

$$J^{-1} = 1 - \frac{\zeta_g^*}{f_0} + \frac{1}{f_0^2}\frac{\partial(u_g, v_g)}{\partial(X, Y)}, \tag{2.5.97}$$

where

$$\zeta_g^* = \frac{\partial v_g}{\partial X} - \frac{\partial u_g}{\partial Y}. \tag{2.5.98}$$

The inverse transformation is possible only if

$$J^{-1} \neq 0. \tag{2.5.99}$$

From Eqs. (2.5.87), (2.5.93), (2.5.94), (2.5.96), and (2.5.97), it may be verified that

$$J^{-1} = \frac{1}{J}. \tag{2.5.100}$$

That is, the reciprocal of the Jacobian of the transformation is identical to the Jacobian of the inverse transformation. When J^{-1} is zero, J is infinity, which occurs when ζ_g is infinite. Thus, the inverse transformation is not possible when geostrophic vorticity (in real space) is extremely large and cyclonic, for example, near extremely intense fronts. In geostrophic space, the inverse transformation becomes impossible when ζ_g^* approaches f_0; note, however, that ζ_g^* is not *really* vorticity, because Φ^* is not really geopotential height.

After plugging u_g and v_g into Eq. (2.5.85), we find that in geostrophic coordinates

$$\frac{\partial u_g}{\partial t} = \frac{1}{J^{-1}} \frac{\partial u_g}{\partial t^*} + \frac{1}{f_0 J^{-1}} \frac{\partial(u_g, v_g)}{\partial(X, t^*)} \tag{2.5.101}$$

$$\frac{\partial v_g}{\partial t} = \frac{1}{J^{-1}} \frac{\partial v_g}{\partial t^*} + \frac{1}{f_0 J^{-1}} \frac{\partial(u_g, v_g)}{\partial(Y, t^*)}. \tag{2.5.102}$$

In a similar manner, after plugging u_g and v_g into Eqs. (2.5.82) and (2.5.83), we find that in geostrophic coordinates

$$\frac{\partial u_g}{\partial x} = \frac{1}{J^{-1}} \frac{\partial u_g}{\partial X} \tag{2.5.103}$$

$$\frac{\partial u_g}{\partial y} = \frac{1}{J^{-1}} \frac{\partial u_g}{\partial Y} + \frac{1}{f_0 J^{-1}} \frac{\partial(u_g, v_g)}{\partial(X, Y)} \tag{2.5.104}$$

$$\frac{\partial v_g}{\partial x} = \frac{1}{J^{-1}} \frac{\partial v_g}{\partial X} - \frac{1}{f_0 J^{-1}} \frac{\partial(u_g, v_g)}{\partial(X, Y)} \tag{2.5.105}$$

$$\frac{\partial v_g}{\partial y} = \frac{1}{J^{-1}} \frac{\partial v_g}{\partial Y}. \tag{2.5.106}$$

It should be noted that

$$0 = \frac{\partial u_g}{\partial x} + \frac{\partial v_g}{\partial y} = \frac{1}{J^{-1}} \left(\frac{\partial u_g}{\partial X} + \frac{\partial v_g}{\partial Y} \right). \tag{2.5.107}$$

That is, the geostrophic wind field is also quasihorizontally nondivergent in geostrophic coordinates. Finally, after plugging u_g and v_g into Eq. (2.5.84), we find that in geostrophic coordinates

$$\frac{\partial u_g}{\partial p} = \frac{1}{J^{-1}} \frac{\partial u_g}{\partial P} + \frac{1}{f_0 J^{-1}} \frac{\partial(u_g, v_g)}{\partial(X, P)} \tag{2.5.108}$$

$$\frac{\partial v_g}{\partial p} = \frac{1}{J^{-1}} \frac{\partial v_g}{\partial P} + \frac{1}{f_0 J^{-1}} \frac{\partial(u_g, v_g)}{\partial(Y, P)}. \tag{2.5.109}$$

Using the transformations in Eqs. (2.5.82), (2.5.83), (2.5.84), and (2.5.85) and the equations of motion, Eqs. (2.5.64) and (2.5.65), we can show that

$$\frac{D}{Dt} = \frac{\partial}{\partial t} + (u_g + u_a) \frac{\partial}{\partial x} + (v_g + v_a) \frac{\partial}{\partial y} + \omega \frac{\partial}{\partial p}$$

$$= \frac{\partial}{\partial t^*} + u_g \frac{\partial}{\partial X} + v_g \frac{\partial}{\partial Y} + \omega \frac{\partial}{\partial P}. \tag{2.5.110}$$

It therefore follows that

$$\frac{Du_g}{Dt} = \frac{\partial u_g}{\partial t^*} + u_g \frac{\partial u_g}{\partial X} + v_g \frac{\partial u_g}{\partial Y} + \omega \frac{\partial u_g}{\partial P} = f_0 v_a \tag{2.5.111}$$

$$\frac{Dv_g}{Dt} = \frac{\partial v_g}{\partial t^*} + u_g \frac{\partial v_g}{\partial X} + v_g \frac{\partial v_g}{\partial Y} + \omega \frac{\partial v_g}{\partial P} = -f_0 u_a. \tag{2.5.112}$$

If the ageostrophic wind variables in geostrophic coordinates are defined in the following way:

$$u_a^* = u_a + \frac{\omega}{f_0}\frac{\partial v_g}{\partial P} \tag{2.5.113}$$

$$v_a^* = v_a - \frac{\omega}{f_0}\frac{\partial u_g}{\partial P}, \tag{2.5.114}$$

then Eqs. (2.5.111) and (2.5.112) may be written compactly as

$$\frac{\partial u_g}{\partial t^*} + u_g \frac{\partial u_g}{\partial X} + v_g \frac{\partial u_g}{\partial Y} = f_0 v_a^* \tag{2.5.115}$$

$$\frac{\partial v_g}{\partial t^*} + u_g \frac{\partial v_g}{\partial X} + v_g \frac{\partial v_g}{\partial Y} = -f_0 u_a^*. \tag{2.5.116}$$

Thus, the *form* of the horizontal equations of motion in geostrophic coordinates is identical to that of the quasigeostrophic equations of motion, save for the neglect of β and the new ageostrophic wind variable. The parcel derivative on the left-hand sides of Eqs. (2.5.115) and (2.5.116) does not contain ageostrophic and vertical advection: The ageostrophic and vertical velocities are implicit in the transformation to geostrophic coordinates.

The vorticity equation in geostrophic coordinates obtained from Eqs. (2.5.98), (2.5.115), and (2.5.116) is:

$$\left(\frac{\partial}{\partial t^*} + \mathbf{v}_g \cdot \boldsymbol{\nabla}_p\right)\zeta_g^* = -f_0\left(\frac{\partial u_a^*}{\partial X} + \frac{\partial v_a^*}{\partial Y}\right). \tag{2.5.117}$$

This is equivalent in *form* to the frictionless version of the quasigeostrophic vorticity equation on an f plane (i.e., for $\beta = 0$).

The geopotential height variable in geostrophic coordinates is defined as

$$\Phi^* = \Phi + \tfrac{1}{2}(u_g^2 + v_g^2). \tag{2.5.118}$$

This is the Bernoulli function for geostrophic motion, less the C_pT term. Substituting Eq. (2.5.118) into Eqs. (2.5.93), (2.5.94), and (2.5.95), we see that

$$\frac{\partial \Phi^*}{\partial X} = \frac{\partial \Phi}{\partial x} \tag{2.5.119}$$

$$\frac{\partial \Phi^*}{\partial Y} = \frac{\partial \Phi}{\partial y} \tag{2.5.120}$$

$$\frac{\partial \Phi^*}{\partial P} = \frac{\partial \Phi}{\partial p}. \tag{2.5.121}$$

Therefore the geostrophic wind in geostrophic coordinates is given as follows:

$$u_g = -\frac{1}{f_0}\frac{\partial \Phi^*}{\partial Y} = -\frac{1}{f_0}\frac{\partial \Phi}{\partial y} \tag{2.5.122}$$

$$v_g = \frac{1}{f_0}\frac{\partial \Phi^*}{\partial X} = \frac{1}{f_0}\frac{\partial \Phi}{\partial x}. \tag{2.5.123}$$

It follows from Eqs. (2.5.118), (2.5.85), (2.5.122), and (2.5.123) that

$$\frac{\partial \Phi^*}{\partial t^*} = \frac{\partial \Phi}{\partial t}. \tag{2.5.124}$$

It is a consequence of Eqs. (2.5.119), (2.5.120), (2.5.121), and (2.5.124) that the geostrophic coordinate transformation is called a *contact transformation.*

From the hydrostatic equation, Eqs. (2.5.76) and (2.5.121), we see that temperature in geostrophic coordinates is

$$T = -\frac{P}{R}\frac{\partial \Phi^*}{\partial P}. \tag{2.5.125}$$

The thermal-wind relations in geostrophic coordinates are therefore as follows:

$$\frac{\partial u_g}{\partial P} = -\frac{1}{f_0}\frac{\partial}{\partial Y}\left(\frac{\partial \Phi^*}{\partial P}\right) = \frac{R}{f_0 P}\frac{\partial T}{\partial Y} \tag{2.5.126}$$

$$\frac{\partial v_g}{\partial P} = \frac{1}{f_0}\frac{\partial}{\partial X}\left(\frac{\partial \Phi^*}{\partial P}\right) = -\frac{R}{f_0 P}\frac{\partial T}{\partial X}. \tag{2.5.127}$$

The adiabatic form of the thermodynamic equation subject to the geostrophic-momentum approximation, Eq. (2.5.45), is given by the following:

$$\frac{\partial \theta}{\partial t} + u_g\frac{\partial \theta}{\partial x} + u_a\frac{\partial \theta}{\partial x} + v_g\frac{\partial \theta}{\partial y} + v_a\frac{\partial \theta}{\partial y} + \omega\frac{\partial \theta}{\partial p} = 0. \tag{2.5.128}$$

Then from Eq. (2.5.110) we know that

$$\frac{D\theta}{Dt} = \frac{\partial \theta}{\partial t^*} + u_g\frac{\partial \theta}{\partial X} + v_g\frac{\partial \theta}{\partial Y} + \omega\frac{\partial \theta}{\partial P} = 0. \tag{2.5.129}$$

Since

$$\theta = T\left(\frac{p_0}{p}\right)^{R/C_p} = T\left(\frac{P_0}{P}\right)^{R/C_p}, \tag{2.5.130}$$

it follows that

$$\frac{\partial T}{\partial t^*} + u_g\frac{\partial T}{\partial X} + v_g\frac{\partial T}{\partial Y} = \omega\sigma^*\frac{P}{R}, \tag{2.5.131}$$

where

$$\sigma^* = -\frac{RT}{P}\frac{\partial \ln \theta}{\partial P}. \tag{2.5.132}$$

(This σ^* is not to be confused with the static-stability parameter in isentropic coordinates discussed in Chap. 1). Thus, the *form* of the thermodynamic equation in geostrophic coordinates is identical to that of the quasigeostrophic thermodynamic equation.

It is not a hyperbole to say that the derivation of the continuity equation in geostrophic coordinates is a very arduous task! From the geostrophic-momentum approximation form of the equations of motion, Eqs. (2.5.64) and (2.5.65), the geostrophic-momentum approximation itself, and the continuity equation

$$\frac{\partial \omega}{\partial p} = -\left(\frac{\partial u_{\mathrm{a}}}{\partial x} + \frac{\partial v_{\mathrm{a}}}{\partial y}\right), \tag{2.5.133}$$

it may be shown after some tedious algebra that

$$\begin{aligned}\frac{D}{Dt}\left(\frac{1}{f_0}\frac{\partial(u_{\mathrm{g}},v_{\mathrm{g}})}{\partial(x,y)}\right) &= \frac{1}{f_0}\frac{\partial}{\partial t}\left(\frac{\partial(u_{\mathrm{g}},v_{\mathrm{g}})}{\partial(x,y)}\right) + (u_{\mathrm{g}}+u_{\mathrm{a}})\frac{1}{f_0}\frac{\partial}{\partial x}\left(\frac{\partial(u_{\mathrm{g}},v_{\mathrm{g}})}{\partial(x,y)}\right)\\ &\quad + (v_{\mathrm{g}}+v_{\mathrm{a}})\frac{1}{f_0}\frac{\partial}{\partial y}\left(\frac{\partial(u_{\mathrm{g}},v_{\mathrm{g}})}{\partial(x,y)}\right) + \omega\frac{1}{f_0}\frac{\partial}{\partial p}\left(\frac{\partial(u_{\mathrm{g}},v_{\mathrm{g}})}{\partial(x,y)}\right)\\ &= \frac{1}{f_0}\frac{\partial(u_{\mathrm{g}},v_{\mathrm{g}})}{\partial(y,p)}\frac{\partial \omega}{\partial x} + \frac{1}{f_0}\frac{\partial(u_{\mathrm{g}},v_{\mathrm{g}})}{\partial(p,x)}\frac{\partial \omega}{\partial y} + \frac{1}{f_0}\frac{\partial(u_{\mathrm{g}},v_{\mathrm{g}})}{\partial(x,y)}\frac{\partial \omega}{\partial p}.\end{aligned} \tag{2.5.134}$$

Consider the geostrophic-momentum form of the vorticity equation, Eq. (2.5.67), expressed as:

$$\frac{D}{Dt}(\zeta_{\mathrm{g}}+f_0) = \left(-\frac{\partial v_{\mathrm{g}}}{\partial p}\frac{\partial}{\partial x} + \frac{\partial u_{\mathrm{g}}}{\partial p}\frac{\partial}{\partial y} + (f_0+\zeta_{\mathrm{g}})\frac{\partial}{\partial p}\right)\omega. \tag{2.5.135}$$

If we add Eq. (2.5.135) to Eqs. (2.5.134), we obtain:

$$\begin{aligned}&\frac{D}{Dt}\left[f_0 + \left(\frac{\partial v_{\mathrm{g}}}{\partial x} - \frac{\partial u_{\mathrm{g}}}{\partial y}\right) + \frac{1}{f_0}\frac{\partial(u_{\mathrm{g}},v_{\mathrm{g}})}{\partial(x,y)}\right]\\ &= \left\{\left(-\frac{\partial v_{\mathrm{g}}}{\partial p} + \frac{1}{f_0}\frac{\partial(u_{\mathrm{g}},v_{\mathrm{g}})}{\partial(y,p)}\right)\frac{\partial}{\partial x} + \left(\frac{\partial u_{\mathrm{g}}}{\partial p} + \frac{1}{f_0}\frac{\partial(u_{\mathrm{g}},v_{\mathrm{g}})}{\partial(p,x)}\right)\frac{\partial}{\partial y}\right.\\ &\quad \left. + \left[f_0 + \left(\frac{\partial v_{\mathrm{g}}}{\partial x} - \frac{\partial u_{\mathrm{g}}}{\partial y}\right) + \frac{1}{f_0}\frac{\partial(u_{\mathrm{g}},v_{\mathrm{g}})}{\partial(x,y)}\right]\frac{\partial}{\partial p}\right\}\omega.\end{aligned} \tag{2.5.136}$$

The extra Jacobian terms will be an aid to our analysis. It is noted that the Jacobian terms are usually smaller than the other terms. For example,

$$\left|\frac{\partial v_{\mathrm{g}}}{\partial x} - \frac{\partial u_{\mathrm{g}}}{\partial y}\right| \sim 10^{-5}\ \mathrm{s}^{-1}. \tag{2.5.137}$$

while

$$\left|\frac{1}{f_0}\frac{\partial(u_{\mathrm{g}},v_{\mathrm{g}})}{\partial(x,y)}\right| \sim \frac{10^{-10}\ \mathrm{s}^{-2}}{10^{-4}\ \mathrm{s}^{-1}} = 10^{-6}\ \mathrm{s}^{-1}. \tag{2.5.138}$$

It is convenient to define a geostrophic "pseudo"-absolute-vorticity vector

$$\boldsymbol{\eta}_g = \eta_{gx}\hat{\mathbf{i}} + \eta_{gy}\hat{\mathbf{j}} + \eta_{gp}\hat{\mathbf{k}}, \tag{2.5.139}$$

where

$$\eta_{gx} = -\frac{\partial v_g}{\partial p} + \frac{1}{f_0}\frac{\partial(u_g,v_g)}{\partial(y,p)} \tag{2.5.140}$$

$$\eta_{gy} = \frac{\partial u_g}{\partial p} - \frac{1}{f_0}\frac{\partial(u_g,v_g)}{\partial(x,p)} \tag{2.5.141}$$

$$\eta_{gp} = f_0 + \left(\frac{\partial v_g}{\partial x} - \frac{\partial u_g}{\partial y}\right) + \frac{1}{f_0}\frac{\partial(u_g,v_g)}{\partial(x,y)}. \tag{2.5.142}$$

Then Eq. (2.5.136) may be written neatly as

$$\frac{D\eta_{gp}}{Dt} = \left(\eta_{gx}\frac{\partial}{\partial x} + \eta_{gy}\frac{\partial}{\partial y} + \eta_{gp}\frac{\partial}{\partial p}\right)\omega \tag{2.5.143}$$

$$= \boldsymbol{\eta}_g \cdot \boldsymbol{\nabla}\omega. \tag{2.5.144}$$

Now from Eq. (2.5.95) it is easily seen that

$$\frac{\partial u_g}{\partial P} = \frac{1}{J}\frac{\partial u_g}{\partial p} - \frac{1}{f_0 J}\frac{\partial(u_g,v_g)}{\partial(x,p)} \tag{2.5.145}$$

$$\frac{\partial v_g}{\partial P} = \frac{1}{J}\frac{\partial v_g}{\partial p} - \frac{1}{f_0 J}\frac{\partial(u_g,v_g)}{\partial(y,p)}. \tag{2.5.146}$$

Substituting Eqs. (2.5.145) and (2.5.146) into Eq. (2.5.95), we find that

$$\begin{aligned} J\frac{\partial}{\partial P} = &\frac{1}{f_0}\left(-\frac{\partial v_g}{\partial p} + \frac{1}{f_0}\frac{\partial(u_g,v_g)}{\partial(y,p)}\right)\frac{\partial}{\partial x} \\ &+ \frac{1}{f_0}\left(\frac{\partial u_g}{\partial p} + \frac{1}{f_0}\frac{\partial(u_g,v_g)}{\partial(p,x)}\right)\frac{\partial}{\partial y} \\ &+ \frac{1}{f_0}\left[f_0 + \left(\frac{\partial v_g}{\partial x} - \frac{\partial u_g}{\partial y}\right) + \frac{1}{f_0}\frac{\partial(u_g,v_g)}{\partial(x,y)}\right]\frac{\partial}{\partial p}. \end{aligned} \tag{2.5.147}$$

Therefore

$$J\frac{\partial}{\partial P} = \frac{1}{f_0}\boldsymbol{\eta}_g \cdot \boldsymbol{\nabla}. \tag{2.5.148}$$

So from Eqs. (2.5.144) and (2.5.148) we see that

$$\frac{D\eta_{gp}}{Dt} = f_0 J\frac{\partial \omega}{\partial P}. \tag{2.5.149}$$

Since total derivatives are independent of reference frame [see Eq. (2.5.110)],

$$\frac{D\eta_{gp}}{Dt} = \left(\frac{\partial}{\partial t^*} + \mathbf{v}_g \cdot \boldsymbol{\nabla}_P\right)\eta_{gp} + \omega\frac{\partial \eta_{gp}}{\partial P}. \tag{2.5.150}$$

We see from Eqs. (2.5.142) and (2.5.87) that

$$\eta_{gp} = f_0 J. \tag{2.5.151}$$

So from Eqs. (2.5.149), (2.5.150), and (2.5.151) it follows that

$$f_0 J \frac{\partial \omega}{\partial P} = \left(\frac{\partial}{\partial t^*} + \mathbf{v}_g \cdot \nabla_P\right) f_0 J + \omega \frac{\partial}{\partial P}(f_0 J). \tag{2.5.152}$$

Then

$$\left(\frac{\partial}{\partial t^*} + \mathbf{v}_g \cdot \nabla_P\right) J = J^2 \frac{\partial}{\partial P}\left(\frac{\omega}{J}\right). \tag{2.5.153}$$

From Eqs. (2.5.100) and (2.5.153) it follows that

$$\left(\frac{\partial}{\partial t^*} + \mathbf{v}_g \cdot \nabla_P\right) J = -J^2 \left(\frac{\partial}{\partial t^*} + \mathbf{v}_g \cdot \nabla_P\right) J^{-1}. \tag{2.5.154}$$

So from Eqs. (2.5.153) and (2.5.154) we get

$$\left(\frac{\partial}{\partial t^*} + \mathbf{v}_g \cdot \nabla_P\right) J^{-1} = -\frac{\partial}{\partial P}\left(\frac{\omega}{J}\right). \tag{2.5.155}$$

The vorticity equation in geostrophic coordinates, Eq. (2.5.117), may be written as

$$\left(\frac{\partial}{\partial t^*} + \mathbf{v}_g \cdot \nabla_P\right)\left[1 - \frac{1}{f_0}\left(\frac{\partial v_g}{\partial X} - \frac{\partial u_g}{\partial Y}\right)\right] = \frac{\partial u_a^*}{\partial X} + \frac{\partial v_a^*}{\partial Y}. \tag{2.5.156}$$

But from Eqs. (2.5.155) and (2.5.97) it follows that

$$\left(\frac{\partial}{\partial t^*} + \mathbf{v}_g \cdot \nabla_P\right)\left[1 - \frac{1}{f_0}\left(\frac{\partial v_g}{\partial X} - \frac{\partial u_g}{\partial Y}\right) + \frac{1}{f_0^2}\frac{\partial(u_g, v_g)}{\partial(X, Y)}\right]$$
$$= -\frac{\partial}{\partial P}\left(\frac{\omega}{J}\right). \tag{2.5.157}$$

If

$$\left|\frac{1}{f_0}\frac{\partial(u_g, v_g)}{\partial(X, Y)}\right| \ll \left|f_0 + \left(\frac{\partial v_g}{\partial X} - \frac{\partial u_g}{\partial Y}\right)\right|, \tag{2.5.158}$$

then

$$J^{-1} = 1 - \frac{1}{f_0}\left(\frac{\partial v_g}{\partial X} - \frac{\partial u_g}{\partial Y}\right) \tag{2.5.159}$$

$$J = 1 + \frac{1}{f_0}\left(\frac{\partial v_g}{\partial x} - \frac{\partial u_g}{\partial y}\right). \tag{2.5.160}$$

So, equating the right-hand sides of Eqs. (2.5.156) and (2.5.157), we find that

$$\frac{\partial u_a^*}{\partial X} + \frac{\partial v_a^*}{\partial Y} + \frac{\partial}{\partial P}\left(\frac{\omega}{J}\right) = 0. \tag{2.5.161}$$

The vertical velocity in geostrophic coordinates is defined as

$$\omega^* = \frac{\omega}{J} = \left(\frac{f_0}{f_0 + \zeta_g}\right)\omega, \tag{2.5.162}$$

so that

$$\frac{\partial u_a^*}{\partial X} + \frac{\partial v_a^*}{\partial Y} + \frac{\partial \omega^*}{\partial P} = 0. \tag{2.5.163}$$

Thus, the continuity equation in geostrophic coordinates has the same *form* as the quasigeostrophic continuity equation. [It should be noted, however, that owing to the approximation in Eq. (2.5.159), mass is not *exactly* conserved in Eq. (2.5.163).]

It was shown earlier [see Eq. (2.5.107)] that the divergence of the geostrophic wind in geostrophic coordinates is zero. From Eqs. (2.5.103), (2.5.104), (2.5.105), (2.5.106), and (2.5.160), we find that vorticity and deformation in geostrophic coordinates are as follows:

$$\zeta_g^* = \left(\frac{f_0}{f_0 + \zeta_g}\right)\zeta_g \tag{2.5.164}$$

$$D_1^* = \left(\frac{f_0}{f_0 + \zeta_g}\right)D_1 \tag{2.5.165}$$

$$D_2^* = \left(\frac{f_0}{f_0 + \zeta_g}\right)D_2. \tag{2.5.166}$$

Vorticity and deformation in geostrophic coordinates may thus be relatively large when the geostrophic vorticity is highly anticyclonic, and less than f_0 in magnitude.

The set of equations, Eqs. (2.5.115), (2.5.116), (2.5.131), and (2.5.163), which are the equations of motion, thermodynamic equation, and continuity equation, respectively, transformed into geostrophic coordinates and subject to the geostrophic-momentum approximation, are known as the *semigeostrophic equations*. B. Hoskins originally developed these equations and gave them their name in the mid-1970s. Their solution represents a distortion of the quasigeostrophic solution.

In summary, the semigeostrophic equations are as follows:

$$\frac{\partial u_g}{\partial t^*} + u_g\frac{\partial u_g}{\partial X} + v_g\frac{\partial u_g}{\partial Y} = f_0 v_a^* \tag{2.5.167}$$

$$\frac{\partial v_g}{\partial t^*} + u_g\frac{\partial v_g}{\partial X} + v_g\frac{\partial v_g}{\partial Y} = -f_0 u_a^* \tag{2.5.168}$$

$$\frac{\partial T}{\partial t^*} + u_g\frac{\partial T}{\partial X} + v_g\frac{\partial T}{\partial Y} = \omega\sigma^*\frac{P}{R}, \tag{2.5.169}$$

or

$$\frac{\partial \theta}{\partial t^*} + u_g \frac{\partial \theta}{\partial X} + v_g \frac{\partial \theta}{\partial Y} + \omega \frac{\partial \theta}{\partial P} = 0 \tag{2.5.170}$$

$$\frac{\partial u_a^*}{\partial X} + \frac{\partial v_a^*}{\partial Y} + \frac{\partial \omega^*}{\partial P} = 0, \tag{2.5.171}$$

where

$$u_a^* = u_a + \frac{\omega}{f_0} \frac{\partial v_g}{\partial P} \tag{2.5.172}$$

$$v_a^* = v_a - \frac{\omega}{f_0} \frac{\partial u_g}{\partial P} \tag{2.5.173}$$

$$\omega^* = \left(\frac{f_0}{f_0 + \zeta_g} \right) \omega \tag{2.5.174}$$

$$\sigma^* = -\frac{RT}{P} \frac{\partial \ln \theta}{\partial P} \tag{2.5.175}$$

$$X = x + \frac{v_g}{f_0} \tag{2.5.176}$$

$$Y = y - \frac{u_g}{f_0} \tag{2.5.177}$$

$$P = p \tag{2.5.178}$$

$$t^* = t. \tag{2.5.179}$$

The set of all X, Y, P, and t^* is called *semigeostrophic space,* while the set of all x, y, p, and t is referred to as *physical* or *real space.*

The restriction that f is treated as a constant can be removed if f is replaced by its value at the location of the air parcel in geostrophic space. In other words, we let $f = f(x,y)$ in Eqs. (2.5.74) and (2.5.75), the horizontal coordinate-transformation relations. The extension of the semi-geostrophic equations to include a variable Coriolis parameter is referred to as *Salmon's generalization.*

Diabatic heating may be added to the semigeostrophic thermodynamic equation, Eqs. (2.5.169) and (2.5.170). However, friction cannot be added easily to the semigeostrophic momentum equations, Eqs. (2.5.167) and (2.5.168).

The static stability parameter in geostrophic space is different from the static stability parameter in physical space. Recall Eq. (2.5.132), which with Eq. (2.5.130) may be written as

$$\sigma^* = -\frac{R}{P} \left(\frac{P}{P_0} \right)^{R/C_p} \frac{\partial \theta}{\partial P}. \tag{2.5.180}$$

Using Eqs. (2.5.148) and (2.5.151), we see that

$$\frac{\partial \theta}{\partial P} = \frac{\boldsymbol{\eta}_g \cdot \nabla \theta}{\eta_{gp}}. \tag{2.5.181}$$

Therefore the static stability parameter in geostrophic space

$$\sigma^* = -\left[-\frac{R}{P}\left(\frac{P}{P_0}\right)^{R/C_p}\right]\frac{\boldsymbol{\eta}_g \cdot \nabla \theta}{\eta_{gp}} \tag{2.5.182}$$

is proportional to a normalized, three-dimensional form of Ertel's potential vorticity for geostrophic motion in real space, except for the extra small Jacobian terms in $\boldsymbol{\eta}_g$.

The difference between Eq. (2.5.182) and σ in real space is that the "vertical" coordinate P in geostrophic space is not exactly vertical: The effect of the coordinate transformation is contained in $\boldsymbol{\eta}_g$. To demonstrate this, let us consider some dependent variable Q in semigeostrophic space. According to Eq. (2.5.148),

$$\frac{\partial Q}{\partial P} = \frac{1}{f_0 J}\boldsymbol{\eta}_g \cdot \nabla Q. \tag{2.5.183}$$

The isopleths of the variable are oriented normal to the P surfaces if

$$\frac{\partial Q}{\partial P} = 0. \tag{2.5.184}$$

It follows then that

$$\boldsymbol{\eta}_g \cdot \nabla Q = 0, \tag{2.5.185}$$

which means that in real space the isopleths of Q and P are oriented along and normal to, respectively, the local orientation of the geostrophic "pseudo"-absolute-vorticity vector (Fig. 2.66). Because the Jacobian terms in Eqs. (2.5.140), (2.5.141), and (2.5.142) are small, the isopleths of Q and P in real space are oriented along and normal to, respectively, the orientation of the three-dimensional curl of the geostrophic wind added to the vector representing the rotation rate of the Earth about the local vertical.

The Sawyer–Eliassen equation in geostrophic coordinates. Suppose that we consider Eqs. (2.5.167), (2.5.168), and (2.5.170), and reformulate Eq. (2.5.53)

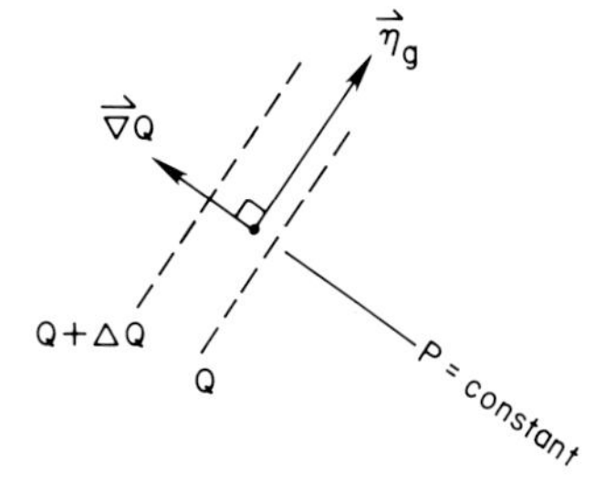

Figure 2.66 Geometrical relationship among isopleths of the arbitrary quantity Q (dashed lines), the gradient of Q (∇Q), isopleths of P (solid lines), and the geostrophic "pseudo"-absolute-vorticity vector ($\boldsymbol{\eta}_g$).

for an adiabatic atmosphere. Then the Sawyer–Eliassen equation in geostrophic space (Y–P) is

$$\frac{\partial^2\psi}{\partial Y^2}\left[-\frac{R}{f_0 P}\left(\frac{P}{P_0}\right)^{\kappa}\frac{\partial\bar{\theta}}{\partial P}\right]+f_0\frac{\partial^2\psi}{\partial P^2}=2\frac{R}{f_0 P}\left(\frac{P}{P_0}\right)^{\kappa}\left(\frac{\partial\theta}{\partial Y}\frac{\partial v_g}{\partial Y}+\frac{\partial\theta}{\partial X}\frac{\partial u_g}{\partial Y}\right), \qquad (2.5.186)$$

in which $\partial\bar{\theta}/\partial P$ is the quasihorizontal average of $\partial\theta/\partial P$. From Eqs. (2.5.181), (2.5.93), and (2.5.94) we see that neglect of quasihorizontal variations of $\partial\theta/\partial P$ in semigeostrophic space is essentially equivalent to the assumption that Ertel's potential vorticity in real space is quasihorizontally uniform. Had we retained the quasihorizontal variation of $\partial\theta/\partial P$, Eq. (2.5.186) would not be a simple Poisson equation: It would have an extra term containing $\partial\psi/\partial Y$ [in which $(\partial/\partial Y)(\partial\theta/\partial P)$ is a factor]. A term of this form does not appear in real space [Eq. (2.5.53)] because it is cancelled by a term representing vertical advection of geostrophic momentum.

We now can compare Eq. (2.5.186), the adiabatic form of the Sawyer–Eliassen equation in semigeostrophic space, to the adiabatic form of Eq. (2.5.53), the Sawyer–Eliassen equation in real space. The forcing function on the right-hand sides have identical forms. However, the operators are different: In semigeostrophic space the $\partial^2\psi/\partial Y\,\partial P$ term and the relative vorticity $-\partial u_g/\partial Y$ factor in the $\partial^2\psi/\partial P^2$ term do not appear. Because the $\partial^2\psi/\partial Y\,\partial P$ term does not appear, the effect of the transformation to geostrophic coordinates (when Ertel's geostrophic potential vorticity is positive and when the forcing function is a constant) is to convert Eq. (2.5.53) into a Poisson equation. The search for a coordination transformation that makes Eq. (2.5.53) a Poisson equation (for constant forcing) is in a sense the inspiration for inventing geostrophic coordinates. In semigeostrophic space the circulation resulting from constant forcing is *not tilted* as it is in real space, even though advection of potential temperature by the ageostrophic part of the wind and advection of geostrophic vorticity by both the ageostrophic and vertical parts of the wind are accounted for.

A. Eliassen first showed this in the early 1960s. Eliassen used

$$m = u_g - f_0 y \qquad (2.5.187)$$

as the transformed, quasihorizontally oriented coordinate rather than Y. Comparing Eq. (2.5.177) to Eq. (2.5.187), we find that m is simply proportional to Y:

$$m = -f_0 Y. \qquad (2.5.188)$$

The quantity m is called the *absolute momentum* because from Eq. (2.5.187) and the equation of motion

$$\frac{Dm}{Dt}=\frac{Du_g}{Dt}-f_0 v = -\frac{\partial\Phi}{\partial x}. \qquad (2.5.189)$$

Hence m is like momentum per unit mass in a nonrotating reference frame. Since m and Y are parallel to each other, the vertical circulation in semigeostrophic space is aligned along the m axis.

The quantity m has also been called the *streamfunction for absolute vorticity* because

$$\hat{\mathbf{k}} \cdot (2\boldsymbol{\Omega} + \boldsymbol{\nabla}_p \times \mathbf{v}_g) = f_0 + \hat{\mathbf{k}} \cdot \boldsymbol{\nabla}_p \times \mathbf{v}_g$$

$$= f_0 + \zeta_g = f_0 - \frac{\partial u_g}{\partial y} = -\frac{\partial m}{\partial y} \tag{2.5.190}$$

$$\hat{\mathbf{j}} \cdot \boldsymbol{\nabla}_p \times \mathbf{v}_g = \frac{\partial u_g}{\partial p} = \frac{\partial m}{\partial p}. \tag{2.5.191}$$

Thus, the vertical component of vorticity is proportional to the horizontal spacing of the m surfaces, and the horizontal temperature gradient (thermal-wind shear) is proportional to the vertical spacing of the m surfaces.

Along an m surface in real space

$$dm = 0 = \frac{\partial m}{\partial y}\, dy + \frac{\partial m}{\partial p}\, dp. \tag{2.5.192}$$

It follows that the m surface and the major axis of the vertical circulation are tilted (Fig. 2.67) with a slope of

$$-\left(\frac{dp}{dy}\right)_m = \frac{\partial m}{\partial y} \Big/ \frac{\partial m}{\partial p} = \left(f_0 - \frac{\partial u_g}{\partial y}\right) \Big/ \left(-\frac{\partial u_g}{\partial p}\right). \tag{2.5.193}$$

Using the thermal-wind relation, we find that

$$-\left(\frac{dp}{dy}\right)_m = \left(f_0 - \frac{\partial u_g}{\partial y}\right) \Big/ \left(-\frac{R}{f_0 p}\frac{\partial T}{\partial y}\right). \tag{2.5.194}$$

It is easy to see that the slope of a Y surface in real space (Fig. 2.67) and the vertical circulation are also tilted as follows:

$$-\left(\frac{dp}{dy}\right)_Y = \left(f_0 - \frac{\partial u_g}{\partial y}\right) \Big/ \left(-\frac{R}{f_0 p}\frac{\partial T}{\partial y}\right), \tag{2.5.195}$$

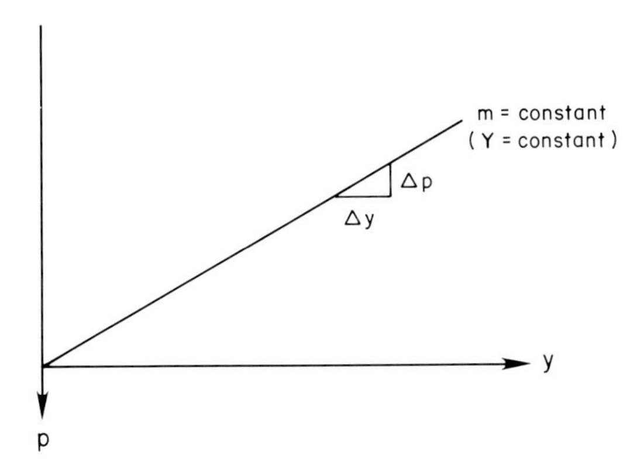

Figure 2.67 A surface of constant m (and Y) in the p–y plane, for constant geostrophic absolute vorticity and constant quasihorizontal meridional temperature gradient. In general there is a *unique* slope to the m (and Y) surface(s) at each point in real space.

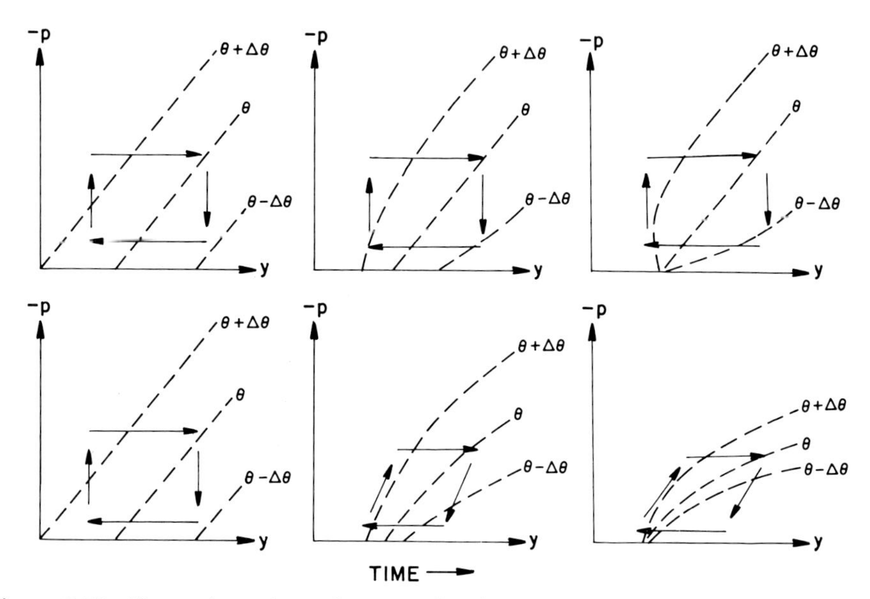

Figure 2.68 Illustration of quasigeostrophic frontogenesis (top) and frontogenesis governed by the geostrophic momentum approximation (bottom). Isotherms of potential temperature (dashed lines); sense of ageostrophic vertical circulation (arrows). In the top row, the vertical circulation does not affect the distribution of surface isotherms; only geostrophic deformation affects the distribution of isotherms. In the bottom row, the ageostrophic circulation does affect the distribution of isotherms (adapted from Bluestein, 1986). (Courtesy of the American Meteorological Society)

which is the same as the slope of a constant-m surface. Thus, when it is cold to the "north" and warm to the "south" (i.e., $\partial T/\partial y < 0$), and absolute vorticity is cyclonic, the vertical circulation tilts to the north (Fig. 2.68 (bottom)): The greater the magnitude the north–south temperature gradient, the smaller the slope of the vertical circulation for a given value of absolute vorticity. The qualitative similarity between Eqs. (2.5.194) and (2.5.195) and Margules' formula (2.2.10) is striking: The latter is based simply upon the kinematics of a temperature discontinuity and the thermal-wind relationship, while the former is based upon semigeostrophic dynamical and thermodynamic principles. The northward slope of the vertical circulation along an east–west-oriented surface frontal zone in the Northern Hemisphere according to Eq. (2.5.194) is typically

$$\left(\frac{dz}{dy}\right)_m = -\frac{1}{\rho g}\left(\frac{dp}{dy}\right)_m$$

$$\sim \frac{1}{(1\ \mathrm{kg\,m^{-3}})(10\ \mathrm{m\,s^{-2}})}\,10^{-4}\ \mathrm{s^{-1}} \Big/ \left(\frac{300\ \mathrm{m^2\,s^{-2}\,K^{-1}}}{10^{-4}\ \mathrm{s^{-1}}\ 100\ \mathrm{kPa}}\right)\left(\frac{10\ \mathrm{K}}{10^5\ \mathrm{m}}\right)$$

$$\sim \frac{1}{300}. \tag{2.5.196}$$

A comparison of quasigeostrophic and semigeostrophic surface frontogenesis. Let us now compare Eq. (2.5.53) to its quasigeostrophic counterpart. If diabatic heating is neglected and $\partial\bar{\theta}/\partial p$ is a function of p only, then the quasigeostrophic form of (2.5.53) is

$$\frac{\partial^2\psi}{\partial y^2}\left[-\frac{\partial\bar{\theta}}{\partial p}\frac{R}{f_0 p}\left(\frac{p}{p_0}\right)^{\kappa}\right]+\frac{\partial^2\psi}{\partial p^2}f_0$$

$$=2\frac{R}{f_0 p}\left(\frac{p}{p_0}\right)^{\kappa}\left(\frac{\partial\theta}{\partial y}\frac{\partial v_{\mathrm{g}}}{\partial y}+\frac{\partial\theta}{\partial x}\frac{\partial u_{\mathrm{g}}}{\partial y}\right). \tag{2.5.197}$$

Equation (2.5.197) has the identical form as the Sawyer–Eliassen equation in geostrophic coordinates, Eq. (2.5.186). The ellipticity condition for Eq. (2.5.197) is that $\partial\bar{\theta}/\partial p<0$, that is, that the atmosphere is statically stable. As in Eq. (2.5.186), $\partial\theta/\partial p$ is quasihorizontally averaged so that (2.5.197) becomes a Poisson equation.

The forcing function on the right-hand side of Eq. (2.5.197) is equivalent to that on the right-hand side of the adiabatic form of the Sawyer–Eliassen equation, Eq. (2.5.53). However, since this equation does not contain a $\partial^2\psi/\partial y\,\partial p$ term on the left-hand side, the circulation does not tilt with height. The terms responsible for the tilt in the geostrophic momentum approximation formulation, Eq. (2.5.50), are the cross-front ageostrophic advection of temperature ($-v_{\mathrm{a}}\,\partial\theta/\partial y$) and the vertical advection of geostrophic momentum ($-\omega\,\partial u_{\mathrm{g}}/\partial p$). Furthermore, the coefficient of the $\partial^2\psi/\partial p^2$ term in Eq. (2.5.197) does not contain the effects of relative vorticity as it does in Eq. (2.5.53). Thus in the quasigeostrophic system the inertial stability has a lower bound (expressed as f_0), and, consequently, v_{a} cannot be much more important than ω unless the static stability is extremely high.

Therefore, one effect of the ageostrophic circulation is to tilt the frontal zone (see bottom of Fig. 2.68). Without tilting, frontogenesis may be accompanied by a production of static instability on the warm side of the front (see top of Fig. 2.68). Furthermore, ageostrophic temperature advection hastens the frontogenetical process at the surface where $\omega\approx 0$. Another way of saying this is that convergence at the surface under the rising branch of the circulation acts frontogenetically, and augments the effects of geostrophic deformation, Eq. (2.3.19). Thus synoptic-scale, geostrophic deformation is usually regarded as initiating frontogenesis, while the completion and maintenance of the front is due to the vertical circulation. (However, coastal frontogenesis and "chinook" frontogenesis are not initiated by geostrophic deformation.)

During quasigeostrophic frontogenesis, regions of large anticyclonic as well as cyclonic vorticity are produced at the surface, since according to the vorticity equation, divergence and convergence act only on f.

A summary comparison of quasigeostrophic frontogenesis and semigeostrophic frontogenesis is found in Table 2.2.

An important result of the semigeostrophic analysis of frictionless, adiabatic frontogenesis is that the frontogenetical process can be considered to be

Table 2.2 Comparison of quasigeostrophic and semigeostrophic surface frontogenesis

Quasigeostrophic surface frontogenesis	Semigeostrophic surface frontogenesis
Slow: only geostrophic deformation acts to increase the quasihorizontal temperature gradient	Fast: convergence accompanying the vertical ageostrophic circulation accelerates the rate of increase in quasihorizontal temperature gradient
Frontal zone is vertically oriented; because quasihorizontal temperature gradient is increased most rapidly at the ground where $\omega \sim 0$, a region of static instability is produced at low levels on the warm side	Frontal zone tilts toward colder air as a result of differential advection by ageostrophic part of vertical circulation; low levels on warm side remain statically stable
Regions of both large anticyclonic and cyclonic vorticity are produced at the surface: convergence and divergence at surface both act only on f	Only regions of large cyclonic vorticity are produced: divergence and convergence at the surface act also on relative vorticity; the effect is therefore weaker when the relative vorticity is anticyclonic

essentially quasigeostrophic even when ageostrophic advection of temperature and geostrophic absolute vorticity, and vertical advection of geostrophic absolute vorticity, are retained, provided that the cross-frontal coordinate is in geostrophic space (or, equivalently, is m). Thus, the smallness of the length scale across a front is not necessarily relevant. Geostrophic coordinates in effect stretch the length scale across the front so that air parcels reside in the frontal zone for a relatively long period of time.

For example, from Eq. (2.5.75) we see that

$$\frac{\partial Y}{\partial y} = 1 - \frac{1}{f_0}\frac{\partial u_g}{\partial y}. \tag{2.5.198}$$

Since the geostrophic shear vorticity $(-\partial u_g/\partial y)$ along the warm side of a frontal zone is cyclonic,

$$\frac{\partial Y}{\partial y} > 1. \tag{2.5.199}$$

Thus, a given distance in semigeostrophic space is associated with a smaller increment in real space. Hence the quasihorizontal resolution in semigeostrophic space is better than it is in real space when the vorticity is cyclonic; when the vorticity is anticyclonic, the resolution is worse. When $\zeta_g = -f_0$, there is a singularity, and the geostrophic horizontal coordinates are identical to the real-space horizontal coordinates [cf. Eqs. (2.5.194) and (2.5.195)]. When the atmosphere is barotropic $(\partial T/\partial y = 0)$, the geostrophic horizontal coordinates are normal to the real-space horizontal coordinates.

The adiabatic, frictionless frontogenetical function, as defined in Eq. (2.3.19) for surface fronts, does not contain a divergence term in geostrophic coordinates. That is, in geostrophic space

$$F = \tfrac{1}{2}\,|\nabla_P \theta|\, D_s \cos 2b, \tag{2.5.200}$$

where D_s is the geostrophic deformation measured in geostrophic coordinates, and b is the angle between the axis of dilatation and the isotherms. The divergence (convergence) effect is implicit in the coordinate transformation.

In addition, the adiabatic, frictionless frontogenetical function as defined in Eq. (2.5.17) in geostrophic coordinates is

$$F = \frac{\sigma^* P}{R} \frac{1}{|\nabla_P \theta|} (\nabla_P \theta \cdot \mathbf{Q}_s), \tag{2.5.201}$$

where $\mathbf{Q}_s$ is the $\mathbf{Q}$ vector expressed in geostrophic coordinates. The effects of ageostrophic and vertical deformation acting on the temperature field are implicit in the coordinate transformation.

Advantages and limitations of semigeostrophic theory. The rapid formation of convective systems may cause the atmosphere to be thrown out of geostrophic balance in an impulsive way, so that the geostrophic momentum approximation is not valid (i.e., $\tau \lesssim 1/f$, where τ is the time scale). Furthermore, if the parcel trajectories are sharply curved, the centripetal acceleration of the ageostrophic wind can be significant. The geostrophic momentum approximation can also fail for straight flow if the parcel accelerations are so strong that the Rossby number is close to 1.

In conclusion, we note that the semigeostrophic equations are useful mainly for the following reasons:

1. They allow us to diagnose easily the vertical motion (and height-tendency) field(s) even when the ageostrophic advection of momentum and temperature and the vertical advection of momentum are significant.
2. They allow us to model features with variable resolution. The resolution is highest in regions of large cyclonic vorticity (i.e., in frontal zones), where we need it.

The major disadvantages in using the semigeostrophic equations are as follows:

1. One has to interpolate data from a Cartesian grid in real space to a Cartesian grid in geostrophic space, and then back again.
2. The coordinate transformations are undefined when $\zeta_g = -f_0$ or when ζ_g is very large.
3. The effects of friction are difficult to handle property.
4. The equations are not entirely accurate for three-dimensional wind fields having substantial curvature (because the advection of ageostrophic momentum by the geostrophic wind is neglected).

Symmetric instability. We found in the preceding section that the Sawyer–Eliassen equation in physical space is *not* elliptic [Eqs. (2.5.58)] if

$$\left(\frac{\partial u_g}{\partial p}\right)^2 + \left[\frac{R}{f_0 p}\left(\frac{p}{p_0}\right)^\kappa\right] \frac{\partial \theta}{\partial p}\left(f_0 - \frac{\partial u_g}{\partial y}\right) \geq 0. \tag{2.5.202}$$

We will now demonstrate that Eq. (2.5.202) is equivalent to the necessary condition for *symmetric instability*. (This terminology was first used by Stone in

1966. The terms *centrifugal, dynamic,* and *inertial* instability are also used.) The formal linear stability analysis of flow in a "symmetric" basic current, which was first done by Eady in 1949, will be bypassed in favor of a "parcel" analysis pioneered by Emanuel during the early 1980s. The latter is more amenable to physical interpretation.

A "symmetric" flow is one whose basic state and perturbations are independent of some horizontal coordinate, which is here designated as the x coordinate. If we included an x variation, we would be analyzing the more general baroclinic instability problem. We therefore seek out the possibility of spontaneously growing vertical circulations in a plane normal to the thermal-wind shear vector. However, it should be recognized that both baroclinic and symmetric instabilities can occur in the atmosphere simultaneously. Baroclinic instability requires a meridional temperature gradient (vertical shear in the geostrophic wind), while symmetric instability depends upon either or both horizontal and vertical shear in the geostrophic wind.

In the parcel theory of convection (Chap. 3), a parcel is displaced *vertically* and the restoring force on the parcel is examined. If the "restoring force" continues to displace the parcel from its initial position, there is instability. In the parcel theory of symmetric flow, a parcel is displaced along a tilted plane, and the restoring force is examined. Because $\partial/\partial x = 0$, the parcel may be viewed as a tube that extends infinitely far off in both directions along the x axis (Fig. 2.69). In the parcel theory of convection it is assumed that the pressure in the air surrounding the parcel is not altered significantly when the parcel is displaced. The same assumption is made for the displaced tube in the parcel theory of symmetric flow. In addition, as in the parcel theory of convection, mixing with the environment is neglected.

Consider the equation of motion in the y direction (with height as the vertical coordinate):

$$\frac{Dv}{Dt} = -fu - \frac{1}{\rho}\frac{\partial p}{\partial y} = f(u_g - u). \qquad (2.5.203)$$

Since the pressure of the surroundings is not altered by a tube-shaped parcel aligned with the x axis, and since the pressure of the tube equals the pressure of the environment to satisfy the dynamic boundary condition, p is equivalent to the pressure of the undisturbed environment. Equation (2.5.203) may be

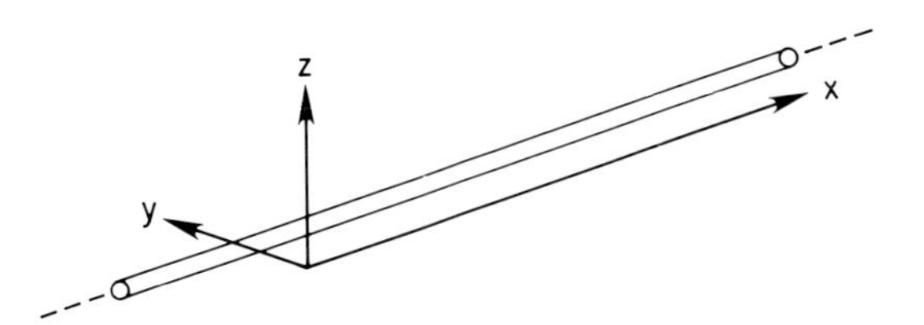

Figure 2.69 A parcel takes the form of an infinitely long tube if there are no variations along the tube; in this illustration, the tube is aligned along the x axis. In the "parcel" theory of symmetric instability, we displace this tube and see whether it is forced back to its original position or it zooms off into oblivion.

expressed as

$$\frac{Dv}{Dt} = f\left\{\left[u_{g0} + \left(\frac{\partial u_g}{\partial z}\right)_0 \delta z + \left(\frac{\partial u_g}{\partial y}\right)_0 \delta y\right] - u\right\} \tag{2.5.204}$$

for small δz and δy, where u_{g0} is the geostrophic-wind component of the tube at its initial position, and $\partial u_g/\partial z$ and $\partial u_g/\partial y$ are evaluated at $y = 0$, $z = 0$.

The equation of motion in the x direction (with height as the vertical coordinate) is as follows:

$$\frac{Du}{Dt} = fv - \frac{1}{\rho}\frac{\partial p}{\partial x} = f(v - v_g) = fv = f\frac{Dy}{Dt}. \tag{2.5.205}$$

(Note that $v_g = 0$.) Integrating (2.5.205) with respect to time, we see that

$$\int_0^t \frac{Du}{Dt}\,dt = \int_0^t f\frac{Dy}{Dt}\,dt. \tag{2.5.206}$$

Then the speed of the tube in the x direction at time t is

$$u(t) = u_{g0} + f(\delta y), \tag{2.5.207}$$

where the initial velocity of the tube (u_{g0}) is geostrophic, and δy is the displacement of the tube in the y direction in a short time δt. (Since the tube is infinitely long, it is more proper to regard the u component of the wind as the speed of parcels within the tube.)

It follows from Eqs. (2.5.207) and (2.5.204) that

$$\frac{Dv}{Dt} = f\left[\frac{\partial u_g}{\partial z}\delta z - \left(f - \frac{\partial u_g}{\partial y}\right)\delta y\right]. \tag{2.5.208}$$

According to the thermal-wind relation (with height as the vertical coordinate)

$$\frac{\partial u_g}{\partial z} = -\frac{g}{fT}\left(\frac{\partial T}{\partial y}\right)_z + \frac{u_g}{T}\frac{\partial T}{\partial z} \approx -\frac{g}{fT}\left(\frac{\partial T}{\partial y}\right)_z. \tag{2.5.209}$$

(The vertical derivative term is usually an order of magnitude less than the horizontal derivative term.) In terms of potential temperature

$$\frac{\partial u_g}{\partial z} = -\frac{g}{f}\frac{1}{\theta}\frac{\partial \theta}{\partial y}. \tag{2.5.210}$$

Along a surface of constant potential temperature

$$d\theta = 0 = \frac{\partial \theta}{\partial y}dy + \frac{\partial \theta}{\partial z}dz. \tag{2.5.211}$$

Then the slope of a θ surface according to Eqs. (2.5.211) and (2.5.210) is

$$\left(\frac{dz}{dy}\right)_\theta = -\frac{\partial \theta}{\partial y}\Big/\frac{\partial \theta}{\partial z} = \left(f\frac{\partial u_g}{\partial z}\right)\Big/\left(\frac{g}{\theta}\frac{\partial \theta}{\partial z}\right). \tag{2.5.212}$$

It follows from Eqs. (2.5.208) and (2.5.212) that

$$\frac{Dv}{Dt} = f\left\{\frac{\partial u_g}{\partial z}\left[\left(f\frac{\partial u_g}{\partial z}\right)\Big/\left(\frac{g}{\theta}\frac{\partial\theta}{\partial z}\right)\right](dy)_\theta - \left(f - \frac{\partial u_g}{\partial y}\right)(dy)_\theta\right\} \quad (2.5.213)$$

for displacements along a constant-θ surface. Since

$$v = \frac{D(\delta y)}{Dt}, \quad (2.5.214)$$

it follows that

$$\frac{D^2}{Dt^2}(\delta y)_\theta + \left\{-\left[f^2\left(\frac{\partial u_g}{\partial z}\right)^2\Big/\left(\frac{g}{\theta}\frac{\partial\theta}{\partial z}\right)\right] + f\left(f - \frac{\partial u_g}{\partial y}\right)\right\}(\delta y)_\theta = 0. \quad (2.5.215)$$

Since the Richardson number for the geostrophic basic state of the environment is

$$\mathrm{Ri} = \left(\frac{g}{\theta}\frac{\partial\theta}{\partial z}\right)\Big/\left(\frac{\partial u_g}{\partial z}\right)^2, \quad (2.5.216)$$

and since the absolute geostrophic vorticity $f + \zeta_g$ is $f - \partial u_g/\partial y$ (remember, $\partial v_g/\partial x = 0$), Eq. (2.5.215) can be written as

$$\frac{D^2(\delta y)_\theta}{Dt^2} + f^2\left(\frac{(\zeta_g + f)}{f} - \frac{1}{\mathrm{Ri}}\right)(\delta y)_\theta = 0. \quad (2.5.217)$$

The solution to Eq. (2.5.217) has the form

$$(\delta y)_\theta \sim e^{i\sigma t}. \quad (2.5.218)$$

Substituting Eq. (2.5.218) into Eq. (2.5.217), we find that

$$\sigma^2 = f^2\left(\frac{(\zeta_g + f)}{f} - \frac{1}{\mathrm{Ri}}\right). \quad (2.5.219)$$

Exponentially growing solutions may exist when σ is imaginary, that is, when

$$\mathrm{Ri} < \frac{f}{\zeta_g + f}. \quad (2.5.220)$$

Since $\zeta_g + f$ is usually on the order of f, symmetric instability may occur when

$$\mathrm{Ri} \lesssim 1. \quad (2.5.221)$$

Thus, symmetric instability, the spontaneous growth of displacements in the plane (in this case along a θ surface; displacements only along isentropic surfaces correspond to adiabatic and hydrostatic conditions, since $D\theta/Dt = 0$ and $Dw/dt = 0$) normal to the axis of symmetry (in this case the x axis) is favored by low static stability, strong vertical geostrophic shear, and weak or anticyclonic geostrophic vorticity. Only the second condition is typically present along frontal zones.

Our parcel analysis of symmetric instability was done only along θ surfaces. What happens if we displace the tube vertically or horizontally? If there is a meridional temperature gradient, $\partial u_g/\partial z \neq 0$, and hence θ surfaces are neither vertical nor horizontal. Only if $\partial u_g/\partial z = 0$ are θ surfaces horizontal; in this case Ri is infinite, and hence the necessary condition for symmetric instability in a barotropic atmosphere is simply that geostrophic absolute vorticity (in the Northern Hemisphere) is anticyclonic (negative) [see Eq. (2.5.219)]:

$$\zeta_g + f < 0. \tag{2.5.222}$$

On the other hand, only if $\partial\theta/\partial z = 0$ are θ surfaces vertical; in this case, however, Ri $= 0$ if the atmosphere is not barotropic, and hence instability can occur for any reasonable midlatitude value of ζ_g.

Let us compute the growth rates (σ) for vertical and horizontal displacements, which are not necessarily along isentropic surfaces. For purely vertical displacements (see Chap. 3)

$$\frac{Dw}{Dt} = \frac{D^2(\delta z)}{Dt^2} = -\frac{g}{\theta}\frac{\partial\theta}{\partial z}\,\delta z. \tag{2.5.223}$$

The effects of *Earth's rotation* are not felt, and it is assumed that the environmental pressure field is in hydrostatic balance, and is not disturbed by the vertically displaced tube.

The solution to Eq. (2.5.223) has the form

$$\delta z \sim e^{i\sigma t}, \tag{2.5.224}$$

where from Eq. (2.5.223) we know that

$$\sigma^2 = \frac{g}{\theta}\frac{\partial\theta}{\partial z} \tag{2.5.225}$$

and hence for typical values of $\partial\theta/\partial z$

$$\sigma \sim \left(\frac{10\ \text{m s}^{-2}}{300\ \text{K}}\frac{50\ \text{K}}{10^4\ \text{m}}\right)^{1/2} \sim 10^{-2}\ \text{s}^{-1}. \tag{2.5.226}$$

Parcel motion is not hydrostatic.

On the other hand, according to Eq. (2.5.208), for purely horizontal displacements

$$\frac{Dv}{Dt} = \frac{D^2}{Dt^2}(\delta y) = -f(\zeta_g + f)\,\delta y. \tag{2.5.227}$$

The effects of *gravity* are not felt directly, and it is assumed that the environmental pressure field is not disturbed by the horizontally displaced tube.

The solution to Eq. (2.5.227) has the form

$$\delta y \sim e^{i\sigma t}, \tag{2.5.228}$$

where from Eq. (2.5.227) we know that

$$\sigma^2 = f(\zeta_g + f); \tag{2.5.229}$$

hence a typical growth rate is

$$\sigma \sim [(10^{-4}\ \mathrm{s}^{-1})(10^{-4}\ \mathrm{s}^{-1})]^{1/2} = 10^{-4}\ \mathrm{s}^{-1}. \tag{2.5.230}$$

The growth rates for unstable, purely vertical displacements are much greater than those for unstable, purely horizontal motions. [If $\partial\theta/\partial z > 0$ and $f(\zeta_g + f) > 0$, the period of oscillation for purely vertical displacements is shorter than that for purely horizontal displacements.]

Suppose, now, that both vertical and horizontal displacements are stable, that is,

$$\frac{\partial\theta}{\partial z} > 0 \qquad \text{and} \qquad f(\zeta_g + f) > 0. \tag{2.5.231}$$

Since the Richardson number may be less than $f/(\zeta_g + f)$, unstable displacements can occur along a θ surface. Thus, the displacement of a tube may be either stable or unstable depending upon the direction of the displacement.

To illustrate this further, consider the "absolute momentum" defined by Eliassen (also called the "pseudoangular momentum" by Emanuel):

$$m_g = u_g - fy. \tag{2.5.232}$$

It represents in a sense the absolute azimuthal velocity about the origin (at $y = 0$).

From Eq. (2.5.232) we see that

$$\frac{\partial m_g}{\partial z} = \frac{\partial u_g}{\partial z} \qquad \text{and} \qquad -\frac{\partial m_g}{\partial y} = f - \frac{\partial u_g}{\partial y}. \tag{2.5.233}$$

Along a constant-m_g surface,

$$dm_g = 0 = \frac{\partial m_g}{\partial y}\,dy + \frac{\partial m_g}{\partial z}\,dz. \tag{2.5.234}$$

Therefore the slope of an m_g surface is

$$\left(\frac{dz}{dy}\right)_{m_g} = -\frac{\partial m_g}{\partial y} \bigg/ \frac{\partial m_g}{\partial z} = \left(f - \frac{\partial u_g}{\partial y}\right) \bigg/ \frac{\partial u_g}{\partial z}. \tag{2.5.235}$$

From Eqs. (2.5.220), (2.5.212), and (2.5.235), we find that the necessary condition for symmetric instability along an isentropic surface is that

$$\left(\frac{dz}{dy}\right)_\theta > \left(\frac{dz}{dy}\right)_{m_g}, \tag{2.5.236}$$

that is, that the slope of θ surfaces in the y direction must be greater than the slopes of m_g surfaces. Since synoptic-scale motions in midlatitudes are along θ surfaces if there is no diabatic heating or friction, Eq. (2.5.236) is the relevant condition for instability.

What can we say about the necessary conditions for symmetric instability with respect to tube displacements along other surfaces? Such displacements must involve heating, friction, orography, or nonhydrostatic dynamics. Consider the v equation of motion, Eq. (2.5.203), expressed in terms of m_g, Eq.

(2.5.232) and

$$m = u - fy, \tag{2.5.237}$$

where m refers to the tube's m. The tube's m is conserved, because

$$\frac{Dm}{Dt} = \frac{Du}{Dt} - fv = -\frac{1}{\rho}\frac{\partial p}{\partial x} = fv_g = 0. \tag{2.5.238}$$

Then

$$\frac{Dv}{Dt} = -fu - \frac{1}{\rho}\frac{\partial p}{\partial y} = -f(u - u_g) = -f(m - m_g). \tag{2.5.239}$$

This means that the horizontal acceleration of a tube is proportional to the difference between its m and the m_g of its environment.

The w equation of motion is (see Chap. 3)

$$\frac{Dw}{Dt} = \frac{g}{\theta}(\theta' - \theta), \tag{2.5.240}$$

where θ' is the potential temperature of the tube and θ is the potential temperature of the environment. The tube's vertical acceleration is proportional to the difference between its potential temperature θ' and the potential temperature of its environment θ.

Refer to Fig. 2.70 and Eqs. (2.5.239) and (2.5.240) in the following discussion.

1. If $\partial\theta/\partial z > 0$ and $\partial m_g/\partial y < 0$, but $(dz/dy)_\theta = (dz/dy)_{m_g}$ [Fig. 2.70(a)], displacements of the tube along the $m_g(\theta)$ surface will result in no accelerations or decelerations, since $Dv/Dt = 0$ and $Dw/Dt = 0$. Displacements of the tube (a) lead to $Dv/Dt > 0$, because $m < m_g$ and $Dw/Dt < 0$, because $\theta' < \theta$; displacements (b) lead to $Dv/Dt < 0$, because $m > m_g$, and $Dw/Dt > 0$, because $\theta' > \theta$. Therefore, for infinitesimal displacements not along the $m_g(\theta)$ surface, the tube will be forced back toward its initial position. Therefore there is either neutral stability or absolute stability when the slope of the m_g surfaces is the same as the slope of θ surfaces.
2. If $\partial\theta/\partial z > 0$ and $\partial m_g/\partial y < 0$, but $(dz/dy)_\theta > (dz/dy)_{m_g}$ [Fig. 2.70(b)], "slantwise" infinitesimal displacements (i.e., along a plane parallel to the x axis, but not parallel to either the y or z axis) of the tube along (a) lead to $Dv/Dt > 0$, because $m < m_g$, and $Dw/Dt > 0$, because $\theta' > \theta$; along (c) $Dv/Dt < 0$, because $m > m_g$, and $Dw/Dt < 0$, because $\theta' < \theta$. Therefore slantwise infinitesimal displacements intermediate in slope between θ and m_g surfaces are symmetrically unstable, since a displaced tube will be forced away from its initial position. On the other hand, slantwise infinitesimal displacements along (b) lead to $Dv/Dt < 0$, because $m > m_g$, and $Dw/Dt > 0$, because $\theta' > \theta$; similarly, slantwise infinitesimal displacements along (d) lead to $Dv/Dt > 0$, because $m < m_g$, and $Dw/Dt < 0$, because $\theta' < \theta$. Therefore infinitesimal slantwise displacements along planes sloping more than θ surfaces are stable.

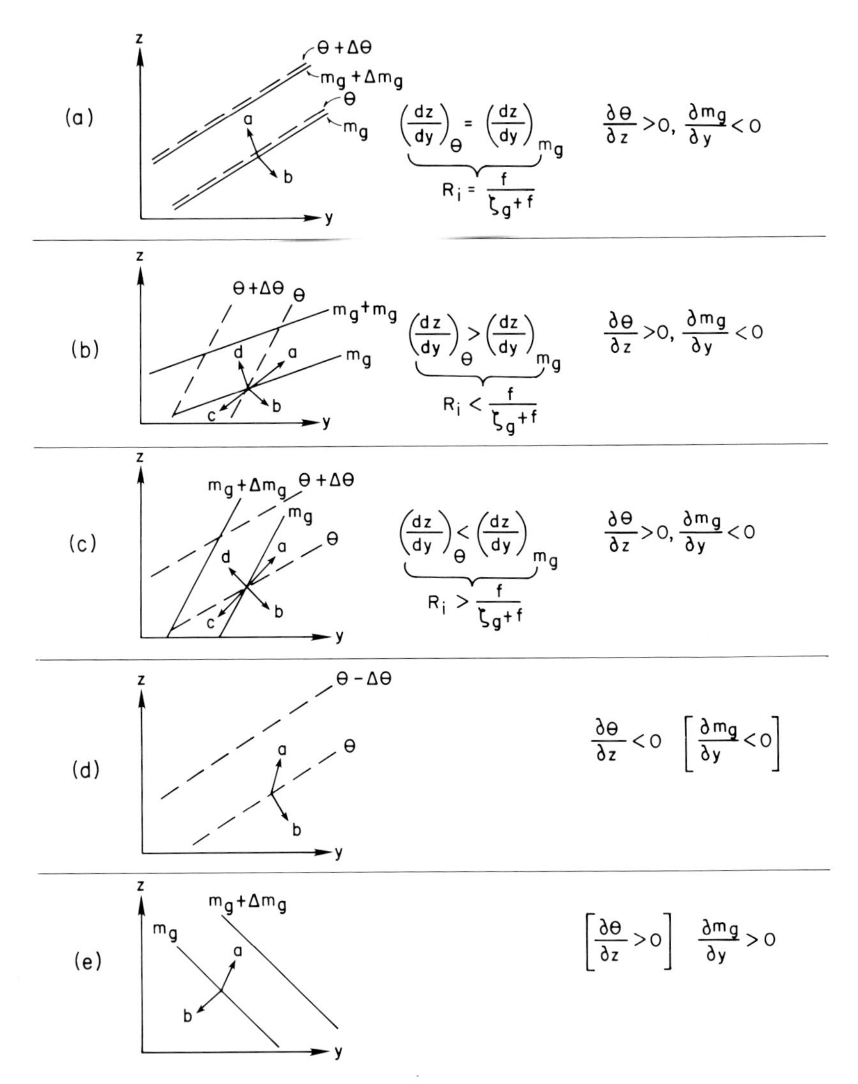

Figure 2.70 Stability criteria for symmetric flow ($\partial/\partial x = 0$) in terms of the slope of the m_g (solid lines) and potential-temperature θ (dashed lines) surfaces. Displacements in the directions *a, b, c,* and *d* are referred to in the text. (a) neutral stability for displacements along the θ and m_g surfaces; otherwise stable for infinitesimal displacements; (b) symmetric instability: unstable for infinitesimal slantwise displacements intermediate in slope between that of θ surfaces and m_g surfaces; otherwise stable; (c) absolute stability: stable for any infinitesimal slantwise displacement; (d) gravitational instability: unstable for any infinitesimal slantwise displacement not along a θ surface; (e) inertial instability: unstable for any infinitesimal slantwise displacement not along an m_g surface.

3. It is left to the reader to see that any infinitesimal slantwise displacements are stable if $\partial\theta/\partial z > 0$ and $\partial m_g/\partial y < 0$, and $(dz/dy)_\theta < (dz/dy)_{m_g}$ [Fig. 2.70(c)]. (It should be noted, however, that *finite* displacements may behave in a more complicated manner. For example, a finite displacement in the $+z$ direction will result in motion down and to the right, until the tube reaches the m_g surface. Then it will oscillate about the m_g surface and progress back to the original location.)
4. It is left to the reader to see that any infinitesimal displacements not along θ surfaces are unstable if $\partial\theta/\partial z < 0$ [Fig. 2.70(d)]. This is a case of pure gravitational instability, and Earth's rotation plays no role.
5. It is left to the reader to see that any infinitesimal displacements not along m_g surfaces are unstable if $\partial m_g/\partial y > 0$ [Fig. 2.70(e)]. This is a case of pure inertial instability, and gravity plays no direct role.

In practice we can check to see if symmetrically unstable vertical circulations are possible in two ways: (a) We can compute Ri and $f/(\zeta_g + f)$, and then identify regions where $\text{Ri} < f/(\zeta_g + f)$. We must be certain that our neglect of diabatic heating and friction are justified, and that the shear vector is essentially unidirectional. In the regions where $\text{Ri} < f/(\zeta_g + f)$, displacements along isentropic surfaces are symmetrically unstable. (b) Another technique for ascertaining the stability of the flow requires a rotation of our coordinate system so that the x axis is parallel to the geostrophic vertical shear vector. (Some synoptic analysts use the y axis as the axis of symmetry, and let $m = v + fx$.) We define the plane through which $y = 0$ arbitrarily, and compute $m_g = u_g - fy$. It does not matter exactly where $y = 0$, because we are only interested in *relative* values of m_g. The surfaces of constant m_g and constant θ are then analyzed. Where the θ surfaces have a greater slope than the m_g surfaces, symmetric instability is possible for displacements intermediate in slope between that of the θ and m_g surfaces.

How wide are symmetrically unstable vertical circulations? Suppose that the slope of a θ surface (Eq. (2.5.212))

$$\left(\frac{dz}{dy}\right)_\theta = \left(f\frac{\partial u_g}{\partial z}\right) \Big/ \left(\frac{g}{\theta}\frac{\partial\theta}{\partial z}\right) \sim \frac{H}{L}, \tag{2.5.241}$$

where H is the depth of the vertical circulation and L is the width of the vertical circulation, and the circulation tilts along the θ surface. The depth of the circulation (H) is limited by the height of the tropopause. If (Eq. (2.5.235))

$$\left(\frac{dz}{dy}\right)_\theta > \left(\frac{dz}{dy}\right)_{m_g} = \left(f - \frac{\partial u_g}{\partial y}\right) \Big/ \frac{\partial u_g}{\partial z}, \tag{2.5.242}$$

then symmetric instability is possible. From Eqs. (2.5.242) and (2.5.241) we see that

$$\frac{H}{L} > \left(f - \frac{\partial u_g}{\partial y}\right) \Big/ \frac{\partial u_g}{\partial z}. \tag{2.5.243}$$

Therefore for typical conditions in midlatitudes,

$$L \lesssim \left[\frac{\partial u_g}{\partial z} \Big/ \left(f - \frac{\partial u_g}{\partial y} \right) \right] H \sim \frac{(50 \text{ m s}^{-1}/10^4 \text{ m})}{10^{-4} \text{ s}^{-1}} (10^4 \text{ m}) = 500 \text{ km}. \quad (2.5.244)$$

In other words, the width of symmetrically unstable vertical circulations is mesoscale, not synoptic scale.

The necessary conditions for symmetric instability can also be expressed in terms of Ertel's potential vorticity (Z) for geostrophic motion. Recall that

$$Z = \frac{(\boldsymbol{\nabla} \times \mathbf{v} + f\hat{\mathbf{k}})}{\rho} \cdot \boldsymbol{\nabla} s, \quad (2.5.245)$$

where $s = C_p \ln \theta$, the specific entropy. In our analysis $v_g = 0$, so that

$$Z = \frac{C_p}{\rho} \left[\frac{\partial u_g}{\partial z} \hat{\mathbf{j}} + \left(f - \frac{\partial u_g}{\partial y} \right) \hat{\mathbf{k}} \right] \cdot \left(\frac{1}{\theta} \frac{\partial \theta}{\partial y} \hat{\mathbf{j}} + \frac{1}{\theta} \frac{\partial \theta}{\partial z} \hat{\mathbf{k}} \right). \quad (2.5.246)$$

If the horizontal temperature field is in thermal-wind balance [Eq. (2.5.210)],

$$Z = \frac{C_p}{\rho} \left[\frac{\partial u_g}{\partial z} \left(-\frac{f}{g} \frac{\partial u_g}{\partial z} \right) + \left(f - \frac{\partial u_g}{\partial y} \right) \frac{1}{\theta} \frac{\partial \theta}{\partial z} \right] \quad (2.5.247)$$

$$= \left(\frac{C_p}{\rho g} \right) f \frac{g}{\theta} \frac{\partial \theta}{\partial z} \left(\frac{\zeta_g + f}{f} - \frac{1}{\text{Ri}} \right). \quad (2.5.248)$$

Thus, according to Eqs. (2.5.248) and (2.5.219) and our analysis of slantwise displacements, the necessary condition for symmetric instability in a gravitationally stable atmosphere may be stated in terms of Ertel's potential vorticity for geostrophic motion; viz. Ertel's potential vorticity must be negative:

$$Z < 0. \quad (2.5.249)$$

Ertel's potential vorticity is expressed more elegantly in isentropic coordinates. For our analysis, the differential of u_g is given by

$$du_g = \frac{\partial u_g}{\partial y} dy + \frac{\partial u_g}{dz} dz. \quad (2.5.250)$$

It follows that

$$\left(\frac{\partial u_g}{\partial y} \right)_\theta = \left(\frac{\partial u_g}{\partial y} \right)_z + \frac{\partial u_g}{\partial z} \left(\frac{dz}{dy} \right)_\theta. \quad (2.5.251)$$

From Eq. (2.5.212) we see that

$$f - \left(\frac{\partial u_g}{\partial y} \right)_\theta = f - \left(\frac{\partial u_g}{\partial y} \right)_z - f \left(\frac{\partial u_g}{\partial z} \right)^2 \Big/ \left(\frac{g}{\theta} \frac{\partial \theta}{\partial z} \right); \quad (2.5.252)$$

hence,

$$(\zeta_g + f)_\theta = (\zeta_g + f)_z - \frac{f}{\text{Ri}}. \quad (2.5.253)$$

Using Eqs. (2.5.253) and (2.5.248), we find that

$$Z = \frac{C_p}{\rho g} \frac{g}{\theta} \frac{\partial \theta}{\partial z} (\zeta_g + f)_\theta. \tag{2.5.254}$$

In other words, the necessary condition for symmetric instability in a gravitationally stable atmosphere is that geostrophic absolute vorticity must be negative on isentropic surfaces. From a practical standpoint, it is easier to calculate Z in isentropic coordinates.

We have looked at Ertel's potential vorticity and the symmetric instability criterion in height coordinates and in isentropic coordinates. Since the Sawyer–Eliassen equation was formulated in pressure coordinates, it would be useful to transform Ertel's potential vorticity into pressure coordinates also. We know that

$$du_g = \frac{\partial u_g}{\partial y} dy + \frac{\partial u_g}{\partial p} dp. \tag{2.5.255}$$

It follows that

$$\left(\frac{\partial u_g}{\partial y}\right)_\theta = \left(\frac{\partial u_g}{\partial y}\right)_p + \frac{\partial u_g}{\partial p}\left(\frac{dp}{dy}\right)_\theta. \tag{2.5.256}$$

Now,

$$d\theta = \frac{\partial \theta}{\partial y} dy + \frac{\partial \theta}{\partial p} dp = 0 \tag{2.5.257}$$

along an isentropic surface, so that

$$\left(\frac{\partial p}{\partial y}\right)_\theta = -\frac{\partial \theta}{\partial y}\bigg/\frac{\partial \theta}{\partial p}. \tag{2.5.258}$$

Using the thermal-wind relation

$$\frac{\partial u_g}{\partial p} = \frac{R}{fp}\left(\frac{p}{p_0}\right)^\kappa \frac{\partial \theta}{\partial y}, \tag{2.5.259}$$

Eqs. (2.5.258) and (2.5.256), we find that

$$-\left(f - \frac{\partial u_g}{\partial y}\right)_\theta \frac{\partial \theta}{\partial p} = -\left(f - \frac{\partial u_g}{\partial y}\right)_p \frac{\partial \theta}{\partial p} - \frac{fp}{R}\left(\frac{p_0}{p}\right)^\kappa \left(\frac{\partial u_g}{\partial p}\right)^2. \tag{2.5.260}$$

Since the left-hand side of Eq. (2.5.260) is proportional to Ertel's potential vorticity, Eq. (2.5.254), the criterion for symmetric instability in a gravitationally stable atmosphere is that

$$\frac{R}{fp}\left(\frac{p}{p_0}\right)^\kappa \frac{\partial \theta}{\partial p}\left(f - \frac{\partial u_g}{\partial y}\right)_p + \left(\frac{\partial u_g}{\partial p}\right)^2 > 0. \tag{2.5.261}$$

This condition is equivalent to the one that ensures that the Sawyer–Eliassen equation is not elliptic (or parabolic), Eq. (2.5.202).

Ertel's potential vorticity is conserved for frictionless, adiabatic motions; it follows that in order for symmetric instability to occur, Z, which is ordinarily positive, must become negative. Vertical gradients of diabatic heating can reduce $\partial\theta/\partial z$, and hence Ri and Z. Z can also be reduced through the turbulent transport of momentum vertically, which can increase $\partial u_g/\partial z$, and hence reduce Ri, or through the turbulent transport of momentum horizontally, which can increase $\partial u_g/\partial y$, and hence reduce ζ_g. The latter two turbulent transports would have to be up the gradient, rather than down the gradient. Diabatic heating is more likely to produce the necessary conditions for symmetric instability, especially above the boundary layer.

Why is symmetric instability important? If there is an ageostrophic, vertical circulation along a front, it could increase in intensity rapidly, and lift air to saturation, resulting in the production of precipitation. The inclusion of moisture into our analysis leads to some interesting results, which will be detailed in Chap. 3. In addition, if an increase in the intensity of the vertical circulation is hastened, frontogenesis may be sped up also as a result of increased surface convergence. However, owing to the high values of cyclonic geostrophic vorticity and static stability along surface fronts, the necessary conditions for symmetric instability should not occur there. Furthermore, the byproducts of an increased rate of frontogenesis owing to symmetric instability, that is, the production of cyclonic vorticity and an increase in static stability through tilting, act to bring the atmosphere back to a neutral state with respect to symmetric instability. It is therefore likely that if we find that $\mathrm{Ri} \approx f/(\zeta_g + f)$, or that the slope of θ surfaces is approximately identical to the slope of m_g surfaces, then symmetrically unstable vertical circulations may have recently occurred. This situation is analogous to that of dry convection: We rarely find that $\partial\theta/\partial z$ is negative, because if it were, vertical mixing would rapidly produce a profile characterized by $\partial\theta/\partial z \approx 0$, that is, one of neutral stability.

2.5.3. The Effects of Friction

On one hand, the direct effects of turbulence on the temperature field are to reduce extrema in the temperature field, and hence to reduce the strength of frontal zones. On the other hand, along the strip of cyclonic geostrophic vorticity produced by convergence along surface fronts (i.e., along the warm edge of the frontal zone), even more convergence is enhanced owing to the effects of turbulent friction on the wind field (Ekman pumping). This results in a narrow jet of rising motion above the surface frontal trough at the top of the boundary layer. The added convergence acts frontogenetically and hastens frontogenesis.

In the boundary layer, vertical shear of the ageostrophic wind field, which is a result of friction, acts to decrease the static stability near the ground on the cold side of the frontal zone and near the top of the boundary layer on the warm side of the frontal zone; it increases the static stability near the ground on the warm side and near the top of the boundary layer on the cold side (Fig. 2.71). This mechanism, referred to by Miller as *vertical deformation* [see Eq.

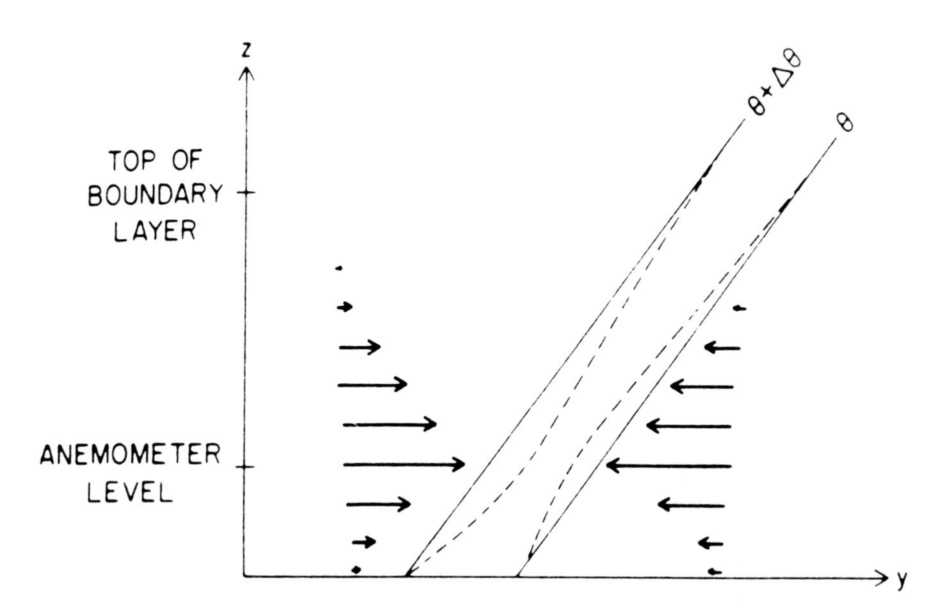

Figure 2.71 Effects of boundary-layer friction on stability within a frontal zone. Frictionally induced ageostrophic wind component normal to the frontal zone (vectors); initial distribution of potential temperature θ (solid lines); later distribution of potential temperature (dashed lines). The vertical gradient of potential-temperature (static stability) changes owing to $-(\partial v/\partial z)(\partial \theta/\partial y)(\partial \theta/\partial z)$, vertical deformation (from Bluestein, 1986). (Courtesy of the American Meteorological Society)

(2.3.21)], may be responsible for the low static stabilities observed near the ground on the cold side of frontal zones, and the concomitant gustiness associated with vertical mixing.

2.5.4. The effects of latent-heat release

Since we sometimes observe the formation of fronts in the absence of cloudiness or precipitation, and since frontlike structures have been produced in laboratory models without any latent-heat release, it is clear that latent-heat release is not necessary, in general, for frontogenesis. However, diabatic heating from latent-heat release could enhance the frontogenetical process if it occurs in the rising branch of the vertical circulation, by producing even more rising motion [see the diabatic heating forcing function in Eq. (2.5.53)], low-level convergence, and therefore even greater frontogenesis.

Evaporative cooling, which occurs as precipitation falls into unsaturated air on the cold side of a front, can also enhance frontogenesis.

Latent-heat release resulting from cumulus convection and friction could act cooperatively along the strip of cyclonic vorticity found in the boundary layer along frontal zones. Ekman pumping could make the atmosphere more susceptible to cumulus convection, while the resultant diabatic heating leads to more vertical motion and low-level convergence. This mechanism may account for the vertical circulations associated with cloud bands in the tropics that form along old frontal zones that still are characterized by strips of cyclonic vorticity, but are no longer accompanied by a strong temperature gradient.

2.5.5. The Relationship between Frontogenesis and Cyclogenesis

Which comes first, cyclogenesis (and anticyclogenesis) or frontogenesis? We have shown how geostrophic deformation, which is associated with the "checkerboard" of surface cyclones and anticyclones in the atmosphere, is often responsible for the initiation of frontogenesis by tightening existing synoptic-scale temperature gradients. On the other hand, the Polar–Front "theorists" or perhaps, more appropriately, "conceptual modelers" (e.g., Bjerknes in 1919), pointed out on the basis of observations that fronts can play an important role in cyclogenesis. We now know, using quasigeostrophic principles, that the strong temperature gradient found along a front can be the site of potentially strong temperature advection, given even a modest amount of cross-isotherm geostrophic flow. Hence, there, seems to be an intimate relationship between surface cyclogenesis and frontogenesis: Each may be responsible for the other.

A. Eliassen has pointed out that synoptic-scale waves in the middle-latitude westerlies can form owing to baroclinic instability, in the absence of fronts. (We also observe waves developing from barotropic processes.) The production of surface anticyclones and cyclones during the amplification of the waves in the westerlies leads to deformation patterns that induce the formation of fronts, and further, to the formation of smaller-scale cyclones along the fronts owing to other instabilities. Thus baroclinic instability may be regarded as a primary process, and frontogenesis a secondary process.

Suppose that baroclinic instability is in fact a primary process. How sensitive is the structure of the fronts that subsequently form to the temperature and height (pressure) structure of the amplifying waves? Consider an atmosphere in which there is uniform (for simplicity only) Ertel's potential vorticity. If there is a channel of concentrated westerly thermal-wind shear (calm winds at the surface, jet aloft), numerical simulations indicate that a cold front forms, but an as-well-developed warm front does not appear. Upward motion is concentrated along a narrow band in the cold front south and southwest of the low, and is more broadly distributed in the weak warm front, east of the low.

If there is a channel of westerly thermal-wind shear (with uniform winds aloft), which is flanked to the north and south by slightly weaker westerly thermal-wind shear and surface cyclonic and anticyclonic geostrophic shear, respectively, a wave in the surface isotherms become sheared in the direction of the initial low-level flow to the north and south; that is, the temperature wave becomes positively tilted on the cyclonic-shear side of the surface geostrophic wind and negatively tilted on the anticyclonic-shear side of the surface geostrophic wind. A strong warm front develops southeast of the low. This warm front is like an occluded front in that the pressure trough is aligned along the thermal ridge. No strong cold front develops south or southwest of the low. A feature like a cold front develops north of the low. This feature looks like the "inverted trough," which is sometimes observed north of cyclones. Upward motion is concentrated in a band along the warm front and in a broader band along the cold-front-like feature to the north.

What is the essential difference between the environment in which the cold front (south of the cyclone) and the warm front, which is like an occlusion, form? In Chapter 1 we found, using quasigeostrophic theory, that in amplifying waves in the baroclinic westerlies, the *waves of geopotential height tilt backward* with height toward the west, against the flow. It can also be shown that the *temperature wave at low levels must tilt forward* with height to the east, with the flow, if the waves amplify. The cold front forms in a region in which the *cold trough* of the temperature wave aloft is displaced to the east, to the warm side, relative to the surface temperature wave. On the other hand, the warm front forms in a region in which the *warm ridge* of the upper temperature wave is displaced to the east, to the warm side, relative to the surface temperature wave. (These results, which are based upon numerical simulations, need to be verified with observations.)

D. Keyser and M. Pecnick have presented numerical evidence that a strong cold front forms when both confluence and horizontal shear act frontogenetically. This happens when there is an along-the-front component to the temperature gradient (as in Figs 2.62 and 2.64), such that both the initial cyclonic shear and that produced along the front further enhances frontogenesis both kinematically and dynamically. This case is also referred to as the *cold-advection case,* owing to the initial cold advection by the along-the-front wind component at low levels in Keyser's simulations. The zones of maximum cyclonic vorticity and strongest potential temperature gradient (not shown) are relatively deep.

When the horizontal shear acts frontolytically (as in Fig. 2.64 along the warm front), the front is also relatively shallow; however, although high values of cyclonic vorticity are produced, the potential-temperature gradient is relatively weak. The vertical circulation in this case is displaced toward the cold air, with a descending branch found also in the warm air (not shown). We refer to this case as the *warm-advection case,* owing to the initial warm advection by the along-the-front wind component at low levels.

Keyser and Pecnick have suggested that the basic-state temperature field in the simulations discussed earlier are such that along-the-front cold advection occurs when there is concentrated westerly thermal wind shear in the initial state, and along-the-front warm advection when there is a channel of westerly thermal-wind shear flanked by weak surface cyclonic and anticyclonic geostrophic shear vorticity in the initial state.

2.5.6. Density-Current Dynamics, Trapped Density Currents, and Barrier Jets

Shallow cold fronts, which behave like "density currents," have been observed in the Southern Plains of the United States, east of the Rocky Mountains, in southeast Australia, and along the east side of the Appalachians. A similar phenomenon, the *alongshore* or *coastal surge,* has been documented along the west coast of the United States. In a density current the boundary between a dense and a lighter fluid (no incendiary pun intended) is forced to move toward the lighter fluid by the horizontal pressure gradient force associated with the increased hydrostatic pressure on the dense side (recall Fig. 2.8). Earth's rotation does not play a fundamental role in the dynamics of a density current unless it is trapped by topography. The horizontal pressure-gradient force (and horizontal temperature gradient) is (are) very large across a density-current boundary. An observer who experiences the passage of such a front notes that the pressure jumps and the temperature falls dramatically. It is noteworthy that such large horizontal temperature gradients can exist in the face of turbulent mixing, which is associated with strong vertical shear along the density-current frontal boundary.

In the reference frame of the frontal surface (Fig. 2.8), air parcels approach from the lighter side and begin to acquire vorticity about a horizontal axis (negative vorticity about the y axis) as they encounter the horizontal gradient in buoyancy along the boundary between dense and lighter air (negative buoyancy in the dense air, neutral buoyancy in the ambient, relatively light air). They then move away from the frontal surface over the denser air.

We make use of the latter analysis to derive a formula for the speed of the density current. Consider a dense fluid at the ground that is H units deep. We simplify the problem by making it two dimensional ($\partial/\partial y = 0$), and by assuming that the dense and light fluids do not mix with each other. The frictionless Boussinesq equations of motion[4] in the x and z directions are as follows:

$$\frac{Du}{Dt} = \frac{\partial u}{\partial t} + u\frac{\partial u}{\partial x} + w\frac{\partial u}{\partial z} = -\frac{1}{\bar{\rho}}\frac{\partial p'}{\partial x} \tag{2.5.262}$$

$$\frac{Dw}{Dt} = \frac{\partial w}{\partial t} + u\frac{\partial w}{\partial x} + w\frac{\partial w}{\partial z} = -\frac{1}{\bar{\rho}}\frac{\partial p'}{\partial z} + B, \tag{2.5.263}$$

where $\bar{\rho}$ is a constant, p' is the "perturbation pressure," and the vertical acceleration induced by buoyancy in the density current

$$B = -g\frac{\rho_{\mathrm{D}} - \rho_{\mathrm{L}}}{\rho_{\mathrm{L}}}, \tag{2.5.264}$$

while

$$B = 0 \tag{2.5.265}$$

elsewhere, where ρ_{D} and ρ_{L} are the densities of the dense and light fluids. We assume that the volume of dense fluid is small compared to the volume of the ambient, light fluid.

Differentiating Eq. (2.5.263) with respect to x and subtracting it from Eq. (2.5.262) differentiated with respect to z, we obtain the following equation for $\partial u/\partial z - \partial w/\partial x$, which is the component of vorticity about the y axis:

$$\frac{\partial}{\partial t}\left(\frac{\partial u}{\partial z} - \frac{\partial w}{\partial x}\right) = -\frac{\partial}{\partial x}\left[u\left(\frac{\partial u}{\partial z} - \frac{\partial w}{\partial x}\right)\right] - \frac{\partial}{\partial z}\left[w\left(\frac{\partial u}{\partial z} - \frac{\partial w}{\partial x}\right)\right] - \frac{\partial B}{\partial x}. \tag{2.5.266}$$

Let us now integrate Eq. (2.5.266) over the area shown in Fig. 2.8; this area encompasses the frontal surface (density-current interface) at the ground, where $w = 0$, and above. We find that

$$\frac{\partial}{\partial t}\int_L^R \int_0^d \left(\frac{\partial u}{\partial z} - \frac{\partial w}{\partial x}\right) dz\, dx = \int_0^d \left[u\left(\frac{\partial u}{\partial z} - \frac{\partial w}{\partial x}\right)\right]_L dz - \int_0^d \left[u\left(\frac{\partial u}{\partial z} - \frac{\partial w}{\partial x}\right)\right]_R dz - \int_L^R \left[w\left(\frac{\partial u}{\partial z} - \frac{\partial w}{\partial x}\right)\right]_d dx + \int_0^H (B_L - B_R)\, dz, \tag{2.5.267}$$

where B_R and B_L are the buoyancies at $x = R$ and L, respectively. The first term on the right-hand side of Eq. (2.5.267) represents the flux of vorticity about the y axis into the area at $x = L$. The second and third terms represent the fluxes of vorticity out from the area at $x = R$ and $z = d$. According to the kinematic lower boundary condition, $w = 0$, and hence there is no flux of vorticity from the ground. The fourth term represents the generation of vorticity by horizontal gradients in buoyancy.

Let us simplify our problem further by considering only density currents that are steady state in the reference frame of the density current. (We therefore are assuming that the acceleration induced by the horizontal pressure-gradient force across the density current is eventually balanced by another force, such as friction or a nonhydrostatic pressure-gradient force.) In addition, let us assume that well away from the density-current boundary, at $x = L$ and R vertical motions are negligible, and hence $\partial w/\partial x$ is also negligible. Air well ahead of the front, which approaches the front, has no buoyancy ($B_R = 0$). It follows that

$$0 = \int_0^d \left(u\frac{\partial u}{\partial z}\right)_L dz - \int_0^d \left(u\frac{\partial u}{\partial z}\right)_R dz - \int_L^R \left(w\frac{\partial u}{\partial z}\right)_d dx + \int_0^H B_L\, dz. \tag{2.5.268}$$

Let us also assume that the density-current boundary moves along with the horizontal wind behind it ($u_{L,0} = 0$) (i.e., the density current is *stagnant*), and that there is no ambient vertical wind shear ($u_{R,d} = u_{R,0}$). It follows that *in the reference frame of the density current*

$$0 = \frac{u_{L,d}^2}{2} + \int_0^H B_L\, dz. \tag{2.5.269}$$

Thus, the speed of the density current $c = -u_{L,d}$, so that

$$c^2 = -2\int_0^H B_L\, dz. \tag{2.5.270}$$

It follows from Eqs. (2.5.270), (2.5.264), and (2.5.265) that

$$c = K\left[g\left(\frac{\rho_D - \rho_L}{\rho_L}\right)H\right]^{1/2} \tag{2.5.271}$$

where

$$K = \sqrt{2}. \tag{2.5.272}$$

(Equation (2.5.271) is proportional (by a factor of $\sqrt{2}$) to the relative phase speed of gravity waves in a two-layer fluid [see Eq. (7.2.3) in Holton (1979)]. However, a density current *moves almost as a material boundary,*[5] while a gravity wave *propagates*; mass bobs up and down, rather than push onward.) Equation (2.5.271) was first derived this way by R. Rotunno et al. in 1988 and is identical to the often-referenced formula derived by Benjamin in 1968. The physical interpretation of Eq. (2.5.271) is that the generation of negative vorticity about the y axis owing to the horizontal buoyancy gradient is balanced by the advection of negative vorticity out from the area at $x = L$. (An alternative interpretation is that a nonhydrostatic horizontal pressure-gradient force slows down an air parcel approaching the density current to a complete stop in the reference frame of the density current. In order to satisfy the dynamic boundary condition (see Vol. I) at the leading edge of the density current, the sum of the nonhydrostatic and hydrostatic pressure just ahead of the density current must be equal to the hydrostatic pressure behind the density current. The speed of the density current is that an air parcel must have in order that its kinetic energy is exhausted, while doing work against the nonhydrostatic pressure-gradient force ahead of the leading edge of the density current.)

We now apply Eq. (2.5.271) to the real atmosphere. Using the ideal gas law, Eq. (2.5.271) can be expressed as

$$c = K\left[gH\left(\frac{(p_D/p_L)\bar{T}_{vw} - \bar{T}_{vc}}{\bar{T}_{vc}}\right)\right]^{1/2}, \tag{2.5.273}$$

where $\bar{T}_{vw}$ and $\bar{T}_{vc}$ are the average virtual temperatures on the warm (light) side and cold (dense) side, and p_D and p_L are the pressures at corresponding heights on the dense (cold) side and light (warm) side. Since

$$\frac{p_D}{p_L} \approx 1 \tag{2.5.274}$$

near the ground,

$$c \approx K\left(gH\frac{\bar{T}_{vw} - \bar{T}_{vc}}{\bar{T}_{vc}}\right)^{1/2}. \tag{2.5.275}$$

Because mixing occurs in the real atmosphere, the actual value of K is somewhat less than $\sqrt{2}$ [mixing was neglected in the formulation of Eqs. (2.5.271) and (2.5.272)]. Estimates of K ranging from 0.7 to 1.0 have been made in thunderstorm-induced density currents. Since the horizontal pressure-gradient force driving the density current is counteracted by friction, mostly near the ground, the density current "noses" forward just above the ground (i.e., is retarded at the ground).

Equation (2.5.275) is somewhat difficult to use unless the vertical temperature profile is known. It follows from the hydrostatic equation that

$$p_{0c} - p_{0w} = \Delta p = gH(\rho_D - \rho_L), \tag{2.5.276}$$

where p_{0c} and p_{0w} are the surface pressures on the cold and warm sides, respectively. (The measurement of p_{0c} is made at the location where the density current levels off at H units in depth.) Substituting Eq. (2.5.276) into Eq. (2.5.271), we find that

$$c = K\left(\frac{\Delta p}{\rho_L}\right)^{1/2}. \tag{2.5.277}$$

In practice ρ_L can be calculated from the surface density with no serious errors.

In the derivation of Eq. (2.5.271) (and subsequent formulations), we assumed that there was no wind ahead of the density current. If there were an ambient wind, the speed of the density current would *not* be simply the sum of the component of the wind normal to the front and Eq. (2.5.277). Laboratory experiments by Simpson and Britter suggest, however, that

$$c = K\left(\frac{\Delta p}{\rho_L}\right)^{1/2} + 0.62\bar{u}, \tag{2.5.278}$$

where $\bar{u}$ is the vertical average of u below H.

What is the source of shallow, cold air in density currents that sometimes appears in cold fronts? In some cold fronts a pool of cool, dense air is produced by evaporative cooling of precipitation falling through unsaturated air behind the front. In others, the source of cool, dense air is sensible heat transfer from relatively warm air to relatively cold ocean water below, or radiative cooling over snow or ice-covered surfaces. Observations indicate that fronts having characteristics of density currents often appear along and near the edges of mountain ranges. It therefore seems as if topography must play an important role.

Topography can "trap" a density current. Part of the evolution of a "coastal surge" on the West Coast of the United States is depicted in Fig. 2.72 for the purpose of illustration. A narrow mesoscale pressure ridge formed at 1200 UTC on May 15, 1985, south of San Francisco, and moved all the way up the coast to British Columbia by May 17 at 1200 UTC. The northern edge of an area of stratus clouds also moved northward with the leading edge of the ridge, which stayed near the coast. As the mesoscale ridge moved northward, a thermal low, initially located over Central California, moved northwestward and off the coast. In general, a wind shift from north to south accompanied the passage of the surge late in its lifetime (after 1800 UTC, May 16).

The surge was triggered when the region of low pressure over California at low levels moved offshore enough to bring the low-level synoptic-scale wind around to a westerly direction, south of the low. At the same time, easterly downslope motion along the coast to the north was accompanied by surface-

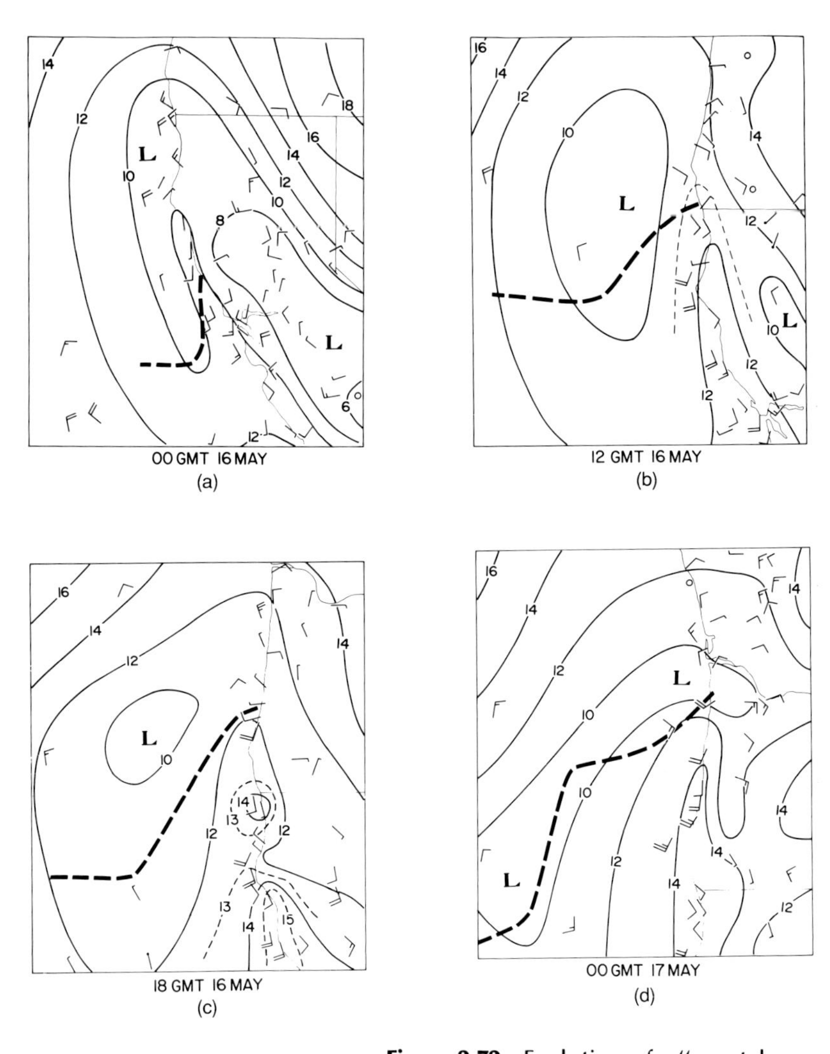

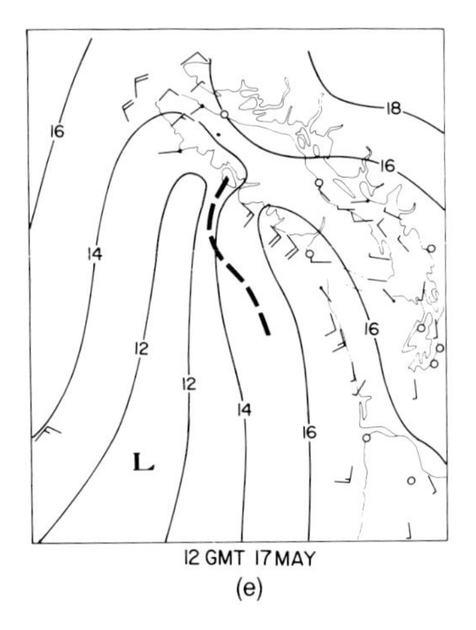

Figure 2.72 Evolution of a "coastal surge" along the west coast of the United States. Isobars of sea-level pressure in mb, with the leading 10 omitted (solid lines); intermediate isobars (thin dashed lines); whole and half wind barbs represent 5 and 2.5 m s^{-1}, respectively; northern boundaries of stratus clouds, based on GOES satellite imagery (thick dashed lines). (a) 0000 UTC (GMT), May 16, 1985, (b) 1200 UTC, May 16, (c) 1800 UTC, May 16, (d) 0000 UTC, May 17, and (e) 1200 UTC, May 17 (from Mass and Albright, 1987). (Courtesy of the American Meteorological Society)

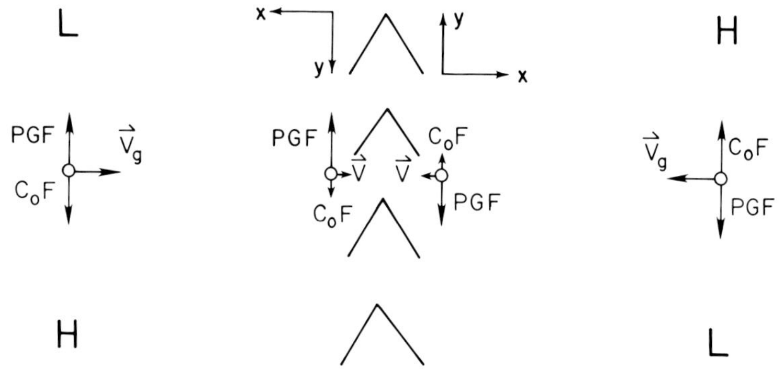

Figure 2.73 Force diagrams for an air parcel in geostrophic balance far away from a mountain range, and not in geostrophic balance near the mountain range. In the example shown the geostrophic wind is normal to the mountain range. As the air parcel approaches the mountains (in this case from either the west or the east), it slows down as it is lifted; if the pressure-gradient force is the same near the mountains as it is far away from the mountains, then the air parcel is no longer in geostrophic balance, and it is deflected in the direction of the pressure-gradient force. Pressure-gradient force (**PGF**), Coriolis force (**CoF**), geostrophic wind ($\mathbf{v}_g$), and total wind ($\mathbf{v}$).

pressure falls and lee-trough formation. Thus, a northward-directed pressure-gradient force appeared along the coast.

Imagine an air parcel in a stable atmosphere at low levels, in near geostrophic balance, approaching the mountain range south of the low. As it hits the mountain range, the advection of lower potential temperature from below acts to increase the hydrostatic pressure along the mountain slope over what it is at the same height above sea level away from the mountains. A pressure-gradient force therefore develops that acts in the direction opposite to the elevation gradient, so that the air parcel slows down. The component of the Coriolis force along elevation contours decreases as the component of wind normal to the mountains decreases; the parcel is therefore deflected to its left by the pressure-gradient force (Fig. 2.73). Because air did not have enough kinetic energy to flow over the mountain range to the east, owing to a very stable lapse rate (cool marine air at the surface surmounted by warm, dry air), there was no north–south Coriolis-force component, and hence the northward-directed pressure-gradient force remained unbalanced, and air accelerated northward. Cool, low-level marine air was then advected northward, and as a hydrostatic consequence, a narrow surface ridge formed. The zone of strong northward pressure gradient at the northern tip of the narrow ridge can be strong enough to behave like a density current.

For southerly flow, the Coriolis force acts toward the east; in the absence of a significant cross-shore wind component, there is a *net force* acting eastward. Hence, the Coriolis force "traps" the air flow along the coastal mountain range. This may also be the mechanism responsible for the "damming" of cold air east of the Appalachians (Figs. 2.74–2.76), which sustains the Baker Ridge

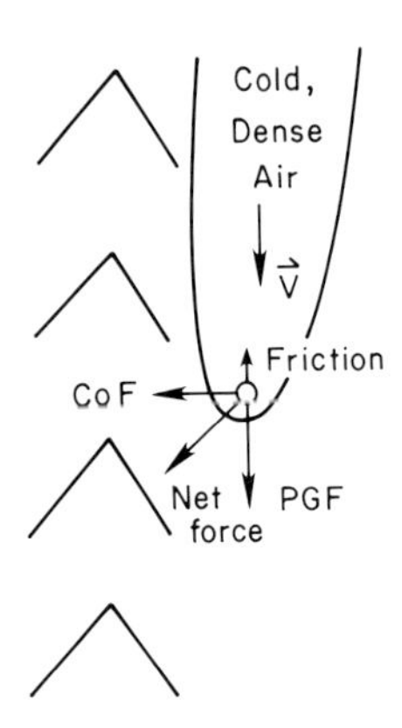

Figure 2.74 Illustration of the "damming" of cold air against a mountain range. Air is accelerated along the mountains, owing to an along-the-mountains pressure gradient force (**PGF**). Since the Coriolis force (**CoF**) acts normal (and to the left in the Northern Hemisphere example shown here) to the motion of the air parcel (**v**), the net force acts to trap the air parcel along the side of the mountain range. If the mountain range were not present, the air parcel would be deflected to the right and geostrophic balance (in the absence of friction) would eventually be attained.

discussed earlier in this chapter. It may also be the mechanism responsible for the southerly propagating cold ridges just east of the Rocky Mountains.

The synoptic-scale flow acts to initiate an along-the-mountain range pressure gradient, while the mountains make it impossible for geostrophic balance to occur in the along-the-range direction: Hence, air accelerates like a trapped density current, owing to the unbalanced[6] pressure-gradient force. Geostrophic balance is not possible within a Rossby radius of deformation $(g\,\Delta\theta\,H/\theta)^{1/2}/f$ of the mountains (this will be explained in the next section).

The trapping of cool air along a mountain range also apparently plays a role in the formation of the Catalina eddy, a mesoscale cyclonic circulation in the low-level winds off the coast of Southern California between spring and early fall. The Catalina eddy is important because the prevailing westerly and northwesterly flow is reversed, and the marine layer deepens so that coastal stratus is more persistent than usual; the air is cooler along the coast, and the air quality improves, owing to the increased depth of the mixed layer.

The eddy is initiated when northerly geostrophic flow created by relative high pressure offshore and relative low pressure inland forces air downslope over a mountain range oriented parallel to the shoreline, which is oriented from west-northwest to east-southeast. Lee troughing occurs south of the mountains, and results in an along-the-shore pressure-gradient force, which has a component from south to north. Cool, humid air is advected northward along the coast and is trapped; as a hydrostatic consequence, a mesoscale ridge forms along the coast, and the moist layer deepens. Thus, an onshore-directed pressure gradient results in an along-the-mountain geostrophic wind that reinforces the ageostrophic wind created by the along-the-mountains pressure gradient. Offshore by a distance in excess of a Rossby radius of deformation, the northerly geostrophic wind is undisturbed. Cyclonic vorticity is generated in the zone separating the trapped southerly flow from the undisturbed northerly flow.

When will the air be forced up and over the mountains? When, on the other hand, will it not have enough kinetic energy to make it over the top of the mountains, and therefore be shunted aside? The amount of potential energy per unit mass needed to be overcome to lift an air parcel up to

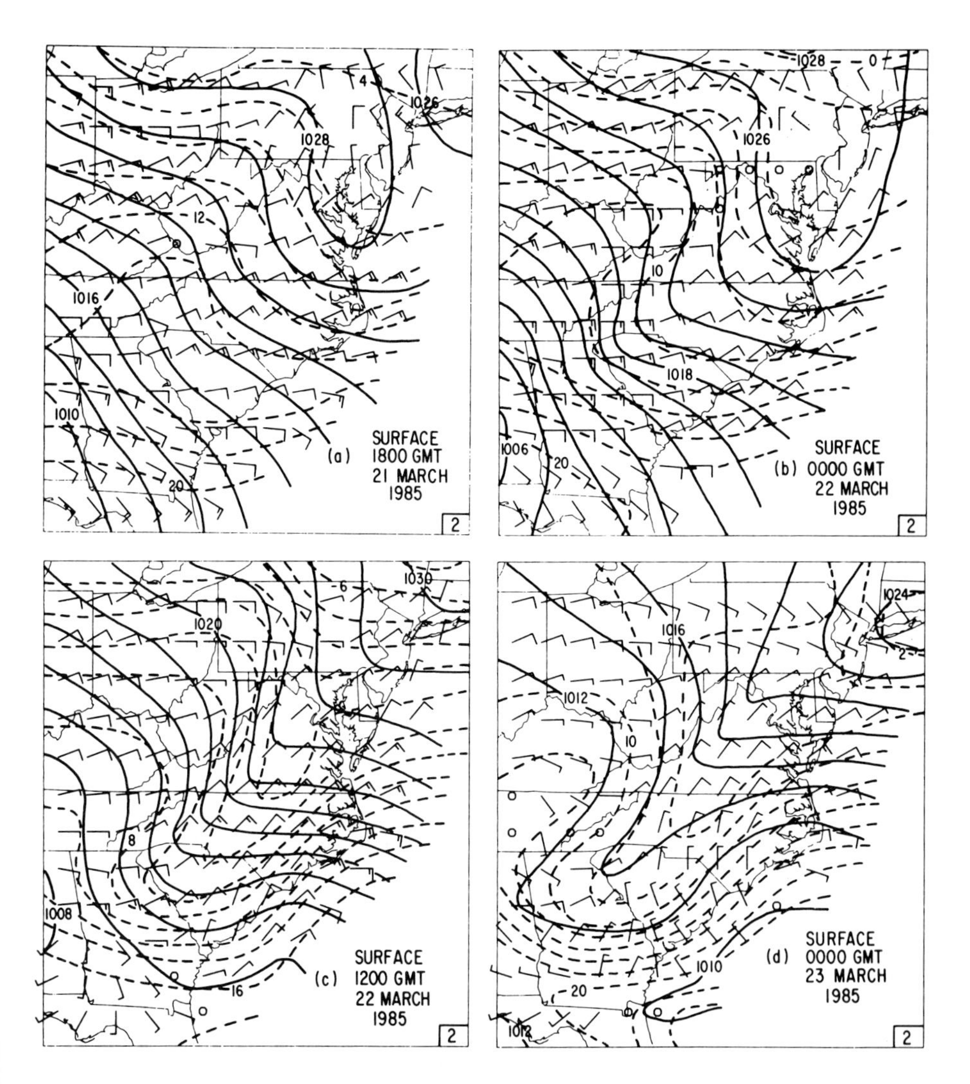

Figure 2.75 Example of cold-air damming along the Appalachians. Altimeter setting in mb (solid lines); surface potential temperature in °C (dashed lines) every 2°C; whole and half wind barbs represent 5 and 2.5 m s^{-1}, respectively. (a) 1800 UTC (GMT), March 21, 1985, (b) 0000 UTC, March 22, 1985, (c) 1200 UTC, March 22, 1985, and (d) 0000 UTC, March 23, 1985. The ridge of high pressure and cold air at the surface are coincident, and build southwestward with time, along the east side of the Appalachians; the winds blow nearly perpendicular to the isobars, along the direction of the pressure gradient (from Bell and Bosart, 1988). (Courtesy of the American Meteorological Society)

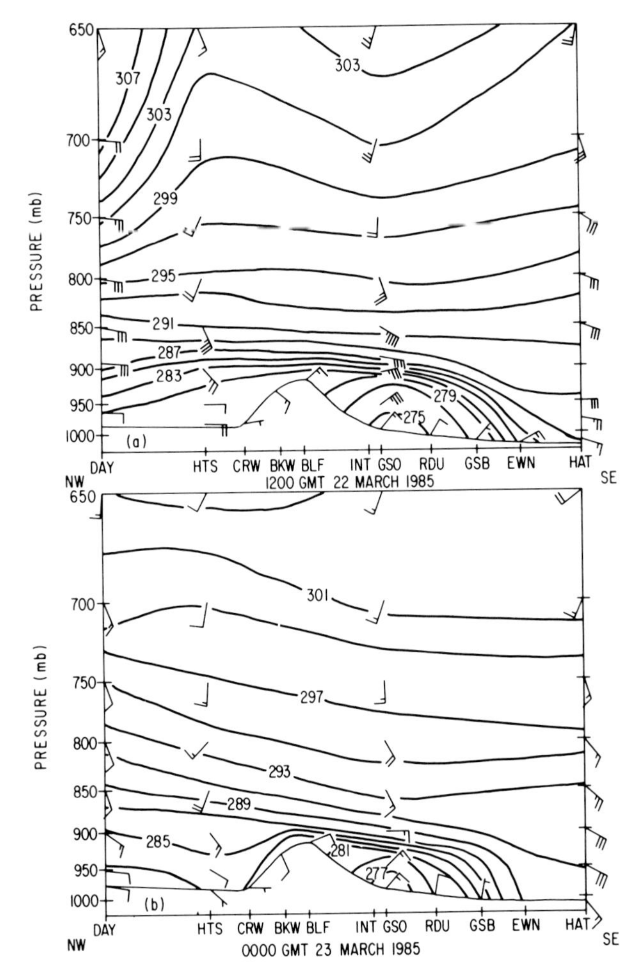

Figure 2.76 Vertical cross sections normal to the mountain range in the cold-air damming example depicted in Fig. 2.75. Potential temperature in K (thick solid line); ground level (thin solid line); winds as in Fig. 2.75; from Dayton, Ohio (DAY), to Cape Hatteras, North Carolina (HAT) (from Bell and Bosart, 1988). (Courtesy of the American Meteorological Society)

mountain-top level (see Eq. (2.5.223)) is proportional to

$$g\frac{\Delta\theta}{\theta}H^2 = N^2H^2, \tag{2.5.279}$$

where N^2 is a constant, and H is the height of the mountain above the surrounding terrain. The kinetic energy per unit mass of the air is $\frac{1}{2}U^2$, where U is the ambient wind speed. The Froude number (F) is often defined as

$$F = \frac{U}{NH}. \tag{2.5.280}$$

The square of F is therefore proportional to the ratio of kinetic energy of the ambient wind to the potential energy that must be overcome to lift an air parcel over the mountains. F^2 may also be interpreted as the ratio of inertial forces $(U(U/H))$ to buoyancy forces $(g\,(\Delta\theta/\theta) = N^2H)$. Low-Froude-number flow is characterized by "blocking," while high-Froude-number flow is characterized by flow over the mountains.

If stable air is already "in place" along and upstream from a mountain range, then a geostrophically balanced air parcel that approaches the mountains can remain in near geostrophic balance, as a result of an upslope-induced increase in the hydrostatic pressure along the mountain slopes; i.e., isobars have turned from a direction normal to the mountains to one more parallel to elevation contours. The along-the-mountain current that develops is called a *barrier jet.* One effect of a barrier jet is to elevate surfaces of potential temperature upstream from a mountain ridge. The barrier jet is different from the trapped density current in that the latter marks the leading edge of the advancing cold air mass, while the former occurs after the initial surge. The height of the top of the cold air mass is shallow and nearly constant in the case of the density current; it is high at high elevation, and drops off with distance away from the mountain ridge in the case of the barrier jet. As the synoptic scale pressure gradient changes, a density current may form; a barrier jet forms after the synoptic-scale pressure gradient has already changed. For example, east of the Rockies a density current may form in between a lee trough, which is characterized by warm downslope air, and a cold high to the north, when the high appears east of the mountains and/or the lee trough forms. After the high and lee trough have moved southward, so that there is an easterly flow of cold, stable air toward the mountains, a barrier jet may form. If the boundary layer is neutrally stratified, then there is no blocking of upslope flow by the mountains, and neither a density current nor a barrier jet form.

2.5.7. Orographically Trapped Gravity Waves

When an air parcel that is initially in geostrophic balance is blocked by topography, geostrophic balance (we neglect friction here for the sake of simplicity) is disturbed, and gravity waves[7] may be forced (Holton, 1979, pp. 155–59). Because gravity waves of this type were first studied in the ocean along coastlines, they are sometimes referred to as *coastally trapped waves.* We will call them *orographically trapped waves,* because orography in the atmosphere plays the role of the coastline in the ocean. The adjective *trapped* refers to the inhibiting effect of the stable stratification and the mountains on the vertical and horizontal propagation of the wave energy. This orographically trapped wave is equivalent to a *Kelvin wave* (Pedlosky, 1979, pp. 75–79).

Suppose that the x axis is oriented normal to a vertical wall (which represents a mountain range or other region of sloping terrain), and the y axis is oriented along the wall, and to the left of the x axis (Fig. 2.73). If there is geostrophic balance in the direction normal to the mountains, the frictionless

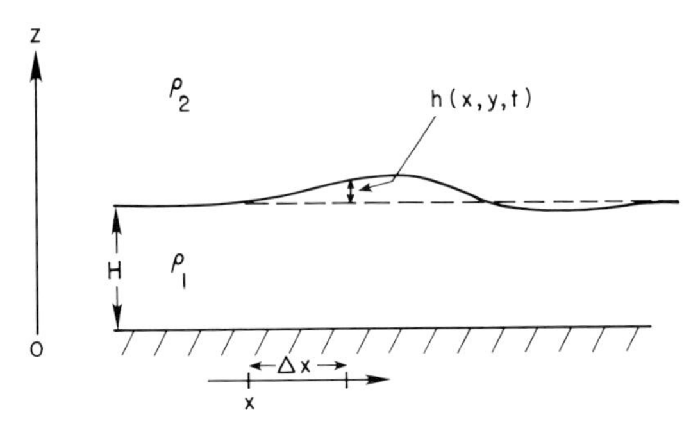

Figure 2.77 Vertical cross section through a two-layer fluid: The densities of the top and bottom fluids are ρ_2 and ρ_1, respectively. The undisturbed depth of the bottom, denser fluid is *H*. The depth by which the interface between the two fluids is disturbed is $h(x,y,t)$.

equations of motion are as follows:

$$0 = fv - \frac{1}{\rho}\frac{\partial p}{\partial x} \tag{2.5.281}$$

$$\frac{Dv}{Dt} = -fu - \frac{1}{\rho}\frac{\partial p}{\partial y}. \tag{2.5.282}$$

To simplify our analysis, we divide the atmosphere into two immiscible layers (Fig. 2.77): Relatively light air (of density ρ_2) overlies relatively dense air (of density ρ_1). The undisturbed height of the interface between the two layers is at $z = H$. Suppose that the interface is perturbed upward by the amount $h(x,y,t)$. Integrating the hydrostatic equation

$$\frac{\partial p}{\partial z} = -\rho g \tag{2.5.283}$$

at x and $x + \Delta x$ from $z = H$ to $H + h$ (h is evaluated at $x + \Delta x$), we find that

$$p(z = H, x) = \rho_2 gh + p(z = H + h, x) \tag{2.5.284}$$

$$p(z = H, x + \Delta x) = \rho_1 gh + p(z = H + h, x + \Delta x). \tag{2.5.285}$$

Subtracting Eq. (2.5.284) from Eq. (2.5.285), and dividing by Δx, we see that

$$\frac{p(z = H, x + \Delta x) - p(z = H, x)}{\Delta x} = \frac{g\,\Delta\rho\,\Delta h}{\Delta x} + \frac{p(z = H + h, x + \Delta x) - p(z = H + h, x)}{\Delta x}, \tag{2.5.286}$$

where $\rho_1 - \rho_2 = \Delta\rho$ and $\Delta h = h - 0$ (h at $x = 0$ is 0). Taking the limit of Eq. (2.5.286) as Δx approaches zero, we obtain the following:

$$\left(\frac{\partial p}{\partial x}\right)_{z=H} = g\,\Delta\rho\frac{\partial h}{\partial x} + \left(\frac{\partial p}{\partial x}\right)_{z=H+h}. \tag{2.5.287}$$

It also follows that

$$\left(\frac{\partial p}{\partial y}\right)_{z=H} = g\,\Delta\rho\,\frac{\partial h}{\partial y} + \left(\frac{\partial p}{\partial y}\right)_{z=H+h}. \tag{2.5.288}$$

At $z=H$, $p_1 \approx p_2$ for relatively small values of h, where p_1 and p_2 are the pressure just below and just above the interface. From the ideal gas law and the definition of potential temperature, it follows that

$$\frac{\Delta\rho}{\rho_1} \approx \frac{\Delta\theta}{\theta_2}, \tag{2.5.289}$$

where $\Delta\theta = \theta_1 - \theta_2$, and θ_1 and θ_2 are the mean potential temperatures in the layers below and above the interface, respectively.

From Eqs. (2.5.287), (2.5.288), (2.5.289), (2.5.281), and (2.5.282), we find that at $z=H$ the equations of motion are as follows:

$$fv = g\frac{\Delta\theta}{\theta_2}\frac{\partial h}{\partial x} + \frac{1}{\rho_1}\left(\frac{\partial p}{\partial x}\right)_{z=H+h} \tag{2.5.290}$$

$$\frac{\partial v}{\partial t} + u\frac{\partial v}{\partial x} + v\frac{\partial v}{\partial y} + w\frac{\partial v}{\partial z} + fu = -g\frac{\Delta\theta}{\theta_2}\frac{\partial h}{\partial y} - \frac{1}{\rho_1}\left(\frac{\partial p}{\partial y}\right)_{z=H+h}. \tag{2.5.291}$$

The pressure-gradient terms involving $\boldsymbol{\nabla}_z h$ are due to the horizontal perturbation pressure gradient, while the pressure-gradient terms involving $\boldsymbol{\nabla}_z p$ are due to the synoptic-scale pressure gradient. We will assume that the pressure-gradient force owing to *synoptic scale* variations in pressure at the top of the interface acts only in the y direction $[-(1/\rho_1)(\partial p/\partial y)_{z=H+h}]$ and is independent of x. Since $h = h(x,y,t)$ and since $g\,\Delta\theta/\theta_2$ is a constant, it follows from Eq. (2.5.290) and the absence of $\partial p/\partial x$ that

$$\frac{\partial v}{\partial z} = 0. \tag{2.5.292}$$

The equations of motion are therefore

$$fv = g\frac{\Delta\theta}{\theta_2}\frac{\partial h}{\partial x} \tag{2.5.293}$$

$$\frac{\partial v}{\partial t} + u\frac{\partial v}{\partial x} + v\frac{\partial v}{\partial y} + fu = -g\frac{\Delta\theta}{\theta_2}\frac{\partial h}{\partial y} - \frac{1}{\rho_1}\left(\frac{\partial p}{\partial y}\right)_{z=H+h}. \tag{2.5.294}$$

The equation of continuity is approximately

$$\frac{\partial u}{\partial x} + \frac{\partial v}{\partial y} + \frac{\partial w}{\partial z} = 0 \tag{2.5.295}$$

if the fractional variation of ρ with respect to height is small. Equation (2.5.295) is the Boussinesq continuity equation, which is equivalent to the equation of continuity for an incompressible fluid. From Eqs. (2.5.294) and (2.5.292), we see that u is independent of height. Integrating Eq. (2.5.295) from $z = 0$, where $w = 0$, to $z = H + h$, the height of the interface, we find that

$$w(z = H + h) = -\left(\frac{\partial u}{\partial x} + \frac{\partial v}{\partial y}\right)(H + h). \tag{2.5.296}$$

The left-hand side of Eq. (2.5.296) may be expanded as follows:

$$w(z - H + h) = \left.\frac{Dz}{Dt}\right|_{H+h} = \frac{DH}{Dt} + \frac{Dh}{Dt} = \frac{\partial h}{\partial t} + u\frac{\partial h}{\partial x} + v\frac{\partial h}{\partial y}. \tag{2.5.297}$$

From Eqs. (2.5.296) and (2.5.297) we find that

$$\frac{\partial h}{\partial t} + u\frac{\partial h}{\partial x} + v\frac{\partial h}{\partial y} + (H + h)\left(\frac{\partial u}{\partial x} + \frac{\partial v}{\partial y}\right) = 0. \tag{2.5.298}$$

Differentiating Eq. (2.5.293) with respect to y, we obtain

$$f\frac{\partial v}{\partial y} = g\frac{\Delta\theta}{\theta_2}\frac{\partial^2 h}{\partial y\,\partial x}, \tag{2.5.299}$$

where f is assumed to be a constant. Differentiating Eq. (2.5.294) with respect to x, and recalling that $(1/\rho_1)(\partial p/\partial y)_{z=H+h}$ is independent of x, we find that

$$\frac{\partial}{\partial t}\left(\frac{\partial v}{\partial x}\right) + \frac{\partial u}{\partial x}\frac{\partial v}{\partial x} + \frac{\partial v}{\partial x}\frac{\partial v}{\partial y} + u\frac{\partial}{\partial x}\left(\frac{\partial v}{\partial x}\right) + v\frac{\partial}{\partial y}\left(\frac{\partial v}{\partial x}\right)$$
$$+ f\frac{\partial u}{\partial x} = -g\frac{\Delta\theta}{\theta_2}\frac{\partial^2 h}{\partial x\,\partial y}. \tag{2.5.300}$$

Adding Eq. (2.5.299) to Eq. (2.5.300), we obtain

$$\frac{\partial}{\partial t}\left(\frac{\partial v}{\partial x}\right) + u\frac{\partial}{\partial x}\left(\frac{\partial v}{\partial x}\right) + v\frac{\partial}{\partial y}\left(\frac{\partial v}{\partial x}\right) + \left(\frac{\partial u}{\partial x} + \frac{\partial v}{\partial y}\right)\left(\frac{\partial v}{\partial x} + f\right) = 0. \tag{2.5.301}$$

Eliminating $\partial u/\partial x + \partial v/\partial y$ from Eq. (2.5.301) by using Eq. (2.5.298), we find that

$$\frac{\partial}{\partial t}\left[\left(f + \frac{\partial v}{\partial x}\right)\Big/(H + h)\right] + u\frac{\partial}{\partial x}\left[\left(f + \frac{\partial v}{\partial x}\right)\Big/(H + h)\right]$$
$$+ v\frac{\partial}{\partial y}\left[\left(f + \frac{\partial v}{\partial x}\right]\Big/(H + h)\right] = 0. \tag{2.5.302}$$

In other words, the quantity $(f + \partial v/\partial x)/(H + h)$, which is a form of potential vorticity, is conserved for horizontal motions.

If initially $v = 0$ and $h = 0$, then

$$\frac{f}{H} = \left(f + \frac{\partial v}{\partial x}\right)\Big/(H + h), \tag{2.5.303}$$

so that

$$fh = H\frac{\partial v}{\partial x}. \tag{2.5.304}$$

From Eq. (2.5.293) and the above, we get the following two differential equations for v and h:

$$\frac{\partial^2 v}{\partial x^2} - \left[f^2 \Big/ \left(g \frac{\Delta\theta}{\theta_2} H\right)\right] v = 0 \tag{2.5.305}$$

$$\frac{\partial^2 h}{\partial x^2} - \left[f^2 \Big/ \left(g \frac{\Delta\theta}{\theta_2} H\right)\right] h = 0. \tag{2.5.306}$$

The solutions to Eqs. (2.5.305) and (2.5.306) in the Northern Hemisphere[8] are as follows:

$$v = v_0 \exp\left[-\left(\frac{f}{[g(\Delta\theta/\theta_2)H]^{1/2}} x \right)\right] \tag{2.5.307}$$

$$h = h_0 \exp\left[-\left(\frac{f}{[g(\Delta\theta/\theta_2)H]^{1/2}} x \right)\right]. \tag{2.5.308}$$

According to Eqs. (2.5.307) and (2.5.308), the component of the wind along the wall that is *not* geostrophic, and the vertical displacement of the interface, are greatest at the wall, and decay to a value of $1/e$ times their value at the wall (v_0 and h_0) at a distance of $[g(\Delta\theta/\theta_2)H]^{1/2}/f$ from the wall. Since $[g(\Delta\theta/\theta_2)H]^{1/2}$ is the speed of a gravity wave in a two-layer fluid (see Holton, 1979, p. 158), $[g(\Delta\theta/\theta_2)H]^{1/2}(1/f)$ may be interpreted as the distance a gravity wave travels in a time of $1/f$: This is called the *Rossby radius of deformation*. Geostrophic balance is not possible within a Rossby radius of deformation of the wall (i.e., the mountains). (The Rossby radius of deformation also appears in the analysis of the **Q**-vector formulation of the quasigeostrophic ω equation in Volume I.) In middle latitudes, for $\Delta\theta = 10$ K and $H = 1$ km, the Rossby radius of deformation is around 200 km; hence significant perturbations in the flow along the wall would be present only within a few hundred kilometers of the wall. The Rossby radius of deformation in this case is therefore mesoscale, not synoptic scale.

From Eqs. (2.5.307), (2.5.308), and (2.5.304) we find that

$$h_0 = -v_0 \frac{H}{[g(\Delta\theta/\theta_2)H]^{1/2}} = v_0 \frac{H}{c_{\text{grav}}}. \tag{2.5.309}$$

In other words, the vertical displacement of the interface at the wall is the distance an air parcel moves along the wall (at the speed v_0) in the time it takes a gravity wave to move, at a speed $c_{\text{grav}} = -[g(\Delta\theta/\theta_2)H]^{1/2}$, a distance equivalent to the mean depth of the dense air (H).

At the wall $u = 0$ (i.e., $u_0 = 0$), and so Eq. (2.5.294) evaluated at $x = 0$, with the aid of Eq. (2.5.309), may be expressed as

$$\frac{\partial v_0}{\partial t} + \left[v_0 - \left(g \frac{\Delta\theta}{\theta_2} H\right)^{1/2} \right] \frac{\partial v_0}{\partial y} = -\frac{1}{\rho_1}\frac{\partial p}{\partial y} (z = H + h,\ x = 0). \tag{2.5.310}$$

This is the y component of the frictionless equation of motion at the wall. Let us include the dissipative effects of turbulent friction near the wall by adding the Rayleigh parameterization of friction to Eq. (2.5.310), so that

$$\frac{\partial v_0}{\partial t} + \left[v_0 - \left(g \frac{\Delta\theta}{\theta_2} H \right)^{1/2} \right] \frac{\partial v_0}{\partial y} + K v_0 = -\frac{1}{\rho_1} \frac{\partial p}{\partial y} (z = H + h, x = 0). \quad (2.5.311)$$

This equation relates the component of wind along the wall (at the wall) to its time rate of change, its along-the-wall derivative, and forcing from the synoptic-scale along-the-wall pressure gradient.

Suppose that

$$p(z = H + h, x = 0, y) = \begin{cases} -\hat{p}\left(\sin \frac{2\pi}{L} y \sin \frac{2\pi}{L} (y - ct) \right), & \text{for } 0 > y > -\frac{L}{2} \quad (2.5.312a) \\ 0, & \text{for other values of } y, \quad (2.5.312b) \end{cases}$$

where

$$c = -500 \text{ km d}^{-1} \sim -6 \text{ m s}^{-1} \quad (2.5.313)$$

$$L = \pi(1000 \text{ km}). \quad (2.5.314)$$

This "forcing" corresponds to that of a typical synoptic-scale system, which in the Northern Hemisphere propagates in the $-y$ direction, with higher terrain to the right. Then it follows that

$$-\frac{1}{\rho_1} \frac{\partial p}{\partial y} (z = H + h, x = 0) = \begin{cases} \frac{\hat{p}}{\rho_1} \frac{2\pi}{L} \sin \frac{2\pi}{L} (2y - ct), & \text{for } 0 > y > -\frac{L}{2} \quad (2.5.315a) \\ 0, & \text{for other values } y. \quad (2.5.315b) \end{cases}$$

Let the amplitude of the pressure variation $\hat{p}$ be about 3 mb. If it were larger, then analytic solutions to Eq. (2.5.311) would not be possible. Thus, the model is somewhat limited, and we must proceed with caution.

Suppose that $K = 1 \text{ d}^{-1}$. This corresponds to an e-folding dissipation time of 1 d.

The qualitative relationship between h_0 and $p(z = H + h_0, x = 0)$, and $p(z = 0, x = 0)$ as a function of time at $y = -L/4$ and $y = -3/8\, L$ is seen in Fig. 2.78. The surface-pressure (dashed lines) trough and the interface-height (solid lines) depresssion occur nearly simultaneously; the surface pressure ridge lags the interface-height crest. The pressure trough at the height of the interface (dotted lines) lags the interface-height depression. The depressed part of the interface-height wave and the surface-pressure trough sharpen up downstream because h_0 and v_0 have opposite signs [cf. Eq. (2.5.309)]; in the vicinity of the interface-height crests (depressions) the air moves in the same (opposite) direction as the gravity wave. The differential motion between crests and depressions results in the sharpening of the wave.

It is not clear how often and under what circumstances significant orographically trapped gravity waves actually occur in the atmosphere. Gravity waves *propagate,* and do not necessarily move cold, surface air along as a

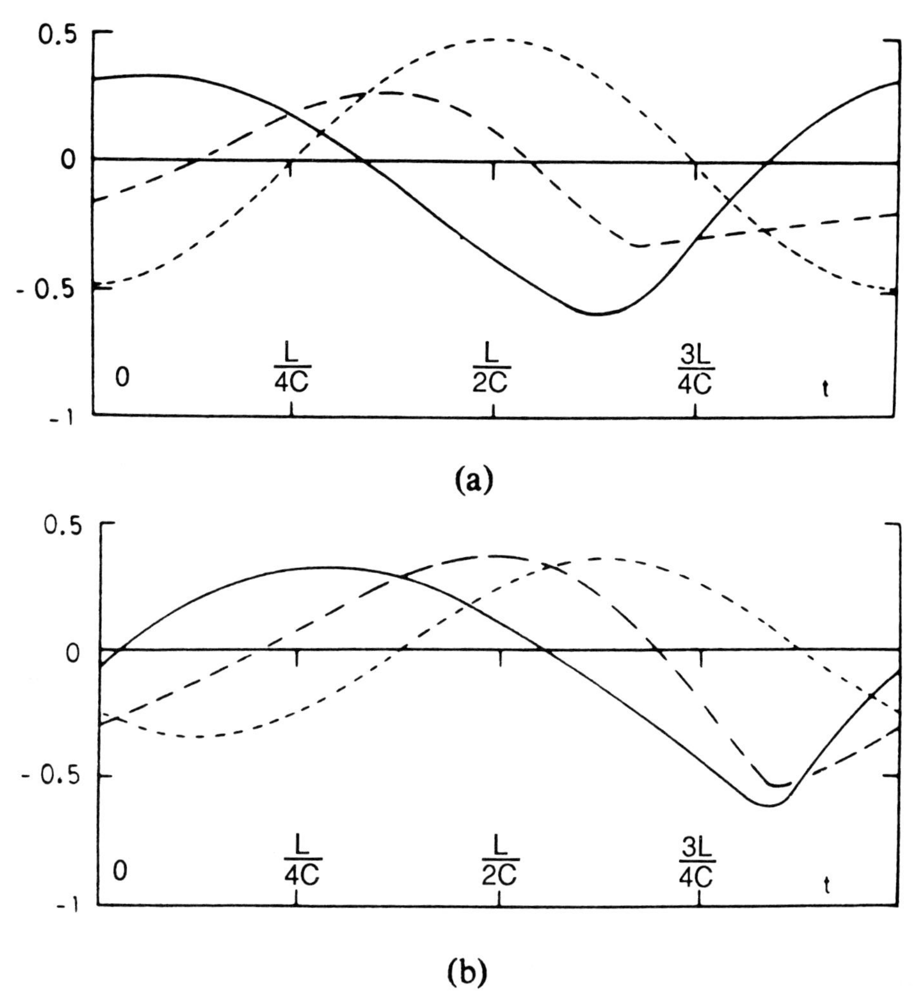

Figure 2.78 Relationship between the disturbance height and the pressure at the interface of the wall boundary ($x = 0$) as a function of time at (a) $y = -L/4$ and (b) farther downstream at $y = -3L/8$. h_0/H (solid line); $[p(z = H + h_0) - \bar{p}(z = H + h_0)]/\rho_1[g(\Delta\rho/\rho_1)H)$ (dotted line); $[p(z = 0) - \bar{p}(z = 0)]/\rho_1[g(\Delta\rho/\rho_1)H]$ (dashed line) (adapted from Gill, 1977).

material boundary. Furthermore, they can exist only if there is a stable layer, a condition that may not necessarily exist *ahead* of a shallow cold front.

2.6. DYNAMICS OF MIDDLE–UPPER TROPOSPHERIC FRONTOGENESIS

Unlike surface frontogenesis, in which vertical motions do not play a direct role (they do play an indirect role through δ) owing to the kinematic lower boundary condition ($\omega = 0$ at the ground, if it is level), middle–upper tropospheric frontogenesis is influenced directly by vertical motion through the tilting term $(\partial\omega/\partial y)(\partial\theta/\partial p)$ in Eq. (2.3.3). The quasihorizontal gradient of the effects of vertical motion on temperature are significant at locations where both vertical motion and static stability are relatively large. The magnitude of vertical motion is usually largest in the middle troposphere, while static stability in general increases with height. The combined effects of vertical motion and static stability are therefore greatest in the middle-to-upper troposphere.

Vertical motions are suppressed at the tropopause in part because static stability is high, and in part because the thermal wind reverses direction. The tropopause therefore acts like an upper boundary, though not a rigid one, to the middle–upper tropospheric front. The tropopause plays to some extent a role in middle–upper tropospheric frontogenesis that is similar to the role the ground plays in surface frontogenesis.

2.6.1. Quasigeostrophic Middle–Upper Tropospheric Frontogenesis

We found in Section 2.4.2 that the middle–upper tropospheric front usually forms in the absence of clouds, in a region of subsidence. Diabatic heating caused by latent-heat release or radiative processes can therefore be ignored.

According to Eq. (2.3.3) and quasigeostrophic theory, the frontogenetical function

$$F = \frac{\partial v_g}{\partial y}\frac{\partial \theta}{\partial y} + \frac{\partial \omega}{\partial y}\frac{\partial \theta}{\partial p}. \tag{2.6.1}$$

Let us for the moment ignore the effect of temperature advection, which is usually relatively small in the middle troposphere, but may be larger near jets, where the tropopause slopes.

Observations indicate that quasihorizontal variations in subsidence (e.g., $\partial\omega/\partial y < 0$) superimposed upon a temperature gradient (e.g., $\partial\theta/\partial y < 0$) in a statically stable atmosphere ($\partial\theta/\partial p < 0$) can initiate middle–upper tropospheric frontogenesis. The quasihorizontal variation in subsidence is due quasigeostrophically to a quasihorizontal variation in vorticity advection becoming more anticyclonic with "height" on the cyclonic-shear side of the flow upstream from a trough or the anticyclonic-shear side of the flow downstream from a ridge (Fig. 2.79). If the static stability and quasihorizontal variation in subsidence remain constant, then frontogenesis occurs at a relatively slow rate according to Eq. (2.6.1), which is as follows:

$$\frac{D}{Dt}\left(-\frac{\partial \theta}{\partial y}\right) = \frac{\partial \omega}{\partial y}\frac{\partial \theta}{\partial p}. \tag{2.6.2}$$

Maximum frontogenesis occurs on the cyclonic-shear side of the flow upstream from the trough and downstream from the ridge.

For typical values of vertical motion and static stability,

$$\frac{D}{Dt}\left(-\frac{\partial \theta}{\partial y}\right) \sim \left(\frac{10^{-4}\ \mathrm{kPa\,s^{-1}}}{10^{6}\ \mathrm{m}}\right)\left(\frac{20\ \mathrm{K}}{40\ \mathrm{kPa}}\right)$$
$$\sim 10^{-11}\text{–}10^{-10}(\mathrm{K\ m^{-1}})\mathrm{s^{-1}}. \tag{2.6.3}$$

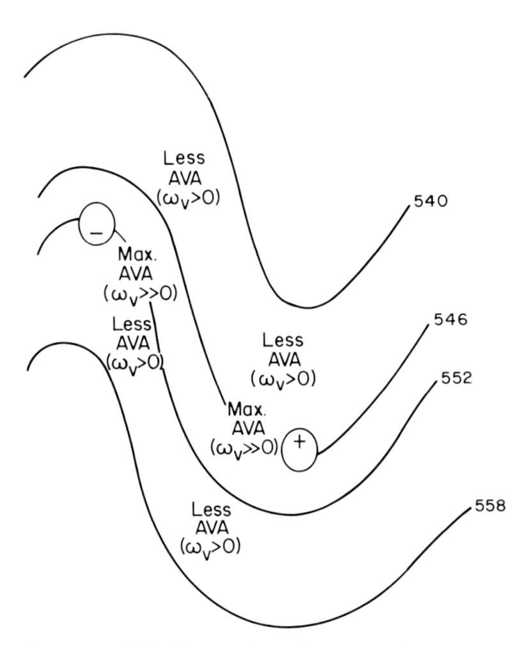

Figure 2.79 Quasigeostrophically induced vertical motion associated with differential vorticity advection (ω_v) in a wavetrain in the middle troposphere in the baroclinic westerlies. 500-mb height in dam (solid lines); regions of maximum and minimum absolute vorticity denoted by + and −, respectively. Maxima in vorticity advection are located just upstream and downstream from troughs and ridges, along the zone of maximum wind current. Strong gradients in vertical motion exist normal to the current, on either side of the strongest current.

The typical temperature gradient of 10 K/1000 km $\sim 10^{-5}$ K m^{-1} only doubles in a time interval of approximately one day ($\sim 10^5$ s) or more. Middle–upper tropospheric frontogenesis usually proceeds more rapidly, and hence a linearized, quasigeostrophic version of Eq. (2.6.2) can explain only part of the frontogenetical process.

Suppose that temperature advection (the effects of $\partial v_g/\partial y$ on $\partial\theta/\partial y$) cannot be ignored. The frontogenetical effect of geostrophic confluence on the temperature gradient is the same as it for surface frontogenesis, Eq. (2.5.2): Frontogenesis through this mechanism is also too slow to explain the observed rates of frontogenesis.

2.6.2. The Geostrophic-Momentum Approximation and Semigeostrophic Middle–Upper Tropospheric Frontogenesis

We saw earlier that quasigeostrophic surface frontogenesis proceeds at a rate much slower than that observed; however, the dynamical equations subject to the geostrophic-momentum approximation can explain rapid surface frontogenesis. It should not be surprising, then, that while quasigeostrophic middle–upper tropospheric frontogenesis proceeds at a rate much slower than that observed, the dynamical equations subject to the geostrophic-momentum approximation can also explain rapid middle–upper tropospheric frontogenesis.

Analysis using the quasigeostrophic ω equation. Consider the effect of the quasihorizontal variation in subsidence associated with quasigeostrophic middle–upper tropospheric frontogenesis. Let us first ignore the effects of temperature advection, if any. The strongest increase with height of anticyclonic vorticity advection, and the strongest subsidence, occur beneath the maximum current, downstream from an upper-level ridge (Fig. 2.79). According to the geostrophic-momentum approximation form of the vorticity equation, Eq. (2.5.67), quasihorizontal variations in vertical velocity act upon vertical shear to change geostrophic vorticity via the tilting effect. Thus, the cross-stream variation in subsidence downstream from an upper-level ridge and upstream from an upper-level trough acts to make the shear vorticity more cyclonic on the cyclonic-shear side of the current and more anticyclonic on the anticyclonic-shear side of the current (Fig. 2.80). Thus, the cross-stream vorticity gradient is increased and a jet, a core of relative maximum in geostrophic wind speed, forms, or if already present, strengthens.

The direct result of this change in vorticity pattern is an increase in the quasihorizontal gradient of anticyclonic vorticity advection aloft and a concomitant *increase* in the gradient of change of anticyclonic vorticity advection with height. It follows, then, according to quasigeostrophic theory, that the quasihorizontal gradient of subsidence is increased in magnitude. In other words, the additional quasihorizontal variation in subsidence and its associated quasihorizontal variation in temperature (via tilting) on the cyclonic-shear side act to produce frontogenesis more rapidly: Mudrick in 1974 first suggested that this positive feedback mechanism (Fig. 2.81) may be responsible for rapid middle–upper tropospheric frontogenesis. Tilting of vorticity in middle–upper tropospheric frontogenesis therefore plays a role similar to that played by convergence (confluence of the ageostrophic wind) acting on the temperature field in surface frontogenesis, since both act to increase the rate of frontogenesis beyond the quasigeostrophic limit and both are due to the secondary circulation.

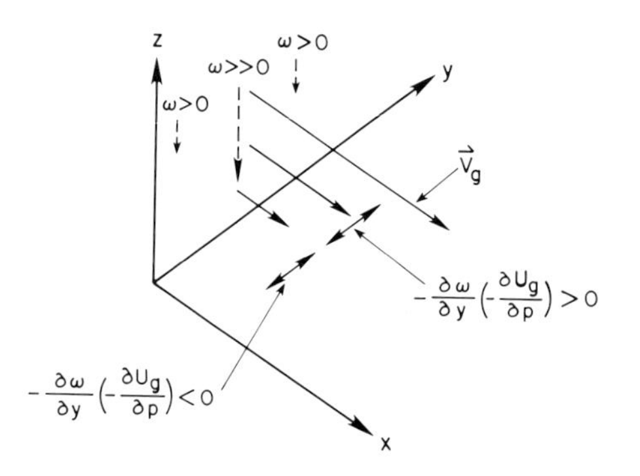

Figure 2.80 The effects of the tilting of vertical shear onto the horizontal as a result of a cross-stream variation in vertical motion. Vertical motion ω (dashed arrows); geostrophic motion $\mathbf{v}_g$ (solid vectors); x axis is aligned along the flow and vertical-shear vector; regions of cyclonic and anticyclonic vorticity generation on the cyclonic and anticyclonic shear sides of the jet as indicated.

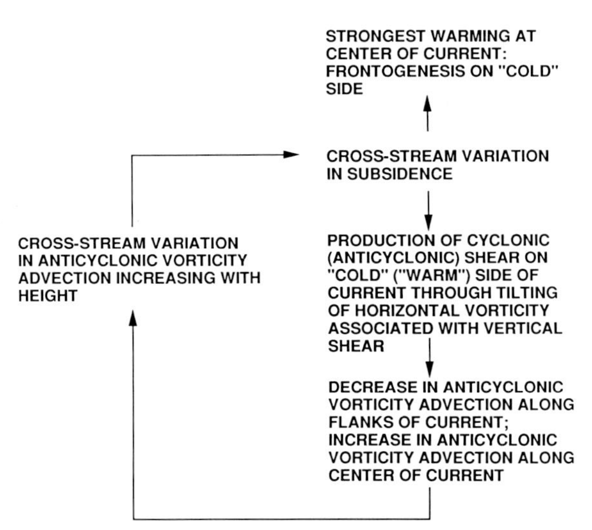

Figure 2.81 Positive feedback mechanism by which a horizontal variation in subsidence is linked to the tilting of horizontal vorticity onto the vertical.

A positive feedback mechanism is not possible downstream from an upper-level trough and upstream from an upper-level ridge. In this case, tilting makes the shear vorticity less cyclonic on the cyclonic-shear side of the current and less anticyclonic on the anticyclonic-shear side of the current. Consequently, the jet becomes less well defined, and hence the quasihorizontal gradient of vorticity advection also decreases. We thus expect middle–upper tropospheric fronts to form in an environment of subsidence upstream from a trough (downstream from a ridge), not in an environment of rising motion, downstream from a trough (upstream from a ridge).

Let us now consider the effects of geostrophic temperature advection. In an amplifying baroclinic wave the temperature field lags the height field: There is cold advection upstream from the trough (Fig. 2.82). According to the quasigeostrophic ω equation, then, there is subsidence upstream from the trough (Fig. 2.83). The subsidence is greatest along the region where the geostrophic current is strongest. The thermodynamic effects of the cross-stream variation in subsidence is frontogenetical on the cyclonic-shear side of the current. Tilting of vertical shear onto the horizontal further increases the quasihorizontal gradient of subsidence (Fig. 2.80), as in the case of vorticity advection.

A positive feedback mechanism does not exist downstream from a trough in an amplifying baroclinic wave, owing to warm advection. Although the cross-stream variation in quasigeostrophically induced rising motion, owing to a maximum in warm advection within the region of strongest geostrophic current, produces a frontogenetical effect on the anticyclonic-shear side, the tilting of vertical shear acts to decrease the quasihorizontal gradient of rising motion.

The combined effects of vorticity advection and temperature advection

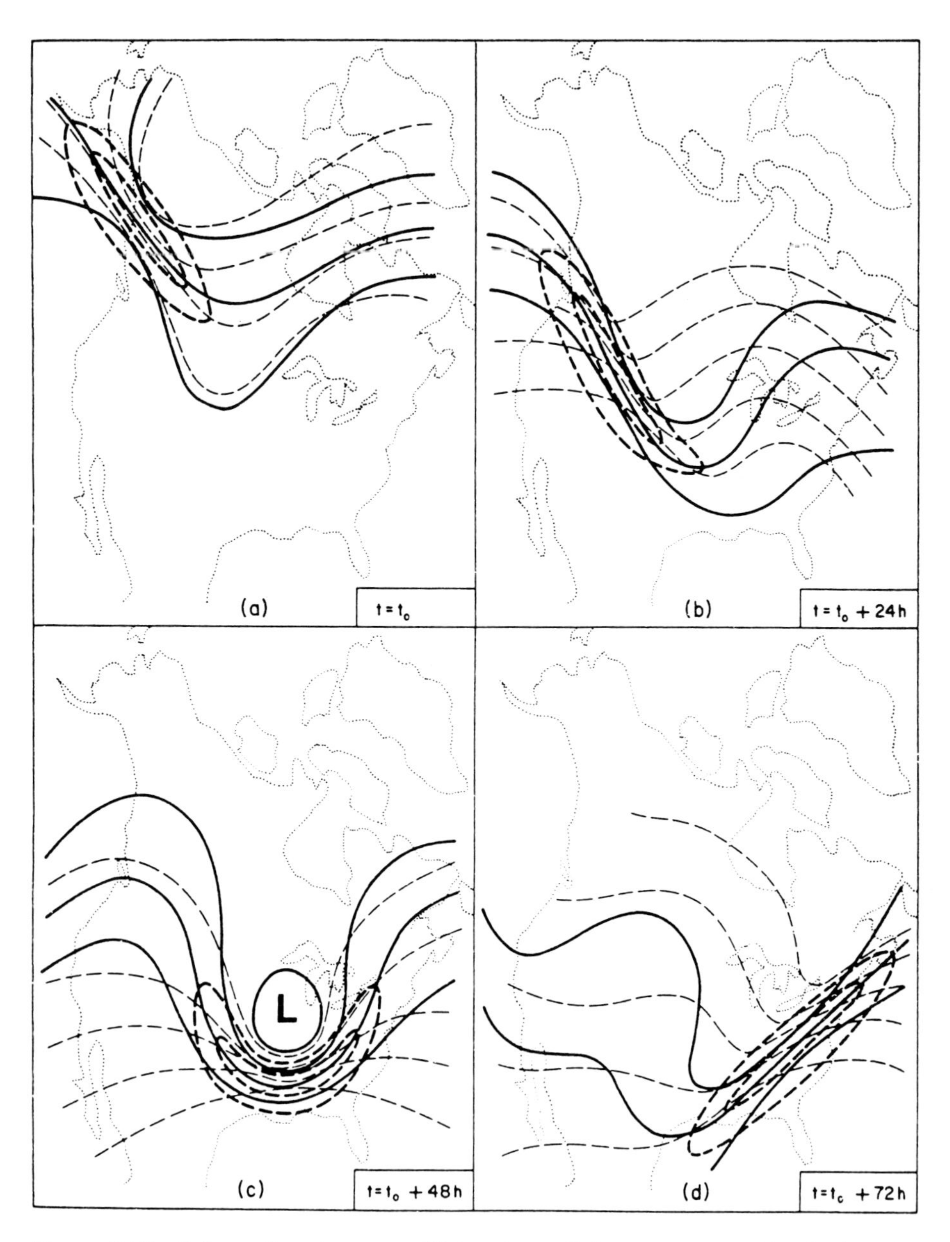

Figure 2.82 Schematic of an amplifying baroclinic wave over North America. Midtropospheric height contours, decreasing toward the north (solid lines), and isotherms, decreasing toward the north (thin dashed lines), as a function of time. In (a) a jet streak (isotach maximum within a stream of strong wind; isotachs indicated by thick dashed lines) enters the region upstream from a trough; (b) the jet streak interacts with the trough, and there is warm advection and cold advection downstream and upstream from the trough, respectively; the trough is negatively tilted and diffluent; (c) the jet streak is in phase with the trough; (d) the jet streak is downstream from the trough; the trough is positively tilted and confluent (cf. also Fig. 1.133; from Shapiro, 1983).

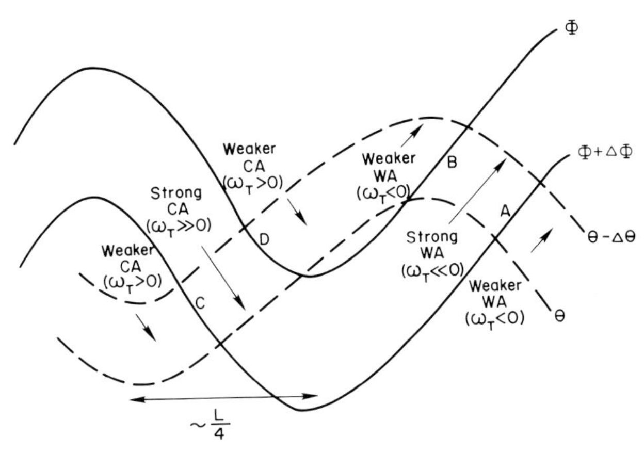

Figure 2.83 Patterns of quasigeostrophically induced vertical motion associated with temperature advection (ω_T) in a baroclinic wave in the westerlies in which there is cold advection (CA) upstream from troughs and warm advection (WA) downstream from troughs. Geopotential height in the middle troposphere (solid lines); potential temperature (dashed lines). A, B, C, and D are regions of anticyclonic shear and warm advection, cyclonic shear and warm advection, anticyclonic shear and cold advection, and cyclonic shear and cold advection, respectively. The wave in the temperature field lags the wave in the geopotential-height field by approximately one-quarter of a wavelength ($L/4$).

therefore act to produce middle–upper tropospheric frontogenesis upstream from baroclinically unstable waves.

Analysis using the Sawyer–Eliassen equation. Shapiro has used the Sawyer–Eliassen equation to diagnose middle–upper tropospheric frontogenesis. This type of analysis takes into account explicitly the action of geostrophic deformation on the temperature field; however, it does not take into account the effects of curvature, which are important in an analysis of vorticity advection; this analysis is therefore basically a two-dimensional one. The Sawyer–Eliassen equation itself is not valid however, if a trough or ridge is so sharply curved that the geostrophic momentum approximation fails.[9] In this case we must be very cautious in interpreting results based upon the Sawyer–Eliassen equation. Recall that the adiabatic forcing function in the Sawyer–Eliassen equation is proportional to the y component (cross-front) of the **Q**-vector,

$$\frac{\partial u_g}{\partial y}\frac{\partial T}{\partial x}+\frac{\partial v_g}{\partial y}\frac{\partial T}{\partial y}.$$

Consider again an amplifying baroclinic wavetrain, in which the temperature wave lags the geopotential-height wave, so that there is cold advection upstream from troughs (Fig. 2.83). If the wavelike current is concentrated along a channel, then the effect of horizontal variations in cold advection, that is, the shear term in Eq. (2.5.53), $(\partial\theta/\partial x)(\partial u_g/\partial y$ [i.e., in the $(\partial T/\partial x)(\partial u_g/\partial y)$ part of the **Q**-vector], may be significant upstream from troughs. (The reader is challenged to consider what happens if the current is *not* concentrated.) In this case (as shown by Keyser and Pecnick using data from numerical simulations), when confluence acts to increase the temperature gradient [e.g., as in Fig. 2.82(a), (b), and (c) upstream from the trough and jet streak], the secondary (vertical) circulation is shifted toward the warm side (as was first suggested by M. Shapiro). The reason for this is that geostrophic shear acts frontolytically and frontogenetically on the cold and warm sides, respectively (Fig. 2.83): Hence the strength of the thermally direct circulation (forced by the confluence) is weakened on the cold side and strengthened on the warm side. Thus, the subsiding branch of the direct circulation shifts over toward the warm side (Fig. 2.84), so that not only is $(\partial v_g/\partial y)(\partial\theta/\partial y)$ positive now, but $(\partial\omega/\partial y)(\partial\theta/\partial p)$ is also positive *locally* [recall Eq. (2.6.1), that $(\partial\omega/\partial y)/(\partial\theta/\partial p) > 0$ contributes to frontogenesis], even though it is negative when viewed from the perspective of the entire wavetrain. The rising branch of the vertical circulation has thus shifted to the anticyclonic-shear side of the current (where it may produce a band of cirrus clouds). The frontogenetical effect of confluence therefore acts to increase the thermal-wind shear, and hence increases or maintains the intensity of the middle–upper tropospheric

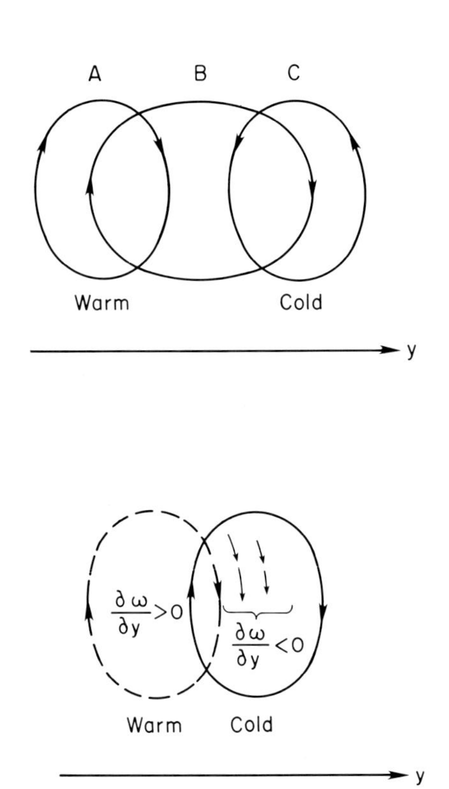

Figure 2.84 Idealized vertical cross sections through a developing middle–upper tropospheric front, upstream from a trough, with the warm, anticylonic-shear side to the left, and the cold, cyclonic-shear side to the right; cold advection is occurring upstream from the trough. The y axis points normal to the geostrophic flow (which is out from the plane of the figure), from the warm side toward the cold side. (top) Thermally direct circulation (B) forced by confluence tightening up the cross-stream temperature gradient; thermally direct circulation (A) forced by anticylonic shear increasing the cross-stream temperature gradient; thermally indirect circulation (C) forced by cyclonic shear decreasing the cross-stream temperature gradient. (bottom) The vertical circulation for confluence only, as B at the top (solid streamline); the vertical circulation for the combined effects of confluence and shear, as A + B + C (dashed streamline). Note that the frontogenetically acting meridional gradient of vertical motion ($\partial\omega/\partial y < 0$) has been shifted to the left, toward the warm side.

current, while *locally* the outer edge of the sinking part of the thermally direct circulation created by the quasihorizontal shear acts frontogenetically.

The ageostrophic vertical circulation associated with the middle–upper tropospheric front can affect the geostrophic forcing function represented by the **Q** vector $[(\partial u_g/\partial y)(\partial T/\partial x) + (\partial v_g/\partial y)(\partial T/\partial y)]$. Although $\partial v_g/\partial y$ is held fixed, $\partial u_g/\partial y$ can be changed. For example, using Eqs. (2.5.44) and (2.5.48), we differentiate Eq. (2.5.43) with respect to y and find that

$$\frac{D}{Dt}\left(\frac{\partial u_g}{\partial y}\right) = \frac{\partial v_a}{\partial y}\left(f_0 - \frac{\partial u_g}{\partial y}\right) - \frac{\partial \omega}{\partial y}\frac{\partial u_g}{\partial p}. \tag{2.6.4}$$

(This is essentially a vorticity equation in which $\partial v_g/\partial x = 0$.) From Fig. 2.84 we see that $\partial\omega/\partial y < 0$, owing to the frontolytical effects of cyclonic shear $(-\partial u_g/\partial y > 0)$ on the "cold" side of the jet. Owing to continuity, $\partial v_a/\partial y = 0$ at the level where $\partial\omega/\partial p = 0$, which is near the level at which $|\partial\omega/\partial y|$ is greatest; $-\partial u_g/\partial p > 0$, and hence $(D/Dt)(\partial u_g/\partial y) < 0$. In other words, the geostrophic shear vorticity $(-\partial u_g/\partial y)$ is made more cyclonic. This in turn results in a further shift of the vertical circulation toward the warmer side, and there is an increase in the frontogenetical horizontal gradient in subsidence: This represents a positive feedback process. (The frontal zone is advected downward, owing to the subsidence.) The feedback process continues until the along-the-front temperature gradient is reduced so much that the shear effect is negligible.

On the other hand, downstream from a trough in a decaying (the temperature field leads the geopotential field, Fig. 2.82) wavetrain having a jetlike current, there is warm advection. In this case the thermally direct vertical circulation, owing to the effect of confluence on the temperature field, is shifted toward the cold, cyclonic-shear side of the jet, as a result of shear-induced frontolysis on the anticyclonic side (Fig. 2.83). (The upper tropospheric front may be connected to the surface front in this case if it has moved around the base of the trough and caught up with the surface front.) The region of maximum upward motion associated with the thermally direct ageostrophic circulation is located nearer to the "cold" air, so that $\partial\omega/\partial y < 0$ on the anticyclonic-shear $(-\partial u_g/\partial y < 0)$ side of the jet. Thus, the vertical circulation acts frontogenetically *locally* $[(\partial\omega/\partial y)(\partial\theta/\partial p) > 0]$. However, according to Eq. (2.6.4), $(D/Dt)(\partial u_g/\partial y) < 0$ on the anticyclonic-shear side. In other words, the geostrophic shear $(-\partial u_g/\partial y)$ vorticity is made more cyclonic. The frontolytical effects of geostrophic shear on the anticyclonic-shear side of the jet consequently are mitigated: A feedback mechanism leading to an increase in horizontal gradient in upward motion is therefore *not* present. The effects of tilting $(\partial\omega/\partial y)$, which were important in the "cold-advection" case, are not important in the "warm-advection" case.

It is emphasized that the aforementioned analyses using the quasigeostrophic ω equation and the Sawyer-Eliassen equation are theories, yet (at the time of this writing) to be confirmed by observations.

2.7. OBSERVATIONAL ASPECTS OF JETS AND JET STREAKS

A jet (from the Latin "to throw") is a region of relatively strong wind that is concentrated into a narrow, quasihorizontal or horizontal stream. The criteria that satisfy the conditions "relatively strong" and "narrow" are subjective.

The three most common types of jets found in midlatitudes are as follows:

1. The Polar-Front jet;
2. The subtropical jet; and
3. The low-level jet.

The term *jet stream* has been used to describe the zonally averaged and time-averaged maximum in the zonal component of the wind in the upper troposphere of the midlatitudes (Fig. 2.85). It has also been used to describe localized, long, narrow regions of high wind speed in the upper troposphere of midlatitudes at any given time.

Just as it is curious that the temperature field has regions of concentrated gradients, frontal zones, it is also fascinating that the wind field has regions of concentrated wind. The discovery of the jet stream during the 1940s at the "Chicago School" by a combination of researchers and forecasters was perhaps as important as the "discovery" of the surface front over 25 years earlier at the Bergen school. The advent of the rawinsonde network, the increased use of high-altitude aircraft, and an effort to investigate the synoptic conditions accompanying severe-storm outbreaks all provided the impetus for the study of jets.

2.7.1. The Polar-Front Jet

The Polar-Front jet is usually found within 50 mb of 250 mb, that is, near the tropopause (Figs. 2.86 and 2.87). (In Section 2.4.2 we noted that the tropopause is located at the level where one or more of the following occur: static stability increases dramatically with height, the wind speed is a maximum, and Ertel's potential vorticity increases to high values.) It is associated with strong quasihorizontal (or horizontal) temperature gradients at low levels and strong vertical wind shear, much of which is associated with an enhanced horizontal temperature gradient according to the thermal-wind relation. The zone of enhanced temperature gradient and vertical shear is often associated with the "Polar" front, the name given to cold fronts that trail cyclones in Polar-Front theory.

Winds in the current of the Polar-Front jet, which are usually from a westerly, southwesterly, or northwesterly direction, may be as high as 75 m s^{-1} or greater in speed. The public's perception of the danger of aircraft reconnaissance in hurricanes having wind speeds of 32 m s^{-1} or greater must be tempered by the realization that commercial aircraft routinely fly safely through the Polar-Front jet! In addition, the public's notion that the meandering of the jet stream is what is responsible for cold and warm

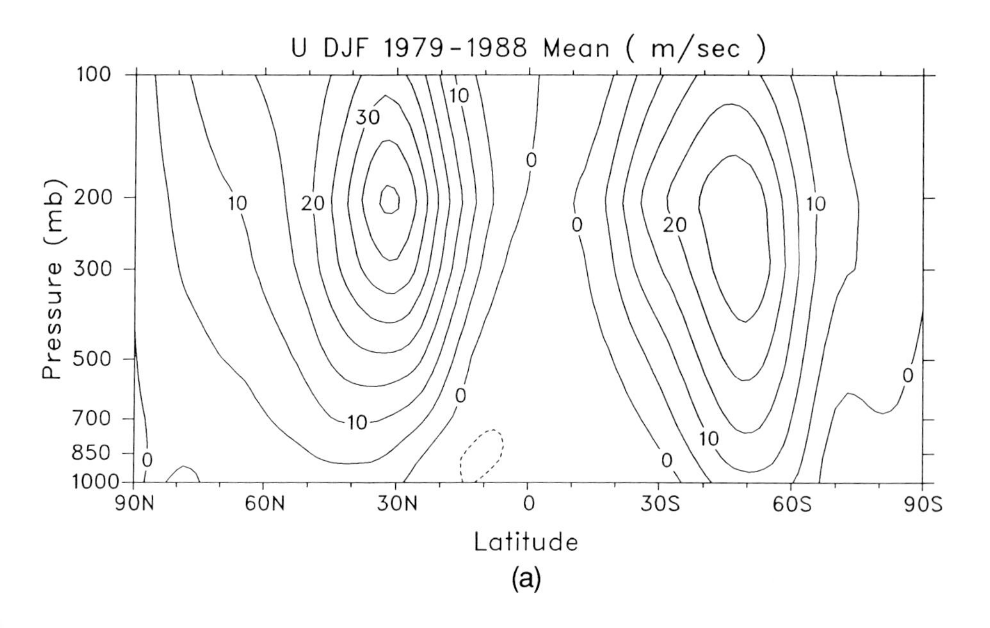

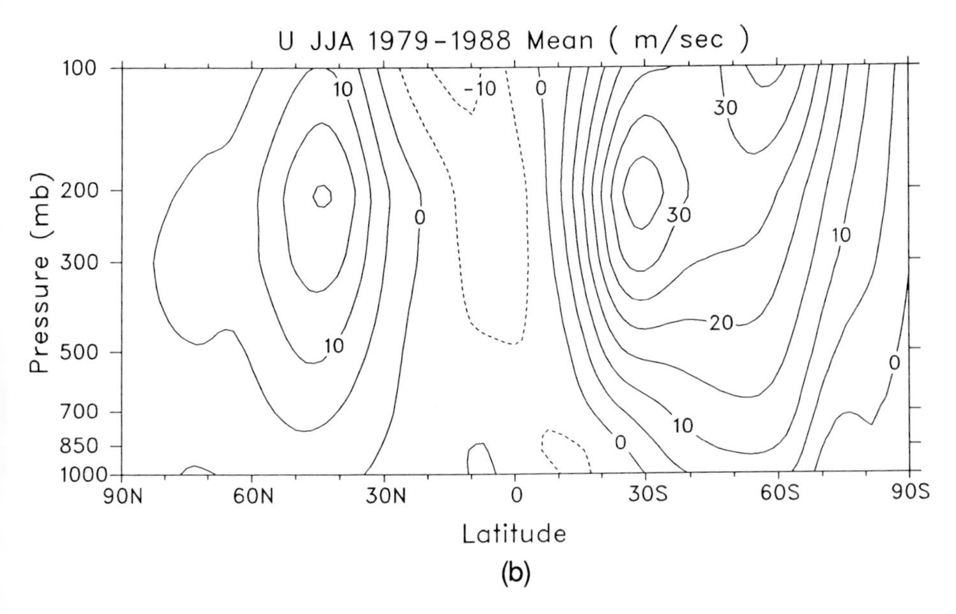

Figure 2.85 Vertical cross-sectional analysis of the average zonal component of the wind in m s^{-1} for (a) the Northern-Hemisphere winter months (Southern Hemisphere summer months) (December, January, and February) and (b) the Northern-Hemisphere summer months (Southern Hemisphere winter months (June, July, and August) (from ECMWF data, 1979–1988; courtesy Kevin Trenberth and Amy Solomon, NCAR).

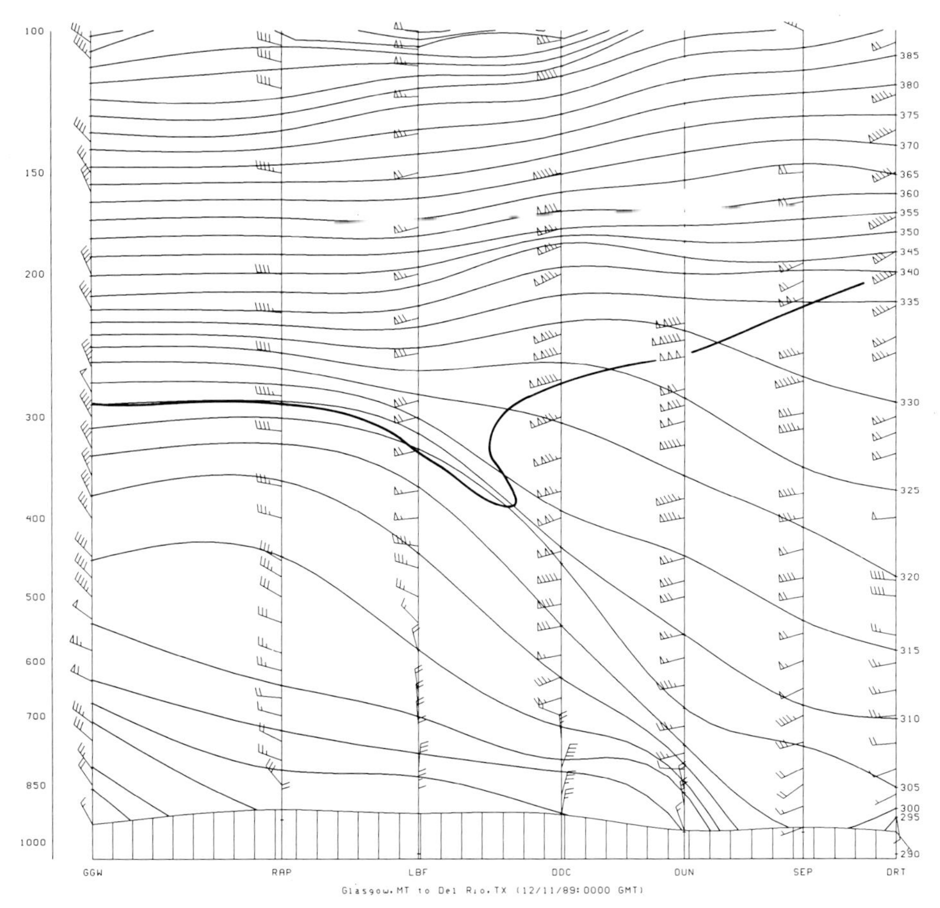

Figure 2.86 Example of the structure of the polar-front jet. Vertical cross section of potential temperature in K (solid lines) and wind (pennants, whole barbs, and half barbs represent 25, 5, and 2.5 m s^{-1}, respectively) for December 11, 1989, from Glasgow, Montana (GGW) southward to Del Rio, Texas (DRT). A surface front is located south of Norman, Oklahoma (OUN); the frontal zone slopes northward, beneath which the winds are northerly, and above which the winds are west-southwesterly. Highest wind speeds associated with the polar-front jet are found near 250 mb, well above the surface frontal zone; highest wind speeds are found at Norman (OUN) (70–75 m s^{-1}), and at Dodge City, Kansas (DDC) (70 m s^{-1}). The tropopause (thick solid line) south of the polar-front jet is around 200 mb; north of the polar-front jet it is lower, at approximately 300 mb (courtesy Tim Hughes, School of Meteorology, University of Oklahoma).

anomalies at the surface is not entirely true: The equatorward and poleward migration of cold and warm air at the surface is to some extent itself responsible for the meandering of the jet stream through the thermal-wind relation.

The height of the tropopause changes across the Polar-Front jet (Fig. 2.88). Consider the thermal-wind relation for the component of the geostrophic wind

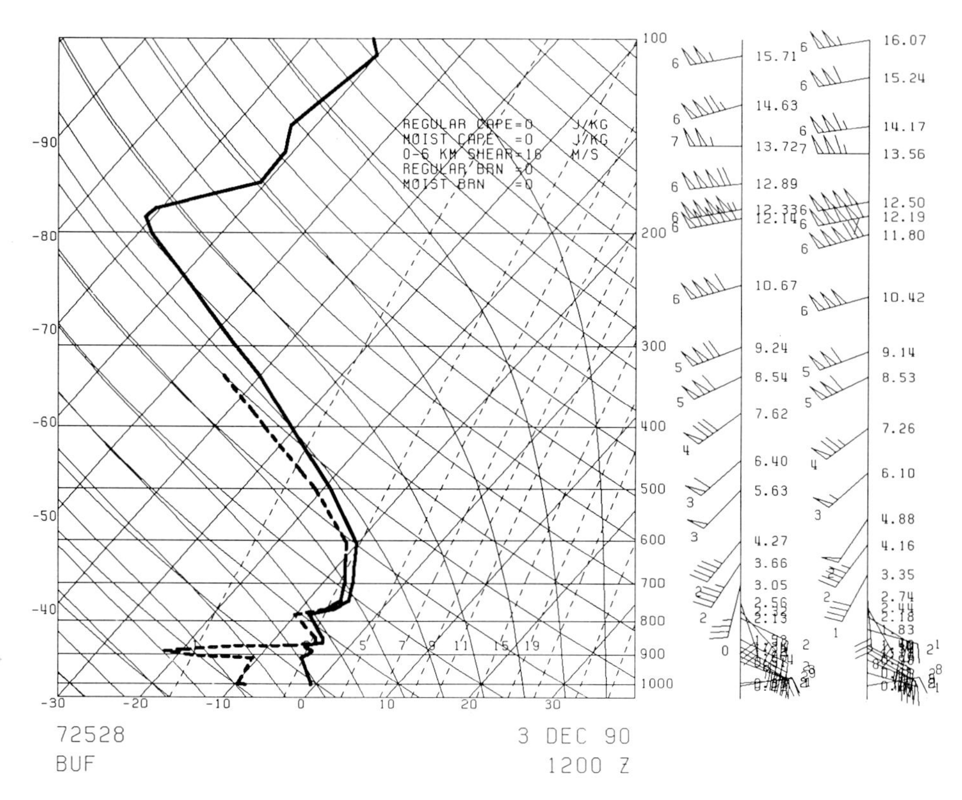

Figure 2.87 Sounding through an intense polar-front jet at Buffalo, New York at 1200 UTC, December 3, 1990. The skewed abscissa is the temperature (°C); the logarithmic ordinate is the pressure (mb). Temperature and dew point plotted as thick solid line and dashed line, respectively. Winds plotted at the right; pennant = 25 m s^{-1}; whole barb = 5 m s^{-1}; half barb = 2.5 m s^{-1}. The winds near the tropopause, which is at 190 mb, are as high as 105 m s^{-1}!

normal and to the left of the temperature gradient:

$$-\frac{\partial u_g}{\partial p} = -\frac{1}{f_0}\frac{\partial}{\partial y}\left(\frac{RT}{p}\right)_p. \tag{2.7.1}$$

Since

$$\theta = T\left(\frac{p_0}{p}\right)^{R/C_p}, \tag{2.7.2}$$

Eq. (2.7.1) may be expressed as

$$-\frac{\partial u_g}{\partial p} = -\frac{1}{f_0}\frac{\partial}{\partial y}\left[\frac{R}{p}\theta\left(\frac{p}{p_0}\right)^{R/C_p}\right]_p. \tag{2.7.3}$$

Then

$$\frac{\partial^2 u_g}{\partial p^2} = \frac{1}{f_0}\frac{\partial}{\partial y}\left(\frac{RT}{p}\frac{1}{\theta}\frac{\partial \theta}{\partial p}\right). \tag{2.1.4}$$

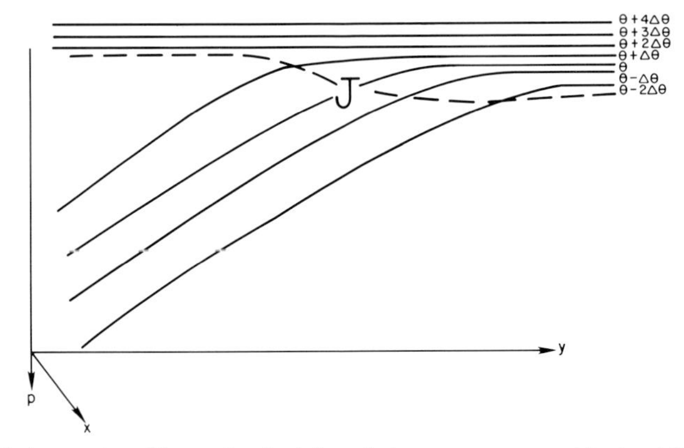

Figure 2.88 Schematic of how the height of the tropopause (dashed line) changes across the polar-front jet (J). Vertical cross section of potential temperature (solid lines) from the subtropics (left) to the polar region (right). Static stability increases with latitude across the jet as a consequence of the thermal-wind relation. The tropical tropopause is higher than the high-latitude tropopause. (cf. Fig. 2.86)

But since

$$\sigma = -\frac{RT}{p}\frac{\partial \ln \theta}{\partial p}, \tag{2.7.5}$$

it follows that

$$\frac{\partial^2 u_g}{\partial p^2} = -\frac{1}{f_0}\frac{\partial \sigma}{\partial y}. \tag{2.7.6}$$

At the "core" of the jet

$$\frac{\partial^2 u_g}{\partial p^2} < 0. \tag{2.7.7}$$

It follows from Eq. (2.7.6) that the static stability parameter σ must increase toward the "cold" side of the jet ($\partial\sigma/\partial y > 0$). This condition is satisfied by a tropopause that slopes such that the "polar" tropopause is lower than the "tropical" tropopause (Fig. 2.88). (This condition, however, is also satisfied by more bizarre tropopause configurations.)

Convective-storm tops are in general lower where the tropopause is low, and hence one cannot conclude that convective storms are not intense simply because their tops are not very high. Thus, although storm tops are higher equatorward of the jet, they are not necessarily more intense than the storms poleward of the jet, which have lower tops. Convective storms will be discussed in detail in Chap. 3. Another consequence of a steeply sloping tropopause is that it is possible that quasihorizontal temperature advection can be significant, owing to the strong quasihorizontal temperature gradient near the jet.

2.7.2. The Subtropical Jet

The *subtropical jet* is a fairly steady jet near 200 mb, and is mainly westerly in direction; it is located at higher levels than the Polar-Front jet. The subtropical jet is a wintertime phenomenon, found between 20° and 35°, the latitude belt of the subtropical anticyclones.

The subtropical jet in the mean appears to be nearly continuous around the globe (Fig. 2.89). The distinction between the Polar-Front jet and the subtropical jet, however, is often not clear on daily weather maps. In fact, it is often difficult to isolate the subtropical jet (as distinct from the Polar-Front jet) on a 200-mb chart on a day-to-day basis in the United States during the winter.

The highest 200-mb wind speeds in the mean are found during the winter in the Northern Hemisphere [Fig. 2.89(a)] off the east coast of Asia near 30°N: Mean wind speeds in excess of 70 m s^{-1} are found southeast of Japan. High wind speeds are also found along the east coast of North America slightly north of 30°N, and in a belt extending from North Africa through Northern India. Weak cyclonic curvature in the mean is found in the subtropics off the west coast of the United States and off the west coast of Africa. During the Southern Hemisphere winter [Fig. 2.89(d)] highest wind speeds are found near 30°S from west of Australia, over Australia, to east of Australia. Highest wind speeds in the mean are in excess of 50 m s^{-1}.

During the Northern Hemisphere summer [Fig. 2.89(c)] only the Polar-Front jet shows up in the mean at 200 mb near 40–45°N. Mean wind speeds of 20–30 m s^{-1} are found, with some in excess of 30 m s^{-1} over Asia. During the Southern Hemisphere summer [Fig. 2.89(b)] there is a belt of winds in excess of 30 m s^{-1} poleward of 40°S in association with the Polar-Front jet between 30°W and 110°E. It is noteworthy that the Polar-Front jet is stronger in the Southern Hemisphere summer than it is in the Northern Hemisphere summer.

The zonally averaged and time-averaged maximum in the zonal component of the wind in the upper troposphere of the midlatitudes (Fig. 2.85) reflects both the Polar-Front jet and the subtropical jet, but more the latter, which is more persistent. Owing to the steadiness and large-scale nature of the subtropical jet, it appears as if planetary-scale processes must play an important role in its maintenance.

A band of cirrus clouds (Fig. 2.90) is often seen on the anticyclonic-shear side of the subtropical jet. Smaller transverse bands are sometimes seen within this main band. (Similar bands sometimes appear in the cirrus canopies of tropical cyclones.) The cirrus may be generated as air rises downstream from a trough in the tropics, and is subsequently transported downstream by the subtropical jet. The sharp poleward edge of the cirrus owes its existence to deformation and a horizontal moisture gradient (Fig. 2.91). There is some observational evidence that, although severe convective storms sometimes develop equatorward of the subtropical jet and its cirrus band, *most* develop poleward of the subtropical jet. The reason for this is not known; however, it is possible that radiative effects from the cirrus may play an important role in suppressing severe convection under the cirrus. It is also possible that subtropical air streams harbor less potential buoyancy.

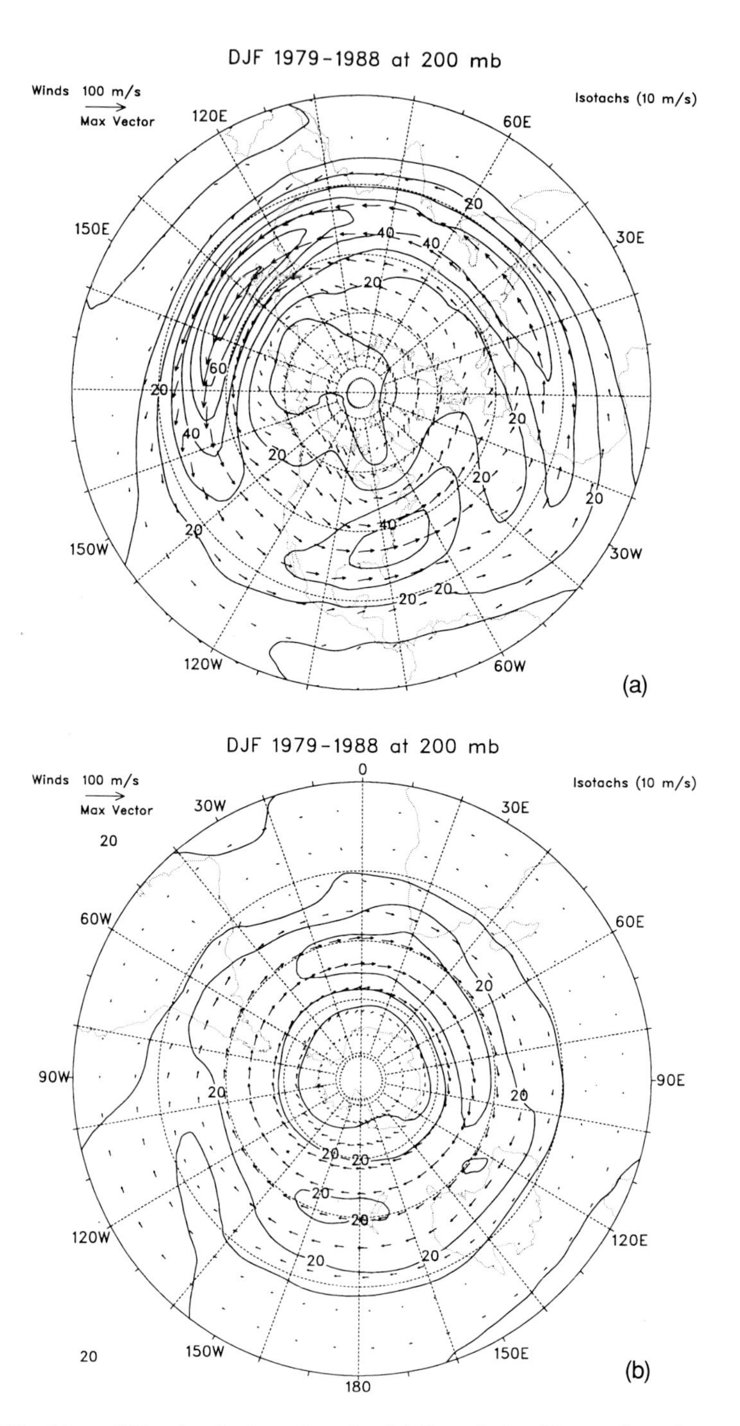

Figure 2.89 Mean 200-mb wind vectors for (a) the winter (December, January, and February) in the Northern Hemisphere, and (b) summer in the Southern Hemisphere; (c) summer (June, July, and August) in the Northern Hemisphere, and (d) winter in the Southern Hemisphere. From ECMWF data for 1979–1988; isotachs in m s^{-1} (solid lines). (Courtesy Kevin Trenberth and Amy Solomon, NCAR)

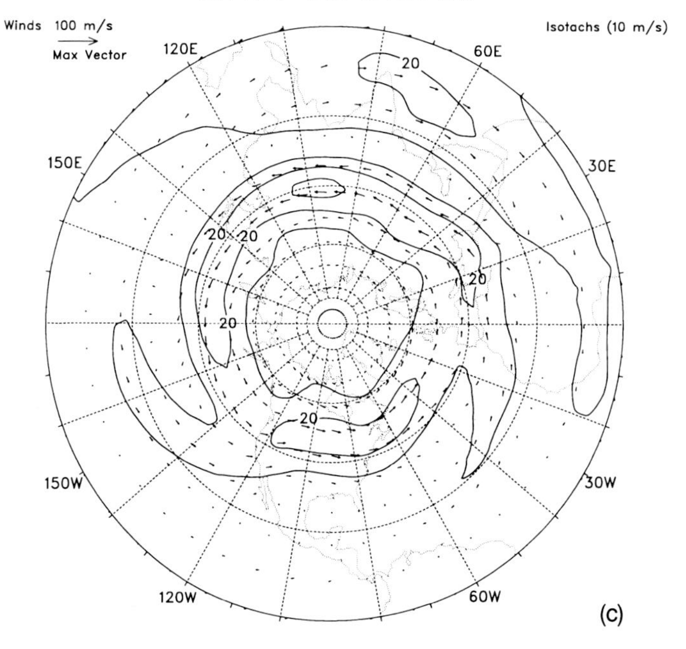

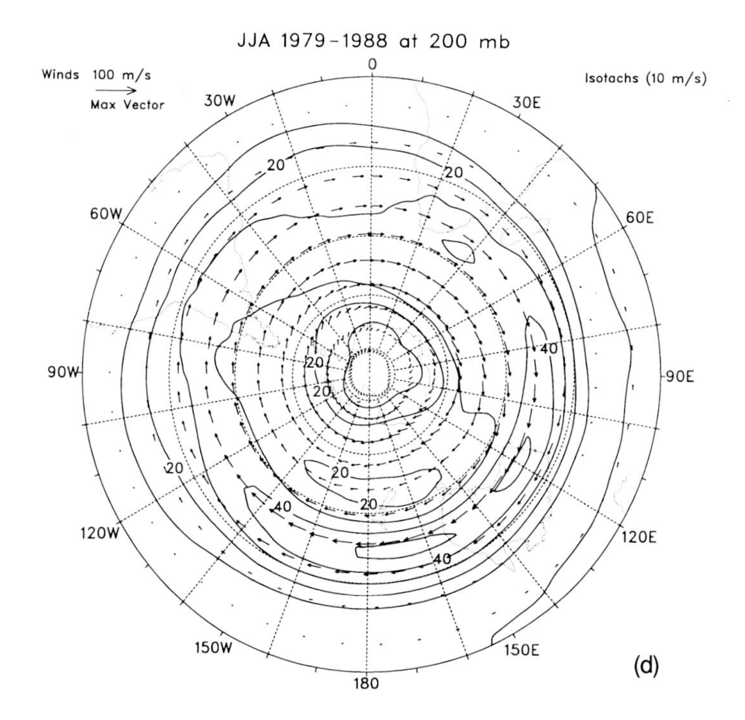

Figure 2.89 (*cont.*)

Figure 2.90 GOES enhanced-infrared satellite photograph at 0130 UTC, May 26, 1980, showing a band of cirrus clouds from the tropics over New Mexico, western Kansas, and central Nebraska, which are associated with the subtropical jet, intruding into a spiraling comma cloud in the midlatitudes. (Also refer to Fig. 1.128.)

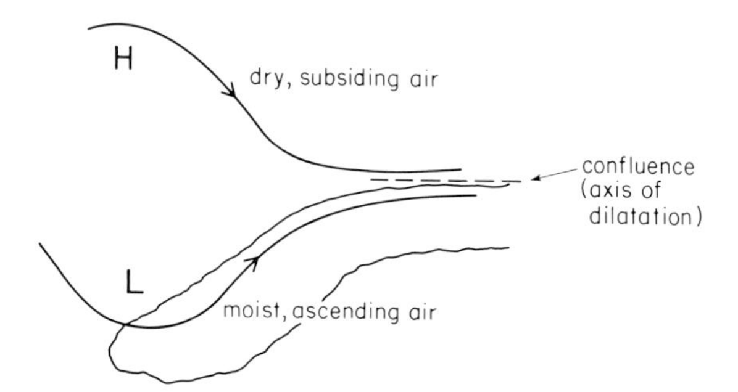

Figure 2.91 Schematic of the relationship between the tropical cirrus cloud shield (solid line) and the axis of dilatation formed between the moist, tropical flow from the southwest and the dry, continental flow from the northwest.

Like the Polar-Front jet, the subtropical jet is associated with a change in height of the tropopause. The meridional temperature gradient associated with the vertical shear under the subtropical jet is concentrated in a shallow layer, not in a relatively deep layer as with the Polar-Front jet (Fig. 2.92). Sometimes the subtropical jet is associated with a portion of the Polar-Front jet that has migrated equatorward into the subtropics and has lost its low-level baroclinicity.

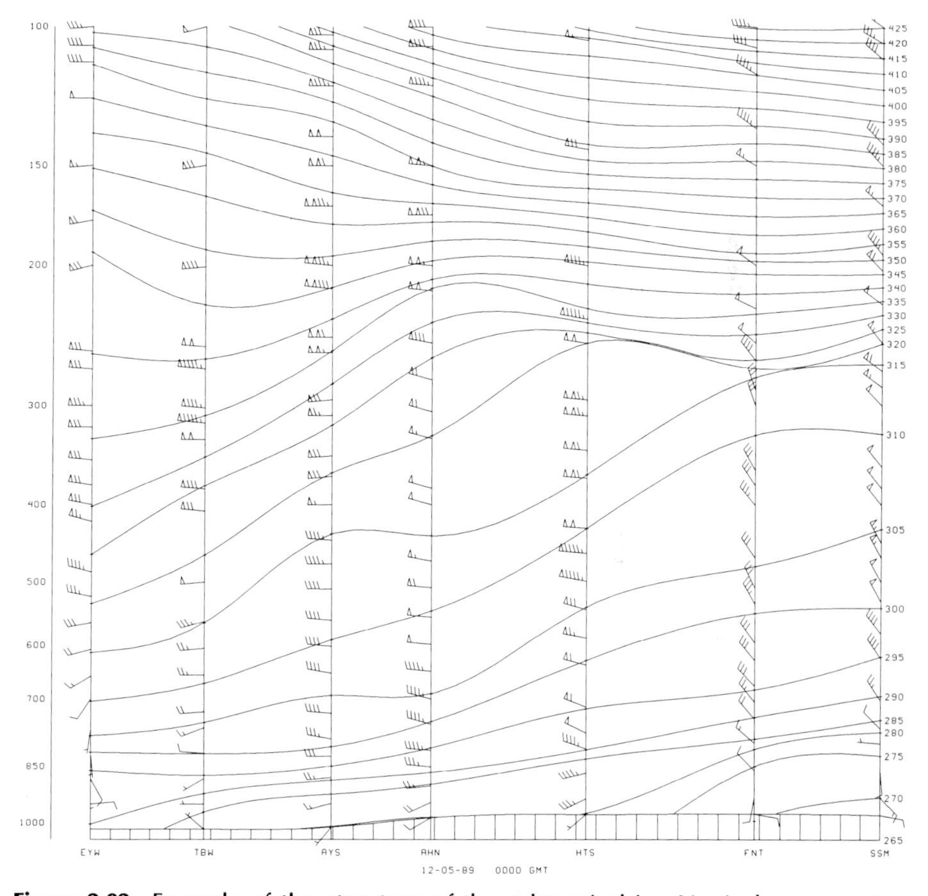

Figure 2.92 Example of the structure of the subtropical jet. Vertical cross section of potential temperature and wind as in Fig. 2.86 at 0000 UTC, December 5, 1989; from Key West, Florida (EYW), through Tampa, Florida (TBW), Waycross, Georgia (AYS), Athens, Georgia (AHN), Huntington, West Virginia (HTS), Flint, Michigan (FNT), and northward to Sault Ste. Marie, Michigan (SSM). Strongest winds are found near 200 mb at AYS (65 m s^{-1}). Below the jet, the vertical wind shear is associated with baroclinicity (zone of relatively tightly packed, sloping potential-temperature isotherms) in the upper troposphere; there is little baroclinicity at low levels, where the winds are southwesterly. An old surface front is dissipating well to the south of Key West, beyond the domain of our analysis. The distance from AHN to HTS is approximately 500 km.

Since the subtropical jet is found at relatively low latitudes, the magnitude of f is small, and hence a relatively small meridional temperature gradient can be associated with relatively large vertical shear. For example, the vertical geostrophic shear associated with a given temperature gradient at 20° latitude is almost twice that at 40° latitude. The zone of strong quasihorizontal (or horizontal) temperature gradient under the subtropical jet is sometimes called the *subtropical front.*

If the flow associated with the jet is sharply curved, then the centripetal acceleration experienced by an air parcel (e.g., passing through a trough or ridge) could be so great that the ageostrophic wind is substantial. The relationship between curvature and wind speed in gradient-wind balance is

$$V_g = V\left(1 + \frac{V}{fR_t}\right), \tag{2.7.8}$$

where V is the actual wind speed, V_g is the geostrophic-wind speed, and R_t is the radius of curvature of the parcel trajectory. Since the geostrophic wind deviates from the actual wind by an amount proportional to the wind speed and curvature $(1/R_t)$, the thermal-wind shear deviates from the actual wind shear in part by an amount proportional to the rate of change of parcel-trajectory curvature with "height" as follows:

$$-\frac{\partial V}{\partial p} - \left(-\frac{\partial V_g}{\partial p}\right) = -\frac{1}{fR_t}\left(-\frac{\partial V}{\partial p}\right) - \frac{V}{f}\left[-\frac{\partial}{\partial p}\left(\frac{1}{R_t}\right)\right]. \tag{2.7.9}$$

If the parcel trajectory is curved anticyclonically and becomes more anticyclonically curved with "height," then the actual shear is more than that expected by the thermal-wind relation. The deviation of the actual shear from the thermal-wind shear is therefore enhanced in a jet stream (where V is large) near a ridge (where $1/R_t$ is negative, and becomes more negative with height below the tropopause in the baroclinic westerlies). Thus, both the anticyclonic curvature of air parcels in ridges in the subtropical jet and the relatively low latitude of the jet dictate that the quasihorizontal temperature gradient associated with the subtropical front need not be very large to explain the vertical shear associated with the subtropical jet.

Computations based upon Eqs. (2.7.8) and (2.7.9) can be simplified in the special case when a wavetrain propagates at a constant speed without changing shape. It was shown in Vol. I that the local time rate of change of an arbitrary quantity Q is related to its rate of change along a trajectory (s_t) and streamline (s), and to the wind speed V as follows:

$$\frac{\partial Q}{\partial t} = V\left(\frac{\partial Q}{\partial s_t} - \frac{\partial Q}{\partial s}\right). \tag{2.7.10}$$

When Q is the wind direction (θ_a), $\partial Q/\partial s_t$ and $\partial Q/\partial s$ are the curvature of the trajectory $(1/R_t)$ and streamline $(1/R_s)$, respectively. Then

$$\frac{\partial \theta_a}{\partial t} = V\left(\frac{1}{R_t} - \frac{1}{R_s}\right), \tag{2.7.11}$$

which is known as *Blaton's formula*. Suppose that a wavetrain propagates at a speed c. If the wavetrain propagates in the direction normal to the ridge and trough axes of the wavetrain, then

$$\frac{D\theta_{\mathrm{a}}}{Dt} = \left(\frac{\partial\theta_{\mathrm{a}}}{\partial t}\right)' + \frac{1}{R_{\mathrm{s}}}\left(V - c\right), \tag{2.7.12}$$

where the prime denotes the reference frame moving along with the wavetrain. It was also shown in Vol. I that

$$\frac{DQ}{Dt} = \frac{\partial Q}{\partial s_{\mathrm{t}}} V, \tag{2.7.13}$$

so that

$$\frac{D\theta_{\mathrm{a}}}{Dt} = \frac{V}{R_{\mathrm{t}}}. \tag{2.7.14}$$

Eliminating $D\theta_{\mathrm{a}}/Dt$ between Eqs. (2.7.12) and (2.7.14), and solving for $1/R_{\mathrm{t}}$, we find that if the wavetrain does not change shape $(\partial\theta_{\mathrm{a}}/\partial t)' = 0$, then

$$R_{\mathrm{t}} = R_{\mathrm{s}}(1 - c/V)^{-1}. \tag{2.7.15}$$

Thus, for this special case, the gradient-wind balance equation and gradient-wind balance thermal-wind equation can be expressed in terms of streamline curvature, which is simpler to compute than trajectory curvature.

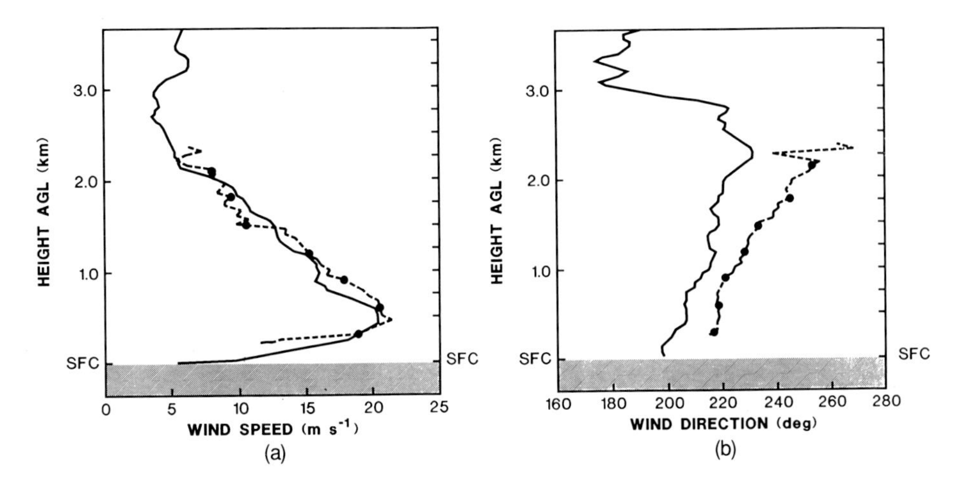

Figure 2.93 Example of a southerly low-level jet in the Southern Plains of the United States (at Norman, Oklahoma) 0930 UTC (0430 CDT), July 29, 1988. (a) Wind speed as a function of height (km AGL) from CLASS rawinsonde data (solid line) and Doppler radar VAD (velocity-azimuth display) data (dashed line); (b) wind direction (degrees) as a function of height (km AGL) as in (a). Note the pronounced maximum in wind speed from the south-southwest at about 400 m AGL (from Stensrud et al., 1990; courtesy R. Maddox, NSSL and the American Meteorological Society).

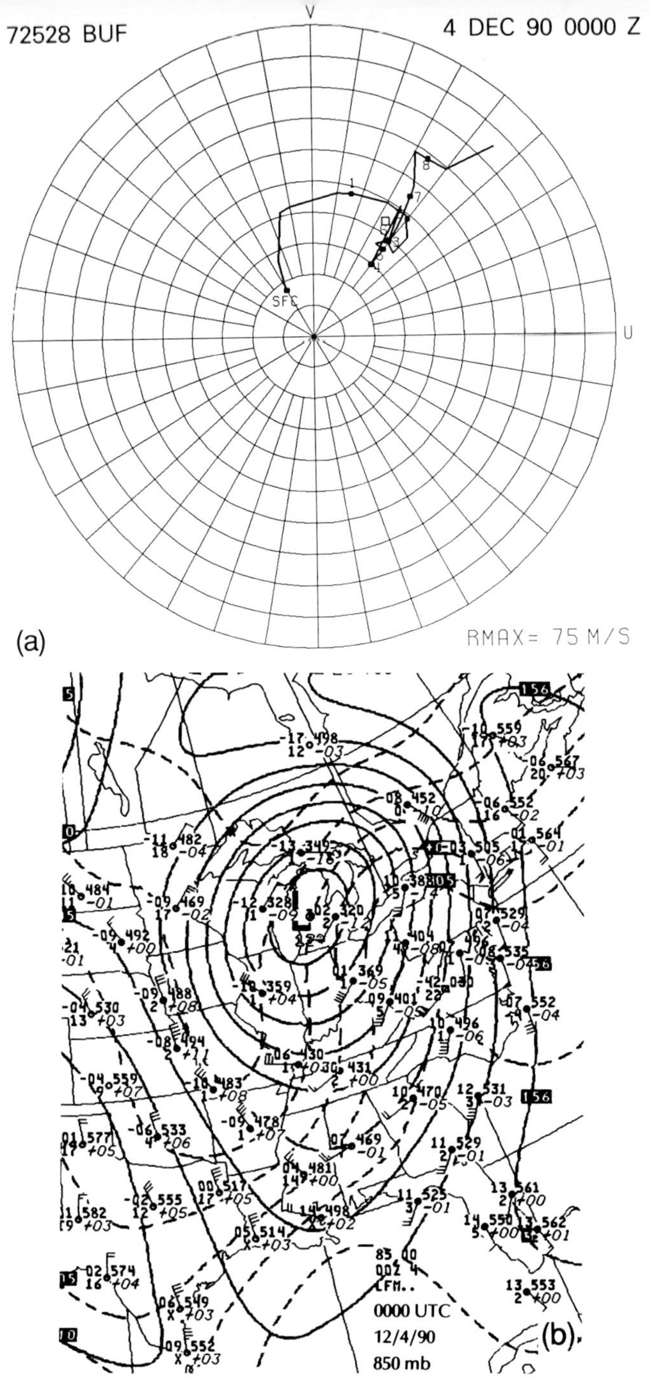

Figure 2.94 Example of a low-level jet ahead of a low-level cyclone. (a) Hodograph for 0000 UTC, December 4, 1990, at Buffalo, New York. Spacing between concentric circles represents 7.5 m s^{-1}; heights given in km MSL. There is a maximum in wind speed from the south-southwest of almost 37.5 m s^{-1} near 1.5 km AGL. The wind speed decreases to 22.5 m s^{-1} at 4 km, and increases above owing to the polar-front jet. (b) NMC 850-mb analysis at 0000 UTC, December 4, 1990. The low-level jet depicted in (a) lies near a warm front just east of the cyclone over Michigan.

2.7.3. The Low-Level Jet

Several types of low-level jets have been identified in the United States. A southerly jet, which is linked to topography, is often found in the Southern Plains of the United States, east of the Rocky Mountains (Fig. 2.93). A low-level jet is sometimes found east or northeast of a mobile low-level cyclone (Fig. 2.94) or ahead of a cold front (not shown). The latter two may be found not only in the southern Plains, but elsewhere. On some occasions northerly low-level jets are found south or southwest of mobile low-level cyclones (Fig. 2.95).

The Southern Plains topographic low-level jet is located at an average height of 800 m AGL. It is often responsible for the rapid advection northward

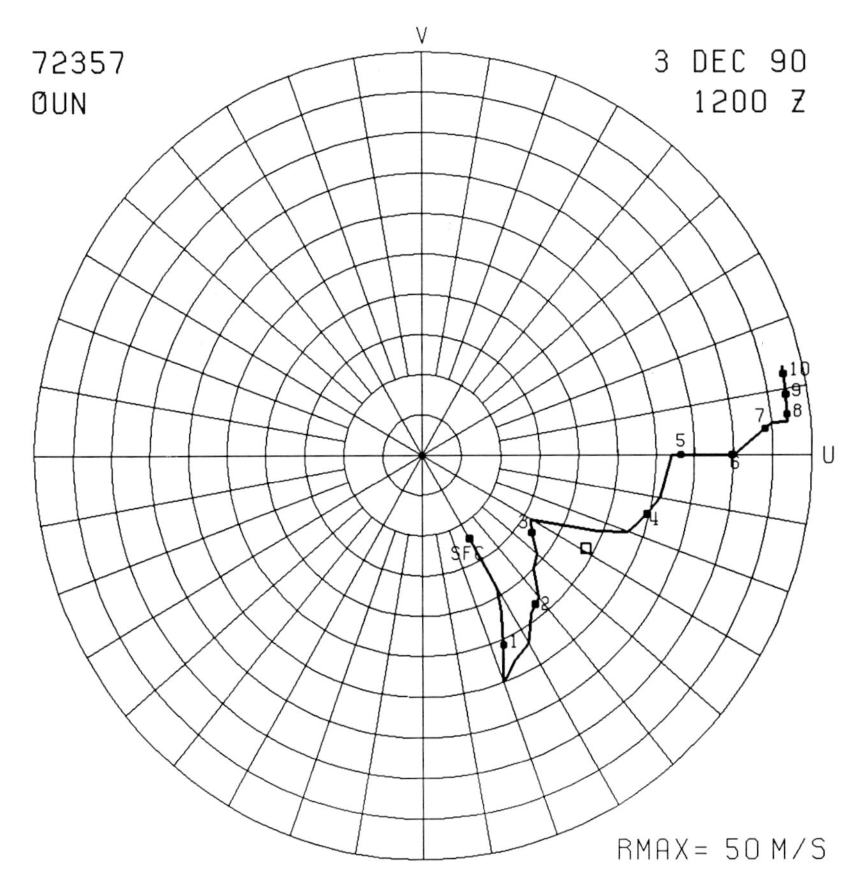

Figure 2.95 Example of a northerly low-level jet to the rear of a mobile low-level cyclone. Hodograph for 1200 UTC, December 3, 1990 at Norman, Oklahoma. Spacing between concentric circles represents 5 m s^{-1}; heights given in km MSL. There is a maximum in wind speed from the north-northwest of 30 m s^{-1} near 1.3 km MSL. The wind speed decreases to slightly over 15 m s^{-1} near 3 km MSL, and increases above owing to the polar-front jet. The low-level jet lies to the rear of a cold front, to the southwest of a cyclone (cf. Fig. 2.27).

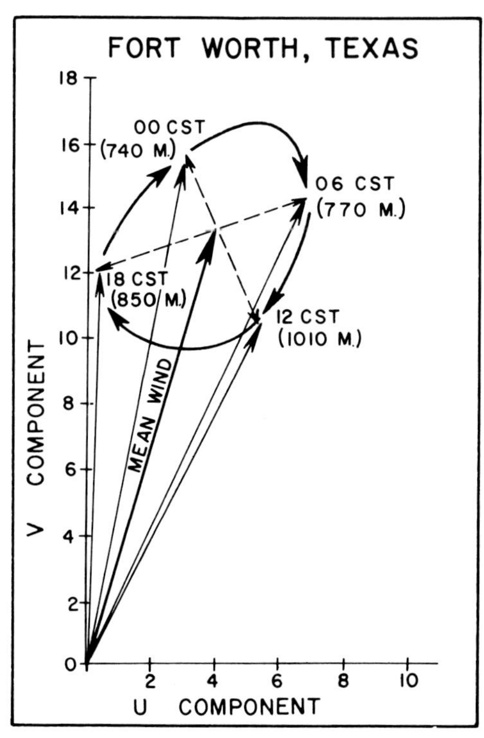

Figure 2.96 Diurnal variation of the Southern Plains low-level jet. Vector mean winds (m s^{-1}) at the level of maximum wind speed on 16 summer days during 1960 at Ft. Worth, Texas. Times (CST) and mean altitudes (m) above the ground at each observation time (from Bonner, 1968). (Courtesy of the American Meteorological Society)

of moisture from the Gulf of Mexico. Undergoing a marked diurnal variation in strength, this low-level jet is strongest at night and weakest during the day (Fig. 2.96). The observed nocturnal maximum in thunderstorm activity over the Plains (see Chap. 3) might be a result of the increase in mositure advection into thunderstorms at cloud-base level at night owing to the low-level jet. The jet is often located just east of a lee trough, and may be located on a surface map where sustained winds are strongest (and the gusts are strongest).

2.7.4. Jet Streaks

An isotach maximum within a jet is called a *jet streak*. The term *jet streak* is usually used to describe wind-speed maxima in the Polar-Front jet or subtropical jet. A jet streak usually propagates at a speed that is slower than the speed of the wind itself. Air parcels consequently accelerate just upstream from the jet streak in the "entrance region," and decelerate just downstream from the jet streak in the "exit region" (Fig. 2.97). If the magnitude of the acceleration or deceleration of an air parcel associated with the wind-speed maximum is on the order of the acceleration induced by the Coriolis force

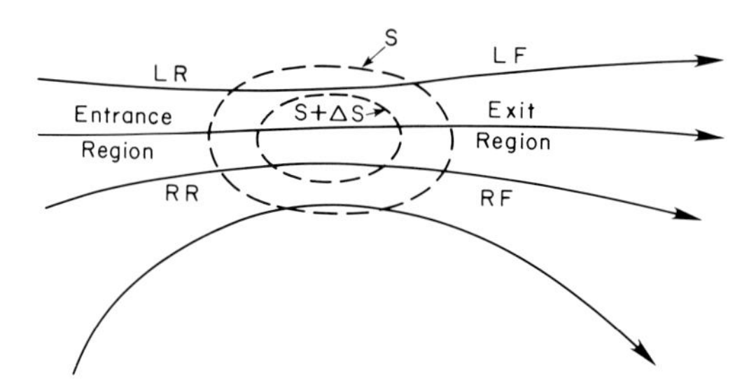

Figure 2.97 Schematic diagram of streamlines (solid lines) and isotachs (S) (dashed lines) in a jet streak. Left-rear quadrant (LR), right-rear quadrant (RR), left-front quadrant (LF), and right-front quadrant (RF).

(i.e., if the Rossby number is of order one), then the jet streak is not a quasigeostrophic phenomenon. We will regard quasigeostrophic jet streaks as minor isotach maxima, and nonquasigeostrophic jet streaks as major isotach maxima. Geostrophic jet streaks are sometimes found in trough-over-ridge features (Fig. 2.98), which are just the opposite of the high-over-low blocks. Ageostrophic jet streaks are found along sharply curved ridges, and may represent portions of an inertial-like oscillation.[9]

The entrance region of a jet streak is subdivided into the left-rear (LR) and right-rear (RR) quadrants, while the exit region is divided into the left-front (LF) and right-front (RF) quadrants (Fig. 2.97). The directions "right" and "left" are given with respect to an observer who is facing downstream (afraid to get wind in his or her face and see what nature has in store upstream); the directions "rear" and "front" refer to the upstream direction, to the rear of the observer, and to the downstream direction, in front of the observer, respectively.

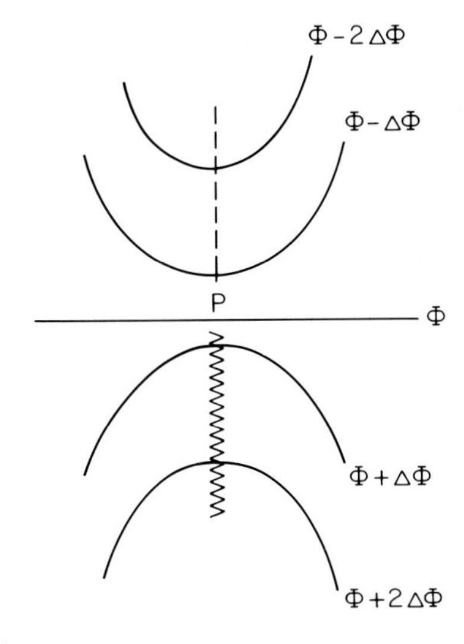

Figure 2.98 Schematic diagram of a jet streak (P) in the geostrophic wind field as a trough (dashed line) -over-ridge (sawtooth line) configuration. Geopotential-height contours (solid lines).

2.8. DYNAMICS OF JETS AND JET STREAKS

Suppose that a jet or jet streak is largely geostrophic (and hydrostatic). Since the formation of a front is accompanied by an increase in thermal-wind shear, the geostrophic wind aloft may also increase and form a jet or jet streak. However, it is also instructive to consider the formation of jets and jet streaks in general, which are not necessarily geostrophic, owing to their small cross-stream scale and associated high wind speeds. Furthermore, the low-level jet is usually in the boundary layer, where turbulent mixing is important.

2.8.1. The Formation of Jets and Jet Streaks

The intensity of a jet flowing in the $+x$ direction may be measured by $\partial^2 u/\partial y^2 + \partial^2 u/\partial z^2$. The intensity of a jet streak, within this jet, may be measured by the three-dimensional Laplacian, $\nabla^2 u = \partial^2 u/\partial x^2 + \partial^2 u/\partial y^2 + \partial^2 u/\partial z^2$. In the center of the core of a jet

$$\frac{\partial^2 u}{\partial y^2} + \frac{\partial^2 u}{\partial z^2} < 0, \tag{2.8.1}$$

$$\frac{\partial u}{\partial y} = 0, \tag{2.8.2}$$

$$\frac{\partial u}{\partial z} = 0. \tag{2.8.3}$$

In the center of the core of a jet streak,

$$\nabla^2 u < 0 \tag{2.8.4}$$

and Eqs. (2.8.2), (2.8.3), and the following condition is satisfied:

$$\frac{\partial u}{\partial x} = 0. \tag{2.8.5}$$

We can regard the rate of change of $\partial^2 u/\partial y^2 + \partial^2 u/\partial z^2$ following air parcel motion as a measure of the increase or decrease in the intensity of the jet. We can similarly regard the rate of change of $\nabla^2 u$ following air parcel motion as a measure of the increase or decrease in the intensity of the jet streak. Let us call the functions

$$J = \frac{D}{Dt}(-\nabla_x^2 u) \tag{2.8.6}$$

where

$$\nabla_x^2 = \frac{\partial^2}{\partial y^2} + \frac{\partial^2}{\partial z^2}, \tag{2.8.7}$$

and

$$J_s = \frac{D}{Dt}(-\nabla^2 u), \tag{2.8.8}$$

the *jetogenetical functions* for jets and jet streaks, respectively. They are analogous to Petterssen's frontogenetical function and are useful for diagnosing frontal processes when only wind data are available. Strictly speaking, Eqs. (2.8.6) and (2.8.8) should be expressed in a semi-Lagrangian framework, that is, one moving with the jet or jet streak rather than one moving along with the air parcel. However, the basic physical mechanisms responsible for the formation of jets and jet streaks should be independent of reference frame, and hence Eqs. (2.8.6) and (2.8.8) should be useful for the analysis of jets, just as the Lagrangian frontogenetical function has been useful for the analysis of fronts.

It follows that

$$J = -\nabla_x^2\left(\frac{Du}{Dt}\right) + \nabla_x^2(\mathbf{v}\cdot\nabla u) - \mathbf{v}\cdot\nabla(\nabla_x^2 u) \tag{2.8.9}$$

$$J_s = -\nabla^2\left(\frac{Du}{Dt}\right) + \nabla^2(\mathbf{v}\cdot\nabla u) - \mathbf{v}\cdot\nabla(\nabla^2 u). \tag{2.8.10}$$

However, if ∇u is uniform[10] in the *neighborhood* of the parcel, then

$$J = -\nabla_s^2\left(\frac{Du}{Dt}\right) \tag{2.8.11}$$

$$J_s = -\nabla^2\left(\frac{Du}{Dt}\right). \tag{2.8.12}$$

The equation of motion along the jet or jet streak is

$$\frac{Du}{Dt} = fv - \frac{1}{\rho}\frac{\partial p}{\partial x} + F_x = fv_a + \beta y v_g + F_x, \tag{2.8.13}$$

where F_x is the component of friction in the direction of the jet. Substituting Eq. (2.8.13) into Eqs. (2.8.11) and (2.8.12), we see that on the jet axis or at the jet streak (where $y = 0$),

$$J = -f_0\,\nabla_x^2 v_a - 2\beta\frac{\partial v}{\partial y} - \nabla_x^2 F_x \tag{2.8.14}$$

$$J_s = -f_0\,\nabla^2 v_a - 2\beta\frac{\partial v}{\partial y} - \nabla^2 F_x. \tag{2.8.15}$$

The jetogenetical function is positive whenever there is a local maximum in the cross-jet component of the ageostrophic wind. Since the ageostrophic wind is perpendicular and to the left of the parcel acceleration vector (in the Northern Hemisphere), "local" parcel accelerations lead to "local" wind maxima. If $-f_0\,\nabla^2 v_a$ is positive somewhere, then by continuity there must be at least one branch of v_a in the opposite direction above or below. Thus, vertical circulations are associated with the formation of jets and jet streaks. These vertical circulations may be driven by frontogenetically or frontolytically induced geostrophic forcing. Thus, jets may indeed be byproducts of frontal processes. The second term, which involves the variation of the Coriolis parameter with latitude, is usually relatively small.

The third term is important for jets in the boundary layer, such as, for example, the "low-level" jet. The diurnal oscillation of the wind vector in the low-level jet is in part related to the diurnal variation in boundary-layer friction. The surface cools at night, and the boundary layer becomes more stable. Vertical mixing is inhibited, and if temperature advection is neglected, the wind speed increases with height up to the top of the boundary layer according to the Ekman profile. On the other hand, during the day the ground heats up, and the boundary layer becomes less stable, or even neutral. Vertical mixing is no longer inhibited, and greater momentum at the top of the boundary layer can be mixed down to the surface. Buajitti and Blackadar in the late 1950s showed that the low-level jet behaves, to some extent, like an inertial oscillation driven by the diurnal oscillation of friction: Maximum (minimum) cross-isobar flow occurs during the day (night), and maximum (minimum) total wind speeds occur at night (during the day).

In the Southern Plains of the United States, during the day, the dry, elevated region to the west heats up more rapidly than the moist, lower elevation region to the east. At night, the west cools off more rapidly than the east. Thus, there is a meridional component to the thermal wind, which is southerly at night, and northerly during the day. It follows that (if friction is ignored) the southerly component of the low-level wind should increase with height at night, and decrease with height during the day.

The diurnal oscillations in surface temperature in the Southern Plains have another effect, because the surface is not level. Cool air near the ground at night "drains" downslope owing to gravity (Fig. 2.99); warm air flows upslope

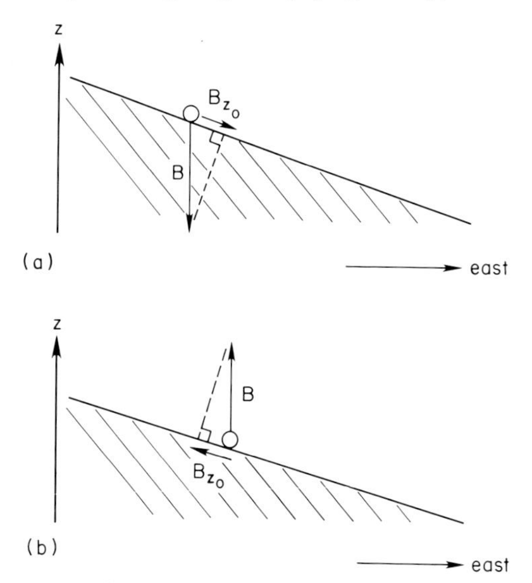

Figure 2.99 Illustration of the diurnal variation in buoyancy of a surface air parcel. **B** is the buoyancy vector; $\mathbf{B}_{z_0}$ is the component of buoyancy along the sloping ground. (a) During the night, the surface air is negatively buoyant, and is accelerated downslope and warms; (b) during the day the surface air is positively buoyant, and is accelerated upslope and cools. The time rate of change of temperature of an air parcel undergoes a diurnal oscillation.

during the day. This effect is thermodynamically mitigated when the atmosphere is stable (e.g., at night), as the cool downslope air is warmed as it is compressed. During the day the warm upslope air cools as it expands; however, owing to decreased static stability, it does not cool too much near the surface. J. Holton has demonstrated that the diurnal oscillation in boundary-layer temperature can in fact produce a diurnal oscillation in the boundary-layer wind over sloping terrain that is similar in some respects to that observed.

We now consider one final process. The Gulf Stream in the Atlantic owes its existence to the presence of North America on the western edge of the Atlantic Ocean. H. Wexler in 1961 proposed that the Southern Plains low-level jet is an atmospheric analog to the oceanic Gulf Stream, and that the Rocky Mountains play the role of the western boundary. It is not clear, however, to what extent the Southern Plains low-level jet is a good analog. The physical mechanism responsible for "western boundary intensification" is the following: A fluid that is deflected northward by a western boundary at constant elevation acquires Earth's vorticity, and therefore must acquire anticyclonic vorticity, if potential vorticity is to be conserved. If this vorticity is mainly manifest as shear, then a high-speed current forms along the western boundary.

2.8.2. The Vertical-Motion Field Near Jets and Jet Streaks

The vertical-motion field can be diagnosed in several ways. The first one we will discuss is the *parcel method.* One calculates the ageostrophic wind field from the field of parcel acceleration. Continuity is then used to determine vertical velocities. The ageostrophic wind may be decomposed into its component parts: the isallobaric component, the inertial–advective component, and the isallotropic component. The specific physical mechanisms responsible for the ageostrophic wind and the vertical-motion fields can thus be diagnosed. Aloft, where wind speeds and gradients of wind speeds are large, the inertial–advective component may be substantial. At low levels, where wind speeds are low, the isallobaric component may be significant because air parcels have a longer period of time to adjust to the changing pressure gradient.

The second method of diagnosis makes use of diagnostic ω equations. Both methods may make use of either quasigeostrophic or semigeostrophic dynamics.

Quasigeostrophic diagnosis

Parcel method. As an air parcel enters a jet streak, its geostrophic wind speed increases. The quasigeostrophic equation of motion (on an f plane) may be expressed as

$$\mathbf{v}_{\mathrm{a}} = \frac{1}{f_0}\hat{\mathbf{k}} \times \frac{D_{\mathrm{g}}\mathbf{v}_{\mathrm{g}}}{Dt} \tag{2.8.16}$$

$$= \frac{1}{f_0}\hat{\mathbf{k}} \times \left(\frac{\partial \mathbf{v}_{\mathrm{g}}}{\partial t} + (\mathbf{v}_{\mathrm{g}} \cdot \nabla)\mathbf{v}_{\mathrm{g}}\right). \tag{2.8.17}$$

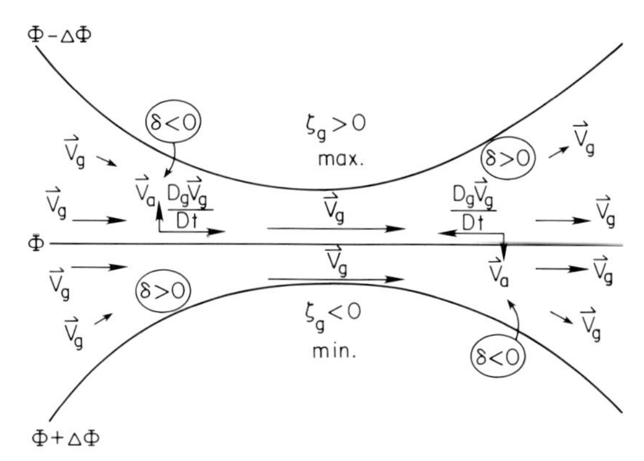

Figure 2.100 Parcel dynamics in a jet streak. Geopotential-height contours Φ (solid lines); $\mathbf{v}_g$ represents the geostrophic wind; $\mathbf{v}_a$ represents the ageostrophic wind, which in this Northern Hemisphere example is oriented perpendicular and to the left of the geostrophic parcel acceleration vector $D_g\mathbf{v}_g/Dt$. There is convergence ($\delta < 0$) in the right-front and left-rear quadrants; there is divergence ($\delta > 0$) in the left-front and right-rear quadrants. A maximum (minimum) in shear and curvature vorticity are found just to the left (right) of the jet streak.

Therefore, as the air parcel accelerates, an ageostrophic wind (in the Northern Hemisphere) blows perpendicular and to the left toward lower height (or pressure); as the air parcel exits the jet streak, its geostrophic wind speed decreases, and an ageostrophic wind blows toward greater height (or pressure). If the accelerations owing to curvature are small, then the jet streak is referred to as a *straight jet streak*. There is convergence in the left-rear and right-front quadrants and divergence in the right-rear and left-front quadrants (Fig. 2.100).

We can deduce this also from the quasigeostrophic vorticity equation

$$\frac{D_g(\zeta_g + f)}{Dt} = -f_0\delta. \tag{2.8.18}$$

As an air parcel enters the left quadrant of a "straight" jet streak, its geostrophic shear vorticity becomes more cyclonic, and hence $\delta < 0$. As it enters the right quadrant, its shear geostrophic vorticity becomes more anticyclonic, and hence $\delta > 0$. As the air parcel exits the left quadrant, its geostrophic shear vorticity becomes less cyclonic, and hence $\delta > 0$; as it exits the right quadrant, its geostrophic shear vorticity becomes less anticyclonic, and hence $\delta < 0$. It has been assumed that f at the parcel changes negligibly (this is valid exactly for zonal flow).

Namias and Clapp, in 1949, first deduced the divergence field from the perspective of energy conservation. An air parcel decelerates in the exit region, loses kinetic energy, and gains potential energy. The latter is consistent with flow from lower to greater heights. Convergence is therefore found to the right and divergence to the left. A similar argument can be made for the entrance region.

If $\omega = 0$ near tropopause level, and the jet streak is located near tropopause level, then according to the continuity equation there must be rising motion below in the right-rear and left-front quadrants, and sinking motion below in the left-rear and right-front quadrants. There is therefore a thermally direct vertical circulation in the entrance region, and a thermally indirect circulation in the exit region. Our analysis is only qualitative, and not quantitative, however, since the Rossby number is often greater than 1. For weaker isotach maxima, however, the quasigeostrophic analysis is quantitatively accurate. If there is substantial curvature in the jet streak, then one must account for centripetal accelerations. One must be careful to consider also the accelerations of an air parcel along its trajectory, not along its streamline, because the jet streak is moving.

The ω-equation method. Referring to Fig. 2.100, we see that there is cyclonic vorticity advection at jet-streak level in the left-front and right-rear quadrants; there is anticyclonic vorticity advection in the left-rear and right-front quadrants. If the jet streak is near the tropopause, then below jet-streak level vorticity advection becomes more cyclonic with height in the left-front and right-rear quadrants, and more anticyclonic with height in left-rear and right-front quadrants. According to the traditional formulation of the quasi-geostrophic ω equation, in the absence of significant temperature advection (remember, $\nabla_p\theta$ reverses at the tropopause, and hence $\nabla_p\theta = 0$ at the tropopause, usually, unless the tropopause slopes, e.g. near a jet), there is rising motion below the jet-streak level in the left-front and right-rear quadrants, and sinking motion in the left-rear and right-front quadrants. A **Q**-vector analysis, of course, gives the same result (Fig. 2.101).

Semigeostrophic diagnosis

Parcel method. The parcel method uses the equation of motion subject to the geostrophic-momentum approximation (on an f plane) as follows:

$$\mathbf{v}_a = \frac{1}{f_0}\hat{\mathbf{k}} \times \frac{D\mathbf{v}_g}{Dt}$$
$$= \frac{1}{f_0}\hat{\mathbf{k}} \times \left(\frac{\partial \mathbf{v}_g}{\partial t} + ((\boldsymbol{v}_g + \boldsymbol{v}_a)\cdot\boldsymbol{\nabla})\boldsymbol{v}_g + \omega\frac{\partial \mathbf{v}_g}{\partial p}\right). \tag{2.8.19}$$

The ω-equation method. The semigeostrophic equations of motion, Eqs. (2.5.167) and (2.5.168), the thermodynamic equation, Eq. (2.5.169), and the continuity equation, Eq. (2.5.171), may be combined to form an ω equation. The derivation of this equation is identical to that of the quasigeostrophic ω equation. The (frictionless and adiabatic) semigeostrophic ω equation in terms of **Q** vectors is as follows:

$$\left(\nabla_P^2 + \frac{f_0^2}{\sigma^*}\frac{\partial^2}{\partial P^2}\right)\omega^* = -2\,\boldsymbol{\nabla}_P\cdot\mathbf{Q}^* \tag{2.8.20}$$

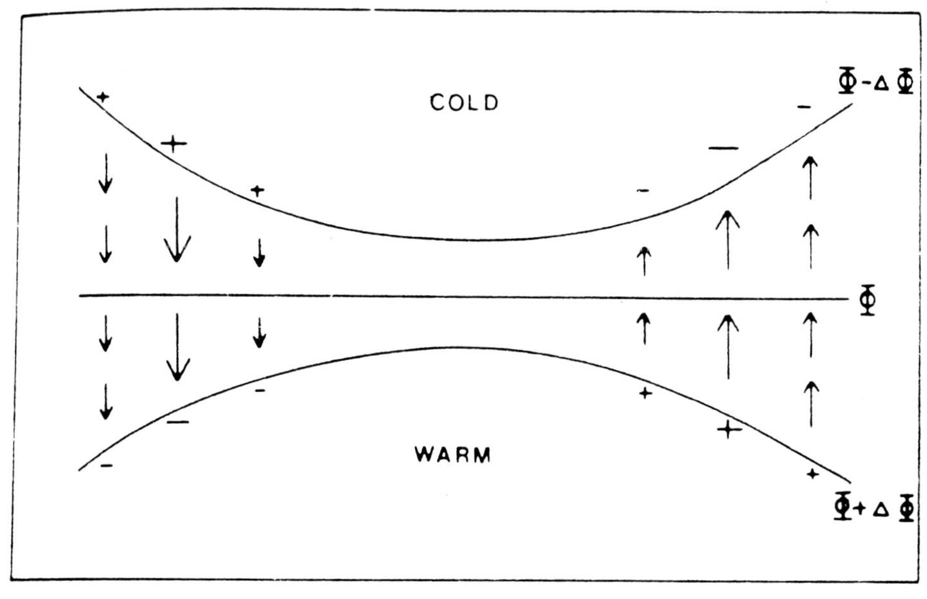

Figure 2.101 Geopotential height contours (solid lines) and **Q** vectors (arrows) for the entrance and exit regions of a jet streak. Signs of $\nabla_p \cdot \mathbf{Q}$ and ω are indicated by − and + (from Bluestein, 1986; adapted from Hoskins et al., 1978). (Courtesy of the American Meteorological Society)

where

$$\mathbf{Q}^* = -\frac{R}{\overline{\sigma^*}P}\begin{pmatrix} \dfrac{\partial \mathbf{v}_g}{\partial X} \cdot \nabla_P T \\ \dfrac{\partial \mathbf{v}_g}{\partial Y} \cdot \nabla_P T \end{pmatrix}, \tag{2.8.21}$$

$$\overline{\sigma^*} = -\frac{R}{P}\left(\frac{P}{P_0}\right)^{R/C_p} \overline{\frac{\boldsymbol{\eta}_g \cdot \nabla\theta}{\eta_{gp}}}, \tag{2.8.22}$$

and the overbar means an average over a pressure surface. The "traditional" formulation of the semigeostrophic ω equation is as follows:

$$\left(\nabla_P^2 + \frac{f_0^2}{\overline{\sigma^*}}\frac{\partial^2}{\partial P^2}\right)\omega^* = -\frac{f_0}{\overline{\sigma^*}}\frac{\partial}{\partial P}[-\mathbf{v}_g \cdot \nabla_P(\zeta_g^* + f)]$$
$$-\frac{R}{\overline{\sigma^*}P}\nabla_P^2(-\mathbf{v}_g \cdot \nabla_P T). \tag{2.8.23}$$

In real space the forcing functions are inversely proportional to Ertel's potential vorticity rather than static stability alone [see Eq. (2.5.182)]: The forcing is therefore enhanced when Ertel's potential vorticity is small.

In order to use Eq. (2.8.20) or (2.8.23) for diagnostic calculations, the geostrophic wind field must be calculated from the observed height field. The gridded height and temperature data in real space then need to be interpolated to semigeostrophic space according to Eqs. (2.5.176) and (2.5.177), so that the forcing functions can be calculated. Computations of ω^* subject to appropriate boundary conditions are then made using relaxation techniques, provided that the Jacobian of the transformation is nonzero. The resulting ω^* field is then transformed back to physical space via Eqs. (2.5.176) and (2.5.177), and then ω^* is transformed back to ω via Eq. (2.5.174).

Differences between the fields of ω computed quasigeostrophically and those computed semigeostrophically are due to the ageostrophic and vertical advection of geostrophic momentum, and to the ageostrophic advection of temperature. Some differences will also result from the interpolations to and from semigeostrophic space, and from the process of averaging σ^* over a pressure surface. An example of a comparison of vertical-velocity fields computed quasigeostrophically, semigeostrophically, and kinematically is shown in Fig. 2.103 for the wind and temperature fields depicted in Fig. 2.102.

Equation (2.8.20), the semigeostrophic ω equation, may itself be transformed into an equation in real space. Unfortunately, the equation is quite complicated, and the physical significance of each term is not easily seen. On the other hand, it is not necessary, using the transformed equation, to interpolate data to and from semigeostrophic space.

The semigeostrophic ω equation must be used cautiously in three-dimensional flows where geostrophic advection of ageostrophic momentum is significant; it is best applied to nearly straight wind fields.

2.8.3. Jet-Streak Propagation

Quasigeostrophic analysis. The rate and direction of propagation of a jet streak can be determined from the field of height tendencies. Using the quasigeostrophic height-tendency equation, and neglecting temperature advection, we find that there are cyclonic vorticity advection and height falls in the left-front and right-rear quadrants; there are anticyclonic vorticity advection and height rises in the left-rear and right-front quadrants. Advection of Earth's vorticity is not significant.

It follows that the jet streak will propagate downstream (from left to right, west to east in Fig. 2.104). Petterssen's formula for the speed of a trough or ridge can be used to compute the actual quasigeostrophic speed, since a jet streak is often defined by the juncture of a trough and a ridge.

The semigeostrophic height-tendency equation. The semigeostrophic equations of motion, Eqs. (2.5.167) and (2.5.168), the thermodynamic equation, Eq. (2.5.169), and the continuity equation, Eq. (2.5.171), may be combined to form a height-tendency equation. The derivation of such an equation is identical to that of the quasigeostrophic height-tendency equation.

The semigeostrophic height-tendency equation is as follows:

$$\left(\nabla_P^2 + \frac{f_0^2}{[\sigma^*]}\frac{\partial^2}{\partial P^2}\right)\chi^* = f_0[-\mathbf{v}_g \cdot \nabla_P(\zeta_g^* + f)] - \frac{f_0^2}{[\sigma^*]}\frac{\partial}{\partial P}\left(\frac{R}{P}(-\mathbf{v}_g \cdot \nabla_P T)\right), \qquad (2.8.24)$$

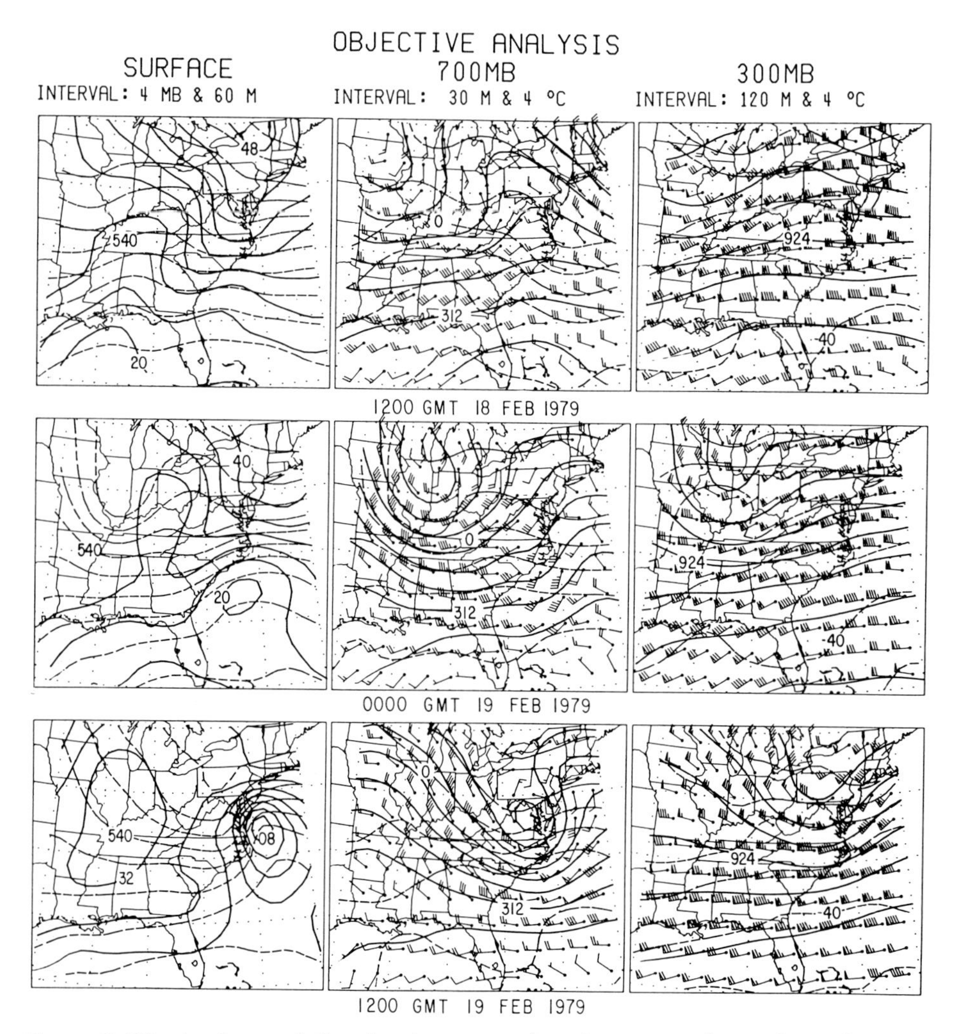

Figure 2.102 Analyses of the development of an intense cyclone along the east coast of the United States at (top) 1200 UTC, February 18, 1979; (middle) 0000, February 19; (bottom) 1200, February 19. Surface analyses of pressure in mb (solid lines) without the leading 10, and 1000–500 mb thickness in dam at 6-dam intervals (dashed lines); 700-mb analyses of height in dam (solid lines) at 30-m intervals, wind (pennants, whole and half barbs represent 25, 5, and 2.5 m s^{-1}, respectively), and temperature in °C (dashed line, at 4°C intervals); 300-mb analyses as at 700 mb (from Bosart and Lin, 1984). (Courtesy of the American Meteorological Society)

where

$$\chi^* = \frac{\partial \Phi^*}{\partial t^*}, \tag{2.8.25}$$

$$[\sigma^*] = \left[-\frac{R}{P}\left(\frac{P}{P_0}\right)^{R/C_p} \frac{\boldsymbol{\eta}_g \cdot \boldsymbol{\nabla}\theta}{\eta_{gp}} \right], \tag{2.8.26}$$

and [ ] means averaged with respect to P.

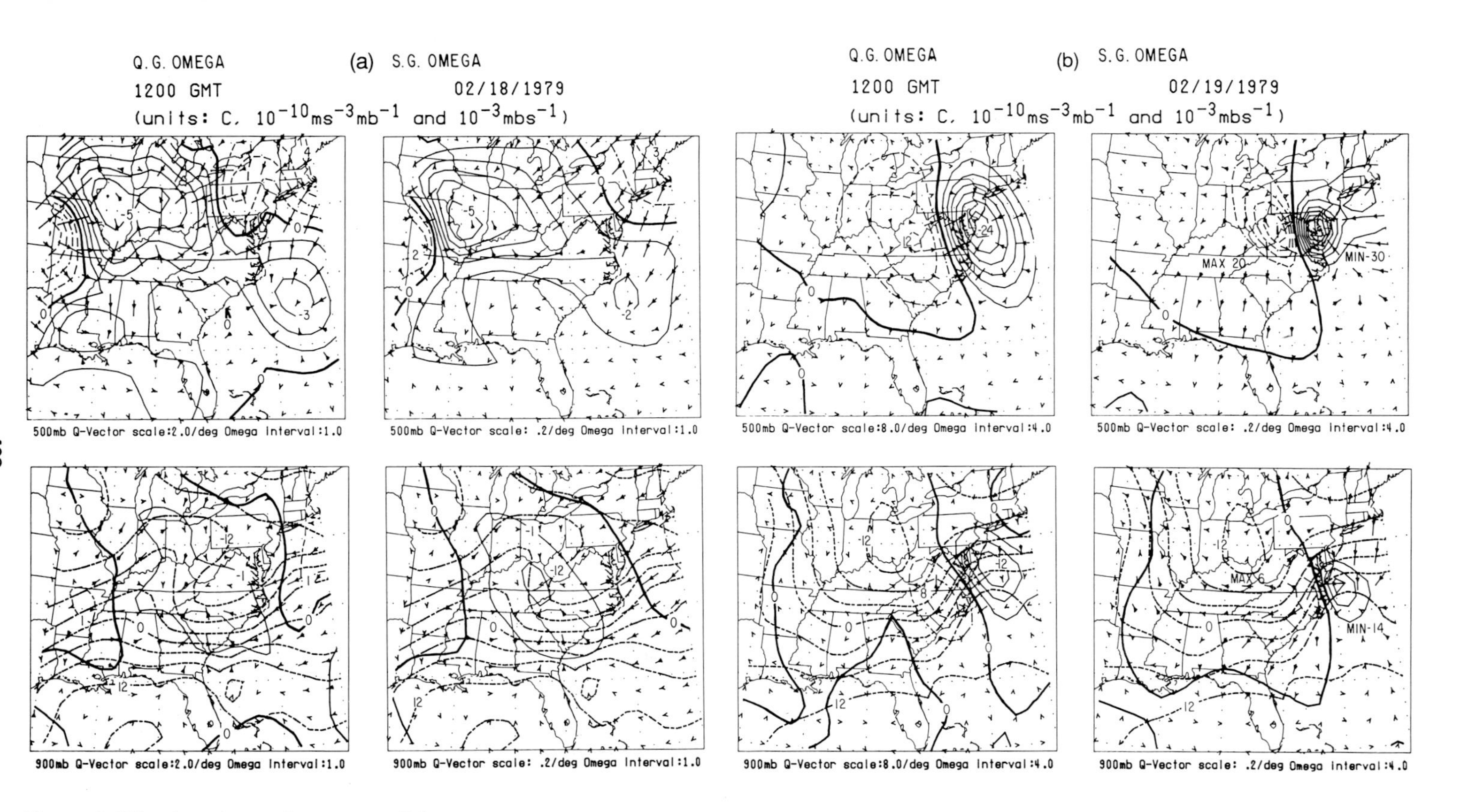

Figure 2.103 (*see legend on page 404*)

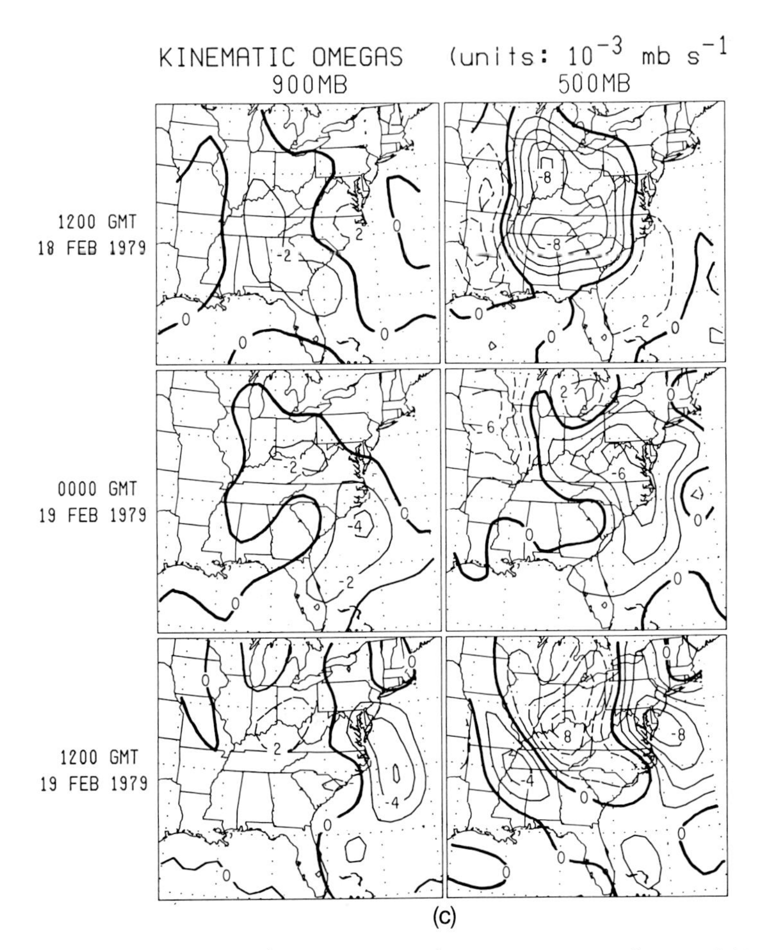

Figure 2.103 A comparison of quasigeostrophic, semigeostrophic, and kinematic computations of vertical velocity ω at 900 and 500 mb for the developing phase of the intense midlatitude cyclone depicted in Fig. 2.102. (a) Vertical-motion diagnoses at (top) 500 mb and (bottom) 900 mb for 1200 UTC, February 18, 1979; quasigeostrophic (Q.G.) and semigeostrophic (S.G.) vertical velocity in 10^{-3} mb s^{-1} (solid, positive, and dashed, negative lines); **Q** vectors scaled in units of 10^{-10} m s^{-3} mb^{-1}. (b) As in (a), but for 1200 UTC, February 19. (c) Kinematically computed vertical velocity ω at (left) 900 mb and (right) 500 mb in 10^{-3} mb s^{-1} (solid, positive and dashed, negative lines) for 1200 UTC, February 18, 0000, February 19 and 1200, February 19 (from Bosart and Lin, 1984). (Courtesy of the American Meteorological Society)

The semigeostrophic height-tendency equation, Eq. (2.8.24), when transformed into real space, is a very complicated equation. Although the physical significance of each term is not obvious, it is not necessary to interpolate data to and from semigeostrophic space.

Like the semigeostrophic ω equation, the semigeostrophic height-tendency equation is best used for nearly straight wind fields.

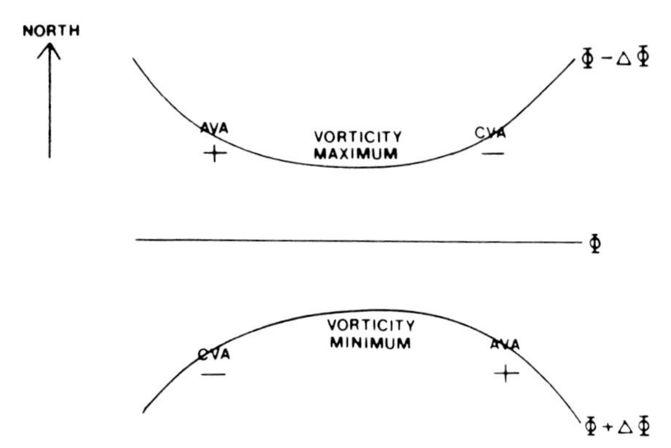

Figure 2.104 Illustration of the propagation of an idealized jet streak in the geostrophic wind field by vorticity advection. Geopotential height Φ (solid lines). CVA and AVA denote cyclonic and anticylonic vorticity advection, respectively; sense of geopotential-height tendencies indicated by + and −. (from Bluestein, 1986). (Courtesy of the American Meteorological Society)

2.8.4. Coupling of an Upper-Level Jet Streak to the Wind Field Below

Uccellini and Johnson in 1979 first suggested that the transverse vertical circulation associated with an upper-level jet streak could become coupled to the wind field at low levels if the circulation extends down to low levels. Underneath the rising branch of the circulation, at the surface, there is convergence; underneath the sinking branch of the circulation, there is divergence. According to the vorticity equation, vorticity will become more cyclonic and surface pressure will fall under the rising branch, while vorticity will become more anticyclonic and surface pressure will rise under the subsiding branch. This pressure-fall/pressure-rise couplet at the surface in the plane of the circulation is associated with an isallobarically induced ageostrophic wind, which is necessarily perpendicular to the flow in the jet streak if the jet streak is straight. The ageostrophic wind exists because the jet streak "feels" the effect of the surface boundary.

In the exit region of a jet streak in westerly flow, the surface-induced ageostrophic flow is toward the pole, while the ageostrophic flow aloft is toward the equator. With cold air on the poleward side, and warm air on the equatorward side, the vertical circulation destabilizes the atmosphere, owing to differential ageostrophic temperature advection; temperature advection becomes more negative with height (warm advection below, cold advection aloft). Furthermore, if there is a supply of low-level moisture on the equatorward side, potential instability is increased owing to the advection of low-level moisture. Fawbush et al., in the early 1950s, in fact observed that one of the synoptic conditions often associated with the development of tornadic storms is the intersection of the low-level projection of an upper-level jet with the axis of a low-level moisture ridge, the latter which Beebe and Bates in 1955 showed is caused by the low-level jet. (There is further significance to these intersecting jets in the formation of tornadic storms, which will be discussed in Chap. 3.)

M. Shapiro, in the early 1980s, suggested that when the thermally indirect circulation associated with the exit region of an upper-level jet streak and upper-level front are situated right over the thermally direct circulation associated with a low-level front, there is subsidence over the warm, moist air mass east of the front, and convection is suppressed since the air aloft is being stabilized (Fig. 2.105). However, when the circulation of the exit region of the

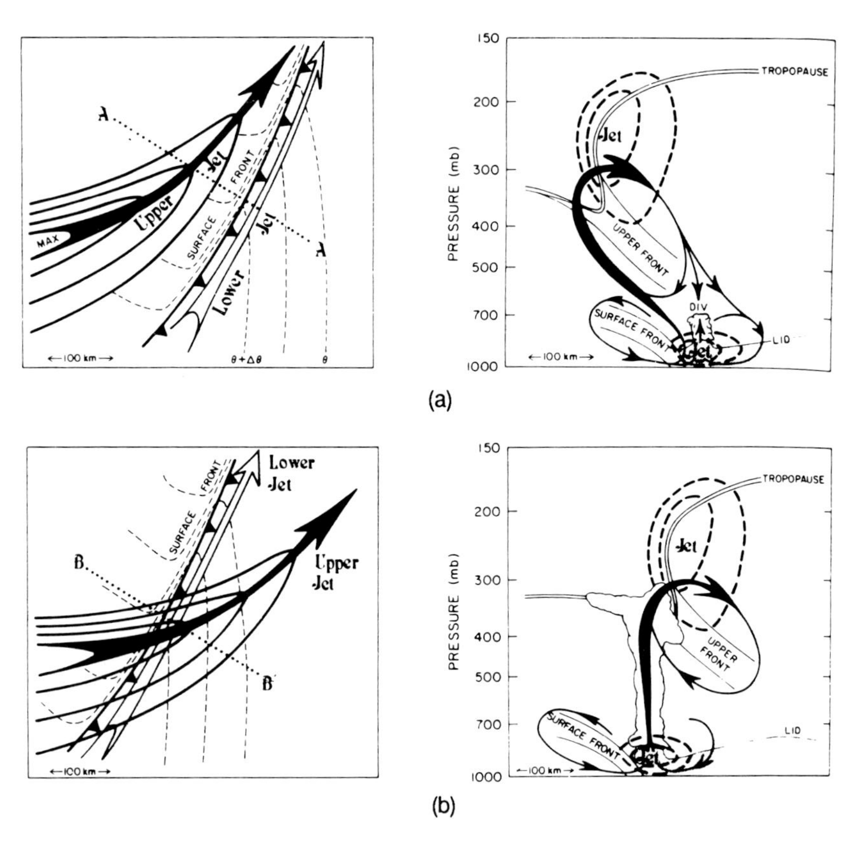

Figure 2.105 (a) Vertically uncoupled upper- and lower-tropospheric jet-front systems and their associated secondary circulation. (Left) Upper jet-front exit region displaced to the west of a surface front and low-level jet. Upper jet isotachs (thick solid lines); upper jet axis (solid arrow); lower jet axis (open arrow); surface potential temperature (thin dashed lines). (Right) Cross section along AA′, left. Upper and lower jet isotachs (thick dashed lines); potential vorticity tropopause (double thin line); moist layer under LID; forced secondary circulation (streamlines with thick arrows). (b) Vertically coupled upper- and lower-tropospheric jet-front systems and their associated secondary circulations. (Left) Upper jet-front exit situated above the surface front and low-level jet. (Right) Cross section along BB′, left. Isopleths and arrows as in (a) (from Shapiro, 1982).

jet streak moves farther east, it becomes "coupled" to the frontal circulation below, and the air mass is destabilized.

In Sec. 2.5.5 we discussed the relationship between surface frontogenesis and surface cyclogenesis. What is the relationship between middle–upper tropospheric frontogenesis and surface cyclogenesis? When a surface cyclone forms downstream from an upper-level trough, a pattern of cold advection is usually formed upstream from the trough. In this case, the environment is conditioned for middle–upper tropospheric frontogenesis upstream from the trough for the reasons discussed in Sec. 2.6.2.

On the other hand, when a middle–upper tropospheric front and its associated jet streak propagate around the base of a trough, the forcing of upward motion is enhanced downstream from the trough; hence, surface cyclogenesis is enhanced. Both middle–upper tropospheric fronts and surface cyclones affect each other, and it may not be correct to generalize that one necessarily *causes* the other.

NOTES

1. "Overrunning" is also used to describe air motion in some warm fronts.
2. If the isotherms are oriented along the axis of contraction, they are not rotated at all; if they are oriented along the axis of dilatation ad initium, then they remain there.
3. A characteristic, we will soon see, that is not much evident in this analysis!
4. See Chap. 3 for a derivation. The Boussinesq equations are valid in an atmosphere in which parcel-density variations are negligible in comparison with density itself, and vertical variations in density are in general negligible. Vertical variations in density, however, are significant in the buoyancy term. We use the Boussinesq equations because they are simpler than the "exact" set of equations.
5. Some of the flow curls upward and backward relative to the density current.
6. Friction may almost balance the pressure-gradient force.
7. The physical mechanism for gravity-wave motion in this case is as follows: The pressure-gradient force causes fluid to flow down the pressure gradient if the magnitude of the Coriolis force is less than that of the pressure-gradient force. If the pressure is relatively high (low) locally, then fluid will diverge from (converge toward) the local pressure anomaly. Thus, the pressure at the high (low) will decrease (increase), owing to the decrease (increase) locally in mass. Eventually the pressure gradients reverse direction as the highs (lows) become lows (highs). Local areas of high and low pressure will thus appear to propagate, as long as the wind field is out of phase with the pressure field.
8. In the Southern Hemisphere, since f is negative, the signs of the exponents in Eqs. (2.5.307) and (2.5.308) are positive.
9. For example, an air parcel that is supergeostrophic coming around a ridge may not attain gradient-wind balance downstream, and will undergo inertial oscillations.
10. The second and third terms in Eqs. (2.8.9) and (2.8.10) are advection terms. In Eqs. (2.8.11) and (2.8.12) we neglect the advection terms.

REFERENCES

General

Hess, S. L., 1959: *Introduction to Theoretical Meteorology*. Holt, Rinehart and Winston, New York.

Palmén, E., and C. W. Newton, 1969: *Atmospheric Circulation Systems*. Academic Press, New York.

Petterssen, S., 1956: *Weather Analysis and Forecasting: Vol. I, Motion and Motion Systems*. McGraw–Hill, New York.

Reiter, E. R., 1963: *Jet-Stream Meteorology*. Univ. of Chicago Press, Chicago.

Section 2.1

Bluestein, H. B., 1986: Fronts and jet streaks: A theoretical perspective. Chap. 9, *Mesoscale Meteorology and Forecasting* (P. Ray, ed.), Amer. Meteor. Soc., Boston, 173–215.

Godson, W. L., 1951: Synoptic properties of frontal surfaces. *Quart. J. Roy. Meteor. Soc.* **77,** 633–53.

Mass, C. F., 1991: Synoptic frontal analysis: Time for a reassessment? *Bull. Amer. Meteor. Soc.* **72,** 348–63.

Palmén, E., 1951: The aerology of extratropical disturbances. *Compendium of Meteorology,* Amer. Meteor. Soc., Boston, 599–620.

Section 2.2.1

Bjerknes, J., 1919: On the structure of moving cyclones. *Geofys. Publ.* **1,** 1–8.

Section 2.2.2

Browning, K. A., and G. A. Monk, 1982: A simple model for the synoptic analysis of cold fronts. *Quart. J. Roy. Meteor. Soc.* **108,** 435–52.

Carbone, R. E., 1982: A severe frontal rainband. Part I: Stormwide hydrodynamic structure. *J. Atmos. Sci.* **39,** 258–79.

Charba, J., 1974: Application of gravity current model to analysis of squall-line gust front. *Mon. Wea. Rev.* **102,** 140–56.

Danielsen, E. F., 1974: The relationship between severe weather, major dust storms and rapid large-scale cyclogenesis (I). *Subsynoptic Extratropical Weather Systems: Observations, Analysis, Modeling and Prediction (Vol. II),* National Center for Atmospheric Research, Boulder, CO, 215–25.

Kreitzberg, C. W., and H. A. Brown, 1970: Mesoscale weather systems within an occlusion. *J. Appl. Meteor.* **9,** 417–32.

Palmén, E., 1948: On the distribution of temperature and wind in the upper westerlies. *J. Meteor.* **5,** 20–27.

Sanders, F., and J. Plotkin, 1966: Detailed analysis of an intense surface cold front. Paper delivered at meeting of Amer. Meteor. Soc., Denver, CO, Jan. 25, 1966.

Shapiro, M. A., 1978: Further evidence of the mesoscale and turbulent structure of upper level jet stream–frontal zone systems. *Mon. Wea. Rev.* **106,** 1100–11.

Shapiro, M. A., T. Hampel, D. Rotzoll, and F. Mosher, 1985: The frontal hydraulic head: A microscale (~1 km) triggering mechanism for mesoconvective weather systems. *Mon. Wea. Rev.* **113,** 1166–83.

Sommers, W. T., 1967: *Mesoscale analysis of complex cold fronts based on surface and tower data.* S.M. Thesis, M.I.T., Cambridge, Mass.

Testud, J., G. Breger, P. Amayenc, M. Chong, B. Nutten, and A. Sauvaget, 1980: A Doppler radar observation of a cold front: Three-dimensional air circulation, related precipitation system and associated wavelike motions. *J. Atmos. Sci.* **37,** 78–98.

Section 2.3.1

Bergeron, T., 1928: Über die dreidimensional verknupfende Wetteranalyse. *Geofys. Publ.* **5,** 1–111.

Bluestein, H. B., 1982: A wintertime mesoscale cold front in the Southern Plains. *Bull. Amer. Meteor. Soc.* **63,** 178–85.

Petterssen, S., 1936: A contribution to the theory of frontogenesis. *Geofys. Publ.* **11,** 1–27.

Section 2.3.2

Miller, J. E., 1948: On the concept of frontogenesis. *J. Meteor.* **5,** 169–71.

Newton, C. W., 1954: Frontogenesis and frontolysis as a three-dimensional process. *J. Meteor.* **11,** 449–61.

Sanders, F., 1955: An investigation of the structure and dynamics of an intense surface frontal zone. *J. Meteor.* **12,** 542–52.

Sanders, F., 1983: Observations of fronts. *Mesoscale Meteorology—Theories, Observations and Models* (D. K. Lilly and T. Gal-Chen, eds.). Reidel, Holland, 175–203.

Section 2.4.1

Baker, D. G., 1970: A study of high pressure ridges to the east of the Appalachian mountains. Ph.D. dissertation, M.I.T., Cambridge, Mass.

Ballentine, R. J., 1980: A numerical investigation of New England coastal frontogenesis. *Mon. Wea. Rev.* **108,** 1479–97.

Beebe, R. G., 1958: An instability line development as observed by the tornado research airplane. *J. Meteor.* **15,** 278–82.

Benjamin, S. G., 1986: Some effects of surface heating and topography on the regional severe storm environment. Part II: Two-dimensional idealized experiments. *Mon. Wea. Rev.* **114,** 330–43.

—, and T. N. Carlson, 1986: Some effects of surface heating and topography on the regional severe storm environment. Part I: Three-dimensional simulations. *Mon. Wea. Rev.* **114,** 307–29.

Bjerknes, J., 1919: On the structure of moving cyclones. *Geofys. Publ.* **1,** 1–8.

—, and H. Solberg, 1922: Life cycle of cyclones and the polar front theory of atmospheric circulation. *Geofys. Publ.* **3,** 1–18.

Bluestein, H. B., 1986: A case study of strong nocturnal convection in Texas. *Preprints, 11th Conf. on Weather Forecasting and Analysis,* Kansas City, MO, Amer. Meteor. Soc., Boston, 223–28.

Bosart, L. F., 1975: New England coastal frontogenesis. *Quart. J. Roy. Meteor. Soc.* **101,** 957–78.

—, 1981: The Presidents' Day snowstorm of 18–19 February 1979: A subsynoptic-scale event. *Mon. Wea. Rev.* **109,** 1542–66.

—, V. Pagnotti, and B. Lettau, 1973: Climatological aspects of eastern United States back-door cold frontal passages. *Mon. Wea. Rev.* **101,** 627–35.

—, C. J. Vaudo, and J. H. Helsdon, Jr., 1972: Coastal frontogenesis. *J. Appl. Meteor.* **11,** 1236–58.

Burgess, D. W., and E. B. Curran, 1985: The relationship of storm type to environment in Oklahoma on 26 April 1984. *Preprints, 14th Conf. on Severe Local Storms,* Indianapolis, IN, Amer. Meteor. Soc., Boston, 208–11.

Carlson, T. N., and F. H. Ludlam, 1968: Conditions for the occurrence of severe local storms. *Tellus* **20,** 203–26.

—, S. G. Benjamin, G. S. Forbes, and Y.-F. Li, 1983: Elevated mixed layers in the regional severe storm environment: Conceptual model and case studies. *Mon. Wea. Rev.* **111,** 1453–73.

Carr, F. H., and J. P. Millard, 1985: A composite study of comma clouds and their association with severe weather over the Great Plains. *Mon. Wea. Rev.* **113,** 370–87.

Carr, J. A., 1951: The East coast "backdoor" front of May 16–20, 1951. *Mon. Wea. Rev.* **79,** 100–5.

Colquhoun, J. R., D. J. Shepherd, C. E., Coulman, R. K. Smith, and K. McInnes, 1985: The Southerly Burster of South Eastern Australia: An orographically forced cold front. *Mon. Wea. Rev.* **113,** 2090–107.

Eldridge, R. H., 1957: A synoptic study of West African disturbance lines. *Quart. J. Roy. Meteor. Soc.* **83,** 303–14.

Fujita, T. T., 1958: Structure and movement of a dry front. *Bull. Amer. Meteor. Soc.* **32,** 1–9.

Hane, C. E., C. L. Ziegler, and H. B. Bluestein, 1992: A summary of dryline experiments conducted in the Southern Plains during COPS-91. *Preprints, Fifth Conf. on Mesoscale Processes,* Atlanta, Amer. Meteor. Soc., Boston, 197–202.

Klemp, J. B., and D. K. Lilly, 1975: The dynamics of wave-induced downslope winds. *J. Atmos. Sci.* **32,** 320–39.

Marks, F. D., and P. M. Austin, 1979: Effects of the New England coastal front on the distribution of precipitation. *Mon. Wea. Rev.* **107,** 53–67.

Mass, C. F., M. D. Albright, and D. J. Brees, 1986: The onshore surge of marine air into the Pacific Northwest: A coastal region of complex terrain. *Mon. Wea. Rev.* **114,** 2602–27.

Matteson, G. T., 1969: The West Texas dry front of June 1967. *M.S. Thesis,* Univ. of Oklahoma, Norman.

McCarthy, J., and S. E. Koch, 1982: The evolution of an Oklahoma dryline. Part I: A meso- and subsynoptic-scale analysis. *J. Atmos. Sci.* **39,** 225–36.

McGuire, E. L., 1962: The vertical structure of three drylines as revealed by aircraft traverses. Rept. No. 7, NSSL, Norman, OK.

Parsons, D. B., M. A. Shapiro, R. M. Hardesty, R. J. Zamora, and J. M. Intrieri, 1991: The finescale structure of a West Texas dryline. *Mon. Wea. Rev.* **119,** 1242–1258.

Penner, C. M., 1955: A three-front model for synoptic analyses. *Quart. J. Roy. Meteor. Soc.* **91,** 89–91.

Rhea, J. O., 1966: A study of thunderstorm formation along drylines. *J. Appl. Meteor.* **5,** 58–63.

Sanders, F., 1955: An investigation of the structure and dynamics of an intense surface frontal zone. *J. Meteor.* **12,** 542–52.

Schaefer, J. T., 1974: The life cycle of the dryline. *J. Appl. Meteor.* **13,** 444–49.

—, 1974: A simulative model of dryline motion. *J. Atmos. Sci.* **31,** 956–64.

—, 1975: Nonlinear biconstituent diffusion: a possible trigger of convection. *J. Atmos. Sci.* **32,** 2278–84.
Shapiro, M. A., 1983: Mesoscale weather systems of the Central United States. *The National STORM Program,* UCAR, Boulder, CO, 3-1–3-77.
—, and D. Keyser, 1990: Fronts, jet streams, and the tropopause. *Extratropical Cyclones* (Chapter 10), Palmén Memorial Volume (C. Newton and E. Holopainen, eds.), American Meteorological Society, Boston, 167–91.
Sun, W.-Y., and Y. Ogura, 1979: Boundary-layer forcing as a possible trigger to a squall-line formation. *J. Atmos. Sci.* **36,** 235–54.
Weston, K. J., 1972: The dry-line of Northern India and its role in cumulonimbus convection. *Quart. J. Roy. Meteor. Soc.* **98,** 519–31.
Ziegler, C. L. and C. E. Hane, 1992: An observational study of the dryline. *Mon. Wea. Rev.* (in review).

Section 2.4.2

Bosart, L. F., 1970: Mid-tropospheric frontogenesis. *Quart. J. Roy. Meteor. Soc.* **96,** 442–71.
Danielsen, E. F., 1968: Stratospheric–tropospheric exchange based on radioactivity, ozone, and potential vorticity. *J. Atmos. Sci.* **25,** 502–18.
Hoskins, B. J., M. E. McIntyre, and A. W. Robertson, 1985: On the use and significance of isentropic potential vorticity maps. *Quart. J. Roy. Meteor. Soc.* **111,** 877–946.
Keyser, D., 1986: Atmospheric fronts: An observational perspective. Chap. 10, *Mesoscale Meteorology and Forecasting* (P. Ray, ed.), Amer. Meteor. Soc., Boston, 216—58.
—, and M. A. Shapiro, 1986: A review of the structure and dynamics of upper-level frontal zones. *Mon. Wea. Rev.* **114,** 452–99.
Reed, R. J., 1955: A study of a characteristic type of upper-level frontogenesis. *J. Meteor.* **12,** 226–37.
—, and F. Sanders, 1953: An investigation of the development of a mid-tropospheric frontal zone and its associated vorticity field. *J. Meteor.* **10,** 338–49.
Sanders, F., L. F. Bosart, and C.-C. Lai, 1991: Initiation and evolution of an intense upper-level front. *Mon. Wea. Rev.* **119,** 1337–67.
Shapiro, M. A., 1980: Turbulent mixing within tropopause folds as a mechanism for the exchange of chemical constituents between the stratosphere and troposphere. *J. Atmos. Sci.* **37,** 994–1004.
Staley, D. O., 1960: Evaluation of potential-vorticity changes near the tropopause and the related vertical motions, vertical advection of vorticity, and transfer of radioactive debris from stratosphere to troposphere. *J. Meteor.* **17,** 591–620.

Section 2.5.1

Eliassen, A., 1959: On the formation of fronts in the atmosphere. *The Atmosphere and the Sea in Motion* (B. Bolin, ed.), Rockefeller Inst. Press, New York, 277–87.
Hoskins, B. J., and M. A. Pedder, 1980: The diagnosis of middle latitude synoptic development. *Quart. J. Roy. Meteor. Soc.* **106,** 707–19.
—, and N. V. West, 1979: Baroclinic waves and frontogenesis. Part II: Uniform potential vorticity jet flows—cold and warm fronts. *J. Atmos. Sci.* **36,** 1663–80.
Keyser, D., M. J. Reeder, and R. J. Reed, 1988: A generalization of Petterssen's

frontogenesis function and its relation to the forcing of vertical motion. *Mon. Wea. Rev.* **116,** 762–80.
Kirk, T. H., 1966: Some aspects of the theory of fronts and frontal analysis. *Quart. J. Roy. Meteor. Soc.* **92,** 374–81.
Namias, J., and P. F. Clapp, 1949: Confluence theory of the high tropospheric jet stream. *J. Meteor.* **6,** 330–36.
Petterssen, S., and J. M. Austin, 1942: Fronts and frontogenesis in relation to vorticity. Papers in Physical Oceanography and Meteorology, M.I.T., and Woods Hole Oceanographic Institution, Cambridge and Woods Hole, Mass.
Stone, P. H., 1966: Frontogenesis by horizontal wind deformation fields. *J. Atmos. Sci.* **23,** 455–65.
Williams, R. T., 1968: A note on quasi-geostrophic frontogenesis. *J. Atmos. Sci.* **25,** 1157–59.
—, and J. Plotkin, 1968: Quasi-geostrophic frontogenesis. *J. Atmos. Sci.* **25,** 201–6.

Section 2.5.2

Blumen, W., 1981: The geostrophic coordinate transformation. *J. Atmos. Sci.* **38,** 1100–5.
Eady, E. T., 1949: Long waves and cyclone waves. *Tellus* **1,** 33–52.
Eliassen, A., 1948: The quasi-static equations of motion. *Geofys. Publ.* **17**(3), 1–44.
—, 1962: On the vertical circulation in frontal zones. *Geofys. Publ.* **24,** 147–60.
—, 1966: Motions of intermediate scale: Fronts and cyclones. *Advances in Earth Science* (P. M. Hurley, ed.), M.I.T. Press, Cambridge, Mass., 111–38.
—, 1984: Geostrophy. *Quart. J. Roy. Meteor. Soc.* **110,** 1–12.
Emanuel, K. A., 1979: Inertial instability and mesoscale convective systems. Part I: Linear theory of inertial instability in rotating viscous fluids. *J. Atmos. Sci.* **36,** 2425–49.
—, 1983: Symmetric instability. *Mesoscale Meteorology—Theories, Observations and Models* (D. K. Lilly and T. Gal-Chen, eds.) D. Reidel, Holland, 217–29.
Gidel, L. T., 1978: Simulation of the differences and similarities of warm and cold surface frontogenesis. *J. Geophys. Res.* **83,** 915–28.
Hoskins, B. J., 1971: Atmospheric frontogenesis models: some solutions. *Quart. J. Roy. Meteor. Soc.* **97,** 139–53.
—, 1974: The role of potential vorticity in symmetric stability and instability. *Quart. J. Roy. Meteor. Soc.* **100,** 480–82.
—, 1975: The geostrophic momentum approximation and the semi-geostrophic equations. *J. Atmos. Sci.* **32,** 233–42.
—, 1982: The mathematical theory of frontogenesis. *Ann. Rev. Fluid Mech.* **14,** 131–51.
—, and F. P. Bretherton, 1972: Atmospheric frontogenesis models: Mathematical formulation and solution. *J. Atmos. Sci.* **29,** 11–37.
—, and I. Draghici, 1977: The forcing of ageostrophic motion according to the semi-geostrophic equations and in an isentropic coordinate model. *J. Atmos. Sci.* **34,** 1859–67.
McWilliams, J. C., and P. R. Gent, 1980: Intermediate models of planetary circulations in the atmosphere and ocean. *J. Atmos. Sci.* **37,** 1657–78.
Salmon, R., 1985: New equations for nearly geostrophic flow. *J. Fluid Mech.* **153,** 461–77.
Sawyer, J. S., 1956: The vertical circulation at meteorological fronts and its relation to frontogenesis. *Proc. Roy. Soc. London* **A234,** 346–62.

Snyder, C., W. Skamarock, and R. Rotunno, 1991: A comparison of primitive equation and semi-geostrophic simulations of baroclinic waves. *J. Atmos. Sci.* **48,** 2179–94.
Stone, P. H., 1966: On non-geostrophic baroclinic stability. *J. Atmos. Sci.* **23,** 390–400.
Williams, R. T., 1967: Atmospheric frontogenesis: A numerical experiment. *J. Atmos. Sci.* **24,** 627–41.
—, 1972: Quasi-geostrophic versus non-geostrophic frontogenesis. *J. Atmos. Sci.* **29,** 3–10.
—, 1974: Numerical simulation of steady-state fronts. *J. Atmos. Sci.* **31,** 1286–96.
Yudin, M. I., 1955: Invariant quantities in large-scale atmospheric processes. *Tr. Glav. Geogiz. Obser.,* No. 55, 3–12.

Section 2.5.3

Bluestein, H. B., 1986: Fronts and jet streaks: A theoretical perspective. (Chap. 9), *Mesoscale Meteorology and Forecasting* (P. Ray, ed.) Amer. Meteor. Soc., Boston, 173–215.
Keyser, D., and R. A. Anthes, 1982: The influence of planetary boundary layer physics on frontal structure in the Hoskins–Bretherton horizontal shear model. *J. Atmos. Sci.* **39,** 1783–802.

Section 2.5.4

Bluestein, H. B., 1977: Synoptic-scale deformation and tropical cloud bands. *J. Atmos. Sci.* **34,** 891–900.
Browning, K. A., and C. W. Pardoe, 1973: Structure of low-level jet streams ahead of mid-latitude cold fronts. *Quart. J. Roy. Meteor. Soc.* **99,** 619–38.
Faller, A. J., 1956: A demonstration of fronts and frontal waves in atmospheric models. *J. Meteor.* **13,** 1–4.
Hoskins, B. J., 1972: Discussions. *Quart. J. Roy. Meteor. Soc.* **98,** 862.
Ross, B. B., and I. Orlanski, 1978: The circulation associated with a cold front. Part II: Moist case. *J. Atmos. Sci.* **35,** 445–65.
Williams, R. T., L. C. Chou, and C. J. Cornelius, 1981: Effects of condensation and surface motion on the structure of steady-state fronts. *J. Atmos. Sci.* **38,** 2365–76.

Section 2.5.5

Bjerknes, J., 1919, op. cit.
Eliassen, A., 1966: Motions of intermediate scale: Fronts and cyclones. *Advances in Earth Science* (P. M. Hurley, ed.), M.I.T. Press, Cambridge, Mass., 111–38.
Hoskins, B. J., and N. V. West, 1979: Baroclinic waves and frontogenesis. Part II: Uniform potential vorticity jet flows—cold and warm fronts. *J. Atmos. Sci.* **36,** 1663–80.
—, and W. A. Heckley, 1981: Cold and warm fronts in baroclinic waves. *Quart. J. Roy. Meteor. Soc.* **107,** 79–90.
Keyser, D., and M. J. Pecnick, 1985: A two-dimensional primitive equation model of frontogenesis forced by confluence and horizontal shear. *J. Atmos. Sci.* **42,** 1259–82.

Section 2.5.6

Bell, G. D., and L. F. Bosart, 1988: Appalachian cold-air damming. *Mon. Wea. Rev.* **116,** 137–61.

Benjamin, J. B., 1968: Gravity current and related phenomena. *J. Fluid, Mech.* **31,** 209–48.

Bosart, L. F., 1983: Analysis of a California Catalina Eddy event. *Mon. Wea. Rev.* **111,** 1619–33.

Dunn, L. B., 1992: Evidence of ascent in a sloped barrier jet and an associated heavy-snow band. *Mon. Wea. Rev.* **120,** 914–24.

Forbes, G. S., R. A. Anthes, and D. W. Thomson, 1987: Synoptic and mesoscale aspects of an Appalachian ice storm associated with cold-air damming. *Mon. Wea. Rev.* **115,** 564–91.

Hobbs, P. V., and P. O. G. Persson, 1982: The mesoscale and microscale structure and organization of clouds and precipitation in midlatitude cyclones. Part V: The substructure of narrow cold-frontal rainbands. *J. Atmos. Sci.* **39,** 280–95.

Holton, J. R., 1979: *An Introduction to Dynamic Meteorology.* Academic Press, New York.

Manins, P. C., and B. L. Sawford, 1982: Mesoscale observations of upstream blocking. *Quart. J. Roy. Meteor. Soc.* **108,** 427–34.

Mass, C. F., and M. D. Albright, 1987: Coastal southerlies and alongshore surges of the West Coast of North America: Evidence of mesoscale topographically trapped response to synoptic forcing. *Mon. Wea. Rev.* **115,** 1707–38.

—, and M. O. Albright, 1989: Origin of the Catalina Eddy. *Mon. Wea. Rev.* **117,** 2406–36.

Rotunno, R., J. B. Klemp, and M. L. Weisman, 1988: A theory for squall lines. *J. Atmos. Sci.* **45,** 463–85.

Seitter, K. L., 1986: A numerical study of atmospheric density current motion including the effects of condensation. *J. Atmos. Sci.* **43,** 3068–76.

Simpson, J. E., and R. E. Britter, 1980: A laboratory model of an atmospheric mesofront. *Quart. J. Roy. Meteor. Soc.* **106,** 485–500.

Wakimoto, R. M., 1982: The life cycle of thunderstorm gust fronts as viewed with Doppler radar and rawinsonde data. *Mon. Wea. Rev.* **110,** 1050–82.

—, 1987: The Catalina Eddy and its effect on pollution over Southern California. *Mon. Wea. Rev.* **115,** 837–55.

Section 2.5.7

Dorman, C. E., 1985: Evidence of Kelvin waves in California's marine layer and related eddy generation. *Mon. Wea. Rev.* **133,** 827–39.

Gill, A. E., 1977: Coastally trapped waves in the atmosphere. *Quart. J. Roy. Meteor. Soc.* **103,** 431–40.

Holton, J. R., 1979: *An Introduction to Dynamic Meteorology,* Academic Press, New York.

Pedlosky, J., 1979: *Geophysical Fluid Dynamics.* Springer-Verlag, New York.

Section 2.6.2

Buzzi, A., A. Trevisan, and G. Salustri, 1981: Internal frontogenesis: A two-dimensional model in isentropic, semi-geostrophic coordinates. *Mon. Wea. Rev.* **109,** 1053–60.

Keyser, D., and M. J. Pecnick, 1985a: A two-dimensional primitive equation model of frontogenesis forced by confluence and horizontal shear. *J. Atmos. Sci.* **42,** 1259–82.

—, and M. J. Pecnick, 1985b: Diagnosis of ageostrophic circulations in a two-dimensional primitive equation model of frontogenesis. *J. Atmos. Sci.* **42,** 1283–305.
—, and M. A. Shapiro, 1986: A review of the structure and dynamics of upper-level frontal zones. *Mon. Wea. Rev.* **114,** 452–99.
Mudrick, S. E., 1974: A numerical study of frontogenesis. *J. Atmos. Sci.* **31,** 869–92.
Shapiro, M. A., 1970: On the applicability of the geostrophic approximation to upper-level frontal-scale motions. *J. Atmos. Sci.* **27,** 408–20.
—, 1981: Frontogenesis and geostrophically forced secondary circulations in the vicinity of jet stream–frontal zone systems. *J. Atmos. Sci.* **38,** 954–73.
—, 1983, op. cit.

Section 2.7

Berggren, R., W. J. Gibbs, and C. W. Newton, 1958: Observational characteristics of the jet stream. A survey of the literature. WMO, TN No. 19.
Lorenz, E. N., 1967: *The Nature and Theory of the General Circulation of the Atmosphere,* WMO, Geneva.
Shapiro, M. A., T. Hample, and D. W. van de Kamp, 1984: Radar wind profiler observations of fronts and jet streams. *Mon. Wea. Rev.* **112,** 1263–66.

Section 2.7.1

Newton, C. W., and A. V. Persson, 1962: Structural characteristics of the subtropical jet stream and certain lower-stratospheric wind systems. *Tellus* **14,** 221–41.
Palmén, E., 1948: On the distribution of temperature and wind in the upper westerlies. *J. Meteor.* **5,** 20–27.
—, and C. W. Newton, 1948: A study of the mean wind and temperature distribution in the vicinity of the polar front in winter. *J. Meteor.* **5,** 220–26.

Section 2.7.2

Defant, F., and H. Taba, 1957: The threefold structure of the atmosphere and the characteristics of the tropopause. *Tellus* **9,** 259–74.
Durran, D. R. and D. B. Weber, 1988: An investigation of the poleward edges of cirrus clouds associated with midlatitude jet streams. *Mon. Wea. Rev.* **116,** 702–14.
Krishnamurti, T. N., 1961: The subtropical jet stream in winter. *J. Meteor.* **18,** 172–91.
Newton, C. W., and A. V. Persson, 1962: Structural characteristics of the subtropical jet stream and certain lower-stratospheric wind systems. *Tellus* **14,** 221–41.
Riehl, H., M. A. Alaka, C. L. Jordan, and R. J. Renard, 1954: The jet stream. *Meteor. Mono.,* No. 7, Amer. Meteor. Soc., Boston.
Whitney, L. F., Jr., 1977: Relationship of the subtropical jet stream to severe local storms. *Mon. Wea. Rev.* **105,** 398–412.
—, A. Timchalk, and T. I. Gray, Jr., 1966: On locating jet streams from TIROS photographs. *Mon. Wea. Rev.* **94,** 127–38.

Section 2.7.3

Bonner, W. D., 1968: Climatology of the low level jet. *Mon. Wea. Rev.* **96,** 833–50.
Browning, K. A., and C. W. Pardoe, 1973: Structure of low-level jet streams ahead of mid-latitude cold fronts. *Quart. J. Roy. Meteor. Soc.* **99,** 619–38.

Stensrud, D. J., M. H. Jain, K. W. Howard, and R. A. Maddox, 1990: Operational systems for observing the lower atmosphere: Importance of data sampling and archival procedures. *J. Atmos. and Oceanic Tech.* **7,** 930–48.

Section 2.7.4

Newton, C. W., 1981: Lagrangian partial-inertial oscillations, and subtropical and low-level monsoon jet streaks. *Mon. Wea. Rev.* **109,** 2474–86.

Section 2.8.1

Bluestein, H. B., 1986: Fronts and jet streaks: A theoretical perspective. (Chap. 9), *Mesoscale Meteorology and Forecasting* (P. Ray, ed.), Amer. Meteor. Soc., Boston, 173–215.

Buajitti, K., and A. K. Blackadar, 1957: Theoretical studies of diurnal wind structure variations in the planetary boundary layer. *Quart. J. Roy. Meteor. Soc.* **83,** 486–500.

Holton, J. R., 1967: The diurnal boundary layer wind oscillation above sloping terrain. *Tellus* **19,** 199–205.

Wexler, H., 1961: A boundary layer interpretation of the low-level jet. *Tellus* **13,** 369–78.

Section 2.8.2

Beebe, R. G., and F. C. Bates, 1955: A mechanism for assisting in the release of convective instability. *Mon. Wea. Rev.* **83,** 1–10.

Bluestein, H. B., 1986: Fronts and jet streaks: A theoretical perspective. (Chap. 9), *Mesoscale Meteorology and Forecasting* (P. Ray, ed.), Amer. Meteor. Soc., Boston, 173–215.

—, and K. W. Thomas, 1984: Diagnosis of a jet streak in the vicinity of a severe weather outbreak in the Texas Panhandle. *Mon. Wea. Rev.* **112,** 2501–22.

Bosart, L. F., and S. C. Lin, 1984: A diagnostic analysis of the Presidents' Day storm of February 1979. *Mon. Wea. Rev.* **112,** 2148–77.

Hoskins, B. J., and I. Draghici, 1977: The forcing of geostrophic motion according to the semi-geostrophic equations in an isentropic coordinate model. *J. Atmos. Sci.* **34,** 1859–67.

Murray, R., and S. M. Daniels, 1953: Transverse flow at entrance and exit to jet streams. *Quart. J. Roy. Meteor. Soc.* **79,** 236–41.

Namias, J., and P. F. Clapp, 1949: Confluence theory of the high tropospheric jet stream. *J. Meteor.* **6,** 330–36.

Newton, C. W., and A. Trevisan, 1984a: Clinogenesis and frontogenesis in jet-stream waves. Part I: Analytical relations to wave structure. *J. Atmos. Sci.* **41,** 2717–34.

—, 1984b: Clinogenesis and frontogenesis in jet-stream waves. Part II: Channel model numerical experiments. *J. Atmos. Sci.* **41,** 2735–55.

Section 2.8.3

Bluestein, H. B., 1986: Fronts and jet streaks: A theoretical perspective. (Chap. 9), *Mesoscale Meteorology and Forecasting* (P. Ray, ed.), Amer. Meteor. Soc., Boston, 173–215.

Section 2.8.4

Beebe, R. G., and F. C. Bates, 1955: A mechanism for assisting in the release of convective instability. *Mon. Wea. Rev.* **83,** 1–10.

Fawbush, E. J., R. C. Miller, and L. G. Starrett, 1951: An empirical method of forecasting tornado development. *Bull. Amer. Meteor. Soc.* **32,** 1–9.

Kocin, P. J., L. W. Uccellini, and R. A. Petersen, 1986: Rapid evolution of a jet streak circulation in a pre-convective environment. *Meteor. Atmos. Phys.* **35,** 103–38.

Shapiro, M. A., 1982: *Mesoscale Weather Systems of the Central United States.* CIRES, Univ. of Colo./NOAA, Boulder, Colo.

Uccellini, L. W., and D. R. Johnson, 1979: The coupling of upper and lower tropospheric jet streaks and implications for the development of severe convective storms. *Mon. Wea. Rev.* **107,** 682–703.

—, P. J. Kocin, R. A. Petersen, C. H Wash, and K. F. Brill, 1984: The Presidents' Day cyclone of 18–19 February 1979: Synoptic overview and analysis of the subtropical jet streak influencing the pre-cyclogenetic period. *Mon. Wea. Rev.* **112,** 31–55.

PROBLEMS

2.1. Suppose that $\partial\theta/\partial y = -a$ at $t = 0$, $u = cx$, and $v = -cy$, where a is a positive constant and c is a constant. How long will it take $-\partial\theta/\partial y$ to increase an order of magnitude (i.e., increase by a factor of 10) (a) if $c > 0$? (b) if $c < 0$? neglect vertical motions and diabatic effects.

2.2. What are the differences among the polar-front jet, the subtropical jet, and the low-level jet? Consider the location, height, behavior, strength, and mechanism of formation.

2.3. Consider the following wind and potential-temperature (θ) fields shown in the figure. (a) From what direction does the ageostrophic wind blow aloft if the atmosphere is in quasigeostrophic equilibrium? Why? (b) Same as (a), but below? (c) Where is vorticity created at the surface? Why?

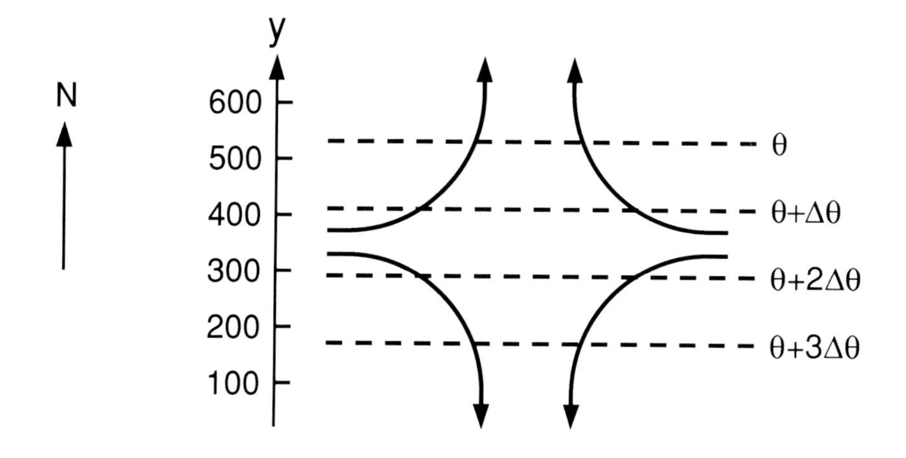

2.4. An east–west-oriented cold front moves southward through stations A, B, and C. Station A is 100 km north of station B, and station B is 100 km north of station C. Thermograph traces at locations A, B, and C at 850 mb on a *huge* tower are displayed in the figure. (a) What is the strength [$K\ (100\ \mathrm{km})^{-1}$] of the front? (b) Note that the strength of the front remains *constant* in the reference frame of the front. What must the meridional gradient of ω [$\mu \mathrm{b\ s}^{-1}\ (100\ \mathrm{km}^{-1}$] be at 850 mb if the flow is adiabatic and if the wind at all stations before the change in temperature is 10 knots from the south, and after the change is 20 knots from the north? Assume that the potential temperature at 500 mb is 15°C greater than the potential temperature at 850 mb, and that the potential-temperature lapse rate ($\partial\theta/\partial p$) is constant. (c) If $\omega = 0$ at 850 mb in the middle of the frontal zone, what is w ($\mathrm{cm\ s}^{-1}$) along the edges of the frontal zone? Use the result from part (b).

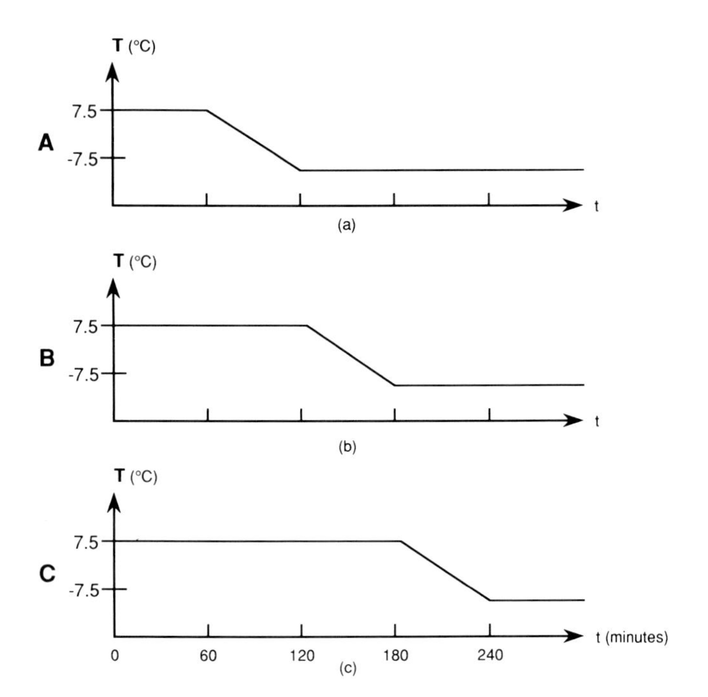

2.5. Estimate the ageostrophic wind speed ($\mathrm{m\ s}^{-1}$) and direction at point A in the figure (page 419) if the flow is steady state and horizontal. Isotachs in $\mathrm{m\ s}^{-1}$ (dashed lines).

2.6. What three physical effects contribute to frontolysis?

2.7. Why are low-level fronts stronger at the surface than aloft?

2.8. Name three characteristics that distinguish low-level (surface) fronts from upper-(mid-) level fronts. Consider their location in the horizontal, their vertical circulation, and their moisture content.

2.9. Explain how horizontally nondivergent synoptic-scale deformation acting on a synoptic-scale horizontal temperature gradient can lead to

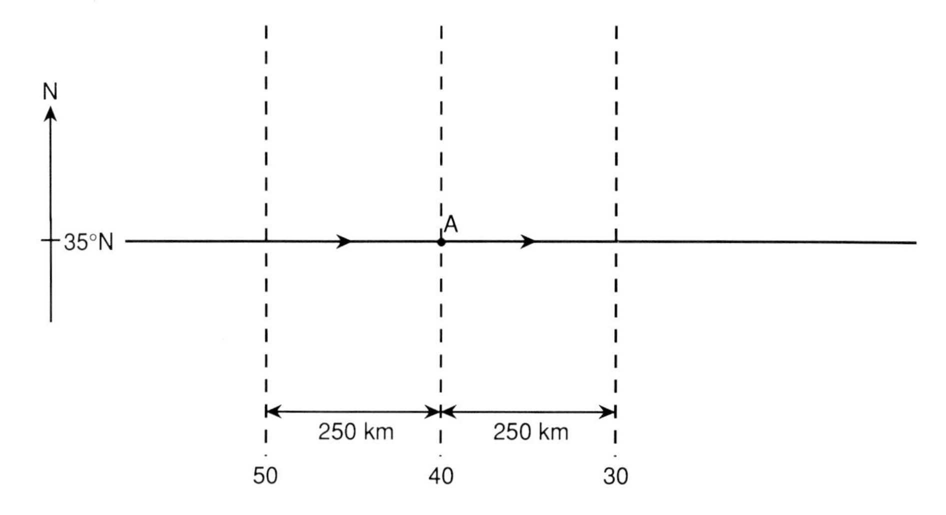

rapid surface frontogenesis, and can result in the production of surface cyclonic vorticity and surface convergence.

2.10. Why are jets associated with fronts?

2.11. How long (days) will it take the horizontal temperature gradient following an air parcel to increase an order of magnitude if there is confluence of $10^{-4}\,\mathrm{s}^{-1}$ acting along the isotherms? Neglect adiabatic temperature changes associated with vertical motions and diabatic heating.

2.12. Show that if the surface isotherms are parallel to the axis of contraction, a circulation will be set up in which the cold air rises and warm air sinks.

2.13. Name three physical mechanisms that contribute to frontogenesis.

2.14. (a) Draw a "balance-of-force" diagram at the exit region of an anticyclonically curved jet streak. Neglect friction. (b) Sketch the geostrophic, ageostrophic, and total wind vectors. Be consistent with part (a); that is, show your vectors in relation to the force vectors.

2.15. Suppose that air at the surface flows uniformly from the west (as indicated in the figure) and has the same temperature at the ground

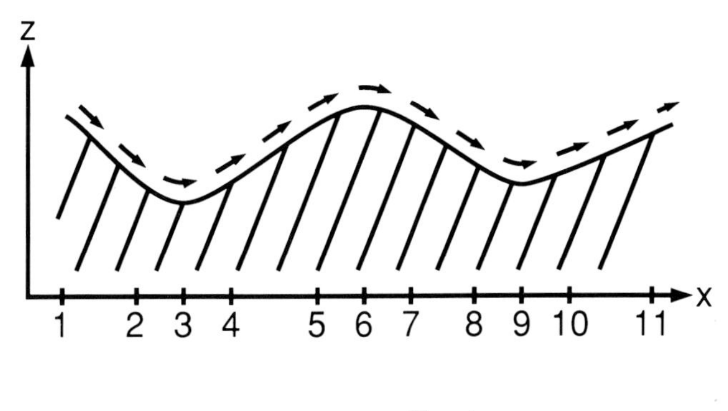

initially. You may assume that the atmsophere is statically stable. Neglect friction. Assume that $v = 0$. (a) For what values of x will there be frontogenesis at the surface? Why? (b) For what values of x will there be a production of cyclonic vorticity at the surface as a result of the frontogenetical process? Why?

2.16. (a) Consider the distributions of quasigeostrophic vertical velocity ω (left) and of potential temperature θ (middle and right) shown in the figure. In the absence of deformation and diabatic heating, what is the rate of change of horizontal potential temperature gradient following air-parcel motion [K (100 km)$^{-1}$ d^{-1}]? (b) Describe qualitatively the vertical circulation as a function of x; that is, for what values of x is there rising motion? Sinking motion? What does the ageostrophic wind field look like?

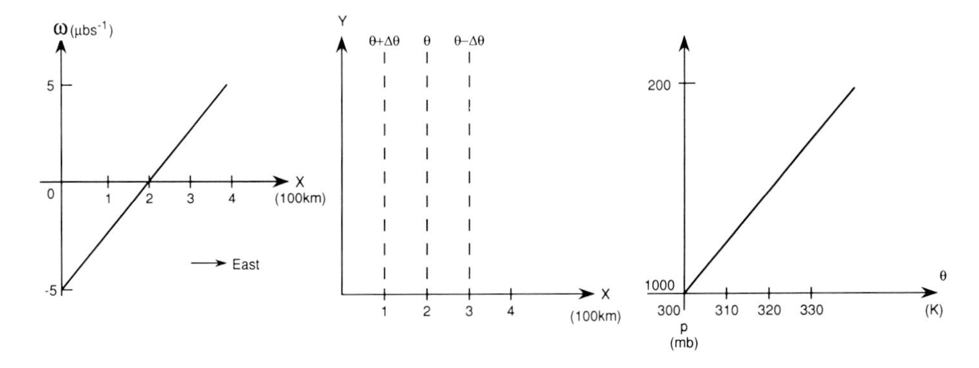

2.17. Why, according to theory, should the surface wind-shift line accompanying a frontal zone shift from the "warm" edge of the frontal zone to the "cold" edge of the frontal zone if the frontal zone suddenly experiences frontolysis?

2.18. Suppose that $u = Dx$, $v = -Dy$, and the potential temperature

$$\theta = Ax + \theta_0, \qquad \text{for } |x| \leq R$$

$$\theta = \begin{cases} \theta_1, & \text{for } x \leq -R \\ \theta_2, & \text{for } x \geq R, \end{cases}$$

where D, A, θ_0, θ_1, θ_2, and R are positive constants. Consider only values of y in the midlatitude region of the Northern Hemisphere. Assume the x axis points toward the east. (a) Neglect vertical motion, friction, and diabatic heating. What is the frontogenetical function in terms of D, R, θ_1, and θ_2 for $|x| < R$? (b) Describe qualitatively the ensuing vertical circulation. Why does it develop? (c) Where is anticyclonic vorticity generated? Why?

2.19. Consider an intense, steady-state, stationary, circular, upper-level cyclone in the Northern Hemisphere in midlatitudes in which the isobars are evenly spaced concentric circles. Draw a qualitative sketch of the streamlines and isotachs of the ageostrophic windfield. Justify your sketch. (Assume the Rossby number is on the order of 1.) Neglect the latitudinal variation in f.

2.20. Consider the geopotential-height and temperature fields shown in the figure. Assume the wind field is in gradient-wind balance. (a) Describe the vertical velocity field as a function of x in a qualitative manner (i.e., where is $\omega < 0$, $\omega > 0$?). (b) Where will anticyclonic vorticity be generated? (c) Is the vertical circulation thermally direct or indirect? Why?

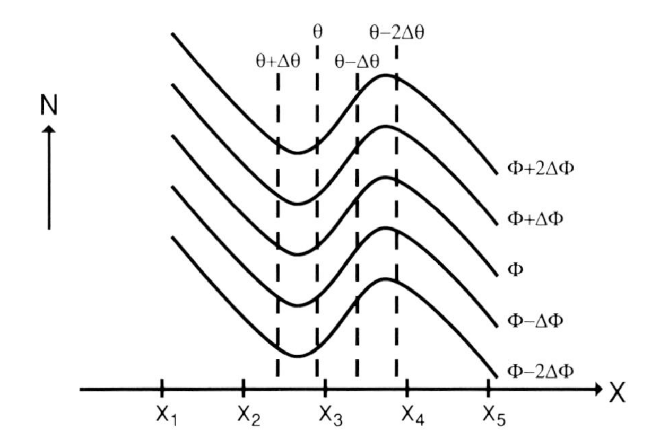

2.21. Why does the dryline move eastward during the day and westward at night in the absence of strong synoptic-scale forcing?

2.22. Consider a zonally oriented frontal zone on a level suface. Suppose that the two-dimensional frontogenetical function is zero. If the axis of contraction is parallel to the potential isotherms, what is the zonal gradient of the zonal wind if the resultant deformation is $3 \times 10^{-5}\ \mathrm{s}^{-1}$? Be sure to indicate the correct units. Neglect diabatic heating.

2.23. Imagine that at 500 mb the wind field is one of pure deformation, the axis of dilatation is oriented northeast–southwest, and isotherms are zonally oriented. How much is the meridional gradient in vertical velocity ($\partial\omega/\partial y$) at 500 mb if the meridional temperature gradient following air-parcel motion increases threefold (i.e., triples) in a day from a starting value of 10 K/500 km? Express your answer in units of $\mu\mathrm{b\ s}^{-1}\ (1000\ \mathrm{km})^{-1}$. Assume the potential temperature of the air parcel is 308 K, the lapse rate is zero (i.e., $\partial T/\partial z = 0$), and that there is no diabatic heating.

2.24. Briefly explain why tropopause folds may be associated with clear-air turbulence.

2.25. (a) Derive an analytic expression that describes the necessary condition for symmetric instability in the atmosphere described by Sanders' analytic model. (b) Find a set of parameters (a, L, $\hat{T}$, etc.) for which the atmosphere is in fact symmetrically unstable.

2.26. Suppose the slope of an m_g surface at 30°N is 1/400. If there is no geostrophic vorticity and the temperature is 5°C, what is the magnitude of the temperature gradient [K (100 km)$^{-1}$] at 700 mb?

2.27. At 40°N, what is the geostrophic shear across a front in which the geostrophic coordinate across the front stretches space by a factor of 5?

2.28. Consider a coordinate system moving along with the *ageostrophic* component of the wind. What is the relationship between X in the ageostrophic coordinate system and y in real space?

2.29. (a) Suppose that the atmosphere is barotropic. Use the Sawyer–Eliassen equation to find an analytic solution to the ageostrophic vertical circulation induced by a diabatic heating source $\overline{dQ}/dt = -A_0(p/R)\cos[(2\pi/L)y]\sin[(2\pi/p_0)p]$, where A_0 is a positive constant, and $-(\partial\theta/\partial p)(R/f_0 p)(p/p_0)^{\kappa} = a = \text{constant} > 0$; $f_0 - \partial u_g/\partial y = c = \text{constant} > 0$. Assume that $v_a = 0$ at $y = \pm L/2$, and $\omega = 0$ at $p = p_0$ and $p_0/2$. (b) Repeat (a), but for the quasigeostrophic version of the Sawyer–Eliassen equation. Contrast the solutions found in (a) and (b).

2.30. (a) List the successes and failures of quasigeostrophic theory in describing surface fronts. (b) Which failures does the geostrophic-momentum approximation correct? Why?

2.31. What synoptic features favor: (a) symmetric instability (b) baroclinic instability?

2.32. Suppose that air having a temperature of -10°C blows out over the Gulf Stream from North America with a speed of 15 m s^{-1}. Suppose the sea surface temperature varies across the Gulf Stream as shown in the figure. (a) If the drag coefficient is 1.2×10^{-3}, the sea-level pressure

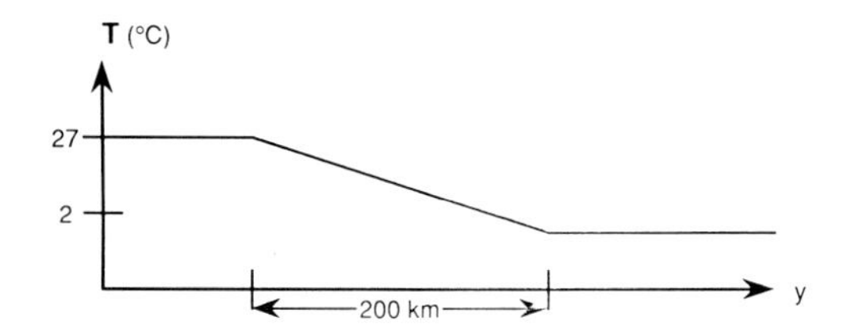

is 1000 mb, and the frontogenetical function is zero, how much confluence or diffluence is acting across the zone of sea-surface temperature gradient? Assume the air temperature at the surface is the same as the sea-surface temperature, and any diabatic heating from the ocean surface is evenly distributed in the lowest 100 mb, and zero above. (b) Is the confluence or diffluence computed in (a) consistent with quasigeostrophic theory? Why or why not?

2.33. (a) Explain why a shallow cold front is more likely to behave like a density current in the presence of a mountain range. (b) Explain "cold-air damming." When and why does it occur?

2.34. (a) Why does a dryline retreat westward at night under quiescent synoptic conditions? Why may it behave like a density current? (b) Consider the surface map shown in the figure. Using the formula $c = K[gH(T_{vw} - T_{vc})/T_{vc}]^{1/2}$, where $K = 1$, estimate the westward speed of the dryline near Abilene, TX (ABI) as a density current. Abilene, TX is at 542 m MSL, and the moist layer extends up to 2 km MSL east of the dryline. MAF is at 874 m MSL, and SEP is at 398 km MSL. (c) What factors other than errors in estimating K, T_{vw}, T_{vc}, and H would introduce errors into your estimate of c?

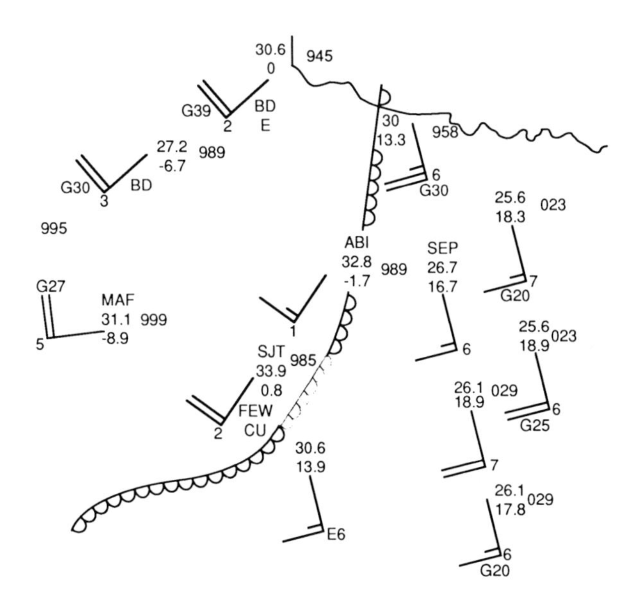

2.35. Within what approximate distance from a midlatitude mountain range (3 km higher than the surrounding ground) may Kelvin waves be excited in an air mass near the ground which is 20°C cooler than the air mass 1 km above?

2.36. In quasigeostrophic frontogenesis, static *instability* can be generated. If the temperature gradient of an air parcel is 10°C/100 km, and static stability does not change following air-parcel motion, what must the vertical shear of the ageostrophic wind [m s^{-1} (100 kPa)$^{-1}$] be if 10^{-6} s^{-1} units convergence act on typical values of static stability at the ground? The action of the ageostrophic wind field on the temperature field, which prevents the generation of static instability, is not included in quasigeostrophic theory.

2.37. What is the Richardson number at 35°N in an isothermal ($T = -21$°C) layer at 500 mb if there is a quasihorizontal temperature gradient of 10°C/100 km? Neglect the vertical shear of the ageostrophic wind.

2.38. Suppose that the frontogenetical function at the ground, where the pressure is 850 mb, is -10 K (100 km)$^{-1}$ d^{-1} and the magnitude of the temperature gradient is 25 K (100 km)$^{-1}$. If there is *convergence* of 5×10^{-5} s^{-1}, and the axis of dilatation is normal to the isotherms, what is $\partial u/\partial x$? Assume the x axis is oriented along the axis of dilatation. Neglect friction and diabatic heating.

2.39. Where does quasigeostrophic surface frontogenesis produce a region of static instability? Why? Why does semigeostrophic frontogenesis *not* produce the region of static stability?

2.40. Derive the vorticity equation from the frictionless equations of motion subject to the geostrophic-momentum approximation.

2.41. Show that

$$\frac{\partial}{\partial t} + (u_g + u_a)\frac{\partial}{\partial x} + (v_g + v_a)\frac{\partial}{\partial y} + \omega\frac{\partial}{\partial p}$$

$$= \frac{\partial}{\partial t^*} + u_g\frac{\partial}{\partial X} + v_g\frac{\partial}{\partial Y} + \omega\frac{\partial}{\partial P},$$

where X, Y, P, and t^* are the independent variables in geostrophic coordinates.

2.42. Express the ageostrophic wind in terms of the nth order of the parcel acceleration, that is, $D^n\mathbf{v}/Dt^n$.

2.43. Consider the height (solid lines; dam) and temperature (dashed lines; °C) fields at 500 mb shown in the figure. Is the necessary condition for symmetric instability satisfied at point P? Assume that the potential temperature increases 2°C km^{-1} with respect to height. Show all your work.

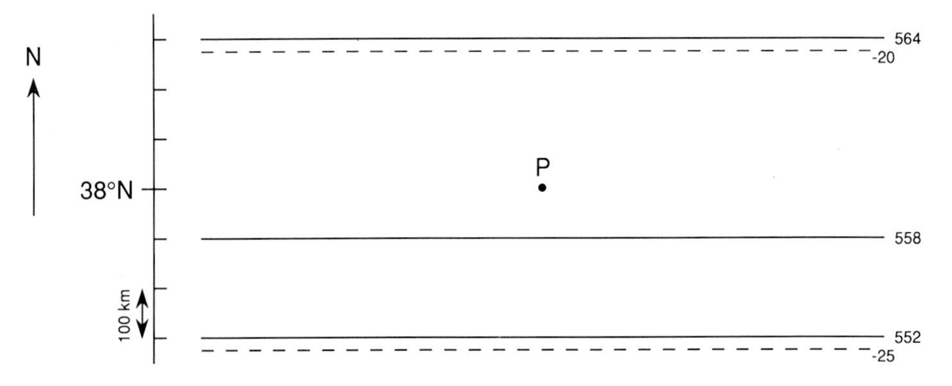

2.44. Consider the vertical velocity (dashed lines; μb s^{-1}), temperature (solid

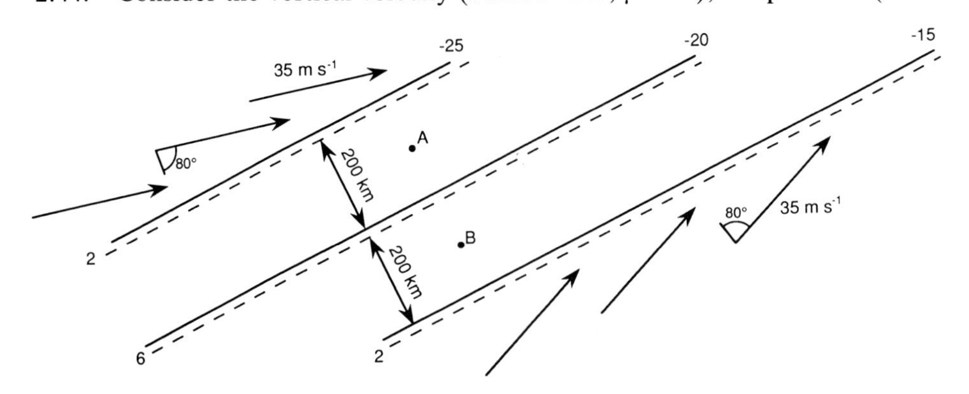

lines; °C), and horizontal wind fields at 500 mb shown in the figure. The temperature at 700 mb is 20°C warmer than it is at 500 mb, and $\partial\theta/\partial p$ is independent of height. Assume there is no diabatic heating. What is the frontogenetical function [K (100 km)$^{-1}$) d^{-1}] (a) at point A, and (b) at point B?

3

Precipitation Systems in the Midlatitudes

It struck me—every Day—
The Lightning was as new
As if the Cloud that instant slit
And let the Fire through—

It burned Me—in the Night—
It Blistered to My Dream—
It sickened fresh upon my sight—
With every Morn that came—

I thought that Storm—was brief—
The Maddest—quickest by—
But Nature lost the Date of This—
And left it in the Sky—

EMILY DICKINSON

3.1. INTRODUCTION

Of all the meteorological phenomena observed on Earth, precipitation probably has the greatest overall impact on our lives. Precipitation affects the clothing we wear, outdoor recreation, crops, and travel. Excessive precipitation results in damaging floods, while lack of precipitation leads to drought conditions and fire danger. The latent heat of condensation associated with the formation and maintenance of precipitation systems, and the radiative transfer properties of the clouds that spawn these systems can have a significant effect on larger-scale motions. In addition, precipitation has had wide-ranging effects in art, music, and poetry. The simple notion of a gaseous substance changing phase, becoming dense enough to fall down to the ground, and then changing phase again, only to return to the air aloft, is itself fascinating!

The most poorly forecast and one of the least well understood of all atmospheric phenomena, a precipitation system is related to various aspects of the larger-scale environment in which it is embedded. The main purpose of this chapter is to relate the character of precipitation systems to their larger-scale environment. By doing so, we can make forecasts of precipitation once we have forecast the environmental features. A secondary goal of this chapter is to describe the detailed phenomenology of precipitation systems.

3.2. TYPES OF PRECIPITATION

Precipitation can take many forms (Table 3.1). Rainfall in 24 h is generally less than 2 cm. However, rates in excess of 12 cm in 24 h can occur locally. Flooding is dependent not only upon rainfall rate but also local topography. Heaviest rainfall rates occur in "convective" showers and thunderstorms. Snowfall in 24 h is generally less than 15–30 cm. However, rates in excess of 50 cm in 24 h can occur.

"Wet snowflakes" are packed very densely with ice crystals. "Dry snowflakes" are packed relatively sparsely with ice crystals. The Cascades of

Table 3.1 Forms of precipitation

Types of precipitation	Characteristics
Drizzle drops (mist)	Drops of water large enough to have enough weight to fall; 0.2–0.5 mm in diameter
Raindrops	Drops of water $\gtrsim$0.5 mm in diameter
"Very light" rain	Surface not completely wet by rain
"Light" rain	Rainfall rate $\leq$0.28 cm h^{-1}
"Moderate" rain	Rainfall rate 0.28–0.75 cm h^{-1}
"Heavy" rain	Rainfall rate >0.75 cm h^{-1}
Snow grains	Small, flattened, and elongated opaque ice particles that have enough weight to fall, but do not shatter or bounce when they hit the ground; analogous to drizzle (called "snizzle" by some television forecasters); $\lesssim$1 mm diameter
Snow	Ice crystals often clumped together in intricate geometric patterns
"Very light"	Surface not completely covered or wet by snow
"Light"	Visibility[a] $>\frac{5}{8}$ mi.
"Moderate"	Visibility[a] $\frac{5}{16}$–$\frac{5}{8}$ mi.
"Heavy"	Visibility[a] $<\frac{5}{16}$ mi.
"Wet snow"	Depth on ground divided by the melted liquid accumulation $\lesssim$10
"Dry snow"	Depth on ground divided by the melted liquid accumulation $\gtrsim$10
Ice pellets (also called "sleet" in the U.S.)	Raindrops or melted snowflakes that have fallen through a subfreezing layer of air and have frozen before hitting the ground; they usually bounce and make an audible sound when they hit the ground
"Sleet" (in the U.K.)	Mixture of rain and snow
Hail	Ice pellets that have grown to a diameter $\gtrsim$5 mm
Snow pellets (also called "soft hail" or "graupel")	Opaque, nearly round ice particles 2–5 mm in diameter; they often break up when hitting the ground if the surface is hard; they are easily crushed
Freezing rain, freezing drizzle	Supercooled raindrops and drizzle drops freezing when hitting a subfreezing surface

[a] The weather observer must be careful to determine when poor visibility is due to fog, rather than to significant snowfall rates.

the northwestern United States often have "wet" snowfall, and relatively warm (albeit subfreezing), moist air at mountain level. On the other hand, the higher Utah and Colorado mountains often have "dry" snowfall, and much colder air with less water vapor content. Snow may be blown by strong winds and reduce visibilities several meters or more above the ground. One of the characteristics of "blizzards" is winds of 16 m s^{-1} or higher, with enough snow in the air to reduce visibilities to about 150 m or less.

Hailstones as large as 8 cm occur in severe thunderstorms. Strong upward motions are necessary to sustain the weight of the hailstone as liquid water is accreted to the hailstone. Damage by hail depends upon its size and the wind speed. Hail will be discussed again in the following in relation to severe thunderstorms. Snow pellets form when supercooled water droplets are collected onto an ice crystal. Snow pellets are common in showers and thunderstorms during the summer in some mountain areas. Freezing rain and freezing drizzle are responsible for ice storms.

It is often difficult to forecast the type of precipitation (i.e., discriminate among ice pellets, freezing rain and freezing drizzle, snow, snow pellets, snow grains, rain, and drizzle) when the temperature is near freezing. Typical soundings for snow, ice pellets, and freezing rain are shown in Fig. 3.1. Drizzle, snow grains, and freezing drizzle can occur in shallow stratus when there is no significant rising motion. Even if the entire atmospheric column is below freezing, freezing drizzle rather than snow may occur, as supercooled drizzle drops reach the ground.

The thickness between two pressure surfaces is often used as a crude indicator of whether precipitation will be liquid or frozen. For example, snow events often occur when the 1000–500 mb thickness is 534 dam or less along East Coast regions of the United States, 540 dam or less inland, and is, for example, as high as 552 dam over the foothills of the Rockies. However, snow can occur even when the thickness is relatively high, for example, if the air in general is relatively warm, but is relatively cool near the bottom of the air column; it may rain even if the thickness is relatively low, for example, if the air is relatively cold aloft, and warm near the surface. Forecasters should refer to the climatology of thickness (for example 1000–500, or 1000–700, or 850–700 mb) versus precipitation type as a function of month for their location. The "critical" rain–snow thickness may change as a function of the date owing to nearby ocean or lake temperature seasonal variations and to surface solar-heating seasonal variations.

When rain or snow falls into a layer of relatively dry air, the precipitation evaporates. Since heat is extracted from the air to evaporate the precipitation, the temperature decreases. It is possible that snowflakes that would otherwise melt before reaching the surface actually reach the surface as a result of evaporative cooling and moistening from previous precipitation.

Freezing rain occurs typically when cold, subfreezing air is present at the surface, (often) owing to a shallow, cold surface anticyclone. Warm advection concurrently occurs above the shallow cold layer. When there is strong rising motion and there is an above-freezing saturated layer of air aloft, freezing rain is likely. Sometimes, however, the cooling that results from the rising motion

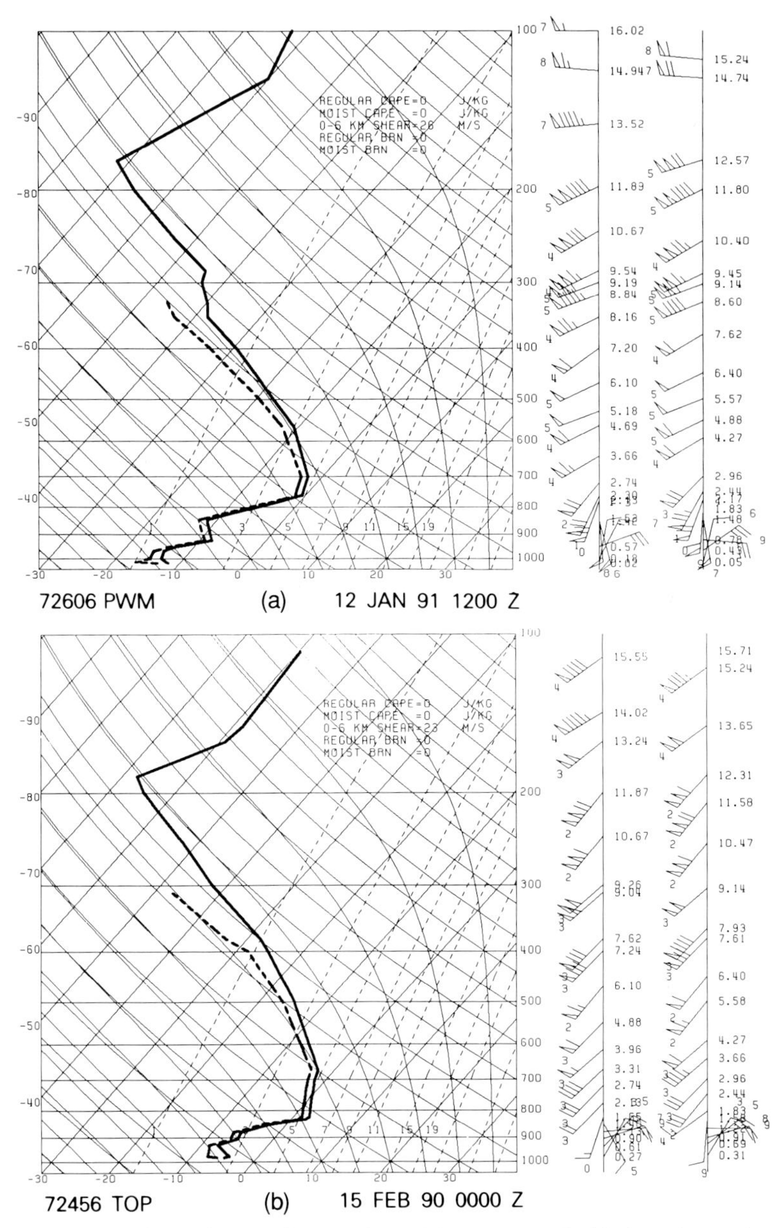

Figure 3.1 Examples of soundings when (a) snow and (b) freezing rain or ice pellets are occurring. In (a) (Portland, Maine at 1200 UTC, January 12, 1991, while moderate-to-heavy snow is being reported) the entire sounding is below freezing, while in (b) (Topeka, Kansas, 0000 UTC, February 15, 1990) there is a layer of above-freezing temperatures between 850 and 750 mb. In (b) some raindrops freeze on the way down to the ground and become ice pellets, while other raindrops reach the ground and freeze upon contact as freezing rain. The above-freezing layer is caused by the advection of warm air from the southwest; the low-level subfreezing layer is due to cold air from the north. Skewed abscissa and logarithmic ordinate are temperature (°C) and pressure (mb); temperature and dew-point plots are thick solid and dashed lines, respectively. Winds plotted to the right; pennant = 25 m s^{-1}; whole barb = 5 m s^{-1}; half barb = 2.5 ms^{-1}.

overwhelms the warming owing to warm advection, and snow may fall instead. When the surface cold-air layer is relatively deep, the raindrops may freeze and ice pellets may form. It is not unusual for freezing rain, rain, snow, and ice pellets to coexist or change from one form into another!

The reader is referred to texts on cloud physics for a detailed treatment of precipitation formation.

3.3. THE CLASSIFICATION OF PRECIPITATION SYSTEMS

Precipitation systems may be classified according to their phenomenology, or according to their physics. It seems logical to expect that the former should be related to the latter. Their phenomenology is determined from radar observations, visual and satellite observations of clouds, and precipitation records.

The horizontal dimensions of precipitation systems range from about 1 km in narrow convective clouds up to roughly 100–400 km in mesoscale systems. The upper limit on the space scale, which is ordinarily mesoscale, is constrained through the effect of latent-heat release in the thermodynamic equation. Precipitation systems, however, are embedded within synoptic-scale systems and usually bear some relationship to them. The time scales of precipitation systems range from about 20 min (10^3 s) to days (10^5 s). Precipitation systems can be as shallow as several kilometers or as deep as 15 km.

Precipitation may be classified as *convective* or *stratiform*. Phenomena characterized by *turbulent* vertical fluxes of heat and momentum such as showers and thunderstorms are convective. The dynamical laws governing their motion may or may not be hydrostatic. For example, showers and thunderstorms are usually nonhydrostatic. However, some precipitation bands in extratropical cyclones and frontal zones are hydrostatic. Convective-precipitation regions are relatively narrow, and precipitation from them tends to be intermittent and intense.

Stratiform systems are characterized by *relatively gentle* vertical fluxes of heat and momentum. Stratiform precipitation, for example, is often found poleward of warm fronts in a broad region of warm advection. The dynamical laws governing their motion are always hydrostatic. Stratiform-precipitation regions are relatively wide, and precipitation from them tends to be steady. Some systems such as squall lines are hybrids in that during portions of their lifetime they have both a convective region and a stratiform region. Many convective precipitation systems evolve into stratiform precipitation systems before they decay. On some occasions convective precipitation develops within an area of stratiform precipitation. An example of this is the "embedded areal" squall line (see Sec. 3.4.9 of this chapter).

The physics of precipitation systems depends upon the following environmental properties:

1. Water-vapor distribution;
2. Vertical velocity;

3. Temperature profile;
4. Microphysics;
5. Vertical wind shear; and
6. Horizontal wind shear.

Our subsequent discussion of precipitation systems will be concerned with the influence of the above on the phenomenology of precipitation systems.

3.4. CONVECTIVE SYSTEMS

3.4.1. The Dynamical Equations for Cumulus Convection

The speeds of air parcels entering or leaving convective clouds can change on the order of 10 $\mathrm{m\,s^{-1}}$ in 10 min or less; that is, the air parcels accelerate (or decelerate) at least $10^{-2}\ \mathrm{m\,s^{-2}} = 1\ \mathrm{cm\,s^{-2}}$. The parcel acceleration is therefore much larger than the Coriolis acceleration, which is on the order of $10^{-3}\ \mathrm{m\,s^{-2}}$. Thus, for time scales on the order of an hour or less, the Coriolis acceleration can be neglected.

The relevant frictionless equations of motion are therefore

$$\frac{Du}{Dt} = -\frac{1}{\rho}\frac{\partial p}{\partial x} \tag{3.4.1}$$

$$\frac{Dv}{Dt} = -\frac{1}{\rho}\frac{\partial p}{\partial y} \tag{3.4.2}$$

$$\frac{Dw}{Dt} = -\frac{1}{\rho}\frac{\partial p}{\partial z} - g. \tag{3.4.3}$$

Although these equations are simpler than those that describe synoptic-scale systems because the effect of the Coriolis force is not included, they are *more* complicated because the hydrostatic approximation cannot be made: The vertical scale of convective clouds is *not* much smaller than the horizontal scale. Let us see why the hydrostatic approximation is not valid from another perspective, and also derive a simplified set of equations valid even when the atmosphere is nonhydrostatic.

Suppose that density is mostly a function of height. Then

$$\rho = \bar{\rho}(z) + \rho'(x,y,z,t), \tag{3.4.4}$$

where the "perturbation" density ρ' is very small compared to $\bar{\rho}$, the density of the "environment." Also, suppose the pressure has a hydrostatically varying part which varies as a function of height alone $[\bar{p}(z)]$ and a much smaller "perturbation" part (p'), so that

$$p = \bar{p}(z) + p'(x,y,z,t), \tag{3.4.5}$$

where

$$\frac{\partial \bar{p}}{\partial z} = -\bar{\rho} g. \tag{3.4.6}$$

The magnitude of p' is on the order of 10 mb or less, while $\bar{p}$ is on the order of 1000 mb. Part of p' is hydrostatic, and the other part is nonhydrostatic.

Substitution of Eqs. (3.4.4)–(3.4.6) into Eqs. (3.4.1)–(3.4.3), and recognition that $(1+a)^{-1} \approx 1-a$ for small a, yields the following:

$$\frac{Du}{Dt} = -\frac{1}{\bar{\rho}}\left(1+\frac{\rho'}{\bar{\rho}}\right)^{-1}\frac{\partial p'}{\partial x} \approx -\frac{1}{\bar{\rho}}\frac{\partial p'}{\partial x} + \frac{1}{\bar{\rho}}\left(\frac{\rho'}{\bar{\rho}}\right)\frac{\partial p'}{\partial x} \tag{3.4.7}$$

$$\frac{Dv}{Dt} = -\frac{1}{\bar{\rho}}\left(1+\frac{\rho'}{\bar{\rho}}\right)^{-1}\frac{\partial p'}{\partial y} \approx -\frac{1}{\bar{\rho}}\frac{\partial p'}{\partial y} + \frac{1}{\bar{\rho}}\left(\frac{\rho'}{\bar{\rho}}\right)\frac{\partial p'}{\partial y} \tag{3.4.8}$$

$$\begin{aligned}\frac{Dw}{Dt} &= -\frac{1}{\bar{\rho}}\left(1+\frac{\rho'}{\bar{\rho}}\right)^{-1}\left(-\bar{\rho} g + \frac{\partial p'}{\partial z}\right) - g \\ &\approx -\frac{1}{\bar{\rho}}\frac{\partial p'}{\partial z} + \frac{1}{\bar{\rho}}\left(\frac{\rho'}{\bar{\rho}}\right)\frac{\partial p'}{\partial z} - \frac{\rho'}{\bar{\rho}} g.\end{aligned} \tag{3.4.9}$$

Since $\rho'/\bar{\rho} \ll 1$, it follows that

$$\frac{Du}{Dt} = -\frac{1}{\bar{\rho}}\frac{\partial p'}{\partial x} \tag{3.4.10}$$

$$\frac{Dv}{Dt} = -\frac{1}{\bar{\rho}}\frac{\partial p'}{\partial y} \tag{3.4.11}$$

$$\frac{Dw}{Dt} = -\frac{1}{\bar{\rho}}\frac{\partial p'}{\partial z} + B, \tag{3.4.12}$$

where the acceleration induced by the buoyancy force

$$B = -\frac{\rho'}{\bar{\rho}} g, \tag{3.4.13}$$

and $\partial p'/\partial x$, $\partial p'/\partial y$, and $\partial p'/\partial z$ are components of $\nabla p'$, the perturbation-pressure gradient vector. The buoyancy force is upward if an air parcel is less dense ($\rho' < 0$) than its environment (Fig. 3.2): If the pressure field in the air parcel is identical to that in the nearby environment, the upward-directed pressure gradient force is greater than the downward-directed force of gravity, owing to the lesser density of the air parcel and hydrostatic balance.

As the air parcel accelerates upward owing to the buoyancy force, air must move out of the way in the volume just above the air parcel to make room for it, while air must move back to fill the volume vacated just below the air parcel (Fig. 3.2). These lateral accelerations above and below the air parcel must be associated with horizontal perturbation-pressure gradients that act outward above the air parcel and inward below the air parcel [Eqs. (3.4.10) and (3.4.11)]. If the pressure field far from the air parcel is not disturbed by the air parcel, then there is a perturbation high-pressure (low-pressure) area above

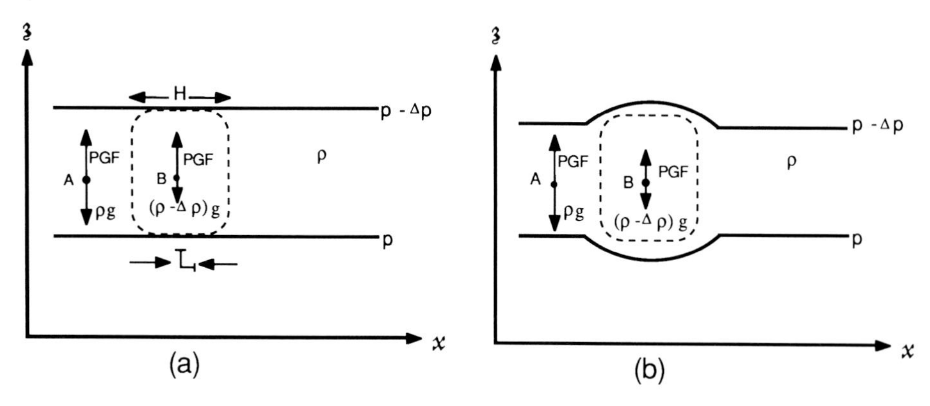

Figure 3.2 The effect on the pressure field of vertical accelerations induced by buoyancy. The environment is in hydrostatic balance at A: The upward-directed pressure-gradient force (**PGF**) is balanced by the downward-directed force of gravity ($-\rho g$). Vertical cross sections of isobars (solid lines) shown in figure. In (a) there is a positively buoyant air parcel (enclosed dashed line) at B of density $\rho - \Delta\rho$, where the upward-directed pressure-gradient force is greater in magnitude than the downward-directed gravitational force. Air above the air parcel is moved away to make room for the air parcel as a result of the perturbation high-pressure region (H); air below the air parcel is moved toward the place where the air parcel had been as a result of the perturbation low-pressure region (L). In (b) the isobars represent the *total* pressure field, that is, the sum of the environmental, hydrostatic, and perturbation (hydrostatic *or* nonhydrostatic) pressure fields. Note that the upward-directed pressure-gradient force is less in magnitude than that shown in (a) for the environmental, hydrostatic pressure field.

(below) the air parcel. In effect there is a *downward*-directed perturbation pressure gradient force, which counteracts the upward-directed buoyancy force.

If an air parcel is very wide (i.e., its depth is small compared to its width), then the volume of air moved above and below the air parcel owing to an upward-directed buoyancy force is relatively large for a given amount of buoyancy. Hence, the magnitudes of the perturbation pressures above and below the air parcel are relatively large, and the downward-directed pressure gradient force essentially cancels the effect of the buoyancy: The atmosphere is nearly hydrostatic (i.e., "quasi-hydrostatic"). On the other hand, if the air parcel is not wide (e.g., its depth is comparable to its width), then the volume of air moved above and below the air parcel is relatively small for a given amount of buoyancy. Hence, the magnitudes of the perturbation pressures above and below the air parcel are relatively small, and the downward-directed pressure gradient force is not large enough to cancel the effect of the buoyancy: The atmosphere is not hydrostatic, and the magnitude of vertical accelerations may be on the order of the buoyancy. We thus see physically why the hydrostatic approximation is scale dependent. [An alternative way of looking at the scale dependence of the hydrostatic approximation is to consider the baroclinic generation of horizontal vorticity around the edge of the air parcel, which we will assume for simplicity is spherical. The vortex line representing the vorticity generated along the edge of the buoyant air parcel is a circle in a horizontal plane around the parcel; the upward (downward) branch of the circulation induced is just inside (just outside) the parcel. If the parcel is relatively wide, then the upward part of the circulation acts mainly along the inner edges of the air parcel, so that the overall effect of the upward motion is weak. However, if the parcel, given the same density deficit, is narrow, then the upward part of the circulation on opposite sides of the parcel reinforce each other, so that the effect of the upward motion is relatively strong.]

Let us rewrite the vertical equation of motion as follows:

$$\frac{Dw}{Dt} = \left(-\frac{1}{\bar{\rho}}\frac{\partial \bar{p}}{\partial z}\right)_1 - (g)_2 - \left(\frac{1}{\bar{\rho}}\frac{\partial p'}{\partial z}\right)_3 + (B)_4. \tag{3.4.14}$$

Terms 1 and 2 are of equal magnitude and opposite sign if the environment is exactly hydrostatic [Eq. (3.4.6)]. If the atmosphere everywhere is exactly hydrostatic, then terms 3 and 4 also are of equal magnitude and opposite sign, there is no vertical motion, and

$$\frac{\partial p'}{\partial z} = -\rho' g; \tag{3.4.15}$$

that is, the perturbation part of the pressure field is also hydrostatic.

If the atmosphere is quasihydrostatic, then terms 1 and 2 cancel each other exactly, while terms 3 and 4 nearly cancel each other. The vertical acceleration Dw/Dt is small, but not necessarily zero. Thus, if the atmosphere is quasihydrostatic, vertical accelerations are small not only compared to the acceleration of gravity, *but also compared to the buoyancy.* How large is the buoyancy in a convective cloud?

The difference between the density of an air parcel (ρ) and that of its environment ($\bar{\rho}$) is

$$\rho' = \rho - \bar{\rho}. \tag{3.4.16}$$

The difference between the temperature of an air parcel (T) and that of its environment ($\bar{T}$) is

$$T' = T - \bar{T}. \tag{3.4.17}$$

From Eqs. (3.4.5), (3.4.17), (3.4.16), and the ideal gas law, it follows that if the pressure of the parcel is the same as that of the environment ($p' = 0$), then

$$-g\frac{\rho'}{\bar{\rho}} = -g\left[\frac{1}{\bar{T}}\frac{\bar{p}}{1+T'/\bar{T}} - \left(\overline{\frac{1}{\bar{T}}\frac{\bar{p}}{1+T'/\bar{T}}}\right)\right]\left(\overline{\frac{1}{\bar{T}}\frac{\bar{p}}{1+T'/\bar{T}}}\right)^{-1}. \tag{3.4.18}$$

For small $T'/\bar{T}$,

$$B = -g\frac{\rho'}{\bar{\rho}} = g\frac{T'}{\bar{T}} = g\left(\frac{T-\bar{T}}{\bar{T}}\right). \tag{3.4.19}$$

The buoyancy of an air parcel is therefore proportional to its "temperature excess" over the environmental temperature. For a parcel-temperature excess of 3°C, $B \sim 0.1\ \mathrm{m\,s^{-2}}$. The density of water vapor may be accounted for if virtual temperature is substituted for temperature; B is then the "virtual" buoyancy. Water loading may be accounted for if the cloud virtual temperature is substituted for virtual temperature.

The continuity equation in a *shallow* atmosphere can be written approximately as

$$\frac{1}{\rho}\frac{D\rho}{Dt} = 0 = \nabla \cdot \mathbf{v} = \frac{\partial u}{\partial x} + \frac{\partial v}{\partial y} + \frac{\partial w}{\partial z}. \qquad (3.4.20)$$

Density varies more rapidly in the vertical than it does in the horizontal; fractional changes in density following air-parcel motion are therefore relatively small for typical vertical velocities in a shallow atmosphere, where vertical variations in density are relatively small. Thus, the atmosphere behaves as if it were incompressible. However, vertical density variations in the environment [$\bar{\rho}(z)$] are retained in the vertical equation of motion, Eqs. (3.4.12) and (3.4.13). Equations (3.4.10)–(3.4.12) and (3.4.20) are known as the *Boussinesq equations*. These equations do not permit sound waves: We do not consider sound waves to be important dynamically for cumulus convection, just as we do not consider gravity waves to be important dynamically for synoptic-scale systems. However, the Boussinesq equations do permit gravity waves.

In a "deep" atmosphere, density does vary substantially with height, and fractional changes in density can be relatively large for deep air-parcel excursions in the vertical. In this case the continuity equation is expressed as:

$$\frac{1}{\rho}\frac{D\rho}{Dt} = \frac{1}{\bar{\rho}(z)} w \frac{\partial \bar{\rho}}{\partial z} = w \frac{\partial \ln \bar{\rho}}{\partial z} = \nabla \cdot \mathbf{v} \qquad (3.4.21)$$

or

$$\nabla \cdot \bar{\rho}(z)\mathbf{v} = 0. \qquad (3.4.22)$$

The latter two are expressions for the "anelastic" continuity equation. They are applicable to the dynamics of deep convection, while the Boussinesq continuity equation is applicable only to shallow convection. The anelastic equations also do not permit sound waves.

The pressure (and density) variable

$$\pi = C_p \theta_0 \left(\frac{p}{p_0}\right)^{\kappa} \qquad (3.4.23)$$

and the potential temperature θ are often used instead of pressure and temperature; in this case density does not appear explicitly in the dynamical equations, which may be written as:

$$\frac{D\mathbf{v}}{Dt} = -\nabla_z \pi' \qquad (3.4.24)$$

$$\frac{Dw}{Dt} = -\frac{\partial \pi'}{\partial z} + B, \qquad (3.4.25)$$

where

$$B = g\frac{\theta - \bar{\theta}}{\bar{\theta}}. \tag{3.4.26}$$

The anelastic continuity equation is written as

$$\nabla \cdot \rho_a \mathbf{v} = 0, \tag{3.4.27}$$

where

$$\rho_a = \rho_a(z) = \rho_{a0}\left(1 - \frac{gz}{C_p\theta_o}\right)^{1/\kappa - 1}. \tag{3.4.28}$$

The latter expresses the vertical variation of density in an adiabatic atmosphere (i.e., one with no variation of θ with height). The Boussinesq continuity equation is still the same as Eq. (3.4.20).

3.4.2. A Brief History of Thunderstorm Research

Pioneering research on thunderstorms in the United States was conducted in Florida and Ohio in the late 1940s under the direction of Horace Byers and Roscoe Braham. Their field program, the Thunderstorm Project, made extensive use of radar and instrumented aircraft. The National Severe Local Storms Research Project began in Kansas City in 1955 after some thunderstorm-related airplane accidents and devastating tornado strikes in populated areas such as Waco, Texas, and Worcester, Massachusetts.[1] The adjective *local* was used to distinguish convective storms from synoptic-scale cyclones. Now, the word *local* is implicit. The National Severe Local Storms Research Project was renamed the National Severe Storms Project (NSSP) in 1961. (The trend of simplifying project names seems to have been dramatically reversed during the "acronym era" of the 1970s and 1980s!) After moving in 1964 to Norman, Oklahoma, a more suitable location for continuing field observations, the NSSP became the National Severe Storms Laboratory (NSSL) under the direction of Edwin Kessler.

Our understanding of thunderstorms and related phenomena has increased dramatically over the past 20 years, much owing to the efforts of NSSL, and also the National Center for Atmospheric Research (NCAR), universities, the National Aeronautics and Space Administration (NASA), and the Air Force. Major projects included the National Hail Research Experiment (NHRE) in 1972–1976, SESAME (Severe Environmental Storms and Mesoscale Experiment) in 1979, CCOPE (Cooperative Convective Precipitation Experiment) in 1982, and the Oklahoma–Kansas PRE (Preliminary Regional Experiments) for STORM (Stormscale Operational and Research Meteorology) in 1985.

During the 1950s, 1960s, and early 1970s conventional 5- and 10-cm radars were used to study the characteristics of the rain and hail regions in thunderstorms. Wind, temperature, and humidity measurements and visual observations were made aboard aircraft flying around and sometimes dangerously inside storms, and by rawinsondes launched around and occasionally into storms. Major breakthroughs occurred during the 1970s when pulsed

Doppler radars and their networks allowed us to determine for the first time the three-dimensional wind field inside storms without actually making *in situ* measurements. Methods of locating lightning channels with VHF (very high frequency) sensors were first used. Visible and infrared photographs from geostationary satellites afforded us a bird's-eye view (from a high-flying bird, indeed!) of the tops of storms and provided us with information concerning the relationship between storms and their neighboring clouds. Visual and instrumented, ground-based observations made by "storm-intercept teams" have been used in conjunction with simultaneous Doppler–radar observations to discover the relationship between radar "signatures" and actual storm phenomena, to discover features too small to be resolved by radar, and to investigate the wind, pressure, temperature, and moisture fields in locations unlikely to be sampled by fixed observation networks (i.e., where "man" and "woman" have never "boldly" gone before). Sophisticated three-dimensional computer models were developed so that convective storms could be studied in a controlled "environment."

3.4.3. The onset of deep convection

We usually refer to convective clouds as *deep* if their depth is a substantial fraction of the depth of the troposphere. Deep cumulus convection may be "based" (i.e., where most of the potentially buoyant air originates) either in the boundary layer or aloft. Cumulus convection usually begins as air parcels are lifted upward to their lifting condensation level (LCL) or as surface air is heated up to "convective temperature" and air parcels are turbulently transported upward to their convective condensation level (CCL). The relative importance of each process varies from case to case. Although the nature of cloud initiation may vary, the physics of the ensuing convective clouds is the same.

We will refer to convection based aloft (i.e., above the boundary layer) as "elevated convection." Elevated convection is usually manifest as *altocumulus castellanus* clouds that grow into cumulonimbi having high bases. Surface heating does not play a major role in producing elevated convection.

"Boundary-layer convection" is manifest as cumulus clouds that grow into cumulonimbi having bases in or at the top of the boundary layer. The source of air at cloud base is from the boundary layer. Surface heating plays an important role in many cases of this type of convection. Elevated surfaces such as mountain peaks often require less total heating to reach convective temperature than the surrounding lower surfaces (Fig. 3.3), and hence can be expected to be the sites of the earliest convection.

The intensity of the initial updraft in boundary-layer convection is to some extent indicated by the height of the "first (radar) echo." A strong updraft results in a higher first echo than a weak updraft because it takes a fixed amount of time for precipitation to form: In the time interval necessary for precipitation formation, the fastest-rising air parcel will be highest when cloud droplets have grown to precipitation size.

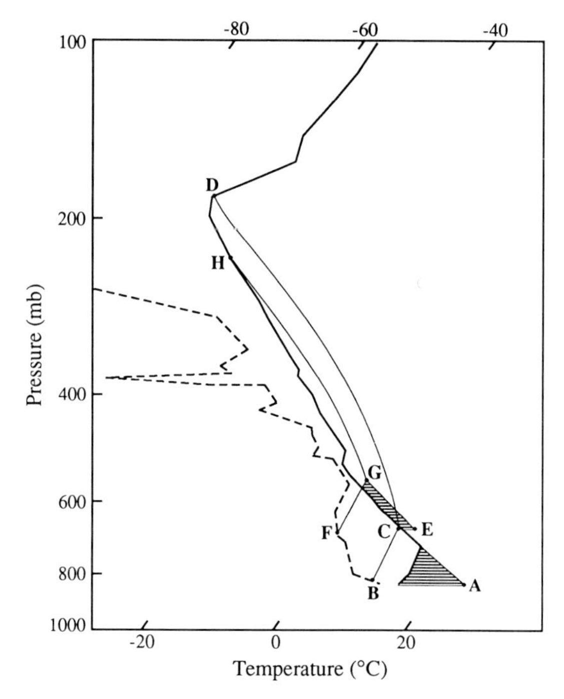

Figure 3.3 Illustration of how less total heating is required to reach convective temperature over high-elevation mountainous terrain than at lower elevations. Sounding for Denver, Colorado, 1200 UTC, June 3, 1981; skewed abscissa is temperature (°C); logarithmic ordinate is pressure (mb). Temperature and dew point plotted as solid and dashed lines, respectively. AC represents the dryadiabat connecting the surface convective temperature at Denver to the surface saturation mixing-ratio isopleth BC; C is at the convective condensation level and the level of free convection for an undilute surface parcel; the parcel's equilibrium level is at D. However, an air parcel beginning at higher elevation (e.g., 700 mb) requires less solar heat (hatched region) to reach convective temperature at E and CCL and LFC at G; FG represents the saturation mixing-ratio isopleth for the parcel at 700 mb; the parcel's equilibrium level is at H. Although the convection based at Denver can be more vigorous than the convection based at 700 mb, the latter begins earlier, because less heat is needed to reach convective temperature. On this day there was a tornado outbreak in the Denver area (adapted from Szoke et al., 1984). (Courtesy of the American Meteorological Society)

If the boundary layer is deep, then the cloud base may be relatively high, and therefore it is sometimes difficult to distinguish elevated convection from boundary-layer convection visually. For example, boundary-layer "rooted" convective cloud bases in the dry, elevated High Plains may be as high as 3–4 km AGL, while cloud bases in convective clouds based in the boundary layer in Florida may be as low as 400–500 m AGL. In general, the most intense convection is rooted in the boundary layer, because usually the highest possible parcel temperature excesses over the environment may be realized when boundary-layer air is lifted.

The nature of the large-scale dynamical lift that brings air parcels to their LCLs is varied. The production of precipitation itself is a highly nonlinear process: It is either precipitating, or it is not; air is saturated and condensation occurs, or it is unsaturated and no precipitation occurs. The lifting mechanism responsible for initiating the sudden onset of this change in state (i.e., no rain to rain, no liquid water content to some liquid water content) is aptly referred to as the *triggering mechanism.*

Triggering mechanisms come in many sizes. On the synoptic scale, quasigeostrophic ascent associated with geostrophic warm advection or geostrophic vorticity advection becoming more cyclonic with height may result in widespread convection. The latter is often associated with the region downstream from short-wave troughs. The former is often associated with the region eastward and poleward of surface cyclones. Although most convective outbreaks occur in broad southwesterly flow aloft, downstream from upper-level troughs, some occur in northwesterly flow (NWF), especially during the summer over the North Central United States when a quasipermanent ridge is parked aloft over the continent (Fig. 3.4). The southwest-flow case (not

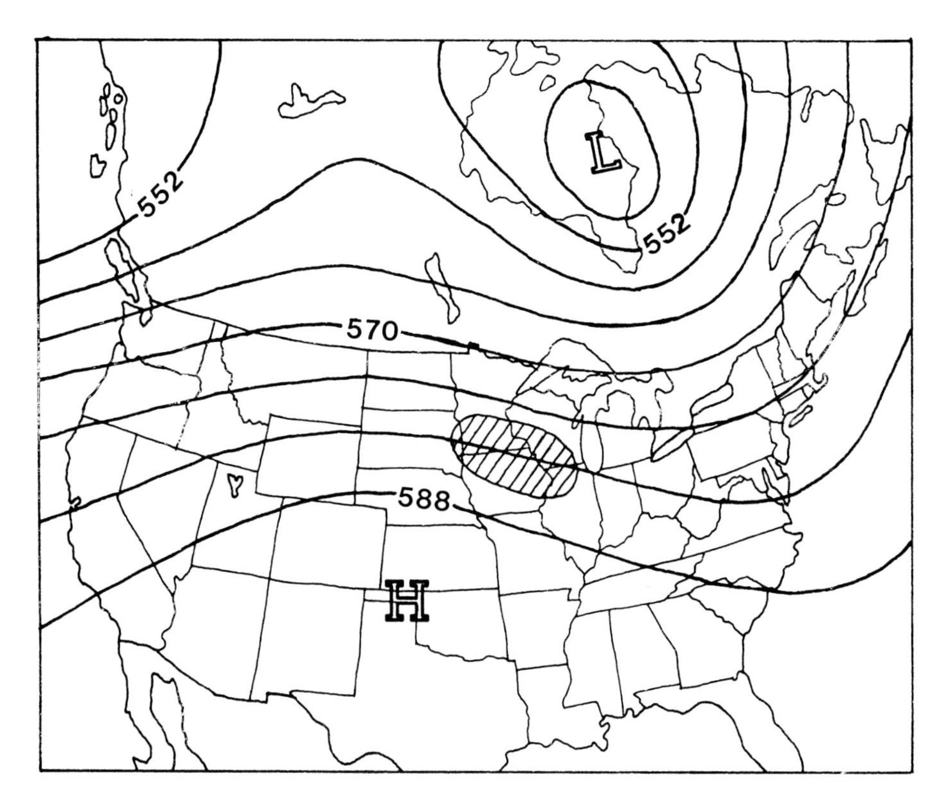

Figure 3.4 Example of 500-mb height contours in dam (solid lines) for a "northwesterly-flow" case (in the United States) of a weak short-wave trough embedded in the basic flow. Severe-thunderstorm outbreak area indicated by area of hatched lines (from Johns, 1984). (Courtesy of the American Meteorological Society)

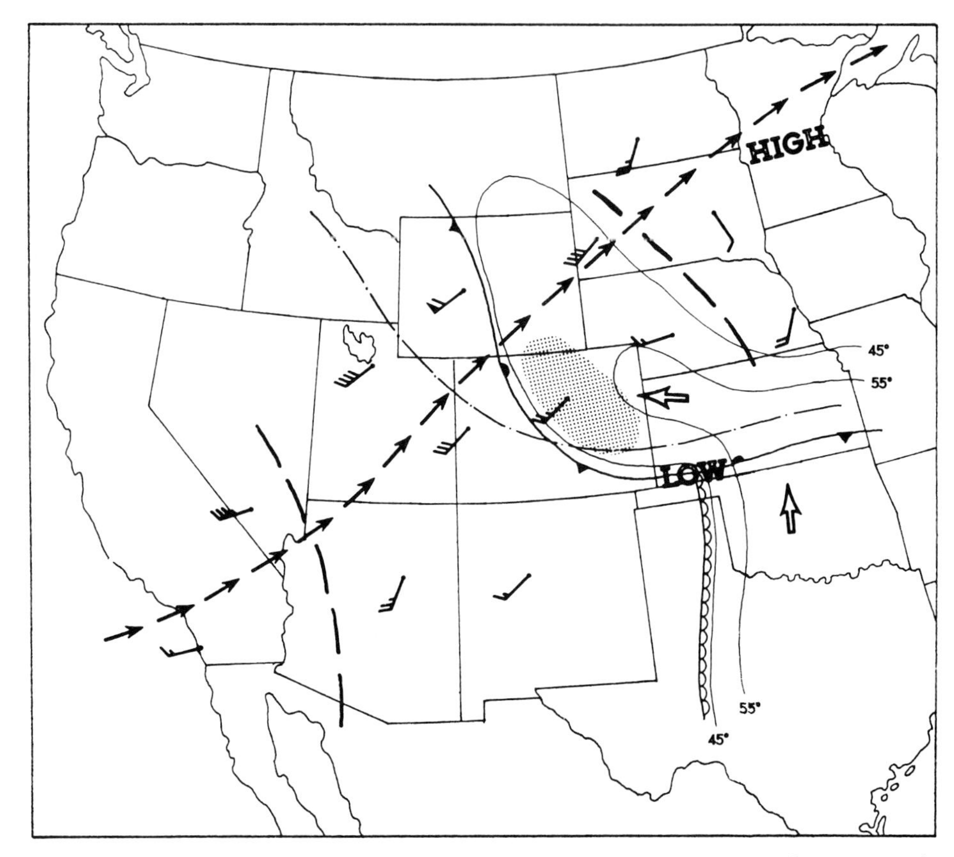

Figure 3.5 Composite High Plains severe thunderstorm parameter chart. Frontal symbols are conventional; surface isodrosotherms in °F (fine lines); dryline (scalloped line); surface flow (large arrows); 700-mb thermal ridge (dashed-dotted line); surface pressure centers (HIGH and LOW); 500-mb winds (flag, full barb, and half barb represent 25, 5, and 2.5 m s^{-1}, respectively); short-wave trough axes (heavy dashed lines); core of strong high-level winds above 500 mb (chain of arrows); region of expected severe thunderstorms (stippled area) (from Doswell, 1980). (Courtesy of the American Meteorological Society)

shown) is often associated with the convective outbreaks in the region eastward and poleward of surface cyclones. Upslope over gently sloping terrain, such as northeasterly or easterly flow in eastern Colorado (Fig. 3.5), may also set the large-scale stage for convection *after* the passage of a back-door cold front.

Quasigeostrophic triggering mechanisms are relatively slow in producing precipitation. For example, if air must be lifted 1 km to trigger convection, then air rising at 1 cm s^{-1} would take approximately an entire day (10^5 s) to reach its LCL. We observe, however, that thunderstorms are not triggered everywhere over the very broad areas experiencing quasigeostrophically related ascent.

Zones of rising motion associated with frontal circulations, upward motion along drylines, upward motion in the left-front and right-rear quadrants of jet streaks, the rising branches of solenoidally induced sea-breeze and land-breeze circulations (or circulations solenoidally induced over horizontal gradients of surface vegetation, soil, or wetness), the rising branch of mountain–valley circulations, upward motion induced by gravity waves, upward motion associated with a myriad of (mesoscale) dynamic instabilities, and lift forced over mesoscale outflow boundaries have also been associated with the onset of deep convection. Since the magnitude of the mesoscale ascent associated with these circulations is on the order of 10 cm s^{-1}, the time required to lift air 1 km to its LCL is only on the order of 3 h (10^4 s).

Intersecting boundaries (e.g., along fronts, outflow boundaries, the dryline) have also been identified as possible loci for convective development, since surface convergence and hence rising motion may be enhanced at these intersections. On even smaller scales, local lifting over gravity currents such as gust fronts and upward accelerations owing to vertical pressure gradients (see Sec. 3.4.6) can also act to trigger convective growth if they are intense enough and last long enough. The precise predictability of cumulus convection can often be limited severely because our ability to pinpoint mesoscale outflow boundaries and storm-scale gust fronts is not good if the convective system responsible for the outflow boundaries/gust fronts has not been triggered yet.

3.4.4. Continuity of water vapor

Although quasigeostrophic lifting is relatively weak, it can modify the environment for subsequent convection that is triggered by the other, smaller-scale, more powerful lifting mechanisms. For example, low-level convergence associated with midtropospheric ascent contributes to destabilization below the LND. In the previous section we considered how air can be brought to its LCL through lifting. In this section we consider how air can be brought closer to its LCL through an increase in water-vapor content.

Let us now consider the effects of the vertical-motion field on the specific humidity. Suppose that the specific humidity in the boundary layer is increased; then the degree of (no pun intended) potential instability is increased, the convective temperature is lowered, and the CCL and LCL are lowered (Fig. 3.6).

The rate at which the specific humidity (q) changes following air-parcel motion is given by the difference between the evaporation (E) and condensation (C) rates per unit mass of moist air

$$\frac{Dq}{Dt} = E - C. \tag{3.4.29}$$

In an Eulerian framework

$$\frac{\partial q}{\partial t} = -\boldsymbol{\nabla} \cdot q\mathbf{v} + q\boldsymbol{\nabla} \cdot \mathbf{v} + E - C. \tag{3.4.30}$$

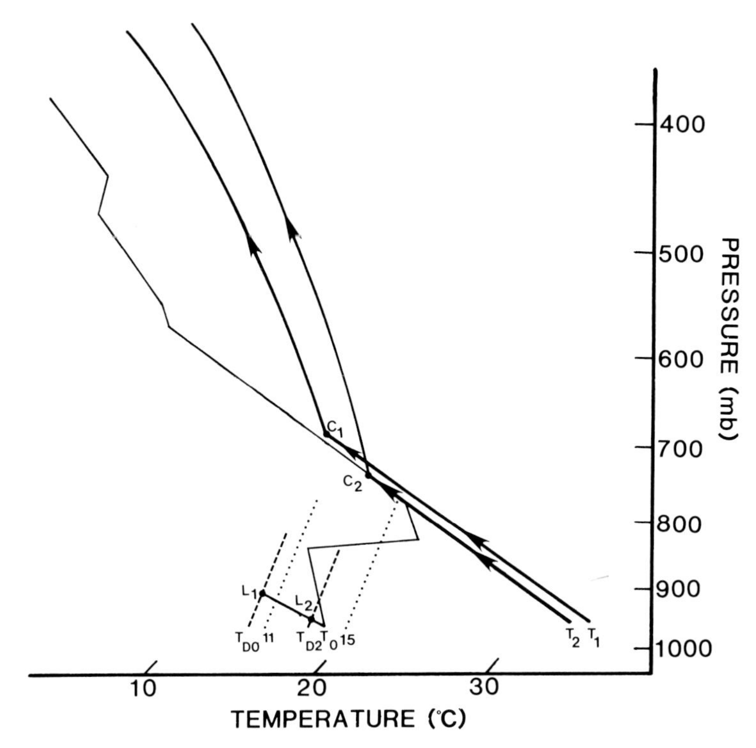

Figure 3.6 Idealized illustration of the effect of an increase in specific humidity on convective temperature, convective condensation level (CCL), and lifting condensation level (LCL). Skewed abscissa and logarithmic ordinate are the temperature (°C) and pressure (mb). Arrows denote paths in skewed temperature–logarithmic-pressure space taken by ascending air parcels. Lines T_2C_2, T_1C_1, and $T_0L_2L_1$ are along dry adiabats. Lines of constant saturation mixing ratio in g kg^{-1} (dotted lines); lines of constant (i.e., well mixed in the moist boundary layer) water-vapor saturation mixing ratio associated with surface dewpoints of T_{D0} and T_{D2} (dashed lines). Air parcel at the surface having temperature of T_0 and dew point of T_{D0} has an LCL of L_1; air parcel having a higher surface dew point of T_{D2} has a lower LCL of L_2. Air parcel having a surface dewpoint of T_{D0} has a convective temperature of T_1 and a CCL of C_1; air parcel having a higher surface dew point of T_{D2} has a lower convective temperature of T_2 and a lower CCL of C_2.

If the variables q and $\mathbf{v}$ are decomposed into mean and eddy components, and pressure is used as the vertical coordinate, then Eq. (3.4.30) may be written as

$$\frac{\partial q}{\partial t} = -\mathbf{v} \cdot \nabla_p q - \omega \frac{\partial q}{\partial p} - \nabla_p \cdot \overline{(q'\mathbf{v}')} - \frac{\partial}{\partial p} \overline{(q'\omega')}$$
$$+ E - C, \tag{3.4.31}$$

where the overbar notation for the mean variables has been dropped and the primed variables represent the eddy components. Local changes in water-vapor content (specific humidity) per unit mass according to Eq. (3.4.31) occur as a result of quasihoriozntal advection ($-\mathbf{v} \cdot \nabla_p q$), vertical advection

$(-\omega\, \partial q/\partial p)$, turbulent horizontal water-vapor flux convergence $(-\nabla_p \cdot \overline{q'\mathbf{v}'})$, turbulent vertical water-vapor flux convergence $[-(\partial/\partial p)(\overline{q'\omega'})]$, and evaporation of liquid water or sublimation of ice into the air, less the loss owing to condensation or the fusion of water vapor into liquid water or ice.

When the water-vapor mixing ratio decreases with height, rising motion contributes to an increase in q (and r) through vertical advection. However, if the boundary layer is well mixed or ω is small in the boundary layer, then vertical advection is negligible.

When water vapor is transported upward (and dry air downward) in the boundary layer through convective eddies [$q' > 0$ when $\omega' < 0$, and $q' < 0$ when $\omega' > 0$; $\overline{(q'\omega')} < 0$], and $\overline{q'\omega'} = 0$ at the surface, then

$$-\frac{\partial}{\partial p}(\overline{q'\omega'}) < 0, \tag{3.4.32}$$

just above the surface; that is, convective currents tend to "mix out" the low-level moisture: Hence the surface dewpoint drops. On the other hand, when a strong "capping inversion" (or "lid") inhibits convective transports out from the boundary layer, so that $\overline{q'\omega'} = 0$ at the top of the boundary layer, then

$$-\frac{\partial}{\partial p}(\overline{q'\omega'}) > 0 \tag{3.4.33}$$

in the boundary layer: The specific humidity increases, the dew point increases, and the convective temperature and the LCL are lowered.

Sometimes the restraining effects of a capping inversion in the presence of evaporation from the soil and vegetation, and horizontal advection, can lead to very high dew points. This effect, which is referred to as the *pooling* or *piling up* of moisture, sometimes occurs along frontal zones and east of drylines.

Equations such as Eqs. (3.4.31) are used in water-vapor budget studies. Often, the difference between the evaporation and condensation rates in a volume is determined from the observed wind and water-vapor fields. Suppose the volume is chosen so that it can be assumed that there is no eddy flux across the sides and top of the volume. Suppose also that the variables are integrated over a long enough period of time or over a large enough volume so that it can be assumed that the water-vapor field is in a steady state. Then integrated over the volume, Eq. (3.4.31) can be expressed as follows:

$$E - P = \overline{\nabla \cdot q\mathbf{v}} + E_0, \tag{3.4.34}$$

where the overbar denotes a time average, E is the total amount of evaporation in the volume, P is the total precipitation that reaches the ground and the net condensate suspended in the air as cloud material, and E_0 is the evaporation from the surface, which is a function of surface wind, humidity, temperature, and water temperature or soil wetness. The net latent heating in the volume is found by multiplying Eq. (3.4.34) by $-L$.

The evaporation, condensation, and conversion of cloud droplets into rain can be parameterized in terms of the water-vapor, temperature, and vertical velocity fields. Edwin Kessler was the first to suggest methods of representing the microphysical processes responsible for cloud droplet growth and the transformation of cloud droplets into rain drops, etc. The reader is referred to his monograph for more details (see references). Kessler's "parameterizations" for evaporation, condensation, and rain formation have been used in many numerical cloud models.

3.4.5. The temperature and moisture stratification in the environment

Convective available potential energy. The acceleration induced by the buoyancy force in moist air, in the absence of liquid water or ice, is

$$B = g\frac{T_v(z) - \bar{T}_v(z)}{\bar{T}_v(z)}. \tag{3.4.35}$$

The dynamical effect of B [Eq. (3.4.12)] alone when w is steady state and independent of x and y is as follows:

$$\frac{Dw}{Dt} = w\frac{dw}{dz} = \frac{d}{dz}(\tfrac{1}{2}w^2) = B. \tag{3.4.36}$$

(We are not including here the effects of the vertical, perturbation pressure-gradient force, which opposes the buoyancy force.) Integrating Eq. (3.4.36) from the level of free convection (LFC) to the equilibrium level (EL), we find that if the air parcel is at rest at the LFC, and if the air parcel does not "entrain" any cooler, drier environmental air, then the updraft at the equilibrium level, the maximum vertical velocity (w_{max}), is

$$w_{max} = \sqrt{2\,\text{CAPE}}, \tag{3.4.37}$$

where CAPE, the *convective available potential energy* (also known as the *buoyant energy*), is given by

$$\text{CAPE} = \int_{\text{LFC}}^{\text{EL}} g\frac{T_v(z) - \bar{T}_v(z)}{\bar{T}_v(z)}\,dz. \tag{3.4.38}$$

In the above expression $T_v(z)$ is the virtual temperature profile of an air parcel ascending moist-adiabatically from the LFC, and $\bar{T}_v(z)$ is the virtual temperature profile of the environment. CAPE represents the area on a thermodynamic diagram enclosed by the environmental temperature profile and the moist adiabat connecting the LFC to the EL (Fig. 3.7). If liquid water is suspended in the updraft and freezes above some level, then the additional effects of the latent-heat release associated with fusion increases the CAPE. Since CAPE is the buoyancy force integrated with respect to height, it represents the work done on the parcel by the environment as the parcel is accelerated upward.

Typical values of CAPE for days when moderate to strong convection occurs range from 1000 to 3000 $m^2\,s^{-2}$ ($J\,kg^{-1}$). Maximum observed values of CAPE are around 5000–7000 $m^2\,s^{-2}$ ($J\,kg^{-1}$). For a CAPE of 2500 $J\,kg^{-1}$,

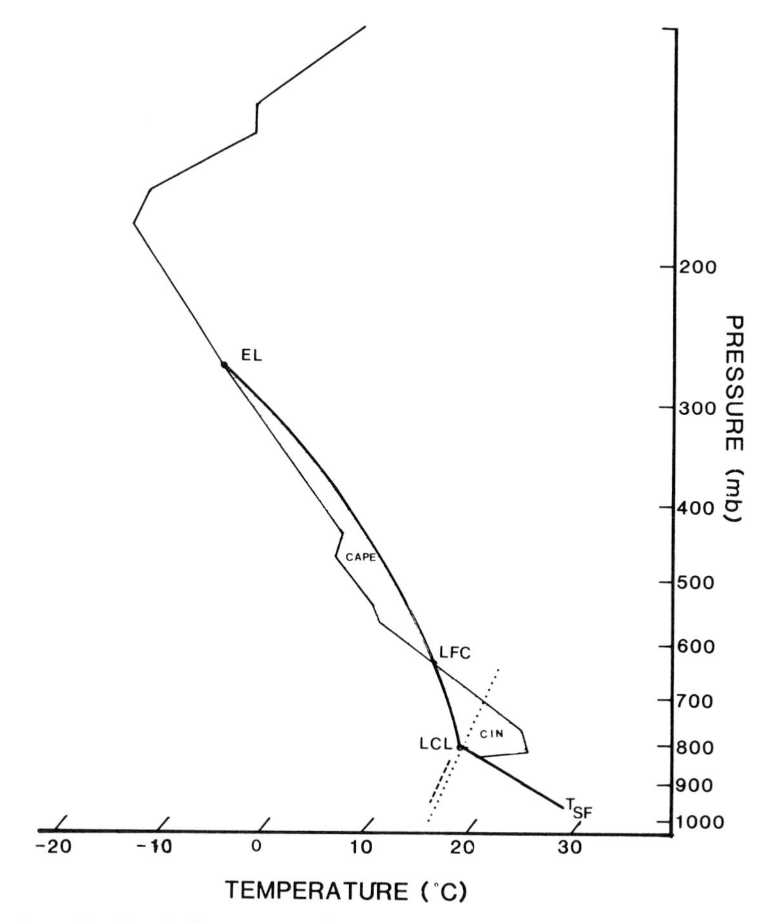

Figure 3.7 Idealized illustration of convective available potential energy (CAPE) and convective inhibition (CIN). Skewed abscissa and logarithmic ordinate are temperature (°C) and pressure (mb), respectively. Surface temperature (T_{SF}); lifting condensation level (LCL); level of free convection (LFC); equilibrium level (EL). Temperature profile (thin solid line); dew point in the moist, well-mixed boundary layer (dashed line); line of constant saturation water-vapor mixing ratio (dotted line); CAPE proportional to the area (hatched area) formed by the temperature curve and the moist adiabat passing through the LFC (thick solid line); thick solid line between the surface temperature and the LCL represents a dryadabiat; CIN is proportional to the area (hatched area) formed by the temperature curve and (above the LCL) the moist adiabat passing through the LCL and (below the LCL) the dry adiabat connecting the LCL to T_{SF} (thick solid line). In this example the lapse rate between the surface and a little below the LCL is dryadiabatic.

w_{max} is approximately 70 m s^{-1} [Eq. (3.4.37)]. Observations in the updrafts of severe storms suggest that Eq. (3.4.37) yields estimates that can be close to the actual values (Fig. 3.8). In numerical simulations, however, the combined effects of the vertically directed perturbation pressure gradient force, water and ice loading, and mixing act to decrease the maximum vertical velocity computed through Eq. (3.4.37) by as much as a factor of two.

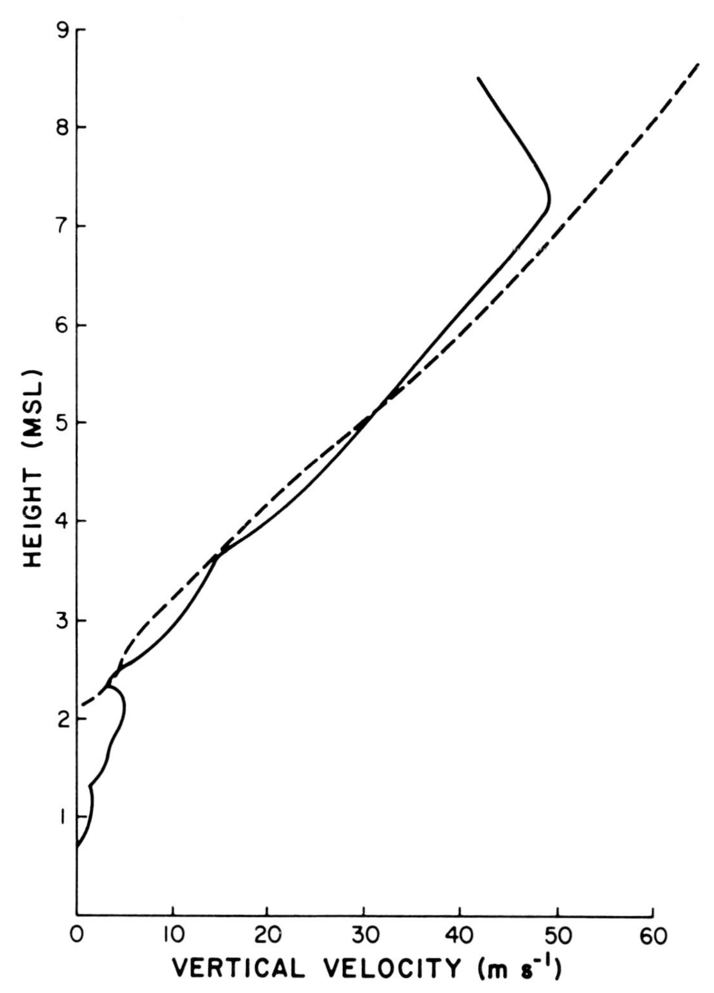

Figure 3.8 Comparison of updraft velocity, in a tornadic storm, estimated from balloon ascent rate (solid line) and from parcel buoyancy $[2\text{CAPE}(z)]^{1/2}$ (dashed line). Height in km MSL (from Bluestein et al., 1988). (Courtesy of the American Meteorological Society)

The computation of CAPE is very sensitive to the surface dew point (or, mean mixing ratio in the lowest 500 m). An increase in mixing ratio of only 1 g kg^{-1} can increase CAPE by 20 percent. This can change the estimate of maximum vertical velocity by as much as 10 percent. Water loading may be accounted for by substituting the cloud virtual temperature $[(1+1.609r_v - r_l)T]$ for the virtual temperature, where r_v and r_l are the water vapor and liquid water mixing ratios.

The work needed to lift an air parcel whose temperature and water-vapor mixing ratio are the means in the lowest 500 m, from rest at the surface to the LFC, is called the *convective inhibition* (CIN). It is the negative of the area enclosed by the environmental temperature profile and the dry and moist adiabats connecting the "surface" to the LFC on a thermodynamic diagram

(Fig. 3.7). When a stable layer or "capping" inversion is present above a moist layer in the boundary layer, the CIN is a measure of the strength of the "cap" or "lid." Using Eq. (3.4.36) below the LFC, we see that the upward motion (push) necessary at the surface so that an air parcel will just reach and come to a rest at the LFC is $\sqrt{2\,\mathrm{CIN}}$. In this case, B is negative, and the air parcel decelerates as it moves upward. For a CIN of only 200 $\mathrm{m^2\,s^{-2}}$ ($\mathrm{J\,kg^{-1}}$), an initial upward push of 20 $\mathrm{m\,s^{-1}}$ is needed.

It is generally assumed that the intensity of a convective storm is a monotonic function of the updraft speed. The latter, we have just shown, is related to the parcel buoyancy, a measure of the parcel's virtual temperature excess over its environment, less loading from liquid water and ice, any downward-directed perturbation pressure gradient force, and loss of buoyancy owing to entrainment of unsaturated environmental air into the updraft.

Stability indices. There are a number of different *stability indices* that represent approximately a parcel's virtual temperature excess. It is difficult to represent the other effects (e.g., pressure forces) that act in opposition to the temperature excess. The most commonly used indices are the *lifted index,* the *totals–totals index,* the *Showalter index,* and the *K index.*

The *Showalter index* is the temperature excess at 500 mb of the environment with respect to an air parcel at 850 mb that is lifted to its LCL (with respect to the moisture at 850 mb), and is then lifted moist-adiabatically above the LCL to 500 mb (Fig. 3.9). Low values (negative values) indicate the possibility of convection. It is not a good index when the moist layer in the boundary layer does not extend as high as 850 mb and in the case of elevated convection when the "roots" of the convective clouds are above 850 mb.

The *totals–totals index* is the sum of the 850 mb temperature and dewpoint less twice the 500 mb temperature. Values in excess of 50 are considered to be associated with very intense convection. The totals–totals index does not require the computation of dry- or moist-adiabatic lapse rates, and humidity. For the purpose of illustration, suppose the temperature and dew point at 850 mb are 15 and 10°C, respectively, and the temperature at 500 mb is −15°C; then the totals–totals index is 55.

The *lifted index* is the temperature excess at 500 mb of the environment with respect to an air parcel in the "moist" layer (i.e., in the boundary layer, where the relative humidity is greater than 60–65 percent) lifted to its LCL, and then lifted moist-adiabatically above the LCL. Usually the air parcel in the moist layer has the estimated mean potential temperature and mixing ratio of the lowest 500 m (Fig. 3.9). (Sometimes only the surface θ and r are used.) A negative lifted index indicates the possibility of convection; a lifted index of less than −6°C indicates the possibility of very intense convection. Lifted indices as low as −10 to −15°C are sometimes observed. The lifted index is not a good indicator of convective activity when the 500-mb temperature is not truly representative of the sounding (for example, if there is a shallow but sharp inversion at or adjacent to 500 mb), or when convection is not rooted in the lowest 500 m. The lifted index is often computed based on the forecast surface temperature and the dew point. Early-morning lifted indices are not

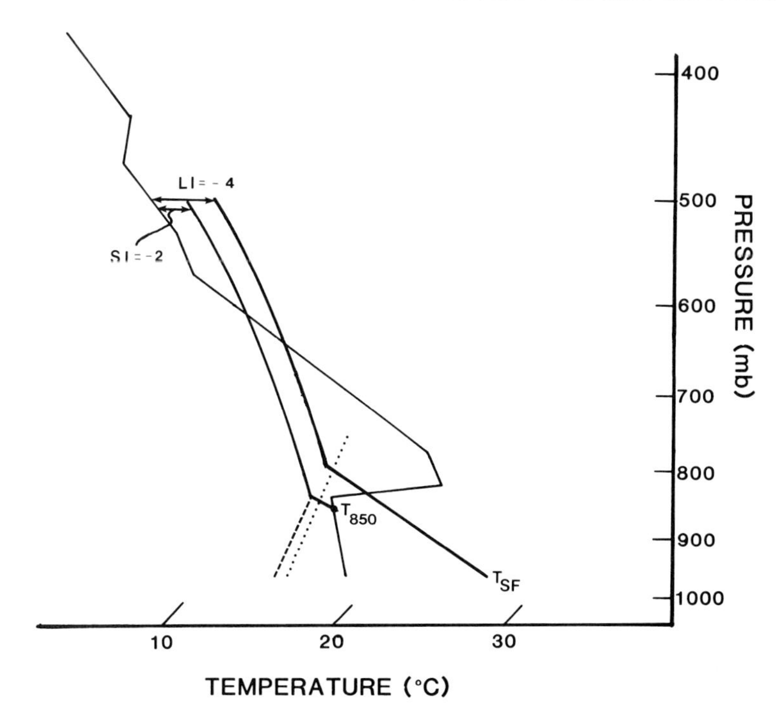

Figure 3.9 Idealized illustration of the computation of the Showalter index (SI) and the lifted index (LI). Skewed abscissa and logarithmic ordinate are temperature in °C and pressure in mb, respectively. Temperature profile (thin solid line); dew point in the moist, well-mixed boundary layer (dashed line); line of constant water-vapor saturation mixing ratio (dotted line); paths taken by ascending air parcel beginning at 850 mb and the surface, where the temperatures are T_{850} and T_{SF}, respectively (thick solid lines).

good indicators of late-afternoon convection, since diurnal surface heating has not been taken into account.

The *K index* is the 850 mb temperature less the 500 mb temperature, added to the 850 mb dew-point temperature less the 700 mb temperature–dew-point depression. It is an indication of potential instability in the lower half of the atmosphere, availability of moisture in the boundary layer, and reduction of buoyancy through entrainment of dry air (near 700 mb). It is most useful for forecasting thunderstorms in the absence of strong, larger-scale forcing.

In the Plains region of the United States, the difference in temperature between 700 and 500 mb is also a good indicator of potential buoyancy. When it is 20–25°C or greater, the lapse rate is a large fraction of the dryadiabatic lapse rate and high CAPE is possible.

Air masses associated with convective precipitation. Convective storms are classified subjectively by forecasters into one of two categories: severe or nonsevere. *Severe* thunderstorms are those that produce tornadoes, strong straight-line winds (wind gusts greater than 25 m s^{-1}), or large hail (greater than 1.9 cm in diameter). The key element of the severe thunderstorm is *its* *ability to inflict damage.* Since some structures are more resistant to damage than others, the officially accepted severe criteria are somewhat arbitrary: Wind gusts of 25 $m\,s^{-1}$ may blow over a mobile home, but not a house; baseball-size hail may break automobile windows, but no damage will be sustained by a strong roof; a small tornado may destroy a small house, but not a strong, steel-reinforced one. Not included in the severe category are intense rainfall-producing storms, which can lead to damaging flash floods, lightning, which can kill people and set fires, and funnel-cloud-producing storms, whose funnel clouds may be tornadoes not yet in contact with the ground. We generally accept the notion that most severe thunderstorms are more "intense" than nonsevere thunderstorms. However, many nonsevere thunderstorms are identical generically to their damaging brother and sisters, even though the strength of the winds and size of the hail produced are just below the "severe" threshold. Although the distinction between severe and nonsevere thunderstorms is useful for forecasting, it has little use in a discussion about the physics of convective storms.

As early as the mid-1880s, J. P. Finley of the Army Signal Service dared to suggest that tornadoes were predictable. It was not until the early 1940s that forecasting methods, based upon upper-air observations, were proposed, in large part in response to military needs. The U.S. Air Force began forecasting severe weather in 1948 in response to an "unwarned" damaging assault by a tornado on Tinker Air Force Base in Oklahoma City. In 1952 the Severe Local Storms Forecasting Unit (SELS) of the U.S. Weather Bureau in Washington, D.C. started to forecast severe weather. The SELS unit moved to Kansas City in 1954 and became the National Severe Storms Forecasting Center (NSSFC) in 1966. As a result of the efforts of the forecasters, four characteristic soundings associated with strong convection have been identified. We will refer to them as *Miller's composite soundings* in honor of the pioneer Col. R. Miller, whose Air Force Manual has been for many years the reference source of severe-storm forecasting techniques.

Type I sounding. The Miller *Type I sounding* (Fig. 3.10) is characterized by a moist, nearly well-mixed (i.e., both the mixing ratio and potential temperature are nearly constant) layer separated from a very dry layer above by a stable layer or inversion. The lapse rate above the "cap" is often nearly dryadiabatic. Although the median depth of the moist layer when intense convection occurs is about 1500 m, only 1000 m (100 mb) of "moisture" is usually sufficient to sustain strong storms. Dewpoints at the surface are usually in excess of 50°F (10°C), while dewpoints at the top of the moist layer (near 850 mb) are usually in excess of 8°C.

Type I soundings have a relatively high lapse rate of θ_e and θ_w owing to the high temperatures and dew points in the boundary layer and low temperatures and dew points at middle and high levels in the troposphere. These soundings are therefore associated with a lot of potential instability. They are sometimes referred to as *loaded-gun soundings.* The CAPE and totals–totals index associated with these soundings are usually large; the lifted index and Showalter index are usually low. These values of the indices indicate the

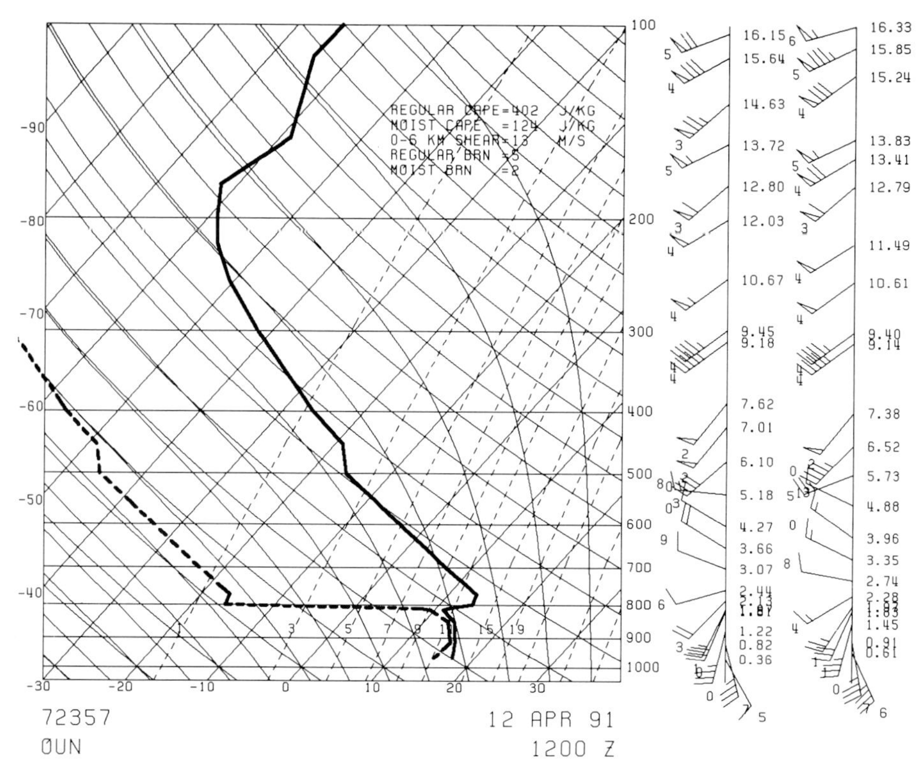

Figure 3.10 Example of a typical Miller "Type I" sounding in the United States. Sounding is at Norman, Oklahoma, at 1200 UTC, April 12, 1991 (0600 LST). Skewed abscissa and logarithmic ordinate are temperature (°C) and pressure (mb), respectively. The plot of temperature and dew point are given by the thick solid and dashed lines, respectively. Winds plotted at the right; pennant = 25 m s^{-1}; whole barb = 5 m s^{-1}; half barb = 2.5 m s^{-1}. The moist air in the boundary layer (surface to 810 mb) comes from the Gulf of Mexico; the mixing ratio from the surface to 850 mb is nearly constant. The layer of low static stability above the moist layer comes from an elevated, deep, dry boundary layer over Mexico. The air above the Mexican layer (above 450 mb) comes from the western United States. The latter is separated from the former by a 50-mb deep stable layer. A stable layer, which in this case is an inversion, caps the moist layer around 800 mb. Several tornadic thunderstorms occurred in north-central Oklahoma later on in the day.

potential for convective storms having intense updrafts. The presence of dry air aloft suggests that the potential for negatively buoyant, evaporatively cooled downdrafts is high, provided that liquid precipitation mixes with the environmental air.

These soundings are common over the Plains region of the United States during the spring. The moist layer is a result of the advection of air northward that has had a history of contact with the Gulf of Mexico. Ordinarily it takes several days for evaporation from the sea surface and the turbulent transfer of water vapor upward to produce a moist layer when dry continental air flows over the Gulf. The cap or lid is a result of relatively warm, dry air aloft from the southwest, west, or northwest. This air may originate in the middle and upper troposphere upstream from a trough in the baroclinic westerlies. This air mass has had a history of descent in part because vorticity advection had been increasingly anticyclonic with height upstream from the trough. The source of this air may also be the tropics; subsequently the air may have had a history of contact with the heated, elevated terrain of Mexico and the Southwest United States. Sometimes both sources of air (Fig. 3.11) above the cap can be identified on a sounding (Fig. 3.10) with the aid of isentropic analysis.

When the cap is very strong, lift may not be strong enough or surface heating may not be sufficient to trigger any convective systems. On days when the strong cap prevents convection from becoming widespread, isolated storms that do form have little competition for water vapor from other storms. Furthermore, if there are few other storms nearby, the probability of cirrus material from nearby anvils reducing the surface heating, and consequently

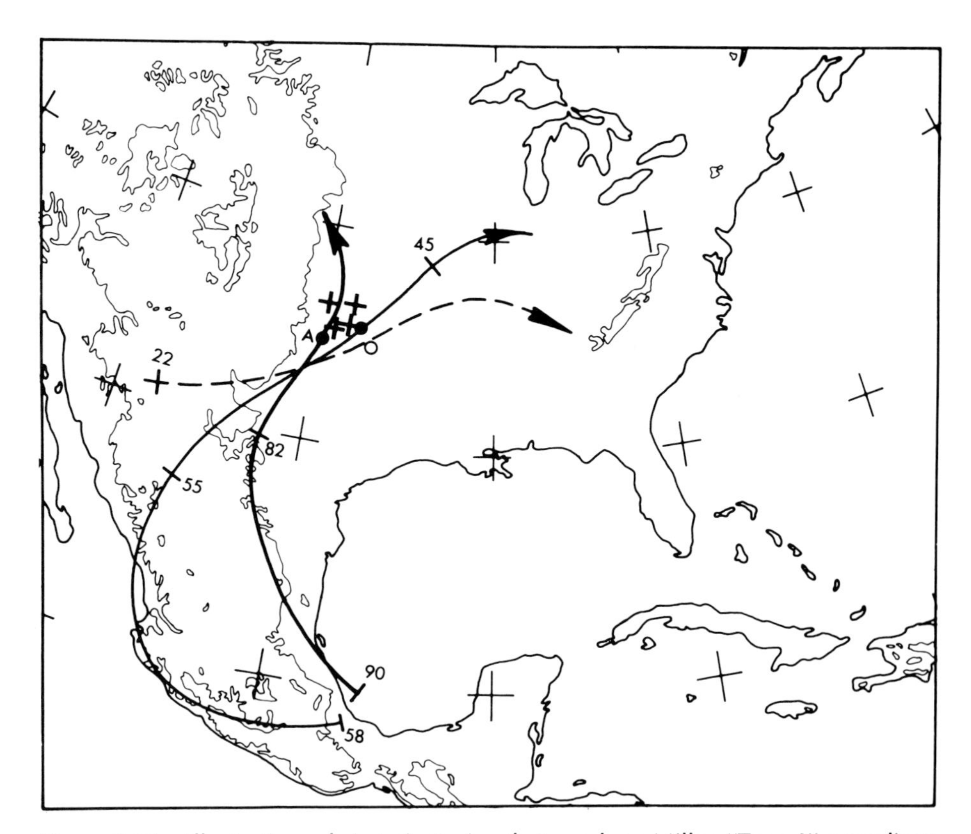

Figure 3.11. Illustration of air trajectories that produce Miller "Type I" soundings in the United States. Composite chart for May 4, 1961. Limiting trajectories for the moist, trade-wind flow (thick line; height of top of layer indicated in tens of mb) and the top of the Mexican air (thin line; height marked in tens of mb). Flow in the high troposphere, at about 220 mb (dashed line). Altus (A) and Oklahoma City (O) denoted by full circles (from Carlson and Ludlam, 1963; courtesy Toby Carlson).

reducing the potential for high surface temperatures, is decreased. A common forecasting problem when the strength of the capping inversion varies spatially is determining where the limits of convective activity will be.

Type II sounding. The Miller *Type II sounding* (Fig. 3.12) is common in the tropics and may be found at one time or another over much of the United States to the east of the Continental Divide. It is common along the Gulf coast and southeastern U.S. coast during the summer. This sounding, sometimes called the *tropical sounding,* is characterized by a deep moist layer (i.e., the relative humidity is greater than 60–65 percent up to at least 7 km AGL) without a capping inversion. The lapse rate is usually less than dryadiabatic but greater than moist adiabatic (i.e., the sounding is conditionally unstable). Without a capping inversion, but with a deep moist layer, there is the possibility of widespread convection.

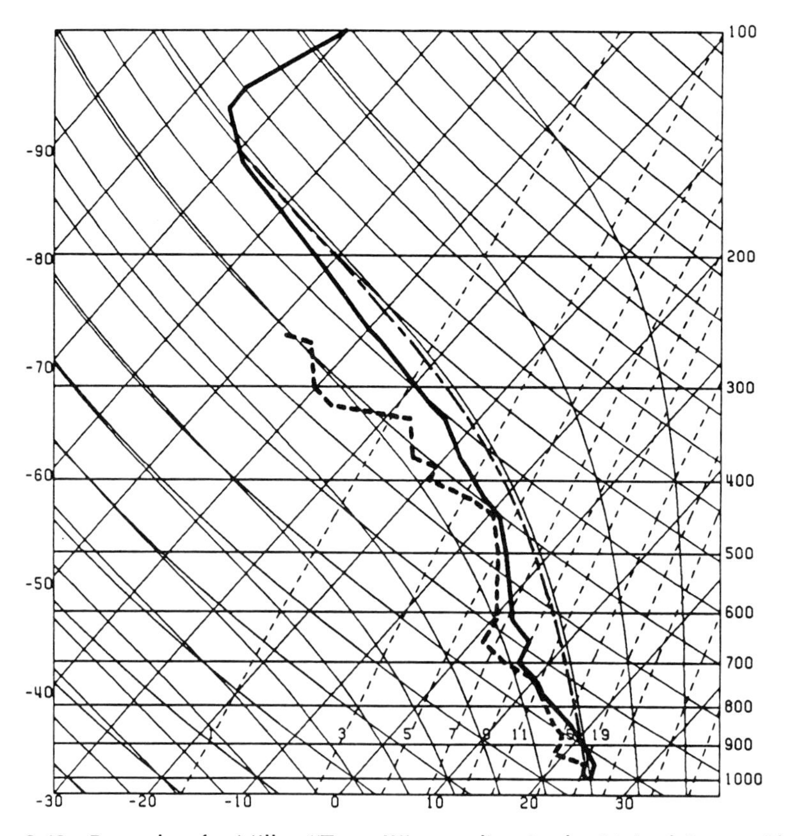

Figure 3.12 Example of a Miller "Type II" sounding in the United States. Skewed abscissa and logarithmic ordinate are temperature (°C) and pressure (mb). Temperature (solid line); dew point (dashed line); moist-adiabat along which surface air parcel ascends (dot-dashed line). For Centerville, Alabama, 0000 UTC, August 17, 1985. This sounding was associated with a tornado outbreak and the remains of Hurricane Danny (from McCaul, 1987). (Courtesy of the American Meteorological Society)

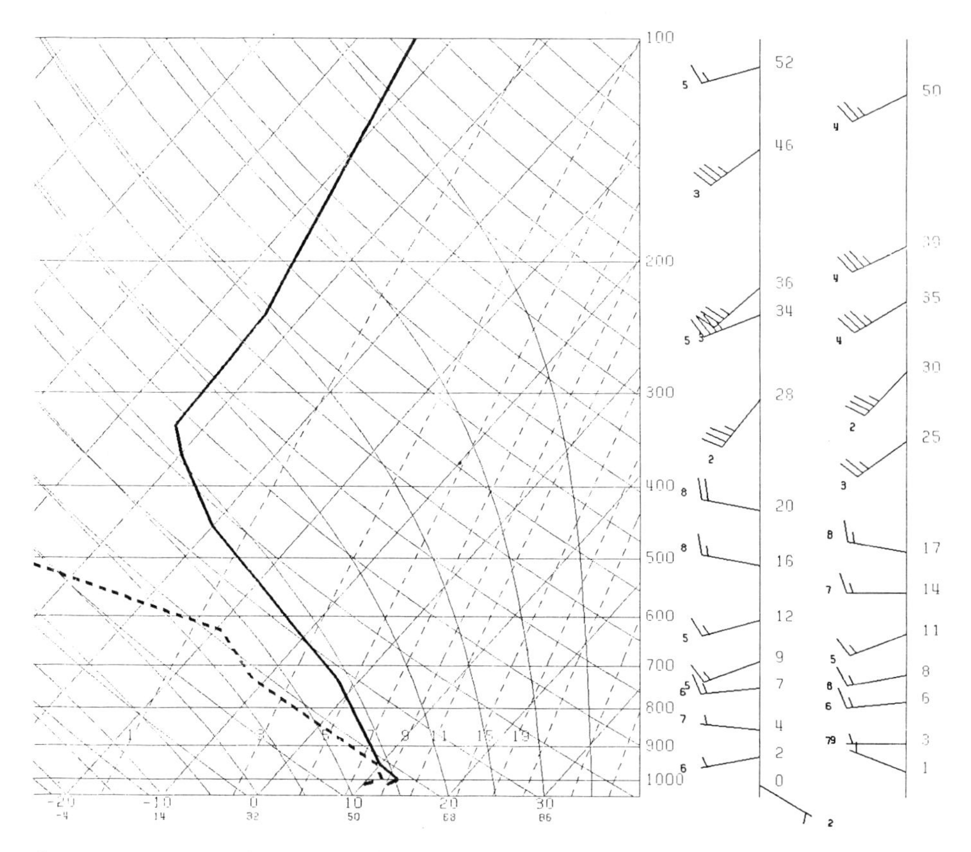

Figure 3.13 Example of a Miller "Type III" sounding in the United States. Sounding for Oakland, California, 1200 UTC, September 10, 1985. Skewed abscissa and logarithmic ordinate are temperature (°C) and pressure (mb), respectively. The plot of temperature and dew point are given by the thick solid and dashed lines, respectively. Winds plotted at the right; whole barb = 5 m s^{-1}; half barb = 2.5 m s^{-1}. A waterspout was reported over nearby San Francisco Bay near the time of this sounding. A cold upper-level low was situated over Northern California. Note the relatively cold −25°C temperature at 500 mb, and the relatively low tropopause (about 325 mb); also note the weak vertical wind shear and light land breeze at the surface from the southeast.

Type III sounding. The Miller *Type III sounding* (Fig. 3.13) is similar to the Type II sounding. However, the temperatures are approximately 10–15°C cooler. These soundings are often characteristic of the region near cold-core upper-level cyclones and troughs, and hence are sometimes referred to as *cold-air soundings.*

Type IV sounding. Finally, the Beebe *Type IV sounding* (Fig. 3.14) does not have a low-level moist layer. Relative humidity generally increases with height in the lower troposphere, and the surface temperatures is high. Because the well-mixed boundary layer is very deep, nearly all the lower half of the sounding, when plotted on a skew T–log p diagram, looks like an inverted

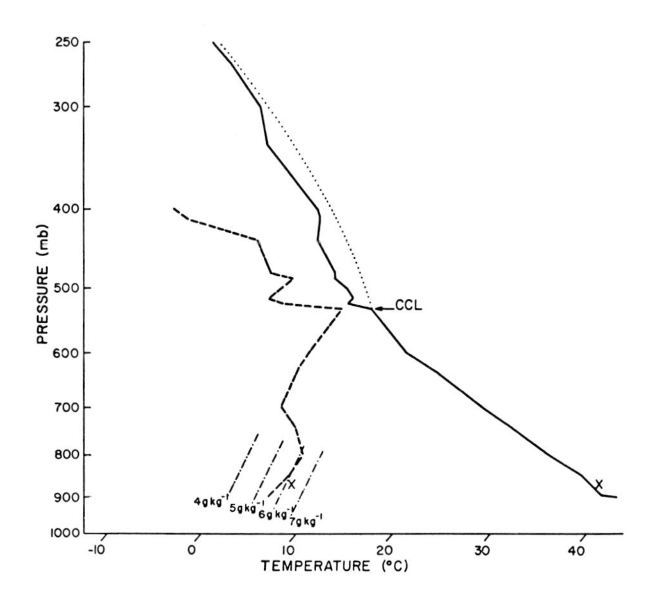

Figure 3.14 Example of a Beebe "Type IV" sounding in the United States (Desert Rock Airport, Mercury, Nevada at 0000 UTC, August 8, 1978). Skewed abscissa and logarithmic ordinate are temperature (°C) and pressure (mb). Temperature (solid line); dew point (dashed line); lines of constant saturation water-vapor mixing ratio (dash-dotted lines); moist adiabat at and above CCL (dotted line). The author observed a tornado in the Sierra Nevada region of California in an environment that was probably similar to this sounding (from Bluestein, 1979). (Courtesy of the American Meteorological Society)

"V": The dew-point profile, which leans to the right with height, is parallel to lines of constant mixing ratio, while the temperature profile, which leans to the left with height and intersects the dew-point profile at the convective condensation level, is parallel to dry adiabats. The Type IV sounding is therefore sometimes referred to as the *inverted-V sounding*. It is common over the High Plains and the western mountain and plateau regions of the United States during the summer. This sounding is produced when dry, low-level, continental tropical air lies beneath moist, maritime, polar air. Convection that occurs in an environment characterized by a Type IV sounding is usually high based. Precipitation falls out through the deep, dry subcloud layer and evaporates, cooling the air and creating vigorous downdrafts that can produce strong surface winds.

The CAPE is one of the most important features of a sounding. Although the common feature in all four types of soundings is conditional instability, the amount of CAPE varies from sounding to sounding. In the Type I sounding, the rapid decrease of temperature with height above the cap implies that CAPE can be quite large; however, the rapid decrease of water vapor with height does not significantly affect the CAPE, even though it does enhance the amount of potential instability through the lapse rates of θ_w and θ_e. The Type II (and III) soundings usually have relatively low values of CAPE, owing to much less steep lapse rates, which often are only slightly greater than pseudomoist adiabatic. The forecaster should note that individual soundings must be analyzed because they may differ substantially from climatology: A Type I sounding can have small CAPE, while a type II sounding can have large CAPE.

The degree of conditional instability (and CAPE) can change owing to differential temperature and moisture advection, differential heating, and vertical motion. The reader is referred back to the static-stability tendency equation (Vol. I) for a more detailed discussion. The existence of CAPE is only a necessary condition for convection; the *amount* of CAPE does not affect the probability that convection will be triggered. It does affect the character of the convection that occurs if it is in fact initiated.

The relationship between soundings, and hail and strong surface winds. Since very intense updrafts are required to suspend large hailstones in the air as they grow into huge spheroids, which have high terminal velocities (e.g., a spherical 7.5 cm hailstone may have a terminal velocity as high as 30 $m\,s^{-1}$), it is reasonable to expect that maximum possible hail size is a function of CAPE. Large values of CAPE are usually found in Type I soundings. Hail is least likely to be associated with the relatively warm Type II sounding. However, hail size is a function of the trajectory of the hailstone through the storm, and a large CAPE does not by itself ensure that the hailstone will grow to a large size.

Some techniques for forecasting peak wind gusts at the surface depend upon the difference in temperature between surface air underneath and air in advance of a thunderstorm. Strong surface winds are possible in storms formed in Type IV environments, since precipitation has plenty of time to evaporate and cool the downdraft, and thereby produce large values of negative buoyancy, before the downdraft finally hits the ground. Type I soundings can also be the environment for strong wind-producing storms owing to the dry air aloft and warm temperatures in the boundary layer.

3.4.6. The Role of Vertical Shear

Gust-front–environment interaction

Ordinary cells. Water droplets (and at high levels, ice crystals) are produced in a growing convective cloud (cumulus congestus) in which there is an updraft on the order of 1–10 $m\,s^{-1}$; this portion of a convective storm's life is called the *cumulus stage.*

The water and ice particles grow and subsequently become too heavy to be held up by the updraft; that is, they become negatively buoyant, and fall out when their terminal velocity becomes larger than the updraft velocity; this portion of a convective storm's life is called the *mature stage.* On radar the region of precipitation is on the order of 100 km^2 in area, and usually extends up to 5 km AGL at least, and often as high as 3 km above the tropopause. The precipitation region is called a *cell.* (Sometimes the updraft itself is referred to as a "cell.") As the precipitation falls, it creates a downdraft that, if there is no vertical shear, *destroys* the updraft *underneath* it, and then mixes with some of the unsaturated air it encounters on the way down. As some of the precipitation evaporates, the air cools and becomes even more negatively buoyant. The downdraft eventually reaches the ground, where in the reference frame of the cloud, it spreads out uniformly in all directions like a pancake. The advancing surface boundary of relatively dense, evaporatively cooled air that originated in the downdraft and is marked by a shift in wind direction and gustiness is called a *gust front.*

Eventually the cloud is characterized entirely by downdraft air and only an ice-crystal anvil, sometimes referred to as an *orphan anvil* (cirrus spissatus cumulonimbogenitus), remains: this final portion of a convective storm's life is called the *dissipating stage.* The anvil sublimates very slowly, and hence may persist for a long time.

The idealized life cycle of such an "ordinary" cell, which occurs under conditions of very weak[2] (or nonexistent) shear, is depicted in Figs. 3.15 and 3.16. The former figure is based upon the model proposed by Byers and Braham after the *Thunderstorm Project.* The entire life cycle of a cell ordinarily lasts about 30 to 50 min. When the layer of air into which the precipitation falls is dry and deep, very strong downdrafts and surface winds may be produced.

Multicell storms. The gust front from an ordinary cell may lift air to its LFC along the periphery of the cell. New, secondary cells may thus form adjacent to an old cell. The secondary cells may trigger more cells over and over again in this manner, until a "multicell" storm complex is formed (Fig. 3.17). The gust fronts from all the cells may merge and produce a large gust front, which is referred to also as an *outflow boundary.* However, much of the environmental air is not lifted to its LFC after the gust front has spread out substantially because the vertical displacement of air becomes less as the layer of cool air spreads out and becomes shallower. As a result, new cells are no longer likely to form.

The magnitude of low-level (i.e., over the depth of the outflow, which is usually 2 km or less) vertical shear affects the ability of a gust front to trigger new cells (Fig. 3.18). In the absence of shear, air is lifted only for a short period of time as it passes over the gust front [Fig. 3.18(a)]. If the low-level shear in the direction of gust-front motion is increased, air parcels are lifted above the gust front more vigorously and for a longer period of time, thus improving the chances that the LFC is reached and that all the CAPE in the environment is used [Fig. 3.18(b)]. If the low-level shear is increased even more, air is lifted less vigorously and for a shorter period of time [Fig. 3.18(c)]. Thus, for certain values of shear (given a "cold pool" of air of certain intensity and depth), a continuous growth of new cells may result.

This can be understood in terms of the horizontal vorticity associated with the circulation about the gust front. Suppose that the environment is neutrally buoyant, while the air behind the gust front is negatively buoyant. When the

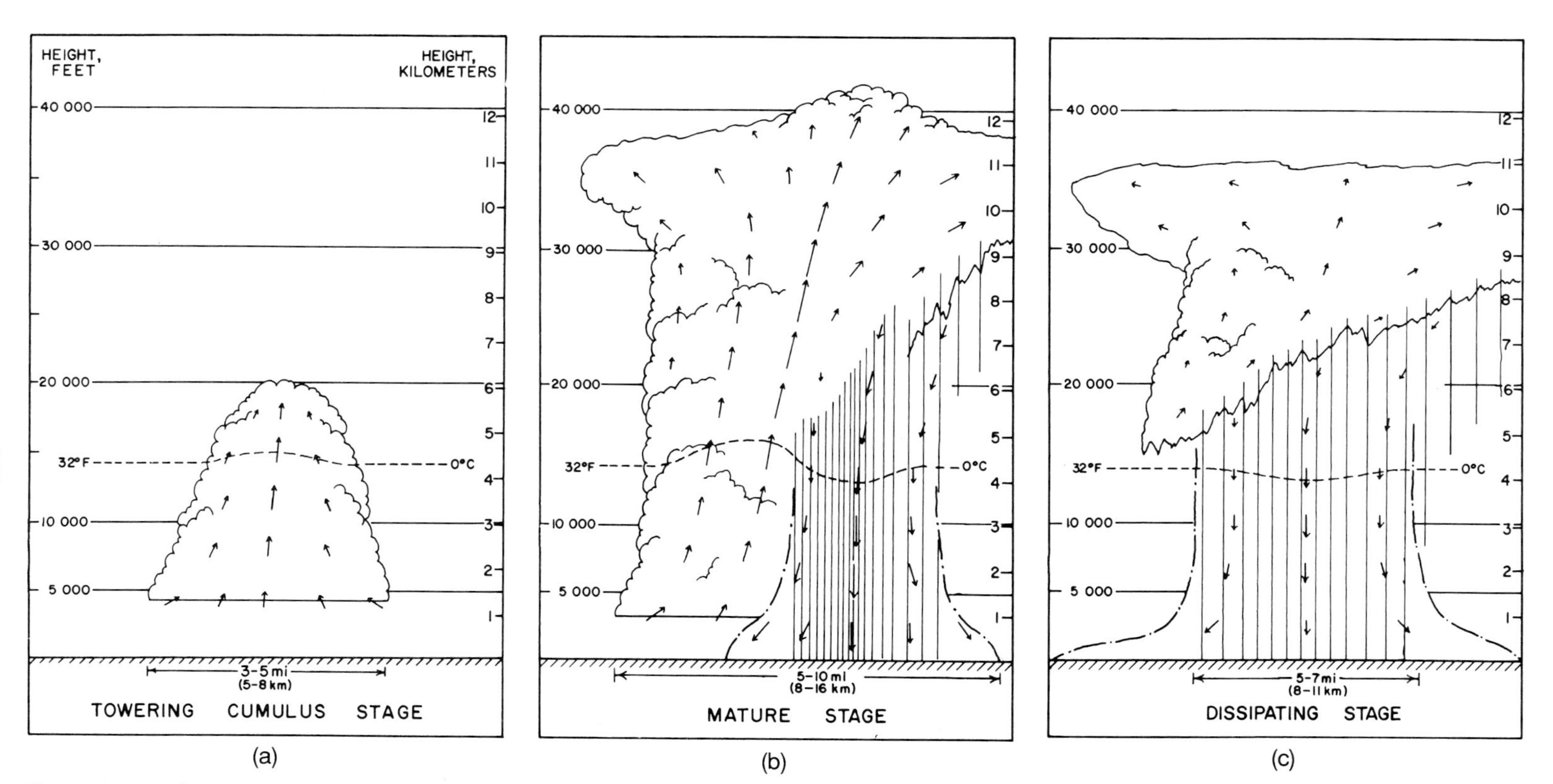

Figure 3.15 The Byers–Braham model of the three stages in the life of a thunderstorm: (a) towering cumulus stage, (b) mature stage, and (c) dissipating stage. Arrows indicate the sense of air motion (from Doswell, 1985).

(a)

(b)

Figure 3.16 Photographs of the (a) towering cumulus, (b), (c) mature, and (d) dissipating stages of a thunderstorm. (a)–(c) were taken August 28, 1971, off the southeast coast of Florida, looking westward; (d) was taken August 28, 1973, near Naples, Florida, looking westward over the Gulf of Mexico (photographs copyright Howard B. Bluestein).

(c)

(d)

Figure 3.16 (*cont.*)

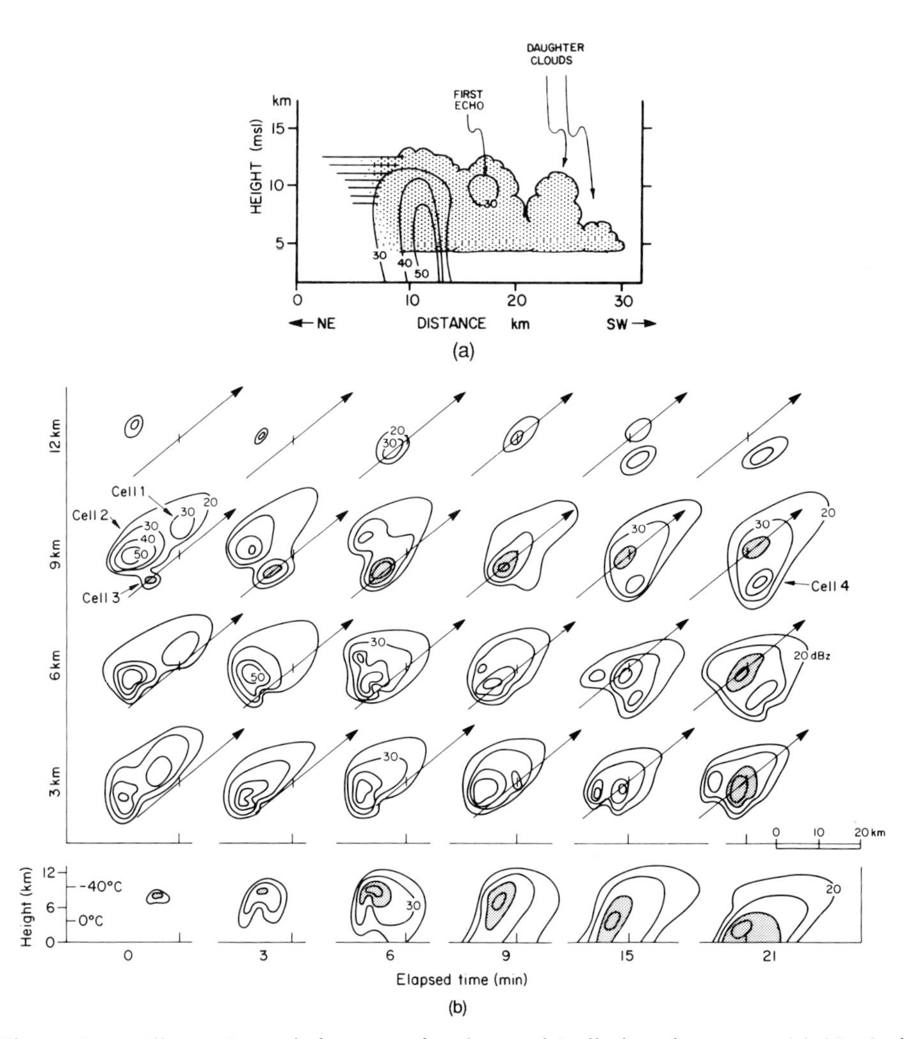

Figure 3.17 Illustration of the growth of a multicell thunderstorm. (a) Vertical cross section from the northeast–southwest of a typical eastward-moving multicell hailstorm in South Dakota showing "daughter" clouds on the storm's right flank. New cells form to the southwest, develop radar-detectable precipitation aloft, and then a deep radar echo; figure is in a sense a look at different stages in the life of a thunderstorm cell. Cloud (stippled area); radar reflectivity in dBZ (solid lines). (after Dennis *et al.*, 1970; from Browning, 1977). (b) Schematic horizontal (above) and vertical (below) radar reflectivity contours in dBZ (solid lines) for an ordinary multicell storm at various stages during its evolution. Cell motion (arrow); cell 3 is shaded to emphasize the history of an individual cell (from Chisholm and Renick, 1972; Browning, 1977). (Courtesy of the American Meteorological Society)

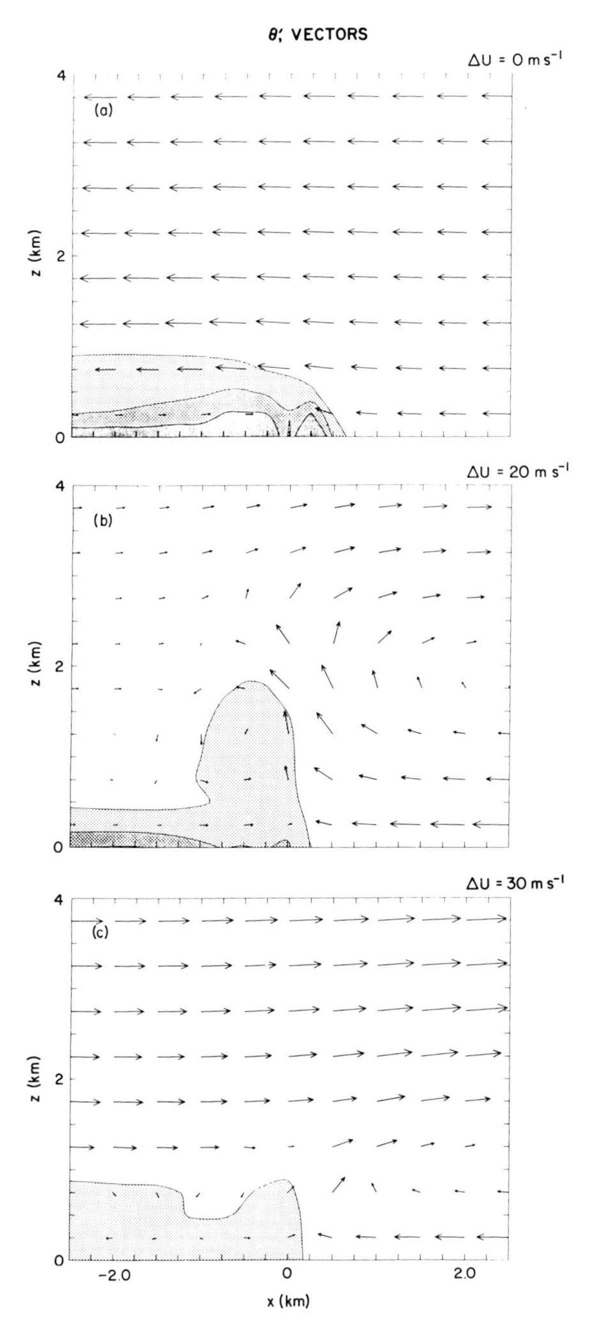

Figure 3.18 The behavior of a spreading cold pool at the ground in a numerically simulated thunderstorm for (a) no vertical shear in the lowest 2 km, (b) 20 m s^{-1} shear in the lowest 2 km, and (c) 30 m s^{-1} shear in the lowest 2 km. In (a)–(c) cold-pool-relative vectors plotted every other grid point (2 grid lengths represents 15 m s^{-1}). Negative potential temperature perturbations shaded at 2 K intervals, beginning at −1 K (from Rotunno et al., 1988). (Courtesy of the American Meteorological Society)

net flux of low-level environmental vertical shear into the gust-front region has sufficient horizontal vorticity to counterbalance the production of horizontal vorticity of the opposite sign by the buoyancy gradient at the gust front (Fig. 3.19), air parcels are lifted the most. Thus, low-level shear is required to keep the gust front from tilting in the upshear direction, and becoming so shallow that air parcels are no longer lifted to their LFC right next to the parent updraft. Keeping the air-parcel trajectory as erect as possible also minimizes the detrimental effects of mixing with the cooler outflow air, whose momentum is opposite to that of the air flowing over the gust front. However, if the low-level shear is too strong, the air parcel trajectories lean *too much* in the downshear direction. Refer to Eq. (2.5.268) for a quantitative assessment of the effects of low-level shear.

Suppose that two gust fronts collide in the absence of strong low-level vertical shear. Then the horizontal vorticity produced by one outflow may counterbalance the horizontal vorticity produced by the other, so that air parcels are displaced upward more than if only one outflow were present (Fig. 3.20). One outflow plays the role of low-level environmental vertical shear. It

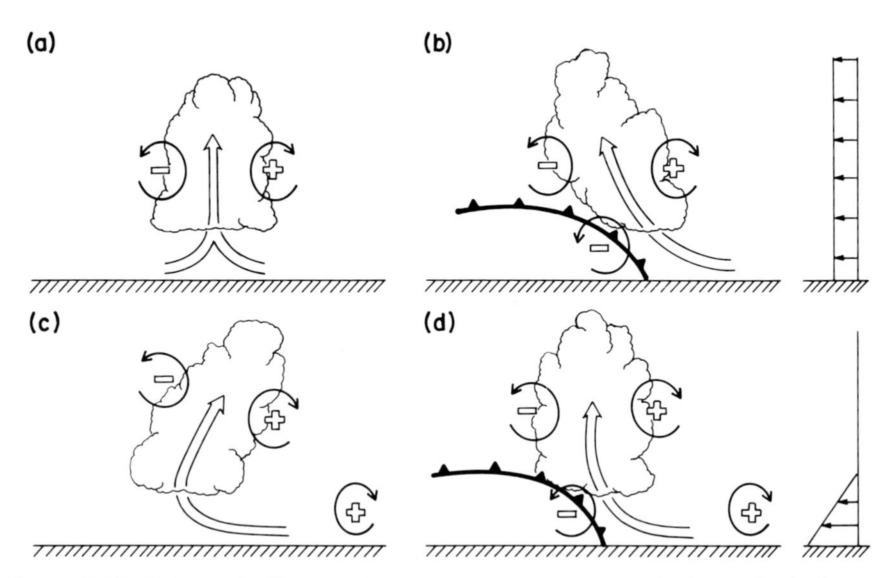

Figure 3.19 Schematic diagram showing how a buoyant updraft may be influenced by vertical wind shear and/or a cold pool. (a) With no vertical wind shear and no cold pool, the axis of the updraft (thick arrow) produced by the thermally created, symmetric vorticity (sense of rotation indicated by thin arrows and plus and minus signs) distribution is vertical. (b) With a cold pool (underneath cold-front symbol) and no shear, the distribution is biased by the negative vorticity of the underlying cold pool and causes the updraft to lean as shown. (c) With shear and no cold pool, the distribution is biased toward positive vorticity, and this causes the updraft to lean as shown. (d) With both a cold pool and shear, the two effects may negate each other, and allow an erect updraft. Environmental wind profile indicated at the right for (a) and (b) (top) and for (c) and (d) (bottom) (from Rotunno et al., 1988). (Courtesy of the American Meteorological Society)

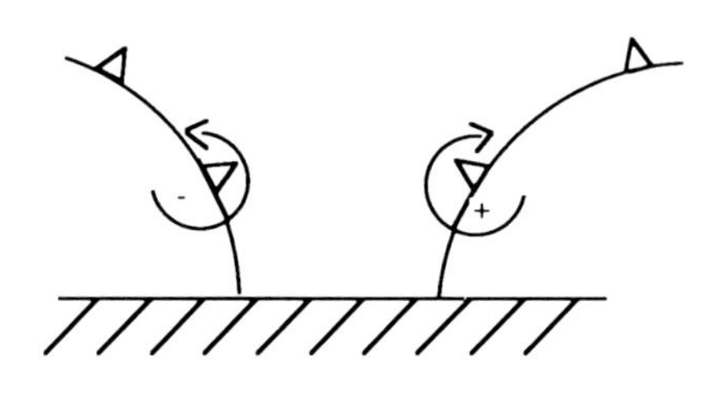

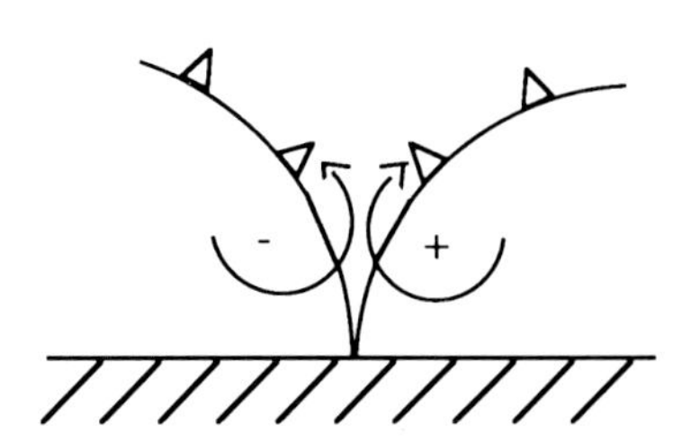

Figure 3.20 Effect of two colliding outflow boundaries (underneath cold-front symbols) on horizontal vorticity (sense of rotation indicated by thin arrows and plusses and minuses). (top) Before collision, (bottom) after collision.

has been observed that the locations where gust fronts collide are often indeed the sites of new convective growth.

The development and dynamical consequences of rotation in supercells. Suppose now that the vertical shear in the environment is increased to a value of approximately $5 \times 10^{-3}\ \mathrm{s}^{-1}$ [$25\ \mathrm{m\,s}^{-1}\ (5\ \mathrm{km})^{-1}$] or greater. If the updraft is relatively weak, owing to low CAPE, the cell will drop its precipitation away from the updraft on its downshear side, into the inflow region of the storm; the original updraft will thus be cut off (blocked) from warm surface inflow from the downshear side. The effect of the wind shear is to make the updrafts lean in the downshear direction and advect kinetic energy associated with the updraft away aloft.

However, if the updraft is localized and strong, owing to high CAPE, then the effect of vertical shear is to *enhance* the intensity and *increase* the longevity of the storm. It is not, however, the *tilt* of the updraft that is of fundamental importance; rather, it is the shear, which is necessary for rotation, that sustains the updraft. To begin a discussion of this interesting process, we first ask the following question: What is the effect of a circularly symmetric updraft on the pressure field in an environment of vertical shear? To answer this, we consider the set of Boussinesq equations, Eqs. (3.4.10)–(3.4.12) and (3.4.20).

We differentiate Eqs. (3.4.10) and (3.4.11) with respect to x and y, respectively, and add them together, and then add Eq. (3.4.12) differentiated with respect to z to them, with $\bar{\rho}$ treated as a constant, to obtain the following divergence equation:

$$\frac{D}{Dt}(\nabla \cdot \mathbf{v}) + \frac{1}{\bar{\rho}}\nabla^2 p' = -\left[\left(\frac{\partial u}{\partial x}\right)^2 + \left(\frac{\partial v}{\partial y}\right)^2 + \left(\frac{\partial w}{\partial z}\right)^2 + 2\frac{\partial u}{\partial y}\frac{\partial v}{\partial x} + 2\frac{\partial u}{\partial z}\frac{\partial w}{\partial x} + 2\frac{\partial v}{\partial z}\frac{\partial w}{\partial y}\right] + \frac{\partial B}{\partial z}. \tag{3.4.39}$$

Substituting for Eq. (3.4.20) in this equation, we can now derive a diagnostic relation between perturbation pressure and gradients in the wind field and the vertical derivative of buoyancy.

Let the wind field be expressed as

$$u = \bar{u}(z) + u'(x,y,z,t) \tag{3.4.40}$$

$$v = \bar{v}(z) + v'(x,y,z,t) \tag{3.4.41}$$

$$w = w'(x,y,z,t), \tag{3.4.42}$$

where $\bar{u}(z)$ and $\bar{v}(z)$ represent the environmental horizontal wind profile, $\bar{w} \approx 0$, and u', v', w' is the wind-field perturbation associated with the storm. The environmental winds $\bar{u}(z)$ and $\bar{v}(z)$ are usually represented graphically as a hodograph. Vertical velocity in the environment is much less than the horizontal velocity, so that the vertical velocity can be ignored. The diagnostic pressure equation for the Boussinesq system is then

$$\nabla^2 p' = -\bar{\rho}\left[\left(\frac{\partial u'}{\partial x}\right)^2 + \left(\frac{\partial v'}{\partial y}\right)^2 + \left(\frac{\partial w'}{\partial z}\right)^2 + 2\frac{\partial u'}{\partial y}\frac{\partial v'}{\partial x} + 2\frac{\partial u'}{\partial z}\frac{\partial w'}{\partial x}\right.$$
$$\left. + 2\frac{\partial v'}{\partial z}\frac{\partial w'}{\partial y}\right] - \bar{\rho}\left(2\frac{\partial \bar{u}}{\partial z}\frac{\partial w'}{\partial x} + 2\frac{\partial \bar{v}}{\partial z}\frac{\partial w'}{\partial y}\right) + \bar{\rho}\frac{\partial B}{\partial z}. \tag{3.4.43}$$

The perturbation pressure associated with gradients in the wind field is called the *dynamic* pressure (p'_{dyn}), while the perturbation pressure associated with the vertical derivative of buoyancy is called the *buoyancy* pressure (p'_{B}). We further partition the dynamic pressure into linear and nonlinear parts.

The total perturbation pressure

$$p' = p'_{\text{dyn}} + p'_{\text{B}} = p'_{\text{L}} + p'_{\text{NL}} + p'_{\text{B}}, \tag{3.4.44}$$

where p'_{L} and p'_{NL} are the perturbation dynamic pressures associated with the linear part of the wind field,

$$-\bar{\rho}\left(2\frac{\partial \bar{u}}{\partial z}\frac{\partial w'}{\partial x} + 2\frac{\partial \bar{v}}{\partial z}\frac{\partial w'}{\partial y}\right) = -2\bar{\rho}\frac{\partial \bar{\mathbf{v}}}{\partial z}\cdot \nabla_z w' \tag{3.4.45}$$

and the nonlinear part of the wind field,

$$-\bar{\rho}\left[\left(\frac{\partial u'}{\partial x}\right)^2 + \left(\frac{\partial v'}{\partial y}\right)^2 + \left(\frac{\partial w'}{\partial z}\right)^2 + 2\frac{\partial u'}{\partial y}\frac{\partial v'}{\partial x}\right.$$
$$\left. + 2\frac{\partial u'}{\partial z}\frac{\partial w'}{\partial x} + 2\frac{\partial v'}{\partial z}\frac{\partial w'}{\partial y}\right],$$

respectively. The terms proportional to

$$\left(\frac{\partial u'}{\partial x}\right)^2 + \left(\frac{\partial v'}{\partial y}\right)^2 + \left(\frac{\partial w'}{\partial z}\right)^2$$

are called the *fluid-extension terms* [they represent the (square of the) magnitude of the gradient of the wind field associated with the storm]; the terms proportional to

$$2\frac{\partial u'}{\partial y}\frac{\partial v'}{\partial x}+2\frac{\partial u'}{\partial z}\frac{\partial w'}{\partial x}+2\frac{\partial v'}{\partial z}\frac{\partial w'}{\partial y}$$

are called the *shear terms.*

The diagnostic Poisson equation for the linear, dynamic term alone is

$$\nabla^2 p'_{\mathrm{L}} = -2\bar{\rho}\frac{\partial \bar{\mathbf{v}}}{\partial z}\cdot \nabla_z w'. \tag{3.4.46}$$

The linear, dynamic perturbation-pressure field depends upon the way in which the vertical-velocity field associated with the updraft is related to the environmental vertical-shear vector. Since a Laplacian tends to change the sign of the variable on which it operates,

$$p'_{\mathrm{L}} \sim \frac{\partial \bar{\mathbf{v}}}{\partial z}\cdot \nabla_z w'. \tag{3.4.47}$$

At a given level there is thus a negative perturbation pressure on the downshear side of an updraft, and a positive perturbation pressure on the upshear side of an updraft (Fig. 3.21). The physical interpretation of Eq. (3.4.47) is that the vertical advection by an updraft of horizontal momentum associated with the environmental vertical shear $(-w'\partial\bar{\mathbf{v}}/\partial z)$ is balanced to a first approximation by the pressure-gradient force $[-(1/\bar{\rho})\nabla_z p']$: The pressure-gradient force is changed in conjunction with the change in momentum advected up from below by the updraft.

If the vertical-shear profile is unidirectional, and the horizontal gradient of w is greatest at midlevels, where w is the largest, then there is a perturbation low-pressure area downshear and perturbation high-pressure area upshear of the updraft at midlevels. As a result of the perturbation–low-pressure area aloft, there is an upward-directed perturbation pressure-gradient force on the downshear flank of the storm (Fig. 3.22); as a result of the perturbation–high-pressure area aloft, there is a downward-directed perturbation pressure-gradient force on the upshear flank of the storm (Fig. 3.22). If the upward-directed pressure-gradient force is strong enough in the layer extending up to the LFC, air will be forced upward and new cells will be triggered on the downshear side. If the downward-directed perturbation pressure-gradient force is strong enough, new cell growth is suppressed on the upshear side. Strong shear from the surface to midlevels in the face of an intense, localized updraft is therefore significant: It promotes new cell growth on the forward, downshear side of the updraft, and suppresses new cell growth on the upshear side.

If the hodograph is curved in a clockwise manner, upward-directed perturbation pressure gradients are favored on the right flank (Fig. 3.21). On the other hand, if the hodograph has counterclockwise curvature, upward-directed perturbation pressure gradients are favored on the left flank (not

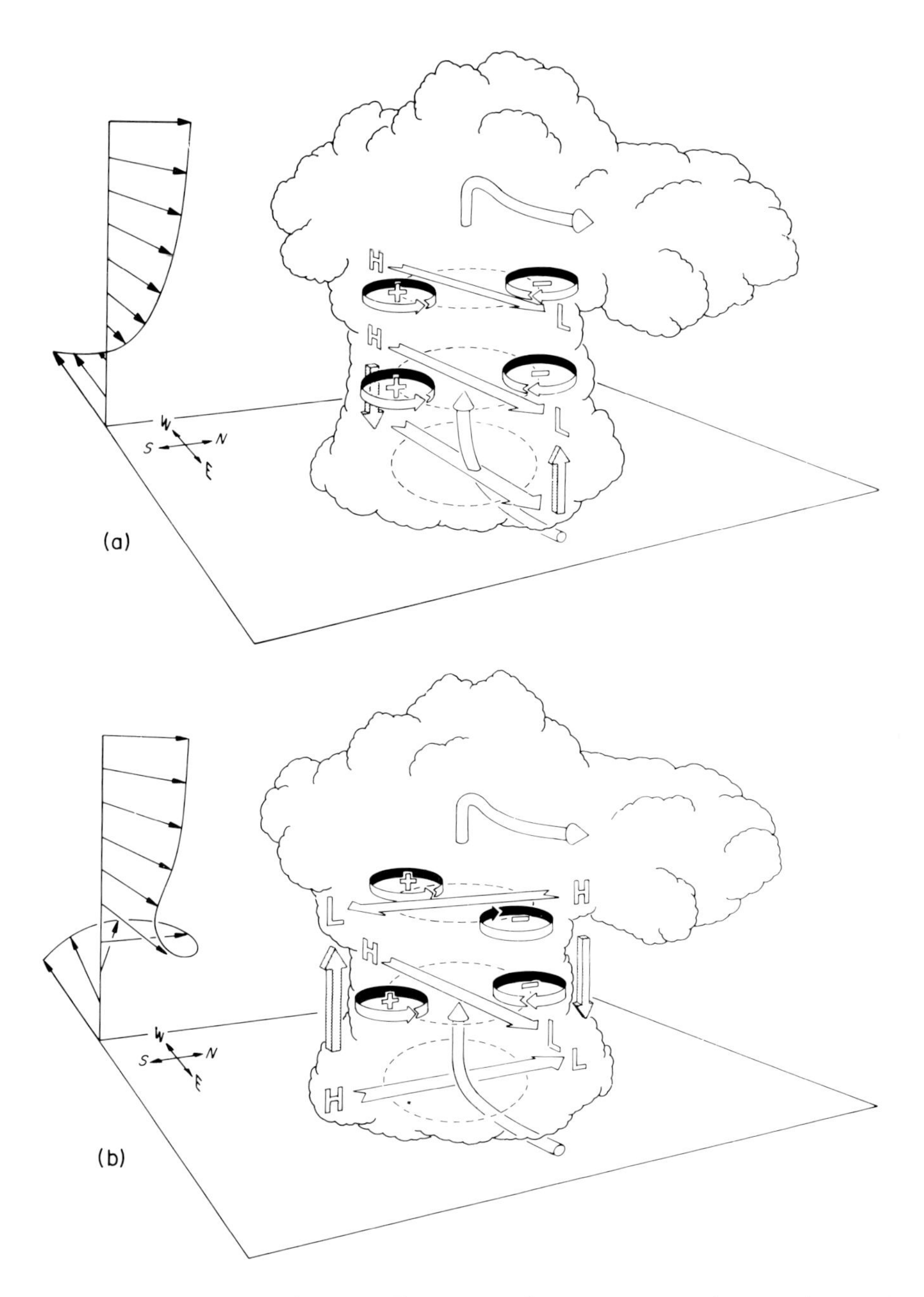

Figure 3.21 Schematic diagram illustrating the pressure and vertical vorticity perturbations arising as an updraft interacts with an environmental vertical wind shear vector that (a) does not change direction with height and (b) turns clockwise with height. The high (H) to low (L) horizontal pressure-gradient forces parallel to the shear vectors (flat arrows) are labeled along with the preferred location of cyclonic (+) and anticyclonic (−) vorticity. Shaded arrows depict orientation of resulting vertical pressure-gradient forces (from Klemp, 1987; adapted from Rotunno and Klemp, 1982). (Courtesy of the American Meteorological Society)

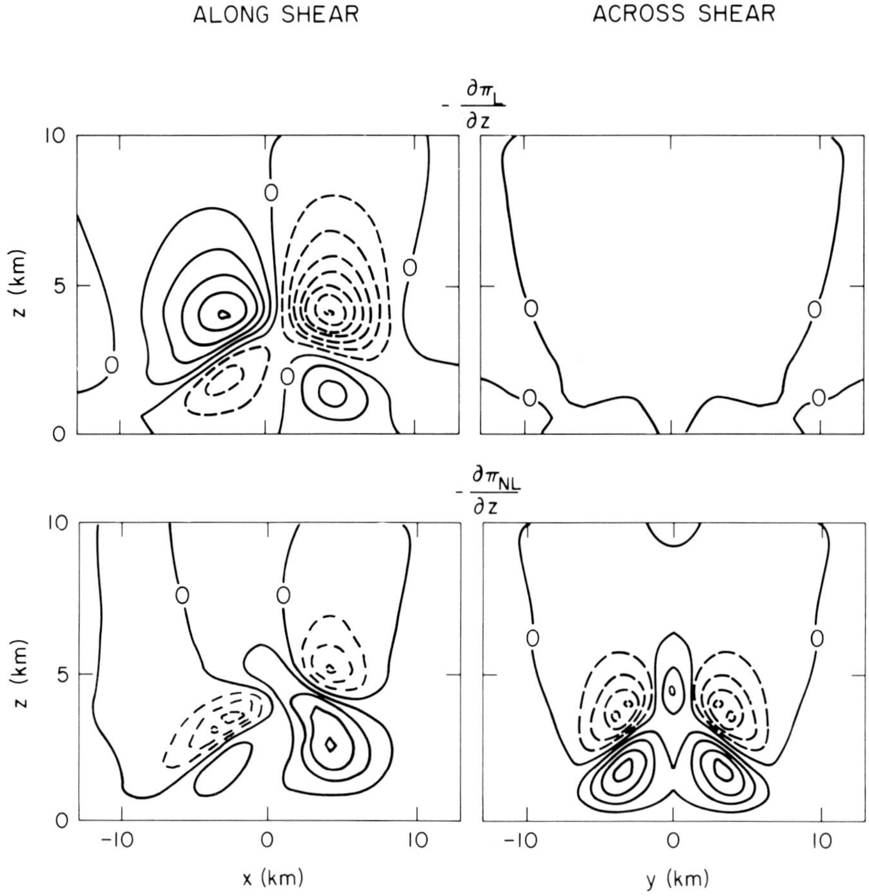

Figure 3.22 Vertical cross section of acceleration induced by vertical perturbation-pressure force for the linear part of the wind field ($-\partial\pi_L/\partial z$ (top) and the nonlinear part of the wind field ($-\partial\pi_{NL}/\partial z$) (bottom) in a numerical simulation 10 min after storm initiation. The environmental wind profile has a straight-line hodograph. Contours plotted every 0.004 m s^{-2} (from Rottuno and Klemp, 1982). (Courtesy of the American Meteorological Society)

shown). Thus, new cell growth is promoted preferentially on the right flank of storms growing in an environment of clockwise hodograph curvature, while new cell growth is promoted preferentially on the left flank of storms growing in an environment of counterclockwise hodograph curvature. Most hodographs are actually curved in a generally clockwise manner, owing to boundary-layer friction and low-level warm advection, and hence "right-flank storms" are more common in high-shear and high-CAPE environments.

Let us now consider the effects of the nonlinear terms on the right-hand side of Eq. (3.4.43). The nonlinear "shear" terms may be expressed as follows:

$$\begin{aligned}\frac{1}{2}\Bigg[&\left(\frac{\partial v'}{\partial x}+\frac{\partial u'}{\partial y}\right)^2-\left(\frac{\partial v'}{\partial x}-\frac{\partial u'}{\partial y}\right)^2+\left(\frac{\partial u'}{\partial z}+\frac{\partial w'}{\partial x}\right)^2-\left(\frac{\partial u'}{\partial z}-\frac{\partial w'}{\partial x}\right)^2\\&+\left(\frac{\partial w'}{\partial y}+\frac{\partial v'}{\partial z}\right)^2-\left(\frac{\partial w'}{\partial y}-\frac{\partial v'}{\partial z}\right)^2\Bigg]\\&=2\frac{\partial v'}{\partial x}\frac{\partial u'}{\partial y}+2\frac{\partial w'}{\partial x}\frac{\partial u'}{\partial z}+2\frac{\partial v'}{\partial z}\frac{\partial w'}{\partial y}.\end{aligned} \tag{3.4.48}$$

The tilting effect in a unidirectionally sheared environment produces a cyclonic–anticyclonic couplet at midlevels (Fig. 3.23); suppose that each member of the couplet is composed of a field of pure rotation about the vertical (z axis). Then the deformation terms $\partial v'/\partial x+\partial u'/\partial y$, $\partial u'/\partial z+\partial w'/\partial x$, and $\partial w'/\partial y+\partial v'/\partial z$ are zero, and the horizontal vorticity terms $\partial u'/\partial z-\partial w'/\partial x$ and $\partial w'/\partial y-\partial v'/\partial z$ are also zero. If we furthermore ignore the fluid-extension terms, it follows that

$$\nabla^2 p'_{\mathrm{NL}}=\frac{\bar{\rho}}{2}\left(\frac{\partial v'}{\partial x}-\frac{\partial u'}{\partial y}\right)^2=\frac{\bar{\rho}}{2}\zeta'^2. \tag{3.4.49}$$

Then the nonlinear perturbation pressure is proportional and opposite in sign to the square of the perturbation vorticity.

$$p'_{\mathrm{NL}}\sim-\zeta'^2. \tag{3.4.50}$$

In other words, the nonlinear shear terms act to produce *low* perturbation pressure in the vicinity of the midlevel cyclonic and anticyclonic vortices. (If the fluid-extension terms were included, we would find, on the other hand, that stretching or shrinking in three dimensions, i.e. divergence and convergence, contribute toward *high* perturbation pressure.) Therefore an upward-directed perturbation pressure gradient develops below the level at which the vortices are most intense (usually at midlevels) on each flank of the old updraft, along a line normal to the vertical-shear vector (Figs. 3.22 and 3.24). It is the *nonlinear-shear effect* that *promotes new or continued cell growth on the flanks* of (alongside) the old cell, while the *linear effect of tilting* biases the cell movement toward the right (left) if the environmental hodograph is curved in a clockwise (counterclockwise) manner; unidirectional shear promotes storms that split, with each member of the split pair having components of motion normal to the shear vector and opposite to each other (Fig. 3.25).

New buoyant updrafts form off the axis of the shear because upward-directed perturbation pressure gradients induce upward accelerations there and lift air to its LFC (Fig. 3.23). Owing to the low-level convergence associated with upward-moving air, vorticity increases through the stretching of the existing vorticity and is advected upward by the updraft: Right movers tend to develop cyclonic rotation, while left movers tend to develop anticyclonic rotation. It is interesting to note that the cyclonic vorticity produced in storms that grow in an environment of clockwise turning shear is *not* due to the Earth's rotation.

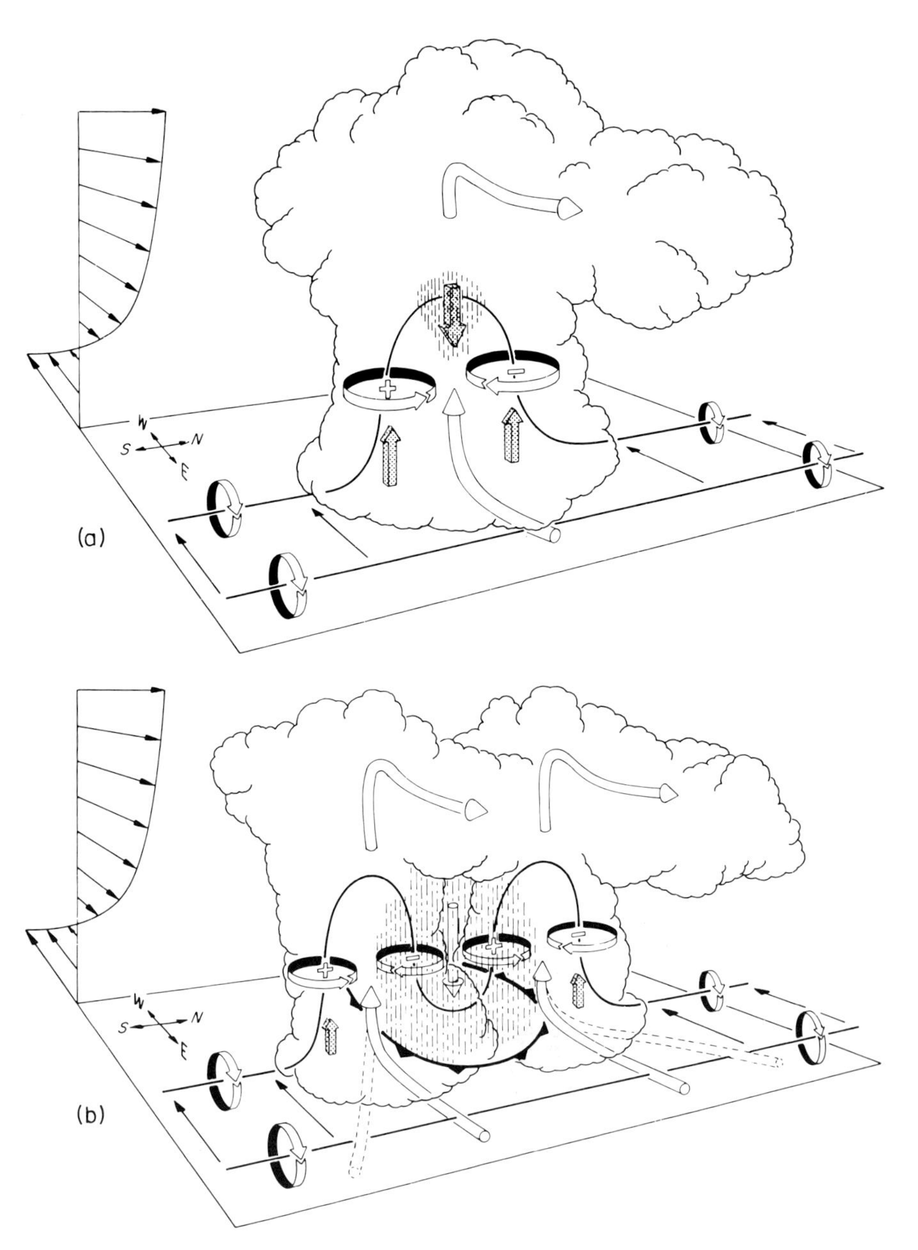

Figure 3.23 Schematic diagram depicting how a typical vortex line (streamline of three-dimensional vorticity vector) contained within (westerly) environmental shear is deformed as it interacts with a convective cell (viewed from the southeast). Direction of cloud-relative airflow (cylindrical arrows); vortex lines (solid lines), with the sense of rotation indicated by circular arrows; the forcing influences that promote new updraft and downdraft growth (shaded arrows); regions of precipitation (vertical dashed lines). (a) Initial stage: Vortex line loops into the vertical as it is swept into the updraft. (b) Splitting stage: Downdraft forming between the splitting updraft cells tilts vortex line downward, producing two vortex pairs. Boundary of the cold air spreading out beneath the storm (cold-front symbol at the surface) (from Klemp, 1987; adapted from Rotunno, 1981). (Courtesy of the American Meteorological Society)

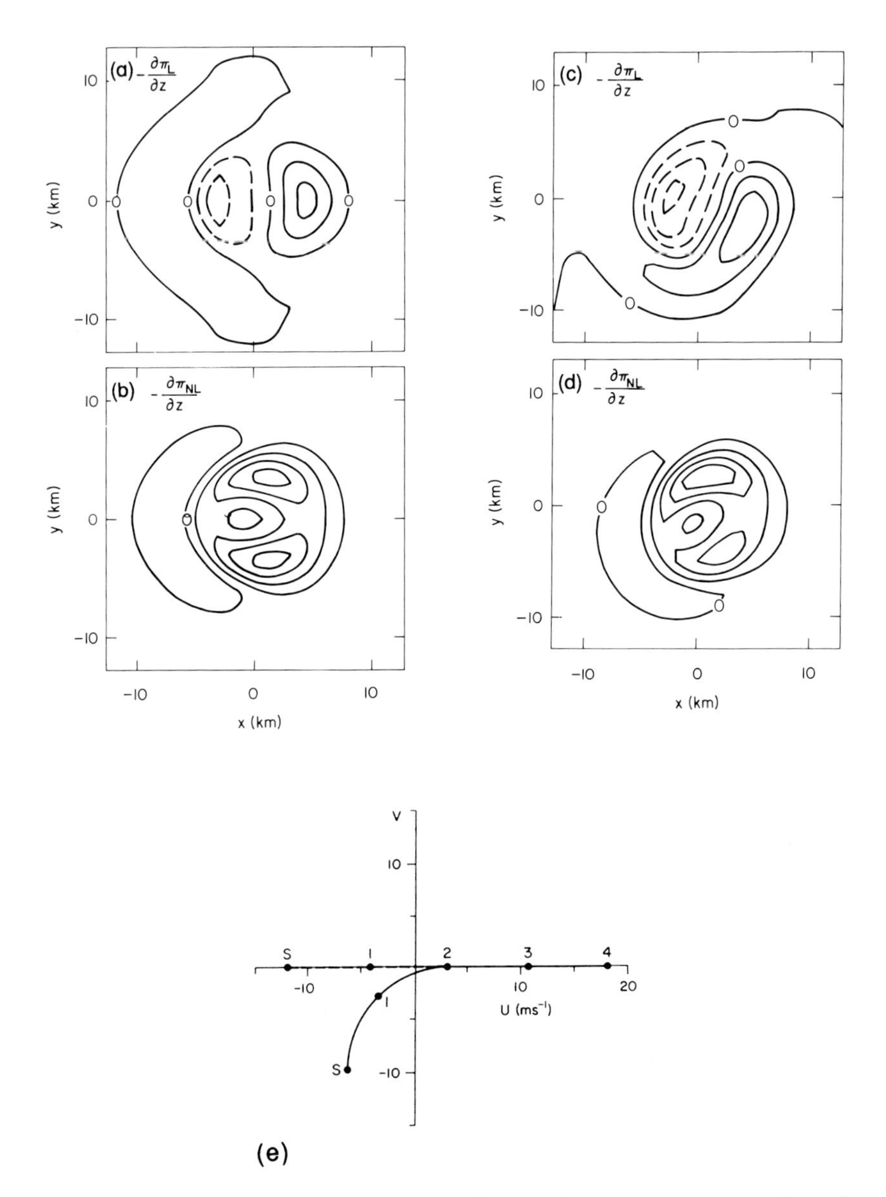

Figure 3.24 Acceleration induced by vertical perturbation-pressure gradient force at 1.5 km in a numerical simulation 10 min after storm initiation for (a) the linear part of the wind field for a straight-line hodograph; (b) the nonlinear part of the wind field for a straight-line hodograph; (c) the linear part of the wind field for a clockwise-turning hodograph; (d) the nonlinear part of the wind field for a clockwise-turning hodograph. Contours plotted every 0.004 m s^{-2}; (e) hodographs used in the simulations; clockwise-turning hodograph indicated by solid line; heights in km AGL (from Rotunno and Klemp, 1982). (Courtesy of the American Meteorological Society)

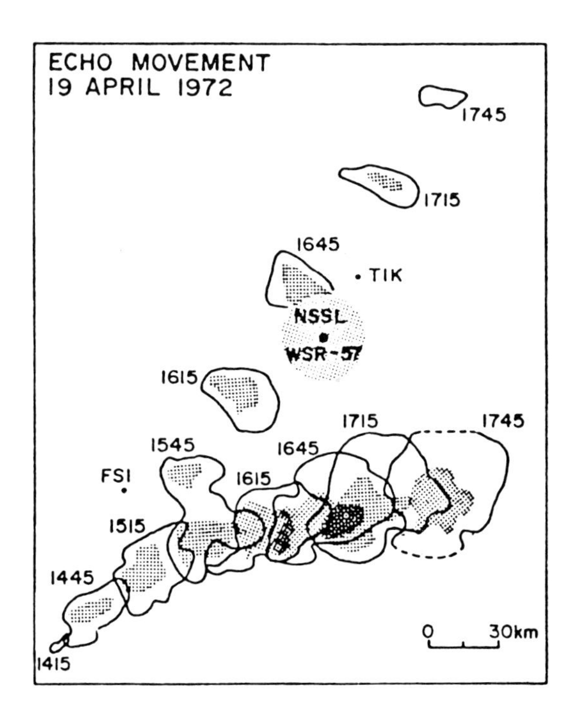

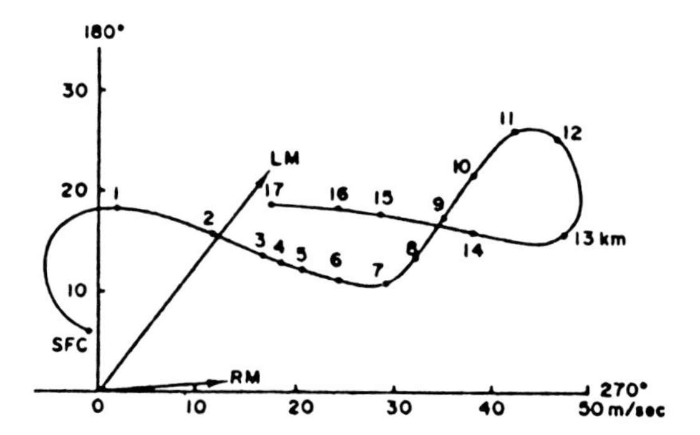

Figure 3.25 (Top) Radar-echo history of a splitting storm observed in south-central Oklahoma. Radar reflectivity of 10 dBZ (solid lines); radar reflectivity in excess of 40 dBZ (stippled regions). Times adjacent to each outline are CST. (Bottom) Hodograph representative of the storm's environment. Heights in km AGL. Motion of the right-moving (RM) and left-moving (LM) cells (from Weisman and Klemp, 1986; adapted from Burgess, 1974). (Courtesy of the American Meteorological Society)

The tendency for the right and left movers to rotate cyclonically and anticyclonically can be demonstrated also using the concept of potential vorticity. Suppose that below cloud base potential temperature is conserved, while above cloud base equivalent potential temperature is conserved: Isentropic surfaces, which are also surfaces of constant θ and θ_e, therefore behave like material surfaces. In the absence of friction and radiative

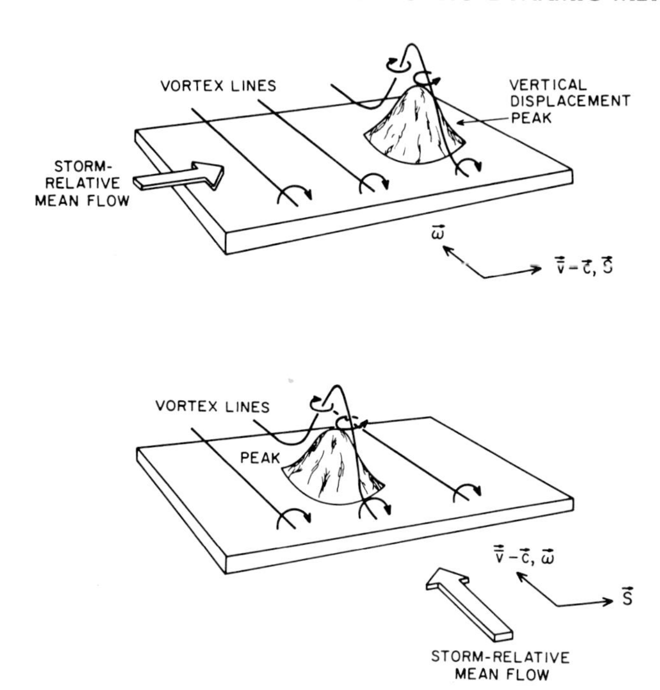

Figure 3.26 (Top) Effect of localized vertical displacement "peak" (i.e., "hump" in isentropic surface) on vortex lines when mean vorticity and mean storm-relative flow $\bar{\mathbf{v}} - \mathbf{c}$ are perpendicular (purely crosswise vorticity). The peak draws up loops of vortex lines (shown slightly above, instead of in, the surface for clarity), giving rise to cyclonic vorticity on the right side of the peak (relative to shear vector **S**) and anticyclonic vorticity on the left side. [Solenoidal effects (not included in figure) actually divert the vortex lines around the right (left) side of a warm (cool) peak, relative to the mean vorticity vector, but do not alter the result.] Environmental flow across expanding peak produces maximum updraft upstream (storm-relative frame) of peak owing to upslope component of vertical velocity. In this case, there is no correlation between vertical velocity and vertical vorticity since centers of the w' field lie in different quadrants of peak from those of ζ' field. (Bottom) As in top panel, but for other extreme when vorticity is purely streamwise (i.e., $\boldsymbol{\omega}$ is parallel to $\bar{\mathbf{v}} - \mathbf{c}$). Here, the upslope (downslope) side of the peak is also the cyclonic (anticyclonic) side, and vertical velocity and vertical velocity are positively correlated (from Davies-Jones, 1984). (Courtesy of the American Meteorological Society)

effects, Ertel's potential vorticity is conserved. In Fig. 3.26 (top), "storm"-relative flow is in the same direction as the vertical-shear vector (**S**), that is, normal to the vorticity vector ($\boldsymbol{\omega}$) associated with **S**. Since the gradient of θ (or θ_e) points vertically, while $\boldsymbol{\omega}$ is directed horizontally, Ertel's potential vorticity is zero. It therefore must *always* be zero. If an updraft deforms an isentropic surface by bulging it upward (think of it as the high θ_e-air in a storm updraft), the vorticity vector develops an upward-directed component (i.e., cyclonic vertical vorticity) on the side of the updraft that lies to the right of **S**, and a downward-directed component (i.e., anticyclonic vertical vorticity) on the opposite side of the updraft. Upward motion in the reference frame of the storm occurs on the upstream side of the bulge, while downward motion occurs on the downstream side. Vertical velocity and vertical vorticity are not correlated with each other.

However, in Fig. 3.26 (bottom), "storm"-relative flow is normal to the vertical-shear vector, that is, parallel to the horizontal vorticity vector associated with **S**. Vertical velocity and vertical vorticity *are* correlated with each other: On the upstream side of the updraft, with respect to the relative flow, there are components of upward vertical velocity *and* positive vertical vorticity; on the downstream side of the updraft, there are components of downward vertical velocity *and* negative vertical vorticity. Therefore low-level convergence occurs under the cyclonically rotating side of the updraft, and the vortex may increase in intensity to that of a "mesocyclone" and be advected upward. Based on Doppler radar observations, a mesocyclone has at least $5 \times 10^{-3}\ \mathrm{s}^{-1}$ units of vorticity, temporal continuity, and continuity in the vertical. Low-level divergence occurs under the anticyclonically rotating side of the updraft, and the anticyclonic vortex weakens.

Can we use the aforementioned theory to assess the potential of mesocyclone formation in right movers on the basis of the environmental hodograph obtained from a sounding or a Doppler wind profiler? *Streamwise vorticity* (ω_s) is defined as the component of the three-dimensional vorticity vector in the direction of the storm-relative flow:

$$\omega_s = \frac{(\mathbf{v} - \mathbf{c}) \cdot \boldsymbol{\nabla} \times \mathbf{v}}{|\mathbf{v} - \mathbf{c}|}, \tag{3.4.51}$$

where **c** is the storm-motion vector. Unlike vorticity, streamwise vorticity is not Galilean invariant. The *helicity* (H) (sometimes called *helicity density*) is the dot product of the relative-flow vector with the vorticity vector:

$$H = (\mathbf{v} - \mathbf{c}) \cdot \boldsymbol{\nabla} \times \mathbf{v}. \tag{3.4.52}$$

It is thought that large positive horizontal helicity in the storm-relative environmental wind field is converted into positive vertical helicity in the storm: R. Davies-Jones has shown how the correlation coefficient for storm vertical vorticity (ζ') and storm vertical velocity, $\overline{\zeta' w'}/(\overline{\zeta'^2 w'^2})^{1/2}$, where the overbar denotes averaging in space, under idealized circumstances is approximately proportional to the *coefficient of streamwise vorticity,* the proportion of vorticity in the streamwise sense,

$$\frac{\mathbf{v} - \mathbf{c} \cdot \boldsymbol{\nabla} \times \mathbf{v}}{|\mathbf{v} - \mathbf{c}|\ |\boldsymbol{\nabla} \times \mathbf{v}|}.$$

The coefficient of streamwise vorticity is also known as *relative helicity*. The latter, H, and ω_s are usually integrated over a portion of the lower troposphere.

Hodographs for which the storm-relative winds are strong and turn rapidly with height are therefore associated with rotating thunderstorms (Fig. 3.27). In order to estimate the potential for mesocyclone formation in right movers, we must be able to estimate the *storm-relative* winds. A crude estimate of storm motion may be made from the pressure-weighted mean wind in the lowest 5 or 6 km. When there is strong vertical shear, this estimate may be refined if we

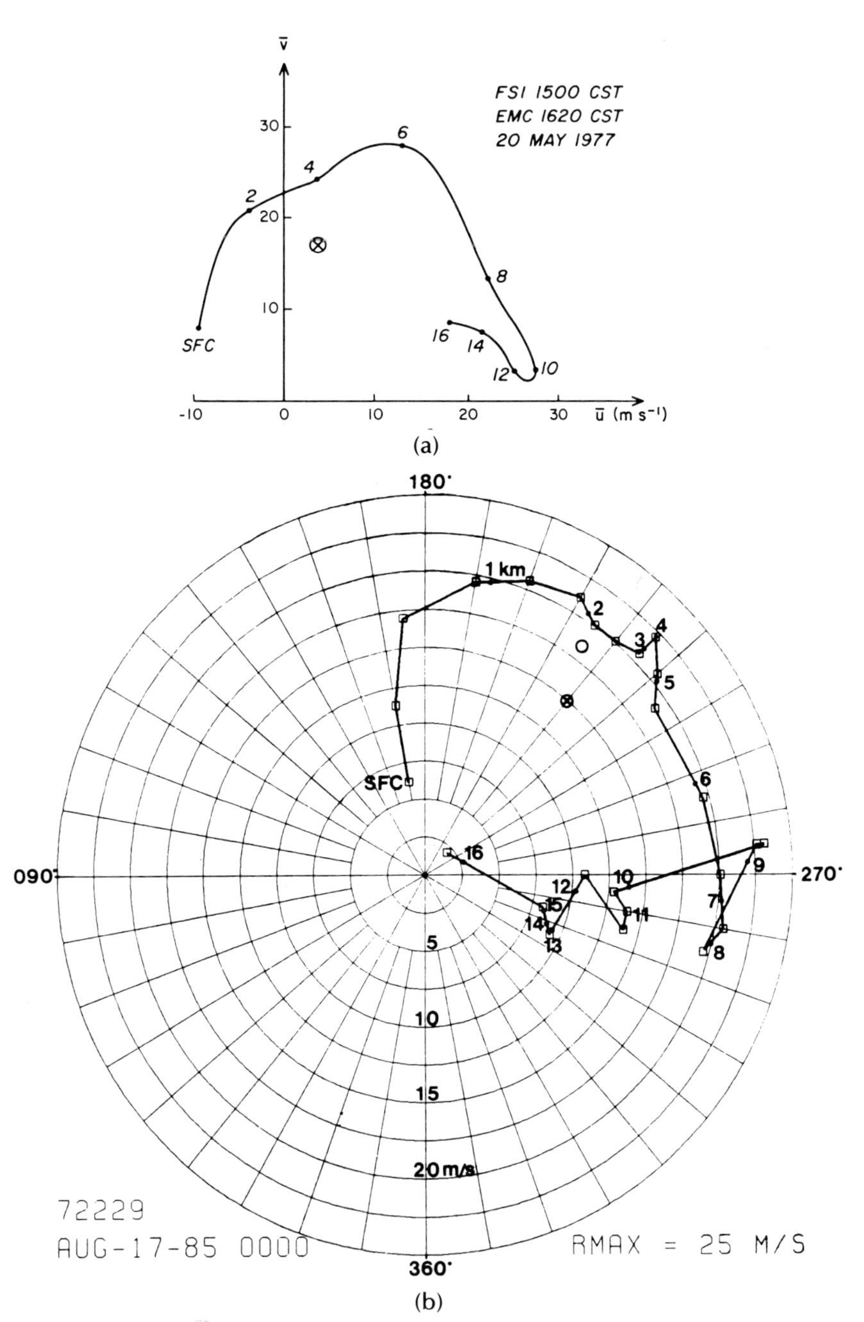

Figure 3.27 Example of a clockwise-turning hodograph on (a) an Oklahoma tornado-outbreak day and (b) a tornado-outbreak day in the remains of a hurricane in Alabama. Storm's translation velocity (circled X) and in (b), pressure-weighted mean winds in the lowest 6 km (open circle). Storm-relative winds (vectors whose tails are at circled X, and whose heads are on hodograph) veer rapidly with height. Heights indicated are km MSL. [(a) from Davies-Jones, 1984; (b) from McCaul, 1987.] [(a) and (b) courtesy of the American Meteorological Society]

account for vertical perturbation pressure-gradient-induced deviant motion by adding a vector of 5–10 $m\,s^{-1}$ to the right of the mean wind.

Storms in which the vertical perturbation pressure-gradient force induced by rotation produces single, vigorously rotating updrafts, sometimes as fast as 40–50 $m\,s^{-1}$, and single precipitation regions (cells) that persist for much longer than the time required for air to circulate completely through them (i.e., longer than about one hour), and that have a component of propagation normal to the vertical-shear vector, are called *supercells*. The supercell is most likely to form in an environment of high CAPE and strong vertical shear. Since hodographs are usually curved in a clockwise manner, supercells tend to rotate cyclonically and move to the right of the mean wind (left in the Southern Hemisphere). They are often distinguished from multicells on radar by their "deviant" motion. Numerical simulations indicate that storms that form in an environment of strong shear and high CAPE, such as supercells, develop more slowly, but last longer and have stronger updrafts than storms that form in an environment of weaker shear.

Supercells usually produce large hail. While tornadoes do not always form in supercells, those that do form tend to be major ones (i.e., large, intense, and relatively long lived). Supercells often develop from multicells (or ordinary cells). As a result, they move with the mean wind early in their life, to the right (or left) of the mean wind during their supercell phase (Fig. 3.28), and with the mean wind late in their life as they weaken. Their path therefore

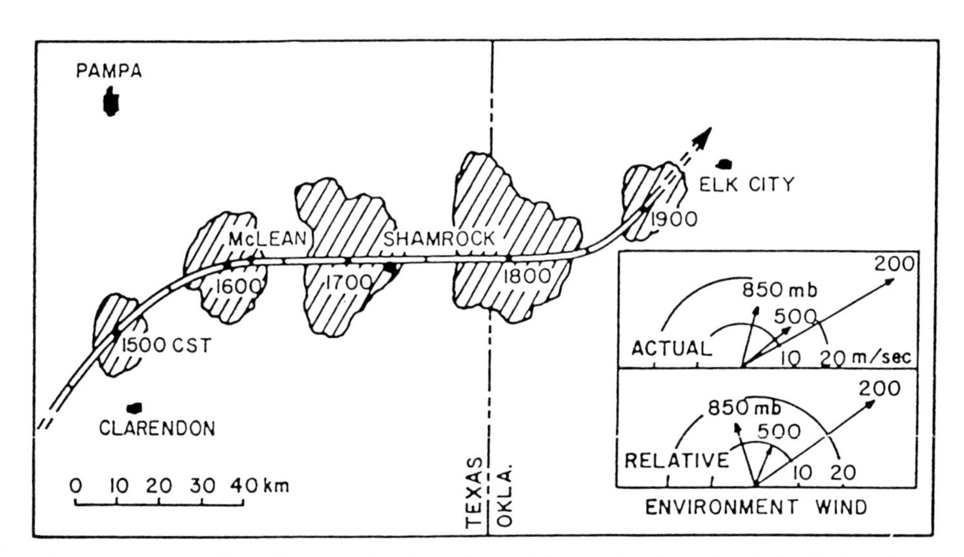

Figure 3.28 Track and successive hourly positions of radar echo (stippled regions indicate radar reflectivity in excess of 30 dBZ) associated with a right-moving supercell on June 1, 1965. Wind hodographs for 1800 CST, relative to the ground and relative to the storm, are shown in the inset. From 1600 to 1830 the storm propagated continuously to the right of the winds during which time it produced heavy rain and hail damage (after Fankhauser, 1971; from Browning, 1977). (Courtesy of the American Meteorological Society)

follows a reversed "S"-like trajectory. This variation in movement as a function of the age of the storm must be taken into account by forecasters who issue warnings to the public. It usually takes an hour or more for a multicell (or ordinary cell) to develop into a supercell. Strong shear apparently inhibits storm development early on; however, stronger, longer-lasting storms eventually develop.

Doug Lilly first suggested that the *longevity* of supercells might be due to their large helicity. Consider the three-dimensional vorticity equation expressed in the following form:

$$\frac{\partial}{\partial t}(\nabla \times \mathbf{v}) - \nabla \times [\mathbf{v} \times (\nabla \times \mathbf{v})] = -\nabla \times \frac{1}{\rho}\nabla p + \nabla \times \mathbf{F}_{\text{fric}}. \tag{3.4.53}$$

If the storm-relative wind points in the same direction as the three-dimensional vorticity vector, then the *Lamb vector*

$$\mathbf{v} \times (\nabla \times \mathbf{v}) = 0. \tag{3.4.54}$$

Then using Eqs. (3.4.54), (3.4.53), and the vector identity

$$\nabla \times \mathbf{a} \times \mathbf{b} = \mathbf{a}(\nabla \cdot \mathbf{b}) - \mathbf{b}(\nabla \cdot \mathbf{a}) - (\mathbf{a} \cdot \nabla)\mathbf{b} + (\mathbf{b} \cdot \nabla)\mathbf{a}, \tag{3.4.55}$$

we find that

$$-(\mathbf{v} \cdot \nabla)(\nabla \times \mathbf{v}) = -[-(\nabla \times \mathbf{v})\nabla \cdot \mathbf{v} + (\nabla \times \mathbf{v} \cdot \nabla)\mathbf{v}]. \tag{3.4.56}$$

The left-hand side of Eq. (3.4.56) represents advection. The terms on the right-hand side represent the divergence effect and tilting. In other words, in purely helical flow advection is balanced by stretching and tilting. Thus, if there is large streamwise vorticity or helicity, it is possible that the effects of (nonlinear) advection are to a great degree cancelled, and hence the cascade of turbulence down to smaller scales, which is caused by nonlinear advection, is suppressed. Hence the dissipative effects of mixing are mitigated.

The base of the updraft in a supercell is marked by a lowered, slowly rotating, nearly precipitation-free cloud base called a *wall cloud* [Fig. 3.29(a)]. (The wall cloud is sometimes also referred to as a *shelf cloud* or *pedestal cloud.*) The adjective *wall* was originally suggested by T. Fujita to describe the downshear (often northeastern in the United States) side of the wall cloud, whose dropoff is often "precipitous." The wall cloud forms when some relatively cool, but more humid air enters the updraft from an adjacent precipitation region, and condenses at a lower level than the ambient air.[3] Wall clouds form when scud clouds appear under cloud base and attach themselves to the cloud base above. If the updraft does not ingest the cooler, more humid air, then the cloud base tends to appear dark and flat. Sometimes a *tail cloud* is observed feeding into the wall cloud from the precipitation region [Fig. 3.29(b)]. When relatively stable air is forcibly lifted to its LFC, the sides and top of the wall cloud take on a laminar, striated appearance, like altocumulus lenticularis (wave clouds) in mountainous areas, when the environment is stable to small vertical displacements of air [Fig. 3.29(c)].

In a storm having a "moderate" updraft, the radar echo is quasisymmetrical and erect (Fig. 3.30). In the strong updraft of a supercell, however,

(a)

(b)

(c)

Figure 3.29 Photograph of (a) a wall cloud; (b) a multiple-vortex wall cloud with tail clouds; (c) a wall cloud having a laminar, striated base. (a) was taken on May 11, 1978, in north-central Oklahoma; view is to the east; baseball-sized hail was reported to the northeast of the wall cloud. (b) was taken May 26, 1978, in West Texas just before a major flood and a small tornado; the view is to the northwest. Both wall clouds were rotating cyclonically, and each was rotating about the other in a cyclonic manner. (c) was taken on May 27, 1985, in northwest Texas; the parent storm for this wall cloud was the right-moving member of a splitting storm; the view is to the northwest [photographs by H. Bluestein; (a) and (b) were taken for NSSL]. [(b) and (c) courtesy of the American Meteorological Society]

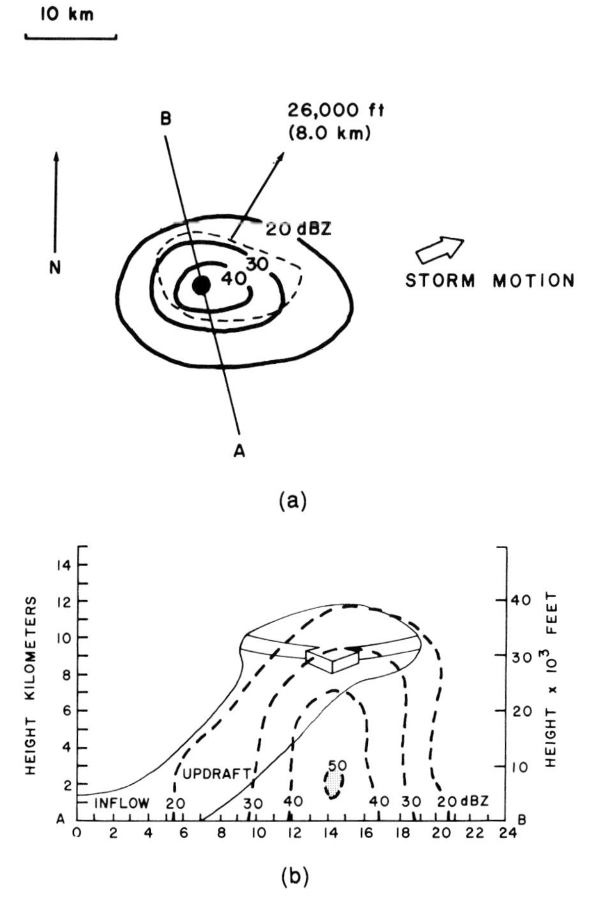

Figure 3.30 (a) Schematic diagram of a thunderstorm having a moderate updraft as seen on a radar PPI during a tilt sequence. Solid lines are low-level reflectivity contours. Dashed line outlines the echo in excess of 20 dBZ derived from the mid-level scan. Black dot is the location of the maximum echo top from the high-level scan. (b) Schematic diagram of vertical cross section [through line AB in (a)]; range-height indicator (RHI) radar display of a thunderstorm with the low-level inflow, a moderate updraft, and outflow aloft (solid lines) superimposed. Radar reflectivity (dashed lines) with reflectivities greater than 50 dBZ stippled (from Lemon, 1977).

precipitation does not have enough time to form until air has reached relatively high levels. On a radar PPI display there is thus a "weak-echo region (WER)" or "vault" at low and middle levels (Fig. 3.31). Visually, the vault sometimes takes on a green or green-blue hue, owing to scattering properties of the cloud droplets. Aloft there may be a "bounded weak-echo region" (BWER) if the updraft is intense (Fig. 3.32). Precipitation particles may flow around the updraft to produce a V-shape to the radar reflectivity pattern at high levels. Precipitation can be advected away from and around (in the Northern

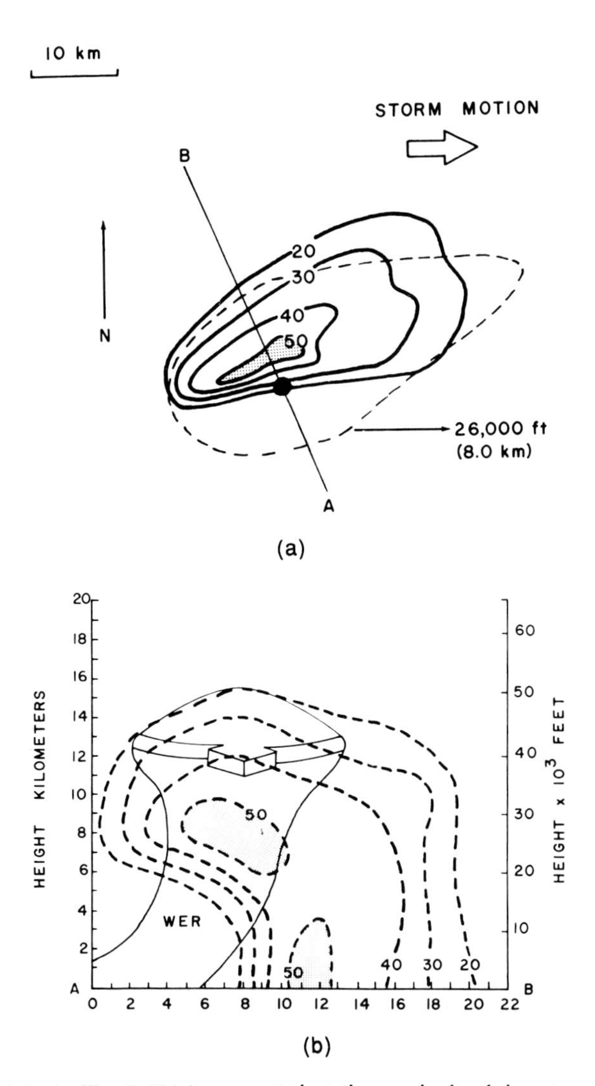

Figure 3.31 (a) As in Fig. 3.30(a), except that the updraft of the storm is strong. (b) As in Fig. 3.30(b), except that the updraft is strong. The WER is the weak-echo region. The cross section is along the line AB shown in (a) (from Lemon, 1977).

Hemisphere westward, southwestward, southward, eastward, and even northeastward from) the precipitation "core" by the mesocyclone to form a "hook echo" (Fig. 3.33). In its early stages, the juxtaposition of a developing hook and a WER looks more like a "notch" of weaker echo indented into the echo core. Hook echoes, however, are not always evident, owing to inadequate radar resolution, or are short lived: Sustained deviant cell motion (not with the mean wind), a WER or BWER, a Doppler-radar-observed mesocyclone, and a strong divergence signature aloft are reliable indicators of a supercell storm.

The visual characteristics of a supercell are depicted in Fig. 3.34. This model is based on observations by severe-storm interceptors ("storm chasers")

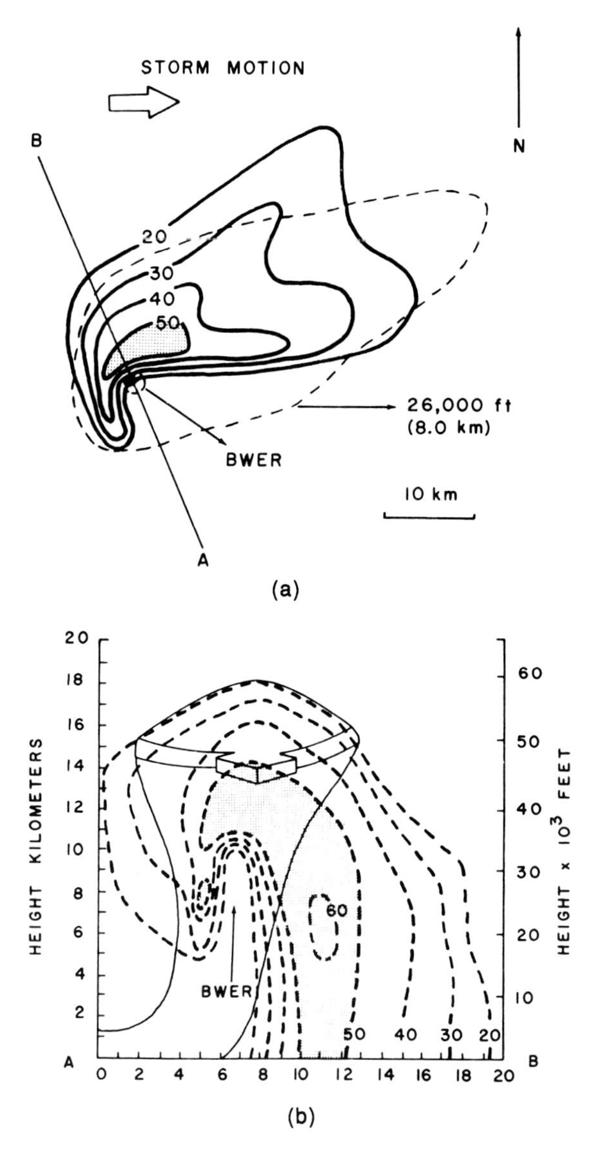

Figure 3.32 (a) As in Fig. 3.30(a), except that the updraft of the storm is intense. BWER is the location of the bounded weak-echo region. (b) As in Fig. 3.30(b), except that the updraft is intense. The cross section is along the line AB shown in (a) (from Lemon, 1977).

in the Central United States. The wall cloud is found under the main tower of the storm. The *flanking line* extends outward from the core of the storm, usually to the west, southwest, or south. It is usually coincident with the gust front that forms to the rear of the storm. Flanking-line cumuli grow and merge with the storm as they take their turn in becoming the *main tower. Penetrating tops* are indicative of buoyant convective elements punching through the stable

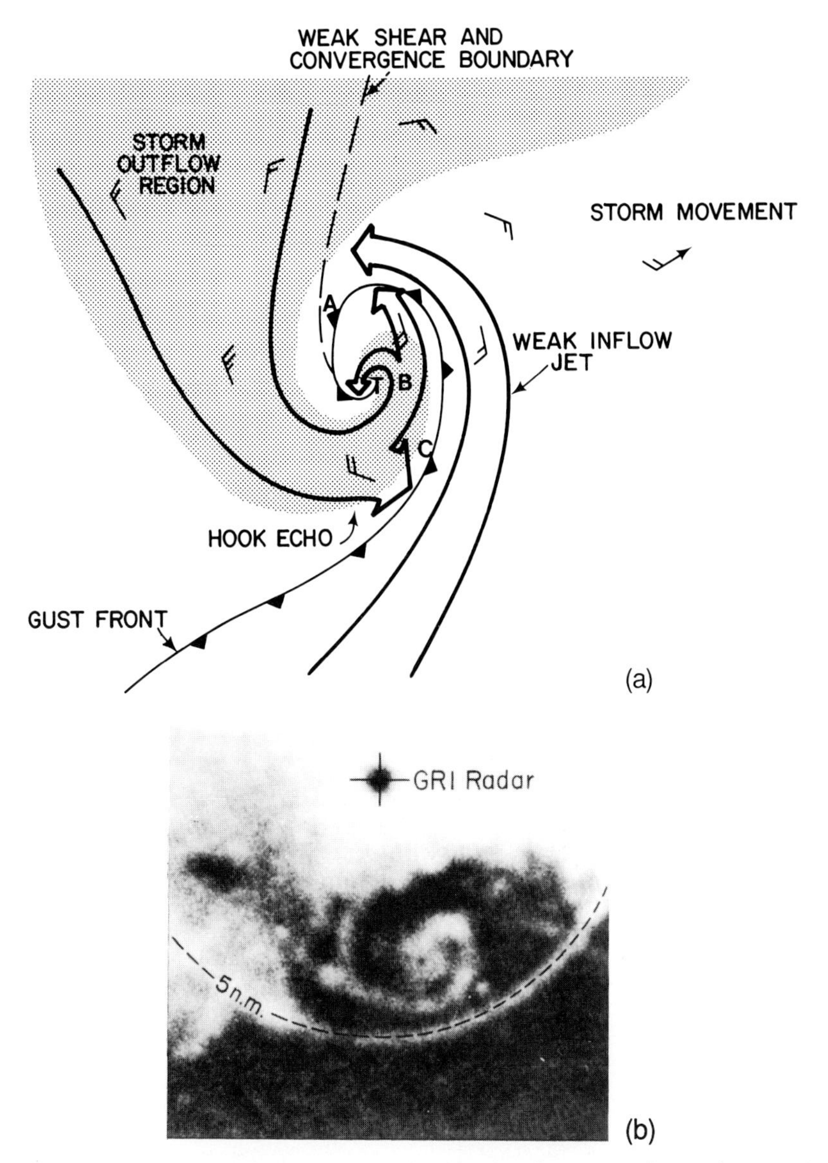

Figure 3.33 (a) Idealized illustration of the low-level mesocyclone characteristics during the tornadic phase of a supercell. Key features include the tornado (T), principal updraft and region of primary tornadogenesis (A), downdraft within the mesocyclone core (B), and possible genesis region of gust front tornadoes (C). A full wind barb is 10 m s^{-1}. Stippled area represents radar echo/precipitation (from Brandes, 1978). (b) Spiral-shaped hook echo in a tornadic supercell in Nebraska (from Fujita, 1981). [(a) and (b) courtesy of the American Meteorological Society]

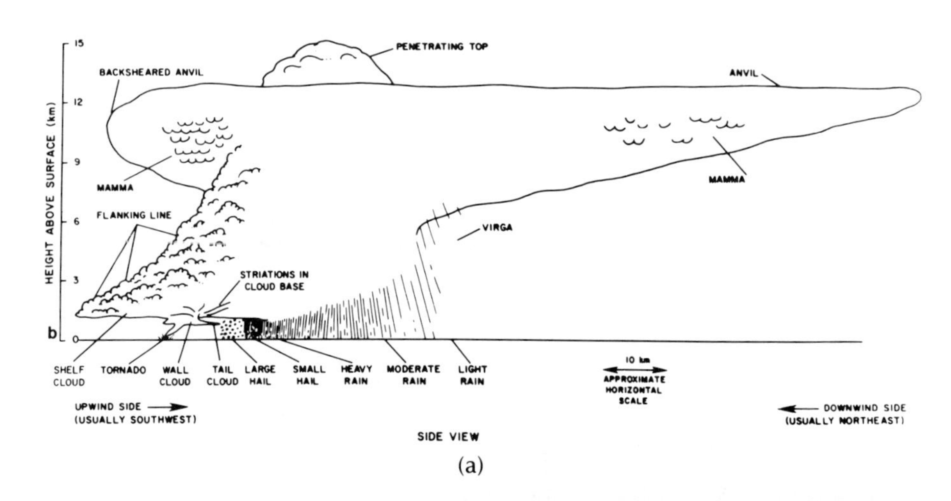
PENETRATING TOP
BACKSHEARED ANVIL
ANVIL
HEIGHT ABOVE SURFACE (km)
15
12
9
6
3
0
MAMMA
FLANKING LINE
MAMMA
VIRGA
STRIATIONS IN CLOUD BASE
b
SHELF CLOUD
TORNADO
WALL CLOUD
TAIL CLOUD
LARGE HAIL
SMALL HAIL
HEAVY RAIN
MODERATE RAIN
LIGHT RAIN
10 km
APPROXIMATE HORIZONTAL SCALE
UPWIND SIDE
(USUALLY SOUTHWEST)
DOWNWIND SIDE
(USUALLY NORTHEAST)
SIDE VIEW

(a)

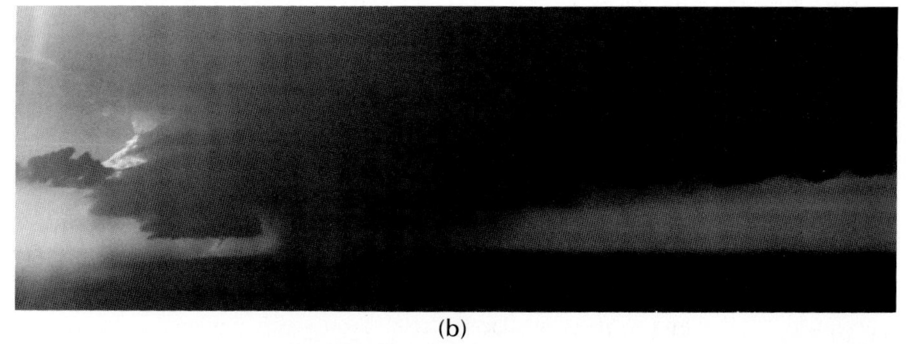

(b)

(c)

Figure 3.35 Photograph of mammatus under the anvil of a thunderstorm complex. View to the west in the Texas Panhandle near Childress on May 4, 1989 (photograph copyright H. Bluestein).

tropopause. Collapsing tops, which occur after an updraft has lost its buoyancy in the stable, stratospheric air, are sometimes referred to as *splashing cirrus*. The main precipitation core is located in the downshear direction with respect to the wall cloud. The anvil of a supercell often is "back-sheared": Its updraft is so strong that some anvil material "shears off" in the upwind direction relative to the storm owing to intense divergence aloft.

Many of the cloud features observed in supercell storms are also observed in ordinary-cell or multicell storms. For example, the latter can have a flanking line, an anvil, a penetrating top, and even a short-lived wall cloud. *Mammatus* (or simply mamma) are often visible underneath the anvil of most storms (Fig.

Figure 3.34 (a) Visual model of an idealized supercell storm in the Northern Hemisphere (from Bluestein and Parks, 1983; after Doswell, 1985). (Courtesy of the American Meteorological Society). (b) Photograph of a supercell storm in south-west Kansas; view toward the west northwest at 2330 UTC, May 26, 1991, from an altitude of approximately 4.9 km AGL aboard a P-3 research aircraft operated by NOAA. The backsheared anvil, flanking line, wall cloud, tornado, precipitation core, forward-flank anvil and mammatus are all visible. Refer to Fig. 3.34(a) to identify the features. (photograph copyright Howard B. Bluestein). (c) Developing supercell in western Oklahoma on April 20, 1985; view toward the northwest. The flanking line and its precipitation-free cloud base are visible (photograph copyright H. Bluestein).

3.35). These are smooth, slowly varying pouches of negatively buoyant "upside-down" convection, and are not necessarily indicators of severe weather or supercell structure. The importance of all these features in the visual supercell model seen in Fig. 3.34(a) is that they *persist* in the same relative locations.

In nature we often observe only one or two supercells within a broad area of high shear and high CAPE. It is a significant forecasting challenge to pinpoint which convective cells may develop into supercells. The interaction among growing cells also has an influence on whether or not a cell develops into a mature supercell.

Low-precipitation storms. Although most supercells have large, intense radar echoes, some do not. The latter have been called *low-precipitation* (LP) *storms* because they produce very little rain. Much of the precipitation consists of hail or a few large raindrops, so that the region under cloud base is nearly translucent (Fig. 3.36). Owing to the lack of precipitation, the LP storm has the appearance of the "skeleton" of a supercell. Often forming in the vicinity of the dryline in the Southern Plains and in the High Plains, LP storms sometimes develop into ordinary supercells. Tornadoes and funnel clouds are sometimes spawned by LP storms; funnel clouds are frequently observed as the LP storm dissipates. The absence of a precipitation-driven, cold downdraft is a hallmark of the LP storm. LP storms require less vertical shear than ordinary supercells, owing to the lack of cold outflow. (On the other hand, supercells may have so much precipitation that the view to the west under the flanking line is completely opaque. Moller and Doswell call these storms *high-precipitation* (HP) *supercells.*)

Factors that can play a role in determining the precipitation efficiency of a storm (i.e., the ratio of the rainfall rate to the total condensation in an air column) include updraft strength and diameter, environmental shear, and the type and density of cloud condensation nuclei (CCN), and the particle trajectories.

If a few large drops of rain or a few large hailstones are produced, then their radar echoes are intense, even though the accompanying rainfall rates are small; if many small drops of rain are produced, then their radar echoes are weak, while the rainfall rates are large. The latter is often the case of precipitation in tropical cyclones, while the former is often the case in supercells.

Storm movement and its consequences. There are two major modes of cell movement: A cell may move passively along with the pressure-weighted mean wind defined over its depth (merrily, merrily, downstream). On the other hand, new cells may form and decay adjacent to older cells in a preferred direction. This apparent movement is called *propagation.* Propagation may be "discrete" or "continuous." Sometimes the difference between discrete propagation and continuous propagation is difficult to discern. Let us make this distinction in the following way: If the time interval between the appearance of successive cells is at least 20 min, the duration of the beginning stage to mature

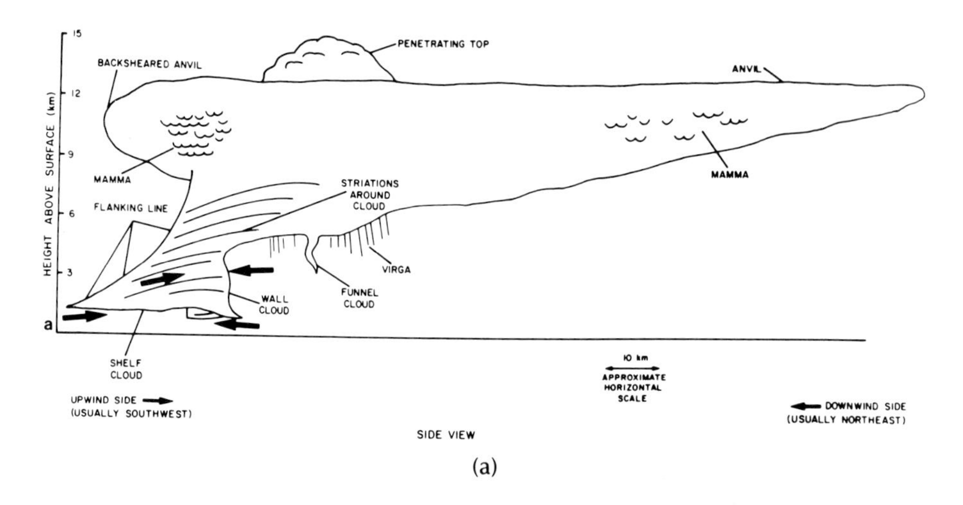

(a)

(b)

Figure 3.36 (a) Visual model of an idealized low-precipitation supercell in the Northern Hemisphere (from Bluestein and Parks, 1983). (Courtesy of the American Meteorological Society). (b) Photograph of a low-precipitation supercell in the Texas Panhandle near the dryline. View is to the west on May 31, 1990; note the tornado under the northern portion of the striated cylindrically shaped tower, which was rotating cyclonically; north (to the right) of the tower, which is usually dark and opaque, it is light! This was one of a series of tornadic LP storms (photograph copyright H. Bluestein).

stage of an isolated, ordinary cell, then the propagation is discrete; otherwise the propagation is continuous.

There are a number of mechanisms for cell propagation, which were discussed earlier. These mechanisms depend upon the vertical shear, CAPE, and behavior of density currents.

The net storm motion is the vector sum of cell motion, discrete propagation, and continuous propagation. When the sum of all three is small, *and* convection is long lived, then there may be local flooding. Generally speaking, since ordinary cells have a small lifetime, flooding can occur either when cell movement is exactly counteracted by propagation (either discrete or continuous), or when cells repeatedly move over the same area. The latter occurs when there is a preferred area for cell development. Flooding is dependent not only on the amount of rain that falls in a given time interval, but also on topography. A given rainfall rate may lead to flooding in a location where runoff is too slow; in another where runoff is fast, the same rainfall rate may not cause flooding.

The level at which storm motion is identical to the environmental wind is called the *steering level* (Fig. 3.37). A steering level may or may not exist. When the environmental vertical shear and CAPE are both large, storms that form in this kind of environment are not likely to have a steering level (Fig. 3.38). If the environmental hodograph is curved in a clockwise manner, storm motion is usually to the right of the mean wind in the lowest 6 km: The storms are called *right movers*. If the hodograph is curved in a counterclockwise manner, storm motion is usually to the left of the mean wind in the lowest 6 km: These storms are called *left movers*. Also, the effects of propagation owing to moving, intersecting boundaries or orography may be to produce a net storm movement that is different from that of the mean wind.

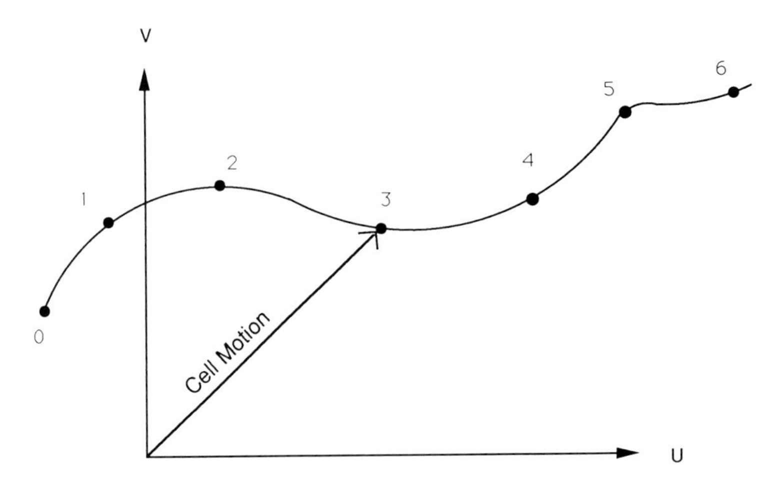

Figure 3.37 Idealized illustration of a hodograph (heights indicated in km AGL) of the environment of a storm for which the "steering level" is 3 km AGL.

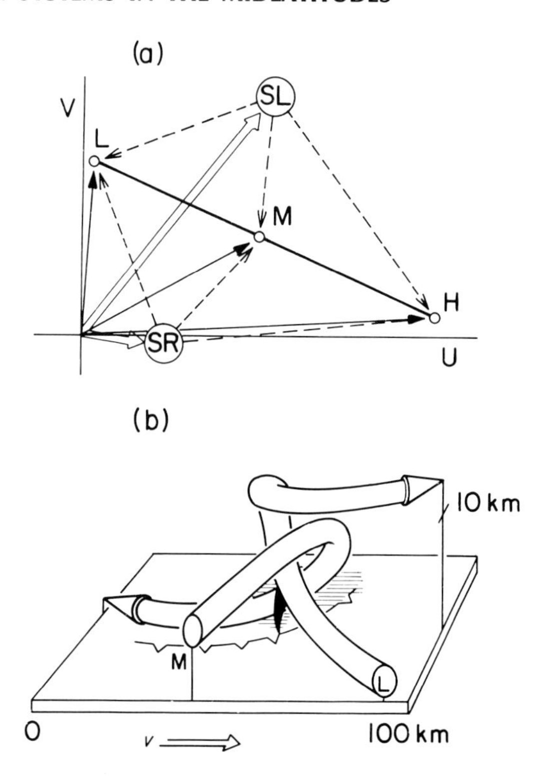

Figure 3.38 Browning's conceptual model for a right-moving supercell (SR). (a) Wind plot illustrating low- (L), middle- (M), and high- (H) level winds relative to the ground (solid arrows) and relative to the storm (dashed arrows). Motion of the SR storm is shown with an open arrow. The motion and relative wind vectors for a possible left-moving supercell (SL) are also shown. (b) Airflow trajectories at selected starting points at low (L) and middle levels (M) (from Klemp, 1987; adapted from Browning, 1964). (Courtesy of the American Meteorological Society)

3.4.7. Summary of the Effects of Vertical Shear on Storm Type

The dynamics of a convective storm[4] (Table 3.2) depend upon the environmental CAPE, density–current behavior of cool, surface outflow, and vertical perturbation pressure-gradient forces. When the environmental vertical shear is relatively weak, only the CAPE and density–current behavior are dynamically significant, and *ordinary* cells occur. When the low-level vertical shear is larger and the CAPE is not too small, a succession of ordinary cells, a multicell storm, can form. If the CAPE is high and the vertical shear is strong, then vertical perturbation pressure-gradient forces owing to rotation become dynamically important, and a *supercell* forms. The effects of the vertical-shear profile in numerical simulations of storms in an environment of CAPE of 2200 J kg^{-1} are illustrated in Fig. 3.39.

In nature, there is a continuum of storm behavior, from the ordinary cell up to the supercell (Fig. 3.40). While the multicell model has *discretely*

Table 3.2 Storm type as a function of vertical shear and CAPE

Vertical shear[a] CAPE	Weak ≲15 m s^{-1}	Moderate ~15–25 m s^{-1}	Strong ≳25 m s^{-1}
Low (500–1000 J kg^{-1})	Ordinary cell	Ordinary cell/supercell	Ordinary cell/supercell
Moderate (~1000–2500 J kg^{-1})	Ordinary cell	Ordinary cell/supercell[b]	Supercell[b]
High (≳2500 J kg^{-1})	Ordinary cell[b]	Ordinary cell[b]/supercell[b]	Supercell[b]

[a] Over lowest 6 km.

[b] Storms in which severe weather is likely. Vertical shear is measured by the length of the hodograph of the environmental winds from the surface to 6 km AGL (small-scale curves and loops are not counted). Supercells can occur even in environments of low CAPE if there is low CIN and if the environment is so moist that entrainment of environmental air does not weaken the updraft significantly. Severe weather is likely in storms produced in an environment of moderate–high CAPE regardless of storm type because the updrafts can be strong (based upon numerical simulations by M. Weisman, NCAR).

evolving updrafts ("strong evolution"), and the supercell model has a *quasisteady* updraft, some storms exhibit "weak evolution," with *gradual changes* in updrafts that are *connected* to each other.

Since the behavior of convective storms is highly sensitive to the environmental CAPE and vertical shear, forecasters must be careful to account for *changes* in environmental CAPE and vertical shear. On the mesoscale, the shear can change owing to the passage of shortwaves or jet streaks or fronts and the response of the boundary-layer winds to developing surface cyclones or to mountain–valley or sea-breeze circulations. On the storm scale, shear can change owing to the environmental response to nearby convection. CAPE can change owing to changes in cloud cover and differential advection of temperature and moisture.

Since storm behavior is a function of both shear *and* updraft strength, and maximum updraft strength is a function of CAPE, it has been useful to consider an empirical quantity, the "*bulk*" *Richardson number* (R), the ratio of CAPE to a quantity proportional to the square of the mean vertical shear integrated over height (S^2) as an indicator of storm type:

$$R = \text{CAPE}/S^2, \tag{3.4.57}$$

where

$$S^2 = \tfrac{1}{2}(\bar{u}_{6000} - \bar{u}_{500})^2, \tag{3.4.58}$$

and $\bar{u}_{6000}$ and $\bar{u}_{500}$ are the pressure-weighted mean-vector wind speeds in the lowest 6 km and 500 m, respectively. Suppose an undilute (i.e., without mixing) air parcel having characteristics of the lowest 500 m rises inside an updraft. Then S^2 represents the difference between the mean environmental kinetic energy associated with horizontal motions in the lowest 6 km and the mean kinetic energy of the horizontal component of the updraft, that is, the kinetic energy extracted from the mean flow into the updraft. Recall that the CAPE represents the maximum possible amount of environmental potential

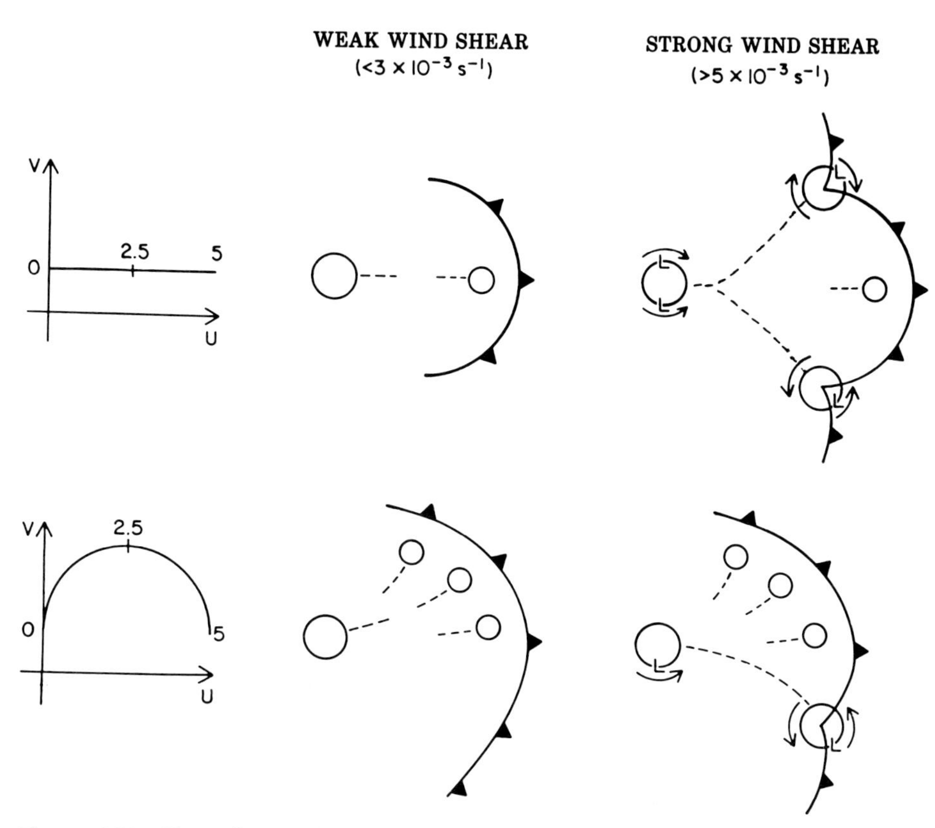

Figure 3.39 The effects of vertical wind shear on the behavior of numerically simulated convective storms. Updraft evolution in (middle) weak and (right) strong wind-shear conditions for unidirectional (top) and clockwise-curved wind (bottom) shear profiles. Hodographs on the far left define the wind shear type; 0-, 2.5-, and 5-km levels are indicated. Large and small circles represent relatively strong and weak updrafts, respectively; the path of each updraft cell is indicated by a dotted line. Updraft structure is depicted at the early and mature phases of each storm; surface gust fronts (cold front symbols) are included at the mature phase. L is the approximate position of significant middle-level mesolow features. The direction of the updraft rotation (if any) is indicated by arrows (from Weisman and Klemp, 1986). (Courtesy of the American Meteorological Society)

energy that can be converted into that portion of the updraft's kinetic energy that is associated with vertical motion.

A measure of the mean relative inflow into the base of a storm is given by $\bar{u}_{6000} - \bar{u}_{500}$ if the storm moves approximately with the mean wind in the lowest 6 km, and the mean wind below 500 m accounts for inflow into the cloud base. The kinetic energy of the mean relative inflow is therefore $\frac{1}{2}(\bar{u}_{6000} - \bar{u}_{500})^2$. Suppose that the kinetic energy of a downdraft is proportional to the CAPE (i.e., what goes up quickly comes down quickly—the harder they come, the harder they fall). Then high values of R may be associated with downdrafts that move away from their parent cells, and cut off the supply of low-level,

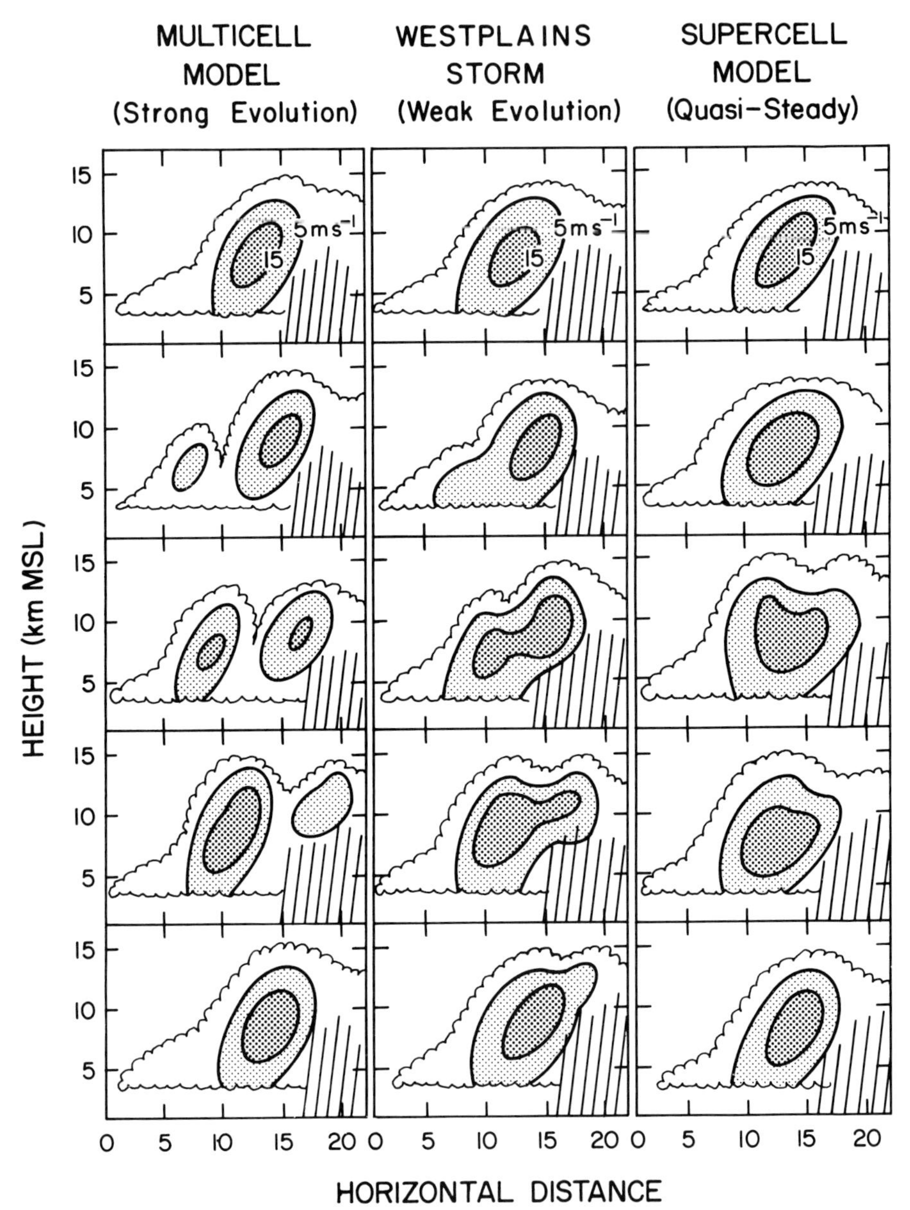

Figure 3.40 Schematic diagram showing the updraft evolution for three different storm models. The contours represent isotachs of vertical wind speed. The left panel depicts the cellular evolution according to the multicell model, involving the formation of discrete updrafts. In the supercell model, on the right, the updraft is shown as being quasisteady. In the model deduced for the "Westplains" storm, shown in the middle, the large updraft undergoes gradual changes but remains singly connected. This is termed *weak evolution,* in contrast to the strong evolution of the multicell case. The time between successive frames, moving down the figure, is 3–5 min (from Foote and Frank, 1983). (Courtesy of the American Meteorological Society)

warm, moist inflow into them. If R falls within a certain range, then there may be a balance between relative inflow and outflow, and hence cells may persist for longer periods of time. In addition, if $\bar{u}_{6000} - \bar{u}_{500}$ is large, the vertical shear is large; if CAPE is large, updrafts and vertical velocity gradients are large. Therefore it is likely that storms developing in this environment will rotate and be long lived.

Observational and numerical work suggest that ordinary cells form exclusively when R is greater than 30–40: The vertical shear is too "weak" for the updraft. When R is smaller, supercells or long-lived multicell storms develop. (Short-lived ordinary cells, however, often form along the downshear flanks of supercells and elsewhere where the gust front triggers new convective cells.)

If R is less than 30–40 *while both the CAPE and vertical shear are small,* long-lived cells are still possible, but they are not severe because updrafts are too weak to support large precipitation particles, too weak to induce strong

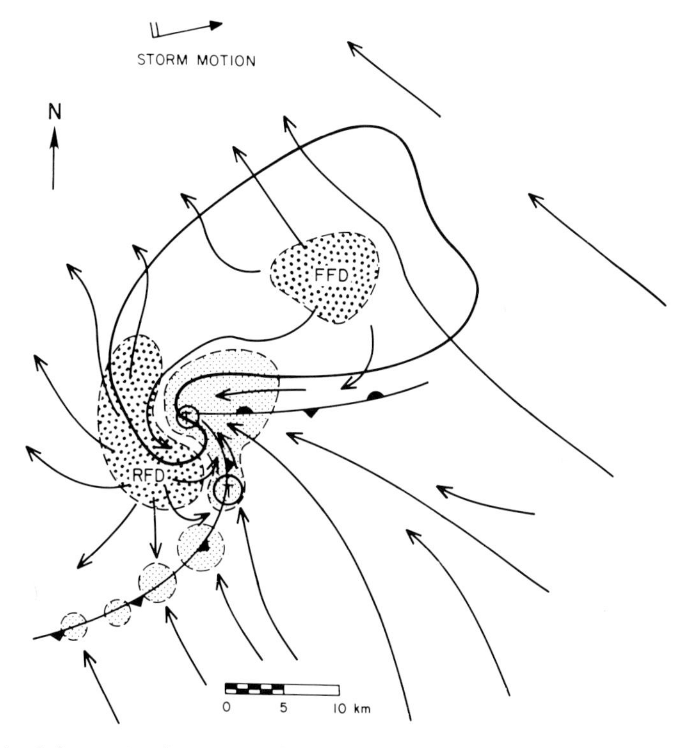

Figure 3.41 Schematic plan view of a tornadic thunderstorm near the surface. The thick line encompasses the radar echo. The cold-front symbol denotes the boundary between the warm inflow and cold outflow and illustrates the occluding gust front. Low-level position of the updraft is finely stippled, while the forward-flank (FFD) and rear-flank (RFD) downdrafts are coarsely stippled. Storm-relative surface flow is shown along with the likely location of tornadoes (encircled T's) (from Lemon and Doswell, 1979). (Courtesy of the American Meteorological Society)

downdrafts, and the tilting effect is not strong enough to develop significant rotation (and upward-directed perturbation pressure gradients). However, since the important effect of gust fronts is not *explicitly* included in the empirical formulation of R, R as a predictor of storm type must be used with some caution.

The forecaster must be careful when forecasting the character of convection on the basis of soundings; convective activity itself can change the "environment" and thus affect the character of subsequent convection.

The subject of the next section is severe weather phenomena associated with convective storms. Tornadoes, hail, and strong winds occur in both ordinary cells and in supercells; it is therefore not practical to classify convective storms in terms of the severe weather they produce.

3.4.8. Severe-Weather Phenomena

Tornadoes and other small-scale vortices. Mesocyclones, which in supercells are 10–15 km across, often spawn tornadic circulations only 10–1000 m across. The tornado vortex detected by a single Doppler radar is called a *tornadic vortex signature* (TVS).

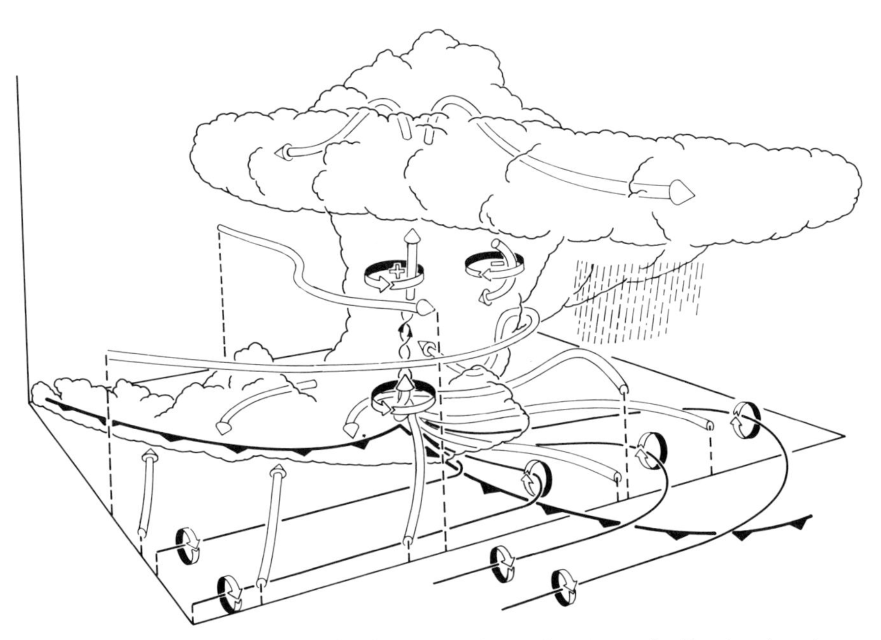

Figure 3.42 Three-dimensional schematic view of a numerically simulated supercell at a stage when the low-level rotation is intensifying. The storm is evolving in an environment of westerly wind shear and is viewed from the southeast. The cylindrical arrows depict the storm-relative flow in and around the storm. The thin lines show the low-level vortex lines, with the sense of rotation indicated by the circular-ribbon arrows. The cold-front symbol marks the boundary of the cold air beneath the storm (from Klemp, 1987). (Courtesy of J. Klemp)

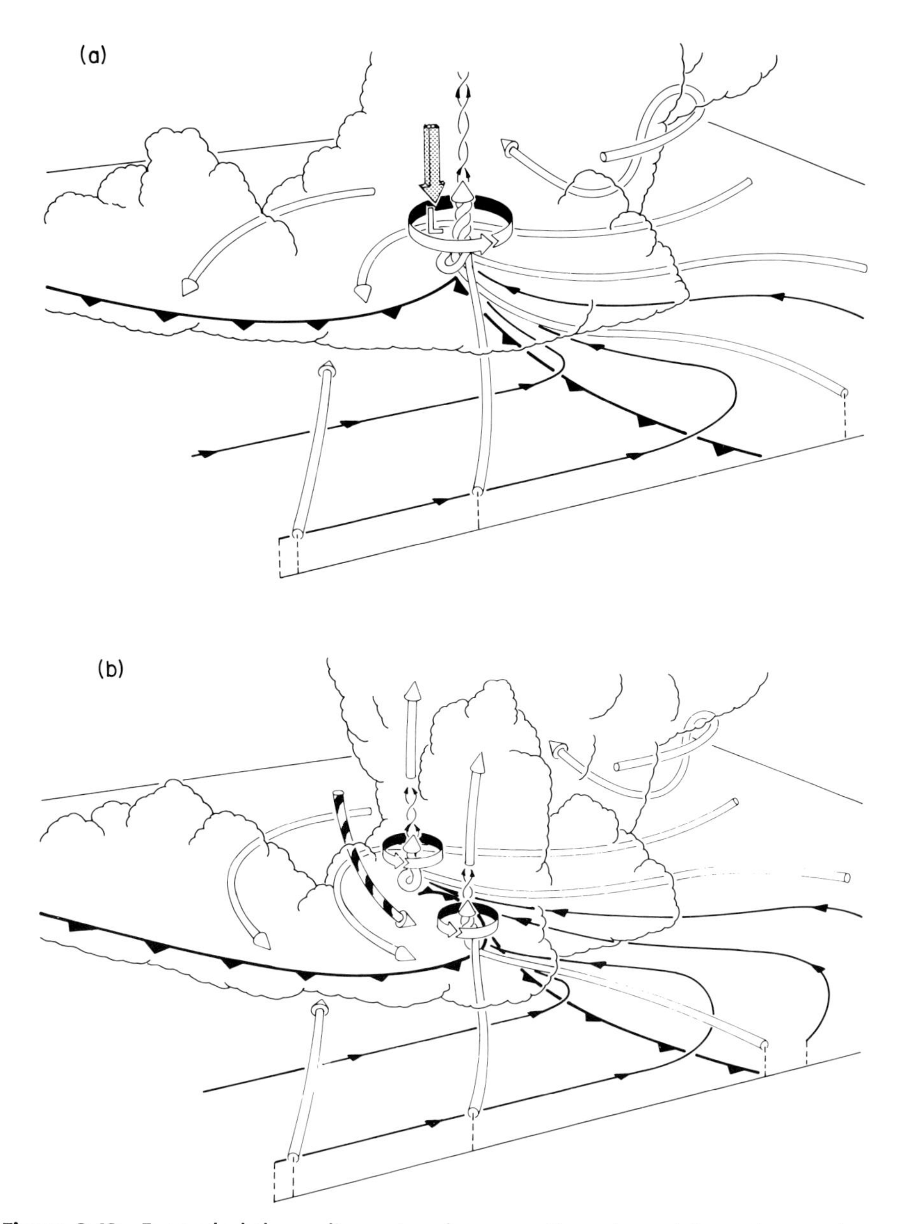

Figure 3.43 Expanded three-dimensional perspective, viewed from the southeast, of the low-level flow (a) at the time depicted in Fig. 3.42, and (b) about 10 min later after the rear-flank downdraft has intensified. Features are drawn as described in Fig. 3.42 except that the vector direction of vortex lines is indicated by arrows along the lines. The shaded arrow in (a) represents the rotationally induced vertical pressure gradient, and the striped arrow in (b) denotes the rear-flank downdraft, which is shifted slightly from the updraft (from Klemp, 1987). (Courtesy of J. Klemp)

Tornadogenesis in supercells is often preceded by the development of a downdraft on the upshear side of the updraft. The downdraft on the downshear side of the updraft (on the forward side of the storm), the *forward-flank downdraft* (FFD) (Fig. 3.41), also plays a role in tornadogenesis; it is due mainly to precipitation loading and evaporative cooling. Along the edge of the FFD, vorticity about the horizontal is generated baroclinically along the boundary between the potentially buoyant (warm) air and negatively buoyant (evaporatively cooled) air. Air parcels gain horizontal vorticity along the edge of the FFD and are tilted upward into the updraft associated (but not exactly colocated) with the mesocyclone, and low-level vertical vorticity is amplified as parcels are stretched when they go up into the updraft (Fig. 3.42). The pressure near the surface falls more rapidly than the pressure aloft in the mesocyclone, and a downward-directed perturbation pressure gradient develops that forces a downdraft (Fig. 3.43). This downdraft spreads out on the rear (with respect to storm motion) side of the storm and becomes the *rear-flank downdraft* (RFD). The boundaries between the RFD and the warm inflow and the FFD and the warm inflow look like storm-scale "fronts"; they intersect at the mesocyclone center and updraft. Sometimes the RFD appears clear, while at other times it is laden with precipitation. Tornadoes form ahead of the RFD, in the updraft, along the edge of the RFD-related gust front (Fig. 3.41). When the RFD "catches up" to the FFD, the mesocyclone occludes: Cold air is now ingested into the updraft and hence it decays.

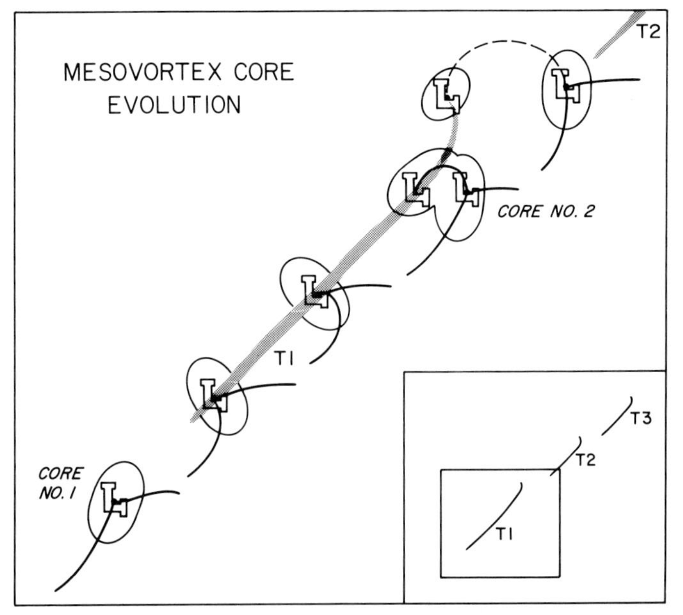

Figure 3.44 Conceptual model of mesocyclone core evolution. Low-level wind discontinuities (thick lines) and tornado tracks (shaded). Inset shows the tracks of the tornado family and the small square is the region expanded in the figure (from Burgess et al., 1982). (Courtesy of the American Meteorological Society)

Figure 3.45 Example of cyclical tornado production. Rope stage of the second tornado in a series and its wall cloud (left), and mature stage of the next tornado (right) and its wall cloud. Looking to the northwest from a position 12.2 km south of Canadian, Texas, 2236 UTC, May 7, 1986 (photograph by H. Bluestein). (Courtesy of the American Meteorological Society)

New mesocyclones may form at the occlusion, the point at which the RFD and FFD now intersect. The periodic development of new wall clouds (and mesocyclones) that spawn tornadoes has been referred to as *cyclical tornadogenesis* (Figs. 3.44 and 3.45). Some long damage paths are usually (but not always) the result of cyclical tornadogenesis.

The tornado frequently first becomes visible as a dust whirl on the ground under the wall cloud (Fig. 3.46a). The condensation funnel soon appears aloft (Fig. 3.46b). Late in life, it may bend at the ground while it stretches in length (Fig. 3.46c) and narrows in width. This final decaying stage in the tornado's life history is called the *rope stage* owing to the ropelike appearance of the condensation funnel. The gust front associated with the RFD may tilt the tornado so that it becomes nearly horizontal.[5] The aforementioned tornado life cycle of 10–30 min is common, but not exclusive. For example, tornadoes sometimes occur in the absence of any condensation funnel, always look like a rope, or never go through a rope stage. Some long damage paths are associated with tornadoes that last much longer than 30 min.

Most tornadoes in supercells are cyclonic. Anticyclonic tornadoes are rare. They have been documented, however, along the RFD gust front away from the mesocyclone (Fig. 3.47).

Although tornadoes frequently consist of a single vortex, they occasionally exist as two or more smaller "suction vortices" that rotate around the center of the wall cloud (Fig. 3.48). Laboratory model and numerical model evidence suggests that multiple vortices are associated with high "swirl ratio," that is, relatively large azimuthal flow compared to radial flow. Suction vortices often have a lifetime of only one revolution around the wall cloud; new ones form and dissipate in the identical location relative to the mesocyclone. Sometimes multiple-vortex tornadoes evolve into single-vortex tornadoes, and vice versa. Some suction vortices extend all the way to cloud base; others are visible only

(a) (b)

(c)

Figure 3.46 A series of photographs depicting the life cycle of a tornado: (a) a dust whirl appears near the ground, underneath the wall cloud; (b) a condensation funnel builds downward from cloud base; (c) the tornado becomes tilted and stretches out just before it dissipates (from Bluestein, 1983). This tornado occurred near Cordell, Oklahoma on May 22, 1981 (photographs copyright H. Bluestein). [(c) courtesy of the American Meteorological Society]

near the ground underneath a broader rotating cloud base. The multiple-vortex phenomenon was first postulated by T. Fujita on the basis of cycloidal damage swaths. It was first filmed in nature in the early 1970s. Multiple-vortex wall clouds have also been observed [cf. Fig. 3.29(b)].

Horizontal wind speeds in tornadoes can be as high as 100–150 $\mathrm{m\,s^{-1}}$. There is some theoretical evidence that even stronger upward vertical velocities can occur near center of a tornado, near the ground. The

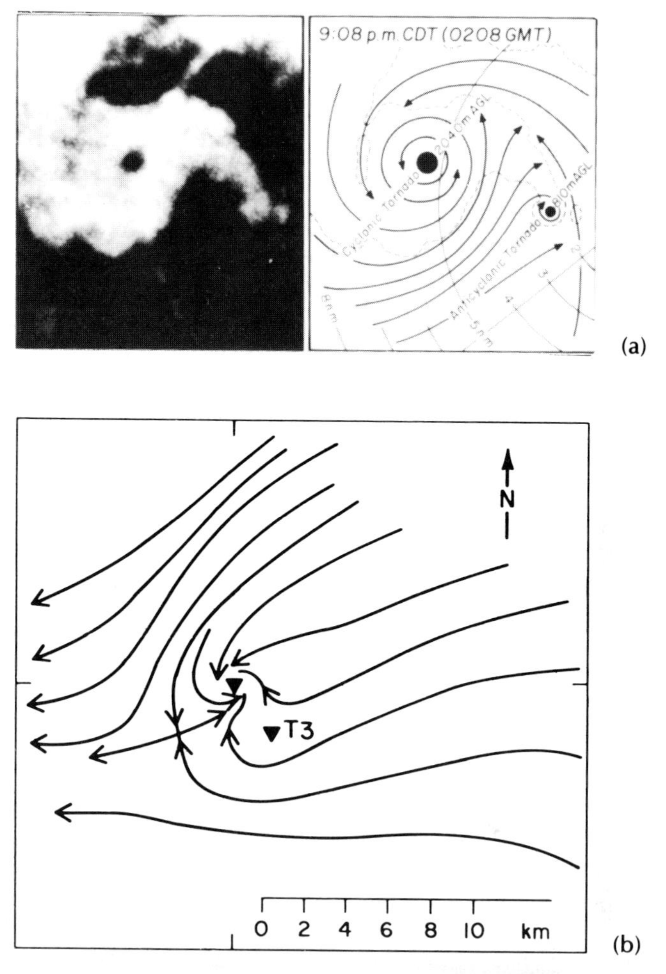

Figure 3.47 (a) (Left) Radar echo (detected by National Weather Service radar at Grand Island, Nebraska) asssociated with cyclonic (hole in echo) and anticyclonic (tip of hook echo bent back along gust front associated with rear-flank downdraft) tornadoes on June 3, 1980, to the northwest of Grand Island. (Right) Wind field (arrows) in relation to hook echo (dashed lines) seen at left (from Fujita, 1981). (b) Smoothed streamlines of the wind field relative to a coordinate system moving with cyclonic tornado denoted by shaded northwestern triangle; T_3 is the location of an anticyclonic tornado. The tornadoes occurred in Iowa on June 13, 1976 (from Brown and Knupp, 1980). [(a) and (b) courtesy of the American Meteorological Society]

"freight-train" roar of some tornadoes might be evidence of even higher wind speeds. The maximum possible tornadic wind speed is a function not only of the hydrostatic-pressure deficit owing to latent-heat release in the updraft, which is a function of the CAPE, but also to the dynamic-pressure deficit, which is a function of the character of the wind field itself [Eq. (3.4.43)], especially in the surface boundary layer.

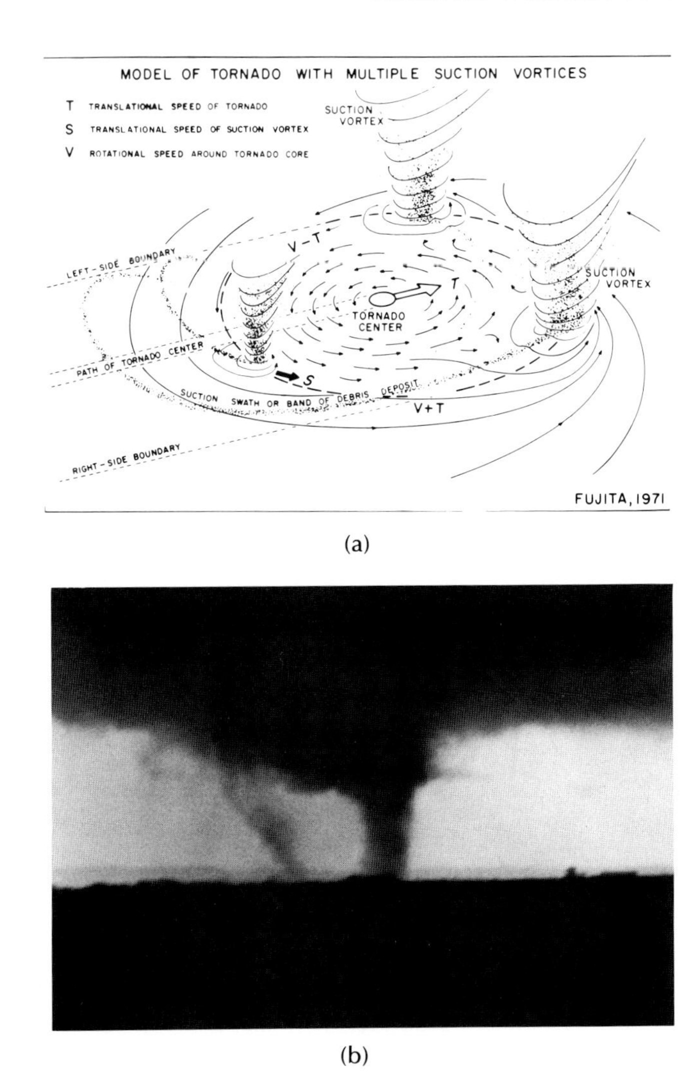

Figure 3.48 (a) Model of tornado with multiple vortices (from Fujita, 1981; proposed by Fujita in 1971). (b) Photograph of a multiple-vortex tornado; view to the south on May 2, 1979, in north-central Oklahoma near Orienta, (photograph taken by H. Bluestein for NSSL). Also see the multiple-vortex tornado in Vol. I, Fig. 1.6. [(a) courtesy of the American Meteorological Society]

Tornadoes do not occur exclusively in supercells. Multicell storms occasionally spawn tornadoes. Growing cumulus congestus in a type III (cold-air) environment sometimes produce weak tornadoes known as *cold-air funnels.*

Lines of growing cumulus congestus, over water, adjacent to precipitation shafts associated with cumulonimbi, sometimes produce waterspouts; some of these may last for as long as 20–30 min. (Some waterspouts, however, may be supercell tornadoes over the water.) Similar phenomena over the land have

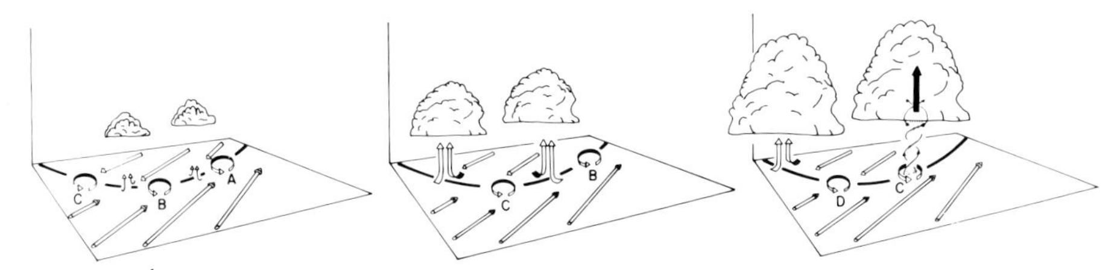

Figure 3.49 Schematic model of the life cycle of the nonsupercell tornado. The black line is the radar detectable convergence boundary. Low-level vortices are labeled with letters. At the left, clouds begin to form over the convergence zone, along which there are pre-existing vortices. In the middle, strong updrafts develop beneath the growing cumulus congestus clouds. At the right, a strong updraft becomes superimposed on one of the pre-existing vortices, and a tornado forms. The tornado dissipates when precipitation falls out of the updraft and the cell collapses (from Wakimoto and Wilson, 1989). (Courtesy of the American Meteorological Society)

been referred to as *landspouts*. They are common in eastern Colorado along the Denver convergence–vorticity zone, and have even been observed over mountainous terrain. These waterspouts (and landspouts) form within 30 min of the first indications of a radar echo, and are difficult to detect by radar unless the radar is close to the parent storm. Sometimes two or more waterspouts or landspouts can exist simultaneously along a line of clouds. Landspouts form when convergence under a rapidly growing cumulus congestus acts on a pre-existing vortex along the convergence line that originally had spawned the cloud (Fig. 3.49). Many landspouts appear to differ from supercell tornadoes in that a boundary and vortex exist even before the cloud begins; supercells, on the other hand, make their own boundary and parent vortex. The parent storms of landspouts form in an environment of weak wind and vertical shear, i.e. in an environment conducive to ordinary-cell (and multicell) development.

Tornadoes can also occur when hurricanes make landfall. These tornadoes usually occur in the right-front quadrant of the hurricane in the face of strong low-level shear.

Gustnadoes are short-lived vortices that occur along gust fronts. They usually do not inflict heavy damage, and typically do not have condensation funnels (Fig. 3.50). Funnels are sometimes observed, however, along gust fronts, pendant from an arcus cloud. Tubular shaped clouds sometimes are also seen along the leading edge of gust fronts, extending upward from roll clouds (Fig. 3.51); these features, however, are not tornadoes.

Tornadoes on rare occasions also form within areas of stratiform precipitation and along frontal rainbands (Fig. 3.52). They may be associated with the spin-up of a pre-existing vortex, like that preceding the landspout. Tornadoes also form on occasion in bow echoes and downbursts, which will be discussed in the next section.

Small, short-lived (1–2 min lifetimes) funnel clouds have been observed

Figure 3.50 Photograph of a gustnado (rotating area of dust) in southwest Oklahoma (do not be fooled by the mountains!) on April 9, 1978. View is toward the east, from a location behind the gust front (photograph for NSSL by H. Bluestein).

Figure 3.51 Photograph of tubular-shaped cloud along a gust front in central Oklahoma on June 14, 1986 (photograph copyright H. Bluestein).

under the ragged bases of cumuli well above the boundary layer (Fig. 3.53) and from ragged bases on the back side of supercells. Visually similar phenomena are observed as LP storms dissipate. These "high-based" funnel clouds do not make contact with the ground below, and are not considered a serious damage threat.

Dust devils are commonly observed over desert areas of fields of dirt under clear skies (Fig. 3.54). They are associated with "dry" convection, and are also not usually considered to be dangerous. Walking or driving through a dust devil can result in the disruption of the vortex. This is not the case for the average tornado! Sometimes dust devils occur under growing cumuli: These may actually be small tornadoes. Steam devils (Fig. 3.55) occur over water when the air temperature is at least 20°C colder than the water. A few final notes of meteorological vortex exotica are as follows: Ash devils have been seen on the volcano Mount St. Helens in Washington, and waterspouts have been observed in the convection forced by the Surtsey volcano.

Straight-line winds. Gust fronts usually produce strong but not necessarily damaging winds. The leading edge of gust fronts may be heralded by a menacing-looking "arcus" or "shelf cloud" (Fig. 3.56), formed when stable air is forced over the cool outflow. *Roll clouds,* which look like long tail clouds, are also seen (Fig. 3.57).

Strong, concentrated downdrafts, however, may be intense enough to produce damage or to make aircraft landings dangerous as planes experience rapid changes in lift as they cross the center of the downdraft (Fig. 3.58). Intense, damage-producing downdrafts on the order of 10 km across are called *downbursts* (Fig. 3.59). T. Fujita pioneered the early research on downbursts. Even smaller-scale downbursts are called *microbursts.* Microbursts may be "dry" or "wet". Dry microbursts are common in the High Plains, when precipitation falls out from a high cloud base and has ample time to evaporate and effect rapid cooling, which produces negatively buoyant air parcels that accelerate downward. Wet microbursts are the result of the entrainment of dry environmental air into a precipitation-bearing cloud. Downbursts have been documented in both ordinary cells and in the RFD of supercells.

Mesoscale convective systems (which are discussed in the following section) that produce widespread (i.e., within an area whose major axis is a least 400 km long) straight-line wind gusts in excess of 26 m s^{-1} are called *derechos.* Derechos produce damaging wind events at least once every 3 h. They frequently occur in the upper Midwest in the United States.

The reader must be cautioned that the classification of damaging winds into two distinct categories, tornadoes and straight-line winds, can be somewhat misleading. Tornadic damage can be inflicted along one side of a weak vortex, when, for example, it translates along at a moderate speed. On the other hand, a microburst may contain a rotating or "twisting" component: Vorticity can be generated, for example, when convergence above a downdraft acts upon ambient vorticity. The vorticity may then be advected downward (but dissipated at the surface in a region of divergence). Vorticity may also be produced along the sides of a downdraft through the tilting of vertical shear.

On very rare occasions bursts of air as hot as 35°C are observed at the

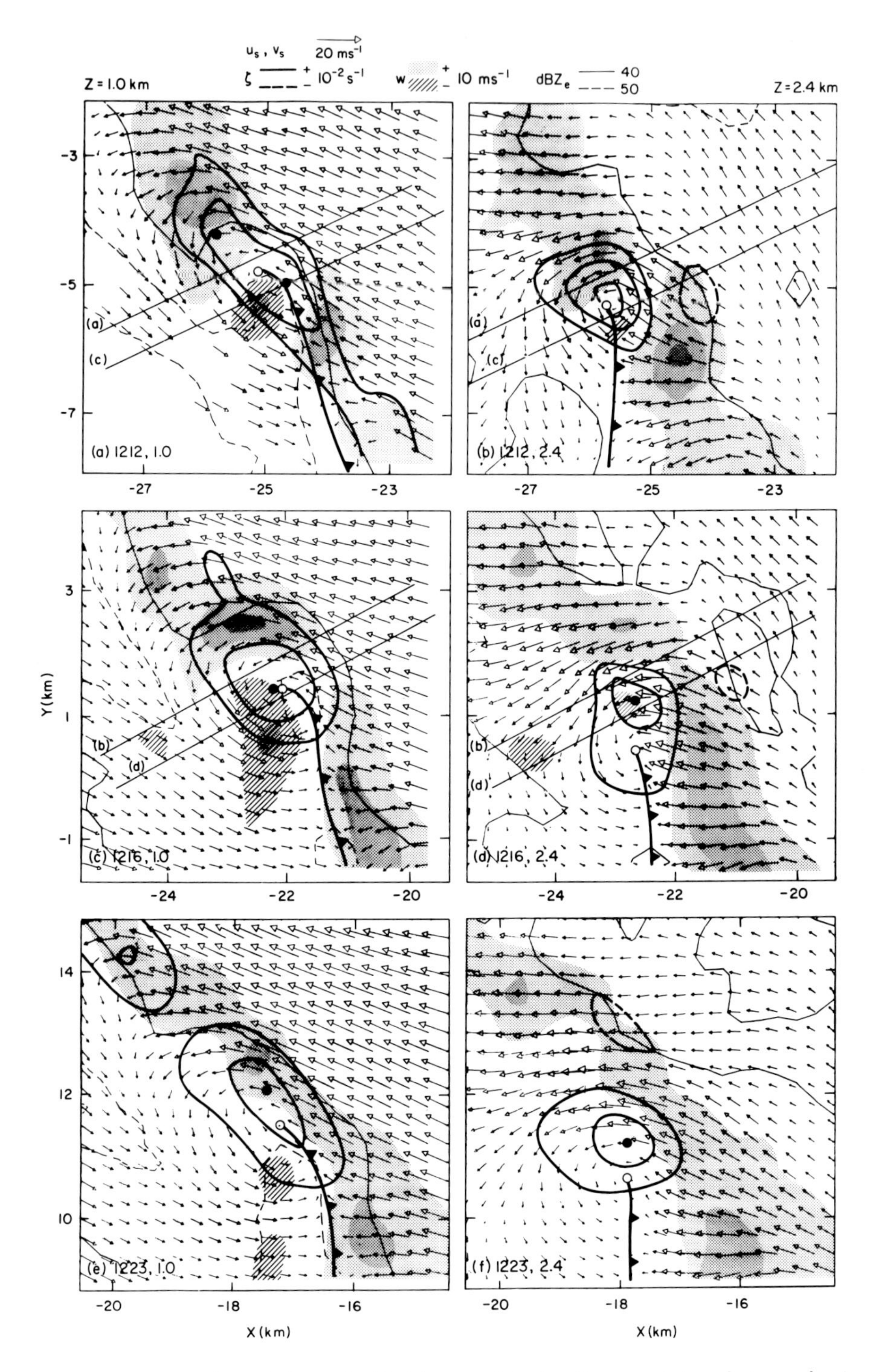

Figure 3.52 Example of the windfield associated with a rainband tornado in California. Horizontal storm-relative airflow at 1 and 2.4 km AGL together with vertical air motion, vorticity, and reflectivity factor. Vector scale is 10 m s^{-1} per grid point. Updraft regions of 10, 20, and 30 m s^{-1} (shaded areas); downdraft regions of 10 and 20 m s^{-1} (hatched areas). Vorticity (heavy solid lines for 1, 2, and

Figure 3.53 Photograph of a high-based funnel cloud pendant from a small cumulus cloud in the Texas Panhandle east of Amarillo on May 25, 1987 (photograph copyright H. Bluestein).

surface near storms. This hot air is the result of an intense, dry downdraft that had been triggered by a negatively buoyant air parcel created through evaporative cooling, and then followed by overshooting of the equilibrium level. This phenomenon is called a *heat burst* (Fig. 3.60).

Hail. The necessary conditions for the formation of hail are the availability of "embryos" of graupel and large frozen raindrops, and supercooled cloud droplets that the former accrete. If the updraft in a supercell is very strong, and an embryo enters it, the embryo will likely not reach hail size before it exits the storm (Fig. 3.61) through the anvil (Fig. 3.62; trajectory 0). If it enters the side of the updraft, it may have time to grow to hail size and fall partially out, but be carried back into the updraft again (i.e., "recycled") where it can grow even more (Fig. 3.62; trajectories 1,2,3). Since strong

---

Figure 3.52 (*cont.*)
$3 \times 10^{-2}\,s^{-1}$). Reflectivity isopleths [fine solid lines (40 dBZ) and dashed lines (30 and 50 dBZ)]. Circulation centers (open dots); vorticity centers, where not otherwise obvious (solid dots): (a) and (b) at 1212 PST, February 5, 1978; (c) and (d) at 1216 while tornado is on the ground; (e) and (f) at 1223 (from Carbone, 1983). (Courtesy of the American Meteorological Society)

Figure 3.54 Photograph of a dust devil in Nevada on August 7, 1978 (photograph copyright H. Bluestein).

Figure 3.55 Photograph of a steam devil over Lake Thunderbird near Norman, Oklahoma, on December 22, 1989 (photograph by H. Bluestein). (Courtesy of the American Meteorological Society)

Figure 3.56 Photograph of an arcus cloud along a gust front in Norman, Oklahoma, on May 27, 1977, early in the morning; looking toward the northwest (photograph copyright H. Bluestein).

Figure 3.57 Photograph of a roll cloud along a gust front in Norman, Oklahoma, on June 2, 1978 (photograph copyright H. Bluestein).

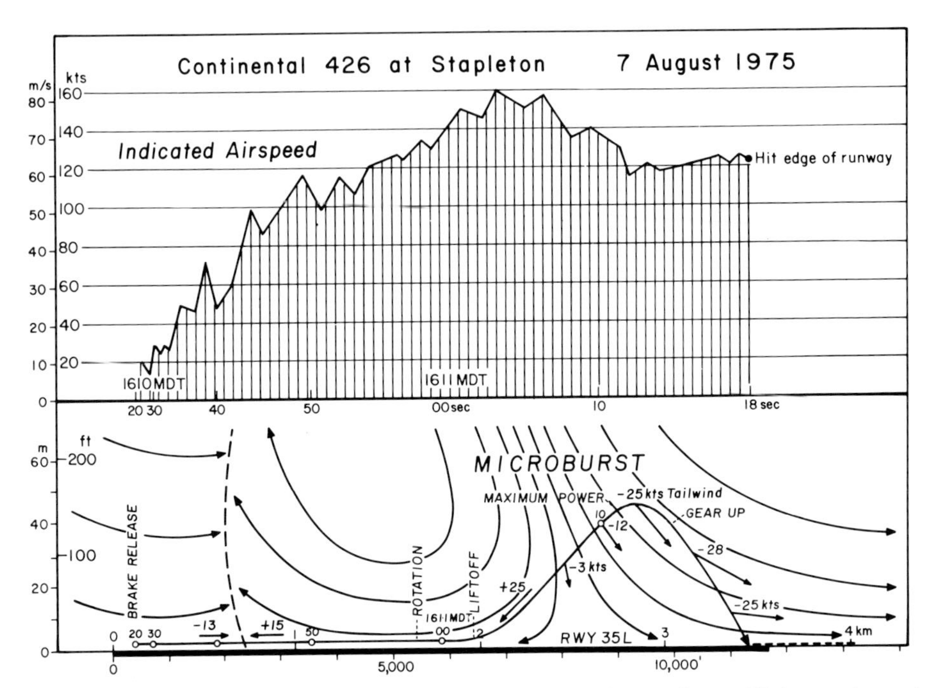

Figure 3.58 Example of a microburst-related aircraft accident. Flight path and indicated airspeed of Continental airlines flight 426 at Stapleton Airport, Denver, Colorado on August 7, 1975 (from Fujita, 1985).

Figure 3.59 Photograph of a downburst over Oklahoma City, Oklahoma, on July 26, 1978 (photograph copyright H. Bluestein).

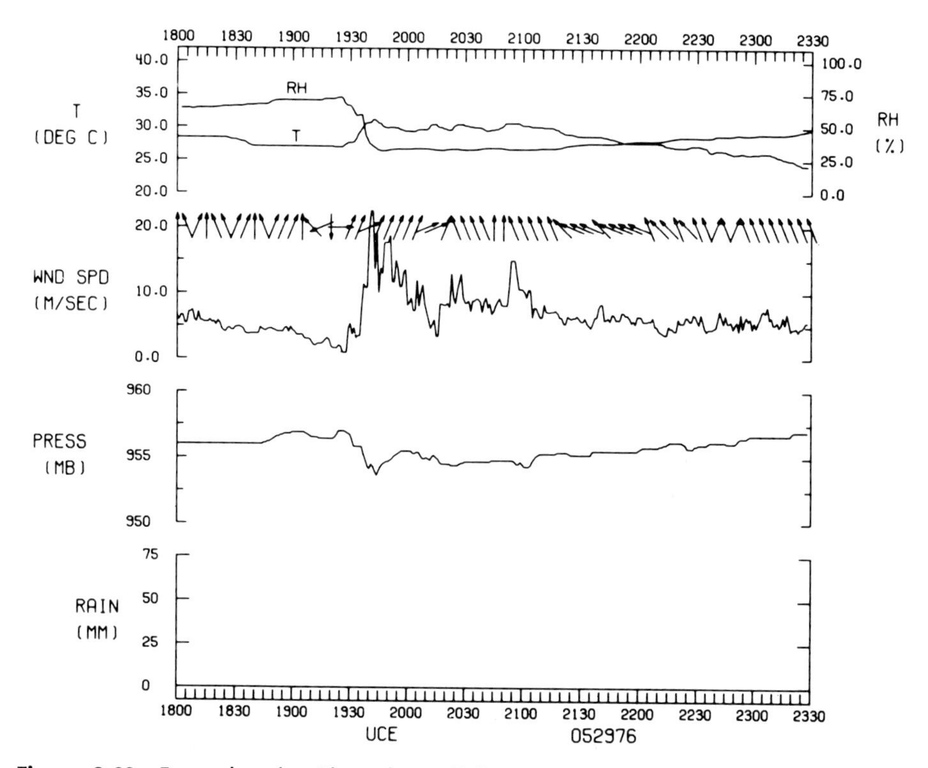

Figure 3.60 Example of a "heat burst." Data on May 29, 1976, for Union City, Oklahoma; time in CST. The heat burst occurred at 1930 (from Johnson, 1983). (Courtesy of the American Meteorological Society)

updrafts can support large, heavy hailstones, the *potential* for producing large hail is great in supercell storms. Supercells are in fact the most prolific producers of large hail. The largest hailstones in supercells are usually at least 1.9 cm in diameter (Fig. 3.63). Hail as large as 10 cm in diameter has been observed. The largest hail usually falls just ahead and to the left of the wall cloud (Fig. 3.64).

Storms composed of one or more ordinary cells are also capable of producing large hail. Some of these storms exist in an environment supporting intense updrafts (high CAPE) and weak vertical shear. The trajectory followed by a growing hailstone, however, may be at least as important as the strength of the updraft. Large hail may be produced even when the environment supports weaker updrafts (smaller CAPE) and there is no "recycling": As long as the hail trajectory is such that it passes through regions of copious supercooled water drops and updrafts strong enough to keep the hail stone from falling to the ground too soon, large hail can be produced. Figure 3.65 shows a conceptual model of a multicell hailstorm in which embryos develop in new, discrete cells on the low-level relative inflow side of the storm, and are swept into the updraft of an older cell.

Hailstones may be spherical, disc-shaped, or ellipsoidal. Sometimes they

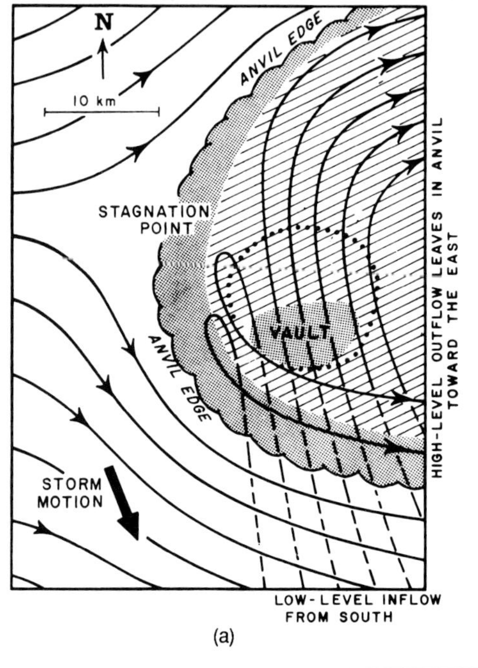

(a)

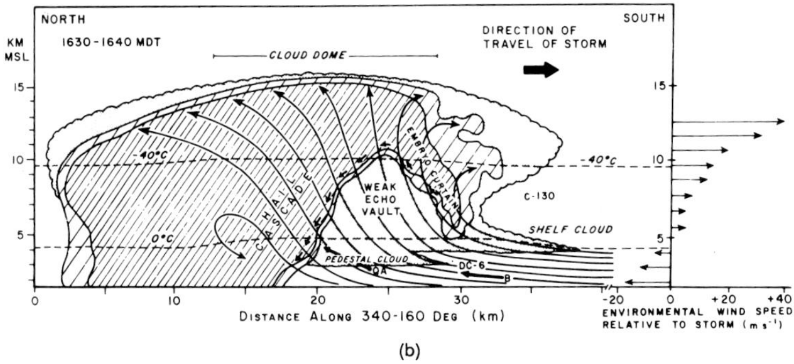

(b)

Figure 3.61 Example of airflow through a supercell. (a) Plan view showing the principal features of the airflow within and around a supercell in Colorado. Regions of radar echo (hatched); areas of cloud devoid of detectable echo (stippled); extent of intense updraft in the middle troposphere (dotted circle); streamlines of airflow relative to the storm (thin lines). Some of the streamlines represent the strong westerly environmental flow at middle levels being diverted around the main updraft; others represent the low-level southerly inflow toward the updraft (dashed lines) and also part of the high-level outflow. (b) Vertical cross section showing features of the visual cloud boundaries of the supercell depicted in (a) superimposed on the radar-echo pattern. The cross section is oriented along the direction of travel of the storm, through the center of the main updraft shown in (a). Two levels of radar reflectivity are represented by different densities of hatched shading. Areas of cloud devoid of detectable echo (stippled). Wind vectors in the plane of the diagram as measured by aircraft (scale is only half that of winds plotted on right side of diagram). Hailstone trajectory (short, thin arrows skirting the boundary of the vault); streamlines of airflow relative to the storm (thin lines). To the right of the diagram is a profile of the wind component along the storm's direction of travel, derived from a sounding 50 km south of the storm (from Browning, 1977). (Courtesy of the American Meteorological Society)

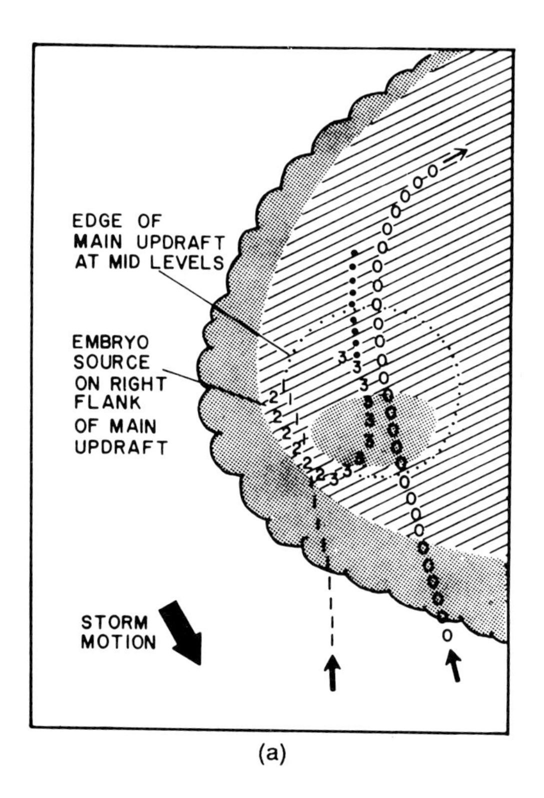

(a)

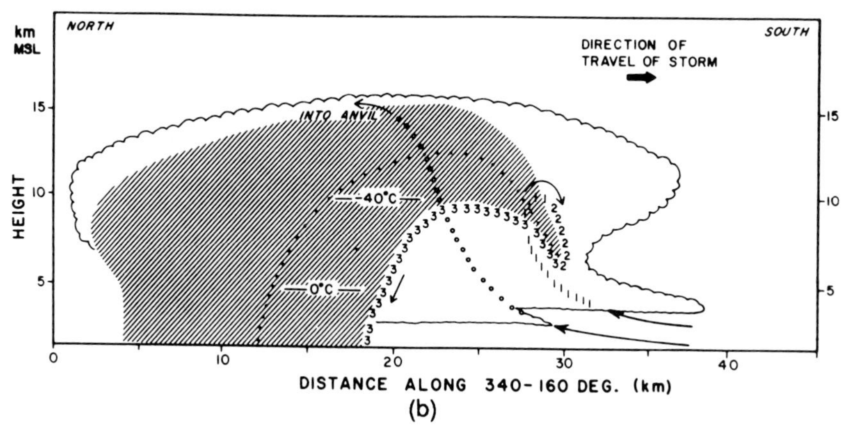

(b)

Figure 3.62 Schematic model of hailstone trajectories within a supercell based upon the airflow model in Fig. 3.61. (a) Plan view corresponding to Fig. 3.61(a). (b) Vertical cross section corresponding to Fig. 3.61(b). The echo distribution and cloud boundaries are as shown in Fig. 3.61. Each of the trajectories 1, 2, and 3 represents a stage in the growth of large hailstones. The transition from 2 to 3 corresponds to the re-entry of a hailstone embryo into the main updraft prior to a final up-and-down trajectory during which the hailstone may grow large, especially if it grows close to the boundary of the vault as in the case of the indicated trajectory 3. Other less favored hailstones will grow a little farther from the edge of the vault and will follow the dotted trajectory. Cloud particles growing "from scratch" within the updraft core are carried rapidly up and out into the anvil along trajectory 0 before they can attain precipitation size (from Browning, 1977, and Browning and Foote, 1976). (Courtesy of the American Meteorological Society)

Figure 3.63 Large hailstones produced in a supercell near El Dorado, Kansas on May 16, 1991 (photograph copyright H. Bluestein).

are smooth; other times they are rough and have spikes. When they are very large (e.g., "baseball" or "softball" or even "grapefruit" size), they are usually widely separated from each other. Smaller hailstones are less widely separated from each other. Very small hailstones (e.g., "pea" size) may be so densely distributed that they pile up into huge drifts. It has been necessary during the heat of the summer to use snowplows to clear away piles of hail. Fog may form over accumulations of hail. Hail can cause extensive damage to crops, windows, roofs, and auto bodies (not to mention exposed human flesh!).

Although it has been said that a "greenish" precipitation shaft is associated with hail, this has never been proved. Greenish areas of rain without hail have also been observed. The task of distinguishing between hail and rain by radar is also difficult. Large radar reflectivities may be indicative of a few large hailstones, many small hailstones, a few large raindrops, *etc.*

Since the shapes of hailstones and raindrops are different, dual-polarization radar techniques can be used to help distinguish between them. A measurement of the difference between the radar reflectivity from two orthogonally

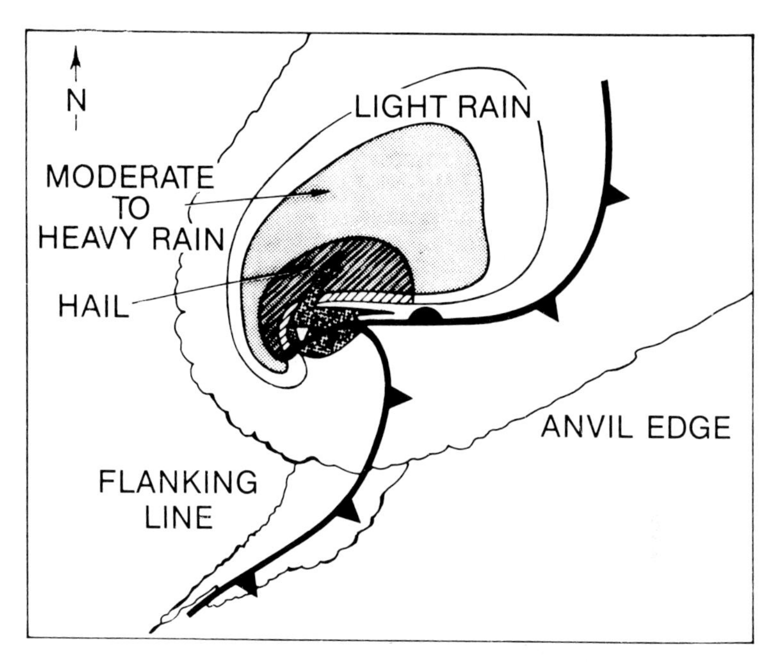

Figure 3.64 Plan view of the distribution of type of precipitation in a typical supercell storm in the Northern Hemisphere. The wall cloud is located at the intersection of the rear-flank downdraft (cold-front symbol) and the forward-flank downdraft (warm-front symbol). In this example the storm is moving toward the northeast (from Doswell, 1985). In a high-precipitation supercell, the area of hail and heavy rain extends farther south and extends around the southern edge of the wall cloud; in a low-precipitation supercell heavy precipitation is found only under the anvil.

polarized beams is called the *differential reflectivity factor,* and is measured in Z_{DR}, where

$$Z_{DR} = 10 \log_{10}(Z_H/Z_V) \tag{3.4.59}$$

and Z_H and Z_V are the horizontally polarized and vertically polarized reflectivity factors, respectively. Since raindrops are flattened as they fall, while hailstones are not, larger values of Z_{DR} are expected from rain; hailstones with irregular shapes tend to tumble and may not produce high values of Z_{DR}. The partial melting of the hailstones, however, makes interpretation more difficult.

Lightning. Although not considered a "severe" criterion for thunderstorms, lightning (all thunderstorms must have lightning!) is very dangerous, and kills more people each year than tornadoes. Lightning may be "cloud-to-ground", intracloud, or "cloud-to-air." Although lightning sometimes appears as a bright, formless flash, it is usually cloud-to-ground or intracloud lightning whose light is diffused inside the cloud. It is also referred to as *sheet lightning.* Cloud-to-ground strikes may be repeated very rapidly a number of times.

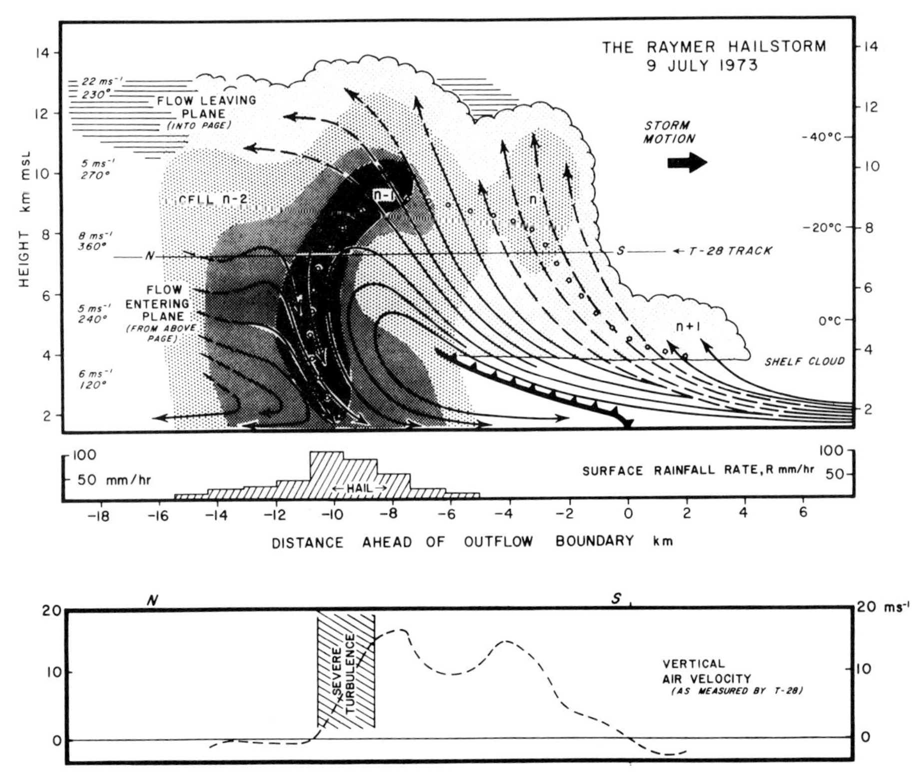

Figure 3.65 Schematic model (top) of a multicell hailstorm in Colorado showing a vertical cross section along the storm's direction of travel through a sequence of evolving cells. Streamlines of flow relative to the storm (solid lines); streamlines on the left side of the figure representing flow into and out of the plane, and on the right side of the figure representing flow remaining within a plane a few kilometers closer to the reader (broken lines); trajectory of a hailstone during its growth from a small droplet at cloud base (open circles); cloud area (light stippled shading); radar reflectivities of 35, 45, and 50 dBZ (three darker grades of stippled shading). The temperature scale on the right side represents the temperature of a parcel lifted from the surface. Winds (m s^{-1}, deg) on the left side are environmental winds relative to the storm based on soundings behind the storm. Surface rainfall rate averaged over 2-min intervals during the passage of the storm (middle). The horizontal line NS through the section at 7.2 km AGL shows the track of a penetrating aircraft (bottom). Vertical air velocity measured by aircraft (from Browning et al., 1976). (Courtesy of the American Meteorological Society)

Lightning viewed from far enough away that thunder is not audible is sometimes referred to as *heat lightning*, since thunderstorms most frequently occur on hot summer days. Sometimes lightning flashes are so frequent (1–2 s^{-1} or more) that at night the sky is illuminated continuously.

The flow of current in cloud-to-ground (CG) lightning may occur either from cloud-to-ground or from ground-to-cloud. It is most common for negative charge to flow from cloud-to-ground. Lightning in which *positive* charge flows from cloud-to-ground is called a "positive" CG flash.

Cloud-to-ground lightning is often observed in areas of intense precipitation and sometimes where precipitation is about to occur. The latter phenomenon has been referred to as the *rain gush*. Lightning is not common in updraft regions and is not often seen in the vicinity of tornadoes. CGs are sometimes observed in the clear air to the east or northeast of wall clouds: These intense, short-lived lightning flashes have been referred to as *staccato flashes*.

The mechanisms that separate charge so that large enough electric potentials can be built up for lightning discharges to occur are not entirely understood. However, it appears that precipitation, small-scale convective motions, and ice crystals play important roles.

Lightning is not common in oceanic areas of the tropics, but it is common over the land. It is sometimes observed in convective snow showers and in precipitation bands in intense, midlatitude, synoptic-scale cyclones.

It is standard practice to identify radar echoes having tops higher than 8 km AGL as thunderstorms, even in the absence of reports of audible thunder at the surface.

The locations where lightning occurs may be "mapped" by measuring the intensity and direction of low-frequency electromagnetic waves emitted by lightning channels. Given that electromagnetic waves travel at the speed of light, the time differences of the arrival of the radio "sferics" (electromagnetic radiation emitted by lightning flash) at an array of sensors may be used to locate the lightning "source." Cloud-to-ground and intracloud lightning flashes produce different wave forms and may therefore be distinguished from each other.

The climatology of thunderstorms in the United States. Thunderstorms in the United States appear most frequently between the afternoon and late-evening hours during the spring and summer months (Fig. 3.66). However, in the Great Plains storms occur mostly at night, while over much of the rest of the United States they occur most frequently during the afternoon.

Solar heating at the Earth's surface plays a major role in producing convective precipitation, since thunderstorms occur most frequently near the time temperatures at the surface have reached their diurnal maximum. The nocturnal maximum over the Great Plains is due in part to the eastward movement of thunderstorms formed over higher terrain to the west late in the day. Some storms, however, are truly nocturnal in that they form at night. These "nocturnal" thunderstorms may be influenced by vertical circulations associated with the low-level jet; they may be associated with an increase in boundary-layer moisture owing to advection by the low-level jet, or with lift associated with the retreating dryline.

Mountain thunderstorm activity "peaks" relatively early in the afternoon in response to intense heating of the sun-facing slopes (and the resulting mountain–valley circulation). They then may move off the mountains and "peak" later on at lower elevation; the latter peak in activity at lower elevation may be due also to the later attainment of convective temperature.

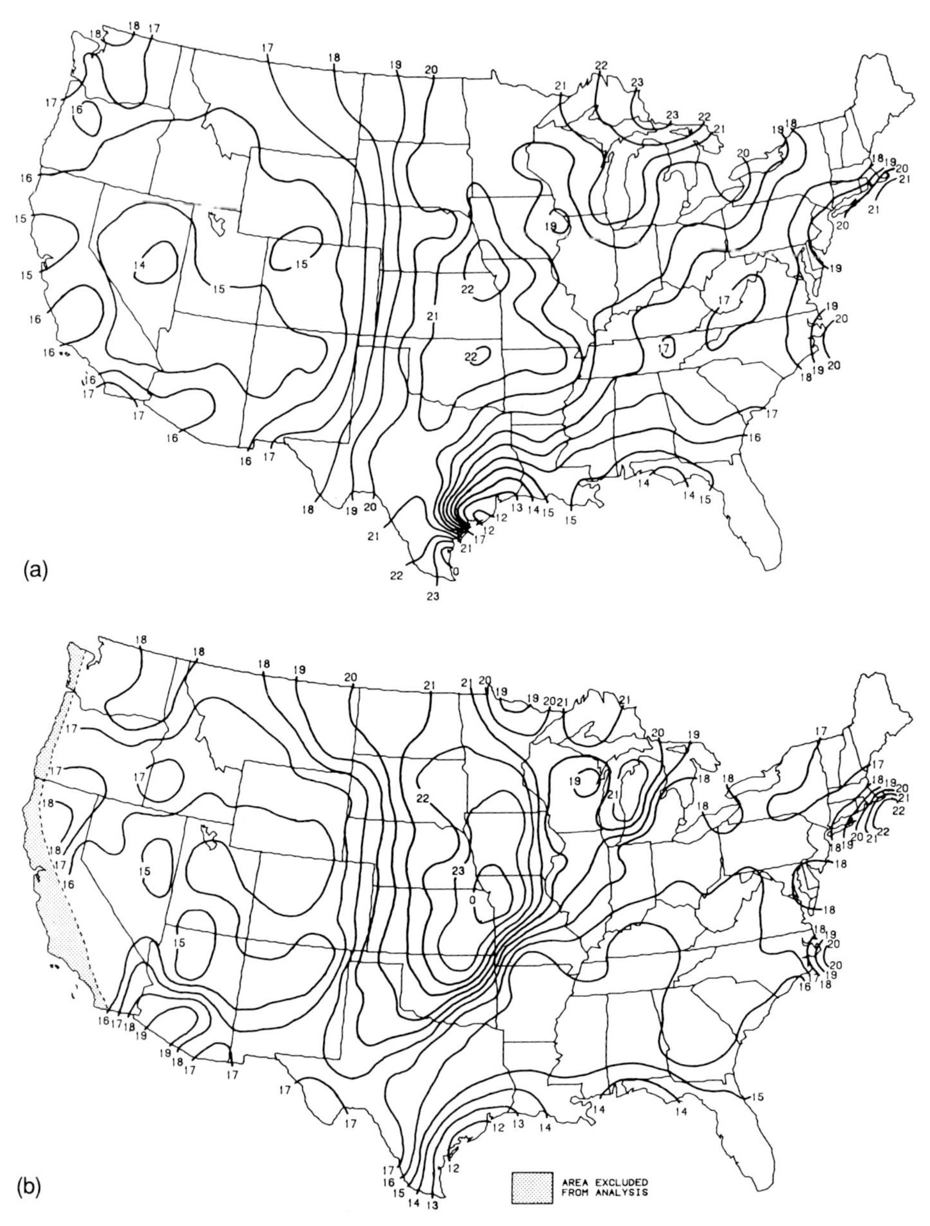

Figure 3.66 Time (LST hour) of maximum thunderstorm activity derived from the first harmonic of hourly thunderstorm activity in the United States. (a) Spring, (b) summer, (c) fall, and (d) winter (from Easterling and Robinson, 1985). (Courtesy of the American Meteorological Society)

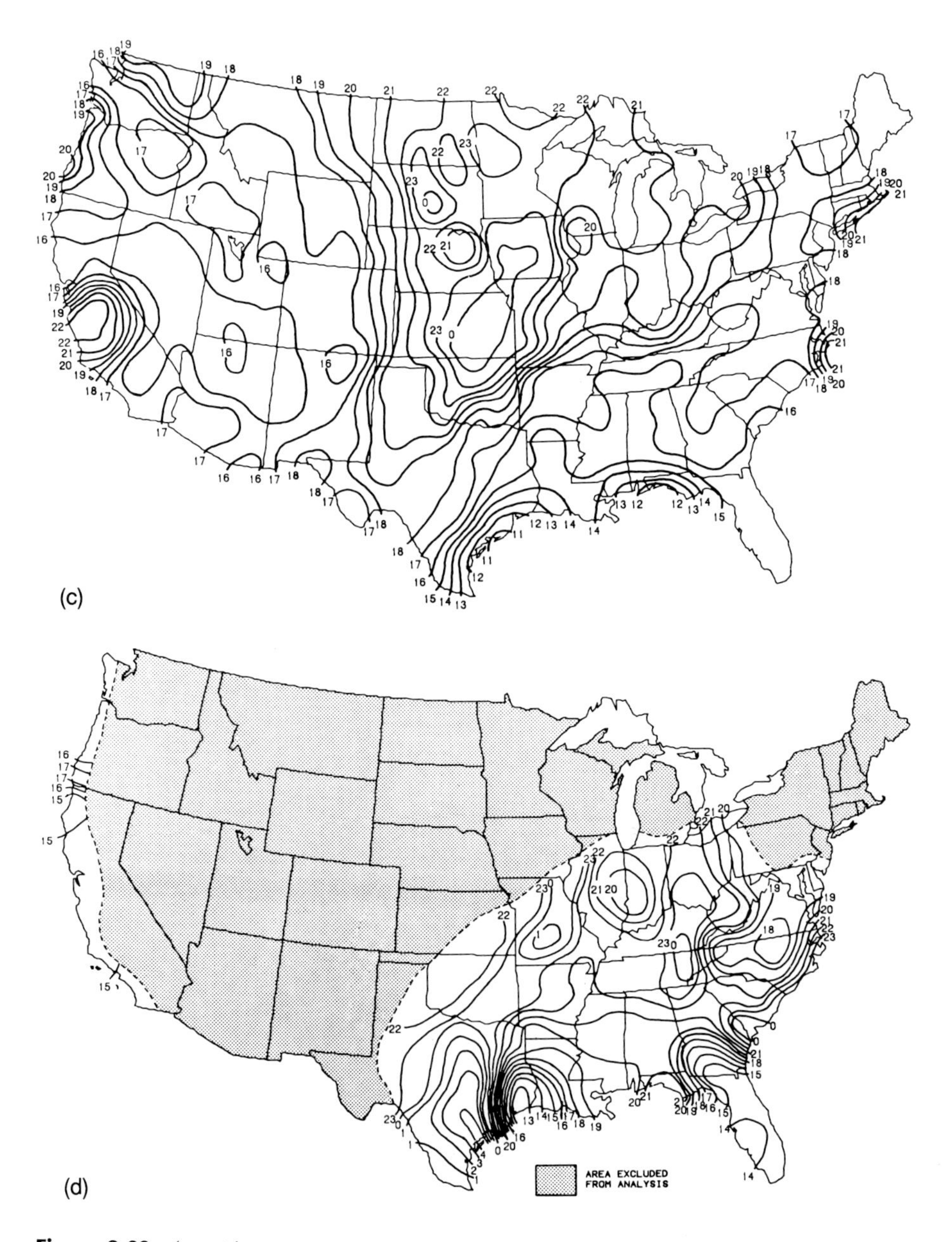

Figure 3.66 *(cont.)*

Thunderstorms are also common near the sea-breeze convergence zone just inland during the day, and near the land-breeze convergence zone just offshore during the early morning hours. Summertime thunderstorm activity over a peninsula responds both to the convergence zones caused by the sea breeze and to the synoptic-scale flow. For example, in Florida, during periods of westerly flow, activity is concentrated during the day on the east coast, while during periods of easterly flow, activity is concentrated on the west coast.

Under weak flow, the "double" sea-breeze convergence zone forces activity inland. In the morning the land-breeze convergence zone keeps activity offshore.

Other features, such as fronts, short-wave troughs, etc., which have vertical circulations associated with them, can trigger thunderstorms at any time of day or night.

The climatology of hail in the United States. In the United States hail falls most frequently in southeastern Wyoming, eastern Colorado, western Kansas, and western Nebraska (Fig. 3.67). Many of the hailstorms in this region are related to the higher terrain and the upslope and advection of moisture from the east or northeast, north of surface cold fronts or stationary fronts. Hail also falls frequently in the Cascades of Washington and over other mountainous areas of the West. Soft hail (graupel) is quite common in mountainous areas. A swath of hail occurrence is found stretching across the Southern Plains. Large hail is confined mostly to the High Plains and Plains regions, even though it is reported occasionally farther east.

The climatology of tornadoes in the United States. Tornadoes in the United States are reported most frequently in a band stretching from West and North Texas through Oklahoma, central and eastern Kansas, and into eastern Nebraska (Fig. 3.68). This region is referred to as *tornado alley*. The elevated

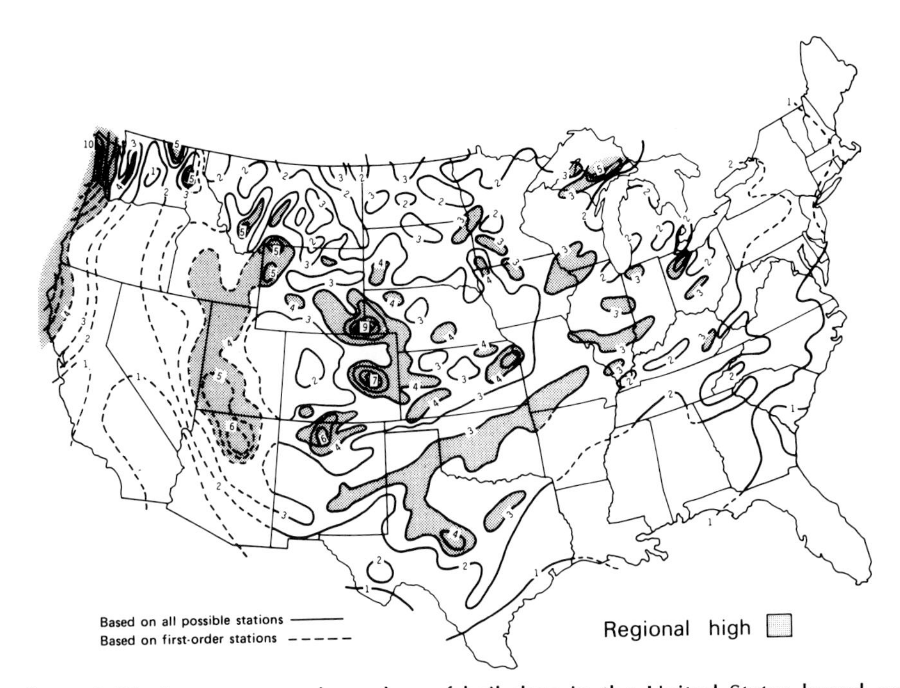

Figure 3.67 Average annual number of hail days in the United States based on point frequencies (from Changnon, 1977). (Courtesy of the American Meteorological Society)

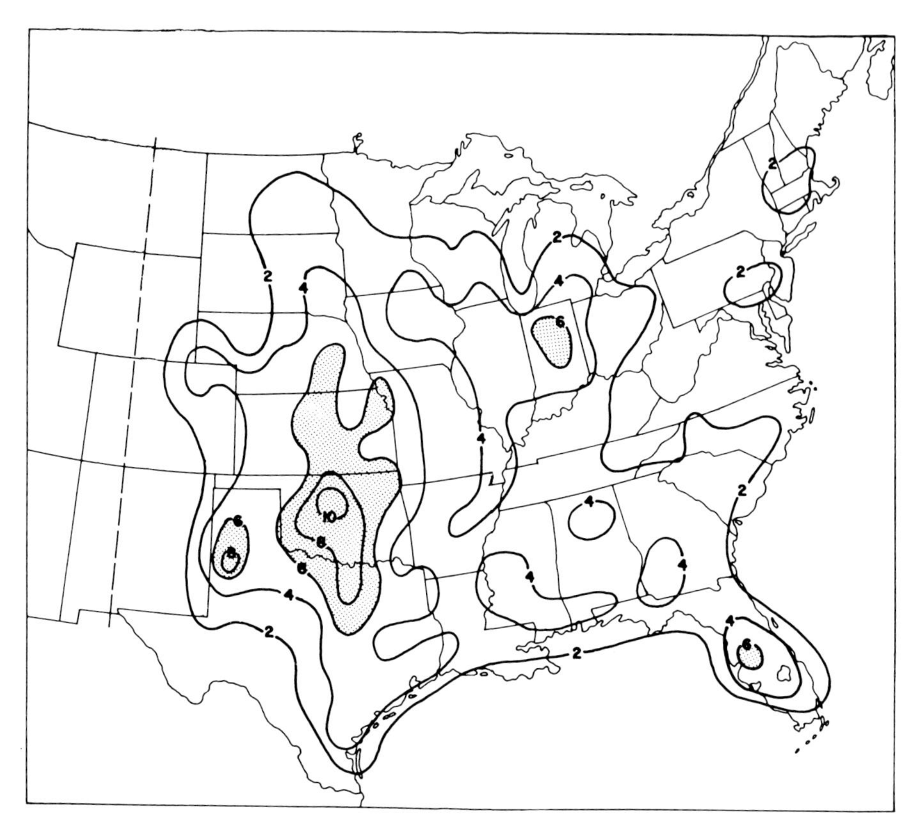

Figure 3.68 Frequency distribution of tornadoes (1950–1976) in the United States. Tornadoes per 2° latitude–longitude overlapping quadrilateral, normalized to 10,000 nm^2 area, per year (from Doswell, 1985; adapted from Kelly et al., 1978). (Courtesy of the American Meteorological Society)

terrain to the west and southwest of the Great Plains and the Gulf of Mexico play important roles in producing Type I soundings having high amounts of CAPE. The sloping terrain is partially responsible for the lee trough and low-level jet, the latter which advects the Gulf marine layer northward, while warm dry air caps the moist layer. The ubiquitous dryline and fronts approaching from the north and northwest are frequent locations of storm formation.

Most tornadoes occur during the late afternoon and evening hours (Fig. 3.69) in the late spring. This suggests that solar heating plays an important role in producing a potentially unstable environment for tornadic storm formation during the afternoon.

Major outbreaks occur primarily during the spring, when strong disturbances in the middle and upper troposphere promote regions of strong lifting and an environment of strong vertical shear and high values of CAPE. Tornadic activity begins along the Gulf States in late winter, and migrates northward, so that summer activity is highest in the Northern Plains (Fig.

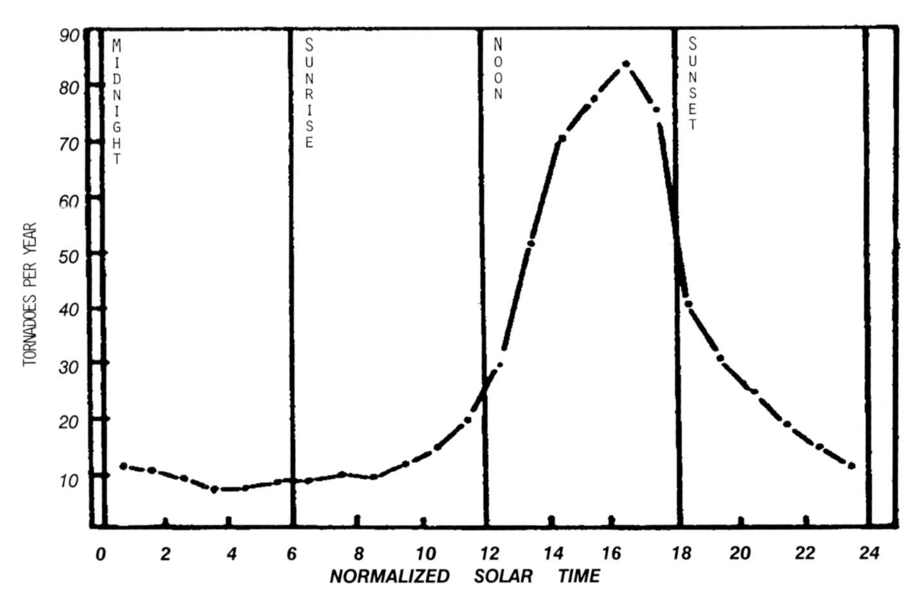

Figure 3.69 Average annual diurnal (NST; normalized solar time) distribution of tornadoes (1950–1976) (from Kelly et al., 1978). (Courtesy of the American Meteorological Society)

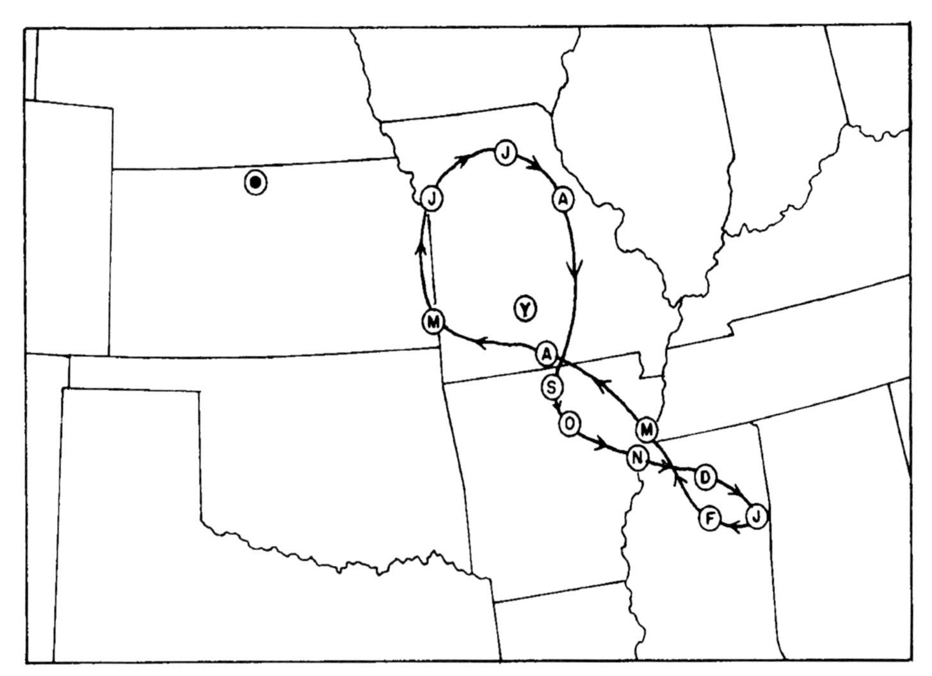

Figure 3.70 Average annual migration of the monthly statistical center (centroid) of tornado activity. The encircled letters indicate the month. The point Y is the centroid for all tornadoes during the year. The solid circle is the geographic center of the continental United States (from Schaefer et al., 1980).

3.70). Environments for tornado outbreaks in the United States have been produced as far east as the East Coast [e.g., Pennsylvania (1985), Worcester, Mass. (1953)]. Tornado outbreaks in late spring and early summer are often associated with short waves moving through northwesterly flow aloft, as a ridge is built up to the west and a trough is ensconced to the east. In spring, on the other hand, outbreaks usually occur in southwesterly flow aloft, downstream from major troughs.

Tornadoes occur along the eastern slopes of the Rockies in Colorado and in Wyoming during the late spring and summer when moist low-level air is advected westward and southwestward owing to an anticyclone to the north and a frontal trough to the south. The *Denver convergence–vorticity zone,* a surface feature that is related to local topography, is a frequent site of tornadic (landspout) storm formation (Fig. 3.71).

Tornadoes along the Gulf states and Florida, which are generally relatively

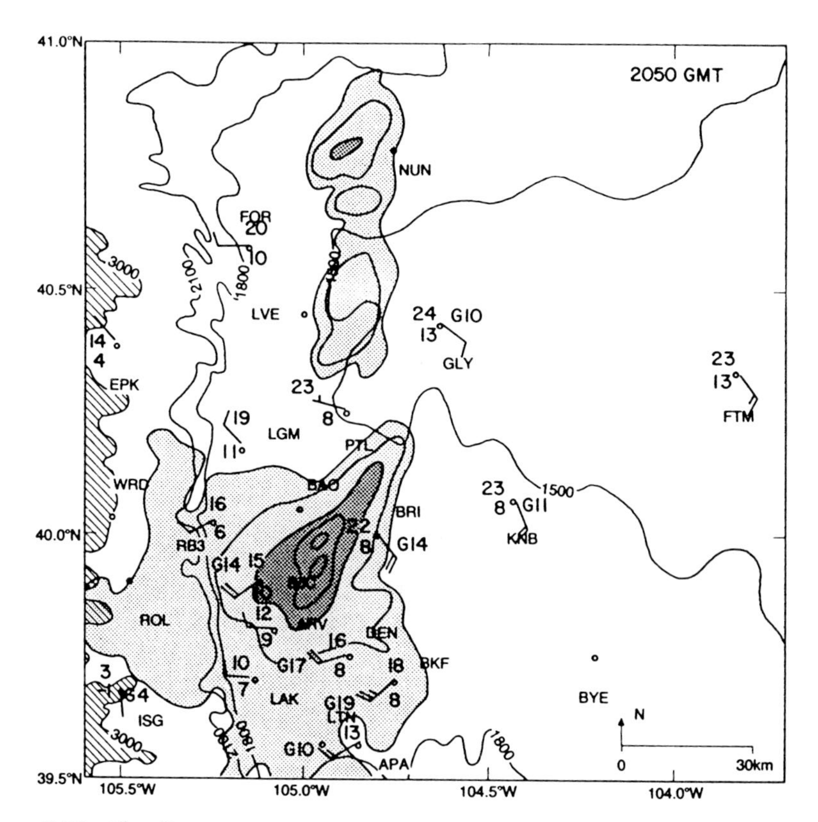

Figure 3.71 The Denver convergence-vorticity zone on June 3, 1981. Whole wind barbs and half wind barbs represent 5 and 2.5 m s^{-1}, respectively. Temperature and dewpoint plotted in °C. Lines of constant elevation in m MSL (solid lines); elevation above 3000 m AGL (hatched); radar reflectivity levels (shaded areas). The winds have a westerly component west of approximately 104.75°W, and an easterly component to the east; strong thunderstorms are located along the convergence zone (from Szoke et al., 1984). (Courtesy of the American Meteorological Society)

weak, are usually associated with Type II soundings. Waterspouts, which rarely do much damage even when they make landfall, are common along the Gulf Coast and southeastern U.S. coasts, but especially around the Florida Keys during the summer.

Some funnel clouds and tornadoes are observed in Type III sounding environments almost anywhere during the spring at lower latitudes and during the summer at higher latitudes. Tornadoes and waterspouts are sometimes reported during the winter along and near the California coast when strong upper-level troughs are offshore.

Thunderstorms may undergo an evolution dictated by changing environmental conditions that are related to topography, rather than those due to actual changing synoptic conditions. For example, mountain thunderstorms may move on to the Plains and encounter a much deeper moist layer. The CAPE might increase just enough to convert a weak, ordinary nontornadic cell into a stronger cell or even a tornadic supercell. On the other hand, a thunderstorm over land may move offshore over cold water and weaken owing to a decrease in CAPE.

3.4.9. The Mesoscale Organization of Convective Cells

Ordinary cells and supercells may be organized in mesoscale patterns, which sometimes include stratiform precipitation. K. Browning and M. Weisman (unpublished conference presentation) have proposed that the two basic types of convective "building blocks," the ordinary cells and supercells, represent a *primary* classification, while their patterns of organization on the mesoscale represent a *secondary* classification.

The basic convective building blocks may be organized into banded or nonbanded areas. A set of convective building blocks that is organized on the mesoscale is called a *mesoscale convective system* (MCS).

Squall lines. Bands of precipitation that are at least partly convective are now called *squall lines.* It is of historical interest that before Polar-Front theory had been introduced, cold fronts were called *squall lines* (*lignes de grain,* from French mariners), while just after Polar-Front theory had been introduced, lines of convection only well ahead of fronts (the *prefrontal squall line* or *instability line*) were considered to be squall lines. The radar echo associated with a precipitation "band" is at least 5 times as long as wide, at least 5–10 km wide, and persists for hours.

Squall lines may appear on radar as solid or "broken" bands of convective precipitation or as bands of stratiform precipitation having an embedded or a leading convective line. Squall lines in the Southern Plains of the United States generally evolve in one of four ways (Fig. 3.72) during the spring.

Squall-line formation. Broken-line formation occurs most frequently along cold fronts, and takes place when a line of cells separated by substantial gaps evolves into a solid line as new cells form in between the old cells, while the cells themselves expand in area. The bulk Richardson number in the

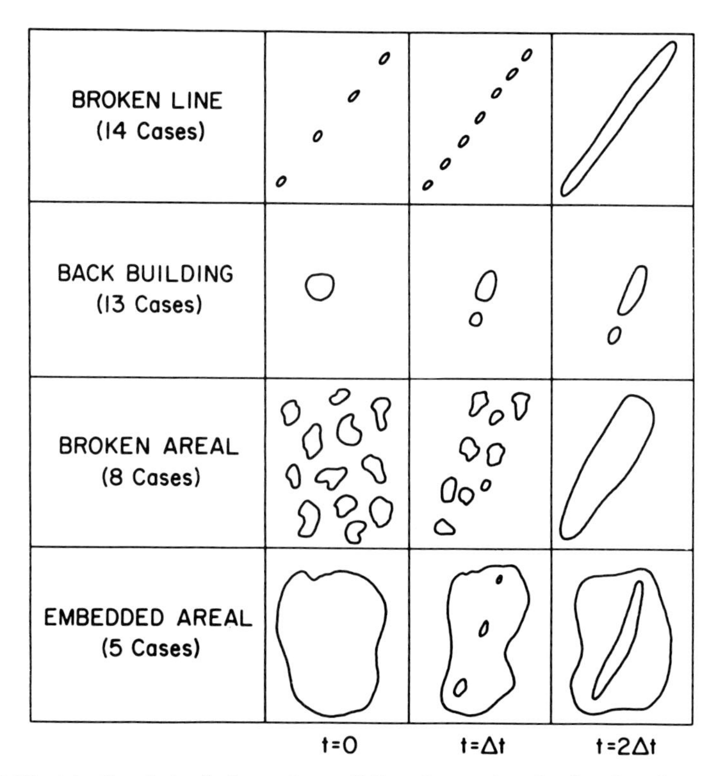

Figure 3.72 Idealized depiction of squall-line formation in the Southern Plains of the United States during the spring (from Bluestein and Jain, 1985). (Courtesy of the American Meteorological Society)

broken-line squall-line environment is relatively high owing to weak vertical shear; relative helicity is low. There is a steering level with respect to cell motion. The cells within a broken-line squall line therefore behave like ordinary cells. The level of free convection is easily attainable through only a little surface heating or lifting. The linear organization is a result of forced ascent along a line.

When a dryline in the Southern Plains intersects a cold front, broken-line squall-line formation often takes place to the northeast of the dryline–front intersection, while supercells tend to form along the dryline–front intersection (Fig. 3.73), at or near the western edge of the squall line.

Backbuilding formation takes place when a new cell forms upstream (with respect to cell motion) from an old cell, grows in areal extent, and moves into and merges with the old cell (Fig. 3.72). Eventually a line segment is produced. Sometimes the backbuilding process occurs separately in several line segments until the line segments join up to form a long line. The bulk Richardson number in the backbuilding squall-line environment is relatively low owing to high CAPE and strong vertical shear; relative helicity is high. Since there is no steering level with respect to cell motion, the effects of propagation normal to the vertical shear vector (i.e., hodograph) are sig-

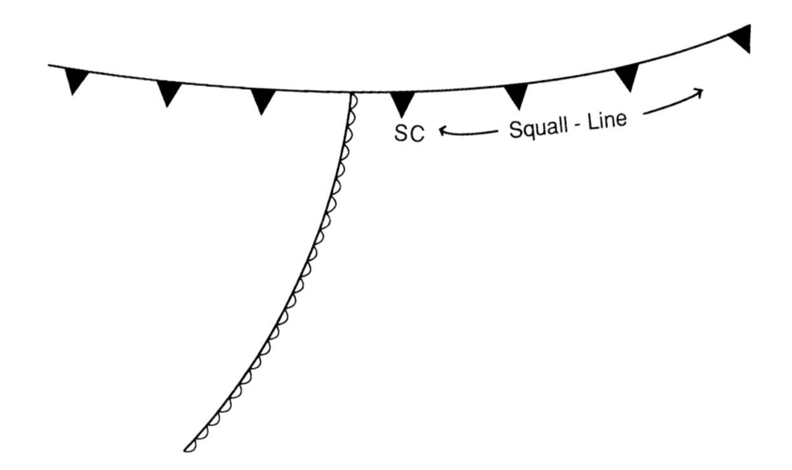

Figure 3.73 Idealized depiction of a squall line along a cold front; a supercell (SC) is found at the western edge of the squall line, near the intersection of the dryline (scalloped line) with the front.

nificant. The parent cells of a backbuilding squall line therefore behave to some extent like supercells. The linear organization is a result of "internal" cell propagation, not "external" forcing such as lift along a cold front. A few "forward-building" squall lines have been observed, but not enough to make general statements about their structure and environment.

Broken-areal formation is the gradual consolidation of an area (which may be a wide zone) of irregularly spaced echoes into a line (Fig. 3.72). The merging of gust fronts in the downshear direction and the resultant triggering of new convective cells along the new line of gust fronts downshear from the old convective cells may explain the broken-areal formation phenomenon. Like broken-line squall lines, broken-areal squall lines are composed of cells that behave like ordinary cells.

Embedded–areal formation is the appearance of a solid band of moderate-to-intense radar echo within a larger area of stratiform precipitation (Fig. 3.72). The low levels in the embedded-areal squall-line environment are markedly cooler and more stable than low levels for other types of squall-line formation owing to the evaporative cooling of stratiform precipitation in unsaturated air below cloud base. Owing to the stable low-level environment and potentially unstable stratification above, "ducted" gravity-wave motion (see Sec. 3.5.4) might be responsible for triggering the band. Ducted gravity waves are those that are trapped when an unstable or neutral layer overlies a stable layer. Dispersion of the waves is limited if there is a "steering level" for the waves in the potentially unstable or neutral layer above the moist layer (i.e., the wind at some level is nearly identical to the wave velocity). Symmetric instability is another possible triggering mechanism of bands of convection within a stratiform area of precipitation (see Sec. 3.5.2 of this chapter). Lines should be triggered along the geostrophic vertical-shear vector.

The composite environmental sounding and hodograph for spring-time

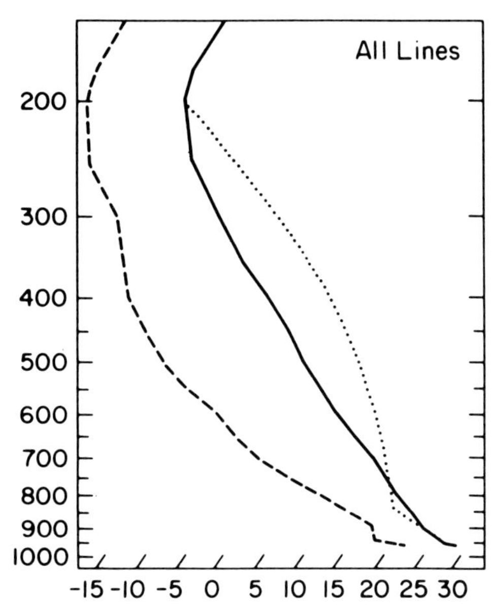

Figure 3.74 Composite sounding for severe squall lines in the Southern Plains of the United States during the spring. Skewed abscissa and logarithmic ordinate are the temperature (°C) and pressure (mb), respectively. Temperature and dew-point profiles plotted as solid lines and dashed lines, respectively. Path taken in temperature–pressure space by surface air parcel as it ascends (dotted line) (from Bluestein and Jain, 1985). (Courtesy of the American Meteorological Society)

squall lines in the Southern Plains of the United States are shown in Figs. 3.74 and 3.75. The right-handed coordinate system of the hodographs is oriented such that the u component of the wind represents flow normal to the squall line, from rear to front, and the v component of the wind represents flow along the squall line. This "natural" coordinate system for lines is called *squall-line coordinates.* Nonsevere squall lines are associated with an environment of weaker CAPE than severe squall lines. Both severe and nonsevere squall lines have some CIN, with an LFC of 750–700 mb. The mean vertical-shear vector is aligned approximately 40–50 degrees to the right of the line, while the low-level shear vector is oriented along the line. There is thus a component of vertical shear normal to the squall line in the lowest 6 km [of approximately 15 m s^{-1} (6 km^{-1})]. The magnitude of the vertical shear in the lowest 6 km is greater in the environment of severe squall lines [22 m s^{-1} (6 $km)^{-1}$] than in the environment of nonsevere squall lines [12 m s^{-1} (6 $km)^{-1}$]. The steering level of squall lines is about 6–7 km AGL; cells move along, and toward the rear of, the squall line.

Once squall lines have formed, they may continue to evolve into other patterns. For example, "solid" squall lines may break up into individual supercells (Fig. 3.76) if the environmental shear increases. Supercells may split, with the right movers interacting with the left movers of adjacent storms. Squall lines may be solid and break up into individual multicells, or vice versa.

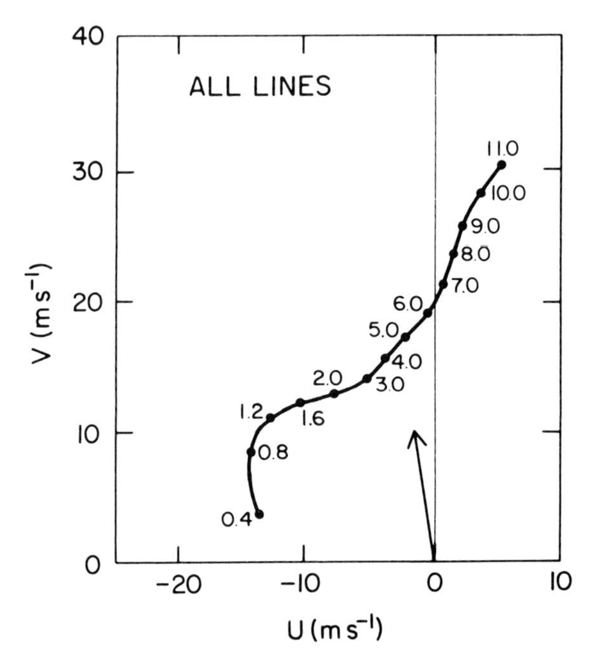

Figure 3.75 Composite hodograph, in a coordinate system moving along with the line, for severe squall lines in the Southern Plains of the United States during the spring. The cell motion is indicated by the vector (from Bluestein and Jain, 1985). (Courtesy of the American Meteorological Society)

Rotation in squall lines. Large tornadoes are common only in squall lines that are composed of supercells. Tornadoes are not common in solid squall lines. Gustnadoes, however, do occur along the gust front of squall lines. In lines of multicells, tornadoes are most likely to occur along the upshear edge of the squall line, where the squall-line's gust front has not cut off the surface inflow of the newest cell. Sometimes the newest cell on the upshear end is a supercell, while the other cells in the line are multicells. Hail is found frequently in squall lines.

Mesocyclones may form aloft within squall lines, but infrequently build down to the surface. A mesocyclone or a larger-scale circulation in the squall line may distort the line into a *line-echo wave pattern* (LEWP) (Fig. 3.77). Although the LEWP is often associated with severe weather, there is evidence that it may be a pattern visible *after* rather than *before* a tornado has occurred, and therefore may not be a useful tornado forecasting signature.

The "bow echo" (Fig. 3.78) is a 60–100 km long curved line of cells, which is associated with long swaths of damaging surface winds. It was first discussed in detail by T. Fujita, and first explained dynamically by M. Weisman. The bow echo is sometimes seen as an isolated entity, and sometimes as part of an LEWP in a squall line. Bow echoes develop 3–4 h after the initiation of an MCS, and can persist for several hours. They can occur when the environmental CAPE is at least $2000\,\mathrm{J\,kg^{-1}}$, and the environmental

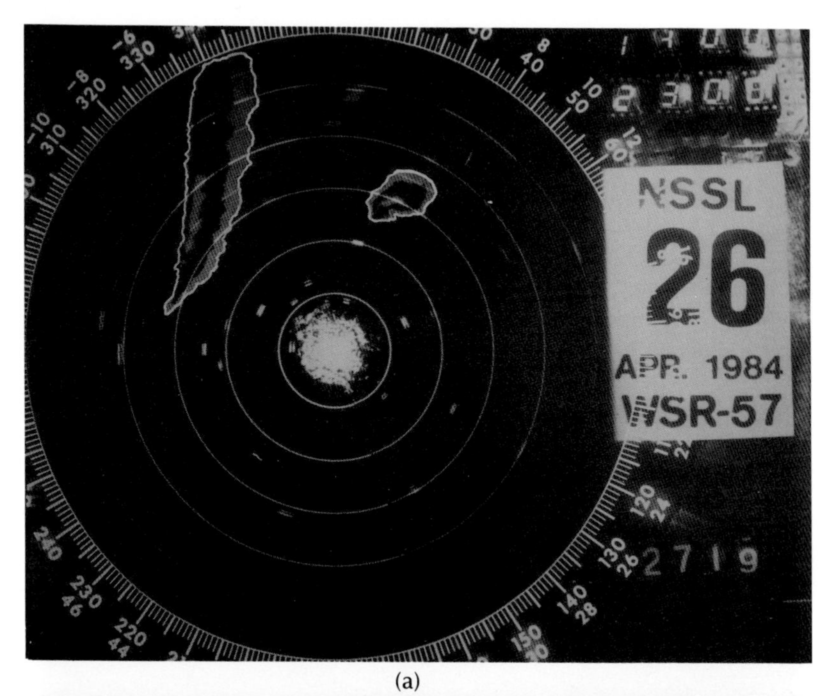

(a)

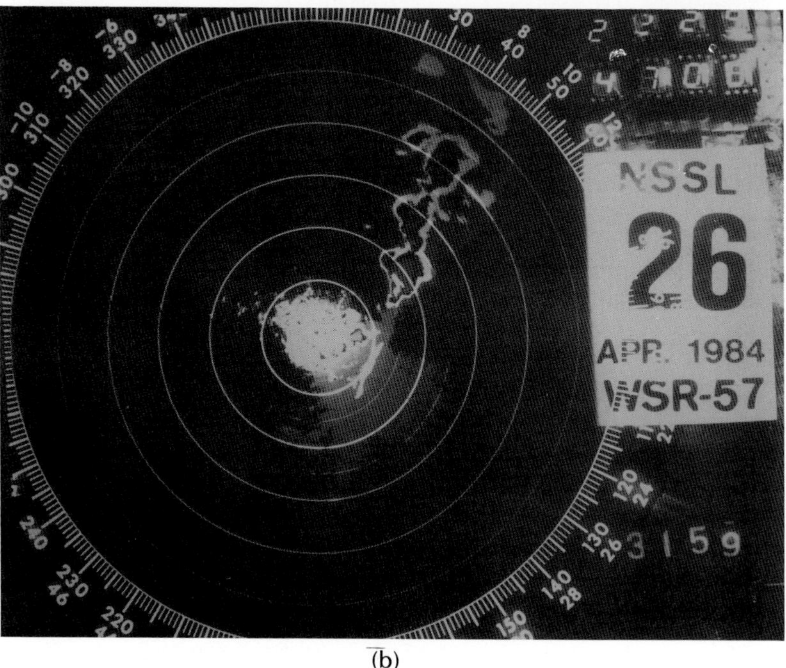

(b)

Figure 3.76 Example of a "solid" squall line, as it appears in plan view on a radar, breaking up into individual supercell storms. Elevation angle is 0°. Contours are in 10-dBZ steps beginning at approximately 10 dBZ. (a) Solid squall line northwest of the National Severe Storms Laboratory, 1900 CST, April 26, 1984; (b) squall line having broken up into individual storms at 2229 CST (from Burgess and Curran, 1985; photographs courtesy D. Burgess, NSSL). (Courtesy of the American Meteorological Society)

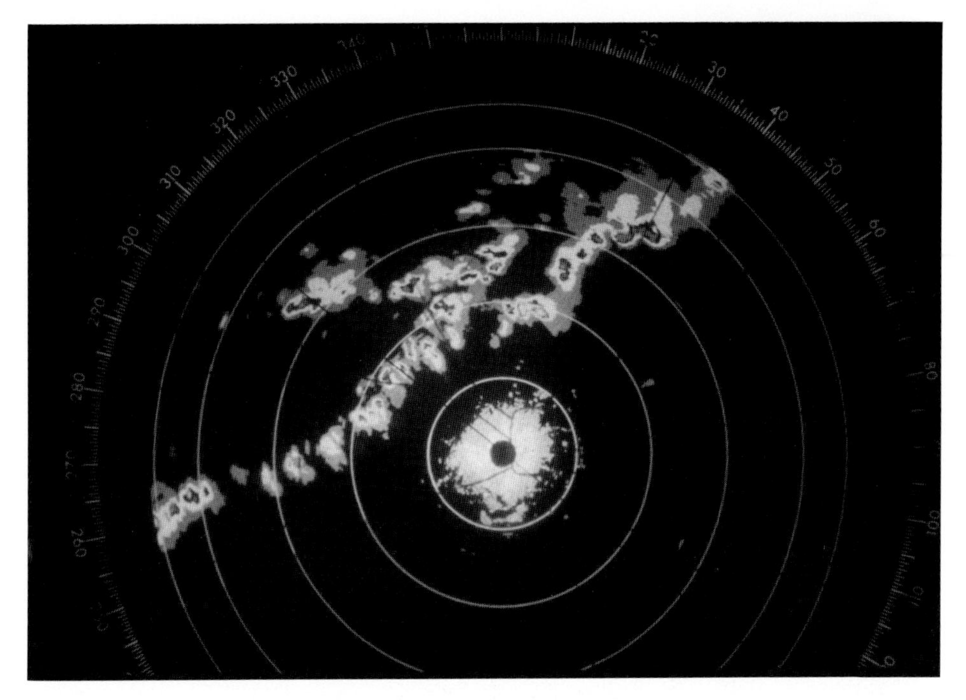

Figure 3.77 Example of a line-echo wave pattern (LEWP) at 1714 CST, July 25, 1986, as viewed from the National Weather Service 5-cm radar at Indianapolis, Indiana (photograph courtesy Ron Przybylinski, National Weather Service, Indianapolis).

vertical shear is greater than 15 m s^{-1} in the lowest 2.5–5 km AGL. (Supercells also occur in such an environment.) A jet that tracks from the rear of the MCS to the leading edge of the bow echo at 2–3 km AGL is found, along with cyclonic and anticyclonic vortices ("book-end" vortices) on the northern and southern flanks of the radar echo, respectively. The book-end vortices, which are produced primarily by tilting (through horizontal gradients of downdraft at the ends of the convective system) of environmental horizontal vorticity, act to enhance the jet. The effect of the Coriolis force acting over several hours, however, eventually makes the cyclonic vortex dominant.

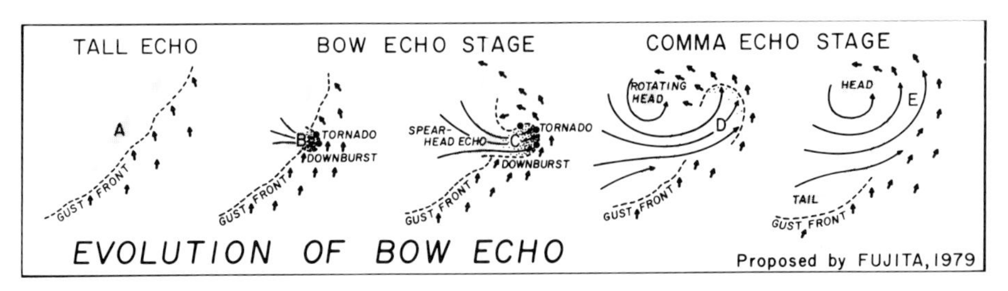

Figure 3.78 Idealized depiction of the evolution of a bow echo (from Fujita, 1981; originally proposed by T. Fujita in 1979). (Courtesy of the American Meteorological Society)

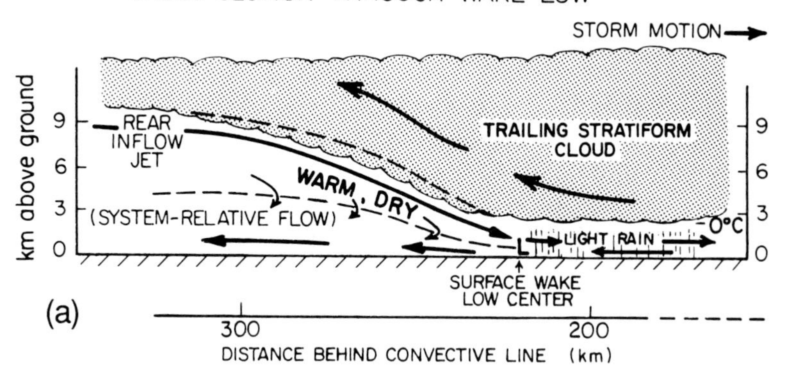

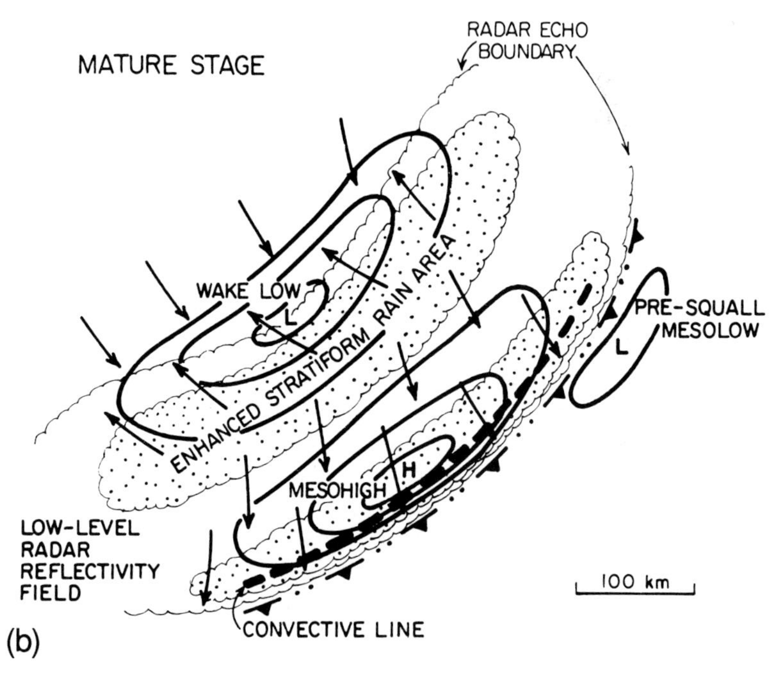

Figure 3.79 (a) Schematic vertical cross section through a wake low and (b) surface pressure and wind fields and precipitation distribution during the mature stage of a squall line. Winds in (a) are system-relative with the dashed line denoting zero relative wind. Arrows indicate streamlines, not trajectories, with those in (b) representing actual winds. Horizontal scales differ in the two schematics (from Johnson and Hamilton, 1988). (Courtesy of the American Meteorological Society)

Meso-highs and meso-lows. The evaporatively cooled air that results from the accumulation of outflow from many cells produces hydrostatically an area of high pressure, a "meso-high" (Figs. 3.79 and 3.80). An observer overtaken by a gust front along a squall line notes a rapid pressure "jump," a wind shift, and a temperature drop as the gust front passes by. A mesoscale area of stratiform precipitation is usually co-located with the surface meso-high. The pressure in the meso-high may be as much as 5 mb or more higher than the ambient pressure. (For example, if the temperature in the lowest 1 km is cooled by 10 K, then the surface pressure rises by approximately 3 mb.)

A mesoscale area of low pressure is sometimes found at the surface (Figs. 3.79 and 3.80) in dry air to the rear (with respect to squall-line motion) of the precipitation area. This "meso-low" or "wake depression" is caused by warming that has occurred in association with subsidence forced outside the squall line.

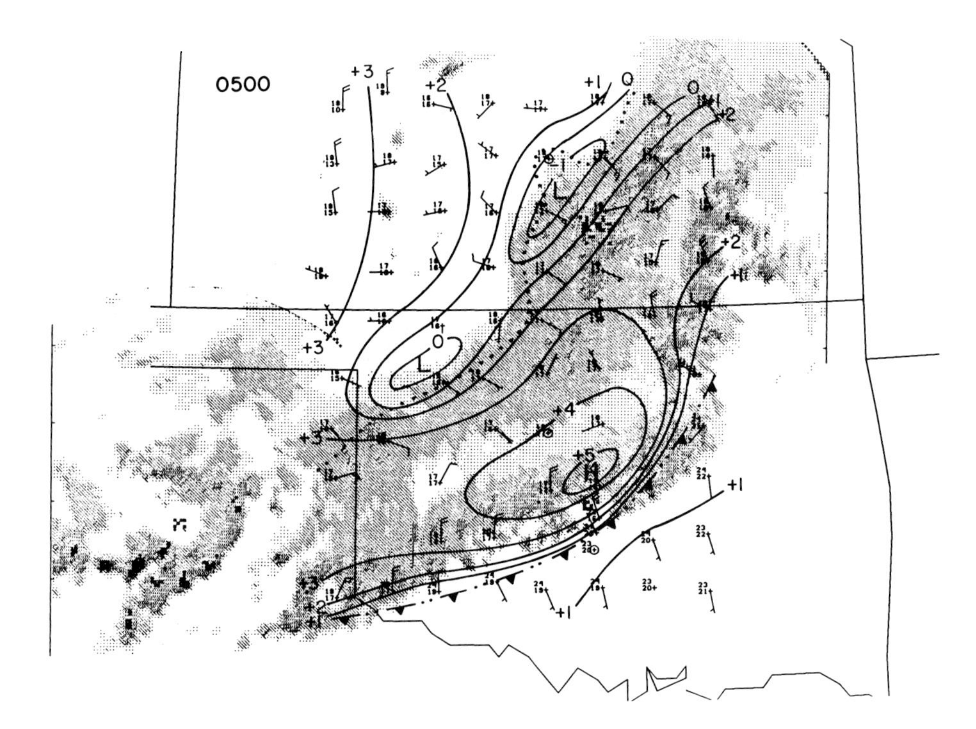

Figure 3.80 Observations of the pressure, temperature, dewpoint, and wind field in a meso-high and a wake low associated with a squall line. Pressure (solid lines) at 518 m indicated as departures, in mb, from 950 mb at 0500 UTC, June 11, 1985 over Kansas and Oklahoma. Radar reflectivity thresholds for the shading are 15, 25, 35, and 50 dBZ. Gust fronts or outflow boundaries (dashed double-dotted lines); termination of surface rainfall based on surface meso-network data (dotted line). Temperature and dew point in °C; whole and half wind barbs represent 5 and 2.5 m s^{-1}, respectively (from Johnson and Hamilton, 1988). (Courtesy of the American Meteorological Society)

Squall-line morphology. Squall lines sometimes have a trailing area of stratiform precipitation, which is separated from the leading[6] band of convective precipitation by a *transition zone* (Fig. 3.81). The region of stratiform precipitation has a "bright band" in the "melting layer" (i.e., where $T = 0°C$). The bright band is caused by enhanced radar return from "wet" ice particles. Relative to the squall line, convective clouds grow, produce precipitation, and decay as they merge with other dying cells in the stratiform-precipitation area. The stratiform-precipitation area thus represents a mortuary for dying cells within the squall line. A time- and space-averaged cross section through a squall line having a trailing area of stratiform precipitation is to some extent a depiction of the life history of a cell; the early part of a cell's life is depicted at the leading edge of the line, while the decay of a cell is depicted at the rear of the line.

Squall lines, whose convective leading edge and trailing stratiform-precipitation pattern is relatively symmetric with respect to an axis normal to and passing through the midpoint of the line are called *symmetric* [Fig. 3.82(a)]. In backbuilding squall lines, however, the life history of a cell is depicted *along* the squall line, for example, from south or southwest to north

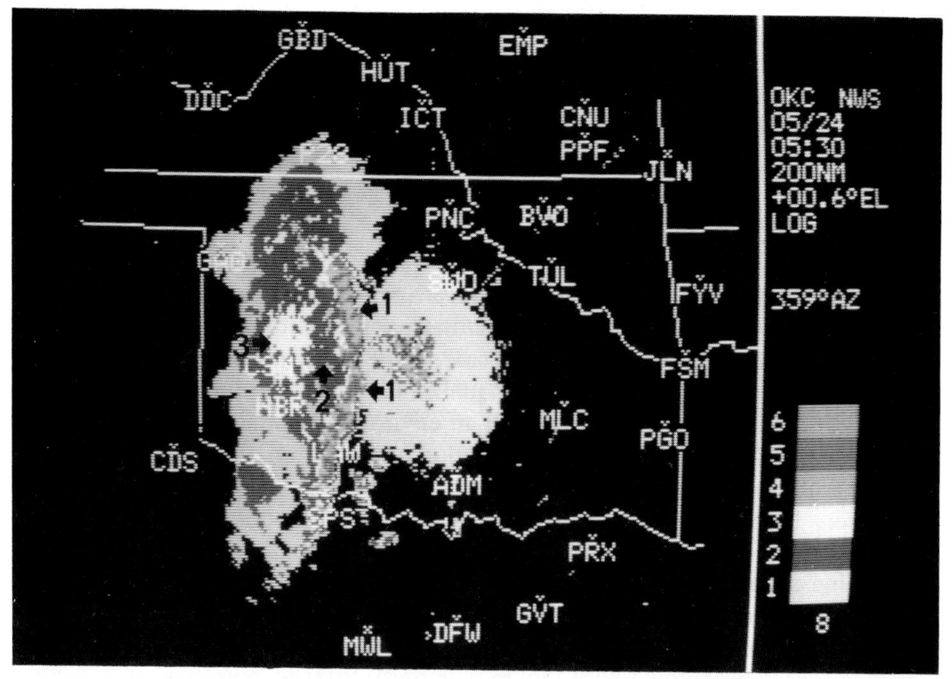

Figure 3.81 Example of the subdivision of the radar echo associated with a squall line into (1) convective, (2) transition zone, and (3) stratiform regions. Radar reflectivity from the National Weather Service WSR-57 radar at 0530 UTC, May 24, 1991; gray shades indicated by levels 1–6 at lower right, with level 6 the most intense, etc. Radar echo region around center of Oklahoma is ground clutter and false return (photograph by H. Bluestein).

the relative minimum in reflectivity found in the transition zone. The jet has been attributed to accelerations induced by a horizontal pressure gradient, which owes its existence to a convergent, mesoscale low-pressure area at midlevels. The "meso-low" is a hydrostatic response to the net latent-heat release in convective cells, and is accompanied by a divergent, "meso-high" in the upper troposphere. As the meso-low and meso-high develop, rising motion on the mesoscale develops, and may maintain or initiate more stratiform precipitation. However, below the mesolow, below the melting layer, there is a mesoscale downdraft. Convergence in the stratiform-precipitation area at midlevels under the region of mesoscale ascent, but above the mesoscale downdraft, may act on Earth's vorticity over the lifetime of the squall line; if convergence persists long enough, and air parcels remain in the area long enough, a mesoscale cyclone or trough is produced. This cyclone or trough is sometimes referred to as a *Neddy eddy,* after Ned Johnston, who first identified it in satellite photographs (Fig. 3.84).

Figure 3.84 Spiral pattern in clouds, in visible satellite photograph, associated with a cyclone in the remnants of a mesoscale convective system over eastern Kansas 2000 UTC, May 27, 1980 (courtesy D. Bartels).

Not all the airflow relative to the squall line is from front to rear. Some ice crystals are carried ahead of the squall line to become part of the leading anvil. In addition, there is some relative flow from the rear to the front of the squall line in a narrow corridor extending from its rear edge (Fig. 3.83). This feature is called the *rear-inflow jet.*

Horizontal buoyancy gradients at the rear edge of the MCS (rear-to-front at the rear edge of the warm pool aloft due to latent heat release, and front-to-rear at the rear edge of the low-level cold pool) generate the rear-inflow jet (Fig. 3.83b and c). Horizontal vorticity below the jet acts *opposite* to the vorticity associated with the leading edge of the cold pool, and helps maintain strong rising motion at the leading edge. When the vorticity generated above the jet is less than that generated below the jet owing to a relatively weak warm region (which is *not* a function of CAPE alone), the jet descends, and the horizontal vorticity above the jet acts in the same sense as the vorticity associated with the leading edge of the cold pool, and rising motion at the leading edge weakens. The relative effects of environmental shear, the strength of the cold pool, and the character of the rear-inflow jet may be examined quantitatively using Eq. (2.5.268)

Although we have described many of the two-dimensional, average aspects of squall lines, most are inherently three dimensional. Lines of supercells, of course, are three dimensional; lines of ordinary cells are also three dimensional. Embedded-areal squall lines, however, may be truly two dimensional.

Because multicells go through a well-defined life cycle as they are incorporated into the squall line, the instantaneous cross-sectional flow pattern relative to the squall line may vary as a function of the lifetime of the multicells. Air that at one instant is flowing toward the squall line and is incorporated into it may be expelled back *ahead* of the squall line as a gust front or forced along the squall line a short time later. Squall lines can thus undergo pulsations. Eventually the pool of evaporatively cooled air becomes so deep that the surface-pressure gradient force acting in the direction of squall-line motion becomes strong enough to accelerate cool, outflow air well ahead of the squall line and "cut-off" warm, surface inflow from the squall line. The reader is referred back to Fig. 3.19 in Sec. 3.4.6 for a discussion of this process from the perspective of vorticity dynamics.

Squall lines usually have a lifetime of approximately 6–12 h. There is some evidence that an inertial gravity-wave oscillation in low-level convergence associated with an old, decaying squall line can trigger another squall line ahead of the old squall line about 12 h later. New convection may also be initiated near a Neddy eddy.

Slowly moving squall lines that have extensive stratiform-precipitation regions can result in heavy rainfall and subsequent flooding. Backbuilding squall lines in which cell motion and propagation counteract each other can also result in locally heavy rain and flooding.

Amorphous mesoscale convective systems and mesoscale convective complexes. Convective building blocks are sometimes distributed randomly in mesoscale convective systems. The term *cloud cluster* was coined in the late

1960s and early 1970s to describe the mesoscale cirrus canopies of some MCS's in the tropics, regardless of the radar-echo pattern associated with the precipitation. Similar systems in the midlatitudes are called *mesoscale convective complexes* (MCC's). The MCC is defined in terms of its appearance in satellite photographs, not in terms of its radar-echo pattern. The cirrus canopy of an MCC above a given height must be large and not too eccentric in shape (Table 3.3, Fig. 3.85). Some segments of squall lines fit into the category of MCC's, even though they are associated with bands of radar echoes. The original intent of the definition of MCC by R. Maddox in 1980 was to identify MCS's that have mesoscale regions of ascending motion in the upper half of the troposphere. We further subdivide MCC's into squall lines and "amorphous mesoscale convective systems." Not all squall lines and amorphous mesoscale convective systems, however, are MCC's.

Amorphous mesoscale convective systems are probably dynamically and thermodynamically similar in many respects to squall lines. A mesoscale region of stratiform precipitation, organized mesoscale ascent aloft, a mesoscale downdraft at low levels, a warm-core low aloft, and high near the tropopause (i.e., anticyclone outflow) resulting from latent-heat release are all common characteristics. Many clusters result from mergers of cells, and in effect are the amorphous equivalents to broken-areal squall lines. Some systems may have evolved from broken-line and broken-areal squall lines that have extensive areas of stratiform precipitation. Others may have evolved from backbuilding squall lines.

MCC's in the United States are most common during the spring from the Gulf States into the Plains and Mississippi Valley, and in the summer in the north-central states. They are rare along the East Coast and West Coasts of the United States. Many develop over the Rocky Mountains and produce during the night much of the growing-season rainfall over the corn and wheat belts. From a global perspective, there is a tendency for MCC's to form downstream, with respect to the upper-level flow, of mountainous areas. Regardless of where an MCC forms, however, it is basically a nocturnal phenomenon, having been born late in the late afternoon or evening. Many MCC's are associated with severe weather as they become organized. They

Table 3.3 Mesoscale convective complex definition, physical characteristics

Size	a. Cloud shield with IR temperature $\leq -32°C$, must have an area $\geq$ 100,000 km^2
	b. Interior cold cloud region with temperature $\leq -52°C$, must have an area $\geq$50,000 km^2
Initiate	Size definitions a. and b. are first satisfied
Duration	Size definitions a. and b. must be met for a period $\geq$6 h
Maximum extent	Contiguous cold cloud shield (IR temperature $\leq -32°C$) reaches maximum size
Shape	Eccentricity (ratio of minor axis to major axis) ≥ 0.7 at time of maximum extent
Terminate	Size definitions a. and b. no longer satisfied

Source: From Maddox (1983).

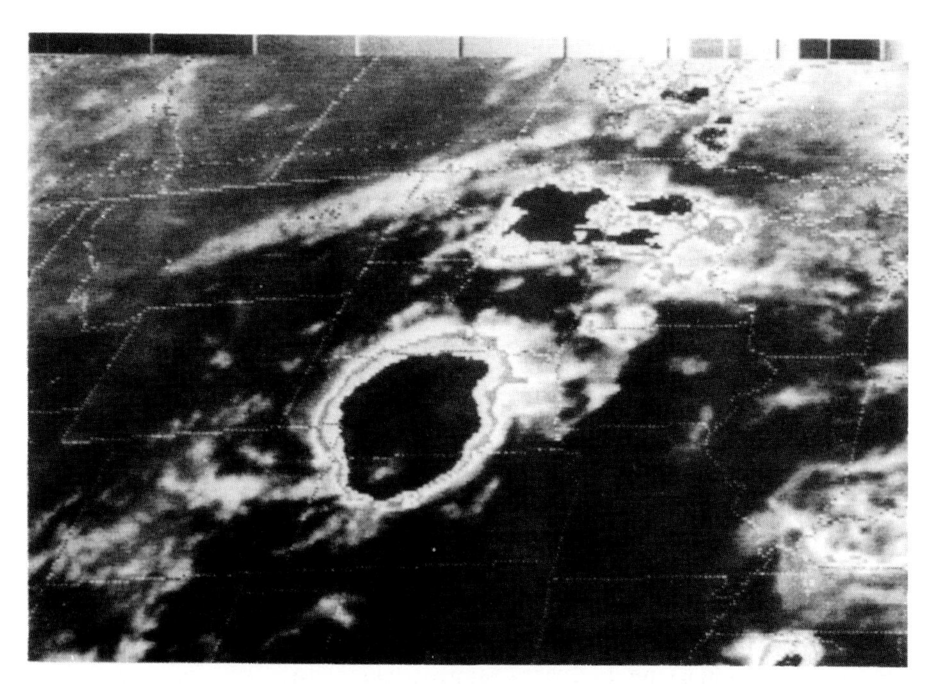

Figure 3.85 Mesoscale convective complex (MCC) over northwest Kansas and central Nebraska as seen in an infrared satellite photograph 0530 UTC (2330 CST), July 15, 1988. The dark areas indicate cold temperatures (i.e., high altitudes); the tiny light area in the southern portion of the cloud mass indicates even colder temperatures and higher altitudes. Note also the tail of lower cloud extending to the southwest of the MCC, which probably extends along an outflow boundary of the complex (courtesy D. Bartels).

often form near east–west-oriented surface fronts under the anticyclonic-shear side of an upper-level jet. Low-level warm advection associated with a low-level jet is often present. There is a warm-core low at midlevels, and strong outflow aloft. The latter can produce strong winds near the tropopause.

The demise of both amorphous MCS's and squall lines is on rare occasions associated with very strong surface winds and a traveling pressure disturbance. The nature of this phenomenon is not understood. However, evidence suggests it is probably a gravity-wave phenomenon.

Other structures

Warm-sector bands. Warm-sector bands are approximately 50-km-wide convective bands located in the "warm" air mass, east and parallel to a surface cold front. They are aligned approximately parallel to the vertical shear associated with the horizontal temperature gradient. The stronger specimens have been compared to "pre-frontal" squall lines. They often appear in series. New clouds are convective, and contain a lot of supercooled water droplets. Older, weaker bands, on the other hand, have a relatively high concentration of ice particles.

Post-frontal bands. Post-frontal bands form in the "cold" air on the western side of a cold front. This environment is characterized by strong subsidence, which is associated quasigeostrophically with cold advection. Post-frontal bands are easily visible on satellite photographs owing to the lack of clouds aloft in the "dry slot."

Open- and closed-cell convection. To the west of post-frontal bands over the ocean, we also often find regions of cumulus convection, which are not necessarily precipitating, however. Convective clouds are arranged as hexagonal cells of roughly 40 km in diameter separated by regions of clear air approximately 20–50 km wide. This cloud pattern is called *open-cell convection* (or *open mesoscale cellular convection*) and refers to a group of hexagonal cells whose centers are free of cloud (Fig. 3.86). *Closed-cell convection* (or *closed mesoscale cellular convection*), on the other hand, refers to a group of hexagonal cells whose centers are cloudy (Fig. 3.86). Open-cell convection patterns form in a boundary layer driven by heating from below (type I). They are mainly found over relatively warm water, downstream from continents during the winter. Closed-cell convection patterns form in a boundary layer driven by cooling (radiative cooling at cloud top) from above (type II). Type II cloud-topped boundary layers (CTBL's) are found mainly over relatively cool water west of continents during the summer. The area covered by open- (and closed-) cell connection is sometimes as great as 10^6 km^2. On satellite photographs these vast areas of cloud are truly spectacular.

Lake-effect convection. Convection occurs frequently over the Great Lakes of the United States during the early winter when cold air in the wake of a cold front flows over relatively warm lake water. The convective clouds produce *lake-effect snowstorms.* When the lake water freezes over, the lake-effect snowstorms do not occur. Necessary conditions for lake-effect snowstorms generally include the following:

1. A nearly dryadiabatic boundary layer at least 1 km deep, capped by an inversion;
2. A fetch (length of area in which waves are generated by the wind, in the direction of the wind) of at least 80 km over the lake water;
3. Geostrophic wind speeds normal to the lake shore in excess of 5 m s^{-1}.

The type of organization of the convective clouds associated with lake-effect snowstorms depends upon the static stability of the upstream air and the wind speed. When the static stability upstream is strong, and the wind speeds are high, long, narrow, parallel bands aligned with wind at the height of the capping inversion are found. These bands do not produce much snowfall. When the static stability is weak and the wind speeds are moderate, bands are found parallel to and on the downwind (lee) side of the lake. These bands are deep and relatively broad; they can produce heavy snowfall. Bands are also found over the center of the lake when the synoptic-scale pressure gradient is so weak that wind speeds are weak, for example, in the center of an

(a)

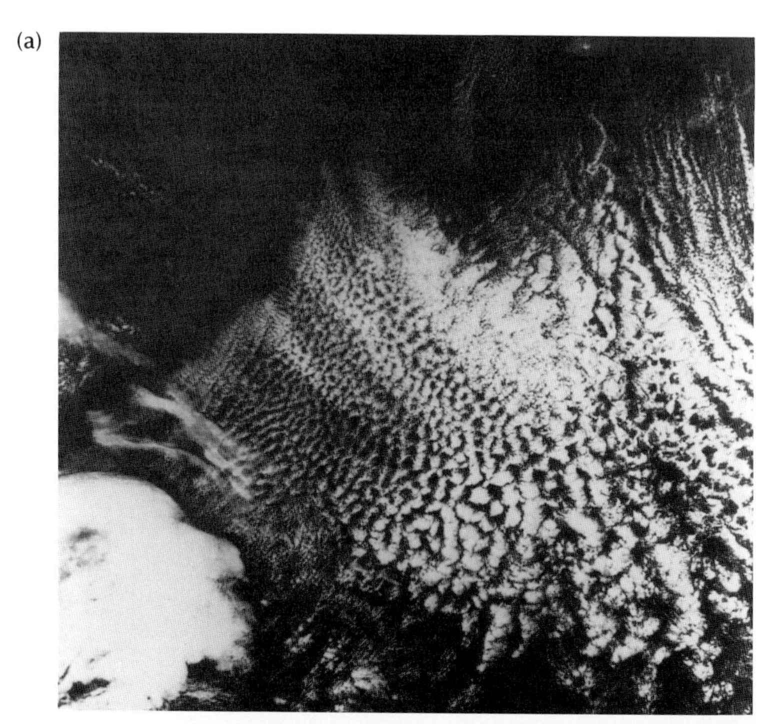

(b)

Figure 3.86 (a) Visible DMSP satellite photograph showing "open-cell" (left) and "closed-cell" (center) convection over the ocean off the Asian continent (over the East China Sea) during a cold-air outbreak at 1154 JST, February 16, 1975 (courtesy E. Agee). (b) Visible satellite photograph showing "closed-cell" (left) and "open-cell" (center) convection over the Pacific Ocean, west of a comma cloud at 2245 UTC, February 12, 1979.

anticyclone. These bands occur when a land breeze from both sides of the lake converges over the center of the lake.

3.5. NONCONVECTIVE SYSTEMS

Precipitation areas are often highly structured even in the absence of convective instability. The organization of the nonconvective precipitation regions is usually also on the mesoscale. The structure of some precipitation areas is related to mesoscale forcing along frontal or other boundaries, or to ascent associated with ducted gravity waves. Other areas of precipitation might be forced by the rising branch of vertical circulations associated with hydrodynamic instabilities such as symmetric instability and conditional symmetric instability.

3.5.1. Characteristics of Nonconvective Systems

Much of what we know about precipitation systems in the absence of convective instability comes from detailed studies by K. Browning and his co-workers in the United Kingdom, P. Hobbs and his co-workers in the Pacific Northwest, a group at M.I.T. in the northeast United States, a group in Illinois, and researchers in Japan. Some research has also been done on upslope snowstorms in Colorado and snowstorms in Oklahoma.

Northwest Pacific and U.K. extratropical surface cyclones are associated with a number of types of precipitation patterns, only some of which may be observed at a given time within a given cyclone (Fig. 3.87). The structure of cyclones off the west coasts of the United States and Europe, however, is usually different from the structure of cyclones over the continental United States. Hence all the precipitation patterns now described cannot necessarily be generalized to *all* cyclones nor are they exhaustive.

Frontal bands. The *narrow cold-frontal band* (Fig. 3.88) (also known as *line convection*) is located along the wind-shift line of the cold front; the band is a boundary-layer phenomenon. Precipitation "cores," ellipsoidal areas of relatively heavy precipitation oriented about 30–35° from the cold front, are sometimes found in the band (Fig. 3.89). The narrow cold-frontal band is only about 5 km wide, and may be the nonconvective counterpart (i.e., actually resulting from *forced* convection rather than *free* convection) to the broken-line squall line. The passage of a narrow cold-frontal band is marked by a jump in pressure and a wind shift. The leading edge of this band thus behaves like a density current. Evaporative cooling and melting of precipitation above the frontal boundary may be the sources of the cold surface air. The jet of rising motion present along fronts, owing to boundary-layer friction, may also be responsible for forcing the narrow cold-frontal band.

Sometimes the narrow cold-frontal band develops a pattern similar to the convective "bow echo"; this pattern is called a *boomerang echo* (Fig. 3.90). Strong horizontal shear is associated with this feature. On rare occasions

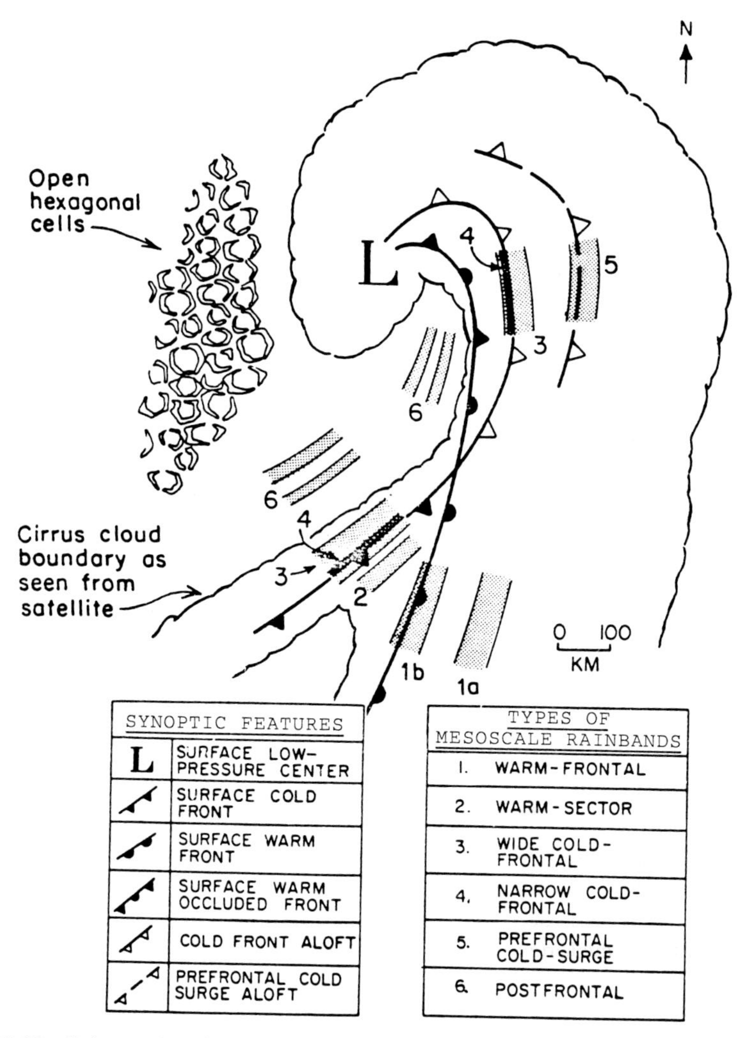

Figure 3.87 Schematic depiction of the types of rainbands (numbers 1–6) observed in extratropical cyclones (from Houze and Hobbs, 1982; originally in Hobbs, 1981).

tornadoes may be associated with it. These are unusual because tornadoes are nearly always associated with convective instability. The breakup of a "solid" precipitation band into cores (with gap regions in between) might be a consequence of the dynamics of a high-density fluid (the "cold" air) overtaking a lower-density fluid. Horizontal shear instabilities might also explain the breakup into cores and gap regions.

Wide cold-frontal bands are about 50 km in width, and are parallel to and along or behind a rearward-sloping surface cold front (Fig. 3.91) within the warm conveyor belt of a cyclone in the middle troposphere.

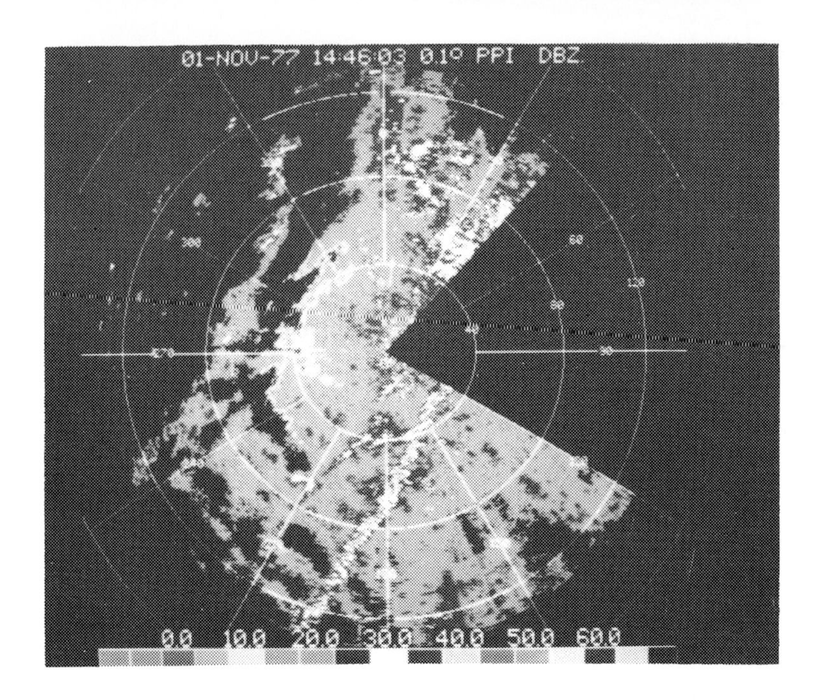

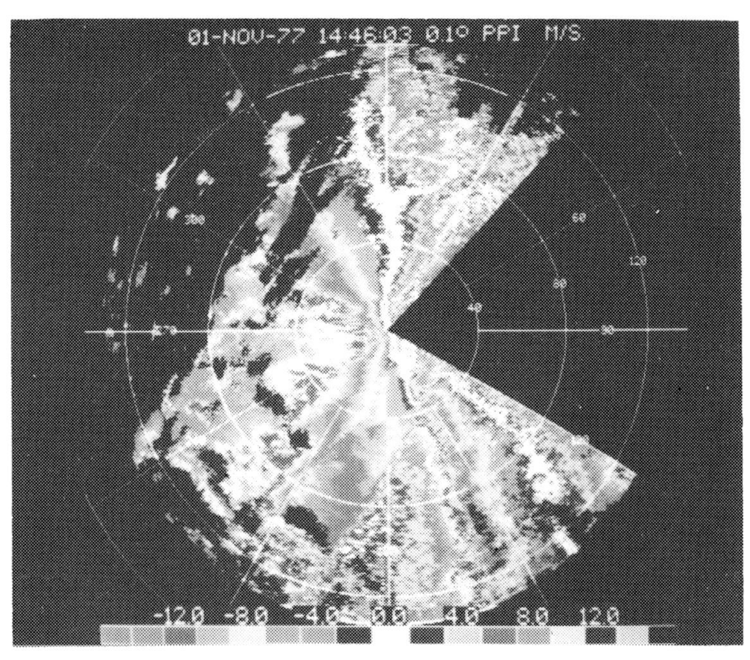

Figure 3.88 Example of an intense, narrow cold-frontal rainband embedded in a broad band of lighter cold-frontal precipitation on November 1, 1977, in the state of Washington. (Top) Plan view of radar reflectivities at 0° elevation angle; shaded scale shown at bottom in dBZ. (Bottom) Single-Doppler velocities at 0° elevation angle; shaded scale shown at bottom in m s^{-1} toward (−) or away (+) from the radar; wind shift evident approximately 40 km to the southeast of the radar (center of the display) marks the position of the surface cold front (from Matejka and Hobbs, 1981; courtesy Peter V. Hobbs, University of Washington).

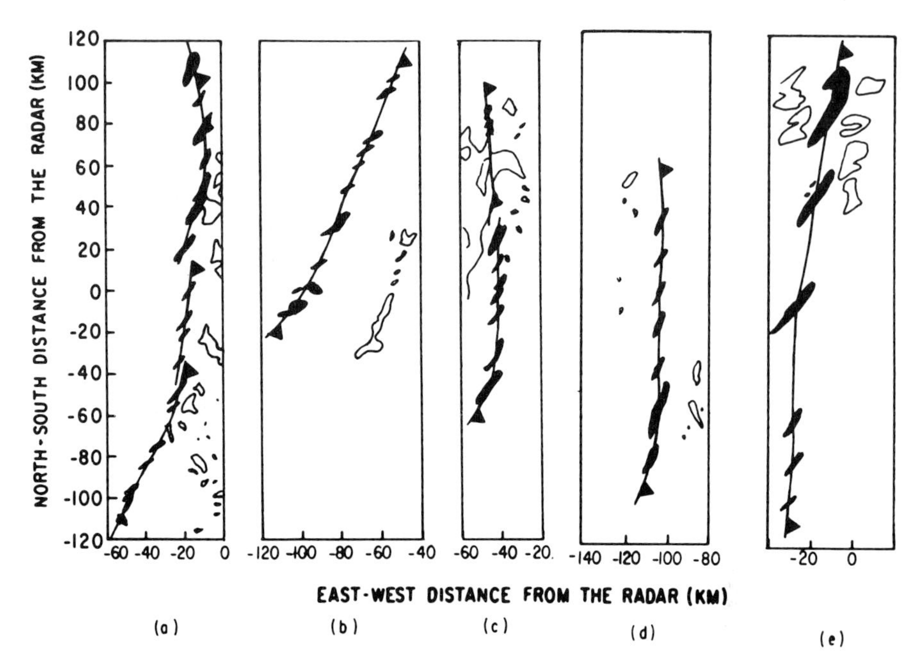

Figure 3.89 Precipitation cores in narrow cold-frontal rainbands. A series of radar reflectivity PPI displays at 0° elevation angle from NCAR's CP-3 radar. In each case the precipitation cores of the narrow cold-frontal rainbands are shaded, with other areas left unshaded. (a) 1733 PST, November 14, 1976, contoured at 23 dBZ level; (b) 0530 PST, November 17, contoured at 39 dBZ level; (c) 0544 PST, November 21, contoured at 29 dBZ level; (d) 0531 PST, December 8, contoured at the 34 dBZ level; (e) 1206 PST, December 17, contoured at the 34 dBZ level (from Parsons and Hobbs, 1983). (Courtesy of the American Meteorological Society)

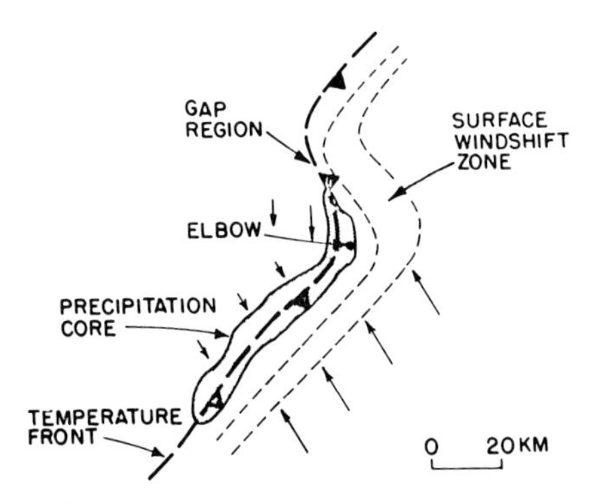

Figure 3.90 Schematic diagram of a "boomerang" echo that Hobbs and Biswas (1979) associated with narrow cold-frontal rainbands (from Houze and Hobbs, 1982).

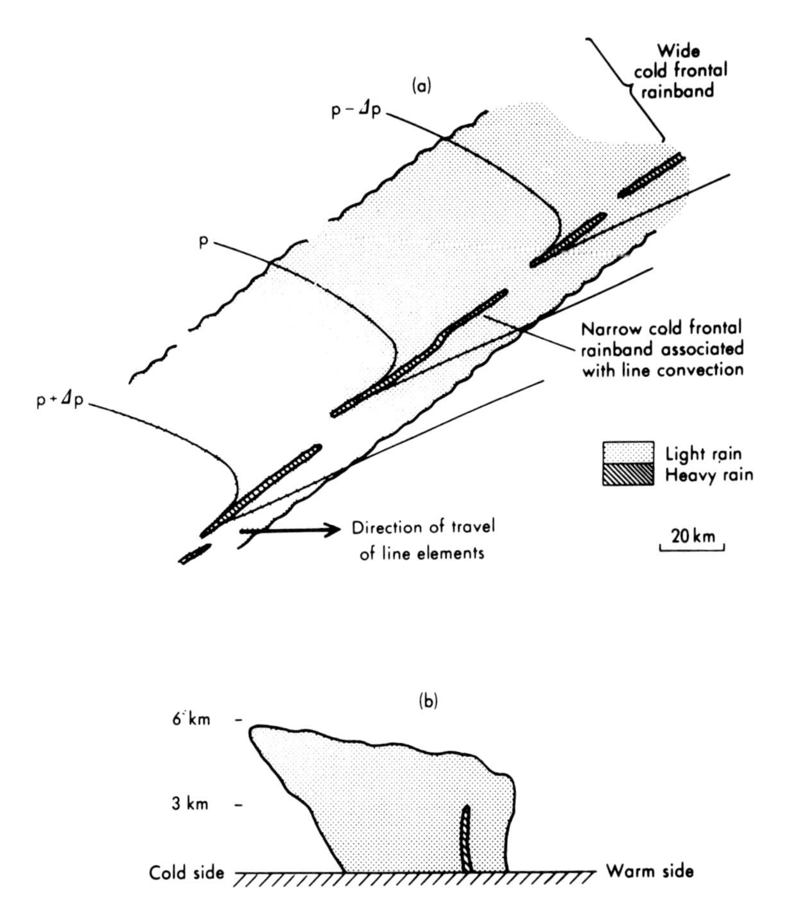

Figure 3.91 Schematic diagram of the wide cold-frontal rainband, and its relationship to the narrow cold-frontal rainband. (a) Plan view and (b) vertical cross section. Intensity of precipitation indicated by shading (from Browning, 1986). (Courtesy of the American Meteorological Society)

Warm-frontal bands are roughly 50 km across, and occur within a region of deep warm advection south or southeast or east of the cyclone within the warm conveyor belt in the middle troposphere. A well-defined warm front, however, may or may not be present. If one is present, the bands tend to be parallel to the warm front, and either along or ahead of it. Warm-frontal bands are embedded within a larger region of stratiform precipitation, like embedded-areal squall lines. The warm-frontal rainbands may be due to the "feeder–seeder" process, ducted gravity waves, symmetric instability, or conditional symmetric instability (all to be discussed later). These bands are similar in structure and perhaps in dynamics to *wide cold-frontal precipitation bands.*

Pre-frontal, cold-surge bands are also similar to wide cold-frontal bands, except that they are associated with weak pulses of low-θ_e (dry, but not necessarily colder) air aloft, east of a forward-sloping surface cold front, that

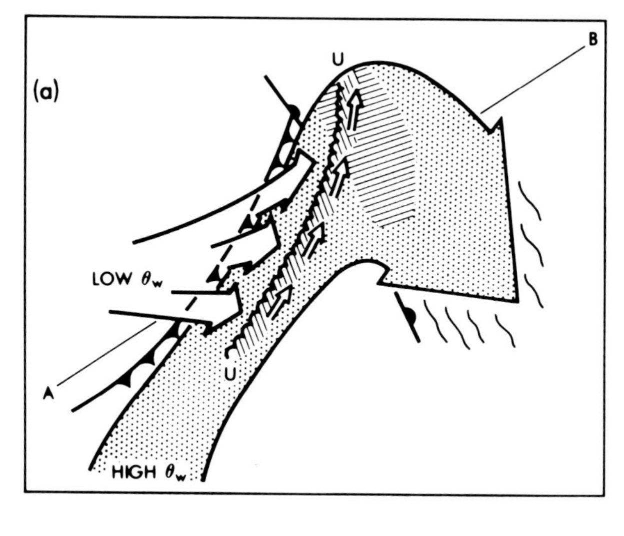

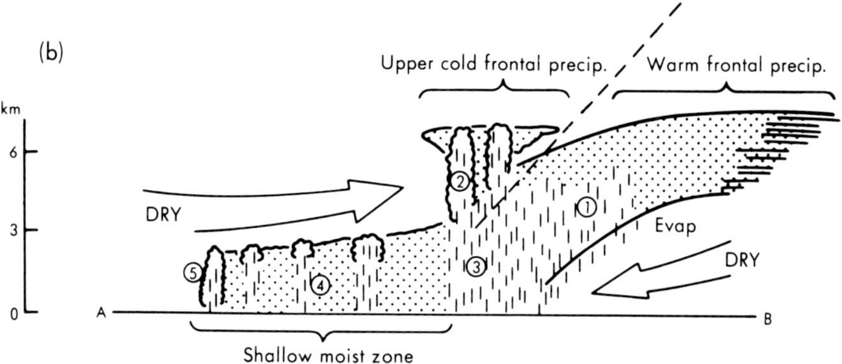

Figure 3.92 Schematic diagrams of a "split-front" or "pre-frontal cold-surge" rainband. (a) Plan view and (b) vertical cross section along the line AB: edge of dry slot aloft (UU); precipitation associated with the "upper-level cold front" and warm front (hatched shading along UU and ahead of the warm front); (1) warm-frontal precipitation, (2) convective-precipitation-generating cells associated with the upper cold front, (3) precipitation from the upper cold-frontal convection descending through an area of warm advection, (4) shallow moist zone between the upper and surface cold fronts characterized by warm advection and scattered outbreaks of mainly light rain and drizzle, and (5) shallow precipitation at the surface cold front itself (from Browning, 1986; after Browning and Monk, 1982). (Courtesy of the American Meteorological Society)

is, ahead of the surface wind shift (Fig. 3.92). The pre-frontal cold-surge band is parallel to or just ahead of the *over-running, upper cold front* in the Browning and Monk *split-front model*; the western edge of the pre-frontal cold-surge band marks the location of the leading edge of the "dry slot" aloft.

Orographic effects. Orography exerts a major influence on nonconvective precipitation systems. For example, the annual precipitation along the western slopes of the Olympic Mountains in Washington state, where moist, marine air flows upslope, is very high; just downstream, on the lee slopes, there is a *rain shadow*. Shallow upslope conditions, however, are not by themselves always sufficient to produce a significant precipitation event; an upper-level trough (i.e., synoptic-scale dynamical lift) is also a necesary synoptic component. Midtroposphere ascent creates "seeder" clouds, which seed the "feeder" clouds below, formed in response to a shallow layer of upslope.

Orography can have other more subtle effects on nonconvective precipitation systems. If cold air is dammed up against a mountain range, and the layer of cold air is very shallow and is not characterized by strong winds (i.e., is stagnant), then the cold-air layer may behave like a density current; consequently a band of heavy precipitation may form along the leading edge of the cold pool, which intersects the ground away from the mountains, where ascent is forced. For example, if during the fall, winter, or spring, shallow, stagnant cold air is pooled against the eastern slopes of the Rockies, which are oriented from north to south, then easterly flow impinging upon the eastern edge of the cold pool may hold it quasi-stationary, and a band of heavy snow may be forced just to the west of the surface boundary of cold air.

On the other hand, if there is strong along-the-mountain flow within the cold-air layer, then the cold-air mass may be behaving like a "barrier jet" as described in Sec. 2.5.6; the cold-air layer is very deep against the mountains, with the top of the cold-air layer dropping off sharply away from the mountains (see Eq. 2.5.308) so that there is no local surface boundary. In this case the isentropic surfaces bulge upward over the cold pool, and force any air that is riding up over the mountains to be lifted before it actually arrives at the mountains; in other words, the effect of the sloping cold-air boundary is to create "upslope" motion away from the mountains! In this way the presence of the mountainous terrain in effect has been broadened. Precipitation gradients in the terrain-normal direction in the case of a barrier jet are weaker than those in the case of a stagnant cold pool because the air is lifted over a broader region, with less vigor. When warm air is advected over the shallow colder air, as often happens in the winter east of the Appalachians, freezing rain may occur and there is an ice storm.

If the boundary layer is not shallow, but is deep and well-mixed and moist adiabatic, then the pattern of precipitation follows the component of true upslope more closely

Orographic precipitation forms "in place," rather than being advected in from elsewhere. This makes forecasting of orographic precipitation a real challenge.

The diurnal variation in precipitation. The diurnal variation in precipitation frequency in the United States during the winter, when a greater fraction of precipitation events is nonconvective, is shown in Fig. 3.93. Precipitation (both convective and nonconvective) tends to be more frequent around sunrise in the central and eastern United States, with the exception of Florida, where afternoon thunderstorms dominate. The early-morning maximum may be related to the diurnal variation in the boundary-layer wind profile. If a

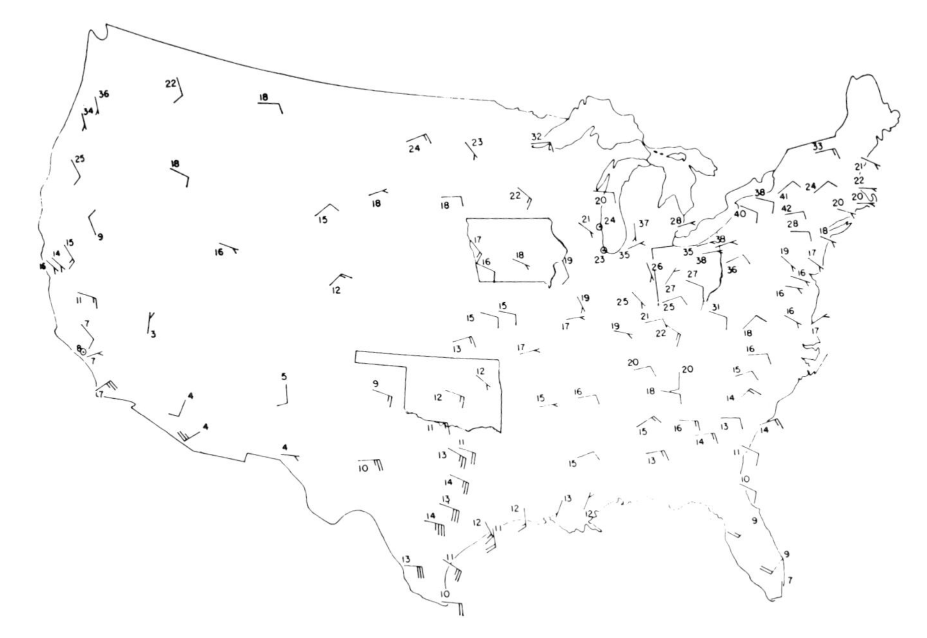

Figure 3.93 Normalized amplitude and phase of the diurnal cycle in the total frequency of precipitation, including trace events, for the winter season (November–March). Normalized amplitude indicated by the configuration of barbs on the tails of the arrows, where each half barb represents 5% and each full barb 10%. Phase indicated by orientation of arrows. Arrow pointing from north indicates a midnight (local time) maximum; one pointing from east indicates an 0600 maximum, etc. Numbers plotted next to station represent 24-h mean frequency in terms of % h with precipitation (from Wallace, 1975). (Courtesy of the American Meteorological Society)

low-level inversion forms at night in the warm-advection region east of a cyclone, then the wind speeds may increase at the top of the boundary layer; thus, warm advection will increase, and synoptic-scale lift should also increase. This hypothesis, however, remains to be tested.

3.5.2. Conditional Symmetric Instability

Conditional symmetric instability (CSI) may be responsible for forcing wide precipitation bands that are aligned along the thermal-wind vector in the saturated layer in which they are generated; we will now show how.

Parcel-method analysis. In Sec. 2.5.2 of Chapter 2 we discussed the dynamics of symmetric instability from the perspective of two-dimensional parcels or "tubes." We will now extend this discussion to include the effects of moisture. Emanuel in 1983 pioneered the use of Lagrangian parcel dynamics in addressing this problem, while Bennetts and Hoskins in 1979 employed a more classical approach. We will as before emphasize Emanuel's approach because it is more amenable to physical interpretation.

In the dry case considered earlier we needed to know the slope of θ surfaces and m_g surfaces. The equation of motion for the lateral component of the tube displacement is Eq. (2.5.239)

$$\frac{Dv}{Dt} = -f(m - m_g), \tag{3.5.1}$$

where $m = u - fy$ and $m_g = u_g - fy$; the equation of motion for the vertical component of tube displacement is Eq. (2.5.240)

$$\frac{Dw}{Dt} = \frac{g}{\theta}(\theta' - \theta), \tag{3.5.2}$$

where θ' is the potential temperature of the tube and θ is the potential temperature of the environment. If the environment is moist, but unsaturated, then θ and θ' are replaced by θ_v and θ'_v; if the environment is saturated, then θ and θ' are replaced by θ_{ev} and θ'_{ev} or θ_{wv} and θ'_{wv}. If the tube is saturated, but the environment is unsaturated, then θ is replaced by θ_v, and θ' is replaced by θ'_{ev} or θ'_{wv}.

We define a surface along which a particular tube is neutrally buoyant as an S surface. If the environment is saturated, then the S surface is one of constant θ_{ev} or θ_{wv}. If the environment is unsaturated, then it is one of constant θ_v only if the tube is unsaturated.

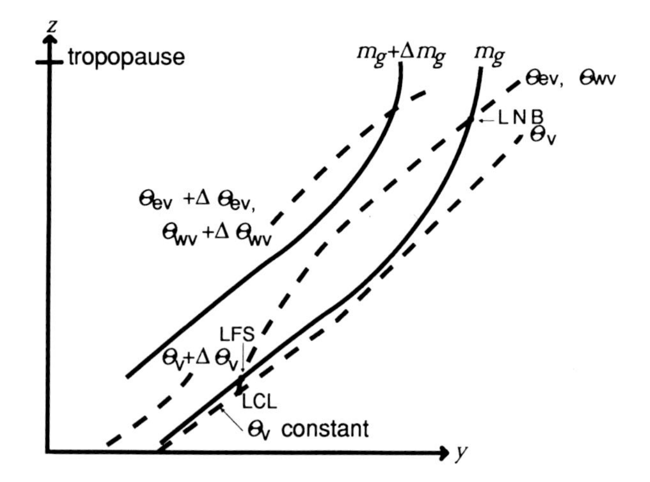

Figure 3.94 Idealized example of a vertical cross section in the Northern Hemisphere, normal to the thermal-wind shear vector, showing surfaces of constant m_g (solid lines), and constant θ_v, θ_{ev}, and θ_{wv}. In this example θ_v, θ_{ev}, and θ_{wv} increase with height (gravitational and conditional stability), m_g decreases with increasing y (inertial stability), and θ_v decreases with y (baroclinic atmosphere, with colder air at larger values of y). Lifting condensation level (LCL); level of free slantwise convection (LFS); level of neutral buoyancy (LNB). Below the LCL the slope of the θ_v surface is less than that of the m_g surface. Note that the slope of θ_{ev} and θ_{wv} surfaces is greater than the slope of θ_v surfaces because the lapse rate of a θ_v surface is greater than that of a θ_{ev} or θ_{wv} surface, and θ_v, θ_{ev}, and θ_{wv} decrease with y.

Refer to Fig. 3.94 for the following discussion. Suppose that an unsaturated tube at the surface is displaced in a slantwise manner with an upward component of motion along its S surface. Assume that the atmosphere is both gravitationally and inertially stable. If the m_g surface that passes through the tube initially has a greater slope than the S surface that passes through the tube, then the atmosphere is symmetrically stable with respect to tube displacements.

When the tube reaches its lifting condensation level (LCL), its S surface becomes a surface of constant θ_{ev} (or θ_{wv}) if the environment is saturated. Along its S surface $Dw/Dt = 0$ [see Eq. (3.5.2)]; however $Dv/Dt < 0$ [see Eq. (3.5.1)] because $m > m_g$ (the parcel's initial m, which is conserved, is m_g; it moves into a region in which the environmental value of m_g is less than the initial m), and hence the tube will be forced back to the left, and then back downward because the parcel moves to the left of the θ_v surface, where $\theta_v' < \theta_v$ [see Eq. (3.5.2)].

However, the tube may eventually be lifted along its S surface to a level at which not only is $Dw/Dt = 0$, but also $Dv/Dt = 0$, if $m = m_g$, that is, if it crosses the original m_g surface again. If above this level the S surface has a greater slope than the m_g surface, then this level is called the *level of free slantwise convection* (LFS). Above the LFS, any further slantwise displacement along the S surface will lead to $Dv/Dt > 0$, since $m < m_g$, and $Dw/Dt > 0$, since the parcel moves to the right of the θ_{ev} surface, where $\theta_{ev}' > \theta_{ev}$ or θ_v; hence the tube will be accelerated toward the right and upward.

Near the tropopause, however, the vertical shear reverses sign (with height), and hence the vertical shear is zero. The m_g surfaces therefore become more steeply sloped [see Eq. (2.5.235)]. In addition, S surfaces become less steeply sloped owing to the strong static stability at the tropopause. Therefore m_g surfaces become more steeply inclined than S surfaces, and hence the atmosphere is symmetrically stable near the tropopause.

The level above the LFS at which $Dv/Dt = 0$ owing to $m = m_g$ (Dw/Dt is still zero along the S surface by definition) is called the *level of neutral buoyancy* (LNB) for slantwise convection (analogous to the equilibrium level for upright convection). Above the LNB the atmosphere is symmetrically stable; below the LNB, but above the LFS, it is symmetrically unstable. Note that, if the tube were lifted from the surface and never became saturated, then displacements along its S surface would be stable. Because the tube becomes saturated, its S surface becomes a θ_{ev} or θ_{wv} surface, which is steeper than a θ_v surface, and hence there may be symmetric instability. If the atmosphere is symmetrically stable with respect to dry, slantwise displacements, but symmetrically unstable with respect to saturated, slantwise displacements, then we say that the atmosphere has *conditional symmetric instability* (CSI). A necessary condition for CSI is that θ_{ev} or θ_{wv} surfaces have greater slopes than m_g surfaces.

How much kinetic energy is required to lift a tube to its LFS? The restoring force $\mathbf{F}$ on the tube is

$$\mathbf{F} = -f(m - m_g)\hat{\mathbf{j}} + \frac{g}{\theta_v}(\theta_v' - \theta_v)\hat{\mathbf{k}}. \tag{3.5.3}$$

The curl of the restoring force is therefore given by

$$\nabla \times \mathbf{F} = \hat{\mathbf{i}}\left(\frac{\partial}{\partial y}\frac{g}{\theta_v}(\theta'_v - \theta_v) + \frac{\partial}{\partial z} f(m - m_g)\right)$$
$$- \hat{\mathbf{j}}\left(\frac{\partial}{\partial x}\frac{g}{\theta_v}(\theta'_v - \theta_v)\right) - \hat{\mathbf{k}}\left(\frac{\partial}{\partial x} f(m - m_g)\right). \qquad (3.5.4)$$

Since $\partial/\partial x = 0$, $\theta'_v = \text{constant}$, and $(\partial/\partial z) f(m - m_g) = -f(\partial u_g/\partial z) = (g/\theta_v)(\partial\theta_v/\partial y)$, it follows that $\nabla \times \mathbf{F} = 0$. (Above the LCL, θ_v is replaced by θ_{ev}.) The restoring-force vector is therefore conservative, and hence a line integral of $\mathbf{F}$ is path independent. The amount of kinetic energy per unit mass (KE) required, the slantwise convective inhibition (SCIN), can be computed as the negative of the work done by the environment on the tube as it is lifted to its LFS on a surface of constant S, where $(g/\theta_v)(\theta'_v - \theta_v) = 0$:

$$\text{SCIN} = \int_{y(z=0)}^{y(z=\text{LFS})} f(m - m_g)(dy)_S. \qquad (3.5.5)$$

Above the LFS the amount of potential energy in the environment that is converted into parcel kinetic energy is called the *slantwise convective available potential energy* (SCAPE). The SCAPE contained between the LFS and the LNB, which is also called the *slantwise positive area* (SPA), is given by

$$\text{SCAPE} = -\int_{y(z=\text{LFS})}^{y(z=\text{LNB})} f(m - m_g)(dy)_S. \qquad (3.5.6)$$

Since the integral of the restoring force is path independent, however, we may express Eq. (3.5.6) in general as

$$\text{SCAPE} = -\int_{y(z=\text{LFS})}^{y(z=\text{LNB})} f(m - m_g)\, dy$$
$$+ \int_{z=\text{LFS}}^{\text{LNB}} \frac{g}{\theta_{ev}}(\theta'_{ev} - \theta_{ev})\, dz. \qquad (3.5.7)$$

Along an m_g surface,

$$\text{SCAPE} = \int_{z=\text{LFS}}^{\text{LNB}} \frac{g}{\theta_{ev}}(\theta'_{ev} - \theta_{ev})(dz)_{m_g}. \qquad (3.5.8)$$

This is a convenient form because it is similar to the expression for CAPE used in the parcel theory of convection. Since m_g surfaces are parallel to the geostrophic coordinate Y, SCAPE is equivalent to the CAPE in geostrophic coordinates. This finding is particularly interesting, because it ties CSI to semigeostrophic theory. If the atmosphere is barotropic, m_g surfaces are vertical and SCAPE = CAPE.

Consider the differential of SCAPE (in general):

$$d(\text{SCAPE}) = -f(m - m_g)\, dy + \frac{g}{\theta_{ev}}(\theta'_{ev} - \theta_{ev})\, dz. \qquad (3.5.9)$$

For a fixed value of dz (dz = constant),

$$d(\text{SCAPE}) = \left(-f(m - m_g)\frac{dy}{dz} + \frac{g}{\theta_{ev}}(\theta'_{ev} - \theta_{ev})\right) dz \tag{3.5.10}$$

$$= \left[-f\frac{\partial m_g}{\partial y}\left(\frac{dy}{dz}\right) dy + \frac{g}{\theta_{ev}}\frac{\partial \theta_{ev}}{\partial y} dy\right] dz, \tag{3.5.11}$$

where

$$m = m_g + \frac{\partial m_g}{\partial y} dy \tag{3.5.12}$$

$$\theta'_{ev} = \theta_{ev} + \frac{\partial \theta_{ev}}{\partial y} dy. \tag{3.5.13}$$

SCAPE is maximized for a fixed vertical displacement (dz) only if

$$\frac{\partial\, \text{SCAPE}}{\partial y} = -f\frac{\partial m_g}{\partial y}\left(\frac{dy}{dz}\right) + \frac{g}{\theta_{ev}}\frac{\partial \theta_{ev}}{\partial y} = 0. \tag{3.5.14}$$

Let us consider the following form of the thermal-wind relation:

$$\frac{g}{\theta_{ev}}\frac{\partial \theta_{ev}}{\partial y} = -f\frac{\partial u_g}{\partial z} = -f\frac{\partial m_g}{\partial z}. \tag{3.5.15}$$

It follows from Eqs. (3.5.14) and (3.5.15) that

$$\left(\frac{dz}{dy}\right) = -\frac{\partial m_g}{\partial y}\Big/\frac{\partial m_g}{\partial z} = \left(\frac{dz}{dy}\right)_{m_g}; \tag{3.5.16}$$

that is, the slope of the displacement is equivalent to the slope of an m_g surface. SCAPE furthermore is maximized [see Eq. (3.5.10)] for a displacement having the slope dz/dy only if

$$\frac{\partial\, \text{SCAPE}}{\partial(dy/dz)} = -f\frac{\partial m_g}{\partial y} dy = -f(m - m_g) = 0, \tag{3.5.17}$$

that is, if $m = m_g$. Therefore *the SCAPE measured along a constant m_g surface is the greatest that can be realized for a given vertical displacement.* Although the SCAPE computed between the LFS and LNB does not depend upon the path of integration, the SCAPE computed up to some particular height below the LNB (but above the LFS) does depend upon the path of integration, because y is not specified even though z is given; the SCAPE is greatest when computed along the m_g surface.

Since SCAPE can be computed by integrating the tube buoyancy $[(g/\theta_{ev})(\theta'_{ev} - \theta_{ev})]$ along an m_g surface [Eq. (3.5.8)], we can ascertain the susceptibility of the atmosphere to slantwise convection simply by considering the susceptibility of the atmosphere to "ordinary" convection *along an m_g surface.* It is necessary to analyze a vertical cross section of m_g, temperature, dew point, θ_v and θ_{ev} in a plane normal to the thermal-wind shear (Fig. 3.95). If θ_{ev} decreases with height along an m_g surface or changes very little (and $\partial\theta_{ev}/\partial z > 0$), then conditional symmetric instability is possible. The sounding

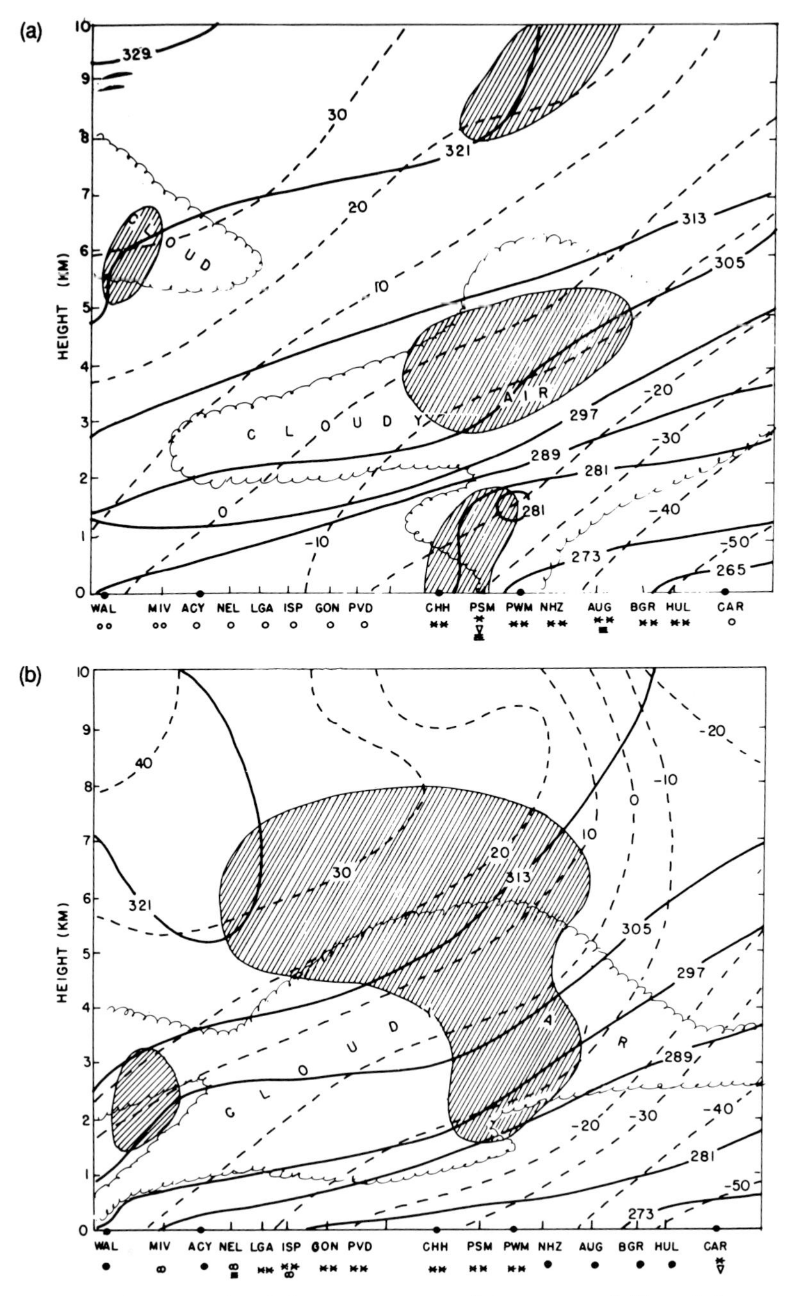

Figure 3.95 An example of the assessment of the susceptibility of the atmosphere to conditional symmetric instability on December 11, 1982, in New England. Vertical cross sections of θ_e in K (solid lines) and m_g in m s^{-1} (dashed lines) in a plane approximately normal to the thermal-wind shear vector at (a) 0000 UTC and (b) 1200 UTC. Clouds are indicated with scalloped lines and regions of possible conditional symmetric instability are shaded. Upper air stations are indicated on the abscissa with large dots. Weather observed at surface stations is indicated below the axis ($* *$ = light snow, $\circ\circ$ = light rain, ∇ = showers, $\equiv$ = fog, ● = skies overcast, ○ = skies broken and ∞ = haze) (from Wolfsberg et al., 1986). (Courtesy of the American Meteorological Society)

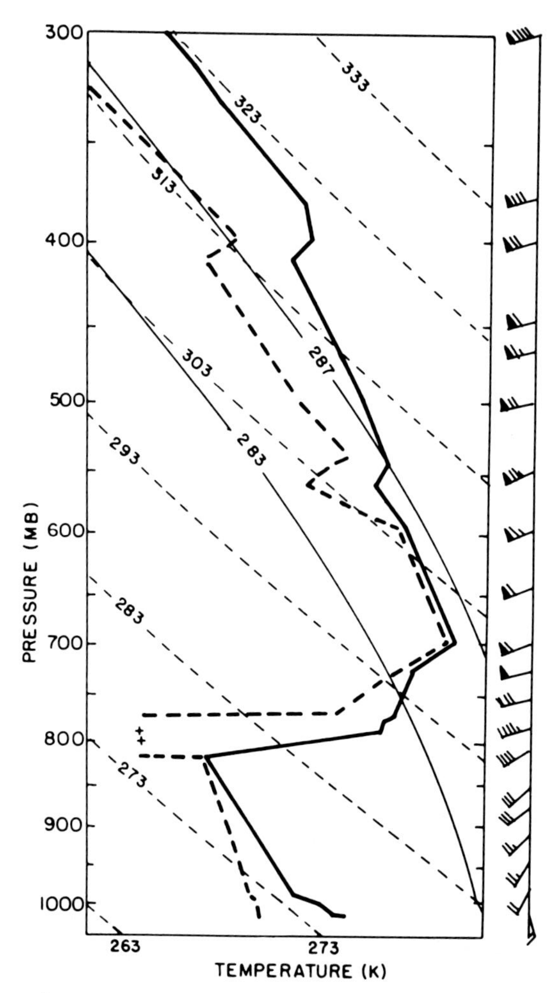

Figure 3.96 Sounding at Chatham, Massachusetts (CHH) at 0000 UTC, December 11, 1982. Temperature (heavy solid line) and dew point (heavy dashed line); reference moist-adiabats (light solid lines) and reference dryadiabats (light dashed lines). Winds plotted along the side; flag, whole wind barbs, half wind barbs represent 25, 5, and 2.5 m s^{-1}, respectively (from Wolfsberg et al., 1986). (Courtesy of the American Meteorological Society)

at Chatham, Mass. (CHH) shown in Fig. 3.96 indicates that the lapse rate in the midtroposphere is nearly moist-adiabatic, or even less than moist-adiabatic. Thus, the sounding is essentially conditionally stable, or neutrally stable at best, with respect to saturated air. However, near 3–3.5 km AGL θ_e decreases with height along the $m_g = -10$ m s^{-1} surface (Fig. 3.95(a)). The sounding plotted along the $m_g = -10$ m s^{-1} surface is shown in Fig. 3.97. A tube lifted from 630 mb would reach its LFS near 580 mb and be unstable above, realizing a buoyancy of about 3°C at 500 mb. A very long horizontal excursion, however, is necessary to achieve this slantwise buoyancy. An air

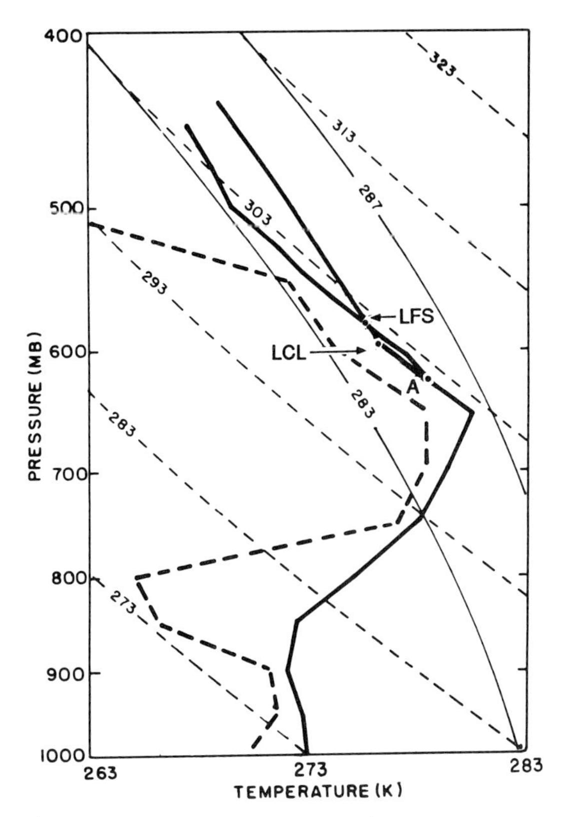

Figure 3.97 Sounding along the $m_g = -10$ m s^{-1} surface at 0000 UTC, December 11, 1982, for the cross section shown in Fig. 3.95(a). Level of free slantwise convection (LFS) and lifting condensation level (LCL) shown for parcel at A. Positive area can be interpreted as potential energy for displacements along the m_g surface (after Wolfsberg et al., 1986). (Courtesy of the American Meteorological Society)

parcel ascending in a slantwise manner may thus find itself in a colder environment to the north, than that directly overhead.

It is not always easy, especially in real-time forecasting situations, to construct very quickly a cross-sectional analysis of m_g, θ_v and θ_{ev}. Therefore, a method for calculating SCAPE using only a single sounding would be useful. Suppose that we make the following approximation (remember, $\partial m_g/\partial y < 0$ if the flow is inertially stable):

$$\frac{\Delta m_g}{L} \approx -\frac{\partial m_g}{\partial y}, \tag{3.5.18}$$

where L is a horizontal distance measured along the y axis (from $y = 0$ to the m_g surface). It follows that

$$L \sim -\Delta m_g \Big/ \frac{\partial m_g}{\partial y} = \Delta m_g \Big/ \left(f - \frac{\partial u_g}{\partial y}\right) \tag{3.5.19}$$

if the geostrophic absolute vorticity is nearly constant over the distance L. If a tube is displaced vertically to a higher m_g surface, at $y=0$,

$$\Delta m_g = \Delta(u_g - fy) = u_g - u_{g0}, \tag{3.5.20}$$

where u_{g0} is the initial geostrophic velocity of the tube. Following Eq. (3.5.18), we can make the following approximation:

$$\frac{\Delta\theta_{ev}}{L} \approx \frac{\partial\theta_{ev}}{\partial y} \tag{3.5.21}$$

if the temperature gradient is nearly constant over the distance L. Using the thermal-wind relation we see that

$$\frac{\Delta\theta_{ev}}{L} \approx -\frac{f\theta_{ev}}{g}\frac{\partial u_g}{\partial z}. \tag{3.5.22}$$

From Eq. (3.5.21) we see that

$$[\theta_{ev}(z)]_{m_g} \approx (\theta_{ev})_{y=0} + \left(\frac{\partial\theta_{ev}}{\partial y}\right)_z L. \tag{3.5.23}$$

Using Eqs. (3.5.23) and (3.5.22), we can derive the following relation:

$$\theta'_{ev} - (\theta_{ev})_{m_g} = \theta'_{ev} - (\theta_{ev})_{y=0} - \frac{f\theta_{ev}}{g}\frac{\partial u_g}{\partial z} L. \tag{3.5.24}$$

Substituting Eq. (3.5.24) into Eq. (3.5.8), we find that

$$\text{SCAPE} = \int_{z=\text{LFS}}^{\text{LNB}} \frac{g}{\theta_{ev}}(\theta'_{ev} - \theta_{ev})\,dz + \int_{z=\text{LFS}}^{\text{LNB}} f\frac{\partial u_g}{\partial z} L\,dz \tag{3.5.25}$$

at $y=0$. But from (3.5.19) and (3.5.20) we see that

$$\begin{aligned} f\frac{\partial u_g}{\partial z} L &= f\frac{\partial u_g}{\partial z}\Delta m_g \Big/ \left(f - \frac{\partial u_g}{\partial y}\right) \\ &= f\frac{\partial u_g}{\partial z}(u_g - u_{g0}) \Big/ \left(f - \frac{\partial u_g}{\partial y}\right) \\ &= \tfrac{1}{2}\left[f \Big/ \left(f - \frac{\partial u_g}{\partial y}\right)\right]\frac{\partial}{\partial z}(u_g - u_{g0})^2. \end{aligned} \tag{3.5.26}$$

Substituting Eq. (3.5.26) into Eq. (3.5.25), we get the following equation for SCAPE:

$$\begin{aligned} \text{SCAPE} = \int_{z=\text{LFS}}^{\text{LNB}} \frac{g}{\theta_{ev}} \Bigg\{ & \left[\theta'_{ev} + \frac{\theta_{ev}}{g}\frac{1}{2}\left[f \Big/ \left(f - \frac{\partial u_g}{\partial y}\right)\right]\frac{\partial}{\partial z}(u_g - u_{g0})^2\right] \\ & - \theta_{ev}\Bigg\}\,dz. \end{aligned} \tag{3.5.27}$$

This is a particularly revealing form of SCAPE, since it is identical to the CAPE for an air parcel, plus the addition of the following term in the integral:

$$\frac{1}{2}\left[f\Big/\left(f-\frac{\partial u_g}{\partial y}\right)\right]\frac{\partial}{\partial z}(u_g-u_{g0})^2.$$

The additional term, which accounts for inertial effects and is proportional to the vertical shear, is substantial for low values of absolute vorticity and high values of vertical shear. In order to use the single-sounding method, the geostrophic shear *across* the thermal wind vector must be computed from constant level maps or from cross-sectional analyses. For typical synoptic-scale values found along a surface frontal zone

$$\frac{\theta_{ev}}{g}\frac{1}{2}\left[f\Big/\left(f-\frac{\partial u_g}{\partial y}\right)\right]\frac{\partial}{\partial z}(u_g-u_{g0})^2$$

$$\sim\frac{300\text{ K}}{10\text{ m s}^{-2}}\frac{10^{-4}\text{ s}^{-1}}{10^{-4}\text{ s}^{-1}}\frac{(50\text{ m s}^{-1})^2}{10^4\text{ m}}\sim 1\text{–}10\text{ K}. \qquad (3.5.28)$$

Therefore slantwise displacements may acquire enough kinetic energy to overcome negative CAPE (i.e., CIN) equivalent to parcel temperature deficits on the order of degrees.

How is SCAPE generated and used up? Since $Dm/Dt=0$ and $D\theta_{ev}/Dt=0$ in frictionless, adiabatic flow, m and θ_{ev} surfaces are simply advected along by

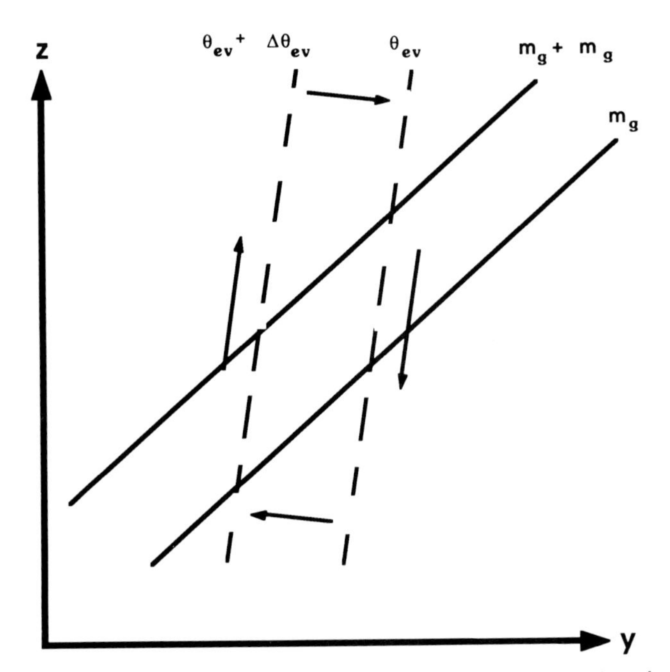

Figure 3.98. Demonstration that the vertical circulation associated with fronts cannot change the distribution of θ_{ev} along m_g surfaces. Idealized vertical cross section of m_g surfaces (solid lines) and virtual equivalent potential temperature (dashed lines); arrows indicate sense of frontal vertical circulation.

the flow. The SCAPE changes only if the distance between θ_{ev} surfaces (along m_g surfaces) changes. The vertical circulation associated with fronts cannot change the *distribution* of θ_{ev} along m_g surfaces, even though it tips the m_g and θ_{ev} surfaces over even further; hence SCAPE is unaltered (Fig. 3.98). Horizontal gradients in diabatic heating can change SCAPE if, for example, the existing horizontal temperature gradient is increased (Fig. 3.99). The SCAPE is decreased by turbulent momentum transports only if the transports act to increase the slope of m_g surfaces relative to the slope of θ_{ev} surfaces: Vertical mixing, which in general acts to make both m and θ_{ev} surfaces vertical ($\partial u_g/\partial z$ and $\partial \theta_{ev}/\partial z$ diminish) acts to decrease SCAPE. Horizontal mixing, which in general acts to make m_g surfaces vertical and θ_{ev} surfaces horizontal [$\partial \theta_{ev}/\partial y$ diminishes, and hence according to the thermal-wind relation, Eq. (3.5.15), $\partial m_g/\partial z$ diminishes], also acts to decrease SCAPE (in the final state, θ_{ev} increases along m_g surfaces).

Observations indicate that conditional symmetric instability may be responsible for some mesoscale wide precipitation bands that are oriented along the vertical-shear vector poleward of warm fronts and stationary fronts, and along cold fronts. It is not believed that precipitation loading can destroy the vertical

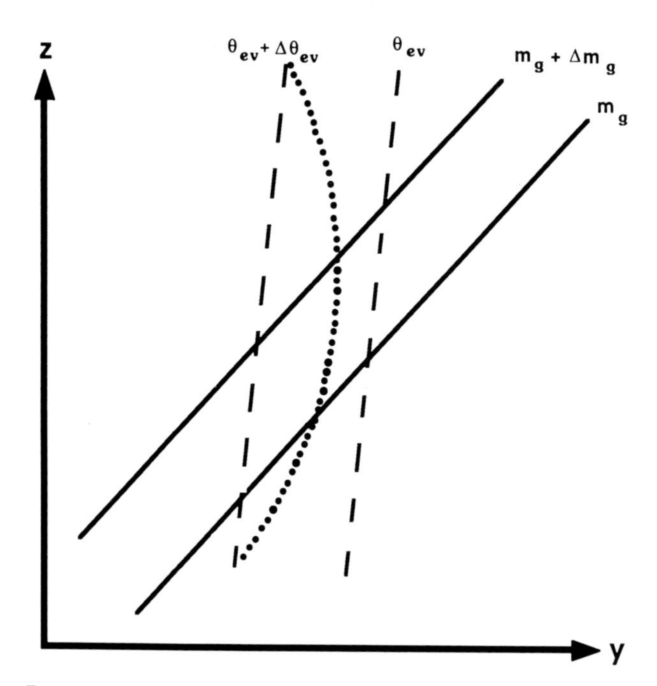

Figure 3.99 Demonstration that diabatic heating can change the slantwise convective available potential energy (SCAPE). Idealized vertical cross section of m_g (solid lines) and θ_{ev} (dashed lines). The dotted line represents the $\theta_{ev} + \Delta\theta_{ev}$ contour after there has been diabatic heating in the middle troposphere. In this example, diabatic heating in the middle troposphere increases the lapse rate of θ_{ev} along the m_g surfaces in the middle troposphere.

circulation associated with the bands because the circulation is sloped so much that precipitation falls away from the region it is being generated.

It is instructive to compute an actual trajectory of a tube as it is displaced along an S surface in a symmetrically unstable (or conditionally symmetrically unstable) environment, and to see what magnitude of vertical velocity can be attained. Consider the idealized S and m_g surfaces shown in Fig. 3.100. The m_g surfaces below $z = H$ are planes that slope toward the right according to

$$m_g(y,z) = \bar{u}_{gz}z - (f - \bar{u}_{gy})y, \tag{3.5.29}$$

where

$$\bar{u}_{gz} = \frac{\partial u_g}{\partial z} = \text{constant} > 0. \tag{3.5.30}$$

$$\bar{u}_{gy} = \frac{\partial u_g}{\partial y} = \text{constant} < 0. \tag{3.5.31}$$

It follows that

$$\frac{-\partial m_g}{\partial y} = f - \bar{u}_{gy} \tag{3.5.32}$$

$$\frac{\partial m_g}{\partial z} = \bar{u}_{gz} = -\frac{g}{f\theta_{ev}}\frac{\partial \theta_{ev}}{\partial y} \tag{3.5.33}$$

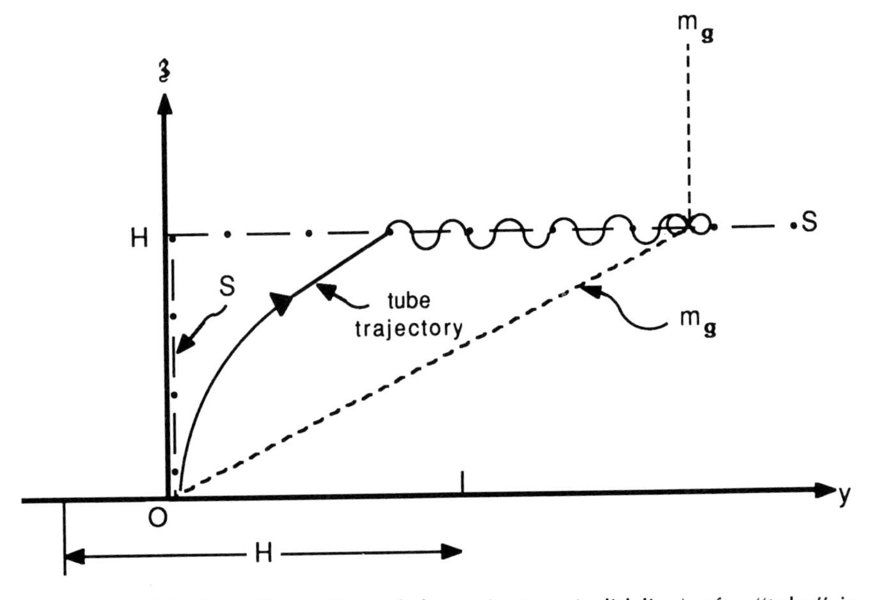

Figure 3.100 Idealized illustration of the trajectory (solid line) of a "tube" in a symmetrically unstable atmosphere. Vertical cross section; S surface (dash-dotted line); m_g surface (dashed line). The tube is given a push initially upward and to the right; we neglect, for simplicity, friction. It accelerates upward and to the right until it reaches the S surface again at $z = H$; it continues to accelerate to the right, but oscillates about the S surface at $z = H$, until it reaches the m_g surface again. It now oscillates about the point defined by the intersection of the S surface and the m_g surface.

according to the thermal-wind relation. Above $z = H$ the atmosphere is barotropic; therefore m_g surfaces are oriented vertically. The S surfaces are oriented vertically below $z = H$, and horizontally above $z = H$. In other words, the atmosphere is symmetrically unstable below $z = H$, and stable above $z = H$. The virtual equivalent potential temperature decreases linearly with y as a result of Eqs. (3.5.30) and (3.5.33), so that

$$\theta_{ev} = \theta_{ev_0} + \frac{\partial \theta_{ev}}{\partial y} y = \theta_{ev_0} - f\frac{\theta_{ev}}{g}\bar{u}_{gz}\, y \qquad (3.5.34)$$

where θ_{ev_0} is the θ_{ev} at the tube's initial location ($y = 0$, $z = 0$). If the tube is initially at rest, then m is zero initially at $y = 0$, $z = 0$. According to Eq. (3.5.29), m_g is also zero at the tube's location initially, so that $u_g = 0$ at $y = 0$, $z = 0$ also.

Substituting Eq. (3.5.29) into the equation of motion in the cross-shear direction, Eq. (3.5.1), given that $m = 0$ initially, we find that at $t = 0$

$$\frac{Dv}{Dt} = -f[-\bar{u}_{gz}z + (f - \bar{u}_{gy})y]. \qquad (3.5.35)$$

Substituting Eq. (3.5.34) into Eq. (3.5.2), the vertical equation of motion, with θ (θ') replaced by θ_{ev} (θ'_{ev}), we find that since the tube's value of θ_{ev}, θ'_{ev}, is θ_{ev_0},

$$\frac{Dw}{Dt} = \frac{g}{\theta_{ev}}(\theta_{ev_0} - \theta_{ev}) = f\bar{u}_{gz}\, y. \qquad (3.5.36)$$

The equations of motion in terms of position (y,z) are therefore as follows:

$$\frac{D^2y}{Dt^2} = f[\bar{u}_{gz}z - (f - \bar{u}_{gy})y] \qquad (3.5.37)$$

$$\frac{D^2z}{Dt^2} = f\bar{u}_{gz}\, y. \qquad (3.5.38)$$

Differentiating Eq. (3.5.37) twice with respect to y and substituting for D^2z/Dt^2 using Eq. (3.5.38), we get the following fourth-order equation for y, the cross-shear component of the tube's displacement from its initial position at $y = 0$, $z = 0$:

$$\frac{D^4y}{Dt^4} + f(f - \bar{u}_{gy})\frac{D^2y}{Dt^2} - f^2\bar{u}_{gz}^2 y = 0. \qquad (3.5.39)$$

Solutions of the form

$$y \sim e^{\sigma t} \qquad (3.5.40)$$

exist, so that upon substitution of Eqs. (3.5.40) into Eq. (3.5.39) we find that

$$\sigma^4 + f(f - \bar{u}_{gy})\sigma^2 - f^2\bar{u}_{gz}^2 = 0. \qquad (3.5.41)$$

Solving Eq. (3.5.41) for σ, we find that

$$\sigma^2 = -\frac{f}{2}\{(f - \bar{u}_{gy}) \mp [(f - \bar{u}_{gy})^2 + 4\bar{u}_{gz}^2]^{1/2}\}. \qquad (3.5.42)$$

In the vicinity of a typical, midlatitude frontal zone,

$$\bar{u}_{gz} \sim \frac{10 \text{ m s}^{-1}}{10^3 \text{ m}} = 10^{-2} \text{ s}^{-1}, \tag{3.5.43}$$

while

$$f - \bar{u}_{gy} \sim 10^{-4} \text{ s}^{-1}. \tag{3.5.44}$$

Therefore

$$\bar{u}_{gz} \gg f - \bar{u}_{gy}, \tag{3.5.45}$$

so that to a good approximation

$$\sigma^2 \approx +f\bar{u}_{gz} \tag{3.5.46}$$

for unstable motions.

The solution to Eq. (3.5.39), given that the initial displacement velocity components are

$$v(t=0) = v_0 \tag{3.5.47}$$

$$w(t=0) = w_0 \tag{3.5.48}$$

and Eq. (3.5.46), is as follows:

$$y = \frac{v_0 + w_0}{2\sigma} \sinh \sigma t + \frac{v_0 - w_0}{2\sigma} \sin \sigma t. \tag{3.5.49}$$

From Eqs. (3.5.49) and (3.5.38) we find that the equation for z is only second order:

$$\frac{D^2 z}{Dt^2} = f\bar{u}_{gz}\left(\frac{v_0 + w_0}{2\sigma} \sinh \sigma t + \frac{v_0 - w_0}{2\sigma} \sin \sigma t\right). \qquad ((3.5.50)$$

The solution to Eq. (3.5.50), given Eqs. (3.5.47) and (3.5.48), is

$$z = \frac{w_0 + v_0}{2\sigma} \sinh \sigma t + \frac{w_0 - v_0}{2\sigma} \sin \sigma t. \tag{3.5.51}$$

For

$$t \gg \frac{1}{\sigma} = \frac{1}{(f\bar{u}_{gz})^{1/2}} \sim \frac{1}{[(10^{-4} \text{ s}^{-1})(10^{-2} \text{ s}^{-1})]^{1/2}} = 10^3 \text{ s} \tag{3.5.52}$$

we see from Eqs. (3.5.49) and (3.5.51) that $y \approx z$. Therefore after about 10^4 s, that is, approximately 3 h later, the tube's trajectory is upward and to the right with a slope of one (Fig. 3.100). The slope of the S surface, however, is infinity, while the slope of the m_g surface is only

$$\left(\frac{dz}{dy}\right)_{m_g} = \left(f - \frac{\partial u_g}{\partial y}\right) \bigg/ \frac{\partial u_g}{\partial z} \sim \frac{10^{-4} \text{ s}^{-1}}{10^{-2} \text{ s}^{-1}} = \frac{1}{100}. \tag{3.5.53}$$

The tube's trajectory thus lies in between the S and m_g surfaces.

Differentiating Eqs. (3.5.49) and (3.5.51) with respect to time, we see that

$$v = \frac{Dy}{Dt} = \frac{v_0 + w_0}{2} \cosh \sigma t + \frac{v_0 - w_0}{2} \cos \sigma t \tag{3.5.54}$$

$$w = \frac{Dz}{Dt} = \frac{v_0 + w_0}{2} \cosh \sigma t + \frac{w_0 + v_0}{2} \cos \sigma t. \tag{3.5.55}$$

For $t \gg 1/\sigma$ it follows from Eqs. (3.5.54) and (3.5.49) that

$$v \approx \sigma y. \tag{3.5.56}$$

Furthermore, from Eqs. (3.5.55) and (3.5.49) it follows that

$$w \approx \sigma y. \tag{3.5.57}$$

The slope of the m_g surfaces in Fig. 3.95 is

$$\left(\frac{dz}{dy}\right)_{m_g} \sim \frac{H}{L} \sim 10^{-2}. \tag{3.5.58}$$

If the length scale L is on the order of 100 km, it follows from Eq. (3.5.58) that H is on the order of 1 km. From Eq. (3.5.52) we see that $\sigma = 10^{-3}\ \text{s}^{-1}$. Then from Eq. (3.5.57) we find that

$$w \sim (10^{-3}\ \text{s}^{-1})(10^{3}\ \text{m}) = 1\ \text{m s}^{-1} \tag{3.5.59}$$

at $z = H$. Therefore vertical velocities on the order of m s^{-1} may be attained in about 3 h or so in a symmetrically unstable atmosphere. This is to be compared with vertical velocities on the order of cm s^{-1} found on the synoptic scale and 10 m s^{-1} found in cumulonimbus convection. The vertical velocity of a tube can accelerate to speeds as high as m s^{-1} in an environment that is unstable with respect to "slantwise" convection, even though it is stable with respect to upright convection.

The increase of vertical velocity with height below the level of maximum upward vertical velocity associated with the vertical circulation itself acts to decrease gravitational and convective stability below by spreading θ_v and θ_{ev} surfaces farther apart. Furthermore, the differential advection of moisture by the ageostrophic circulation associated with the vertical circulation might, in some instances, reduce $\partial\theta_{ev}/\partial z$ to the point where conditional instability of the traditional kind is produced. If conditional instability is present, we expect that vertical circulations associated with it will dominate over any vertical circulations associated with symmetric instability, because buoyant accelerations are much greater in magnitude than inertial accelerations.

In the absence of mixing, the tube continues to experience an upward buoyancy force and an inertial force to the right until it reaches $z = H$, after which it no longer has a buoyancy force acting upon it. In the continued presence of inertial force, though, it overshoots the S surface and then oscillates about it; it reaches an equilibrium point at the intersection of the S and m_g surfaces, about which it oscillates in both the y and z directions.

The relationship of CSI to potential vorticity. In Chap. 2 we discussed symmetric instability in a dry (or moist, but always unsaturated) atmosphere.

We found that a necessary condition for symmetric instability (when the temperature field is in thermal-wind balance) is that Ertel's potential vorticity must be negative. From Eq. (2.5.246) we see that this implies that

$$(\nabla \times \mathbf{v}_g + f\hat{\mathbf{k}}) \cdot \nabla\theta < 0. \tag{3.5.60}$$

However, if the atmosphere is moist, then θ_w (or θ_e) is substituted for θ.

Suppose we consider the Boussinesq equations, Eqs. (3.4.10), (3.4.11), (3.4.12), (3.4.26), and (3.4.20), and now include the Coriolis force. The three-dimensional vorticity equation may then be expressed as follows:

$$\frac{D}{Dt}(\nabla \times \mathbf{v} + f\hat{\mathbf{k}}) = [(\nabla \times \mathbf{v} + f\hat{\mathbf{k}}) \cdot \nabla]\mathbf{v} - \nabla \times \frac{1}{\rho}\nabla p' + \nabla \times \mathbf{F}_{\text{fric}} + \nabla \times g\left(\frac{\theta - \bar{\theta}}{\bar{\theta}}\right)\hat{\mathbf{k}}. \tag{3.5.61}$$

The term $\nabla \times g[(\theta - \bar{\theta})/\bar{\theta}]\hat{\mathbf{k}}$ represents the generation of horizontal vorticity resulting from horizontal buoyancy gradients. The rate of change of Ertel's "wet-bulb" (moist) potential vorticity following air-parcel motion can be expressed as follows:

$$\frac{D}{Dt}\left(\frac{C_p}{\rho\theta_w}(\nabla \times \mathbf{v} + f\hat{\mathbf{k}}) \cdot \nabla\theta_w\right) = \frac{C_p}{\rho\theta_w}\nabla \times \mathbf{F}_{\text{fric}} \cdot \nabla\theta_w + \frac{1}{\rho}(\nabla \times \mathbf{v} + f\hat{\mathbf{k}}) \cdot \nabla\left(\frac{1}{T}\frac{dQ}{dt}\right) + \frac{C_p}{\rho\theta_w}\nabla \times g\left(\frac{\theta - \bar{\theta}}{\bar{\theta}}\right)\hat{\mathbf{k}} \cdot \nabla\theta_w. \tag{3.5.62}$$

The first two terms on the right-hand side of Eq. (3.5.62) represent the frictional and diabatic generation of Ertel's potential vorticity. The additional potential-vorticity generation term involving buoyancy [the last one on the right-hand side in Eq. (3.5.62)] represents the generation of horizontal vorticity, from horizontal buoyancy gradients, in the direction of the wet-bulb potential temperature gradient. It may be expressed, with the aid of a vector identity, as $(C_p/\rho\theta_w\bar{\theta})\hat{\mathbf{k}} \cdot (\nabla\theta_w \times \nabla\theta)$. When $\nabla\theta_w$ is oriented within 90° of $\nabla\theta$ and to its left (Fig. 3.101), Ertel's potential vorticity may be reduced to the point where it becomes negative; hence conditional symmetric instability can be produced even when the atmosphere is inertially stable $(\zeta + f > 0)$ and convectively stable $(\partial\theta_w/\partial z > 0)$. Since (in the Northern Hemisphere) the thermal-wind vector is oriented perpendicular and to the left of $\nabla\theta$, and θ_w is a monotonically increasing function of water-vapor content, it follows that Ertel's "wet-bulb" potential vorticity may be reduced when water-vapor content increases in the direction of the thermal wind.

This can happen when there is a "moist tongue" of high dew points east of a surface cyclone and drier air to the southwest; water-vapor content increases in the same direction as the thermal-wind vector points: Thus Ertel's potential vorticity is decreased by the buoyancy term.

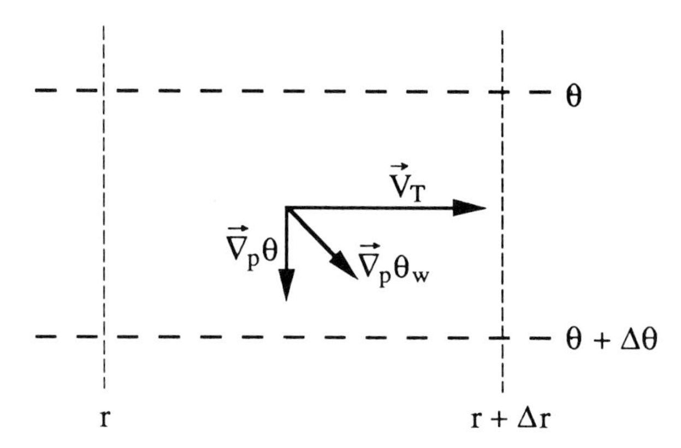

Figure 3.101 An idealized distribution of potential temperature (θ) (long dashed lines) and water-vapor mixing ratio r (short dashed lines) for which there is a generation of *negative* potential vorticity: The thermal wind ($\mathbf{v}_T$) is westerly (from left to right); the moisture gradient points toward the east. However the gradient of θ points toward the south, and the gradient of θ_w points toward the southeast.

Since a necessary condition for CSI is that "moist" potential vorticity is negative, and vorticity is a factor of Ertel's potential vorticity, regions in which vorticity is cyclonic and large in magnitude are not regions where CSI is likely. For example, we would expect to find that the region far downstream from an upper-level trough is likelier to harbor CSI than the region closer to the trough axis. (It has also been suggested that weak warm advection, as evidenced by slight veering of the winds with height, and little directional shear and flow curvature also favor CSI bands. The former is conducive to quasigeostrophic ascent leading to saturation; the latter assures "symmetry.")

If a symmetrically unstable circulation acts to bring the atmosphere back towards a state of neutral stability or very small stability, then Ertel's potential vorticity also becomes very small according to Eq. (2.5.248). Static stability in geostrophic coordinates is proportional to Ertel's potential vorticity in real space according to Eq. (2.5.182). The relative strength of the vertical branch of the circulation in geostrophic coordinates, furthermore, which is equivalent to the branch of the vertical circulation that tilts along the m_g surface in physical space, depends upon the relative magnitude of the static stability in geostrophic coordinates to the Coriolis parameter [see Eq. (2.5.186)]. If there is a frontogenetical circulation in the presence of a small Ertel's potential vorticity on the warm side of a frontal zone, then the upward branch of the tilted circulation is enhanced relative to the horizontal branch (Fig. 3.102). Through this mechanism, precipitation bands formed as a result of conditional symmetric instability may persist in the presence of frontogenetical forcing long after the SCAPE has been used up.

3.5.3. The "Feeder–Seeder" Process

Microphysical processes may play a role in organizing some precipitation bands. Precipitation within a stratiform area may be enhanced when ice

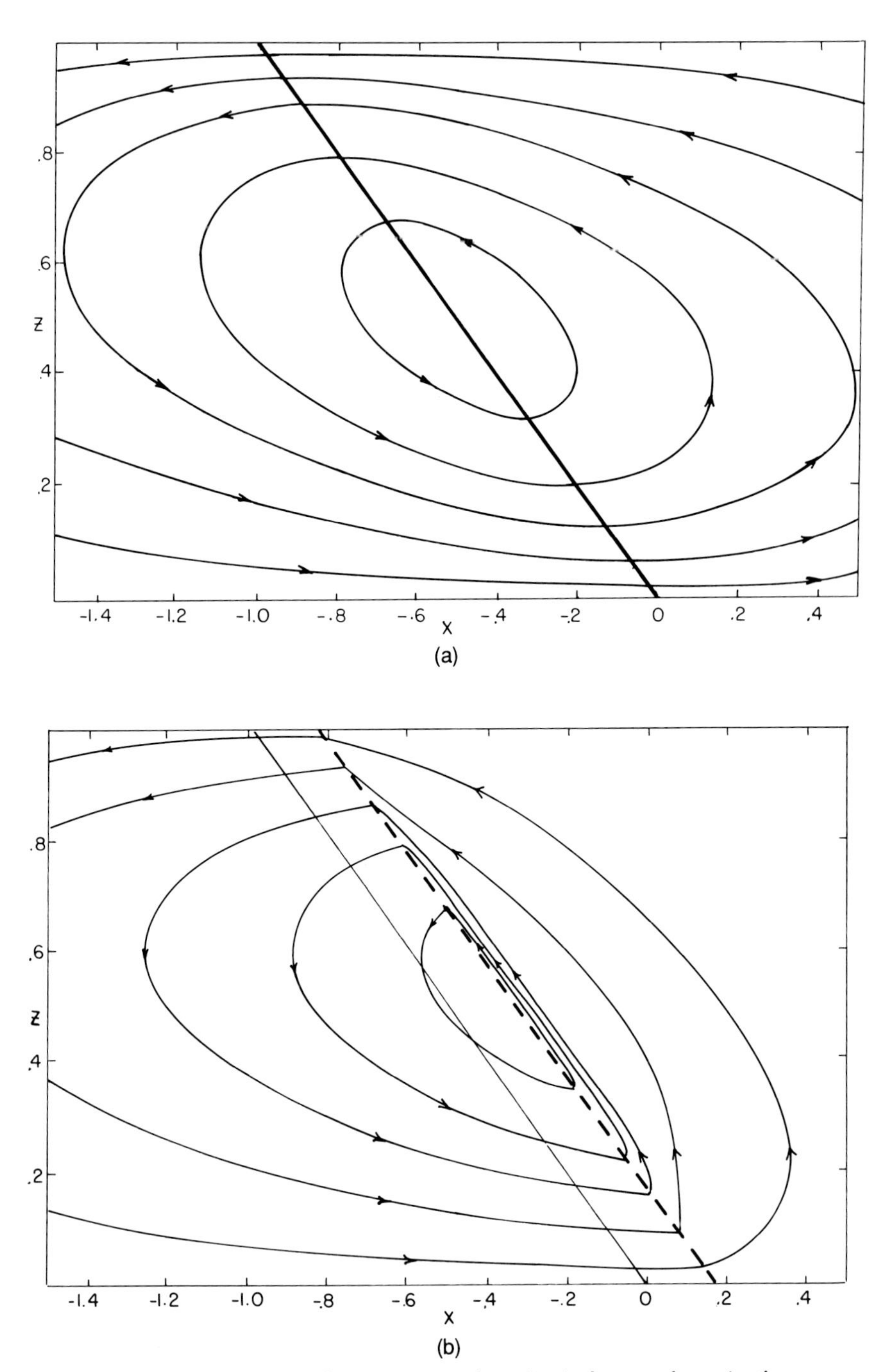

Figure 3.102 (a) Vertical circulation (streamfunction) about a front in the presence of typical values of Ertel's potential vorticity. (b) Vertical circulation about a front in the presence of very low values of Ertel's potential vorticity (based upon solutions to the Sawyer–Eliassen equation; from Emanuel, 1985). (Courtesy of the American Meteorological Society)

particles fall out from "generating cells" aloft and "seed" the stratiform cloud below. The bulk of precipitation that reaches the ground, however, comes from the stratiform cloud, the "feeder" cloud. The formation of precipitation through seeding of the stratiform feeder cloud is called the "seeder–feeder" process.

The generating cells (Fig. 3.103) are often located in layers of convective instability above a warm-frontal zone. If a generating cell travels along with the mean wind in its layer, then a band of precipitation may form in the stratiform cloud below. The "feeder–seeder" process is thus hybrid in that both a layer of convective instability and stability are present. It is one of the few significant mesoscale precipitation processes that is essentially entirely microphysical in nature.

3.5.4. Ducted Gravity Waves

Figure 3.104 illustrates the idealized structure of a propagating gravity wave. The locations of greatest rising (sinking) motion are a quarter of a wavelength downstream (with respect to propagation) of the locations of highest (lowest) surface pressure. Wind speeds in the direction of wave propagation are greatest (smallest) at the locations of highest (lowest) surface pressure.

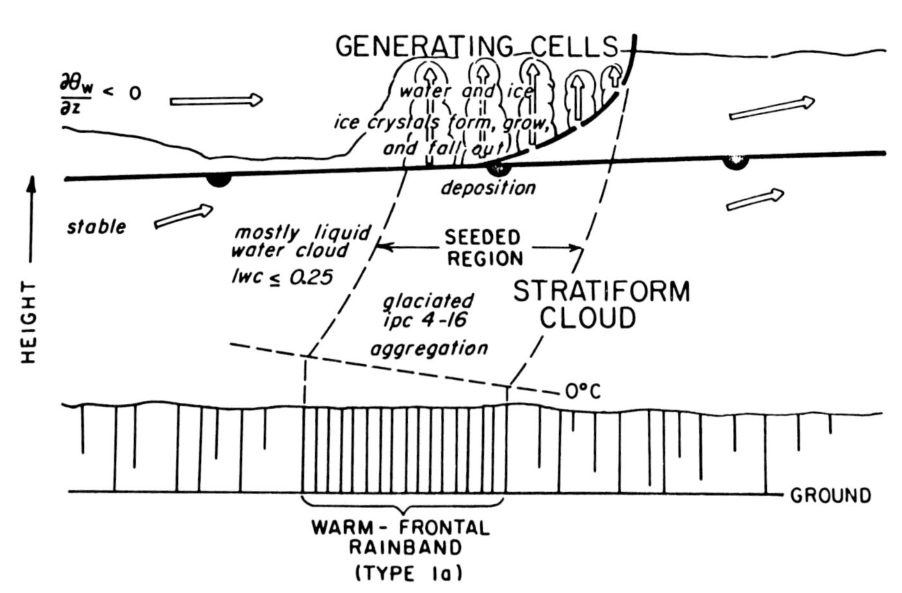

Figure 3.103 Illustration of the feeder–seeder process: Conceptual model of the vertical cross section of a warm-frontal rainband. Vertical hatching below cloud base represents precipitation: The density of the hatching corresponds qualitatively to the precipitation rate. The heavy broken line branching out from the front is a warm-frontal zone with convective ascent in the generating cells. Ice-particle concentrations (ipc) are given in number per liter; cloud liquid water contents (lwc) are in g m^{-3}. The motion of the rainband is from left to right. The seeder zone is above the warm front; the feeder cloud is below the warm front (from Houze and Hobbs, 1982; originally from Hobbs, 1978, and Majejka et al., 1980).

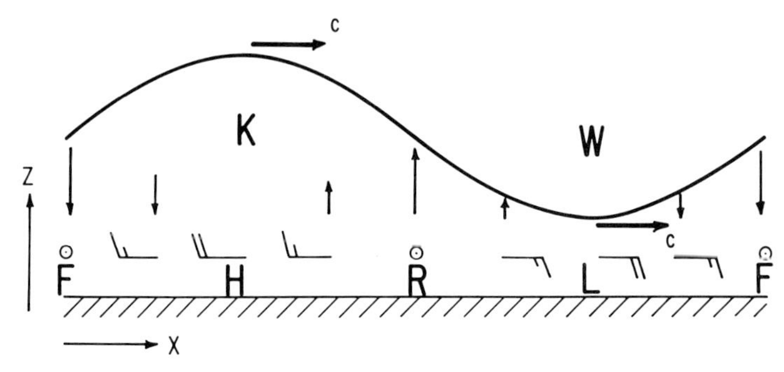

Figure 3.104 Idealized vertical cross section of a linear plane gravity wave, with no basic current, propagating toward the right at speed c. Representative isentropic surface or temperature inversion (heavy sinusoidal line); surface-pressure extrema (H and L); cold and warm temperature anomalies (K and W, respectively); falling and rising motion (F and R, respectively) (from Bosart and Sanders, 1986). (Courtesy of the American Meteorological Society)

Although convective storms can generate gravity waves, for example, when a downdraft hits the ground, and stable air is lifted over the gust front, or as a buoyant air parcel penetrates the tropopause, gravity waves themselves sometimes can trigger convective systems if their amplitude is large enough that air parcels can be displaced upward to their LCL and LFC. In the absence of convective instability or conditional symmetric instability, gravity waves may also force nonconvective bands. Gravity waves may be generated orographically when stable air is lifted over mountains; inertial-gravity waves may be generated through the frontogenetical process and as air pracels move through jet streaks, and quasigeostrophic equilibrium is disturbed.

It can be shown that since air density decreases monotonically in the atmsophere, gravity waves can propagate vertically, and hence wave energy is lost to the atmosphere above. Gravity waves cannot travel very far from their source before their amplitude becomes too small to lift air to its lifting condensation level. Richard Lindzen and K.-K. Tung, however, showed in 1976 that a stable layer ($\partial\theta/\partial z > 0$ in an unsaturated atmosphere, $\partial\theta_e/\partial z > 0$ in a saturated atmosphere), under the following conditions, can act like a "duct" and reflect upward-propagating waves, so that gravity waves can travel away from their source with less loss of energy:

1. The vertical wavelength of the nth mode of "long" gravity waves is approximately

$$L_{zn} \approx \frac{2\pi(c_n - \bar{u})}{N}, \tag{3.5.63}$$

where $c_n - \bar{u}$ is the phase speed relative to the mean flow ($\bar{u}$) in the layer for the nth mode, and the Brunt–Väisälä frequency

$$N = \left(g \frac{\partial \ln \theta}{\partial z}\right)^{1/2} \tag{3.5.64}$$

if the atmosphere is unsaturated, and

$$N = \left(g \frac{\partial \ln \theta_e}{\partial z}\right)^{1/2} \qquad (3.5.65)$$

if the atmosphere is saturated. The relative phase speed $(c_n - \bar{u})$ is given by the following relation:

$$c_n - \bar{u} \approx \frac{NH}{\pi(\frac{1}{2} + n)}, \qquad n = 0, 1, 2, \ldots, \qquad (3.5.66)$$

where H is the depth of the stable layer, the "duct." The phase speed is not a function of the *horizontal scale*. The longest mode, the $n = 0$ mode, is expected to dominate, because the slower, shorter modes have weakened more after traveling a given horizontal distance than the faster, longer modes. From Eqs. (3.5.66) and (3.5.63) we see that the vertical wavelength of the longest mode

$$L_{z0} = 4H. \qquad (3.5.67)$$

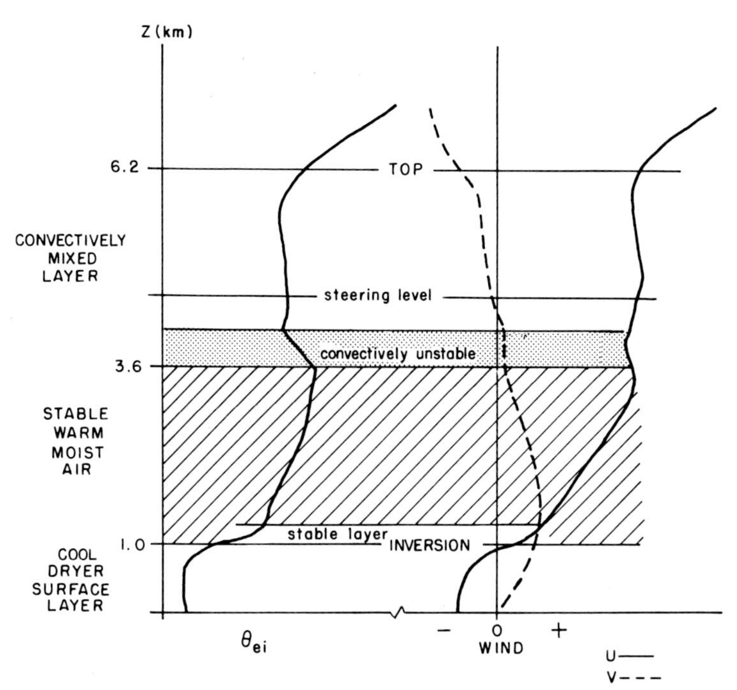

Figure 3.105 Schematic representation of typical environmental conditions for ducted gravity-wave (banded) disturbances. Equivalent potential temperature (θ_{ei}) is on the left, and the mean velocity (relative to band movement) profiles are on the right; *u* component parallel to the band; *v* component normal to the band (from Marks, 1975).

For other modes,

$$L_{zn} \approx \frac{2H}{\frac{1}{2} + n} \tag{3.5.68}$$

and hence L_{zn} is shorter. From Eqs. (3.5.67) and (3.5.68) we conclude that the depth of the stable layer must accommodate one-quarter of the vertical wavelength that corresponds to the phase speed.

2. There cannot be any level ("critical level" or "steering level") in the stable layer where the wind speed is equal to the phase speed of the gravity wave ($c = \bar{u}$), or else wave energy would be completely absorbed.
3. Just above the stable layer there is a layer in which the Richardson number is less than 0.25. This may be due to a nearly dryadiabatic lapse rate ($N \approx 0$) in the presence of nonzero vertical shear if the atmosphere is unsaturated, or $\partial\theta_e/\partial z \approx 0$ if the atmosphere is saturated, or to very strong vertical shear.
4. There must be a critical level in the unstable layer or the phase speed of the gravity wave must be *almost* the same as the flow speed ($c = \bar{u}$).

Ducted gravity waves may be associated with bands of upward motion strong enough to lift air to saturation. In a saturated environment, they may

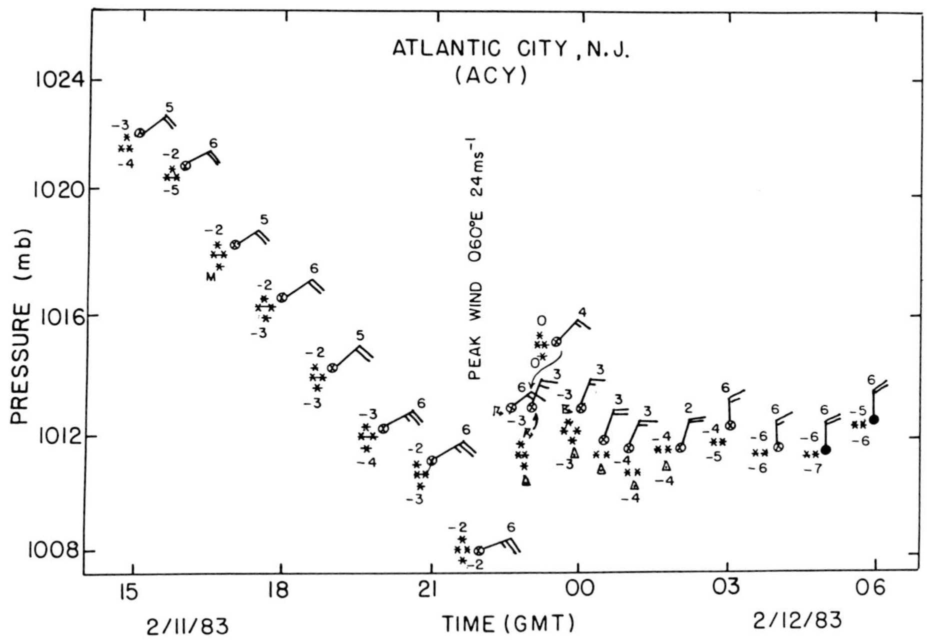

Figure 3.106 Time section of hourly and special observations as a function of mean sea-level pressure for Atlantic City, New Jersey, during the "Megalopolitan" snowstorm of 1983 in the northeast United States. Temperature and dew point plotted in °C. Whole wind barb and half wind barb represent 5 and 2.5 m s^{-1}, respectively; tens unit of wind direction plotted at the end of each wind barb (from Bosart and Sanders, 1986). (Courtesy of the American Meteorological Society)

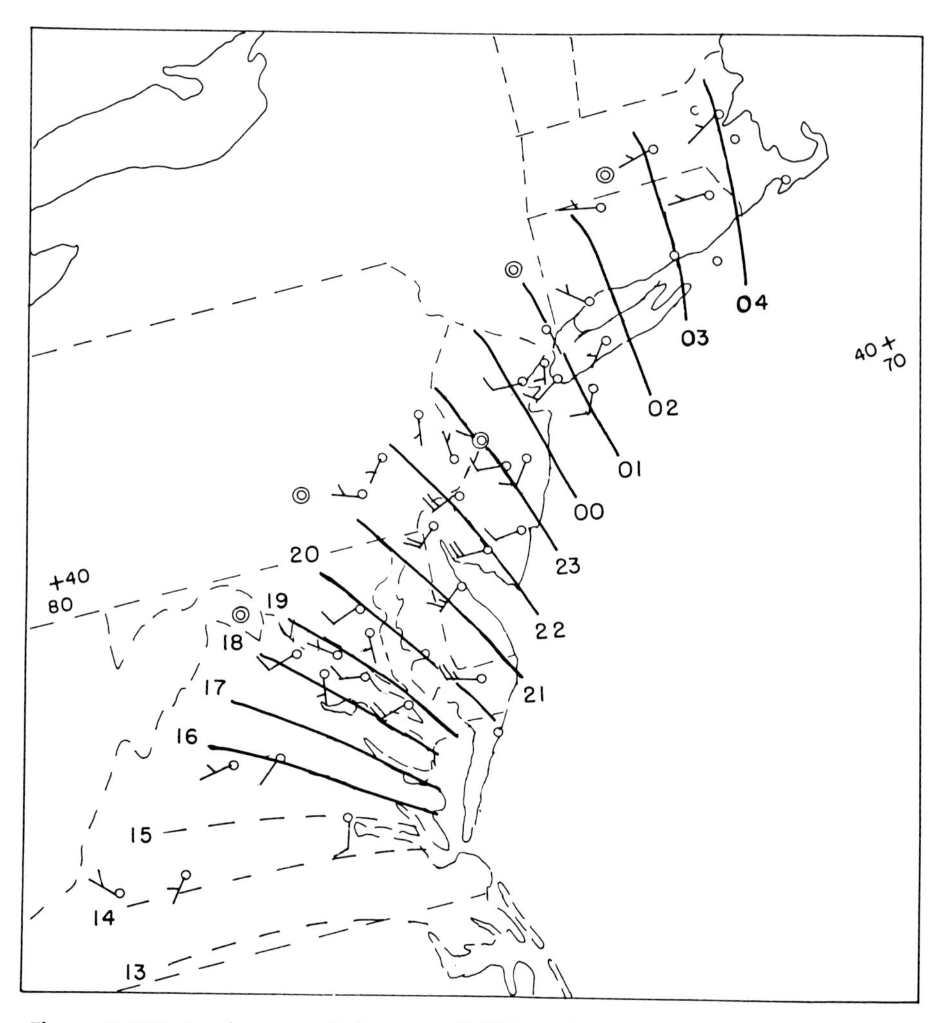

Figure 3.107 Isochrones of the time (UTC) at the end of the hour of greatest vector wind shift during the "Megalopolitan" snowstorm of 1983. Dashed line indicates the ill-organized phase. One full (half) wind barb represents 5 m s^{-1} (2.5 m s^{-1}) magnitude of vector shift (from Bosart and Sanders, 1986). (Courtesy of the American Meteorological Society)

enhance precipitation within an existing area of stratiform precipitation. Figure 3.105 depicts the typical environmental conditions for mesoscale precipitation bands in New England; we note that conditions (2), (3), and (4) are present.

Figure 3.106 shows evidence of solitary, large-amplitude gravity-wave motion north and northeast of an extratropical cyclone. The pressure falls, rises, and falls again (at a number of surface observing sites). (Pressure changes as rapid as 11 mb in 15 min have been documented!) The progression of a band of enhanced precipitation and wind shift are depicted in Figure 3.107.

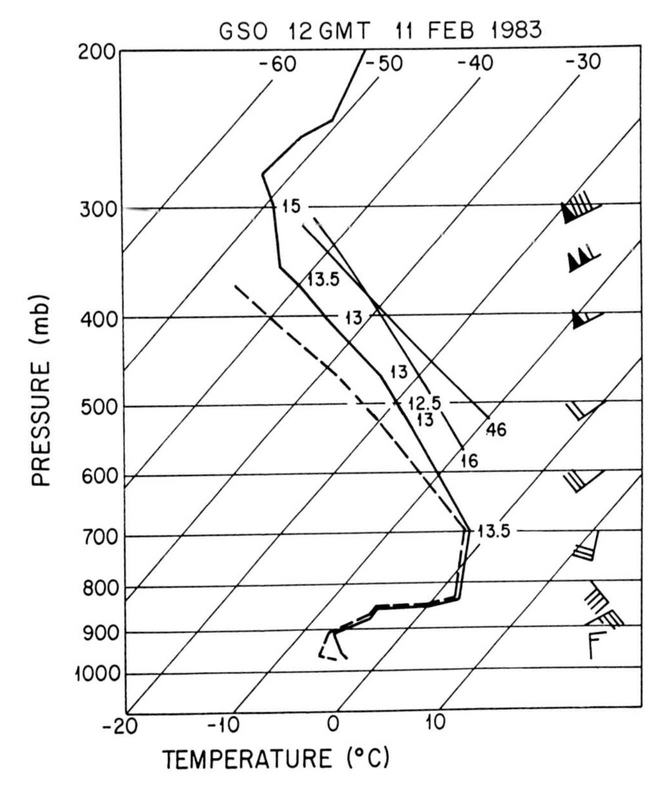

Figure 3.108 Assessment of a sounding [skew T–log p diagram for Greensboro, North Carolina (GSO)] for the possibility of ducted gravity-wave propagation during the "Megalopolitan" snowstorm of 1983. Temperature and dew point plotted as thick solid and dashed lines, respectively. Winds plotted at the right; pennant = 25 m s^{-1}; whole barb = 5 m s^{-1}; half barb = 2.5 m s^{-1}. The Richardson number is below 0.25 between 360 and 650 mb, where there is a possible steering level. In the low-level stable layer, between 700 and 900 mb, $N = [9.8(300 - 274)/287(2050)]^{1/2}$ s^{-1} = 0.021 s^{-1}; $c_0 - \bar{u} = 27.4$ m s^{-1} = (0.021 s^{-1})(2050 m)/(π/2). Since $\bar{u} \approx 1$ m s^{-1} (toward 050°), $c_0 \approx 26$ m s^{-1}. The observed value of c was a little less than 15 m s^{-1} (from Bosart and Sanders, 1986). (Courtesy of the American Meteorological Society)

An illustration comparing Lindzen and Tung's theoretical estimate of gravity-wave speed with the observed speed is found in Fig. 3.108. The reader is cautioned that it is sometimes difficult to identify clearly the precise vertical extent of the stable duct; it is also not always possible to find a representative sounding. In Fig. 3.108 we also note that the stable duct is elevated above the ground; it is not required that it be bounded below by the ground.

The reader should note that the "feeder-seeder" process may also be operating in an environment supportive of ducted gravity waves: A region of convective instability supportive of generating cells lies over a stable layer in which the "feeder" cloud is located.

NOTES

1. The author, who was 4 years old and living 40 miles to the east in Boston when Worcester was hit, well remembers that sultry, hazy, tragic afternoon in June.
2. The discussion will be more quantitative later.
3. Lowered cloud bases can form whenever relatively humid air gets ingested into an updraft. However, in ordinary cells these "lowerings" do not persist for long and do not rotate; they disappear as precipitation falls out and the updraft decays.
4. *Not* a storm *system,* which is composed of more than one convective storm.
5. It thus takes on the appearance of the classic tornado (simulated with the aid of a windsock and an automobile) in the motion-picture version of *The Wizard of Oz.*
6. In convective cells aligned along a warm front, however, the stratiform-precipitation area may *lead* the convective line.

REFERENCES

General

Atlas, D., D. R. Booker, H. Byers, R. H. Douglas, T. Fujita, D. C. House, F. H. Ludlam, J. S. Malkus, C. W. Newton, Y. Ogura, R. A. Schleusener, B. Vonnegut, and R. T. Williams, 1963: *Severe Local Storms,* Meteor. Mono. **5**(27), Amer. Meteor. Soc., Boston.

Browning, K. A., 1977: The structure and mechanisms of hailstorms. *Hail: A Review of Hail Science and Hail Suppression* (B. Foote and C. Knight, eds.). Meteor. Mono. **16,** Amer. Meteor. Soc., Boston, 1–43.

Doswell, C. A., III, 1982: *The Operational Meteorology of Convective Weather (Vol. I): Operational Mesoanalysis.* NOAA Tech. Memo. NWS NSSFC-5, U.S. Dept. of Commerce, NOAA, NWS, Kansas City, MO.

—, 1985: *The Operational Meteorology of Convective Weather (Vol II): Storm Scale Analysis.* NOAA Tech. Memo. ERL ESG-15, NOAA, ERL, ESG, Boulder, CO.

Houze, R. A., and P. V. Hobbs, 1982: Organization and structure of precipitating cloud systems. *Adv. in Geophysics* **24,** 225–315.

Kessler, E. (ed.), 1981: *Thunderstorms: A Social, Scientific, and Technological Documentary (Vol. 1), The Thunderstorm in Human Affairs.* U.S. Dept. of Commerce, NOAA, ERL, Boulder, CO.

—, 1986: *Thunderstorms: A Social, Scientific, and Technological Documentary (Vol. 2), Thunderstorm Morphology and Dynamics.* Univ. of Oklahoma Press, Norman, Okla.

—, 1982: *Thunderstorms: A Social, Scientific, and Technological Documentary (Vol. 3), Instruments and Techniques for Thunderstorm Observation and Analysis.* U.S. Dept. of Commerce, NOAA, ERL, Boulder, CO.

Ludlam, F. H., 1980: *Clouds and Storms: The Behavior and Effect of Water in the Atmosphere.* Penn State Univ. Press, Univ. Park, PA.

Miller, R. C., 1972: *Notes on Analysis and Severe-Storm Forecasting Procedures of the Air Force Global Weather Central.* Tech. Rep. 200 (Rev. 1975), U.S. Air Force, Air Weather Service.

Schaefer, J. T., 1986: Severe thunderstorm forecasting: A historical perspective. *Weather and Forecasting* **1,** 164–89.

Section 3.3

Smull, B. F., and R. A. Houze, 1987: Dual-Doppler radar analysis of a midlatitude squall line with a trailing region of stratiform rain. *J. Atmos. Sci.* **44,** 2128–48.

Section 3.4

Dutton, J. A., and G. H. Fichtl, 1969: Approximate equations of motion for gases and liquids. *J. Atmos. Sci.* **26,** 241–54.

Ogura, Y., and N. A. Phillips, 1962: Scale analysis of deep and shallow convection in the atmosphere. *J. Atmos. Sci.* **19,** 173–79.

Phillips, N. A., 1970: Unpublished course notes. M.I.T., Cambridge, Mass.

Section 3.4.3

Bluestein, H. B., and K. W. Thomas, 1984: Diagnosis of a jet streak in the vicinity of a severe weather outbreak in the Texas Panhandle. *Mon. Wea. Rev.* **112,** 2499–520.

Doswell, C. A., III, 1980: Synoptic-scale environments associated with High Plains severe thunderstorms. *Bull. Amer. Meteor. Soc.* **99,** 1388–400.

—, 1987: The distinction between large-scale and mesoscale contribution to severe convection: A case study example. *Wea. and Forecasting* **2,** 3–16.

Johns, R. H., 1982: A synoptic climatology of northwest flow severe weather outbreaks. Part I: Nature and significance. *Mon. Wea. Rev.* **110,** 1653–63.

—, 1984: A synoptic climatology of northwest-flow severe weather outbreaks. Part II: Meteorological parameters and synoptic patterns. *Mon. Wea. Rev.* **112,** 449–64.

Koch, S. E., 1984: The role of an apparent mesoscale frontogenetical circulation in squall line initiation. *Mon. Wea. Rev.* **112,** 2090–111.

McGinley, J., 1986: Nowcasting mesoscale phenomena (Chap. 28). *Mesoscale Meteorology and Forecasting* (P. Ray, ed.), Amer. Meteor. Soc., Boston, 657–88.

Pielke, R. A., and M. Segal, 1986: Mesoscale circulations forced by differential terrain heating (Chap. 22). *Mesoscale Meteorology and Forecasting* (P. Ray, ed.), Amer. Meteor. Soc., Boston, 516–48.

Purdom, J., 1976: Some uses of high resolution GOES imagery in the mesoscale forecasting of convection and its behavior. *Mon. Wea. Rev.* **104,** 1474–83.

Rhea, J. O., 1966: A study of thunderstorm formation along drylines. *J. Appl. Meteor.* **5,** 58–63.

Szoke, E. J., M. L. Weisman, J. M. Brown, F. Caracena, and T. W. Schlatter, 1984: A subsynoptic analysis of the Denver tornadoes of 3 June 1981. *Mon. Wea. Rev.* **112,** 790–808.

Uccellini, L. W., 1975: A case study of apparent gravity wave initiation of severe convective storms. *Mon. Wea. Rev.* **103,** 497–513.

Wilson, J. W., and W. E. Schreiber, 1986: Initiation of convective storms at radar-observed boundary-layer convergence lines. *Mon. Wea. Rev.* **114,** 2516–36.

Section 3.4.4

Kessler, E., 1969: *On the Distribution and Continuity of Water Substance in Atmospheric Circulations.* Meteor. Mono. **10** (32), Amer. Meteor. Soc., Boston.

Section 3.4.5

Beebe, R. G., 1955: Types of airmasses in which tornadoes occur. *Bull. Amer. Meteor. Soc.* **36,** 349–50.

Bluestein, H. B., 1979: A mini-tornado in California. *Mon. Wea. Rev.* **107,** 1227–29.

—, E. W. McCaul, Jr., G. P. Byrd, and G. R. Woodall, 1988: Mobile sounding observations of a tornadic storm near the dryline: The Canadian, Texas storm of 7 May 1986. *Mon. Wea. Rev.* **116,** 1790–1804.

Carlson, T. N., and F. H. Ludlam, 1968: Conditions for the occurrence of severe local storms. *Tellus* **20,** 203–26.

Colby, F. P., Jr., 1983: Convective inhibition as a predictor of the outbreak of convection in AVE-SESAME II. *Preprints, 13th Conf. on Severe Local Storms,* Tulsa, Amer. Meteor. Soc., Boston, 324–27.

Fawbush, E. J., and R. C. Miller, 1953: A method for forecasting hailstone size at the Earth's surface. *Bull. Amer. Meteor. Soc.* **34,** 235–44.

—, and R. C. Miller, 1954a: A basis for forecasting peak wind gusts in non-frontal thunderstorms. *Bull. Amer. Meteor. Soc.* **35,** 14–19.

—, and R. C. Miller, 1954b: The types of airmasses in which North American tornadoes form. *Bull. Amer. Meteor. Soc.* **35,** 154–65.

Galway, J. G., 1956: The lifted index as a predictor of latent instability. *Bull. Amer. Meteor. Soc.* **37,** 528–29.

George, J. J., 1960: *Weather Forecasting for Aeronautics.* Academic Press, New York.

McCaul, E. W., Jr., 1987: Observations of the Hurricane "Danny" tornado outbreak of 16 August. 1985. *Mon. Wea. Rev.* **115,** 1206–23.

Showalter, A. K., 1953: A stability index for thunderstorm forecasting. *Bull. Amer. Meteor. Soc.* **34,** 350–52.

Section 3.4.6

Bluestein, H. B., 1986: Visual aspects of the flanking line in severe thunderstorms. *Mon. Wea. Rev.* **114,** 788–95.

—, and C. R. Parks, 1983: A synoptic and photographic climatology of low-precipitation severe thunderstorms in the Southern Plains. *Mon. Wea. Rev.* **111,** 2034–46.

Brandes, E. A., 1978: Mesocyclone evolution and tornadogenesis: Some observations. *Mon. Wea. Rev.* **106,** 995–1011.

Browning, K. A., 1964: Airflow and precipitation trajectories within severe local storms which travel to the right of the winds. *J. Atmos. Sci.* **21,** 634–39.

—, and R. J. Donaldson, Jr., 1963: Airflow and structure of a tornadic storm. *J. Atmos. Sci.* **20,** 533–45.

Burgess, D., 1974: Study of a right-moving thunderstorm utilizing new single Doppler evidence. M.S. Thesis, Dept. of Meteorology, Univ. of Oklahoma, Norman.

—, R. J. Donaldson, T. Sieland, and J. Hinkelman, 1979: Part I—Meteorological Applications. *Final Report on the Joint Doppler Operational Project* (JDOP), NOAA Tech. Memo. ERL NSSL-86, NSSL, Norman, OK, 12–15.

Byers, H. R., and R. R. Braham, Jr., 1949: *The Thunderstorm.* Supt. of Documents, U.S. Government Printing Office, Washington, D.C.

Chisholm, A. J., and J. H. Renick, 1972: The kinematics of multicell and supercell Alberta hailstorms. "Alberta Hail Studies, 1972," Research Council of Alberta Hail Studies, Report No. 72-2, 24–31.

Davies-Jones, R., 1984: Streamwise vorticity: The origin of updraft rotation in supercell storms. *J. Atmos. Sci.* **41,** 2991–3006.

Dennis, A. S., C. A. Schock, and A. Koscielski, 1970: Characteristics of hailstorms of western South Dakota. *J. Appl. Meteor.* **9,** 127–35.

Fujita, T., 1959: *A detailed analysis of the Fargo tornadoes of June 20, 1957.* Tech. Rep. No. 5 to U.S. Wea. Bur., Univ. of Chicago.

Klemp, J. B., 1987: Dynamics of tornadic thunderstorms. *Ann. Rev. Fluid Mech.* **19,** 1–33.

—, and R. Rotunno, 1983: A study of the tornadic region within a supercell thunderstorm. *J. Atmos. Sci.* **40,** 359–77.

Lazarus, S. M., and K. K. Droegemeier, 1988: Simulation of convective initiation along gust fronts. *Preprints, 15th Conf. on Severe Local Storms,* Baltimore, Amer. Meteor. Soc., Boston, 241–44.

Lemon, L. R., 1977: *New Severe Thunderstorm Radar Identification Techniques and Warning Criteria: A Preliminary Report.* NOAA Tech. Memo. NWS NSSFC-1, TDU, NSSFC, Kansas City.

Lilly, D. K., 1986: The structure, energetics and propagation of rotating convective storms. Part II: Helicity and storm stabilization. *J. Atmos. Sci.* **43,** 126–40.

McCaul, E. W., Jr., 1987: Observations of the Hurricane "Danny" tornado outbreak of 16 August 1985. *Mon. Wea. Rev.* **115,** 1206–23.

Moller, A. R., and C. A. Doswell, 1988: A proposed advanced storm spotter's training program. *Preprints, 15th Conf. on Severe Local Storms,* Baltimore, Amer. Meteor. Soc., Boston, 173–77.

Rotunno, R., 1981: On the evolution of thunderstorm rotation. *Mon. Wea. Rev.* **109,** 577–86.

—, and J. B. Klemp, 1982a: The influence of the shear-induced pressure gradient on thunderstorm motion. *Mon. Wea. Rev.* **110,** 136–51.

—, and J. B. Klemp, 1982b: On the rotation and propagation of simulated supercell thunderstorms. *J. Atmos. Sci.* **42,** 271–92.

—, J. B. Klemp, and M. L. Weisman, 1988: A theory for strong, long-lived squall lines. *J. Atmos. Sci.* **45,** 463–85.

Stout, G. E., and F. A. Huff, 1953: Radar records Illinois tornadogenesis. *Bull. Amer. Meteor. Soc.* **34,** 281–84.

Weisman, M. L., and J. B. Klemp, 1982: The dependence of numerically simulated convective storms on vertical wind shear and buoyancy. *Mon. Wea. Rev.* **110,** 504–20.

—, and J. B. Klemp, 1984: The structure and classification of numerically simulated convective storms in directionally varying wind shears. *Mon. Wea. Rev.* **112,** 2479–98.

—, and J. B. Klemp, 1986: Characteristics of isolated convective storms (Chap. 15). *Mesoscale Meteorology and Forecasting* (P. Ray, ed.), Amer. Meteor. Soc., Boston, 331–58.

Wilhelmson, R. B., and C.-S. Chen, 1982: A simulation of the development of successive cells along a cold outflow boundary. *J. Atmos. Sci.* **39,** 1466–83.

Section 3.4.7

Foote, G. B., and H. W. Frank, 1983: Case study of a hailstorm in Colorado. Part III: Airflow from triple-Doppler measurements. *J. Atmos. Sci.* **40,** 686–707.

Section 3.4.8

Bluestein, H. B., 1983: Surface meteorological observations in severe thunderstorms. Part II: Field experiments with TOTO. *J. Clim. Appl. Meteor.* **22,** 919–30.

—, 1984: Photographs of the Canyon, Texas storm of 26 May 1978. *Mon. Wea. Rev.* **112,** 2521–23.
—, 1985: The formation of a "landspout" in a "broken-line" squall line in Oklahoma. *Preprints, 14th Conf. on Severe Local Storms,* Indianapolis, Amer. Meteor. Soc., Boston, 267–70.
—, 1988: Funnel clouds pendant from high-based cumulus clouds. *Weather* **43,** 220–21.
Brown, J. M., and K. R. Knupp, 1980: The Iowa cyclonic–anticyclonic tornado pair and its parent thunderstorm. *Mon. Wea. Rev.* **108,** 1626–46.
Browning, K. A. and G. B. Foote, 1976: Airflow and hail growth in supercell storms and some implications for hail suppression. *Quart. J. Roy. Meteor. Soc.* **102,** 499–534.
—, J. C. Fankhauser, J.-P. Chalon, P. J. Eccles, R. G. Strauch, F. M. Merrem, D. J. Musil, E. L. May, and W. R. Sand, 1976: Structure of an evolving hailstorm, Part V: Synthesis and implications for hail growth and hail suppression. *Mon. Wea. Rev.* **104,** 603–10.
Burgess, D., V. Wood, and R. Brown, 1982: Mesocyclone evolution statistics. *Preprints, 12th Conf. on Severe Local Storms,* San Antonio, Amer. Meteor. Soc., Boston, 422–24.
Busk, H. G., 1927: Land waterspouts. *Meteor. Magazine,* 289–91.
Carbone, R. E., 1983: A severe frontal rainband. Part II: Tornado parent vortex circulation. *J. Atmos. Sci.* **40,** 2639–54.
Changnon, S. A., 1977: The scales of hail. *J. Appl. Meteor.* **16,** 626–48.
Cooley, J. R., 1978: Cold air funnel clouds. *Mon. Wea. Rev.* **106,** 1368–72.
Davies-Jones, R., and E. Kessler, 1974: Tornadoes (Chap. 16). *Weather and Climate Modification* (W. Hess, ed.), Wiley, New York, 552–95.
Doviak, R. J., and D. S. Zrnic, 1984: *Doppler Radar and Weather Observations.* Academic Press, New York, 212–23.
Easterling, D. R., and P. J. Robinson, 1985: The diurnal variation of thunderstorm activity in the United States. *J. Clim. Appl. Meteor.* **24,** 1048–58.
Fiedler, B. H., and R. Rotunno, 1986: A theory for the maximum windspeeds in tornado-like vortices. *J. Atmos. Sci.* **43,** 2328–40.
Foote, G. B., and C. A. Knight (eds.), 1977: Hail: A Review of Hail Science and Hail Suppression. *Meteor. Mono.* **16,** Amer. Meteor. Soc., Boston.
Frank, N. L., P. L. Moore, and G. E. Fisher, 1967: Summer shower distribution over the Florida peninsula as deduced from digitized radar data. *J. Appl. Meteor.* **6,** 309–16.
Fujita, T. T., 1981: Tornadoes and downbursts in the context of generalized planetary scales. *J. Atmos. Sci.* **38,** 1511–34.
—, 1985: *The Downburst.* Univ. of Chicago.
Golden, J. H., 1974: On the life cycle of Florida Keys' waterspouts, I. *J. Appl. Meteor.* **13,** 676–92.
—, and D. Purcell, 1978: Life cycle of the Union City, Oklahoma tornado and comparison with waterspouts. *Mon. Wea. Rev.* **106,** 3–11.
Hales, J. E., Jr., 1983: Synoptic features associated with Los Angeles tornado occurrences. *Preprints, 13th Conf. on Severe Local Storms,* Tulsa, Okla., Amer. Meteor. Soc., Boston, 132–35.
Johnson, B. C., 1983: The heat burst of 29 May 1976. *Mon. Wea. Rev.* **111,** 1776–92.
Karr, T. W., and R. L. Wooten, 1976: Summer radar echo distribution around Limon, Colorado. *Mon. Wea. Rev.* **104,** 728–34.
Kelly, D. L., J. T. Schaefer, R. P. McNulty, C. A. Doswell III, and R. F. Abbey, Jr., 1978: An augmented tornado climatology. *Mon. Wea. Rev.* **106,** 1172–83.

Klemp, J. B., 1987: Dynamics of tornadic thunderstorms. *Ann. Rev. Fluid Mech.* **19,** 1–33.

Krider, E. P., R. C. Noggle, and M. A. Uman, 1976: A gated, wideband magnetic direction finder for lightning return strokes. *J. Appl. Meteor.* **15,** 301–6.

Lemon, L. R., and C. A. Doswell III, 1979: Severe thunderstorm evolution and mesocyclone structure as related to tornadogenesis. *Mon. Wea. Rev.* **107,** 1184–97.

Rasmussen, E. N., R. E. Peterson, J. E. Minor, and B. D. Campbell, 1982: Evolutionary characteristics and photogrammetric determination of wind speeds within the Tulia outbreak tornadoes 28 May 1980. *Preprints, 12th Conf. on Severe Local Storms,* San Antonio, Amer. Meteor. Soc., Boston, 301–4.

Rotunno, R., 1984: An investigation of a three-dimensional asymmetric vortex. *J. Atmos. Sci.* **41,** 283–98.

Schaefer, J. T., D. L. Kelly, C. A. Doswell, III, J. G. Galway, R. J. Williams, R. P. McNulty, L. R. Lemon, and B. D. Lambert, 1980: Tornadoes-when, where and how often? *Weatherwise* **33,** 52–9.

Sinclair, P. C., 1969: General characteristics of dust devils. *J. Appl. Meteor.* **8**, 32–45.

Szoke, E. J., M. L. Weisman, J. M. Brown, F. Caracena, and T. W. Schlatter, 1984: A subsynoptic analysis of the Denver tornadoes of 3 June 1981. *Mon. Wea. Rev.* **112,** 790–808.

Uman, M. A., 1969: *Lightning.* McGraw–Hill, New York.

Wakimoto, R., and J. Wilson, 1989: Non-supercell tornadoes. *Mon. Wea. Rev.* **117,** 1113–40.

Wilson, J. W., 1986: Tornadogenesis by nonprecipitation induced wind shear lines. *Mon. Wea. Rev.* **114,** 270–84.

Section 3.4.9

Agee, E. M., 1987: Mesoscale cellular convection over the oceans. *Dyn. Atmos. Oceans* **10,** Elsevier, Amsterdam, 317–41.

Bluestein, H. B., and M. H. Jain, 1985: Formation of mesoscale lines of precipitation: Severe squall lines in Oklahoma during the spring. *J. Atmos. Sci.* **42,** 1711–32.

—, G. T. Marx, and M. H. Jain, 1987: Formation of mesoscale lines of precipitation: Non-severe squall lines in Oklahoma during the spring. *Mon. Wea. Rev.* **115,** 2719–27.

Braham, R. R., Jr., and R. D. Kelly, 1982: Lake-effect snowstorms on Lake Michigan, USA. *Cloud Dynamics* (E. Agee and T. Asai, eds.), D. Reidel, 87–101.

Burgess, D. W., and E. B. Curran, 1985: The relationship of storm type to environment in Oklahoma on 26 April 1984. *Preprints, 14th Conf. on Severe Local Storms,* Indianapolis, Amer. Meteor. Soc., Boston, 208–11.

Crook, N. A., 1987: Moist convection ahead of a surface cold front. *Extended Abstracts, 3rd Conf. on Mesoscale Processes,* Vancouver, B.C., Amer. Meteor. Soc., Boston, 90.

Fujita, T., and H. A. Brown, 1958: A study of mesosystems and their radar echoes. *Bull. Amer. Meteor. Soc.* **39,** 538–54.

—, 1959: Precipitation and cold air production in mesoscale thunderstorm systems. *J. Meteor.* **16,** 454–66.

—, 1981: Tornadoes and downbursts in the context of generalized planetary scales. *J. Atmos. Sci.* **38,** 1511–34.

Hane, C. E., 1986: Extratropical squall lines and rainbands. (Chap. 16) *Mesoscale Meteorology and Forecasting* (P. Ray, ed.), Amer. Meteor. Soc., Boston, 359–89.

—, C. J. Kessinger, and P. S. Ray, 1987: The Oklahoma squall line of 19 May 1977. Part II: Mechanisms for maintenance of the region of strong convection. *J. Atmos. Sci.* **44,** 2866–83.

Hjelmfelt, M. R., 1990: Numerical study of the influence of environmental conditions on lake-effect snowstorms over Lake Michigan. *Mon. Wea. Rev.* **118,** 138–50.

Houze, R. A., Jr., B. F. Smull, and P. Dodge, 1990: Mesoscale organization of springtime rainstorms in Oklahoma. *Mon. Wea. Rev.* **118,** 613–54.

Johnson, R. H., and P. J. Hamilton, 1988: The relationship of surface pressure features to the precipitation and air flow structure of an intense midlatitude squall line. *Mon. Wea. Rev.* **116,** 1444–72.

Johnston, E. C., 1982: Mesoscale vorticity centers induced by mesoscale convective complexes. *Preprints, 9th Conf. on Weather Forecasting and Analysis,* Seattle, Amer. Meteor. Soc., Boston, 196–200.

Kelly, R. D., 1986: Mesoscale frequencies and seasonal snowfalls for different types of Lake Michigan snowstorms. *J. Climate Appl. Meteor.* **25,** 308–21.

Kessinger, C. J., P. S. Ray, and C. E. Hane, 1987: The Oklahoma squall line of 19 May 1977. Part I: A multiple Doppler analysis of convective and stratiform structures. *J. Atmos. Sci.* **44,** 2840–64.

Maddox, R. A., 1980: Mesoscale convective complexes. *Bull. Amer. Meteor. Soc.* **61,** 1374–87.

—, 1983: Large-scale meteorological conditions associated with midlatitude, mesoscale convective complexes. *Mon. Wea. Rev.* **111,** 1475–93.

Nolen, R. H., 1959: A radar pattern associated with tornadoes. *Bull. Amer. Meteor. Soc.* **40,** 277–79.

Rutledge, S. A., and R. A. Houze, Jr., 1987: A diagnostic modeling study of the trailing stratiform region of a midlatitude squall line. *J. Atmos. Sci.* **44,** 2640–56.

Smull, B. F., and R. A. Houze, Jr., 1987: Dual-Doppler radar analysis of a midlatitude squall line with a trailing region of stratiform rain. *J. Atmos. Sci.* **44,** 2128–48.

Weisman, M. L., 1992: The genesis of severe long-lived bow echoes. *J. Atmos. Sci.* (in press).

—, 1992: The role of convectively generated rear-inflow jets in the evolution of long-lived meso-convective systems. *J. Atmos. Sci.* (in press).

Section 3.5.1

Browning, K. A., 1986: Conceptual models of precipitation systems. *Weather and Forecasting* **1,** 23–41.

Byrd, G. P., 1989: A composite analysis of winter season overrunning precipitation bands over the Southern Plains of the United States. *J. Atmos. Sci.* **46,** 1119–32.

Dunn, L., 1987: Cold air damming by the Front Range of the Colorado Rockies and its relationship to locally heavy snows. *Wea. and Forecasting* **2,** 177–89.

Forbes, G. S., R. A. Anthes, and D. W. Thomson, 1987: Synoptic and mesoscale aspects of an Appalachian ice storm associated with cold-air damming. *Mon. Wea. Rev.* **115,** 564–91.

Hobbs, P. V., 1981: Mesoscale structure in midlatitude frontal systems. *Proc., IAMAP Symposium on Nowcasting: Mesoscale Observation and Short-Range Prediction,* Hamburg, Eur. Space Agency Publ. SP-165, 29–36.

—, and K. R. Biswas, 1979: The cellular structure of narrow cold-frontal rainbands. *Quart. J. Roy. Meteor. Soc.* **105,** 723–27.

Houze, R. A., Jr., and P. V. Hobbs, 1982: Organization and structure of precipitating cloud systems. *Adv. in Geophys.* **24,** 225–315.

James, P. K., and K. A. Browning, 1979: Mesoscale structure of line convection at surface cold fronts. *Quart. J. Roy. Meteor. Soc.* **105,** 371–82.

Matejka, T. J., and P. V. Hobbs, 1981: The use of a single Doppler radar in short-range forecasting and real-time analysis of extratropical cyclones. *Proc., IAMAP Symposium on Nowcasting: Mesoscale Observation and Short-Range Prediction,* Hamburg, Eur. Space Agency Publ. SP-165, 177–81.

Parsons, D. B., and P. V. Hobbs, 1983: The mesoscale and microscale structure and organization of clouds and precipitation in midlatitude cyclones. VIII: Formation, development, interaction and dissipation of rainbands. *J. Atmos. Sci.* **40,** 559–79.

Wallace, J. M., 1975: Diurnal variations in precipitation and thunderstorm frequency over the conterminous United States. *Mon. Wea. Rev.* **103,** 406–19.

Section 3.5.2

Bennetts, D. A., and B. J. Hoskins, 1979: Conditional symmetric instability—a possible explanation for frontal rainbands. *Quart. J. Roy. Meteor. Soc.* **105,** 945–62.

Emanuel, K. A., 1983a: On assessing local conditional symmetric instability from atmospheric soundings. *Mon. Wea. Rev.* **111,** 2016–33.

—, 1983b: The Lagrangian parcel dynamics of moist symmetric instability. *J. Atmos. Sci.* **40,** 2368–76.

—, 1985: Frontal circulations in the presence of small moist symmetric stability. *J. Atmos. Sci.* **42,** 1062–71.

Wolfsberg, D. G., K. A. Emanuel, and R. E. Passarelli, 1986: Band formation in a New England winter storm. *Mon. Wea. Rev.* **114,** 1552–69.

Section 3.5.3

Hobbs, P. V., 1978: Organization and structure of clouds and precipitation on the mesoscale and microscale in cyclonic storms. *Rev. Geophys. Space Phys.* **16,** 741–55.

Houze, R. A., Jr., and P. V. Hobbs, 1982, op. cit.

Matejka, T. J., R. A. Houze, Jr., and P. V. Hobbs, 1980: Microphysics and dynamics of clouds associated with mesoscale rainbands in extratropical cyclones. *Quart J. Roy. Meteor. Soc.* **106,** 29–56.

Section 3.5.4

Bosart, L. F., and J. P. Cussen, Jr., 1973: Gravity wave phenomena accompanying East Coast cyclogenesis. *Mon. Wea. Rev.* **101**, 446–54.

Bosart, L. F., and F. Sanders, 1986: Mesoscale structure in the megalopolitan snowstorm of 11–12 February 1983. Part III: A large-amplitude gravity wave. *J. Atmos. Sci.* **43,** 924–39.

—, and A. Seimon, 1988: A case study of an unusually intense atmospheric gravity wave. *Mon. Wea. Rev.* **116,** 1857–86.

Eom, J, 1975: Analysis of the internal gravity wave occurrence of April 19, 1970 in the Midwest. *Mon. Wea. Rev.* **103,** 217–26.

Ley, B. E., and W. R. Peltier, 1978: Wave generation and frontal collapse. *J. Atmos. Sci.* **35,** 3–17.

Lindzen, R. S., and K. K. Tung, 1976: Banded convective activity and ducted gravity waves. *Mon. Wea. Rev.* **104,** 1602–17.

Marks, F. D., 1975: A study of the mesoscale precipitation patterns associated with the New England coastal front. M.S. Thesis, Dept. of Meteorology, M.I.T., Cambridge, Mass.

Pecnick, M. J., and J. A. Young, 1984: Mechanics of a strong sub-synoptic gravity wave deduced from satellite and surface observations. *J. Atmos. Sci.* **41,** 1850–62.

Schneider, R. S., 1990: Large-amplitude mesoscale wave disturbances within the intense Midwest extratropical cyclone of 15 December 1987. *Wea. and Forecasting* **5,** 533–58.

Uccellini, L. W., and S. E. Koch, 1987: The synoptic setting and possible energy sources for mesoscale wave disturbances. *Mon. Wea. Rev.* **115,** 721–29.

PROBLEMS

3.1. (a) Name six types of mesoscale rainbands (that are *not* topographically excited) found in extratropical cyclones. Where (in the conceptual model of an extratropical cyclone) is each found? How is each oriented (with respect to features in the conceptual model)? (b) Name five physical mechanisms that have been proposed to explain them.

SELECTED ANSWERS TO PROBLEMS

1.1. From northwest to southeast.

1.2. Northerly at B, southerly at A.

1.5. (a) AVA everywhere, CVA nowhere; (b) CVA for $B<y<C$, AVA for $C<y<D$ and $A<y<B$.

1.6. $\omega_T<0$ southeast of low, $\omega_T>0$ northwest of low, $\omega_V>0$ near low.

1.8. Low will elongate in the northeast–southwest direction.

1.9. (a) CVA west of troughs, east of ridges; AVA east of troughs, west of ridges; (b) same as in (a).

1.10. $\omega_T<0$ southwest of high; $\omega_T>0$ northeast of high.

1.11. Westward (very slowly unless low has long wavelength).

1.12. If there is less CVA above 500 mb than there is below 500 mb.

1.13. Eastward.

1.17. 5620 km.

1.18. (a) No motion; (b) nowhere.

1.19. (a) At B and C; (b) at none of the points shown.

1.20. Upward at D; downward at B; none at A or C.

1.21. $7.11\ \mathrm{m\,s^{-1}}$.

1.23. (a) χ_V at A (0), B (−), C (0), D (+), E (+), F (0), G (−), H (−), I (0), J (+), K (0), L (0), M (−), N (0), O (+), P (+), Q (0), R (−), S (−), T (0), U (+), V (0); χ_T at A–K (+), L–V (−); (b) χ_V at A–K (−), L–V (+), χ_T at A–K (+), L–V (−).

1.24. $-2.35\ \mu\mathrm{b\,s^{-1}}$.

1.25. (a) $\delta>0$ at B, $\delta<0$ at *D*, $\delta=0$ at A and C; (b) toward the west.

1.26. (a) Rise at 1, fall at 3 and 14, no change elsewhere; (b) rise at 1, 11, and 14, fall at 3 and 4, no change elsewhere.

1.28. East of troughs and west of ridges.

1.31. Rising motion where $L/4 < x < 3L/4$, sinking motion where $0 < x < L/4$ and $3L/4 < x < L$.
1.32. (a) $10^{-12}\ \mathrm{s^{-2}\,kPa^{-1}}$; (b) $10^{-10}\ \mathrm{s^{-2}}$; (c) $10^{-10}\ \mathrm{s^{-2}}$.
1.33. $20\ \mathrm{m^2\,s^{-2}\,kPa^{-2}}$.
1.34. $0 < x < 200$ km.
1.35. (a) Eastward; (b) westward.
1.41. (a) 29.7°C $\mathrm{d^{-1}}$; (b) −17.6°C $\mathrm{d^{-1}}$; effect of secondary circulation less than geostrophic "forcing;" (c) 16.6°C $\mathrm{d^{-1}}$; (d) −29.8°C $\mathrm{d^{-1}}$; effect of secondary circulation greater than geostrophic "forcing."
1.44. (a) Surface cyclogenesis induced east of negative-IPV anomaly aloft; surface anticyclogenesis induced west of negative-IPV anomaly aloft; (b) surface cyclogenesis induced west of negative-IPV anomaly aloft; surface anticyclogenesis induced east of negative-IPV anomaly aloft.
1.45. (a) Cyclone; (b) $\omega > 0$ west of positive θ anomaly, $\omega < 0$ east of positive θ anomaly.
1.46. $\mathrm{IPV} = -g(\zeta_\theta + f)\,\partial\theta/\partial p \sim 10^{-5}\ \mathrm{m^2\,K\,kg^{-1}\,s^{-1}}$.
1.47. (b) Propagation acts to move jet streak westward, while basic current advects it eastward; if horizontal scale of jet streak is small, the effect of propagation is negligible, and the basic current moves jet streak eastward; if horizontal scale of jet streak is large, the effect of propagation is dominant, and the jet streak propagates westward.
1.48. (a) Decrease; (b) eastward; (c) increase; (d) westward.
2.1. (a) $(\ln 10)/c$; (b) never!
2.3. (a) From the north; (b) from the south; (c) near $y = 500$ km, where there is surface convergence.
2.4. (a) 15 K $(100\ \mathrm{km})^{-1}$; (b) 54.1 $\mu\mathrm{b\,s^{-1}}$ $(100\ \mathrm{km})^{-1}$; (c) 25.4 $\mathrm{cm\,s^{-1}}$ at the southern edge, −25.4 $\mathrm{cm\,s^{-1}}$ at the northern edge.
2.5. 4.8 $\mathrm{m\,s^{-1}}$ from the north.
2.11. 0.267 d.
2.14. See the following figure.

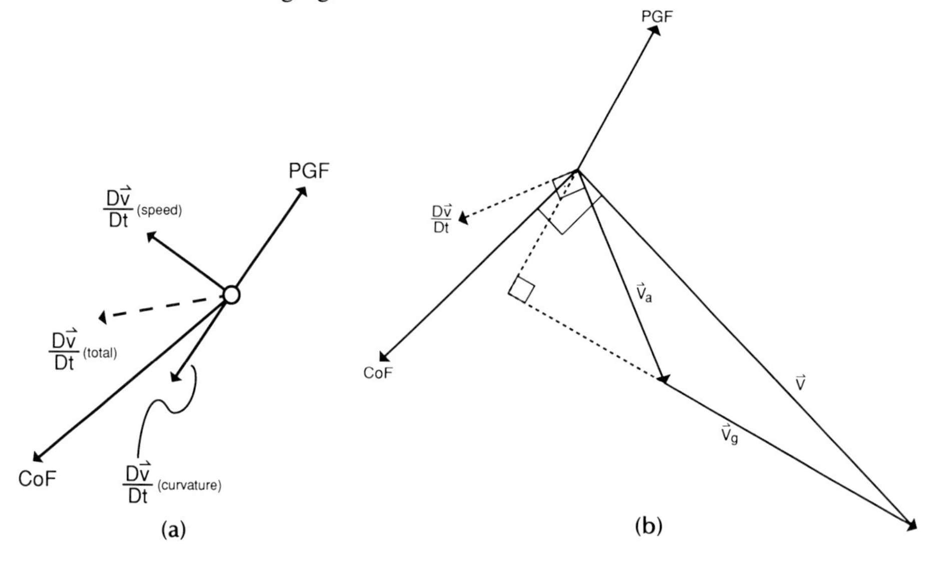

2.15. (a) $2<x<4$, $5<x<7$, $8<x<10$; (b) $x\sim 2$, 7, 8.
2.16. (a) 8.1 K $(100\text{ km})^{-1}\text{ d}^{-1}$; (b) rising motion at $x\sim 300$ km, sinking motion at $x\sim 100$ km; westerly ageostrophic wind below, easterly ageostrophic wind above.
2.18. (a) $-(D/2R)(\theta_2-\theta_1)$; (b) westerly ageostrophic wind aloft, easterly ageostrophic wind below, rising motion at $x\sim -R$, sinking motion at $x\sim R$; (c) below near $x=R$, aloft near $x=-R$, where $\delta>0$.
2.19. The streamlines are anticyclonic circles; the isotachs are concentric circles, with highest values near the center.
2.20. (a) $\omega<0$ at $x=x_4$, $\omega>0$ at $x=x_2$, $\omega\sim 0$ elsewhere; (b) at x_2 below, x_4 aloft; (c) thermally indirect.
2.22. $-1.5\times 10^{-5}\text{ s}^{-1}$.
2.23. $0.752\ \mu\text{b s}^{-1}\ (1000\text{ km})^{-1}$.
2.25. (a) $\alpha^2[\ln(p/1000)]^2+\alpha[2\ln(p/1000)]+[1-f_0^2T_0\gamma(p)/Ra^2]>0$.
2.26. $6.69\text{ K }(100\text{ km})^{-1}$.
2.27. $-3.75\times 10^{-4}\text{ s}^{-1}$.
2.28. $X=-(Dy/Dt)/f$.
2.29. (a)

$$v_a=-\frac{A_0}{f_0C_p}\left[a\left(\frac{2\pi}{L}\right)^2+c\left(\frac{2\pi}{p_0}\right)^2\right]^{-1}\left(\frac{2\pi}{L}\right)\left(\frac{2\pi}{p_0}\right)\sin(2\pi/L)y\cos(2\pi/p_0)p,$$

$$\omega=\frac{A_0}{f_0C_p}\left[a\left(\frac{2\pi}{L}\right)^2+c\left(\frac{2\pi}{p_0}\right)^2\right]^{-1}\left(\frac{2\pi}{L}\right)^2\cos(2\pi/L)y\sin(2\pi/p_0)p,$$

where $c=f_0-\partial u_g/\partial y$; (b) as in (a), but $c=f_0$.
2.32. (a) $-2.14\times 10^{-5}\text{ s}^{-1}$ (confluence); (b) yes.
2.34. (b) 14.4 m s^{-1}.
2.35. ~250 km.
2.36. $\sim 0.5\text{ m s}^{-1}\ (100\text{ kPa})^{-1}$.
2.37. 0.18.
2.38. $4.85\times 10^{-6}\text{ s}^{-1}$.
2.43. no; $\text{Ri}=9.1$, $f_0/(\zeta_g+f_0)=2.6$.
2.44. (a) 13.2 $(100\text{ km})^{-1}\text{ d}^{-1}$; (b) 2.48 K $(100\text{ km})^{-1}\text{ d}^{-1}$.

APPENDIX 1

Solution to Sanders' Analytic Model

A.1. VERTICAL VELOCITY

The quasigeostrophic ω equation, subject to the three-dimensional distribution of Φ, Eq. (1.2.6), and the vertically varying σ, Eq. (1.2.8), is as follows:

$$\left(\nabla_p^2 + \frac{f_0^2 p^2}{RT_0\gamma}\frac{\partial^2}{\partial p^2}\right)\omega = \left\{\frac{4(2\pi/L)^3 Ra\hat{T}}{f_0 T_0 \gamma}\left[p\left(\ln\frac{1000}{p}\right)\left(1-\alpha\ln\frac{1000}{p}\right)\left(1-\frac{\alpha}{2}\ln\frac{1000}{p}\right)\right]\right.$$
$$\left. - \frac{(2\pi/L)\hat{T}\beta}{T_0\gamma}p\left(1-\alpha\ln\frac{1000}{p}\right)\right\}\sin\frac{2\pi}{L}x\cos\frac{2\pi}{L}y$$
$$-\left[\frac{4(2\pi/L)^3 a\hat{\Phi}_0}{f_0T_0\gamma}p\left(1-\alpha\ln\frac{1000}{p}\right)\right]\sin\frac{2\pi}{L}(x+\lambda)\cos\frac{2\pi}{L}y$$
$$+\left\{\frac{2(2\pi/L)^4\hat{\Phi}_0\hat{T}}{f_0T_0\gamma}\sin\frac{2\pi}{L}\lambda\left[p\left(1-\alpha\ln\frac{1000}{p}\right)\right]\right\}\sin\frac{4\pi}{L}y. \quad \text{(A.1.1)}$$

The forcing functions having

$$\sin(2\pi/L)x\cos(2\pi/L)y, \qquad \sin(2\pi/L)(x+\lambda)\cos(2\pi/L)y, \quad \text{and} \quad \sin(4\pi/L)y$$

as factors are associated with solutions ω_1, ω_2, and ω_3, respectively.

The exact, rather complicated-looking, solutions to Eq. (A.1.1) subject to Eqs. (1.2.13) and (1.2.14) are as follows:

$$\omega_1 = (K_{11}^\omega P_{11}^\omega + K_{12}^\omega P_{12}^\omega)\sin\frac{2\pi}{L}x\cos\frac{2\pi}{L}y, \quad \text{(A.1.2)}$$

where

$$K_{11}^\omega = \frac{(2\pi/L)Ra\hat{T}}{f_0T_0\gamma} \quad \text{(A.1.3)}$$

$$K_{12}^\omega = \frac{\hat{T}\beta}{2(2\pi/L)T_0\gamma}, \quad \text{(A.1.4)}$$

are measures of the strength of the response of each forcing function, while

$$P_{11}^\omega = [2k + 6\alpha k(k+1) + 6\alpha^2k^2(k+2)]p\left[1-\left(\frac{p}{1000}\right)^{q-1}\right]$$
$$-[2+6\alpha k+6\alpha^2k(k+1)]p\ln\frac{1000}{p}$$
$$+(3\alpha+3\alpha^2k)p\left(\ln\frac{1000}{p}\right)^2 - \alpha^2p\left(\ln\frac{1000}{p}\right)^3 \quad \text{(A.1.5)}$$

$$P_{12}^\omega = (\alpha k+1)p\left[1-\left(\frac{p}{1000}\right)^{q-1}\right] - \alpha p\ln\frac{1000}{p}, \quad \text{(A.1.6)}$$

where

$$k = \frac{f_0^2}{2(2\pi/L)^2 R T_0 \gamma} \tag{A.1.7}$$

$$q - 1 = \tfrac{1}{2}\left(1 + \frac{4}{k}\right)^{1/2} - \tfrac{1}{2} \tag{A.1.8}$$

determine the shape of the vertical profile.

$$\omega_2 = K_2^\omega P_2^\omega \sin\frac{2\pi}{L}(x + \lambda)\cos\frac{2\pi}{L}y, \tag{A.1.9}$$

where

$$K_2^\omega = \frac{2(2\pi/L)a\hat{\Phi}_0}{f_0 T_0 \gamma}, \tag{A.1.10}$$

while

$$P_2^\omega = (\alpha k + 1)p\left[1 - \left(\frac{p}{1000}\right)^{q-1}\right] - \alpha p \ln\frac{1000}{p}. \tag{A.1.11}$$

Finally,

$$\omega_3 = K_3^\omega P_3^\omega \sin\frac{4\pi}{L}y, \tag{A.1.12}$$

where

$$K_3^\omega = \frac{(2\pi/L)^2\hat{\Phi}_0\hat{T}\sin(2\pi/L)\lambda}{2f_0 T_0 \gamma}, \tag{A.1.13}$$

while

$$P_3^\omega = \alpha p \ln\left(\frac{1000}{p}\right) - \left(\alpha\frac{k}{4} + 1\right)p\left[1 - \left(\frac{p}{1000}\right)^{r-1}\right] \tag{A.1.14}$$

and where

$$r - 1 = \tfrac{1}{2}(1 + 8/k)^{1/2} - \tfrac{1}{2}. \tag{A.1.15}$$

The total solution to Eq. (A.1.1) is

$$\omega = \omega_1 + \omega_2 + \omega_3 \tag{A.1.16}$$

$$= \omega_{11} + \omega_{12} + \omega_{21} + \omega_{22} + \omega_3, \tag{A.1.17}$$

where

$$\omega_{11} = K_{11}^\omega P_{11}^\omega \sin\frac{2\pi}{L}x\cos\frac{2\pi}{L}y \tag{A.1.18}$$

$$\omega_{12} = K_{12}^\omega P_{12}^\omega \sin\frac{2\pi}{L}x\cos\frac{2\pi}{L}y \tag{A.1.19}$$

$$\omega_{21} = \omega_{22} = \frac{K_2^\omega P_2^\omega}{2}\sin\frac{2\pi}{L}(x + \lambda)\cos\frac{2\pi}{L}y \tag{A.1.20}$$

$$\omega_3 = K_3^\omega P_3^\omega \sin\frac{4\pi}{L}y. \tag{A.1.21}$$

Note that the (first) subscripts on the ωs, 1, 2, and 3, are associated with $\sin(2\pi/L)x\cos(2\pi/L)y$, $\sin(2\pi/L)(x+\lambda)\cos(2\pi/L)y$, and $\sin(4\pi/L)y$ distributions in x and y. The profiles of each $K^{\omega}P^{\omega}$ term are shown in Fig. 1.36 for $\lambda = L/4$. This phase lag was chosen because it represents the phase lag for most rapid intensification.

A.2. HEIGHT TENDENCY

The exact solutions to Eq. (1.1.10) using the solutions to ω and the height field, Eq. (1.2.6), are as follows:

$$\chi_1 = (K^{\chi}_{11}P^{\chi}_{11} - K^{\chi}_{12}P^{\chi}_{12} + K^{\chi}_{13}P^{\chi}_{13} - K^{\chi}_{14}P^{\chi}_{14}) \sin\frac{2\pi}{L}x \cos\frac{2\pi}{L}y, \tag{A.1.22}$$

where

$$K^{\chi}_{11} = \frac{R\hat{T}\beta}{2(2\pi/L)} \tag{A.1.23}$$

$$K^{\chi}_{12} = \frac{(2\pi/L)R^2 a\hat{T}}{f_0} \tag{A.1.24}$$

$$K^{\chi}_{13} = \frac{f_0 Ra\hat{T}}{2(2\pi/L)T_0\gamma} \tag{A.1.25}$$

$$K^{\chi}_{14} = \frac{f_0^2\hat{T}\beta}{4(2\pi/L)^3T_0\gamma} \tag{A.1.26}$$

$$P^{\chi}_{11} = \left(\ln\frac{1000}{p}\right)\left(1 - \frac{\alpha}{2}\ln\frac{1000}{p}\right) \tag{A.1.27}$$

$$P^{\chi}_{12} = \left[\left(\ln\frac{1000}{p}\right)\left(1 - \frac{\alpha}{2}\ln\frac{1000}{p}\right)\right]^2 \tag{A.1.28}$$

$$\begin{aligned} P^{\chi}_{13} = {} & [2k + 6\alpha k(k+1) + 6\alpha^2k^2(k+2)]\left[q\left(\frac{p}{1000}\right)^{q-1} - 1\right] \\ & - [2 + 6\alpha k + 6\alpha^2 k(k+1)]\left(1 - \ln\frac{1000}{p}\right) \\ & + (3\alpha + 3\alpha^2 k)\left[2\ln\frac{1000}{p} - \left(\ln\frac{1000}{p}\right)^2\right] \\ & - \alpha^2\left[3\left(\ln\frac{1000}{p}\right)^2 - \left(\ln\frac{1000}{p}\right)^3\right] \end{aligned} \tag{A.1.29}$$

$$P^{\chi}_{14} = \alpha\left(1 - \ln\frac{1000}{p}\right) - (\alpha k + 1)\left[q\left(\frac{p}{1000}\right)^{q-1} - 1\right] \tag{A.1.30}$$

$$\chi_2 = (K^{\chi}_{21}P^{\chi}_{21} - K^{\chi}_{22} + K^{\chi}_{23}P^{\chi}_{23}) \sin\frac{2\pi}{L}(x+\lambda)\cos\frac{2\pi}{L}y, \tag{A.1.31}$$

where

$$K_{21}^{\chi} = \frac{(2\pi/L)Ra\hat{\Phi}_0}{f_0} \tag{A.1.32}$$

$$K_{22}^{\chi} = \frac{\hat{\Phi}_0\beta}{2(2\pi/L)} \tag{A.1.33}$$

$$K_{23}^{\chi} = \frac{f_0 a\hat{\Phi}_0}{(2\pi/L)T_0\gamma} \tag{A.1.34}$$

$$P_{21}^{\chi} = \left(\ln\frac{1000}{p}\right)\left(1 - \frac{\alpha}{2}\ln\frac{1000}{p}\right) \tag{A.1.35}$$

$$P_{23}^{\chi} = (\alpha k + 1)\left[q\left(\frac{p}{1000}\right)^{q-1} - 1\right] - \alpha\left(1 - \ln\frac{1000}{p}\right). \tag{A.1.36}$$

Finally,

$$\chi_3 = K_{31}^{\chi}P_{31}^{\chi}\sin\frac{4\pi}{L}y, \tag{A.1.37}$$

where

$$K_{31}^{\chi} = \frac{f_0\hat{\Phi}_0\hat{T}\sin(2\pi\lambda/L)}{8T_0\gamma} \tag{A.1.38}$$

$$P_{31}^{\chi} = \alpha\left(1 - \ln\frac{1000}{p}\right) - (\alpha k/2 + 1)\left[r\left(\frac{p}{1000}\right)^{r-1} - 1\right]. \tag{A.1.39}$$

The total solution (χ) to Eq. (1.1.10), i.e., to Eq. (1.2.22) is

$$\chi = \chi_1 + \chi_2 + \chi_3 \tag{A.1.40}$$

$$= \chi_{11} + \chi_{11}^{\delta} + \chi_{12} + \chi_{12}^{\delta} + \chi_{21} + \chi_{21}' + \chi_{23}^{\delta} + \chi_3^{\delta}, \tag{A.1.41}$$

where

$$\chi_{11} = K_{11}^{\chi}P_{11}^{\chi}\sin\frac{2\pi}{L}x\cos\frac{2\pi}{L}y \tag{A.1.42}$$

$$\chi_{11}^{\delta} = K_{13}^{\chi}P_{13}^{\chi}\sin\frac{2\pi}{L}x\cos\frac{2\pi}{L}y \tag{A.1.43}$$

$$\chi_{12} = -K_{12}^{\chi}P_{12}^{\chi}\sin\frac{2\pi}{L}x\cos\frac{2\pi}{L}y \tag{A.1.44}$$

$$\chi_{12}^{\delta} = -K_{14}^{\chi}P_{14}^{\chi}\sin\frac{2\pi}{L}x\cos\frac{2\pi}{L}y \tag{A.1.45}$$

$$\chi_{21} = K_{21}^{\chi}P_{21}^{\chi}\sin\frac{2\pi}{L}(x + \lambda)\cos\frac{2\pi}{L}y \tag{A.1.46}$$

$$\chi_{21}' = -K_{22}^{\chi}\sin\frac{2\pi}{L}(x + \lambda)\cos\frac{2\pi}{L}y \tag{A.1.47}$$

$$\chi_{23}^{\delta} = K_{23}^{\chi} P_{23}^{\chi} \sin \frac{2\pi}{L}(x + \lambda) \cos \frac{2\pi}{L} y \tag{A.1.48}$$

$$\chi_{3}^{\delta} = K_{31}^{\chi} P_{31}^{\chi} \sin \frac{4\pi}{L} y. \tag{A.1.49}$$

A.3. VELOCITY OF SURFACE FEATURES

At the center of each surface cyclone

$$x = L/2 - \lambda \pm nL, \qquad n = 0, 1, 2, \ldots, \tag{A.1.50}$$

and so

$$c_x(p = 1000 \text{ mb}) = c_{x1} \cos \frac{2\pi}{L} \lambda + c_{x2} \tag{A.1.51}$$

$$c_y(p = 1000 \text{ mb}) = c_{y3} \sin \frac{2\pi}{L} \lambda, \tag{A.1.52}$$

where

$$\begin{aligned} c_{x1} = &\frac{f_0 \hat{T}}{2(2\pi/L)^2 T_0 \gamma \hat{\Phi}_0} \Big(Ra\{[2k + 6\alpha k(k+1) \\ &+ 6\alpha^2 k^2 (k+2)](q-1) - [2 + 6\alpha k + 6\alpha^2 k(k+1)]\} \\ &+ \frac{f_0 \beta}{2(2\pi/L)^2} [(\alpha k + 1)(q-1) - \alpha] \Big) \end{aligned} \tag{A.1.53}$$

$$c_{x2} = \frac{f_0 a}{(2\pi/L)^2 T_0 \gamma} [(\alpha k + 1)(q-1) - \alpha] - \frac{\beta}{2(2\pi/L)^2} \tag{A.1.54}$$

$$c_{y3} = \frac{f_0 \hat{T}\{(r-1)[\alpha(k/2) + 1] - \alpha\}}{4(2\pi/L) T_0 \gamma}. \tag{A.1.55}$$

Index